UK NATURE CONSERVATION

No. 3

A review of the scarce and
threatened Coleoptera of
Great Britain

Part 1

by

P.S. Hyman

Revised and updated by

M.S. Parsons

Further copies can be obtained from
The UK Joint Nature Conservation Committee
Peterborough

ISBN 1873 701 10 1

Anchonidium unguiculare (Aubé, 1850)
Curculionidae
by T.M. Eccles

1 mm

CONTENTS

1. Introduction 1

2. Scope of this review 2

3. An introduction to the data sheets 2

4. Methods and sources of information. 5

5. Status category definitions and criteria 7

6. The study of beetles in britain 10

7. Abbreviations 12

8. Acknowledgements 13

9. Further reading 13

Appendix 1: Water beetle statuses 17

Appendix 2: Species listed by status category 19

Appendix 3: Taxonomic list of Red Data Book and notable species in part 1 of the
  review 29

Bibliography to the introduction 55

THE DATA SHEETS 59

Aderidae	61	Chrysomelidae	170
Anobiidae	62	Cisidae	210
Anthicidae	67	Clambidae	213
Anthribidae	69	Cleridae	213
Apionidae	72	Coccinellidae	216
Attelabidae	83	Colydiidae	220
Biphyllidae	87	Corylophidae	226
Bostrichidae	87	Cucujidae	227
Bruchidae	87	Curculionidae	230
Buprestidae	88	Dermestidae	311
Byrrhidae	92	Drilidae	313
Cantharidae	94	Elateridae	314
Carabidae	99	Endomychidae	330
Cerambycidae	155	Erotylidae	331
Cerylonidae	169	Eucnemidae	332

Geotrupidae 334
Heteroceridae 335
Histeridae 335
Hypocopridae 344
Lampyridae 344
Limnichidae 344
Lucanidae 345
Lycidae 345
Lyctidae 347
Lymexylidae 347
Melandryidae 348
Meloidae 355
Melyridae 357
Mordellidae 361
Mycetophagidae 366
Mycteridae 368
Oedemeridae 368
Peltidae 371
Phalacridae 371
Phloiophilidae 373
Platypodidae 374

Psephenidae 374
Ptinidae 375
Pyrochroidae 376
Pythidae 377
Rhizophagidae 377
Salpingidae 380
Scaphidiidae 381
Scarabaeidae 382
Scirtidae 396
Scolytidae 398
Scraptiidae 405
Silphidae 409
Silvanidae 411
Sphaeritidae 412
Sphindidae 412
Tenebrionidae 413
Tetratomidae 419
Throscidae 420
Trogidae 420
Trogossitidae 421

References quoted in data sheets 423

Indexes 457

Index to the names of invertebrates mentioned in the text 457
Index to scientific names of plants mentioned in the text 479

1. INTRODUCTION

by Dr R.S. Key

With almost 4,000 species, the Coleoptera comprises the third largest order of British invertebrates after Diptera and Hymenoptera. Species of beetle are found in just about every terrestrial, freshwater and intertidal habitat.

Although many species of beetle are common in Britain, many others are under some degree of threat. Some species have become extinct in historical times and there is good evidence from subfossil material in Bronze Age peaty deposits that the process of extinction has been under way almost since man came to these islands (eg. Buckland, 1979; Girling, 1982). Many more continue to decline alarmingly. To date, however, no autecological conservation studies other than minor surveys have been commissioned, or species recovery programmes implemented, for any species of beetle. The purposes of this review is to identify the scarcer species of beetle in Britain and the ones under threat, and to suggest ways in which these species can be conserved.

Perhaps because of their bewildering diversity and specialism, beetles have been neglected in the evaluation of conservation sites and the planning of their management. Another reason for the neglect of beetles by conservationists has been that, with the possible exception of the ladybirds (Coccinellidae), and perhaps a few highly coloured species, beetles have held little esteem among non-entomologists. Seemingly unpleasant habits of a few species, together with the existence of a number of pests of crops and stored products, has led to their having enjoyed little support with the general public. Lack of knowledge of the ecology, behaviour, distribution and conservation needs of the large majority of beetle species has not helped. Species in some families are difficult to identify reliably, often requiring dissection of their genitalia, while the larvae of many remain undescribed.

Despite all this, beetles have much to offer conservation in addition to having conservation needs of their own. They have a high degree of specialisation, coupled with short, often annual life cycles and sometimes poor powers of dispersal and colonisation. This means that they can be very useful indicators of both habitat quality and continuity of suitable habitat conditions. They frequently respond much more rapidly to environmental changes than does vegetation and may thus be used for early warning of habitat change and degradation. A diverse beetle fauna, typical of the habitat on a site, may be a very good indication of its health.

Although there is still a great deal to learn about beetles, considerable information is already available on many species. This may be used to determine the conservation needs of individual species and assemblages of beetles. Until now, this has been dispersed among the notebooks and minds of entomologists, in articles, papers and books, and on the data labels of specimens in private and museum collections. The aim of this Review is to collate as much of this information as possible and to synthesise it into statements on the distribution, ecology, threats to, and conservation needs of the individual species. It should allow conservation workers to identify important sites for beetles and to take into account their needs in planning site management. Hopefully, entomologists will realise and address the gaps in our knowledge.

Five hundred and forty six species of beetle were identified in Shirt (1987) as being under some degree of threat in Britain (54 of which were considered to be extinct) and some guidelines were given towards their conservation. This Review extends that approach, covering additional species that are somewhat less scarce in Britain but are by no means common. At the same time, it reviews the species of beetle covered by the Red Data Book in the light of additional information that was not available when the Red Data Book was being compiled, suggesting additions, changes and deletions. **This is not, however, a Red Data Book and the changes, additions and deletions should be considered provisional until they are published in a future edition of the Red Data Book.** It is these amended statuses that will be used in future by the statutory nature conservation bodies.

1

INTRODUCTION

2. SCOPE OF THIS REVIEW

This Review is produced in two parts. The first concentrates on readily identifiable and popular families of beetles, including most of those families for which there exist recording schemes co-ordinated by the Institute of Terrestrial Ecology's Biological Records Centre. The families covered in part one are:

Aderidae, Anobiidae, Anthribidae, Anthicidae, Apionidae, Attelabidae, Biphyllidae, Bostrichidae, Bruchidae, Buprestidae, Byrrhidae, Cantharidae, Carabidae, Cerambycidae, Cerylonidae, Chrysomelidae, Cisidae, Clambidae, Cleridae, Coccinellidae, Colydiidae, Corylophidae, Cucujidae, Curculionidae, Dermestidae, Drilidae, Elateridae, Endomychidae, Erotylidae, Eucnemidae, Geotrupidae, Heteroceridae, Histeridae, Hypocopridae, Lampyridae, Limnichidae, Lucanidae, Lycidae, Lyctidae, Lymexylidae, Melandryidae, Meloidae, Melyridae, Mordellidae, Mycetophagidae, Mycteridae, Oedemeridae, Peltidae, Phalacridae, Phloiophilidae, Platypodidae, Psephenidae, Pyrochroidae, Pythidae, Rhizophagidae, Salpingidae, Scaphidiidae, Scarabaeidae, Scirtidae, Scolytidae, Scraptiidae, Silphidae, Silvanidae, Sphaeritidae, Sphindidae, Tenebrionidae, Tetratomidae, Throscidae, Trogidae and Trogossitidae.

Families with no scarce species are therefore omitted. These are the Byturidae, Dascillidae, Derodontidae, Eucinetidae, Leptinidae, Merophysiidae, Nemonychidae, Rhipiphoridae and Urodontidae.

Families to be covered in Part two are: Cryptophagidae, Hydrophilidae (terrestrial species), Lathridiidae, Leiodidae, Nitidulidae, Pselaphidae, Ptiliidae, Scydmaenidae, Sphaeriidae and Staphylinidae.

This Review covers predominantly terrestrial groups of beetles only and excludes the water beetles - members of the families Dytiscidae, Elmidae, Gyrinidae, Haliplidae, Hydraenidae, Hydrophilidae, Hygrobiidae and Noteridae. An analysis of the statuses of these is given by Foster (1986) and Foster and Eyre (in press). A full list of these species statuses is given in Appendix 1.

Site evaluation using water beetle assemblages is more advanced than evaluation using terrestrial beetles and other aquatic groups of insects. Computerised *Two-way Indicator Species Analysis* (TWINSPAN) of the data set of the national recording scheme, organised by the Balfour-Browne Club, has enabled recurrent assemblages to be identified and sites of similar aquatic habitats can be ranked within certain geographical areas of the country using methodology called WETSCORE (Foster 1987b; Foster and Eyre in press).

3. AN INTRODUCTION TO THE DATA SHEETS

Information on each species is given in standardised format. The data sheets are designed to be completely self-contained. This should help reserve managers etc. to compile information on species relating to a single site.

Information on the data sheets is divided into specific sections:

3.1. The species' name and family.

The species' full name, family and synonymy is given, together with any English names specific to that species (otherwise, the English name used for the beetle's family is given if there is one, e.g. "a soldier beetle" for a member of the Cantharidae). The basic checklist used is based on the Royal Entomological Society of London checklist (Kloet and Hinks 1977) and incorporates subsequent changes identified in the Society's bulletin "Antenna" and in other publications. Changes in nomenclature published up to the end of 1990 are incorporated. Full specific and generic synonomy, together with published mis-spellings, misidentifications etc. are included. Species are indexed by all synonyms, generic and specific.

3.2. The species' status.

The definitions and criteria for ascribing species to scarcity/threat categories are given in a separate section below. Species that are considered to be extinct are also included.

3.3. Distribution.

Reference is made to any distribution map that may exist for the species and a list of those areas of Britain from which a species has been recorded is given.

3.3.1. The scarcer species.

With the exception of species in category Nationally Notable B and Nationally Notable, distribution is usually as a list of Watsonian vice-counties (Dandy 1969). Where individual records cannot be ascribed to vice-counties (normally, records from older literature sources), the original, less specific area is given. This is often either a county which comprises more than one vice-county (e.g. "Yorkshire"), or a vague area such as "London district" or "Bristol area", which could refer to more than one vice-county. Vice-counties are listed in numerical order with more vague areas interpolated in the approximate position in the numerically ordered list.

To differentiate between old and relatively recent records, separate lists are given for vice-counties represented by records prior to 1970 and from 1970 onwards. This can give some evidence of change in status, although it must be noted that the 1970 onwards records only cover a 20 year period, whereas pre-1970 records may be from a period dating back as far as the beginning of the previous century. Some species are recorded very infrequently but from a wide area and the two time periods are therefore not directly comparable.

3.3.2. Notable B and Nationally Notable species.

For these species, the distribution that is given is in the form of the old Nature Conservancy Council (NCC) regional structure as it was prior to April 1991. This has changed with the re-organisation of the NCC but the older structure is used for the purposes of this Review. A map of these regions is given in Figure 1. Where a species has been recorded in all of the regions within England, Scotland or Wales, just the countries are listed. Records from the periods pre-1970 and 1970 onwards are not differentiated for Notable B species.

3.4. Habitat and ecology.

Information on habitat associations, larval and adult ecology, behaviour, foodplant(s), phenology etc. is presented. Where ecological information is derived from Continental sources, this is stated, as individual species' ecology may sometimes be different in Britain. For example, the leaf beetle *Chrysolina cerealis* (Chrysomelidae), which is widespread and polyphagous on various labiates in mainland Europe, but in Britain is montane and monophagous on thyme *Thymus drucei*. Latin names are given for the food plants of phytophagous species.

3.5. Status.

The proposed changes and additions to the Red Data Book categories (Shirt 1987) are identified (see notes on page 1). The distribution in Britain is described, giving indications of geographical trends which may not be readily apparent from the list of counties or regions given in the "Distribution" section. For some species, particularly those restricted to a single site and those that have not been recorded for many years, actual localities are sometimes referred to, although the general policy has been to avoid long lists of individual sites. Changes in range, the perceived degree of threat, extent to which populations may be localised and the likelihood of under-recording are also described. Protected species are also identified.

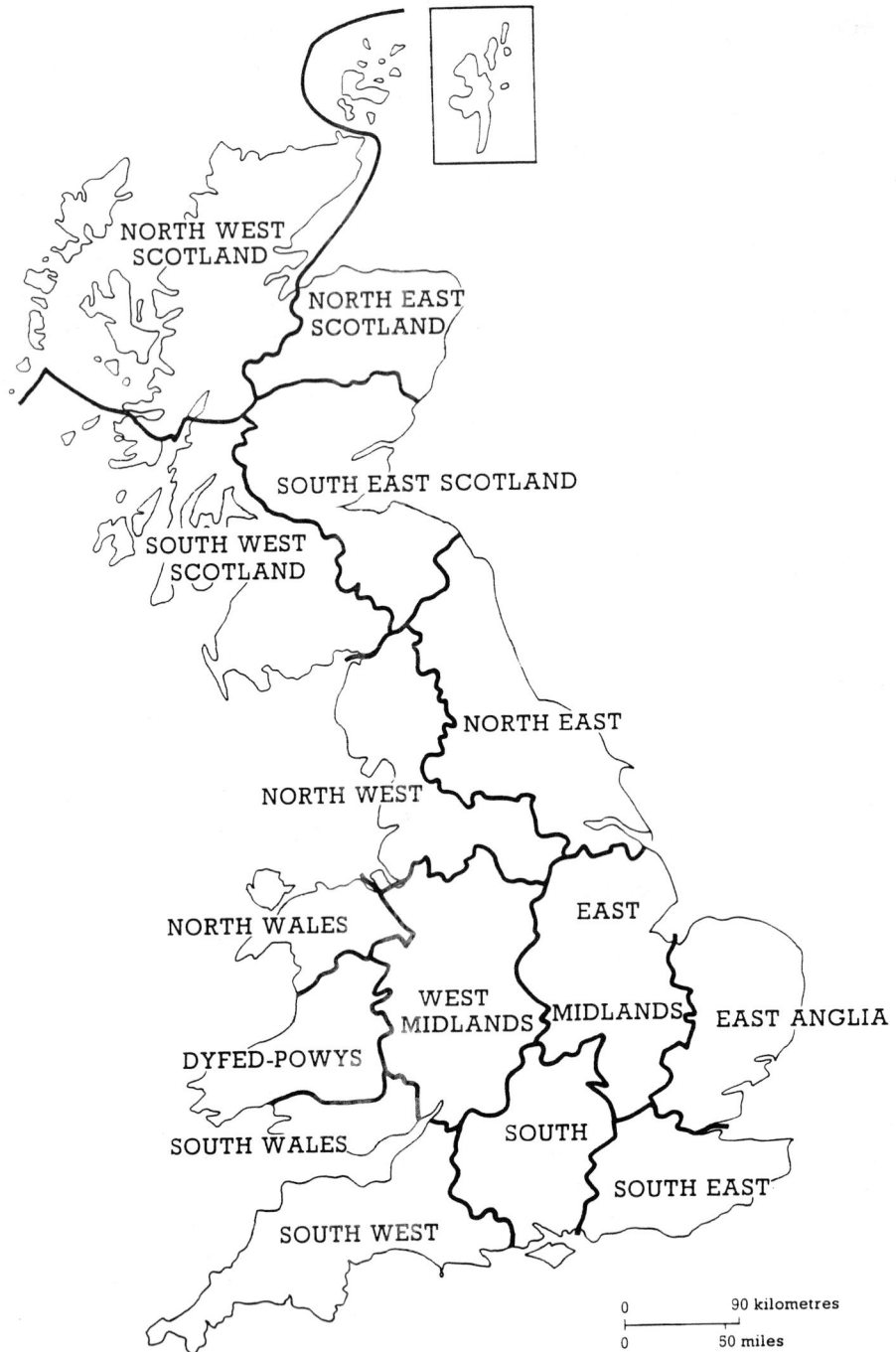

FIGURE 1: Map of NCC regions as at 1 January 1985.

3.6. Indicator status.

Species considered to be indicative of the continuity of mature timber habitat are identified as such, together with their grading. These are the species referred to by Harding and Rose (1986), Garland (1983) and Speight (1989). Statuses in Harding and Rose (1986) supersede those of Allen, Hunter, Johnson and Skidmore *in* Harding (1977) which are not normally listed.

3.7. Threats.

Threats to the overall habitat as well as to the specific microhabitat of the species are included. Known threats to individual populations, or examples of activities that have led to the destruction of populations of the species are given where appropriate. Otherwise, the statement summarises those activities which are perceived as most likely to put individual populations at risk. For some species where the ecology is unknown or uncertain, or for ones not apparently associated with a particular habitat type or niche, this section may be omitted.

3.8. Management and conservation.

Precautionary measures and positive action to maintain populations are suggested. Where active measures have already been taken to safeguard individual populations, these are described. For the majority of species, where specific requirements are unknown, this section may be generalised, identifying management practices that have been found to favour those aspects of the habitat with which the species may be associated. The occurrence of the species on protected sites is sometimes included. For some species where the ecology is unknown or uncertain, or for ones not apparently associated with a particular habitat type or niche, this section may be omitted.

3.9. Published sources.

Literature references pertinent to the species, in categories Nationally Notable A and above, are listed.

4. METHODS AND SOURCES OF INFORMATION.

During the preparation of this Review, as much information as possible was collated within the constraints of the contract period.

4.1. History of the Review.

This Review was commissioned in 1985 by the Terrestrial Invertebrate Zoology Branch of the Chief Scientist's Directorate of the NCC, the government's statutory nature conservation body. Dr Hyman was employed on a two year contract but it proved impossible to complete such a wide ranging review within that timescale and the contract was extended. Dr Hyman left the NCC in 1988 prior to the completion of the Review. The project lapsed until 1990 when Mr M. Parsons was taken on to complete the project. Progress reports were given by Key (1989a & b, 1991). In a review of statutory nature conservation bodies in Britain which came into operation in April 1991, the NCC was split into three separate bodies for England, Scotland and Wales, together with a Joint Nature Conservation Committee. The JNCC was established "for the purposes of nature conservation, and fostering the understanding thereof" (Environmental Protection Act 1990). The JNCC therefore took over responsibility for completing this Review.

4.2. Initial selection of species.

After consultation with a small number of specialists and reference to various standard works on British beetles, a discussion document (Hyman 1985) was put together. This was circulated for comment among a larger number of British coleopterists, using the circulation list of the Coleopterist's Newsletter. Through this

INTRODUCTION

document, opinions were sought on the suitability of species for inclusion. It was also intended to generate a rough guide to the status categories which should be adopted.

4.3. Second consultative stage.

The results of this consultation were collated into a second working document (Hyman 1986). Provisional statuses to all the scarcer British species of beetle were proposed and included suggested changes to the Red Data Book status of some species. The statuses contained in this document were included in Ball (1986) which covered most invertebrate groups.

Provisional data sheets were subsequently devised for each scarce species using information from the literature, the Invertebrate Site Register (see below) and from further limited consultation with experts. These were again circulated to coleopterists with a request for records, biological observations and further comment. These records will be used in future site evaluation and defence.

The returned information was collated and combined, and then re-examined to determine whether further changes in status were necessary in the light of this additional information. Changes at this stage were carried out in consultation with recognised specialists on the families of beetles concerned. Alterations to the statuses in Hyman (1986) and Shirt (1987) are listed in Appendix 3. Table 1 gives a breakdown of changes in the number of species allocated to each status category, within part 1 of this Review and relevant families in Hyman (1986).

TABLE 1: Numbers of species allocated to RDB and Notable Status in this Review and Hyman (1986).

Status	This review	Hyman (1986)
Notable B	398	456
Notable A	222	338
RDB 3	103	157
RDB 2	52	45
RDB 1	135	101
Extinct	53	52
RDB I	23	-
RDB K	57	-
Total	**1043**	**1149**

4.4. Sources of information.

4.4.1. Literature.

We are grateful to Mr A.B. Drane who made available his exhaustive collection of literature on the British beetle fauna. So comprehensive was this that, with the exception of additional reference to the "standard" works on British beetles, it was not considered necessary to conduct a further literature survey. The earliest standard work consulted was that of Fowler (1887 - 1891), which summarised earlier works. Subsequent to Dr Hyman's departure from the project, the major entomological journals were searched for additional information covering the period 1988 to 1990.

4.4.2. Amateur and professional entomologists.

A full list of contributing entomologists is given in the acknowledgements section (page 13).

4.4.3. Institute of Terrestrial Ecology: Biological Records Centre, beetle recording schemes.

The organisers of the majority of the terrestrial Coleoptera recording schemes were contacted and almost all contributed to this Review. Access by the author to the data sets was not sought and the author relied on the recording scheme organisers to consult the actual records.

4.4.4. The NCC's Invertebrate Site Register (ISR).

The ISR was set up by the NCC in 1980. It is a computerised inventory of sites of significance for invertebrate conservation and contains records of significant species of all groups of invertebrates. Information from the literature, professional and amateur entomologists, regional and national museums, biological records centres and taxonomic based recording schemes are all consulted by the ISR. This database was consulted extensively in the preparation of this report. It now resides with the JNCC. The ISR also has access to a computerised entomological bibliography, ENTSCAPE, listing a broad spectrum of journals relevant to the British insect fauna. This was used in the collation of the reference list.

4.4.5. Museums.

Museum collections were not consulted directly by the author owing to the constraints of time. Some contributing coleopterists work in museums with significant collections of beetles and were able to include information from the museum collection on the data sheets circulated by Dr Hyman.

5. STATUS CATEGORY DEFINITIONS AND CRITERIA

Criteria for the selection of species into the Red Data Book categories follow Shirt (1987), with minor modifications which are italicised. Categories RDB K (Insufficiently Known) and RDB I (Indeterminate) are used for the first time for Coleoptera and are based on the criteria used by Wells, Pyle and Collins (1983). Criteria for the selection of Nationally Notable species generally follow Eversham (1983). For the purposes of site evaluation for the selection of Sites of Special Scientific Interest the term Nationally Notable is now replaced by the term Nationally Scarce, but the criteria remain unchanged (NCC 1989).

Red Data Book Category 1. RDB 1 - ENDANGERED

Definition.   Taxa in danger of extinction and whose survival is unlikely if casual factors continue operating.

Included are those taxa whose numbers have been reduced to a critical level or whose habitats have been so dramatically reduced that they are deemed to be in immediate danger of extinction. Also included are *some* taxa that are *possibly* extinct.

Criteria. Species which are known or *believed to occur* as only a single population within one 10km square of the National Grid.

Species which only occur in habitats known to be especially vulnerable.

Species which have shown a rapid or continuous decline over the last twenty years and are now *estimated* to exist in five or fewer 10km squares.

Species which are *possibly* extinct *but have been recorded this century* and if rediscovered would need protection.

Red Data Book Category 2. RDB 2 - VULNERABLE

Definition. Taxa *believed* likely to move into the Endangered category in the near future if the casual factors continue operating.

Included are taxa of which most or all of the populations are decreasing because of *over-exploitation*, extensive destruction of habitat or other environmental disturbance; taxa with populations that have been seriously depleted and whose ultimate security is not yet assured; and taxa with populations that are still abundant but are under threat from serious adverse factors throughout their range.

Criteria. Species declining throughout their range.

Species in vulnerable habitats.

Red Data Book Category 3. RDB 3 - RARE

Definition. Taxa with small populations that are not at present Endangered or Vulnerable, but are at risk.

These taxa are usually localised within restricted geographical areas or habitats or are thinly scattered over a more extensive range.

Criteria. Species which are *estimated to exist* in only fifteen or fewer *post 1970* 10km squares. *This criterion may be relaxed where populations are likely to exist in over fifteen 10m squares but occupy small areas of especially vulnerable habitat.*

Red Data Book Category 5. RDB - ENDEMIC

Definition. Taxa which are not known to occur naturally outside Britain. Taxa within this category may also be in any of the other RDB categories *or not threatened at all.*

There are few truly endemic species of beetle in Britain. Most that have been identified are in fairly obscure groups which are relatively poorly known and may eventually be discovered elsewhere in Europe. Two species merit special mention. The flea beetle *Psylliodes luridipennis* (Chrysomelidae) is only known from Lundy Island in the Bristol Channel, feeding on Lundy Island cabbage *Rhynchosinapis wrightii*, a plant which is also considered endemic. The fern-feeding weevil *Syagrius intrudens* (Curculionidae) is, as yet, known from nowhere else in the world, but is thought to be of Australasian origin and it is assumed that it has been introduced from a part of the world where it has yet to be discovered - "an introduced endemic"! This species is not afforded conservation status here.

Red Data Book Appendix. RDB App. - EXTINCT

Definition.   Taxa which were formerly native to Britain but have not been recorded since 1900.

Red Data Book Category I. RDB I - INDETERMINATE

Definition.   Taxa *considered* to be Endangered, Vulnerable or Rare, but where there is not enough information to say which of the three categories (RDB 1 to 3) is appropriate.

Red Data Book Category K. RDB K - INSUFFICIENTLY KNOWN

Definition.   Taxa that are suspected but not definitely known to belong to any of the above categories, because of lack of information.

Criteria.   Taxa recently discovered or recognised in Britain which may prove to be more widespread in the future (although some recent discoveries may be placed in other categories if the group to which they belong is thought not to be under-recorded).

Taxa with very few or perhaps only a single known locality but which belong to poorly recorded or taxonomically difficult groups such as the Ptiliidae, aleocharine Staphylinidae, Mordellidae etc.

Species with very few or perhaps only a single known locality, inhabiting inaccessible or infrequently sampled but widespread habitats. Examples include some northern moorland species, ones associated with agricultural situations and ones which are adult only during the winter.

Species with very few or perhaps only a single known locality and of questionable native status, but not clearly falling into the category of recent colonist, vagrant or introduction.

Nationally Notable (Scarce) Category A. Na - NOTABLE A

Definition.   Taxa which do not fall within RDB categories but which are none-the-less uncommon in Great Britain and thought to occur in 30 or fewer 10km squares of the National Grid or, for less well-recorded groups, within seven or fewer vice-counties.

Nationally Notable (Scarce) Category B. Nb - NOTABLE B

Definition.   Taxa which do not fall within RDB categories but which are none-the-less uncommon in Great Britain and thought to occur in between 31 and 100 10km squares of the National Grid or, for less-well recorded groups between eight and twenty vice-counties.

Nationally Notable (Scarce). NOTABLE

Definition.   Species which are estimated to occur within the range of 16 to 100 10km squares. The subdividing of this category into Notable A and Notable B has not been attempted for some species in this Review.

INTRODUCTION

5.1. Protected species.

Two species of beetle, the violet click beetle *Limoniscus violaceus* (Elateridae) and the rainbow leaf beetle *Chrysolina cerealis* (Chrysomelidae) are legally protected under part 1.9, Schedule 5 of the 1981 Wildlife and Countryside Act. Under the act, it is an offence to intentionally kill, injure or take, have in possession any live or dead wild animal, part of or anything derived from such animal, or to damage, destroy or obstruct access to any structure or place which the species uses for shelter or protection, or disturb such animal while it is occupying a structure or place which it uses for that purpose. It is an offence to sell, offer or expose for sale or have in possession or transport for the purpose of sale any live or dead wild animal included in Schedule 5, or any part of, or anything derived from such an animal; or publishes or causes to be published any advertisement likely to be understood as the buying or selling or intends to buy or sell, any of these things.

Species on Schedule 5 are reviewed at five year intervals, next due in 1991.

5.2. Species not included.

Certain species may meet one or more of the above criteria but are not considered to warrant consideration in conservation evaluation or are only known from Ireland:

1. Species associated with non-native plants in Great Britain, including species associated with pine and other conifers but not occurring in areas where pine is considered native.

2. Migrant and vagrant/accidental species - ie species with no established population.

3. Obligate synanthropic species.

4. Species which are recorded only from Ireland in the British Isles.

6. THE STUDY OF BEETLES IN BRITAIN

Britain has an excellent history of the study of Coleoptera, dating back to the beginning of the nineteenth century. Its fauna has arguably been the most intensively studied in the world. There has never existed, however, any formal society devoted to the study of all beetles in Britain, although the very active Balfour-Browne Club exists for the study of water beetles. Activities have been co-ordinated to some extent by the national entomological societies such as the British Entomological and Natural History Society and the Amateur Entomologists' Society, which has produced a Coleopterist's Handbook (Walsh and Dibb 1954, Cooter and Cribb 1974, Cooter 1991), and also by various regional entomological societies such as the Lancashire and Cheshire Entomological Society, the entomological section of the Yorkshire Naturalists' Union, the Dyfed Invertebrate Group etc.

In 1981, a major meeting was organised by the Biological Records Centre at Monks Wood to review the future of beetle study in Britain. After that meeting, a quarterly Coleopterist's Newsletter was launched, the readership of which now forms a loose grouping of British entomologists who study the Coleoptera. More or less annual residential field meetings have been organised to study and record the beetle fauna of various parts of Britain and there has been a second large workshop meeting, in 1986, again at Monks Wood. The circulation of the newsletter now stands at 120 and gives a rough guide to the number of active coleopterists in Britain.

6.1. Recording Schemes.

Another focus for beetle study has been the various beetle recording schemes co-ordinated by the Biological Records Centre. There now exist 14 beetle schemes. Most schemes offer an identification service and some have produced newsletters. Preliminary or provisional atlases have been produced for selected ground beetles (Luff 1982), click beetles (Mendel 1988), and various families of water beetles (Foster 1981, 1983, 1984, 1985, 1987a, 1990). Atlases are currently in preparation for the Carabidae, Ptiliidae, Cryptophagidae; Atomariinae and certain subfamilies of Staphylinidae. A separate scheme is operated for the more conspicuous of the ladybirds by the Department of Genetics at the University of Cambridge, and this has again produced distribution maps (Majerus, Forge and Walker 1990).

6.1.1. Schemes in operation in 1991:

Carabidae - ground beetles:
Dr M.L. Luff, Dept. of Agricultural Biology, The University, Newcastle upon Tyne NE1 7RU.

Aquatic Coleoptera - water beetles:
Dr G.N. Foster, Balfour-Browne Club, 3 Eglinton Terrace, Ayr KA7 1JJ.

Elmidae - riffle beetles:
Mr D.G. Holland, N.R.A. North West Region, PO Box 12, Town House, Buttermarket Street, Warrington, Cheshire WA1 2QG.

Staphylinidae - rove beetles:
Mr P. Hammond, Dept. of Entomology, Natural History Museum of London, Cromwell Road, London SW7 5BD.

Scarabaeoidea - dung beetles and chafers:
Dr D.B. Shirt, c/o Biological Records Centre, Monks Wood Experimental Station, Abbots Ripton, Huntingdon, Cambs PE17 2LS.

Atomariinae and Ptiliidae:
Mr C. Johnson, Department of Entomology, Manchester Museum, The University, Manchester M13 9PL.

Lymexyloidea and Heteromera:
Dr R.S. Key, English Nature, Northminster House, Peterborough, PE1 1UA.

Coccinellidae - ladybirds:
Dr J. Muggleton, MAFF, Slough Laboratory, London Road, Slough, Berks SL3 7HJ.

Elateroidea - click beetles:
Mr H. Mendel, c/o The Museum, High Street, Ipswich, Suffolk IP1 3QH.

Cantharoidea and Buprestoidea - soldier beetles, jewel beetles and glow worms:
Dr K.N.A. Alexander, 22 Cecily Hill, Cirencester, Glos GL7 2EF.

Chrysomelidae and Bruchidae -leaf beetles and pulse beetles:
Dr M.L. Cox, Commonwealth Institute of Entomology, Dept. of Entomology, Natural History Museum of London, Cromwell Road, London SW7 5BD.

Cerambycidae - longhorn beetles:
Dr P.F.G. Twinn, Upper Woodlands, Llanover, Abergavenny, Gwent NP7 9EP.

Curculionoidea (part) - Orthocerous weevils:
Dr P.S. Hyman, Luton Museum, Wardown House, Luton, Bedfordshire LU2 7HA.

Scolytidae - bark beetles:
Mr T.G. Winter, Forestry Commission, Alice Holt Lodge, Wrecclesham, Farnham, Surrey GU10 4LH.

6.2. Beetle collecting.

To identify many species of beetle it is necessary to retain and kill specimens for dissection and as comparative or voucher material. While some conservation workers may find this distasteful, or even wish to call for the outlawing of insect collecting, it must be recognised that the collecting of specimens is necessary adjunct to their study and conservation. The collection of specimens rarely poses a threat to the species concerned and for this reason the NCC has in the past recommended only a very few, exceptionally vulnerable species for legal protection.

Individual populations of insects, even of scarce or threatened species, tend to be large, and the depredations of the entomologist are usually slight when compared with natural mortality through, for example, predation or parasitism. Perhaps the greatest threat to populations of scarce or threatened beetles is habitat destruction or degradation. Provided that sensible precautions are taken to avoid damage to habitats while studying and collecting beetles and other invertebrates, the statutory bodies wish to encourage more entomologists to study beetles and communicate the results of their studies to other entomologists and conservationists. A code of conduct for insect collecting was drawn up by the Joint Committee for the Conservation of British Insects in

INTRODUCTION

1972. An addendum to this, more specific to the methods used for beetle collecting can be found in Key *in* Cooter (1991).

6.3. The future.

There are obviously considerable limitations to the use of rarity alone in the assessment of species and sites for conservation. There is a need to measure the degree of "habitat fidelity" of individual species and identify the recurrent assemblages of species associated with particular habitat conditions in different parts of Britain as has been done for the water beetles (Foster and Eyre in press). This will enable the degree of typicality and representativeness of the faunas of habitats to be assessed, these being important features in identification of conservation sites (Ratcliffe 1977) but very much neglected for the invertebrates. This has barely started but promises a more powerful way to assess site's conservation value and trends in its habitat's "health" with time. There is also a need to develop further the "habitat quality indicator" approach in tandem with the above, particularly in the identification of species associated with historical continuity of habitat conditions. An excellent start has been made in this direction for the mature timber habitat (Harding and Rose 1986, Garland 1983) and a tentative start has been made for old calcareous grasslands (Alexander 1985).

It is the intention of the JNCC to publish updates of both the Red Data Book and the National Reviews of groups of British Invertebrates. It is hoped that future reviews will increasingly cover the above points.

Another way forward that has been adopted by the statutory nature conservation bodies is the drawing up of individual Species Recovery Programmes. To date, Recovery Programmes have been drawn up for the protected species listed on Schedule 5 of the 1981 Wildlife and Countryside Act (Whitten 1990, 1991). Targets, priorities and management proposals are set down to bring individual species back from the brink of extinction. Given the number of species of beetle that are endangered, it is unlikely that this approach will be appropriate for the majority of species. If, however, individual Recovery Programmes are carefully selected, the ecological insights gained in research, habitat management techniques developed, and the sites managed for the small number of species for which programmes are implemented will benefit the conservation of a wider variety of beetles and invertebrates as a whole.

The JNCC's Species Conservation Branch is maintaining data files on all British Red Data Book and Nationally Notable species and would be pleased to receive records and biological information concerning any of the species dealt with here, as well as views on the inclusion, exclusion and grading of any species. We would also strongly encourage entomologists to publish new records and observations on the biology of these species. The biology and habitat requirements of many species are insufficiently known and further information is needed to allow the correct management to be defined and implemented for their conservation. Information can now be accepted in machine-readable form via the biological recording package RECORDER, available from English Nature and designed to facilitate the exchange of biological records. Please address correspondence to: *The Invertebrate Site Register, Joint Nature Conservation Committee, Monkstone House, City Road, Peterborough, PE1 1JY.*

7. ABBREVIATIONS

BRC	-	Biological Records Centre	NNR	-	National Nature Reserve
ISR	-	Invertebrate Site Register	RDB 1	-	Red Data Book category 1 Endangered
JCCBI	-	Joint Committee for the Conservation of British Insects	RDB 2	-	Red Data Book category 2 Vulnerable
JNCC	-	Joint Nature Conservation Committee	RDB 3	-	Red Data Book category 3 Rare
Na	-	Notable A			
Nb	-	Notable B	RDB 5	-	Red Data Book category 5 Endemic
NCC	-	Nature Conservancy Council			

RDB App. - Red Data Book Appendix
            Extinct
RDB I   - Red Data Book category I
            Indeterminate

RDB K   - Red Data Book category K
            Insufficiently Known
SSSI   - Site of Special Scientific Interest

8. ACKNOWLEDGEMENTS

During the compilation of this Review, a large number of entomologists gave a considerable amount of advice and information on candidate species. Special thanks are due to Mr A.A. Allen, Mr P. Hodge, Dr M.G. Morris and Prof. J.A. Owen for the exceptional amount of work that they put into this Review.

Other contributors were:

Dr K.N.A. Alexander, Mr D.M. Appleton, Mr D. Atty, Dr R.G. Booth, Mr S. Bowestead, Mr J.H. Bratton, Mr I. Carter, Mr M. Collier, Mr J. Cooter, Dr M.L. Cox, Mr R. Crossley, Dr R.A. Crowson, Mr M. Darby, Mr M.L. Denton, Mr A.B. Drane, Mr A. Duff, Mr T. Eccles, Mr W.A. Ely, Mr J.H. Flint, Mr P. Follett, Mr A.P. Foster, Dr G.N. Foster, Mr A.P. Fowles, Mr S.P. Garland, Mr P.M. Hammond, Mr T.D. Harrison, Mr N.H. Heal, Mr R. Jones, Dr R.S. Key, Dr P. Kirby, Mr A.H. Kirk-Spriggs, Mr D. Lott, Dr M.L. Luff, Mr C. MacKechnie-Jarvis, Dr M.E.N. Majerus, Mr R.J. Marsh, Mr I. McClenaghan, Mr H. Mendel, Mrs J.M. Morgan, Mr R.K.A. Morris, Mr D.R. Nash, Mrs M. Palmer, Mr J. Parry, Mr M.S. Parsons, Mr E.G. Philp, Dr R.D. Pope, Mr J.R.W. Read, Mr M. Russell, Dr D.B. Shirt, Dr M. Sinclair, Mr P. Skidmore, Mr E. Smith, Mr R.R. Uhthoff-Kaufmann, Dr A.C. Warne, Mr R.C. Welch, Mr P. Whitehead, Mr S.A. Williams, Mr T.G. Winter.

Special thanks are due to Dr S.G. Ball for much computer work on the production of this Review, to Mr A.B. Drane for access to and loan of his extensive collection of literature on the British beetle fauna, to Dr P. Kirby and Ms S.J.J. Lambert for considerable help with the revision of the data sheets, to Mr R.K.A. Morris for editing and production of the manuscript, to Mr C.G.N. Faulkner and Mr G.E. Shreeves for help with data input and last, but not least, the long-suffering typists of the NCC.

9. FURTHER READING

9.1. Checklist.

Kloet G.S and Hinks W.D. (1977).

9.2. General reference guides to beetles.

For many years there was a dearth of general texts on beetles, Linssen (1959) being the most useful but also much sought after as a collectors item and consequently very expensive. Recent texts of continental origin and translated for the British market include Bechyne (1956), Bílý (1990) and Harde and Hammond (1984); the latter is extremely well illustrated and both contain much useful biological information. Among recent continental publications, du Chatenet (1990) deserves mention. This is an excellent guide to about half the European families of beetles and includes maps of the European distribution of a number of British species. It is, however, in French.

9.3. Beetle biology/ecology.

This is an immense subject with a great variety of sources. Perhaps the most useful text in the first instance is Crowson (1981), an expensive work which embraces all aspects of beetle biology from physiology, evolution and classification to behaviour and ecology. This is also an excellent source of references. A useful

INTRODUCTION

guide to techniques which also covers sections on beetle biology is Cooter (1991). Many aspects of invertebrate ecology require more specialist treatment and Unwin and Corbet (1991) a guide to insects, plants and microclimate, is one of a series of publications that provide advice on techniques, apparatus design and construction, and suggestions for projects. An example of a detailed account of one facet of insect ecology is illustrated by Donisthorpe's study of Myrmecophilous insects (1927), a most readable account of insect ecology which provides much useful information on beetle behaviour.

In recent years, the use of beetles in site evaluation and as indicators of little modified habitats has been in the ascendancy. Approaches vary from those of Alexander (1985) whose notes on the beetle fauna of calcareous grasslands were intended to stimulate discussion and feedback, to more detailed works in Stork (1990), covering topics such as the use of ground beetles in site evaluation and the response of ground beetles to habitat degradation and regeneration attempts on peat bogs. These topics are still relatively new and there is much scope for original work. The basis for much of this work must be that of the inventory approach such as Hunter (1977) whose work on the ecology of pinewood beetles includes lists of species associated with this habitat.

9.4. Entomological techniques.

Cooter J. (1991) is a manual of entomological techniques aimed at coleopterists. It also contains sections on beetle biology, conservation, biological recording and food plant relationships.

9.5. Beetle/insect conservation.

There is little doubt that the beetle fauna of Great Britain is changing, although detailed accounts are few. Hammond (1974) illustrates the historical change in the period since the last ice age in the county of Essex. To some extent, this means that the use of indicator species is becoming increasingly important and has led to various attempts to identify threatened assemblages such as those inhabiting the dead wood habitat. Amongst the most useful are accounts by Garland (1983) and Harding and Rose (1986). Practical applications of this concept are most advanced in the use of water beetles for assessing the conservation value of wetlands Foster (1987b) and Foster and Eyre (in press). The importance of overlooked habitats has led to attempts to raise the profile of specific problems such as the conservation value of dead wood (Stubbs 1972), a subject much expanded upon by Speight (1989) who examines this subject in a European context and provides useful suggestions for dead wood management. At a more popular level, the argument against removal of dead wood for use in wood burning stoves is explored by Harding (1981).

Conservation of beetles and, indeed, other insects can be promoted in a variety of ways which may involve a partnership between education and legislation. In the first instance, education of the general public and the entomologist must be the priority. There is ample evidence of changing emphasis within the entomological community, illustrated by the recent publication of Fry and Lonsdale (1991) a theoretical guide to insect conservation, Thomas and Collins (1991) a collection of papers on insect conservation world wide, and the emphasis on conservation in the recently revised Coleopterist's Handbook (Key *in* Cooter 1991). That insect conservation is worthwhile, there can be no doubt, but until now, there has been no text to help the conservation manager. This problem will shortly be addressed by Kirby (in press), a lavishly illustrated handbook on practical ways to conserve insects.

It is unfortunate that insects cannot be conserved without legislation, but this is clearly the case. Levels of legislation vary and European approaches are discussed by Collins (1987). This work is, however, rapidly becoming out of date with the development of the Berne Convention and forthcoming European Community Habitat and Species Directives. Identification of scarce and threatened species is well advanced in Great Britain. Commencing with Shirt (1987), the publication of a series of Reviews (eg. Falk 1991)of which this is one, is an attempt to refine the process. On a World level, Wells, Pyle and Collins (1983) is the first attempt to define the most threatened species but as only seven species of beetles are represented, it barely scratches the surface.

14

The question of insect introductions or reintroductions is a contentious issue and one for which guidelines are very necessary. These can be found in the code issued by the Joint Committee for the Conservation of Insects (JCCBI 1986).

9.6. Identification works on British beetles.

There are two classic works on British beetles, both of which are old and out of print. Fowler (1887-1891), and Fowler and Donisthorpe (1913) remain the most comprehensive general guide to the British beetles but little biological information is given and the keys are difficult to use. Published in six volumes, this is a collectors item and very expensive. More easily obtainable is Joy (1932, reprinted 1976) but in addition to unreliable biological and status comments, the keys include some which are difficult to use. A comprehensive addendum to Joy is in preparation by the British Entomological and Natural History Society.

Identification to family level is possible using alternative keys in Crowson (1954, reprinted 1967), Crowson 1956 and Unwin (1984). None of these keys are foolproof and the former two are taxonomic rather than practical, which makes them difficult to use. The latter is well laid out and includes good text illustrations but is hampered by the inclusion of errors in the keys and exceptions to couplets that key out incorrectly. The best key to families remains that of Joy, although a number of genera are now within different families.

9.7. Keys to particular groups.

In addition to keys to individual families, a number of artificial groupings are favoured by students of beetles or are recognised through economic importance. Amongst these, wood boring species and pests of stored products are of considerable economic importance and are covered by Hickin (1975), Hinton (1945) and Munroe (1966).

The water beetles are a popular grouping with a club (The Balfour-Browne Club) devoted to their study, which publishes regular bulletins. The club's namesake is also the author of a monograph on the group (Balfour-Browne 1940-1958)and a Royal Entomological Society key (Balfour-Browne 1953). More recently, the group has been the subject of an excellent key by Friday (1988). Amongst continental literature, Hansen (1987) and Holmen (1987) are also useful. Holland (1972) also provides a key to the larvae and pupae of the Elminthidae.

9.7.1. Keys to particular families.

If a family is not represented here, then the only keys available are in Joy or Fowler. Only keys to whole families are given here. There are innumerable keys to individual genera which are outside the scope of this work.

Aderidae:	Buck (1954).	Cantharidae:	A tabular key to the larger species by K.N.A. Alexander accompanies the introductory material to the recording scheme (unpublished).
Anobiidae:	Hickin (1975), Buck (1958).		
Anthribidae:	Morris (1990).	Carabidae:	Forsyth (1987), Lindroth (1985 and 1986), Lindroth (1974), Trautner and Geigenmuller (1987).
Anthicidae:	Buck (1954).		
Apionidae:	Morris (1990).	Cerambycidae:	Duffy (1952), Hickin (1987),Hickin (1975).
Attelabidae:	Morris (1990).	Choragidae:	Morris (1990).
Bostrichidae:	Hickin (1975).	Clambidae:	Johnson (1966).
Buprestidae:	Bilý (1982), Levy (1977), Hickin (1975).	Cleridae:	Gerstmeir (1991).
Byturidae:	Hinks (1944).		

INTRODUCTION

Coccinellidae: Majerus and Kearns (1989), Moon (1986), Pope (1953).

Colydiidae: Buck (1955).

Curculionidae: Morris (1991), Morris (1990).

Dermestidae: Hinton (1945), Munroe (1966). (some species).

Dytiscidae: Balfour-Browne (1953), Friday (1988), Hansen (1987), Holland (1972), Holmen (1987).

Elmidae: Balfour-Browne (1953), Friday (1988), Hansen (1987), Holland (1972), Holmen (1987).

Geotrupidae: Jessop (1986).

Gyrinidae: Balfour-Browne (1953), Friday (1988), Hansen (1987), Holland (1972), Holmen (1987).

Haliplidae: Balfour-Browne (1953), Friday (1988), Hansen (1987), Holland (1972), Holmen (1987).

Heteroceridae: Clarke (1973).

"Heteromera": Buck (1954).

Histeridae: Halstead (1963).

Hydraenidae: Balfour-Browne (1953), Friday (1988), Hansen (1987), Holland (1972), Holmen (1987).

Hydrophilidae: Balfour-Browne (1953), Friday (1988), Hansen (1987), Holland (1972), Holmen (1987).

Hygrobiidae: Balfour-Browne (1953), Friday (1988), Hansen (1987), Holland (1972), Holmen (1987).

Lathridiidae: Rucker (1991).

Lucanidae: Jessop (1986).

Lyctidae: Hickin (1960), Hickin (1975).

Lymexylidae: Hickin (1975).

Melandryidae: Buck (1954).

Meloidea: Buck (1954).

Monotomidae: Peacock (1977).

Mordellidae: Buck (1954).

Mycetophagidae: Buck (1955).

Mycteridae: Buck (1954).

Nemonychidae: Morris (1990), Morris (1991).

Noteridae: Balfour-Browne (1953), Friday (1988), Hansen (1987), Holland (1972), Homen (1987).

Oedemeridae: Buck (1954).

Phalacridae: Thompson (1958).

Platypodidae: Duffy (1953), Hickin (1975).

Pselaphidae: Pearce (1957).

Pyrochroidae: Buck (1954).

Pythidae: Buck (1954).

Rhipiphoridae: Buck (1954).

Rhizophagidae: Peacock (1977).

Salpingidae: Buck (1954).

Scarabaeoidea: Britton (1956), Jessop (1986).

Scolytidae: Duffy (1953).

Silphidae: Schawaller (1991).

Sphaeritidae: Halstead (1963).

Staphylinidae: Tottenham (1954), only selected families.

Tenebrionidae: Brendell (1975).

Tetratomidae: Buck (1954).

Trogidae: Jessop (1986).

Urodontidae: Morris (1990).

APPENDIX 1: WATER BEETLE STATUSES

Statuses from Ball (1986) are listed together with the revised statuses, which have been updated by Dr G.N.Foster.

Species	Status in Ball (1986)	Revised status	Species	Status in Ball (1986)	Revised status
Haliplidae			*Ilybius aenescens* Thomson,1870	Nb	Nb
			Ilybius fenestratus (F.,1781)	Nb	Nb
Peltodytes caesus (Duftschmid,1805)	Nb	Nb	*Ilybius guttiger* (Gyllenhal,1808)	Nb	Nb
Haliplus apicalis Thomson,1868	Nb	Nb	*Ilybius subaeneus* Erichson,1837	Nb	Nb
Haliplus furcatus Seidlitz,1887	RDB 1	RDB 1	*Rhantus aberratus*	RDB 1	RDB 1
Haliplus heydeni Wehncke,1875	Nb	Nb	Gemminger & von Harold,1868		
Haliplus laminatus (Schaller,1783)	Nb	Nb	*Rhantus frontalis* (Marsham,1802)	Nb	Nb
Haliplus mucronatus Stephens,1828	RDB 3	Na	*Rhantus grapii* (Gyllenhal,1808)	Nb	Nb
Haliplus variegatus Sturm,1834	RDB 3	RDB 3	*Rhantus suturalis* (MacLeay,1825)	Nb	Nb
Haliplus varius Nicolai,1822	RDB 3	RDB K	*Hydaticus seminiger* (Degeer,1774)	Nb	Nb
			Hydaticus transversalis (Pontoppidan,1763)	Na	RDB 3
Noteridae			*Graphoderus bilineatus* (Degeer,1774)	RDB 1	RDB 1
			Graphoderus cinereus (L.,1758)	RDB 3	RDB 3
Noterus crassicornis (Müller,O.F.,1776)	Nb	Nb	*Graphoderus zonatus* (Hoppe,1795)	RDB 1	RDB 1
			Acilius canaliculatus (Nicolai,1822)	Nb	RDB 3
Dytiscidae			*Dytiscus circumcinctus* Ahrens,1811	Na	Na
			Dytiscus circumflexus F.,1801	Nb	Nb
Laccophilus ponticus Sharp,1882	RDB 2	RDB 2	*Dytiscus dimidiatus* Bergstraesser,1778	Na	RDB 3
Hydrovatus clypealis Sharp,1876	RDB 3	Na	*Dytiscus lapponicus* Gyllenhal,1808	Na	Nb
Hydroglyphus geminus (F.,1792)	-	Nb			
Bidessus minutissimus (Germar,1824)	RDB 3	RDB 3	**Gyrinidae**		
Bidessus unistriatus (Schrank,1781)	RDB 1	RDB 1			
Hygrotus decoratus (Gyllenhal,1810)	Nb	Nb	*Gyrinus aeratus* Stephens,1835	Nb	Nb
Hygrotus quinquelineatus (Zetterstedt,1828)	Nb	Nb	*Gyrinus caspius* Ménétriés,1832	Nb	-
Coelambus nigrolineatus (von Steven,1808)	RDB 3	Na	*Gyrinus distinctus* Aubé,1938	Na	RDB 3
Coelambus novemlineatus (Stephens,1829)	Nb	Nb	*Gyrinus minutus* F.,1798	Nb	Nb
Coelambus parallelogrammus (Ahrens,1812)	Nb	Nb	*Gyrinus natator* (L.,1758)	Na	RDB 1
Hydroporus cantabricus Sharp,1882	Na	RDB 3	*Gyrinus paykulli* Ochs,1927	Na	Na
Hydroporus elongatulus Sturm,1835	RDB 3	RDB 3	*Gyrinus opacus* Sahlberg,C.R.,1819	Na	Na
Hydroporus ferrugineus Stephens,1829	Nb	Nb	*Gyrinus suffriani* Scriba,1855	Na	RDB 3
Hydroporus glabriusculus Aubé,1838	RDB 3	RDB 3	*Gyrinus urinator* Illiger,1807	Nb	Nb
Hydroporus longicornis Sharp,1871	Nb	Nb			
Hydroporus longulus Mulsant,1860	-	Nb	**Hydrophilidae**		
Hydroporus marginatus (Duftschmid,1805)	Na	Nb			
Hydroporus neglectus Schaum,1845	Nb	Nb	*Georissus crenulatus* (Rossi,1794)	Na	Na
Hydroporus obsoletus Aubé,1838	Na	Nb	*Spercheus emarginatus* (Schaller,1783)	RDB 1	RDB 1
Hydroporus rufifrons (Müller,O.F.,1776)	RDB 2	RDB 2	*Hydrochus angustatus* Germar,1824	Nb	Nb
Hydroporus scalesianus Stephens,1828	RDB 2	RDB 2	*Hydrochus brevis* (Herbst,1793)	RDB 3	RDB 3
Stictonectes lepidus (Olivier,1795)	Nb	Nb	*Hydrochus carinatus* Germar,1824	RDB 3	RDB 3
Graptodytes bilineatus (Sturm,1835)	RDB 3	RDB 3	*Hydrochus elongatus* (Schaller,1783)	RDB 3	RDB 3
Graptodytes flavipes (Olivier,1795)	RDB 2	RDB 2	*Hydrochus ignicollis* Motschulsky,1860	RDB 3	RDB 3
Graptodytes granularis (L.,1767)	Nb	Nb	*Hydrochus megaphallus*	-	RDB 2
Deronectes latus (Stephens,1829)	Na	Nb	Berge Henegouwen,1988		
Potamonectes depressus depressus (F.,1775)	Nb	Nb	*Hydrochus nitidicollis* Mulsant,1844	RDB 3	RDB 3
Potamonectes griseostriatus (Degeer,1774)	Nb	Nb	*Helophorus alternans* Gené,1836	Na	Na
Oreodytes alpinus (Paykull,1798)	Na	RDB 3	*Helophorus arvernicus* Mulsant,1846	Nb	Nb
Oreodytes davisi (Curtis,1831)	-	Nb	*Helophorus dorsalis* (Marsham,1802)	RDB 3	Nb
Scarodytes halensis (F.,1787)	Nb	Nb	*Helophorus fulgidicollis* Motschulsky,1860	Nb	Nb
Laccornis oblongus (Stephens,18335)	Na	RDB 3	*Helophorus griseus* Herbst,1793	Nb	Nb
Agabus biguttatus (Olivier,1795)	Nb	Nb	*Helophorus laticollis* Thomson,181868	RDB 2	RDB 2
Agabus brunneus (F.,1798)	RDB 2	RDB 2	*Helophorus longitarsis* Wollaston,1864	RDB 3	RDB 3
Agabus chalconatus (Panzer,1796)	Nb	Nb	*Helophorus nanus* Sturm,1836	Na	Nb
Agabus conspersus (Marsham,1802)	Nb	Nb	*Helophorus strigifrons* Thomson,1868	Nb	Nb
Agabus labiatus (Brahm,1790)	Nb	Nb	*Helophorus tuberculatus* Gyllenhal,1808	RDB 3	RDB 3
Agabus melanarius Aubé,1837	Na	Nb			
Agabus striolatus (Gyllenhal,1808)	RDB 2	RDB 2			
Agabus uliginosus (L.,1761)	Na	Nb			
Agabus undulatus (Schrank,1776)	RDB 2	RDB 3			
Agabus unguicularis (Thomson,1867)	Nb	Nb			

Species	Status in Ball (1986)	Revised status	Species	Status in Ball (1986)	Revised status
Cercyon atricapillus (Marsham,1802)	Nb	Nb	*Ochthebius poweri* Rye,1869	RDB 3	RDB 3
Cercyon bifenestratus Küster,1851	RDB 3	Na	*Ochthebius punctatus* Stephens,1829	Na	Nb
Cercyon convexiusculus Stephens,1829	Nb	Nb	*Ochthebius pusillus* Stephens,1835	Na	RDB 3
Cercyon granarius Erichson, 1837	RDB 3	RDB 3	*Ochthebius viridis* Peyron,1858	Nb	Nb
Cercyon sternalis Sharp,1918	Nb	Nb	*Hydraena minutissima* Stephens,1829	Nb	Nb
Cercyon tristis (Illiger,1801)	Nb	Nb	*Hydraena nigrita* Germar,1824	Nb	Nb
Cercyon ustulatus (Preyssler,1790)	Nb	Nb	*Hydraena palustris* Erichson,1837	RDB 2	RDB 2
Paracymus aeneus (Germar,1824)	RDB 1	RDB 1	*Hydraena pulchella* Germar,1824	Na	RDB 3
Paracymus scutellaris (Rosenhauer,1856)	Nb	Nb	*Hydraena pygmaea* Waterhouse,1833	RDB 3	RDB 3
Limnoxenus niger (Zschach,1788)	Na	Nb	*Hydraena rufipes* Curtis,1830	Na	Nb
Anacaena bipustulata (Marsham,1802)	Nb	Nb	*Hydraena testacea* Curtis,1830	Nb	Nb
Laccobius atratus (Rottenberg,1874)	Nb	Nb	*Limnebius aluta* (Bedel,1881)	Na	RDB 3
Laccobius atrocephalus Reitter,1872	Nb	Nb	*Limnebius crinifer* Rey,1884	RDB 3	RDB I
Laccobius simulatrix d'Orchymont,1932	RDB3	RDB I	*Limnebius nitidus* (Marsham,1802)	Nb	Nb
Laccobius sinuatus Motschulsky,1849	Nb	Nb	*Limnebius papposus* Mulsant,1844	Nb	Nb
Helochares lividus (Forster,1771)	Nb	Nb			
Helochares obscurus (Müller,O.F.)	RDB 3	RDB 3	**Dryopidae**		
Helochares punctatus Sharp,1869	Nb	Nb			
Enochrus affinis (Thunberg,1794)	Nb	Nb	*Pomatinus substriatus* (Müller,P.W.J.,1806)	-	Na
Enochrus bicolor (F.,1792)	Nb	Nb	*Dryops anglicanus* Edwards,1909	RDB 3	RDB 3
Enochrus coarctatus (Gredler,1863)	Nb	-	*Dryops auriculatus* (Fourcroy,1785)	Nb	Nb
Enochrus halophilus (Bedel,1878)	-	Na	*Dryops griseus* (Erichson,1847)	RDB 3	RDB 3
Enochrus isotae Hebauer,1981	RDB 3	RDB 3	*Dryops nitidulus* (Heer,1841)	Na	RDB 3
Enochrus melanocephalus (Olivier,1792)	Nb	Nb	*Dryops similaris* Bollow,1936	Na	RDB 3
Enochrus ochropterus (Marsham,1802)	Nb	Nb	*Dryops striatellus*	Na	RDB 3
Enochrus quadripunctatus (Herbst,1797)	-	Nb	(Fairmaire & Brisout,1859)		
Chaetarthria seminulum (Herbst,1797)	Nb	Nb			
Hydrochara caraboides (L.,1758)	RDB 1	RDB 1	**Elmidae**		
Hydrophilus piceus (L.,1758)	RDB 3	RDB 3			
Berosus affinis Brulle,1835	Nb	Nb	*Macronychus quadrituberculatus*	RDB 3	RDB 3
Berosus luridus (L.,1761)	Nb	Nb	Müller,P.W.J.,1806		
Berosus signaticollis (Charpentier,1825)	Nb	Nb	*Normandia nitens* (Müller,P.W.J.,1817)	RDB 2	RDB 2
Berosus fulvus Kuwert, 1888	RDB 3	RDB 3	*Oulimnius major* (Rey,1889)	RDB 3	Na
			Oulimnius rivularis (Rosenhauer,1856)	Na	Na
Hydraenidae			*Oulimnius troglodytes* (Gyllenhal,1827)	Nb	Nb
			Riolus cupreus (Müller,P.W.J.,1806)	Nb	Nb
Ochthebius aeneus Stephens,1835	RDB 1	RDB 1	*Riolus subviolaceus* (Müller,P.W.J.1817)	Nb	Nb
Ochthebius auriculatus Rey,1885	Nb	Nb	*Stenelmis canaliculata* (Gyllenhal,1808)	RDB 2	RDB 2
Ochthebius bicolon Germar,1824	Nb	Nb			
Ochthebius exaratus Mulsant,1844	Na	RDB 3			
Ochthebius exsculptus Germar,1824	Nb	Nb			
Ochthebius lenensis Poppius,1907	RDB 2	RDB 2			
Ochthebius lejolisi Mulsant & Rey,1861	Nb	Nb			
Ochthebius marinus (Paykull,1798)	Nb	Nb			
Ochthebius nanus Stephens,1829	Nb	Nb			

Note: Terrestrial species of *Cercyon* are covered in part 2 of the Review.

APPENDIX 2: SPECIES LISTED BY STATUS CATEGORY

In this appendix the species are listed in checklist order within status categories.

EXTINCT

Carabidae	*Bembidion octomaculatum* (Goeze,1777)
	Agonum sahlbergi (Chaudoir,1850)
	Diachromus germanus (L.,1758)
	Acupalpus elegans (Dejean,1829)
	Lebia marginata (Fourcroy,1785)
	Lebia scapularis (Fourcroy,1785)
Histeridae	*Saprinus subnitescens* Bickhardt,1909
	Hister quadrinotatus Scriba,1790
	Hister illigeri Duftschmid,1805
Lucanidae	*Platycerus caraboides* (L.,1758)
Scarabaeidae	*Aphodius scrofa* (F.,1787)
	Brindalus porcicollis (Illiger,1803)
	Rhyssemus germanus (L.,1767)
	Pleurophorus caesus (Creutzer in Panzer,1796)
	Onthophagus taurus (Schreber,1759)
	Polyphylla fullo (L.,1758)
Elateridae	*Ampedus sanguineus* (L.,1758)
	Cardiophorus gramineus (Scopoli,1763)
	Cardiophorus ruficollis (L.,1758)
	Selatosomus cruciatus (L.,1758)
Lampyridae	*Lamprohiza splendidula* (L.,1767)
Dermestidae	*Anthrenus pimpinellae* (F.,1775)
	Anthrenus scrophulariae (L.,1758)
Bostrichidae	*Bostrichus capucinus* (L.,1758)
Cleridae	*Tilloidea unifasciatus* (F.,1787)
	Trichodes alvearius (F.,1792)
	Trichodes apiarius (L.,1758)
	Tarsostenus univittatus (Rossi,1792)
Melyridae	*Ebaeus pedicularius* (L.,1758)
Cerylonidae	*Murmidius ovalis* (Beck,1817)
Coccinellidae	*Nephus bisignatus* (Boheman,1850)
Mycetophagidae	*Mycetophagus fulvicollis* F.,1792
Colydiidae	*Oxylaemus cylindricus* (Panzer,1796)
Mycteridae	*Mycterus curculioides* (F.,1781)
Scraptiidae	*Scraptia dubia* (Olivier,1790)
	Anaspis septentrionalis Champion,1891
Meloidae	*Meloe variegatus* Donovan,1793
Cerambycidae	*Leptura virens* L.,1758
	Strangalia attenuata (L.,1758)
	Obrium cantharinum (L.,1767)
	Plagionotus arcuatus (L.,1758)
Chrysomelidae	*Clytra laeviuscula* Ratzeburg,1837
	Cryptocephalus violaceus Laicharting,1781
	Hypocassida subferruginea (Schrank,1776)
Attelabidae	*Rhynchites auratus* (Scopoli,1763)
	Rhynchites bacchus (L.,1758)
Curculionidae	*Coniocleonus hollbergi* (Fahraeus,1842)
	Chromoderus affinis (Schrank,1781)
	Hypera arundinis (Paykull,1792)
	Lepyrus capucinus (Schaller,1783)
	Phloeophagus gracilis (Rosenhauer,1856)
	Bagous petro (Herbst,1795)
	Bagous binodulus (Herbst,1795)
Scolytidae	*Trypophloeus granulatus* (Ratzeburg,1837)

ENDANGERED   RDB 1

Carabidae	*Omophron limbatum* (F.,1777)
	Carabus intricatus L.,1761
	Dyschirius extensus Putzeys,1845
	Trechus subnotatus Dejean,1831
	Bembidion humerale Sturm,1825
	Tachys edmondsi Moore,1956
	Tachys walkerianus Sharp,1913
	Pterostichus aterrimus (Herbst,1784)
	Pterostichus kugelanni (Panzer,1797)
	Agonum quadripunctatum (Degeer,1774)
	Amara fusca Dejean,1828
	Harpalus obscurus (F.,1792)
	Harpalus cupreus Dejean,1829
	Harpalus honestus (Duftschmid,1812)
	Harpalus melancholicus Dejean,1829
	Badister anomalus (Perris,1866)
	Panagaeus cruxmajor (L.,1758)
	Chlaenius nitidulus (Schrank,1781)
	Chlaenius tristis (Schaller,1783)
	Callistus lunatus (F.,1775)
	Lebia cruxminor (L.,1758)
	Lebia cyanocephala (L.,1758)
	Dromius quadrisignatus Dejean,1825
	Cymindis macularis (Fischer von Waldheim,1824)
	Drypta dentata (Rossi,1790)
	Brachinus sclopeta (F.,1792)
Histeridae	*Teretrius fabricii* Mazur,1972
	Paromalus parallelepipedus (Herbst,1792)
	Paralister obscurus (Kugelann,1792)
Silphidae	*Thanatophilus dispar* (Herbst,1793)
	Aclypea undata (Müller,1776)
	Silpha carinata Herbst,1783
Trogidae	*Trox perlatus* Goeze,1777
Scarabaeidae	*Aegialia rufa* (F.,1792)
	Aphodius brevis Erichson,1848
	Aphodius lividus (Olivier,1789)
	Aphodius niger (Panzer,1796)
	Aphodius quadrimaculatus (L.,1761)
	Aphodius subterraneus (L.,1758)
	Euheptaulacus sus (Herbst,1783)
	Heptaulacus testudinarius (F.,1775)
	Copris lunaris (L.,1758)
	Onthophagus nutans (F.,1787)
	Gnorimus variabilis (L.,1758)
Byrrhidae	*Curimopsis nigrita* (Palm,1934)
Buprestidae	*Anthaxia nitidula* (L.,1758)
	Aphanisticus emarginatus (Olivier, 1790)
Elateridae	*Lacon querceus* (Herbst,1784)
	Ampedus nigerrimus (Lacordaire,1835)
	Ampedus ruficeps (Mulsant and Guillebeau,1855)
	Megapenthes lugens (Redtenbacher,1842)
	Melanotus punctolineatus (Pelerin,1829)
	Limoniscus violaceus (Müller,1821)

	Anostirus castaneus (L.,1758)
	Elater ferrugineus L.,1758
	Synaptus filiformis (F.,1781)
Eucnemidae	*Eucnemis capucina* Ahrens,1812
	Hylis cariniceps (Reitter,1902)
Cantharidae	*Malthodes brevicollis* (Paykull,1798)
Lampyridae	*Phosphaenus hemipterus* (Goeze,1777)
Dermestidae	*Globicornis nigripes* (F.,1792)
Anobiidae	*Gastrallus immarginatus* (Müller,1821)
Peltidae	*Ostoma ferrugineum* (L.,1758)
Melyridae	*Hypebaeus flavipes* (F.,1787)
	Axinotarsus pulicarius (F.,1777)
Rhizophagidae	*Rhizophagus oblongicollis* Blatch & Horner,1892
Cucujidae	*Laemophloeus monilis* (F.,1787)
	Leptophloeus clematidis (Erichson,1846)
Silvanidae	*Silvanoprus fagi* (Guerin-Meneville,1844)
Coccinellidae	*Clitostethus arcuatus* (Rossi,1794)
Colydiidae	*Cicones undatus* Guerin-Meneville,1829
	Endophloeus markovichianus (Pill. & Mitt.,1783)
	Teredus cylindricus (Olivier,1790)
Tenebrionidae	*Platydema violaceum* (F.,1790)
	Lagria atripes Mulsant and Guillebeau,1855
	Omophlus rufitarsis (Leske,1785)
Melandryidae	*Abdera affinis* (Paykull,1799)
	Melandrya barbata (F.,1787)
Scraptiidae	*Scraptia fuscula* Müller,1821
Oedemeridae	*Chrysanthia nigricornis* (Westhoff,1881)
Meloidae	*Meloe autumnalis* Olivier,1792
	Meloe brevicollis Panzer,1793
	Meloe cicatricosus Leach,1811
	Apalus muralis (Forster,1771)
Anthicidae	*Anthicus tristis* Schmidt,1842
Cerambycidae	*Acmaeops collaris* (L.,1758)
	Strangalia revestita (L.,1767)
	Lamia textor (L.,1758)
	Oberea oculata (L.,1758)
Bruchidae	*Bruchidius olivaceus* (Germar,1824)
Chrysomelidae	*Oulema erichsoni* (Suffrian,1841)
	Labidostomis tridentata (L.,1758)
	Gynandrophthalma affinis (Illiger,1794)
	Cryptocephalus coryli (L.,1758)
	Cryptocephalus exiguus Schneider,1792
	Cryptocephalus nitidulus F.,1787
	Cryptocephalus primarius Harold,1872
	Bromius obscurus (L.,1758)
	Chrysolina cerealis (L.,1767)
	Chrysomela tremula F.,1787
	Galeruca interrupta Illiger,1802
	Longitarsus aeruginosus (Foudras,1860)
	Longitarsus ferrugineus (Foudras,1860)
	Longitarsus nigerrimus (Gyllenhal,1827)
	Apteropeda splendida Allard,1860
	Dibolia cynoglossi (Koch,1803)
	Psylliodes attenuata (Koch,1803)
	Psylliodes hyoscyami (L.,1758)
	Cassida denticollis Suffrian,1844
Apionidae	*Apion brunnipes* Boheman,1839
Curculionidae	*Otiorhynchus auropunctatus* Gyllenhal,1834
	Cathormiocerus attaphilus Brisout,1880
	Cathormiocerus britannicus Blair,1934
	Sitona gemellatus Gyllenhal,1834
	Lixus algirus (L.,1758)
	Lixus paraplecticus (L.,1758)
	Lixus vilis (Rossi,1790)

	Hypera pastinacae (Rossi,1790)
	Limobius mixtus (Boheman,1834)
	Hylobius transversovittatus (Goeze,1777)
	Dryophthorus corticalis (Paykull,1792)
	Bagous brevis Gyllenhal,1836
	Bagous czwalinai Seidlitz,1891
	Bagous diglyptus Boheman,1845
	Bagous longitarsis Thomson,1868
	Bagous nodulosus Gyllenhal,1836
	Bagous lutosus (Gyllenhal,1813)
	Bagous puncticollis Boheman,1854
	Pachytychius haematocephalus (Gyllenhal,1836)
	Ceutorhynchus insularis Dieckmann,1971
	Ceutorhynchus syrites Germar,1824
	Rhinoncus albicinctus Gyllenhal,1837
Scolytidae	*Triotemnus coryli* (Perris,1855)
	Ernoporus caucasicus Lindemann,1876
	Ernoporus tiliae (Panzer,1793)

VULNERABLE   RDB 2

Carabidae	*Cicindela hybrida* L.,1758
	Dyschirius obscurus (Gyllenhal,1827)
	Harpalus froelichi Sturm,1818
	Polistichus connexus (Fourcroy,1785)
Silphidae	*Silpha obscura* L.,1758
Scarabaeidae	*Diastictus vulneratus* (Sturm,1805)
	Gnorimus nobilis (L.,1758)
Buprestidae	*Trachys minuta* (L.,1758)
Elateridae	*Ampedus cardinalis* (Schiödte,1865)
	Ampedus rufipennis (Stephens,1830)
	Ampedus tristis (L.,1758)
	Negastrius pulchellus (L.,1761)
	Negastrius sabulicola (Boheman,1853)
	Cardiophorus erichsoni du Buysson,1901
	Dicronychus equiseti (Herbst,1784)
Anobiidae	*Xyletinus longitarsis* Jansson,1942
Lymexylidae	*Lymexylon navale* (L.,1758)
Coccinellidae	*Nephus quadrimaculatus* (Herbst,1783)
Endomychidae	*Lycoperdina succincta* (L.,1767)
Tenebrionidae	*Diaperis boleti* (L.,1758)
	Prionychus melanarius (Germar,1813)
Melandryidae	*Hypulus quercinus* (Quensel,1790)
Oedemeridae	*Ischnomera cinerascens* (Pandelle,1867)
	Oedemera virescens (L.,1767)
Aderidae	*Aderus brevicornis* (Perris,1869)
Cerambycidae	*Gracilia minuta* (F.,1781)
	Pyrrhidium sanguineum (L.,1758)
Chrysomelidae	*Donacia bicolora* Zschach,1788
	Zeugophora flavicollis (Marsham,1802)
	Cryptocephalus biguttatus (Scopoli,1763)
	Cryptocephalus decemmaculatus (L.,1758)
	Cryptocephalus querceti Suffrian,1848
	Cryptocephalus sexpunctatus (L.,1758)
	Chrysolina crassicornis (Helliesin,1911)
	Psylliodes luridipennis Kutschera,1864
Anthribidae	*Tropideres sepicola* (F.,1792)
	Tropideres niveirostris (F.,1798)
Curculionidae	*Otiorhynchus ligustici* (L.,1758)
	Cathormiocerus socius Boheman,1843
	Polydrusus marginatus Stephens,1831

Strophosoma fulvicorne Walton,J.,1846
Liparus germanus (L.,1758)
Leiosoma pyrenaeum Brisout,1866
Anchonidium unguiculare (Aubé,1850)
Bagous cylindrus (Paykull,1800)
Bagous argillaceus Gyllenhal,1836
Dorytomus affinis (Paykull,1800)
Ceutorhynchus pilosellus Gyllenhal,1837
Ceutorhynchus querceti (Gyllenhal,1813)
Baris analis (Olivier,1790)
Tychius quinquepunctatus (L.,1758)
Rhynchaenus testaceus (Müller,1776)

RARE   RDB 3

Carabidae	*Cicindela germanica* L.,1758
	Pelophila borealis (Paykull,1790)
	Dyschirius angustatus (Ahrens,1830)
	Trechus rivularis (Gyllenhal,1810)
	Bembidion virens Gyllenhal,1827
	Pogonus luridipennis (Germar,1822)
	Amara alpina (Paykull,1790)
	Amara famelica Zimmerman,1832
	Amara strenua Zimmermann,1832
	Harpalus cordatus (Duftschmid,1812)
	Harpalus parallelus Dejean,1829
	Harpalus puncticollis (Paykull,1798)
	Harpalus sabulicola (Panzer,1796)
	Anisodactylus poeciloides (Stephens,1828)
	Dromius vectensis Rye,1872
	Lionychus quadrillum (Duftschmid,1812)
Sphaeritidae	*Sphaerites glabratus* (F.,1792)
Histeridae	*Aeletes atomarius* (Aubé,1842)
	Acritus homoeopathicus Wollaston,1857
	Hypocaccus metallicus (Herbst,1792)
Scarabaeidae	*Aphodius consputus* Creutzer,1799
Psephenidae	*Eubria palustris* Germar,1818
Heteroceridae	*Heterocerus hispidulus* Kiesenwetter,1843
Elateridae	*Ampedus cinnabarinus* (Eschscholtz,1829)
	Procraerus tibialis (Boisdvl. & Lacordaire,1835)
	Athous subfuscus (Müller,1764)
	Selatosomus angustulus (Kiesenwetter,1858)
	Selatosomus nigricornis (Panzer,1799)
	Agriotes sordidus (Illiger,1807)
	Adrastus rachifer (Fourcroy,1785)
Throscidae	*Aulonothroscus brevicollis* (Bonvouloir,1859)
	Trixagus elateroides (Heer,1841)
Eucnemidae	*Dirhagus pygmaeus* (F.,1792)
	Hylis olexai (Palm,1955)
Cantharidae	*Cantharis fusca* L.,1758
	Malthodes crassicornis (Mäklin,1846)
Dermestidae	*Trinodes hirtus* (F.,1781)
Anobiidae	*Dryophilus anobioides* Chevrolat,1832
	Caenocara bovistae (Hoffmann,1803)
Ptinidae	*Ptinus lichenum* Marsham,1802
Trogossitidae	*Nemozoma elongatum* (L.,1761)
Cleridae	*Thanasimus rufipes* (Brahm,1797)
Melyridae	*Malachius aeneus* (L.,1758)
	Malachius barnevillei Puton,1865
	Malachius vulneratus Abeille,1891
Rhizophagidae	*Rhizophagus parvulus* (Paykull,1800)

	Monotoma angusticollis (Gyllenhal,1827)
Erotylidae	*Triplax lacordairii* Crotch,1870
	Triplax scutellaris Charpentier,1825
Corylophidae	*Orthoperus brunnipes* (Gyllenhal,1808)
Coccinellidae	*Coccinella quinquepunctata* L.,1758
Endomychidae	*Lycoperdina bovistae* (F.,1792)
Cisidae	*Rhopalodontus perforatus* (Gyllenhal,1813)
	Cis coluber Abeille,1874
	Cis dentatus Mellié,1848
Colydiidae	*Myrmechixenus vaporariorum* Geur.-Men.,1843
	Synchita separanda (Reitter,1882)
	Langelandia anophthalma Aubé,1843
	Colydium elongatum (F.,1787)
	Oxylaemus variolosus (Dufour,1843)
Tenebrionidae	*Bolitophagus reticulatus* (L.,1767)
	Corticeus unicolor Piller and Mitterpacher,1783
Melandryidae	*Osphya bipunctata* (F.,1775)
Scraptiidae	*Scraptia testacea* Allen,1940
Oedemeridae	*Ischnomera caerulea* (L.,1758)
Meloidae	*Meloe rugosus* Marsham,1802
Anthicidae	*Anthicus scoticus* Rye,1872
Cerambycidae	*Grammoptera ustulata* (Schaller,1783)
	Leptura fulva Degeer,1775
	Leptura sanguinolenta L.,1761
	Leptura sexguttata F.,1775
	Mesosa nebulosa (F.,1781)
Chrysomelidae	*Macroplea appendiculata* (Panzer,1794)
	Donacia aquatica (L.,1758)
	Hydrothassa hannoveriana (F.,1775)
	Phyllodecta polaris Schneider,1886
	Ochrosis ventralis (Illiger,1807)
	Psylliodes sophiae Heikertinger,1914
Attelabidae	*Rhynchites pauxillus* Germar,1824
	Byctiscus populi (L.,1758)
Apionidae	*Apion rubiginosum* Grill,1893
	Apion dispar Germar,1817
	Apion minimum Herbst,1797
	Nanophyes gracilis Redtenbacher,1849
Curculionidae	*Cathormiocerus maritimus* Rye,1874
	Cathormiocerus myrmecophilus (Seidlitz,1868)
	Brachysomus hirtus (Boheman,1845)
	Coniocleonus nebulosus (L.,1758)
	Hypera diversipunctata (Schrank,1798)
	Pissodes validirostris (Sahlberg,1834)
	Bagous collignensis (Herbst,1797)
	Bagous frit (Herbst,1795)
	Smicronyx coecus (Reich,1797)
	Smicronyx reichi (Gyllenhal,1836)
	Ceutorhynchus parvulus Brisout,1869
	Ceutorhynchus unguicularis Thomson,1871
	Ceutorhynchus urticae Boheman,1845
	Ceutorhynchus verrucatus Gyllenhal,1837
	Phytobius olssoni Israelson,1972
	Baris scolopacea Germar,1818
	Anthonomus rufus Gyllenhal,1836
	Miarus micros (Germar,1821)
Scolytidae	*Tomicus minor* (Hartig,1834)
	Pityophthorus lichtensteini (Ratzeburg,1837)

APPENDIX 2

INDETERMINATE · RDB I

Carabidae	*Bradycellus csikii* Laczó,1912
	Badister meridionalis Puel,1925
Histeridae	*Hetaerius ferrugineus* (Olivier,1789)
Scaphidiidae	*Scaphisoma assimile* Erichson,1845
Scirtidae	*Elodes elongata* Tournier,1868
Byrrhidae	*Simplocaria maculosa* Erichson,1847
Lycidae	*Platycis cosnardi* (Chevrolat,1829)
Anobiidae	*Hemicoelus nitidus* (Herbst,1793)
	Caenocara affinis (Sturm,1837)
Hypocopridae	*Hypocoprus latridioides* Motschulsky,1839
Coccinellidae	*Exochomus nigromaculatus* (Goeze,1777)
Colydiidae	*Myrmechixenus subterraneus* Chevrolat,1835
Scraptiidae	*Anaspis schilskyana* Csiki,1915
Chrysomelidae	*Chaetocnema conducta* (Motschulsky,1838)
	Cassida nebulosa L.,1758
Apionidae	*Apion laevigatum* (Paykull,1792)
Curculionidae	*Otiorhynchus morio* (F.,1781)
	Barypeithes pyrenaeus Seidlitz,1868
	Procas armillatus (F.,1801)
	Procas granulicollis Walton,1848
	Ceutorhynchus arquatus (Herbst,1795)
	Rhynchaenus decoratus (Germar,1821)
Platypodidae	*Platypus parallelus* (F.,1801)

INSUFFICIENTLY KNOWN · RDB K

Carabidae	*Bembidion argenteolum* Ahrens,1812
Histeridae	*Halacritus punctum* (Aubé,1842)
	Saprinus virescens (Paykull,1798)
	Onthophilus punctatus (Müller,1776)
	Hister quadrimaculatus L.,1758
	Epierus comptus Erichson,1834
Scaphidiidae	*Scaphium immaculatum* (Olivier,1790)
Scarabaeidae	*Onthophagus fracticornis* (Preyssler,1790)
	Melolontha hippocastani F.,1801
Clambidae	*Clambus pallidulus* Reitter,1911
Melyridae	*Dasytes coeruleus* (Degeer,1774)
	Sphinginus lobatus (Olivier,1790)
Phalacridae	*Olibrus flavicornis* (Sturm,1807)
	Stilbus atomarius (L.,1767)
Cerylonidae	*Anommatus diecki* Reitter,1875
Corylophidae	*Orthoperus aequalis* Sharp,1885
	Rypobius ruficollis (du Val,1854)
Coccinellidae	*Hippodamia tredecimpunctata* (L.,1758)
	Vibidia duodecimguttata (Poda,1761)
Scraptiidae	*Anaspis bohemica* Schilsky,1898
	Anaspis melanostoma Costa,1854
Mordellidae	*Mordella holomelaena* Apfelbeck,1914
	Mordella leucaspis Kuester,1849
	Mordellistena brevicauda (Boheman,1849)
	Mordellistena humeralis (L.,1758)
	Mordellistena neuwaldeggiana (Panzer,1796)
	Mordellistena parvula (Gyllenhal,1827)
	Mordellistena acuticollis Schilsky,1895
	Mordellistena nanuloides Ermisch,1967
	Mordellistena pseudopumila Ermisch,1962

	Mordellistena parvuloides Ermisch,1956
Chrysomelidae	*Agelastica alni* (L.,1758)
	Longitarsus bearei Kevan,1967
	Longitarsus longiseta Weise,1889
	Chaetocnema aerosa (Letzner,1847)
	Psylliodes luteola (Müller,1776)
Apionidae	*Apion lemoroi* Brisout,1880
	Apion armatum Gerstaecker,1854
Curculionidae	*Sitona cinerascens* (Fahraeus,1840)
	Sitona puberulus Reitter,1903
	Lixus scabricollis Boheman,1843
	Hypera ononidis Chevrolat,1863
	Gronops inaequalis Boheman,1842
	Bagous arduus Sharp,1917
	Dorytomus majalis (Paykull,1792)
	Ceutorhynchus hepaticus Gyllenhal,1837
	Ceutorhynchus moelleri Thomson,1868
	Ceutorhynchus thomsoni Kolbe,1900
	Anthonomus chevrolati Desbrochers,1868
	Anthonomus humeralis (Panzer,1795)
	Anthonomus piri Kollar,1837
	Tychius crassirostris Kirsch,1871
	Tychius polylineatus (Germar,1824)
	Miarus plantarum (Germar,1824)
	Rhynchaenus calceatus Germar,1821
	Rhynchaenus populi (F.,1792)
	Rhynchaenus pseudostigma Temperé,1982

NOTABLE A · Na

Carabidae	*Cicindela sylvatica* L.,1758
	Carabus clatratus L.,1761
	Calosoma inquisitor (L.,1758)
	Leistus montanus Stephens,1827
	Nebria complanata (L.,1767)
	Nebria livida (L.,1758)
	Nebria nivalis (Paykull,1790)
	Elaphrus lapponicus Gyllenhal,1810
	Dyschirius nitidus (Dejean,1825)
	Perileptus areolatus (Creutzer,1799)
	Thalassophilus longicornis (Sturm,1825)
	Bembidion nigropiceum (Marsham,1802)
	Bembidion semipunctatum Donovan,1806
	Bembidion ephippium (Marsham,1802)
	Bembidion schueppeli Dejean,1831
	Tachys micros (Fischer von Waldheim,1828)
	Tachys scutellaris Stephens,1828
	Agonum gracilipes (Duftschmid,1812)
	Agonum scitulum Dejean,1828
	Agonum sexpunctatum (L.,1758)
	Amara infima (Duftschmid,1812)
	Amara nitida Sturm,1825
	Amara quenseli (Schoenherr,1806)
	Zabrus tenebrioides (Goeze,1777)
	Harpalus melleti Heer,1837
	Harpalus punctatulus (Duftschmid,1812)
	Harpalus dimidiatus (Rossi,1790)
	Harpalus quadripunctatus Dejean,1829
	Harpalus tenebrosus Dejean,1829
	Harpalus vernalis (Duftschmid,1812)
	Anisodactylus nemorivagus (Duftschmid,1812)

Bradycellus distinctus (Dejean,1829)
Stenolophus skrimshiranus Stephens,1828
Acupalpus brunnipes (Sturm,1825)
Acupalpus flavicollis (Sturm,1825)
Licinus punctatulus (F.,1792)
Badister peltatus (Panzer,1797)
Masoreus wetterhalli (Gyllenhal,1813)
Dromius longiceps Dejean,1826
Dromius sigma (Rossi,1790)
Cymindis axillaris (F.,1794)

Histeridae Abraeus granulum Erichson,1839
Gnathoncus buyssoni Auzat,1917
Hypocaccus rugiceps (Duftschmid,1805)

Silphidae Nicrophorus vestigator Herschel,1807
Aclypea opaca (L.,1758)

Trogidae Trox sabulosus (L.,1758)

Geotrupidae Odonteus armiger (Scopoli,1772)
Geotrupes pyrenaeus (Charpentier,1825)

Scarabaeidae Aphodius nemoralis Erichson,1848
Aphodius sordidus (F.,1775)
Euheptaulacus villosus Gyllenhal,1806
Psammodius asper (F.,1775)
Onthophagus nuchicornis (L.,1758)
Amphimallon ochraceus (Knoch,1801)

Clambidae Clambus nigrellus Reitter,1914

Scirtidae Cyphon kongsbergensis Munster,1924
Cyphon punctipennis Sharp,1872
Scirtes orbicularis (Panzer,1793)

Byrrhidae Curimopsis setigera (Illiger,1798)

Limnichidae Limnichus pygmaeus (Sturm,1807)

Buprestidae Agrilus pannonicus (Piller and Mitterpacher,1783)
Agrilus sinuatus (Olivier,1790)
Agrilus viridis (L., 1758)
Trachys scrobiculatus Kiesenwetter,1857

Elateridae Ampedus elongantulus (F.,1787)
Ampedus sanguinolentus (Schrank,1776)
Ischnodes sanguinicollis (Panzer,1793)
Fleutiauxellus maritimus (Curtis,1840)
Fleutiauxellus quadripustulatus (F.,1792)
Ctenicera pectinicornis (L.,1758)

Drilidae Drilus flavescens (Fourcroy,1785)

Cantharidae Rhagonycha elongata (Fallén,1807)

Lycidae Pyropterus nigroruber (Degeer,1774)

Anobiidae Dorcatoma dresdensis Herbst,1792
Dorcatoma serra Panzer,1795

Ptinidae Ptinus palliatus Perris,1847

Melyridae Aplocnemus nigricornis (F.,1792)
Dasytes niger (L.,1761)
Cerapheles terminatus (Ménétriés,1832)

Rhizophagidae Rhizophagus picipes (Olivier,1790)
Cyanostolus aeneus (Richter,1820)

Cucujidae Uleiota planata (L.,1761)
Pediacus depressus (Herbst,1797)
Cryptolestes spartii (Curtis,1834)
Notolaemus unifasciatus (Latreille,1804)

Erotylidae Tritoma bipustulata F.,1775

Phalacridae Phalacrus brunnipes Brisout,1863

Cerylonidae Anommatus duodecimstriatus (Müller,1821)

Coccinellidae Platynaspis luteorubra (Goeze,1777)
Coccinella magnifica Redtenbacher,1843

Mycetophagidae Mycetophagus populi F.,1798
Mycetophagus quadriguttatus Müller,1821

Colydiidae Cicones variegatus (Hellwig,1792)
Aulonium trisulcum (Fourcroy,1785)

Tenebrionidae Mycetochara humeralis (F.,1787)

Tetratomidae Tetratoma desmaresti Latreille,1807

Salpingidae Lissodema cursor (Gyllenhal,1813)
Rabocerus foveolatus (Ljungh,1824)

Pythidae Pytho depressus (L.,1767)

Pyrochroidae Schizotus pectinicornis (L.,1758)

Melandryidae Anisoxya fuscula (Illiger,1798)
Abdera quadrifasciata (Curtis,1829)
Abdera triguttata (Gyllenhal,1810)
Xylita laevigata (Hellenius,1786)

Scraptiidae Anaspis thoracica (L.,1758)

Mordellidae Tomoxia bucephala Costa,1854

Anthicidae Anthicus bimaculatus (Illiger,1801)
Anthicus salinus Crotch,1866

Cerambycidae Prionus coriarius (L.,1758)
Grammoptera variegata (Germar,1824)
Leptura scutellata F.,1781
Judolia sexmaculata (L.,1758)
Strangalia aurulenta (F.,1792)
Strangalia nigra (L.,1758)
Molorchus umbellatarum (von Schreber,1759)
Saperda carcharias (L.,1758)
Saperda scalaris (L.,1758)

Chrysomelidae Macroplea mutica (F.,1792)
Donacia dentata Hoppe,1795
Donacia impressa Paykull,1799
Donacia obscura Gyllenhal, 1813
Donacia sparganii Ahrens,1810
Plateumaris braccata (Scopoli, 1772)
Zeugophora turneri Power,1863
Cryptocephalus frontalis Marsham,1802
Cryptocephalus punctiger Paykull,1799
Chrysolina graminis (L.,1758)
Chrysolina marginata (L.,1758)
Chrysolina sanguinolenta (L.,1758)
Calomicrus circumfusus (Marsham,1802)
Phyllotreta vittata (F.,1801)
Aphthona nigriceps (Redtenbacher,1842)
Longitarsus absynthii Kutschera,1862
Longitarsus agilis (Rye,1868)
Longitarsus curtus (Allard,1860)
Longitarsus fowleri Allen,1967
Longitarsus ganglbaueri Heikertinger,1912
Longitarsus nigrofasciatus (Goeze,1777)
Longitarsus parvulus (Paykull,1799)
Longitarsus quadriguttatus (Pontoppidan,1763)
Longitarsus rutilus (Illiger,1807)
Altica brevicollis Foudras,1860
Crepidodera impressa (F.,1801)
Podagrica fuscipes (F.,1775)
Mantura chrysanthemi (Koch,1803)
Chaetocnema sahlbergi (Gyllenhal,1827)
Pilemostoma fastuosa (Schaller,1783)
Cassida hemisphaerica Herbst,1799

Anthribidae Anthribus fasciatus Forster,1771
Choragus sheppardi Kirby,1819

Attelabidae Rhynchites olivaceus Gyllenhal,1833

Apionidae Apion affine Kirby,1808
Apion soror Rey,1895
Apion semivittatum Gyllenhal,1833
Apion difficile Herbst,1797
Apion genistae Kirby,1811
Apion cineraceum Wencker,1864
Apion flavimanum Gyllenhal,1833

APPENDIX 2

Curculionidae

Apion curtisi Stephens,1831
Apion astragali (Paykull,1800)
Apion intermedium Eppelsheim,1875
Apion reflexum Gyllenhal,1833
Apion laevicolle Kirby,1811
Apion schoenherri Boheman,1839
Trachyphloeus digitalis (Gyllenhal,1827)
Trachyphloeus laticollis Boheman,1843
Omiamima mollina (Boheman,1834)
Polydrusus sericeus (Schaller,1783)
Tropiphorus obtusus (Bonsdorff,1785)
Rhinocyllus conicus (Froelich,1792)
Hypera meles (F.,1792)
Limobius borealis (Paykull,1792)
Cionus longicollis Brisout,1863
Cionus nigritarsis Reitter,1904
Plinthus caliginosus (F.,1775)
Magdalis barbicornis (Latreille,1804)
Magdalis duplicata Germar,1818
Magdalis phlegmatica (Herbst,1797)
Cossonus linearis (F.,1775)
Phloeophagus truncorum (Germar,1824)
Bagous lutulosus (Gyllenhal,1827)
Bagous subcarinatus Gyllenhal,1836
Dorytomus hirtipennis Bedel,1884
Dorytomus salicis Walton,1851
Notaris aethiops (F.,1792)
Pseudostyphlus pillumus (Gyllenhal,1836)
Mononychus punctumalbum (Herbst,1784)
Rutidosoma globulus (Herbst,1795)
Coeliodes nigritarsis Hartmann,1895
Trichosirocalus horridus (Panzer,1801)
Trichosirocalus rufulus (Dufour,1851)
Ceutorhynchus angulosus Boheman,1845
Ceutorhynchus atomus Boheman,1845
Ceutorhynchus euphorbiae Brisout,1866
Ceutorhynchus pectoralis Weise,1895
Ceutorhynchus pervicax Weise,1883
Ceutorhynchus pulvinatus Gyllenhal,1837
Ceutorhynchus pumilio (Gyllenhal,1827)
Ceutorhynchus quercicola (Paykull,1792)
Phytobius muricatus Brisout,1867
Phytobius quadricorniger Colonnelli,1986
Phytobius quadrinodosus (Gyllenhal,1813)
Tapinotus sellatus (F.,1794)
Baris laticollis (Marsham,1802)
Baris lepidii Germar,1824
Tychius lineatulus Stephens,1831
Tychius parallelus (Panzer,1794)
Tychius tibialis Boheman,1843
Sibinia sodalis Germar,1824
Mecinus janthinus Germar,1821
Gymnetron beccabungae (L.,1761)
Gymnetron collinum (Gyllenhal,1813)
Gymnetron linariae (Panzer,1795)
Gymnetron rostellum (Herbst,1795)
Rhynchaenus foliorum (Müller,1764)

Scolytidae

Dryocoetinus alni (Georg,1856)
Taphrorychus bicolor (Herbst,1793)
Ernoporus fagi (F.,1798)
Trypophloeus asperatus (Gyllenhal,1813)
Pityogenes quadridens (Hartig,1834)
Pityogenes trepanatus (Nördlinger,1848)

NOTABLE B   Nb

Carabidae

Cicindela maritima Latreille et Dejean,1822
Carabus monilis F.,1792
Carabus nitens L.,1758
Notiophilus aesthuans (Motschulsky,1864)
Notiophilus quadripunctatus Dejean,1826
Blethisa multipunctata (L.,1758)
Elaphrus uliginosus F.,1792
Dyschirius impunctipennis Dawson,1854
Miscodera arctica (Paykull,1798)
Patrobus septentrionis Dejean,1828
Aepus marinus (Ström,1783)
Aepus robini (Laboulbène,1849)
Trechus fulvus Dejean,1831
Trechus rubens (F.,1792)
Trechus discus (F.,1792)
Asaphidion pallipes (Duftschmid,1812)
Bembidion litorale (Olivier,1790)
Bembidion nigricorne Gyllenhal,1827
Bembidion bipunctatum (L.,1761)
Bembidion pallidipenne (Illiger,1802)
Bembidion stomoides Dejean,1831
Bembidion obliquum Sturm,1825
Bembidion fluviatile Dejean,1831
Bembidion lunatum (Duftschmid,1812)
Bembidion monticola Sturm,1825
Bembidion saxatile Gyllenhal,1827
Bembidion testaceum (Duftschmid,1812)
Bembidion laterale (Samouelle,1819)
Bembidion quadripustulatum Serville,1821
Bembidion gilvipes Sturm,1825
Bembidion clarki Dawson,1849
Bembidion fumigatum (Duftschmid,1812)
Tachys bistriatus (Duftschmid,1812)
Tachys parvulus (Dejean,1831)
Pogonus littoralis (Duftschmid,1812)
Pterostichus aethiops (Panzer,1796)
Pterostichus angustatus (Duftschmid,1812)
Pterostichus anthracinus (Panzer,1795)
Pterostichus cristatus (Dufour,1820)
Pterostichus gracilis (Dejean,1828)
Pterostichus lepidus (Leske,1785)
Pterostichus longicollis (Duftschmid,1812)
Pterostichus oblongopunctatus (F.,1787)
Calathus ambiguus (Paykull,1790)
Platyderus ruficollis (Marsam,1802)
Agonum ericeti (Panzer,1809)
Agonum livens (Gyllenhal,1810)
Agonum nigrum Dejean,1828
Agonum versutum Sturm,1824
Amara consularis (Duftschmid,1812)
Amara curta Dejean,1828
Amara equestris (Duftschmid,1812)
Amara fulva (Müller,1776)
Amara lucida (Duftschmid,1812)
Amara praetermissa (Sahlberg,1827)
Amara spreta Dejean,1831
Harpalus ardosiacus Lutschnik,1922
Harpalus azureus (F.,1775)
Harpalus rupicola Sturm,1818

	Harpalus schaubergerianus Puel,1937
	Harpalus serripes (Quensel in Schoenherr,1806)
	Harpalus servus (Duftschmid,1812)
	Harpalus smaragdinus (Duftschmid,1812)
	Dicheirotrichus obsoletus (Dejean,1829)
	Stenolophus teutonus (Schrank,1781)
	Acupalpus consputus (Duftschmid,1812)
	Acupalpus exiguus Dejean,1829
	Licinus depressus (Paykull,1790)
	Badister dilatatus Chaudoir,1837
	Badister unipustulatus Bonelli,1813
	Panagaeus bipustulatus (F.,1775)
	Chlaenius nigricornis (F.,1787)
	Oodes helopioides (F.,1792)
	Odacantha melanura (L.,1767)
	Lebia chlorocephala (Hoffmannsegg,1803)
	Demetrias imperialis (Germar,1824)
	Demetrias monostigma Samouelle,1819
	Cymindis vaporariorum (L.,1758)
	Brachinus crepitans (L.,1758)
Histeridae	*Plegaderus dissectus* Erichson,1839
	Saprinus cuspidatus Ihssen,1949
	Saprinus immundus (Gyllenhal,1827)
	Hypocaccus rugifrons (Paykull,1798)
	Baeckmanniolus dimidiatus (Illiger,1807)
	Myrmetes piceus (Paykull,1809)
	Hister bissexstriatus F.,1801
	Grammostethus marginatus (Erichson,1834)
Silphidae	*Nicrophorus interruptus* Stephens,1830
	Dendroxena quadrimaculata (Scopoli,1772)
	Silpha tyrolensis Laicharting,1781
Scaphidiidae	*Scaphisoma boleti* (Panzer,1793)
Lucanidae	*Lucanus cervus* (L., 1758)
Geotrupidae	*Geotrupes mutator* (Marsham,1802)
	Geotrupes vernalis (L.,1758)
Scarabaeidae	*Aegialia sabuleti* (Panzer,1796)
	Aphodius coenosus (Panzer,1798)
	Aphodius conspurcatus (L.,1758)
	Aphodius distinctus (Müller,1776)
	Aphodius fasciatus (Olivier,1789)
	Aphodius paykulli Bedel,1908
	Aphodius plagiatus (L.,1767)
	Aphodius porcus (F.,1792)
	Aphodius putridus (Fourcroy,1785)
	Aphodius zenkeri Germar,1813
	Onthophagus vacca (L.,1767)
	Omaloplia ruricola (F.,1775)
	Cetonia cuprea F.,1775
Scirtidae	*Cyphon pubescens* (F.,1792)
	Prionocyphon serricornis (Müller,1821)
	Hydrocyphon deflexicollis (Müller,1821)
Byrrhidae	*Byrrhus arietinus* Steffahny,1842
	Porcinolus murinus (F.,1794)
Heteroceridae	*Heterocerus fusculus* Kiesenwetter,1843
Buprestidae	*Agrilus angustulus* (Illiger,1803)
	Agrilus laticornis (Illiger,1803)
	Aphanisticus pusillus (Olivier,1790)
Elateridae	*Ampedus nigrinus* (Herbst,1784)
	Ampedus quercicola (du Buysson, 1887)
	Ampedus pomorum (Herbst,1784)
	Cardiophorus asellus Erichson,1840
	Harminius undulatus (Degeer,1774)
	Athous campyloides Newman,1833
	Selatosomus bipustulatus (L.,1767)

	Selatosomus impressus (F.,1792)
Eucnemidae	*Melasis buprestoides* (L.,1761)
Cantharidae	*Ancistronycha abdominalis* (F.,1798)
	Cantharis obscura L.,1758
	Rhagonycha lutea (Müller,1764)
	Rhagonycha translucida (Krynicki,1832)
	Silis ruficollis (F.,1775)
	Malthinus balteatus Suffrian,1851
	Malthinus frontalis (Marsham,1802)
	Malthodes fibulatus Kiesenwetter,1852
	Malthodes guttifer Kiesenwetter,1852
	Malthodes maurus (Castelnau,1840)
Lycidae	*Dictyoptera aurora* (Herbst,1784)
	Platycis minuta (F.,1787)
Dermestidae	*Megatoma undata* (L.,1758)
	Ctesias serra (F.,1792)
Anobiidae	*Ptinomorphus imperialis* (L.,1767)
	Anobium inexspectatum Lohse,1954
	Hadrobregmus denticollis (Crtz. in Panzer,1796)
	Dorcatoma flavicornis (F.,1792)
	Anitys rubens (Hoffmann,1803)
Ptinidae	*Ptinus sexpunctatus* Panzer,1792
	Ptinus subpilosus Sturm,1837
Lyctidae	*Lyctus linearis* (Goeze,1777)
Phloiophilidae	*Phloiophilus edwardsi* Stephens,1830
Peltidae	*Thymalus limbatus* (F.,1787)
Cleridae	*Tillus elongatus* (L.,1758)
	Opilo mollis (L.,1758)
	Korynetes caeruleus (Degeer,1775)
Melyridae	*Aplocnemus pini* (Redtenbacher,1849)
	Dasytes plumbeus (Müller,1776)
	Dasytes puncticollis Reitter,1888
	Dolichosoma lineare (Rossi,1794)
	Malachius marginellus Olivier,1790
Lymexylidae	*Hylecoetus dermestoides* (L.,1761)
Rhizophagidae	*Rhizophagus nitidulus* (F.,1798)
Sphindidae	*Sphindus dubius* (Gyllenhal,1808)
Cucujidae	*Dendrophagus crenatus* (Paykull,1799)
Silvanidae	*Silvanus bidentatus* (F.,1792)
Biphyllidae	*Diplocoelus fagi* Guerin-Meneville,1844
Phalacridae	*Olibrus millefolii* (Paykull,1800)
	Olibrus pygmaeus (Sturm,1807)
Cerylonidae	*Cerylon fagi* Brisout,1867
Corylophidae	*Orthoperus nigrescens* Stephens,1829
Coccinellidae	*Scymnus femoralis* (Gyllenhal,1827)
	Scymnus schmidti Fuersch,1958
	Scymnus limbatus Stephens,1832
	Hyperaspis pseudopustulata Mulsant,1853
	Adonia variegata (Goeze,1777)
Endomychidae	*Symbiotes latus* Redtenbacher,1849
Cisidae	*Sulcacis bicornis* (Mellié,1848)
	Cis festivus (Panzer,1793)
	Cis jacquemarti Mellié,1848
	Cis lineatocribratus Mellié,1848
Mycetophagidae	*Mycetophagus piceus* (F.,1777)
Colydiidae	*Orthocerus clavicornis* (L.,1758)
	Synchita humeralis (F.,1792)
Tenebrionidae	*Opatrum sabulosum* (L.,1758)
	Crypticus quisquilius (L.,1761)
	Eledona agricola (Herbst,1783)
	Scaphidema metallicum (F.,1792)
	Helops caeruleus (L.,1758)
	Cylindrinotus pallidus (Curtis,1830)
	Prionychus ater (F.,1775)

APPENDIX 2

	Pseudocistela ceramboides (L.,1758)
Tetratomidae	*Tetratoma ancora* F.,1791
Salpingidae	*Lissodema quadripustulata* (Marsham,1802)
	Rabocerus gabrieli Gerhardt,1901
Pyrochroidae	*Pyrochroa coccinea* (L., 1761)
Melandryidae	*Hallomenus binotatus* (Quensel,1790)
	Orchesia micans (Panzer,1793)
	Orchesia minor Walker,1836
	Abdera biflexuosa (Curtis,1829)
	Abdera flexuosa (Paykull,1799)
	Phloiotrya vaudoueri Mulsant,1856
	Zilora ferruginea (Paykull,1798)
	Melandrya caraboides (L.,1761)
	Conopalpus testaceus (Olivier,1790)
Mordellidae	*Variimorda villosa* (Schrank,1781)
Oedemeridae	*Ischnomera cyanea* (F.,1787)
	Ischnomera sanguinicollis (F.,1787)
	Oncomera femorata (F.,1792)
Meloidae	*Meloe violaceus* Marsham,1802
Anthicidae	*Anthicus angustatus* Curtis,1838
	Anthicus bifasciatus (Rossi,1792)
Aderidae	*Aderus oculatus* (Paykull,1798)
	Aderus populneus (Creutzer in Panzer,1796)
Cerambycidae	*Rhagium inquisitor* (L.,1758)
	Aromia moschata (L.,1758)
	Phymatodes alni (L.,1767)
	Anaglyptus mysticus (L.,1758)
	Pogonocherus fasciculatus (Degeer,1775)
	Acanthocinus aedilis (L.,1758)
	Stenostola dubia (Laicharting,1784)
	Phytoecia cylindrica (L.,1758)
Bruchidae	*Bruchus atomarius* (L.,1761)
Chrysomelidae	*Donacia cinerea* Herbst, 1784
	Donacia clavipes Fabricius, 1792
	Donacia crassipes F.,1775
	Donacia thalassina Germar,1811
	Plateumaris affinis (Kunze, 1818)
	Orsodacne lineola (Panzer,1795)
	Cryptocephalus aureolus Suffrian,1847
	Cryptocephalus bilineatus (L.,1767)
	Cryptocephalus bipunctatus (L.,1758)
	Cryptocephalus parvulus Müller,1776
	Chrysolina haemoptera (L.,1758)
	Chrysolina oricalcia (Müller,1776)
	Chrysolina violacea (Müller,1776)
	Phaedon concinnus Stephens,1831
	Phytodecta decemnotata (Marsham,1802)
	Luperus flavipes (L.,1767)
	Phyllotreta aerea Allard,1859
	Phyllotreta cruciferae (Goeze,1777)
	Longitarsus anchusae (Paykull,1799)
	Longitarsus ballotae (Marsham,1802)
	Longitarsus brunneus (Duftschmid,1825)
	Longitarsus dorsalis (F.,1781)
	Longitarsus lycopi (Foudras,1860)
	Longitarsus nasturtii (F.,1792)
	Longitarsus ochroleucus (Marsham,1802)
	Longitarsus plantagomaritimus Dollman,1912
	Longitarsus suturalis (Marsham,1802)
	Longitarsus tabidus (F.,1775)
	Altica ericeti (Allard,1859)
	Lythraria salicariae (Paykull,1800)
	Chalcoides nitidula (L.,1758)
	Epitrix atropae Foudras,1860
	Podagrica fuscicornis (L.,1767)
	Mantura obtusata (Gyllenhal,1813)
	Mantura rustica (L.,1767)
	Chaetocnema subcoerulea (Kutschera,1864)
	Apteropeda globosa (Illiger,1794)
	Mniophila muscorum (Koch,1803)
	Psylliodes chalcomera (Illiger,1807)
	Cassida nobilis L.,1758
	Cassida prasina Illiger,1798
Anthribidae	*Platyrhinus resinosus* (Scopoli,1763)
	Platystomos albinus (L.,1758)
	Anthribus nebulosus Forster,1771
Attelabidae	*Rhynchites cupreus* (L.,1758)
	Rhynchites cavifrons Gyllenhal,1833
	Rhynchites interpunctatus Stephens,1831
	Rhynchites longiceps Thomson,1888
	Rhynchites tomentosus Gyllenhal,1839
	Byctiscus betulae (L.,1758)
Apionidae	*Apion limonii* Kirby,1808
	Apion sedi Germar,1818
	Apion vicinum Kirby,1808
	Apion pubescens Kirby,1811
	Apion stolidum Germar,1817
	Apion gyllenhali Kirby,1808
	Apion cerdo Gerstaecker,1854
	Apion difforme Ahrens,1817
	Apion dissimile Germar,1817
	Apion filirostre Kirby,1808
	Apion ryei Blackburn,1874
	Apion varipes Germar,1817
Curculionidae	*Otiorhynchus desertus* Rosenhauer,1847
	Otiorhynchus raucus (F.,1777)
	Otiorhynchus scaber (L.,1758)
	Caenopsis fissirostris (Walton,J.,1847)
	Trachyphloeus alternans Gyllenhal,1834
	Trachyphloeus aristatus (Gyllenhal,1827)
	Trachyphloeus asperatus Boheman,1843
	Trachyphloeus spinimanus Germar,1824
	Phyllobius vespertinus (F.,1792)
	Polydrusus pulchellus Stephens,1831
	Polydrusus confluens Stephens,1831
	Polydrusus flavipes (Degeer,1775)
	Polydrusus mollis (Ström,1768)
	Barypeithes sulcifrons (Boheman,1843)
	Brachysomus echinatus (Bonsdorff,1785)
	Strophosoma faber (Herbst,1784)
	Cneorhinus plumbeus (Marsham,1802)
	Barynotus squamosus Germar,1824
	Tropiphorus elevatus (Herbst,1795)
	Tropiphorus terricola (Newman,1838)
	Tanymecus palliatus (F.,1787)
	Sitona macularius (Marsham,1802)
	Sitona ononidis Sharp,1866
	Sitona waterhousei Walton,1846
	Cleonus piger (Scopoli,1763)
	Larinus planus (F.,1792)
	Hypera dauci (Olivier,1807)
	Hypera fuscocinerea (Marsham,1802)
	Liparus coronatus (Goeze,1777)
	Leiosoma oblongulum Boheman,1842
	Gronops lunatus (F.,1775)
	Magdalis carbonaria (L.,1758)
	Magdalis cerasi (L.,1758)
	Anoplus roboris Suffrian,1840

Mesites tardii (Curtis,1825)
Cossonus parallelepipedus (Herbst,1795)
Pselactus spadix (Herbst,1795)
Trachodes hispidus (L.,1758)
Cryptorhynchus lapathi (L.,1758)
Acalles ptinoides (Marsham,1802)
Acalles roboris Curtis,1835
Bagous limosus (Gyllenhal,1827)
Bagous tempestivus (Herbst,1795)
Bagous glabrirostris (Herbst,1795)
Bagous lutulentus (Gyllenhal,1813)
Hydronomus alismatis (Marsham,1802)
Dorytomus filirostris (Gyllenhal,1836)
Dorytomus ictor (Herbst,1795)
Dorytomus salicinus (Gyllenhal,1827)
Dorytomus tremulae (F.,1787)
Notaris bimaculatus (F.,1787)
Notaris scirpi (F.,1792)
Thryogenes scirrhosus (Gyllenhal,1836)
Grypus equiseti (F.,1775)
Orthochaetes insignis (Aubé,1863)
Orthochaetes setiger (Beck,1817)
Smicronyx jungermanniae (Reich,1797)
Coeliodes erythroleucos (Gmelin in L.,1790)
Coeliodes ruber (Marsham,1802)
Zacladus exiguus (Olivier,1807)
Stenocarus umbrinus (Gyllenhal,1837)
Trichosirocalus barnevillei (Brisout,1866)
Trichosirocalus dawsoni (Brisout,1869)
Ceutorhynchus campestris Gyllenhal,1837
Ceutorhynchus constrictus (Marsham,1802)
Ceutorhynchus geographicus (Goeze,1777)
Ceutorhynchus hirtulus Germar,1824
Ceutorhynchus mixtus Mulsant and Rey,1858
Ceutorhynchus punctiger Sahlberg,1835
Ceutorhynchus rapae Gyllenhal,1837
Ceutorhynchus resedae (Marsham,1802)
Ceutorhynchus terminatus (Herbst,1795)
Ceutorhynchus triangulum Boheman,1845
Ceutorhynchus trimaculatus (F.,1775)
Ceutorhynchus viduatus (Gyllenhal,1813)
Eubrychius velutus (Beck,1817)
Litodactylus leucogaster (Marsham,1802)
Phytobius canaliculatus Fahraeus,1843
Phytobius comari (Herbst,1795)
Phytobius waltoni Boheman,1843
Drupenatus nasturtii (Germar,1824)
Baris picicornis (Marsham,1802)
Anthonomus conspersus Desbrochers,1868
Anthonomus ulmi (Degeer,1775)
Anthonomus varians (Paykull,1792)
Furcipus rectirostris (L.,1758)
Curculio betulae (Stephens,1831)
Curculio rubidus (Gyllenhal,1836)
Curculio villosus F.,1781
Tychius pusillus Germar,1842
Tychius schneideri (Herbst,1795)
Tychius squamulatus Gyllenhal,1836
Sibinia arenariae Stephens,1831
Sibinia potentillae Germar,1824
Sibinia primitus (Herbst,1795)
Ellescus bipunctatus (L.,1758)
Acalyptus carpini (F.,1792)
Miarus graminis (Gyllenhal,1813)

Mecinus circulatus (Marsham,1802)
Mecinus collaris Germar,1821
Gymnetron melanarium (Germar,1821)
Gymnetron veronicae (Germar,1821)
Gymnetron villosulum Gyllenhal,1838
Rhynchaenus iota (F.,1787)
Rhynchaenus pratensis (Germar,1821)

Scolytidae *Scolytus mali* (Bechstein,1805)
Scolytus ratzeburgi Janson,1856
Leperisinus orni (Fuchs,1906)
Kissophagus hederae (Schmitt,1843)
Xyloterus signatus (F.,1792)
Xyleborus dispar (F.,1792)
Xyleborus dryographus (Ratzeburg,1837)

Platypodidae *Platypus cylindrus* (F.,1792)

APPENDIX 3: TAXONOMIC LIST OF RED DATA BOOK AND NOTABLE SPECIES IN PART 1 OF THE REVIEW

Hyman (1986), in addition to the Red Data Book and Nationally Notable statuses, used the following terms:

List 1:   Naturalised species of rare occurrence.
List 2:   Non-established immigrant species and species of doubtful occurrence or status.
List 3:   Rare synanthropic species.
List 4:   Species of rare occurrence in Ireland, not recorded from Great Britain.

Shirt (1987), in addition to the Red Data Book categories, also used the following terms:

+   Category 1 species that are believed to be extinct.
!   Listed on Schedule 5 of the Wildlife and Countryside Act, 1981.
(5)   Also listed in category 5 (Endemic).
*   Taxa which are believed to be rare but are too recently discovered or recognised to be certain of placing.

Species	Status in Hyman (1986)	Status in Shirt (1987)	Revised Status	Page Number
Carabidae				
Cicindela germanica L.,1758	RDB 3	RDB 3	RDB 3	123
Cicindela hybrida L.,1758	RDB 3	RDB 3	RDB 2	123
Cicindela maritima Latreille et Dejean,1822	Nb	-	Nb	123
Cicindela sylvatica L.,1758	Na	-	Na	124
Omophron limbatum (F.,1777)	RDB 1	RDB 1	RDB 1	143
Carabus auratus L.,1761	List 1	-	-	
Carabus cancellatus Illiger, 1798	List 2	-	-	
Carabus clatratus L.,1761	Na	-	Na	120
Carabus intricatus L.,1761	RDB 1	RDB 1	RDB 1	121
Carabus monilis F.,1792	Nb	-	Nb	121
Carabus nitens L.,1758	Nb	-	Nb	121
Calosoma inquisitor (L.,1758)	Na	-	Na	120
Calosoma sycophanta (L.,1758)	List 2	-	-	
Leistus montanus Stephens,1827	RDB 3	RDB 3	Na	139
Pelophila borealis (Paykull,1790)	Na	-	RDB 3	145
Nebria complanata (L.,1767)	Na	-	Na	141
Nebria livida (L.,1758)	Na	-	Na	142
Nebria nivalis (Paykull,1790)	RDB 3	RDB 3	Na	142
Notiophilus aesthuans (Motschulsky,1864)	Nb	-	Nb	142
Notiophilus quadripunctatus Dejean,1826	Nb	-	Nb	143
Blethisa multipunctata (L.,1758)	Nb	-	Nb	118
Elaphrus lapponicus Gyllenhal,1810	Na	-	Na	130
Elaphrus uliginosus F.,1792	Na	-	Nb	130
Dyschirius angustatus (Ahrens,1830)	RDB 3	RDB 3	RDB 3	128
Dyschirius extensus Putzeys,1845	RDB 3	RDB 3	RDB 1	128
Dyschirius impunctipennis Dawson,1854	Nb	-	Nb	129
Dyschirius nitidus (Dejean,1825)	Na	-	Na	129
Dyschirius obscurus (Gyllenhal,1827)	RDB 1	RDB 1	RDB 2	129
Dyschirius politus (Dejean,1825)	Nb	-	-	
Dyschirius thoracicus (Rossi,1790)	Nb	-	-	
Clivina collaris (Herbst,1784)	Nb	-	-	
Miscodera arctica (Paykull,1798)	Nb	-	Nb	141
Patrobus septentrionis Dejean,1828	Nb	-	Nb	145
Perileptus areolatus (Creutzer,1799)	Na	-	Na	145
Aepus marinus (Ström,1783)	Nb	-	Nb	100
Aepus robini (Laboulbène,1849)	Nb	-	Nb	101
Thalassophilus longicornis (Sturm,1825)	Na	-	Na	152
Trechus fulvus Dejean,1831	Nb	-	Nb	153

APPENDIX 3

Species	Status in Hyman (1986)	Status in Shirt (1987)	Revised Status	Page Number
Trechus rivularis (Gyllenhal,1810)	RDB 1	RDB 1	RDB 3	153
Trechus rubens (F.,1792)	Nb	-	Nb	154
Trechus secalis (Paykull,1790)	Nb	-	-	
Trechus subnotatus Dejean,1831	RDB 1	RDB 1	RDB 1	154
Trechus discus (F.,1792)	Nb	-	Nb	153
Trechus micros (Herbst,1784)	Nb	-	-	
Asaphidion curtum (Heyden,1870)	Na	-	-	
Asaphidion pallipes (Duftschmid,1812)	Nb	-	Nb	108
Asaphidion stierlini (Heyden,1880)	Na	-	-	
Bembidion argenteolum Ahrens,1812	List 4	-	RDB K	110
Bembidion litorale (Olivier,1790)	Na	-	Nb	113
Bembidion nigricorne Gyllenhal,1827	Nb	-	Nb	114
Bembidion nigropiceum (Marsham,1802)	Na	-	Na	115
Bembidion bipunctatum (L.,1761)	Nb	-	Nb	111
Bembidion pallidipenne (Illiger,1802)	Nb	-	Nb	115
Bembidion stomoides Dejean,1831	Nb	-	Nb	117
Bembidion obliquum Sturm,1825	Nb	-	Nb	115
Bembidion semipunctatum Donovan,1806	Na	-	Na	117
Bembidion ephippium (Marsham,1802)	Na	-	Na	111
Bembidion virens Gyllenhal,1827	RDB 1	RDB 1	RDB 3	118
Bembidion geniculatum Heer,1838	Nb	-	-	
Bembidion fluviatile Dejean,1831	Nb	-	Nb	112
Bembidion lunatum (Duftschmid,1812)	Nb	-	Nb	114
Bembidion maritimum Stephens,1835	Nb	-	-	
Bembidion monticola Sturm,1825	Nb	-	Nb	114
Bembidion saxatile Gyllenhal,1827	Nb	-	Nb	116
Bembidion stephensi Crotch,1871	Nb	-	-	
Bembidion testaceum (Duftschmid,1812)	Nb	-	Nb	117
Bembidion laterale (Samouelle,1819)	Nb	-	Nb	113
Bembidion humerale Sturm,1825	RDB 1	RDB 1	RDB 1	113
Bembidion quadripustulatum Serville,1821	Nb	-	Nb	116
Bembidion gilvipes Sturm,1825	Nb	-	Nb	112
Bembidion schueppeli Dejean,1831	Na	-	Na	116
Bembidion clarki Dawson,1849	Nb	-	Nb	111
Bembidion fumigatum (Duftschmid,1812)	Nb	-	Nb	112
Bembidion normannum Dejean,1831	Nb	-	-	
Bembidion callosum Kuster, 1847	List 2	-	-	
Bembidion octomaculatum (Goeze,1777)	RDB APPENDIX	RDB APPENDIX	EXTINCT	115
Bembidion iricolor Bedel,1879	Nb	-	-	
Tachys bistriatus (Duftschmid,1812)	Nb	-	Nb	151
Tachys bisulcatus (Nicolai,1822)	List 2	-	-	
Tachys edmondsi Moore,1956	RDB 3	RDB 3 (5)	RDB 1 Endemic	151
Tachys micros (Fischer von Waldheim,1828)	RDB 3	RDB 3	Na	151
Tachys parvulus (Dejean,1831)	Na	-	Nb	152
Tachys quadristriatus (Duftschmid,1812)	List 2	-	-	
Tachys scutellaris Stephens,1828	RDB 3	RDB 3	Na	152
Tachys walkerianus Sharp,1913	Na	-	RDB 1	152
Pogonus littoralis (Duftschmid,1812)	Nb	-	Nb	146
Pogonus luridipennis (Germar,1822)	Na	-	RDB 3	146
Pterostichus aethiops (Panzer,1796)	Nb	-	Nb	147
Pterostichus angustatus (Duftschmid,1812)	Nb	-	Nb	147
Pterostichus anthracinus (Panzer,1795)	Nb	-	Nb	147
Pterostichus aterrimus (Herbst,1784)	RDB 1	RDB 1	RDB 1	148
Pterostichus cristatus (Dufour,1820)	Nb	-	Nb	148
Pterostichus gracilis (Dejean,1828)	Nb	-	Nb	148
Pterostichus kugelanni (Panzer,1797)	Na	-	RDB 1	149
Pterostichus lepidus (Leske,1785)	Nb	-	Nb	149
Pterostichus longicollis (Duftschmid,1812)	Nb	-	Nb	149
Pterostichus macer (Marsham,1802)	Nb	-	-	
Pterostichus oblongopunctatus (F.,1787)	Nb	-	Nb	150

Species	Status in Hyman (1986)	Status in Shirt (1987)	Revised Status	Page Number
Abax parallelus (Duftschmid,1812)	List 2	-	-	
Calathus ambiguus (Paykull,1790)	Nb	-	Nb	119
Sphodrus leucophthalmus (L.,1758)	List 3	-	-	
Laemostenus complanatus (Dejean,1828)	List 1	-	-	
Laemostenus terricola (Herbst,1784)	Nb	-	-	
Platyderus ruficollis (Marsam,1802)	Nb	-	Nb	146
Agonum ericeti (Panzer,1809)	Nb	-	Nb	101
Agonum gracilipes (Duftschmid,1812)	Na	-	Na	101
Agonum livens (Gyllenhal,1810)	Na	-	Nb	101
Agonum lugens (Duftschmid,1812)	List 4	-	-	
Agonum nigrum Dejean,1828	Nb	-	Nb	102
Agonum quadripunctatum (Degeer,1774)	Na	-	RDB 1	102
Agonum sahlbergi (Chaudoir,1850)	RDB 1	RDB 1+	EXTINCT	102
Agonum scitulum Dejean,1828	Na	-	Na	103
Agonum sexpunctatum (L.,1758)	Na	-	Na	103
Agonum versutum Sturm,1824	Nb	-	Nb	103
Perigona nigriceps (Dejean,1831)	List 1	-	-	
Amara alpina (Paykull,1790)	RDB 3	RDB 3	RDB 3	104
Amara consularis (Duftschmid,1812)	Na	-	Nb	104
Amara cursitans (Zimmermann,1832)	Na	-	-	
Amara curta Dejean,1828	Na	-	Nb	104
Amara equestris (Duftschmid,1812)	Nb	-	Nb	105
Amara famelica Zimmerman,1832	Na	-	RDB 3	105
Amara fulva (Müller,1776)	Nb	-	Nb	105
Amara fusca Dejean,1828	RDB 2	RDB 2	RDB 1	105
Amara infima (Duftschmid,1812)	Na	-	Na	106
Amara lucida (Duftschmid,1812)	Nb	-	Nb	106
Amara nitida Sturm,1825	Na	-	Na	106
Amara praetermissa (Sahlberg,1827)	Na	-	Nb	107
Amara quenseli (Schoenherr,1806)	Na	-	Na	107
Amara spreta Dejean,1831	Nb	-	Nb	107
Amara strenua Zimmermann,1832	Na	-	RDB 3	107
Zabrus tenebrioides (Goeze,1777)	Na	-	Na	154
Harpalus ardosiacus Lutschnik,1922	Nb	-	Nb	130
Harpalus azureus (F.,1775)	Nb	-	Nb	130
Harpalus calceatus (Duftschmid,1812)	List 2	-	-	
Harpalus cordatus (Duftschmid,1812)	Na	-	RDB 3	131
Harpalus melleti Heer,1837	Na	-	Na	133
Harpalus obscurus (F.,1792)	Na	-	RDB 1	133
Harpalus parallelus Dejean,1829	Na	-	RDB 3	133
Harpalus punctatulus (Duftschmid,1812)	Na	-	Na	134
Harpalus puncticeps (Stephens,1828)	Nb	-	-	
Harpalus puncticollis (Paykull,1798)	Na	-	RDB 3	134
Harpalus rupicola Sturm,1818	Nb	-	Nb	135
Harpalus sabulicola (Panzer,1796)	Na	-	RDB 3	135
Harpalus schaubergerianus Puel,1937	Na	-	Nb	136
Harpalus cupreus Dejean,1829	RDB 1	RDB 1	RDB 1	131
Harpalus dimidiatus (Rossi,1790)	Na	-	Na	131
Harpalus froelichi Sturm,1818	Na	-	RDB 2	132
Harpalus honestus (Duftschmid,1812)	RDB 1	RDB 1+	RDB 1	132
Harpalus melancholicus Dejean,1829	Na	-	RDB 1	132
Harpalus neglectus Serville,1821	Nb	-	-	
Harpalus quadripunctatus Dejean,1829	Na	-	Na	135
Harpalus serripes (Quensel in Schoenherr,1806)	Nb	-	Nb	136
Harpalus servus (Duftschmid,1812)	Nb	-	Nb	136
Harpalus smaragdinus (Duftschmid,1812)	Nb	-	Nb	137
Harpalus tenebrosus Dejean,1829	Na	-	Na	137
Harpalus vernalis (Duftschmid,1812)	Na	-	Na	137
Anisodactylus nemorivagus (Duftschmid,1812)	Na	-	Na	108
Anisodactylus poeciloides (Stephens,1828)	Na	-	RDB 3	108

APPENDIX 3

Species	Status in Hyman (1986)	Status in Shirt (1987)	Revised Status	Page Number
Scybalicus oblongiusculus (Dejean,1829)	RDB 1	RDB 1+	-	
Diachromus germanus (L.,1758)	RDB APPENDIX	RDB APPENDIX	EXTINCT	126
Dicheirotrichus obsoletus (Dejean,1829)	Nb	-	Nb	126
Bradycellus csikii Laczó,1912	RDB 3	RDB 3	RDB I	119
Bradycellus distinctus (Dejean,1829)	Na	-	Na	119
Stenolophus plagiatus Gorham,1901	List 2	-	-	
Stenolophus skrimshiranus Stephens,1828	Na	-	Na	150
Stenolophus teutonus (Schrank,1781)	Nb	-	Nb	150
Acupalpus brunnipes (Sturm,1825)	Na	-	Na	99
Acupalpus consputus (Duftschmid,1812)	Nb	-	Nb	99
Acupalpus dorsalis (F.,1787)	Nb	-	-	
Acupalpus elegans (Dejean,1829)	RDB 1	RDB 1	EXTINCT	99
Acupalpus exiguus Dejean,1829	Nb	-	Nb	100
Acupalpus flavicollis (Sturm,1825)	Na	-	Na	100
Licinus depressus (Paykull,1790)	Nb	-	Nb	139
Licinus punctatulus (F.,1792)	Na	-	Na	140
Badister anomalus (Perris,1866)	Na	-	RDB 1	109
Badister dilatatus Chaudoir,1837	Na	-	Nb	109
Badister meridionalis Puel,1925	Na	-	RDB I	109
Badister peltatus (Panzer,1797)	Na	-	Na	110
Badister sodalis (Duftschmid,1812)	Nb	-	-	
Badister unipustulatus Bonelli,1813	Nb	-	Nb	110
Panagaeus bipustulatus (F.,1775)	Nb	-	Nb	144
Panagaeus cruxmajor (L.,1758)	RDB 2	RDB 2	RDB 1	144
Chlaenius nigricornis (F.,1787)	Nb	-	Nb	122
Chlaenius nitidulus (Schrank,1781)	RDB 1	RDB 1	RDB 1	122
Chlaenius tristis (Schaller,1783)	RDB 1	RDB 1	RDB 1	122
Callistus lunatus (F.,1775)	RDB 1	RDB 1	RDB 1	120
Oodes helopioides (F.,1792)	Nb	-	Nb	144
Odacantha melanura (L.,1767)	Nb	-	Nb	143
Masoreus wetterhalli (Gyllenhal,1813)	Na	-	Na	141
Lebia chlorocephala (Hoffmannsegg,1803)	Nb	-	Nb	138
Lebia cruxminor (L.,1758)	RDB 1	RDB 1	RDB 1	138
Lebia cyanocephala (L.,1758)	Na	-	RDB 1	138
Lebia marginata (Fourcroy,1785)	RDB APPENDIX	RDB APPENDIX	EXTINCT	139
Lebia scapularis (Fourcroy,1785)	RDB APPENDIX	RDB APPENDIX	EXTINCT	139
Demetrias imperialis (Germar,1824)	Nb	-	Nb	125
Demetrias monostigma Samouelle,1819	Nb	-	Nb	125
Dromius longiceps Dejean,1826	RDB 2	RDB 2	Na	126
Dromius quadrisignatus Dejean,1825	RDB 3	RDB 3	RDB 1	126
Dromius sigma (Rossi,1790)	RDB 2	RDB 2	Na	127
Dromius vectensis Rye,1872	Na	-	RDB 3	127
Lionychus quadrillum (Duftschmid,1812)	RDB 3	RDB 3	RDB 3	140
Cymindis axillaris (F.,1794)	Na	-	Na	124
Cymindis macularis (Fischer von Waldheim,1824)	Na	-	RDB 1	124
Cymindis vaporariorum (L.,1758)	Nb	-	Nb	125
Polistichus connexus (Fourcroy,1785)	RDB 2	RDB 2	RDB 2	146
Drypta dentata (Rossi,1790)	RDB 1	RDB 1	RDB 1	128
Brachinus crepitans (L.,1758)	Nb	-	Nb	118
Brachinus sclopeta (F.,1792)	List 2	-	RDB 1	118

Sphaeritidae

Sphaerites glabratus (F.,1792)	RDB 3	RDB 3	RDB 3	412

Histeridae

Teretrius fabricii Mazur,1972	RDB 1	RDB 1+	RDB 1	343
Plegaderus dissectus Erichson,1839	Nb	-	Nb	342
Plegaderus vulneratus (Panzer,1796)	Nb	-	-	

32

Species	Status in Hyman (1986)	Status in Shirt (1987)	Revised Status	Page Number
Abraeus granulum Erichson,1839	Na	-	Na	335
Aeletes atomarius (Aubé,1842)	RDB 3	RDB 3	RDB 3	336
Acritus homoeopathicus Wollaston,1857	RDB 3	RDB 3	RDB 3	336
Halacritus punctum (Aubé,1842)	Na	-	RDB K	338
Gnathoncus buyssoni Auzat,1917	Na	-	Na	337
Saprinus cuspidatus Ihssen,1949	Nb	-	Nb	342
Saprinus immundus (Gyllenhal,1827)	Nb	-	Nb	342
Saprinus subnitescens Bickhardt,1909	RDB APPENDIX	RDB APPENDIX	EXTINCT	343
Saprinus virescens (Paykull,1798)	Na	-	RDB K	343
Hypocaccus metallicus (Herbst,1792)	RDB 2	RDB 2	RDB 3	339
Hypocaccus rugiceps (Duftschmid,1805)	RDB 2	RDB 2	Na	340
Hypocaccus rugifrons (Paykull,1798)	Nb	-	Nb	340
Baeckmanniolus dimidiatus (Illiger,1807)	Nb	-	Nb	337
Myrmetes piceus (Paykull,1809)	Nb	-	Nb	340
Paromalus parallelepipedus (Herbst,1792)	RDB 1	RDB 1	RDB 1	341
Onthophilus punctatus (Müller,1776)	Na	-	RDB K	341
Hister bissexstriatus F.,1801	Nb	-	Nb	338
Hister quadrimaculatus L.,1758	RDB 2	RDB 2	RDB K	339
Hister quadrinotatus Scriba,1790	RDB APPENDIX	RDB APPENDIX	EXTINCT	339
Hister illigeri Duftschmid,1805	RDB APPENDIX	RDB APPENDIX	EXTINCT	339
Paralister obscurus (Kugelann,1792)	RDB 2	RDB 2	RDB 1	341
Grammostethus marginatus (Erichson,1834)	Nb	-	Nb	338
Epierus comptus Erichson,1834	RDB 3	RDB 3*	RDB K	337
Hetaerius ferrugineus (Olivier,1789)	RDB 3	RDB 3	RDB I	338

Silphidae

Species	Status in Hyman (1986)	Status in Shirt (1987)	Revised Status	Page Number
Nicrophorus germanicus (L.,1758)	List 2	-	-	
Nicrophorus interruptus Stephens,1830	Nb	-	Nb	409
Nicrophorus vestigator Herschel,1807	Na	-	Na	410
Thanatophilus dispar (Herbst,1793)	RDB 3	RDB 3	RDB 1	411
Aclypea opaca (L.,1758)	Nb	-	Na	409
Aclypea undata (Müller,1776)	RDB 3	RDB 3	RDB 1	409
Dendroxena quadrimaculata (Scopoli,1772)	Nb	-	Nb	409
Silpha carinata Herbst,1783	RDB 1	RDB 1	RDB 1	410
Silpha obscura L.,1758	Nb	-	RDB 2	410
Silpha tristis Illiger,1798	Nb	-	-	
Silpha tyrolensis Laicharting,1781	Nb	-	Nb	411

Scaphidiidae

Species	Status in Hyman (1986)	Status in Shirt (1987)	Revised Status	Page Number
Scaphium immaculatum (Olivier,1790)	RDB 1	RDB 1	RDB K	382
Scaphisoma assimile Erichson,1845	Na	-	RDB I	381
Scaphisoma boleti (Panzer,1793)	Nb	-	Nb	382

Lucanidae

Species	Status in Hyman (1986)	Status in Shirt (1987)	Revised Status	Page Number
Lucanus cervus (L., 1758)	-	-	Nb	345
Platycerus caraboides (L.,1758)	RDB APPENDIX	RDB APPENDIX	EXTINCT	345

Trogidae

Species	Status in Hyman (1986)	Status in Shirt (1987)	Revised Status	Page Number
Trox perlatus Goeze,1777	RDB 1	RDB 1	RDB 1	420
Trox sabulosus (L.,1758)	Na	-	Na	421

Geotrupidae

Species	Status in Hyman (1986)	Status in Shirt (1987)	Revised Status	Page Number
Odonteus armiger (Scopoli,1772)	RDB 3	RDB 3	Na	334
Geotrupes mutator (Marsham,1802)	Nb	-	Nb	?
Geotrupes pyrenaeus (Charpentier,1825)	Na	-	Na	?

APPENDIX 3

Species	Status in Hyman (1986)	Status in Shirt (1987)	Revised Status	Page Number
Geotrupes vernalis (L.,1758)	Nb	-	Nb	334

Scarabaeidae

Species	Status in Hyman (1986)	Status in Shirt (1987)	Revised Status	Page Number
Aegialia rufa (F.,1792)	RDB 1	RDB 1	RDB 1	382
Aegialia sabuleti (Panzer,1796)	Nb	-	Nb	383
Aphodius brevis Erichson,1848	RDB 1	RDB 1	RDB 1	383
Aphodius coenosus (Panzer,1798)	Nb	-	Nb	384
Aphodius conspurcatus (L.,1758)	Nb	-	Nb	384
Aphodius consputus Creutzer,1799	Na	-	RDB 3	384
Aphodius distinctus (Müller,1776)	Nb	-	Nb	385
Aphodius fasciatus (Olivier,1789)	Nb	-	Nb	385
Aphodius lividus (Olivier,1789)	RDB 3	RDB 3	RDB 1	385
Aphodius nemoralis Erichson,1848	Nb	-	Na	386
Aphodius niger (Panzer,1796)	RDB 1	RDB 1	RDB 1	386
Aphodius paykulli Bedel,1908	Nb	-	Nb	386
Aphodius plagiatus (L.,1767)	Nb	-	Nb	387
Aphodius porcus (F.,1792)	Nb	-	Nb	387
Aphodius putridus (Fourcroy,1785)	Nb	-	Nb	387
Aphodius quadrimaculatus (L.,1761)	RDB 3	RDB 3	RDB 1	388
Aphodius scrofa (F.,1787)	RDB APPENDIX	RDB APPENDIX	EXTINCT	388
Aphodius sordidus (F.,1775)	Nb	-	Na	388
Aphodius subterraneus (L.,1758)	RDB 3	RDB 3	RDB 1	389
Aphodius zenkeri Germar,1813	Nb	-	Nb	389
Euheptaulacus sus (Herbst,1783)	RDB 3	RDB 3	RDB 1	391
Euheptaulacus villosus Gyllenhal,1806	Na	-	Na	391
Heptaulacus testudinarius (F.,1775)	RDB 3	RDB 3	RDB 1	392
Saprosites mendax Blackburn,1892	List 1	-	-	
Psammodius asper (F.,1775)	Na	-	Na	395
Psammodius caelatus (LeConte,1857)	List 2	-	-	
Brindalus porcicollis (Illiger,1803)	RDB 1	RDB 1+	EXTINCT	389
Rhyssemus germanus (L.,1767)	RDB APPENDIX	RDB APPENDIX	EXTINCT	396
Diastictus vulneratus (Sturm,1805)	RDB 2	RDB 2	RDB 2	390
Pleurophorus caesus (Creutzer in Panzer,1796)	RDB APPENDIX	RDB APPENDIX	EXTINCT	395
Copris lunaris (L.,1758)	RDB 1	RDB 1	RDB 1	390
Onthophagus fracticornis (Preyssler,1790)	Na	-	RDB K	393
Onthophagus nuchicornis (L.,1758)	Na	-	Na	394
Onthophagus nutans (F.,1787)	RDB APPENDIX	RDB APPENDIX	RDB 1	394
Onthophagus taurus (Schreber,1759)	RDB APPENDIX	RDB APPENDIX	EXTINCT	394
Onthophagus vacca (L.,1767)	Nb	-	Nb	395
Omaloplia ruricola (F.,1775)	Nb	-	Nb	393
Amphimallon ochraceus (Knoch,1801)	Na	-	Na	383
Melolontha hippocastani F.,1801	Nb	-	RDB K	393
Polyphylla fullo (L.,1758)	RDB APPENDIX	RDB APPENDIX	EXTINCT	395
Cetonia cuprea F.,1775	Nb	-	Nb	390
Oxythyrea funesta (Poda,1761)	List 2	-	-	
Gnorimus nobilis (L.,1758)	RDB 3	RDB 3	RDB 2	391
Gnorimus variabilis (L.,1758)	RDB 1	RDB 1	RDB 1	392
Trichius fasciatus (L.,1758)	Nb	-	-	
Trichius zonatus Germar,1831	List 2	-	-	

Clambidae

Species	Status in Hyman (1986)	Status in Shirt (1987)	Revised Status	Page Number
Clambus nigrellus Reitter,1914	Na	-	Na	213
Clambus pallidulus Reitter,1911	Na	-	RDB K	213

Eucinetidae

Species	Status in Hyman (1986)	Status in Shirt (1987)	Revised Status	Page Number
Eucinetus meridionalis (Laporte de Castelnau,1836)	RDB 3	RDB 3	-	

34

Species	Status in Hyman (1986)	Status in Shirt (1987)	Revised Status	Page Number
Scirtidae				
Elodes elongata Tournier,1868	RDB 3	RDB 3*	RDB I	397
Elodes minuta (L.,1767)	Nb	-	-	
Cyphon kongsbergensis Munster,1924	Na	-	Na	396
Cyphon pubescens (F.,1792)	RDB 3	RDB 3	Nb	396
Cyphon punctipennis Sharp,1872	Na	-	Na	396
Prionocyphon serricornis (Müller,1821)	RDB 3	RDB 3	Nb	397
Hydrocyphon deflexicollis (Müller,1821)	Nb	-	Nb	397
Scirtes orbicularis (Panzer,1793)	RDB 3	RDB 3	Na	398
Byrrhidae				
Simplocaria maculosa Erichson,1847	RDB 3	RDB 3	RDB I	93
Byrrhus arietinus Steffahny,1842	Nb	-	Nb	92
Porcinolus murinus (F.,1794)	Nb	-	Nb	93
Curimopsis nigrita (Palm,1934)	RDB 1	RDB 1	RDB 1	92
Curimopsis setigera (Illiger,1798)	Na	-	Na	93
Psephenidae				
Eubria palustris Germar,1818	RDB 3	RDB 3	RDB 3	374
Heteroceridae				
Heterocerus fusculus Kiesenwetter,1843	Nb	-	Nb	335
Heterocerus hispidulus Kiesenwetter,1843	RDB 3	RDB 3	RDB 3	335
Limnichidae				
Limnichus pygmaeus (Sturm,1807)	-	-	Na	344
Buprestidae				
Melanophila acuminata (Degeer,1774)	Na	-	-	
Anthaxia nitidula (L.,1758)	RDB 1	RDB 1	RDB 1	90
Agrilus angustulus (Illiger,1803)	Nb	-	Nb	88
Agrilus laticornis (Illiger,1803)	-	-	Nb	88
Agrilus pannonicus (Piller and Mitterpacher,1783)	RDB 2	RDB 2	Na	89
Agrilus sinuatus (Olivier,1790)	RDB 2	RDB 2	Na	89
Agrilus viridis (L., 1758)	RDB 2	RDB 2	Na	90
Aphanisticus emarginatus (Olivier, 1790)	Na	-	RDB 1	90
Aphanisticus pusillus (Olivier,1790)	Nb	-	Nb	91
Trachys minuta (L.,1758)	Na	-	RDB 2	91
Trachys scrobiculatus Kiesenwetter,1857	Na	-	Na	92
Trachys troglodytes Gyllenhal in Schoenherr,1817	Na	-	-	
Elateridae				
Lacon querceus (Herbst,1784)	RDB 1	RDB 1	RDB 1	324
Ampedus cardinalis (Schiödte,1865)	RDB 2	RDB 2	RDB 2	314
Ampedus cinnabarinus (Eschsholtz,1829)	RDB 3	RDB 3	RDB 3	315
Ampedus elongantulus (F.,1787)	Na	-	Na	315
Ampedus nigerrimus (Lacordaire,1835)	RDB 1	RDB 1	RDB 1	316
Ampedus nigrinus (Herbst,1784)	Nb	-	Nb	316
Ampedus quercicola (du Buysson, 1887)	-	-	Nb	317
Ampedus pomorum (Herbst,1784)	Nb	-	Nb	316
Ampedus praeustus (Fabricius,1792)	RDB APPENDIX	-	-	
Ampedus ruficeps (Mulsant and Guillebeau,1855)	RDB 1	RDB 1	RDB 1	317
Ampedus rufipennis (Stephens,1830)	RDB 2	RDB 2	RDB 2	318

Species	Status in Hyman (1986)	Status in Shirt (1987)	Revised Status	Page Number
Ampedus sanguineus (L.,1758)	RDB APPENDIX	RDB APPENDIX	EXTINCT	318
Ampedus sanguinolentus (Schrank,1776)	Nb	-	Na	318
Ampedus tristis (L.,1758)	RDB 3	RDB 3	RDB 2	319
Ischnodes sanguinicollis (Panzer,1793)	Na	-	Na	324
Procraerus tibialis (Boisduval and Lacordaire,1835)	RDB 2	RDB 2	RDB 3	327
Megapenthes lugens (Redtenbacher,1842)	RDB 1	RDB 1	RDB 1	326
Fleutiauxellus maritimus (Curtis,1840)	Na	-	Na	323
Fleutiauxellus quadripustulatus (F.,1792)	Na	-	Na	323
Negastrius pulchellus (L.,1761)	RDB 3	RDB 3	RDB 2	326
Negastrius sabulicola (Boheman,1853)	RDB 3	RDB 3	RDB 2	327
Zorochros flavipes (Aubé,1850)	List 2	-	-	
Panspoeus guttatus Sharp,1877	List 1	-	-	
Cardiophorus asellus Erichson,1840	Nb	-	Nb	320
Cardiophorus erichsoni du Buysson,1901	Na	-	RDB 2	321
Cardiophorus gramineus (Scopoli,1763)	RDB APPENDIX	RDB APPENDIX	EXTINCT	321
Cardiophorus ruficollis (L.,1758)	RDB APPENDIX	RDB APPENDIX	EXTINCT	321
Dicronychus equiseti (Herbst,1784)	Na	-	RDB 2	322
Melanotus punctolineatus (Pelerin,1829)	RDB 3	RDB 3	RDB 1	326
Limoniscus violaceus (Müller,1821)	RDB 1	RDB 1	RDB 1	325
Harminius undulatus (Degeer,1774)	RDB 3	RDB 3	Nb	324
Athous campyloides Newman,1833	Na	-	Nb	320
Athous subfuscus (Müller,1764)	RDB 3	RDB 3	RDB 3	320
Ctenicera pectinicornis (L.,1758)	Nb	-	Na	321
Anostirus castaneus (L.,1758)	RDB 1	RDB 1	RDB 1	319
Selatosomus angustulus (Kiesenwetter,1858)	RDB 3	RDB 3	RDB 3	328
Selatosomus bipustulatus (L.,1767)	Na	-	Nb	328
Selatosomus cruciatus (L.,1758)	RDB APPENDIX	RDB APPENDIX	EXTINCT	328
Selatosomus impressus (F.,1792)	Nb	-	Nb	329
Selatosomus melancholicus (F.,1793)	List 4	-	-	
Selatosomus nigricornis (Panzer,1799)	Nb	-	RDB 3	329
Elater ferrugineus L.,1758	RDB 1	RDB 1	RDB 1	322
Agriotes sordidus (Illiger,1807)	Nb	-	RDB 3	314
Synaptus filiformis (F.,1781)	RDB 3	RDB 3	RDB 1	329
Adrastus rachifer (Fourcroy,1785)	Na	-	RDB 3	314

Cerophytidae

Cerophytum elateroides (Latrielle,1809)	RDB APPENDIX	-	-	

Throscidae

Aulonothroscus brevicollis (Bonvouloir,1859)	RDB 3	RDB 3	RDB 3	420
Trixagus elateroides (Heer,1841)	-	-	RDB 3	420

Eucnemidae

Eucnemis capucina Ahrens,1812	RDB 1	RDB 1	RDB 1	332
Dirhagus pygmaeus (F.,1792)	RDB 3	RDB 3	RDB 3	332
Melasis buprestoides (L.,1761)	Nb	-	Nb	334
Epiphanis cornutus Eschscholtz,1829	List 1	-	-	
Hylis cariniceps (Reitter,1902)	RDB 1	RDB 1	RDB 1	333
Hylis olexai (Palm,1955)	RDB 3	RDB 3	RDB 3	333

Drilidae

Drilus flavescens (Fourcroy,1785)	Nb	-	Na	313

Species	Status in Hyman (1986)	Status in Shirt (1987)	Revised Status	Page Number
Cantharidae				
Ancistronycha abdominalis (F.,1798)	Na	-	Nb	94
Cantharis figurata Mannerheim,1843	Na	-	-	
Cantharis fusca L.,1758	Na	-	RDB 3	94
Cantharis obscura L.,1758	Nb	-	Nb	94
Cantharis pallida Goeze,1777	Nb	-	-	
Cantharis thoracica (Olivier,1790)	Nb	-	-	
Rhagonycha elongata (Fallén,1807)	-	-	Na	97
Rhagonycha lutea (Müller,1764)	Nb	-	Nb	98
Rhagonycha translucida (Krynicki,1832)	Nb	-	Nb	98
Silis ruficollis (F.,1775)	Nb	-	Nb	98
Malthinus balteatus Suffrian,1851	-	-	Nb	95
Malthinus frontalis (Marsham,1802)	-	-	Nb	95
Malthodes brevicollis (Paykull,1798)	RDB 3	RDB 3	RDB 1	95
Malthodes crassicornis (Mäklin,1846)	RDB 3	RDB 3	RDB 3	96
Malthodes fibulatus Kiesenwetter,1852	Nb	-	Nb	96
Malthodes flavoguttatus Kiesenwetter,1852	Nb	-	-	
Malthodes guttifer Kiesenwetter,1852	Nb	-	Nb	97
Malthodes maurus (Castelnau,1840)	Nb	-	Nb	97
Malthodes mysticus Kiesenwetter,1852	Nb	-	-	
Lampyridae				
Lamprohiza splendidula (L.,1767)	-	-	EXTINCT	344
Phosphaenus hemipterus (Goeze,1777)	RDB 1	RDB 1	RDB 1	344
Lycidae				
Dictyoptera aurora (Herbst,1784)	Nb	-	Nb	345
Pyropterus nigroruber (Degeer,1774)	RDB 3	RDB 3	Na	346
Platycis cosnardi (Chevrolat,1829)	RDB 1	RDB 1	RDB I	346
Platycis minuta (F.,1787)	Nb	-	Nb	346
Dermestidae				
Globicornis nigripes (F.,1792)	RDB 1	RDB 1	RDB 1	312
Megatoma undata (L.,1758)	Nb	-	Nb	312
Ctesias serra (F.,1792)	Nb	-	Nb	311
Anthrenus pimpinellae (F.,1775)	RDB APPENDIX	RDB APPENDIX	EXTINCT	311
Anthrenus scrophulariae (L.,1758)	RDB APPENDIX	RDB APPENDIX	EXTINCT	311
Trinodes hirtus (F.,1781)	RDB 3	RDB 3	RDB 3	313
Derodontidae				
Laricobius erichsoni Rosenhauer,1846	List 1	-	-	
Anobiidae				
Ptinomorphus imperialis (L.,1767)	Nb	-	Nb	66
Dryophilus anobioides Chevrolat,1832	Na	-	RDB 3	65
Ernobius angusticollis (Ratzeburg,1837)	Na	-	-	
Ernobius gigas (Mulsant and Rey,1863)	RDB 3	RDB 3	-	
Ernobius nigrinus (Sturm,1837)	Nb	-	-	
Ernobius pini (Sturm,1837)	Na	-	-	
Gastrallus immarginatus (Müller,1821)	RDB 1	RDB 1	RDB 1	65
Hemicoelus nitidus (Herbst,1793)	Na	-	RDB I	66
Anobium inexspectatum Lohse,1954	Nb	-	Nb	62
Hadrobregmus denticollis (Creutzer in Panzer,1796)	Na	-	Nb	65
Xyletinus longitarsis Jansson,1942	Na	-	RDB 2	66

Species	Status in Hyman (1986)	Status in Shirt (1987)	Revised Status	Page Number
Dorcatoma dresdensis Herbst,1792	RDB 1	RDB 1	Na	63
Dorcatoma flavicornis (F.,1792)	Nb	-	Nb	64
Dorcatoma serra Panzer,1795	Na	-	Na	64
Caenocara affinis (Sturm,1837)	RDB 1	RDB 1	RDB I	63
Caenocara bovistae (Hoffmann,1803)	Na	-	RDB 3	63
Anitys rubens (Hoffmann,1803)	Na	-	Nb	62

Ptinidae

Ptinus dubius Sturm,1837	Na	-	-	
Ptinus lichenum Marsham,1802	Nb	-	RDB 3	375
Ptinus palliatus Perris,1847	-	-	Na	375
Ptinus sexpunctatus Panzer,1792	-	-	Nb	375
Ptinus subpilosus Sturm,1837	Nb	-	Nb	376

Bostrichidae

Bostrichus capucinus (L.,1758)	RDB 3	RDB 3	EXTINCT	87

Lyctidae

Lyctus linearis (Goeze,1777)	Nb	-	Nb	347

Phloiophilidae

Phloiophilus edwardsi Stephens,1830	Nb	-	Nb	373

Trogossitidae

Nemozoma elongatum (L.,1761)	RDB 3	RDB 3	RDB 3	421

Peltidae

Ostoma ferrugineum (L.,1758)	RDB 1	RDB 1	RDB 1	371
Thymalus limbatus (F.,1787)	Na	-	Nb	371

Cleridae

Tillus elongatus (L.,1758)	Nb	-	Nb	215
Tilloidea unifasciatus (F.,1787)	RDB 1	RDB APPENDIX	EXTINCT	215
Thaneroclerus buqueti (Lefebvre,1835)	List 2	-	-	
Opilo mollis (L.,1758)	Nb	-	Nb	214
Thanasimus rufipes (Brahm,1797)	Na	-	RDB 3	214
Trichodes alvearius (F.,1792)	RDB APPENDIX	RDB APPENDIX	EXTINCT	215
Trichodes apiarius (L.,1758)	RDB APPENDIX	RDB APPENDIX	EXTINCT	216
Paratillus carus (Newman,1840)	List 1	-	-	
Tarsostenus univittatus (Rossi,1792)	RDB APPENDIX	RDB APPENDIX	EXTINCT	214
Korynetes caeruleus (Degeer,1775)	Nb	-	Nb	213
Necrobia ruficollis (F.,1775)	Nb	-	-	

Melyridae

Aplocnemus nigricornis (F.,1792)	Na	-	Na	357
Aplocnemus pini (Redtenbacher,1849)	Nb	-	Nb	357
Dasytes coeruleus (Degeer,1774)	List 2	-	RDB K	358
Dasytes niger (L.,1761)	Nb	-	Na	358
Dasytes plumbeus (Müller,1776)	Nb	-	Nb	359
Dasytes puncticollis Reitter,1888	Nb	-	Nb	359
Dolichosoma lineare (Rossi,1794)	Nb	-	Nb	359
Ebaeus pedicularius (L.,1758)	RDB APPENDIX	RDB APPENDIX	EXTINCT	359

Species	Status in Hyman (1986)	Status in Shirt (1987)	Revised Status	Page Number
Hypebaeus flavipes (F.,1787)	RDB 1	RDB 1	RDB 1	359
Axinotarsus pulicarius (F.,1777)	RDB 2	RDB 2	RDB 1	357
Malachius aeneus (L.,1758)	RDB 3	RDB 3	RDB 3	360
Malachius barnevillei Puton,1865	RDB 3	RDB 3	RDB 3	360
Malachius marginellus Olivier,1790	Nb	-	Nb	361
Malachius vulneratus Abeille,1891	RDB 3	RDB 3	RDB 3	361
Sphinginus lobatus (Olivier,1790)	List 1	-	RDB K	361
Cerapheles terminatus (Ménétriés,1832)	Na	-	Na	358
Troglops cephalotes (Olivier,1790)	List 2	-	-	

Lymexylidae

Hylecoetus dermestoides (L.,1761)	Nb	-	Nb	347
Lymexylon navale (L.,1758)	RDB 2	RDB 2	RDB 2	348

Rhizophagidae

Rhizophagus nitidulus (F.,1798)	Nb	-	Nb	378
Rhizophagus oblongicollis Blatch and Horner,1892	RDB 1	RDB 1	RDB 1	379
Rhizophagus parvulus (Paykull,1800)	RDB 3	RDB 3	RDB 3	379
Rhizophagus picipes (Olivier,1790)	RDB 3	RDB 3	Na	379
Cyanostolus aeneus (Richter,1820)	RDB 3	RDB 3	Na	377
Monotoma angusticollis (Gyllenhal,1827)	RDB 3	RDB 3	RDB 3	378
Monotoma quadrifoveolata Aubé,1837	RDB 3	RDB 3	-	

Sphindidae

Sphindus dubius (Gyllenhal,1808)	Nb	-	Nb	412

Hypocopridae

Hypocoprus latridioides Motschulsky,1839	Na	-	RDB I	344

Cucujidae

Uleiota planata (L.,1761)	RDB 2	RDB 2	Na	229
Dendrophagus crenatus (Paykull,1799)	Nb	-	Nb	227
Pediacus depressus (Herbst,1797)	Na	-	Na	229
Laemophloeus monilis (F.,1787)	RDB 1	RDB 1	RDB 1	228
Cryptolestes spartii (Curtis,1834)	Na	-	Na	227
Notolaemus unifasciatus (Latreille,1804)	RDB 3	RDB 3	Na	228
Leptophloeus clematidis (Erichson,1846)	RDB 2	RDB 2	RDB 1	228

Silvanidae

Silvanus bidentatus (F.,1792)	RDB 3	RDB 3	Nb	411
Silvanoprus fagi (Guerin-Meneville,1844)	RDB 3	RDB 3	RDB 1	411
Cryptamorpha desjardinsi (Guérin-Méneville,1844)	List 2	-	-	

Biphyllidae

Diplocoelus fagi Guerin-Meneville,1844	Nb	-	Nb	87

Erotylidae

Triplax lacordairii Crotch,1870	RDB 3	RDB 3	RDB 3	331
Triplax scutellaris Charpentier,1825	RDB 3	RDB 3	RDB 3	331
Tritoma bipustulata F.,1775	Na	-	Na	331

APPENDIX 3

Species	Status in Hyman (1986)	Status in Shirt (1987)	Revised Status	Page Number
Phalacridae				
Phalacrus brunnipes Brisout,1863	Nb	-	Na	372
Olibrus affinis (Sturm,1807)	Nb	-	-	
Olibrus flavicornis (Sturm,1807)	Na	-	RDB K	371
Olibrus millefolii (Paykull,1800)	Nb	-	Nb	372
Olibrus pygmaeus (Sturm,1807)	Nb	-	Nb	372
Stilbus atomarius (L.,1767)	Na	-	RDB K	373
Cerylonidae				
Anommatus diecki Reitter,1875	RDB 3	-	RDB K	169
Anommatus duodecimstriatus (Müller,1821)	List 3	-	Na	169
Murmidius ovalis (Beck,1817)	Na	-	EXTINCT	170
Cerylon fagi Brisout,1867	Na	-	Nb	170
Corylophidae				
Orthoperus aequalis Sharp,1885	List 2	-	RDB K	226
Orthoperus atomarius (Heer,1841)	RDB APPENDIX	RDB APPENDIX	-	
Orthoperus brunnipes (Gyllenhal,1808)	RDB 3	RDB 3	RDB 3	226
Orthoperus nigrescens Stephens,1829	Nb	-	Nb	226
Rypobius ruficollis (du Val,1854)	RDB 3	RDB 3	RDB K	227
Coccinellidae				
Coccidula scutellata (Herbst,1783)	Nb	-	-	
Clitostethus arcuatus (Rossi,1794)	RDB 1	RDB 1	RDB 1	216
Scymnus femoralis (Gyllenhal,1827)	Nb	-	Nb	219
Scymnus nigrinus Kugelann,1794	Nb	-	-	
Scymnus schmidti Fuersch,1958	Nb	-	Nb	220
Scymnus limbatus Stephens,1832	Nb	-	Nb	220
Nephus bisignatus (Boheman,1850)	RDB APPENDIX	RDB APPENDIX	EXTINCT	218
Nephus quadrimaculatus (Herbst,1783)	RDB 2	RDB 2	RDB 2	219
Hyperaspis pseudopustulata Mulsant,1853	Nb	-	Nb	218
Platynaspis luteorubra (Goeze,1777)	Na	-	Na	219
Exochomus nigromaculatus (Goeze,1777)	Na	-	RDB I	217
Hippodamia tredecimpunctata (L.,1758)	RDB 3	RDB 3	RDB K	218
Adonia variegata (Goeze,,1777)	Nb	-	Nb	216
Coccinella magnifica Redtenbacher,1843	RDB 3	RDB 3	Na	216
Coccinella quinquepunctata L.,1758	RDB 3	RDB 3	RDB 3	217
Halyzia sedecimguttata (L.,1758)	Na	-	-	
Vibidia duodecimguttata (Poda,1761)	RDB APPENDIX	RDB APPENDIX	RDB K	220
Endomychidae				
Symbiotes latus Redtenbacher,1849	Nb	-	Nb	330
Lycoperdina bovistae (F.,1792)	Nb	-	RDB 3	330
Lycoperdina succincta (L.,1767)	RDB 2	RDB 2	RDB 2	330
Cisidae				
Rhopalodontus perforatus (Gyllenhal,1813)	Na	-	RDB 3	212
Sulcacis bicornis (Mellié,1848)	Nb	-	Nb	213
Cis coluber Abeille,1874	RDB 3	RDB 3	RDB 3	210
Cis festivus (Panzer,1793)	Nb	-	Nb	211
Cis dentatus Mellié,1848	Na	-	RDB 3	211
Cis jacquemarti Mellié,1848	Nb	-	Nb	212
Cis lineatocribratus Mellié,1848	Nb	-	Nb	212
Cis punctulatus Gyllenhal,1827	Nb	-	-	

Species	Status in Hyman (1986)	Status in Shirt (1987)	Revised Status	Page Number
Mycetophagidae				
Mycetophagus fulvicollis F.,1792	Na	-	EXTINCT	366
Mycetophagus piceus (F.,1777)	Nb	-	Nb	366
Mycetophagus populi F.,1798	Na	-	Na	367
Mycetophagus quadriguttatus Müller,1821	List 3	-	Na	367
Colydiidae				
Myrmechixenus subterraneus Chevrolat,1835	Na	-	RDB I	223
Myrmechixenus vaporariorum Guerin-Meneville,1843	Na	-	RDB 3	223
Orthocerus clavicornis (L.,1758)	Nb	-	Nb	223
Synchita humeralis (F.,1792)	Na	-	Nb	224
Synchita separanda (Reitter,1882)	RDB 3	RDB 3	RDB 3	225
Cicones undatus Guerin-Meneville,1829	Na	-	RDB 1	221
Cicones variegatus (Hellwig,1792)	Na	-	Na	221
Endophloeus markovichianus (Piller and Mitter.,1783)	RDB APPENDIX	-	RDB 1	222
Langelandia anophthalma Aubé,1843	Na	-	RDB 3	222
Colydium elongatum (F.,1787)	RDB 3	RDB 3	RDB 3	222
Aulonium ruficorne (Olivier,1790)	List 2	-	-	
Aulonium trisulcum (Fourcroy,1785)	Nb	-	Na	220
Pycnomerus fuliginosus Erichson,1842	List 1	-	-	
Teredus cylindricus (Olivier,1790)	RDB 1	RDB 1	RDB 1	225
Oxylaemus cylindricus (Panzer,1796)	RDB APPENDIX	RDB APPENDIX	EXTINCT	224
Oxylaemus variolosus (Dufour,1843)	RDB 3	RDB 3	RDB 3	224
Tenebrionidae				
Blaps lethifera Marsham,1802	List 3	-	-	
Blaps mortisaga (L.,1758)	RDB APPENDIX	RDB APPENDIX	-	
Opatrum sabulosum (L.,1758)	Nb	-	Nb	416
Crypticus quisquilius (L.,1761)	Nb	-	Nb	413
Bolitophagus reticulatus (L.,1767)	RDB 3	RDB 3	RDB 3	413
Eledona agricola (Herbst,1783)	Nb	-	Nb	414
Diaperis boleti (L.,1758)	RDB 2	RDB 2	RDB 2	414
Scaphidema metallicum (F.,1792)	Nb	-	Nb	418
Platydema violaceum (F.,1790)	RDB 1	RDB 1+	RDB 1	417
Corticeus fraxini (Kugelann,1794)	Nb	-	-	
Corticeus linearis (F.,1790)	Nb	-	-	
Corticeus unicolor Piller and Mitterpacher,1783	RDB 3	RDB 3	RDB 3	413
Helops caeruleus (L.,1758)	Nb	-	Nb	415
Cylindrinotus pallidus (Curtis,1830)	Nb	-	Nb	414
Lagria atripes Mulsant and Guillebeau,1855	Na	-	RDB 1	415
Prionychus ater (F.,1775)	Nb	-	Nb	417
Prionychus melanarius (Germar,1813)	RDB 2	RDB 2	RDB 2	417
Pseudocistela ceramboides (L.,1758)	Na	-	Nb	418
Mycetochara humeralis (F.,1787)	Nb	-	Na	416
Cteniopus sulphureus (L.,1758)	Nb	-	-	
Omophlus rufitarsis (Leske,1785)	RDB 1	RDB 1	RDB 1	416
Tetratomidae				
Tetratoma ancora F.,1791	Nb	-	Nb	419
Tetratoma desmaresti Latreille,1807	Na	-	Na	419
Salpingidae				
Lissodema cursor (Gyllenhal,1813)	Na	-	Na	380
Lissodema quadripustulata (Marsham,1802)	Nb	-	Nb	380
Rabocerus foveolatus (Ljungh,1824)	Nb	-	Na	381

Species	Status in Hyman (1986)	Status in Shirt (1987)	Revised Status	Page Number
Rabocerus gabrieli Gerhardt,1901	Nb	-	Nb	381
Salpingus ater (Paykull,1798)	Nb	-	-	

Mycteridae

Mycterus curculioides (F.,1781)	RDB APPENDIX	RDB APPENDIX	EXTINCT	368

Pythidae

Pytho depressus (L.,1767)	Na	-	Na	377

Pyrochroidae

Pyrochroa coccinea (L., 1761)	Nb	-	Nb	376
Schizotus pectinicornis (L.,1758)	RDB 3	RDB 3	Na	376

Melandryidae

Hallomenus binotatus (Quensel,1790)	Nb	-	Nb	351
Orchesia micans (Panzer,1793)	Nb	-	Nb	353
Orchesia minor Walker,1836	Nb	-	Nb	353
Anisoxya fuscula (Illiger,1798)	RDB 3	RDB 3	Na	350
Abdera affinis (Paykull,1799)	RDB 1	RDB 1	RDB 1	348
Abdera biflexuosa (Curtis,1829)	Nb	-	Nb	349
Abdera flexuosa (Paykull,1799)	Nb	-	Nb	349
Abdera quadrifasciata (Curtis,1829)	Na	-	Na	349
Abdera triguttata (Gyllenhal,1810)	Nb	-	Na	350
Phloiotrya vaudoueri Mulsant,1856	Nb	-	Nb	354
Xylita laevigata (Hellenius,1786)	Na	-	Na	354
Hypulus quercinus (Quensel,1790)	RDB 2	RDB 2	RDB 2	351
Zilora ferruginea (Paykull,1798)	Nb	-	Nb	354
Melandrya barbata (F.,1787)	RDB 1	RDB 1	RDB 1	352
Melandrya caraboides (L.,1761)	Nb	-	Nb	352
Conopalpus testaceus (Olivier,1790)	Nb	-	Nb	351
Osphya bipunctata (F.,1775)	RDB 3	RDB 3	RDB 3	353

Scraptiidae

Scraptia dubia (Olivier,1790)	Na	-	EXTINCT	407
Scraptia fuscula Müller,1821	Na	-	RDB 1	408
Scraptia testacea Allen,1940	Na	-	RDB 3	408
Anaspis bohemica Schilsky,1898	Na	-	RDB K	405
Anaspis melanostoma Costa,1854	RDB 3	RDB 3	RDB K	406
Anaspis schilskyana Csiki,1915	RDB 1	RDB 1	RDB I	406
Anaspis septentrionalis Champion,1891	Na	-	EXTINCT	407
Anaspis thoracica (L.,1758)	Nb	-	Na	407

Mordellidae

Tomoxia bucephala Costa,1854	RDB 3	RDB 3	Na	365
Variimorda villosa (Schrank,1781)	Nb	-	Nb	366
Mordella holomelaena Apfelbeck,1914	Na	-	RDB K	361
Mordella leucaspis Kuester,1849	Na	-	RDB K	362
Mordellistena brevicauda (Boheman,1849)	Nb	-	RDB K	363
Mordellistena humeralis (L.,1758)	Nb	-	RDB K	363
Mordellistena neuwaldeggiana (Panzer,1796)	Nb	-	RDB K	364
Mordellistena parvula (Gyllenhal,1827)	Nb	-	RDB K	364
Mordellistena acuticollis Schilsky,1895	Na	-	RDB K	362
Mordellistena nanuloides Ermisch,1967	Na	-	RDB K	364
Mordellistena pseudopumila Ermisch,1962	Na	-	RDB K	365

Species	Status in Hyman (1986)	Status in Shirt (1987)	Revised Status	Page Number
Mordellistena parvuloides Ermisch,1956	Na	-	RDB K	365
Rhipiphoridae				
Metoecus paradoxus (L.,1761)	Nb	-	-	
Oedemeridae				
Chrysanthia nigricornis (Westhoff,1881)	RDB 1	RDB 1	RDB 1	368
Ischnomera cyanea (F.,1787)	Nb	-	Nb	369
Ischnomera cinerascens (Pandelle,1867)	RDB 2	RDB 2	RDB 2	369
Ischnomera caerulea (L.,1758)	-	-	RDB 3	368
Ischnomera sanguinicollis (F.,1787)	Na	-	Nb	370
Oncomera femorata (F.,1792)	Nb	-	Nb	370
Oedemera virescens (L.,1767)	RDB 3	RDB 3	RDB 2	370
Meloidae				
Lytta vesicatoria (L.,1758)	List 2	-	-	
Meloe autumnalis Olivier,1792	RDB 3	RDB 3	RDB 1	355
Meloe brevicollis Panzer,1793	RDB 3	RDB 3	RDB 1	355
Meloe cicatricosus Leach,1811	RDB 3	RDB 3	RDB 1	356
Meloe rugosus Marsham,1802	RDB 3	RDB 3	RDB 3	356
Meloe variegatus Donovan,1793	RDB 3	RDB 3	EXTINCT	356
Meloe violaceus Marsham,1802	Nb	-	Nb	356
Apalus muralis (Forster,1771)	RDB 1	RDB 1	RDB 1	355
Anthicidae				
Anthicus angustatus Curtis,1838	Nb	-	Nb	67
Anthicus bifasciatus (Rossi,1792)	Nb	-	Nb	67
Anthicus bimaculatus (Illiger,1801)	Na	-	Na	67
Anthicus salinus Crotch,1866	Nb	-	Na	68
Anthicus scoticus Rye,1872	Na	-	RDB 3	68
Anthicus tobias Marseul,1879	Nb	-	-	
Anthicus tristis Schmidt,1842	Nb	-	RDB 1	68
Aderidae				
Aderus brevicornis (Perris,1869)	RDB 1	RDB 3	RDB 2	61
Aderus oculatus (Paykull,1798)	Nb	-	Nb	61
Aderus populneus (Creutzer in Panzer,1796)	Na	-	Nb	61
Cerambycidae				
Prionus coriarius (L.,1758)	Na	-	Na	164
Arhopalus tristis (F.,1787)	Nb	-	-	
Tetropium castaneum (L.,1758)	RDB 3	RDB 3	-	
Rhagium inquisitor (L.,1758)	Nb	-	Nb	165
Acmaeops collaris (L.,1758)	RDB 1	RDB 1	RDB 1	155
Grammoptera holomelina Pool,1905	List 2	-	-	
Grammoptera ustulata (Schaller,1783)	RDB 3	RDB 3	RDB 3	157
Grammoptera variegata (Germar,1824)	Na	-	Na	158
Leptura fulva Degeer,1775	Na	-	RDB 3	159
Leptura rubra L.,1758	RDB 3	RDB 3	-	
Leptura rufa Brulle,1832	RDB 3	-	-	
Leptura sanguinolenta L.,1761	Na	-	RDB 3	159
Leptura scutellata F.,1781	Na	-	Na	160
Leptura sexguttata F.,1775	RDB 3	RDB 3	RDB 3	160
Leptura virens L.,1758	RDB APPENDIX	RDB APPENDIX	EXTINCT	161

APPENDIX 3

Species	Status in Hyman (1986)	Status in Shirt (1987)	Revised Status	Page Number
Judolia cerambyciformis (Schrank,1781)	Nb	-	-	
Judolia sexmaculata (L.,1758)	Na	-	Na	158
Strangalia attenuata (L.,1758)	RDB APPENDIX	RDB APPENDIX	EXTINCT	167
Strangalia aurulenta (F.,1792)	Na	-	Na	167
Strangalia nigra (L.,1758)	Na	-	Na	168
Strangalia quadrifasciata (L.,1758)	Nb	-	-	
Strangalia revestita (L.,1767)	RDB 3	RDB 3	RDB 1	168
Trinophyllum cribratum Bates,1878	List 2	-	-	
Gracilia minuta (F.,1781)	Na	-	RDB 2	157
Obrium brunneum (F.,1792)	Na	-	-	
Obrium cantharinum (L.,1767)	RDB APPENDIX	RDB APPENDIX	EXTINCT	162
Molorchus minor (L.,1758)	Nb	-	-	
Molorchus umbellatarum (von Schreber,1759)	Na	-	Na	162
Aromia moschata (L.,1758)	Na	-	Nb	156
Hylotrupes bajulus (L.,1758)	List 3	-	-	
Callidium violaceum (L.,1758)	RDB 3	RDB 3	-	
Pyrrhidium sanguineum (L.,1758)	RDB 1	RDB 2	RDB 2	165
Phymatodes alni (L.,1767)	Nb	-	Nb	163
Plagionotus arcuatus (L.,1758)	RDB APPENDIX	RDB APPENDIX	EXTINCT	163
Anaglyptus mysticus (L.,1758)	Nb	-	Nb	156
Lamia textor (L.,1758)	RDB 2	RDB 2	RDB 1	159
Mesosa nebulosa (F.,1781)	RDB 3	RDB 3	RDB 3	161
Pogonocherus fasciculatus (Degeer,1775)	Nb	-	Nb	164
Acanthocinus aedilis (L.,1758)	Nb	-	Nb	155
Agapanthia villosoviridescens (Degeer,1775)	Nb	-	-	
Saperda carcharias (L.,1758)	Na	-	Na	166
Saperda scalaris (L.,1758)	Na	-	Na	166
Oberea oculata (L.,1758)	RDB 1	RDB 1	RDB 1	162
Stenostola dubia (Laicharting,1784)	Nb	-	Nb	167
Phytoecia cylindrica (L.,1758)	Nb	-	Nb	163

Bruchidae

Species	Status in Hyman (1986)	Status in Shirt (1987)	Revised Status	Page Number
Bruchus atomarius (L.,1761)	-	-	Nb	88
Bruchus pisorum (L.,1758)	Nb	-	-	
Bruchus rufipes Herbst,1783	Nb	-	-	
Bruchidius cisti (F.,1775)	Nb	-	-	
Bruchidius incarnatus (Boheman,1833)	Na	-	-	
Bruchidius olivaceus (Germar,1824)	Na	-	RDB 1	87

Chrysomelidae

Species	Status in Hyman (1986)	Status in Shirt (1987)	Revised Status	Page Number
Macroplea appendiculata (Panzer,1794)	RDB 3	RDB 3	RDB 3	200
Macroplea mutica (F.,1792)	RDB 3	RDB 3	Na	201
Donacia aquatica (L.,1758)	Na	-	RDB 3	186
Donacia bicolora Zschach,1788	Nb	-	RDB 2	186
Donacia cinerea Herbst, 1784	Nb	-	Nb	187
Donacia clavipes Fabricius, 1792	Nb	-	Nb	187
Donacia crassipes F.,1775	Na	-	Nb	187
Donacia dentata Hoppe,1795	Na	-	Na	187
Donacia impressa Paykull,1799	Na	-	Na	188
Donacia obscura Gyllenhal, 1813	RDB 2	RDB 2	Na	188
Donacia sparganii Ahrens,1810	Na	-	Na	189
Donacia thalassina Germar,1811	Na	-	Nb	189
Plateumaris affinis (Kunze, 1818)	Nb	-	Nb	206
Plateumaris braccata (Scopoli, 1772)	Na	-	Na	206
Orsodacne cerasi (L.,1758)	Nb	-	-	
Orsodacne lineola (Panzer,1795)	Nb	-	Nb	203
Zeugophora flavicollis (Marsham,1802)	RDB 1	RDB 1	RDB 2	210
Zeugophora subspinosa (F.,1781)	Nb	-	-	

Species	Status in Hyman (1986)	Status in Shirt (1987)	Revised Status	Page Number
Zeugophora turneri Power,1863	Na	-	Na	210
Lema cyanella (L.,1758)	Nb	-	-	
Oulema erichsoni (Suffrian,1841)	RDB 3	RDB 3	RDB 1	203
Oulema septentrionis Weise,1880	List 4	-	-	
Lilioceris lilii (Scopoli,1763)	List 2	-	-	
Labidostomis tridentata (L.,1758)	RDB 1	RDB 1	RDB 1	191
Clytra laeviuscula Ratzeburg,1837	RDB APPENDIX	RDB APPENDIX	EXTINCT	180
Clytra quadripunctata (L.,1758)	Nb	-	-	
Gynandrophthalma affinis (Illiger,1794)	RDB 1	RDB 1	RDB 1	190
Cryptocephalus aureolus Suffrian,1847	Nb	-	Nb	180
Cryptocephalus biguttatus (Scopoli,1763)	RDB 2	RDB 2	RDB 2	180
Cryptocephalus bilineatus (L.,1767)	Nb	-	Nb	181
Cryptocephalus bipunctatus (L.,1758)	Nb	-	Nb	181
Cryptocephalus coryli (L.,1758)	RDB 1	RDB 1	RDB 1	181
Cryptocephalus decemmaculatus (L.,1758)	RDB 2	RDB 2	RDB 2	182
Cryptocephalus exiguus Schneider,1792	RDB 1	RDB 1	RDB 1	182
Cryptocephalus frontalis Marsham,1802	Na	-	Na	182
Cryptocephalus hypochaeridis (L.,1758)	Nb	-	-	
Cryptocephalus moraei (L.,1758)	Nb	-	-	
Cryptocephalus nitidulus F.,1787	RDB 1	RDB 1	RDB 1	183
Cryptocephalus parvulus Müller,1776	Na	-	Nb	183
Cryptocephalus primarius Harold,1872	RDB 1	RDB 1	RDB 1	184
Cryptocephalus punctiger Paykull,1799	Na	-	Na	184
Cryptocephalus querceti Suffrian,1848	RDB 2	RDB 2	RDB 2	184
Cryptocephalus sexpunctatus (L.,1758)	RDB 2	RDB 2	RDB 2	185
Cryptocephalus violaceus Laicharting,1781	RDB APPENDIX	RDB APPENDIX	EXTINCT	185
Bromius obscurus (L.,1758)	RDB 1	RDB 1	RDB 1	172
Lamprosoma concolor (Sturm,1807)	Nb	-	-	
Chrysolina brunsvicensis (Gravenhorst,1807)	Nb	-	-	
Chrysolina cerealis (L.,1767)	RDB 1	RDB 1!	RDB 1	176
Chrysolina crassicornis (Helliesin,1911)	RDB 1	RDB 2	RDB 2	176
Chrysolina graminis (L.,1758)	Na	-	Na	177
Chrysolina haemoptera (L.,1758)	Nb	-	Nb	177
Chrysolina marginata (L.,1758)	Na	-	Na	178
Chrysolina menthastri (Suffrian,1851)	Nb	-	-	
Chrysolina oricalcia (Müller,1776)	Nb	-	Nb	178
Chrysolina sanguinolenta (L.,1758)	Na	-	Na	178
Chrysolina varians (Schaller,1783)	Nb	-	-	
Chrysolina violacea (Müller,1776)	Nb	-	Nb	179
Phaedon concinnus Stephens,1831	Na	-	Nb	204
Hydrothassa glabra (Herbst,1783)	Nb	-	-	
Hydrothassa hannoveriana (F.,1775)	RDB 3	RDB 3	RDB 3	191
Chrysomela aenea L.,1758	Nb	-	-	
Chrysomela populi L.,1758	Nb	-	-	
Chrysomela tremula F.,1787	RDB 1	RDB 1	RDB 1	179
Phytodecta decemnotata (Marsham,1802)	Nb	-	Nb	205
Phytodecta viminalis (L.,1758)	Nb	-	-	
Phyllodecta polaris Schneider,1886	RDB 3	RDB 3	RDB 3	204
Galerucella grisescens (Joannis,1865)	List 2	-	-	
Galeruca interrupta Illiger,1802	RDB 1	RDB 1	RDB 1	190
Galeruca tanaceti (L.,1758)	Nb	-	-	
Phyllobrotica quadrimaculata (L.,1758)	Nb	-	-	
Luperus flavipes (L.,1767)	Nb	-	Nb	200
Calomicrus circumfusus (Marsham,1802)	Na	-	Na	173
Agelastica alni (L.,1758)	RDB APPENDIX	RDB APPENDIX	RDB K	170
Phyllotreta aerea Allard,1859	Nb	-	Nb	204
Phyllotreta cruciferae (Goeze,1777)	Nb	-	Nb	204
Phyllotreta flexuosa (Illiger,1794)	Nb	-	-	
Phyllotreta tetrastigma (Comolli,1837)	Nb	-	-	
Phyllotreta vittata (F.,1801)	Na	-	Na	205

Species	Status in Hyman (1986)	Status in Shirt (1987)	Revised Status	Page Number
Aphthona atrovirens Foerster,1849	Nb	-	-	
Aphthona herbigrada (Curtis,1837)	Nb	-	-	
Aphthona nigriceps (Redtenbacher,1842)	Na	-	Na	171
Longitarsus absynthii Kutschera,1862	Na	-	Na	191
Longitarsus aeruginosus (Foudras,1860)	Na	-	RDB 1	192
Longitarsus agilis (Rye,1868)	Na	-	Na	192
Longitarsus anchusae (Paykull,1799)	Na	-	Nb	192
Longitarsus ballotae (Marsham,1802)	Nb	-	Nb	193
Longitarsus bearei Kevan,1967	Na	-	RDB K	193
Longitarsus brunneus (Duftschmid,1825)	Na	-	Nb	193
Longitarsus longiseta Weise,1889	Na	-	RDB K	195
Longitarsus curtus (Allard,1860)	Na	-	Na	193
Longitarsus dorsalis (F.,1781)	Nb	-	Nb	194
Longitarsus ferrugineus (Foudras,1860)	Na	-	RDB 1	194
Longitarsus fowleri Allen,1967	Na	-	Na	195
Longitarsus ganglbaueri Heikertinger,1912	Na	-	Na	195
Longitarsus holsaticus (L.,1758)	Nb	-	-	
Longitarsus lycopi (Foudras,1860)	Na	-	Nb	196
Longitarsus nasturtii (F.,1792)	Na	-	Nb	196
Longitarsus nigerrimus (Gyllenhal,1827)	RDB 1	RDB 1	RDB 1	196
Longitarsus nigrofasciatus (Goeze,1777)	Na	-	Na	197
Longitarsus obliteratus (Rosenhauer,1847)	Nb	-	-	
Longitarsus ochroleucus (Marsham,1802)	Nb	-	Nb	197
Longitarsus parvulus (Paykull,1799)	Na	-	Na	197
Longitarsus pellucidus (Foudras,1860)	Nb	-	-	
Longitarsus plantagomaritimus Dollman,1912	Na	-	Nb	198
Longitarsus quadriguttatus (Pontoppidan,1763)	RDB 3	RDB 3	Na	198
Longitarsus reichei (Allard,1860)	Na	-	-	
Longitarsus rutilus (Illiger,1807)	RDB 2	RDB 2	Na	199
Longitarsus suturalis (Marsham,1802)	Na	-	Nb	199
Longitarsus tabidus (F.,1775)	Nb	-	Nb	199
Altica brevicollis Foudras,1860	Na	-	Na	170
Altica britteni Sharp,1914	Nb	-	-	
Altica ericeti (Allard,1859)	Nb	-	Nb	171
Batophila aerata (Marsham,1802)	Nb	-	Nb	
Lythraria salicariae (Paykull,1800)	Nb	-	Nb	200
Ochrosis ventralis (Illiger,1807)	Nb	-	RDB 3	202
Crepidodera impressa (F.,1801)	Na	-	Na	180
Chalcoides nitidula (L.,1758)	Na	-	Nb	176
Epitrix atropae Foudras,1860	Nb	-	Nb	190
Podagrica fuscicornis (L.,1767)	Nb	-	Nb	207
Podagrica fuscipes (F.,1775)	Na	-	Na	207
Mantura chrysanthemi (Koch,1803)	Na	-	Na	201
Mantura matthewsi (Curtis,1833)	Nb	-	-	
Mantura obtusata (Gyllenhal,1813)	Nb	-	Nb	202
Mantura rustica (L.,1767)	Nb	-	Nb	202
Chaetocnema aerosa (Letzner,1847)	Na	-	RDB K	175
Chaetocnema arida Foudras,1860	Na	-	-	
Chaetocnema aridula (Gyllenhal,1827)	Na	-	-	
Chaetocnema conducta (Motschulsky,1838)	RDB 3	RDB 3	RDB I	175
Chaetocnema confusa (Boheman,1851)	Na	-	-	
Chaetocnema sahlbergi (Gyllenhal,1827)	Na	-	Na	175
Chaetocnema subcoerulea (Kutschera,1864)	Nb	-	Nb	176
Apteropeda globosa (Illiger,1794)	Na	-	Nb	172
Apteropeda splendida Allard,1860	Na	-	RDB 1	172
Mniophila muscorum (Koch,1803)	Nb	-	Nb	202
Dibolia cynoglossi (Koch,1803)	RDB 1	RDB 1	RDB 1	185
Psylliodes attenuata (Koch,1803)	Na	-	RDB 1	207
Psylliodes chalcomera (Illiger,1807)	Na	-	Nb	208
Psylliodes hyoscyami (L.,1758)	RDB 1	RDB 1	RDB 1	208

Species	Status in Hyman (1986)	Status in Shirt (1987)	Revised Status	Page Number
Psylliodes luridipennis Kutschera,1864	RDB 1	RDB 1 (5)	RDB 2 Endemic	208
Psylliodes luteola (Müller,1776)	Na	-	RDB K	209
Psylliodes sophiae Heikertinger,1914	RDB 3	RDB 3	RDB 3	209
Pilemostoma fastuosa (Schaller,1783)	Na	-	Na	206
Hypocassida subferruginea (Schrank,1776)	RDB APPENDIX	RDB APPENDIX	EXTINCT	191
Cassida denticollis Suffrian,1844	RDB 3	RDB 3	RDB 1	173
Cassida hemisphaerica Herbst,1799	Na	-	Na	173
Cassida murraea L.,1767	Na	-	-	
Cassida nebulosa L.,1758	Na	-	RDB I	174
Cassida nobilis L.,1758	Nb	-	Nb	174
Cassida prasina Illiger,1798	Nb	-	Nb	174
Cassida sanguinosa Suffrian,1844	List 4	-	-	
Cassida vittata de Villers,1789	Nb	-	-	

Nemonychidae

Cimberis attelaboides Fabricius,1787	Nb	-	-	

Anthribidae

Platyrhinus resinosus (Scopoli,1763)	Na	-	Nb	70
Tropideres sepicola (F.,1792)	RDB 3	RDB 3	RDB 2	71
Tropideres niveirostris (F.,1798)	RDB 3	RDB 3	RDB 2	71
Platystomos albinus (L.,1758)	Na	-	Nb	70
Anthribus fasciatus Forster,1771	Na	-	Na	69
Anthribus nebulosus Forster,1771	Nb	-	Nb	69
Choragus sheppardi Kirby,1819	Na	-	Na	69

Urodontidae

Bruchela rufipes (Olivier,1790)	RDB 3	RDB 3	-	

Attelabidae

Attelabus nitens (Scopoli,1763)	Nb	-	-	
Apoderus coryli (L.,1758)	Nb	-	-	
Rhynchites auratus (Scopoli,1763)	RDB APPENDIX	RDB APPENDIX	EXTINCT	84
Rhynchites bacchus (L.,1758)	RDB APPENDIX	RDB APPENDIX	EXTINCT	84
Rhynchites cupreus (L.,1758)	Nb	-	Nb	85
Rhynchites cavifrons Gyllenhal,1833	-	-	Nb	84
Rhynchites olivaceus Gyllenhal,1833	Na	-	Na	85
Rhynchites sericeus Herbst,1797	RDB APPENDIX	RDB APPENDIX	-	
Rhynchites interpunctatus Stephens,1831	-	-	Nb	85
Rhynchites longiceps Thomson,1888	Nb	-	Nb	85
Rhynchites pauxillus Germar,1824	Na	-	RDB 3	86
Rhynchites tomentosus Gyllenhal,1839	-	-	Nb	86
Byctiscus betulae (L.,1758)	Na	-	Nb	83
Byctiscus populi (L.,1758)	Na	-	RDB 3	83

Apionidae

Apion affine Kirby,1808	Na	-	Na	72
Apion lemoroi Brisout,1880	RDB 3	RDB 3	RDB K	78
Apion limonii Kirby,1808	Nb	-	Nb	78
Apion sedi Germar,1818	Nb	-	Nb	81
Apion soror Rey,1895	Na	-	Na	81
Apion pallipes Kirby,1808	Nb	-	-	
Apion semivittatum Gyllenhal,1833	Na	-	Na	81
Apion urticarium (Herbst,1784)	Nb	-	-	
Apion difficile Herbst,1797	Na	-	Na	74

Species	Status in Hyman (1986)	Status in Shirt (1987)	Revised Status	Page Number
Apion genistae Kirby,1811	Na	-	Na	76
Apion cruentatum Walton,1844	Nb	-	-	
Apion rubiginosum Grill,1893	Na	-	RDB 3	80
Apion cineraceum Wencker,1864	Na	-	Na	73
Apion flavimanum Gyllenhal,1833	Na	-	Na	76
Apion vicinum Kirby,1808	Nb	-	Nb	82
Apion curtisi Stephens,1831	Nb	-	Na	74
Apion pubescens Kirby,1811	Nb	-	Nb	79
Apion stolidum Germar,1817	Nb	-	Nb	82
Apion brunnipes Boheman,1839	RDB 3	RDB 3	RDB 1	73
Apion armatum Gerstaecker,1854	List 2	-	RDB K	72
Apion dispar Germar,1817	RDB 3	RDB 3*	RDB 3	75
Apion laevigatum (Paykull,1792)	Na	-	RDB I	78
Apion astragali (Paykull,1800)	Na	-	Na	72
Apion afer Gyllenhal,1833	Nb	-	-	
Apion gyllenhali Kirby,1808	Nb	-	Nb	77
Apion intermedium Eppelsheim,1875	Na	-	Na	77
Apion minimum Herbst,1797	Na	-	RDB 3	79
Apion reflexum Gyllenhal,1833	Na	-	Na	79
Apion waltoni Stephens,1839	Nb	-	-	
Apion cerdo Gerstaecker,1854	Nb	-	Nb	73
Apion difforme Ahrens,1817	-	-	Nb	75
Apion dissimile Germar,1817	Na	-	Nb	75
Apion filirostre Kirby,1808	Nb	-	Nb	75
Apion laevicolle Kirby,1811	Na	-	Na	77
Apion ryei Blackburn,1874	List 2	RDB 5	Nb	80
Apion schoenherri Boheman,1839	Na	-	Na	80
Apion varipes Germar,1817	Nb	-	Nb	82
Nanophyes gracilis Redtenbacher,1849	Na	-	RDB 3	83

Curculionidae

Species	Status in Hyman (1986)	Status in Shirt (1987)	Revised Status	Page Number
Otiorhynchus aurifer Boheman,1843	List 1	-	-	
Otiorhynchus auropunctatus Gyllenhal,1834	RDB 1	RDB 1	RDB 1	284
Otiorhynchus crataegi Germar,1824	List 1	-	-	
Otiorhynchus desertus Rosenhauer,1847	Nb	-	Nb	284
Otiorhynchus ligustici (L.,1758)	RDB 2	RDB 2	RDB 2	284
Otiorhynchus morio (F.,1781)	RDB 3	RDB 3	RDB I	285
Otiorhynchus niger (Fabricius,1775)	List 2	-	-	
Otiorhynchus porcatus (Herbst,1795)	Nb	-	Nb	
Otiorhynchus raucus (F.,1777)	Nb	-	Nb	285
Otiorhynchus scaber (L.,1758)	Nb	-	Nb	285
Otiorhynchus uncinatus Germar,1824	List 4	-	-	
Caenopsis fissirostris (Walton,1847)	Nb	-	Nb	244
Caenopsis waltoni (Boheman,1843)	Nb	-	-	
Peritelus sphaeroides Germar,1824	RDB APPENDIX	RDB APPENDIX	-	
Trachyphloeus alternans Gyllenhal,1834	Na	-	Nb	303
Trachyphloeus aristatus (Gyllenhal,1827)	Nb	-	Nb	303
Trachyphloeus asperatus Boheman,1843	Nb	-	Nb	304
Trachyphloeus digitalis (Gyllenhal,1827)	Na	-	Na	304
Trachyphloeus laticollis Boheman,1843	Na	-	Na	304
Trachyphloeus spinimanus Germar,1824	Nb	-	Nb	305
Cathormiocerus attaphilus Brisout,1880	RDB 1	RDB 1	RDB 1	244
Cathormiocerus britannicus Blair,1934	RDB 1	RDB 1	RDB 1 Endemic	244
Cathormiocerus maritimus Rye,1874	RDB 3	RDB 3	RDB 3	245
Cathormiocerus myrmecophilus (Seidlitz,1868)	RDB 3	RDB 3	RDB 3	245
Cathormiocerus socius Boheman,1843	RDB 2	RDB 2	RDB 2	245
Omiamima mollina (Boheman,1834)	RDB 3	RDB 3	Na	283
Phyllobius vespertinus (F.,1792)	Nb	-	Nb	286
Polydrusus chrysomela (Olivier,1807)	Nb	-	-	

Species	Status in Hyman (1986)	Status in Shirt (1987)	Revised Status	Page Number
Polydrusus confluens Stephens,1831	Nb	-	Nb	290
Polydrusus flavipes (Degeer,1775)	Nb	-	Nb	290
Polydrusus marginatus Stephens,1831	Na	-	RDB2	290
Polydrusus mollis (Ström,1768)	Nb	-	Nb	291
Polydrusus pilosus Gredler,1866	Na	-	-	
Polydrusus prasinus (Olivier,1790)	RDB APPENDIX	RDB APPENDIX	-	
Polydrusus pulchellus Stephens,1831	Na	-	Nb	291
Polydrusus sericeus (Schaller,1783)	Na	-	Na	291
Barypeithes curvimanus (Jaquelin du Val,1855)	List 4	-	-	
Barypeithes pyrenaeus Seidlitz,1868	Na	-	RDB I	243
Barypeithes sulcifrons (Boheman,1843)	Nb	-	Nb	243
Brachysomus echinatus (Bonsdorff,1785)	Nb	-	Nb	243
Brachysomus hirtus (Boheman,1845)	RDB 3	RDB 3	RDB 3	243
Strophosoma fulvicorne Walton,J.,1846	RDB 3	RDB 3	RDB 2	301
Strophosoma faber (Herbst,1784)	Nb	-	Nb	301
Cneorhinus plumbeus (Marsham,1802)	-	-	Nb	258
Barynotus squamosus Germar,1824	-	-	Nb	242
Tropiphorus elevatus (Herbst,1795)	Nb	-	Nb	307
Tropiphorus obtusus (Bonsdorff,1785)	Na	-	Na	307
Tropiphorus terricola (Newman,1838)	-	-	Nb	307
Tanymecus palliatus (F.,1787)	Na	-	Nb	302
Sitona ambiguus Gyllenhal,1834	Nb	-	-	
Sitona cambricus Stephens,1831	Nb	-	-	
Sitona cinerascens (Fahraeus,1840)	List 2	-	RDB K	298
Sitona cylindricollis (Fahraeus,1840)	Nb	-	-	
Sitona gemellatus Gyllenhal,1834	RDB 1	RDB 1	RDB 1	298
Sitona lineellus (Bonsdorff,1785)	Nb	-	-	
Sitona macularius (Marsham,1802)	Nb	-	Nb	298
Sitona ononidis Sharp,1866	Na	-	Nb	299
Sitona puberulus Reitter,1903	Na	-	RDB K	299
Sitona waterhousei Walton,1846	Nb	-	Nb	299
Coniocleonus hollbergi (Fahraeus,1842)	RDB APPENDIX	RDB APPENDIX	EXTINCT	259
Coniocleonus nebulosus (L.,1758)	Na	-	RDB 3	260
Chromoderus affinis (Schrank,1781)	RDB 3	RDB 3	EXTINCT	257
Cleonus piger (Scopoli,1763)	Na	-	Nb	258
Lixus algirus (L.,1758)	RDB 1	RDB 1	RDB 1	276
Lixus scabricollis Boheman,1843	-	-	RDB K	276
Lixus elongatus (Goeze,1777)	List 2	-	-	
Lixus iridis Olivier,1807	List 2	-	-	
Lixus paraplecticus (L.,1758)	RDB 1	RDB 1	RDB 1	276
Lixus vilis (Rossi,1790)	RDB 1	RDB 1	RDB 1	277
Larinus planus (F.,1792)	Nb	-	Nb	273
Rhinocyllus conicus (Froelich,1792)	Na	-	Na	293
Hypera arundinis (Paykull,1792)	RDB APPENDIX	RDB APPENDIX	EXTINCT	271
Hypera dauci (Olivier,1807)	Na	-	Nb	271
Hypera diversipunctata (Schrank,1798)	RDB 3	RDB 3	RDB 3	271
Hypera fuscocinerea (Marsham,1802)	Nb	-	Nb	271
Hypera meles (F.,1792)	RDB 3	RDB 3	Na	272
Hypera ononidis Chevrolat,1863	Na	-	RDB K	272
Hypera pastinacae (Rossi,1790)	RDB 1	RDB 1	RDB 1	273
Limobius borealis (Paykull,1792)	Nb	-	Na	274
Limobius mixtus (Boheman,1834)	RDB 2	RDB 2	RDB 1	275
Cionus longicollis Brisout,1863	Na	-	Na	257
Cionus nigritarsis Reitter,1904	Nb	-	Na	257
Alophus triguttatus (F.,1775)	Nb	-	Nb	
Lepyrus capucinus (Schaller,1783)	RDB APPENDIX	RDB APPENDIX	EXTINCT	274
Hylobius transversovittatus (Goeze,1777)	RDB 3	RDB 3	RDB 1	270
Liparus coronatus (Goeze,1777)	Nb	-	Nb	275
Liparus germanus (L.,1758)	RDB 2	RDB 2	RDB 2	275
Leiosoma oblongulum Boheman,1842	Na	-	Nb	273

APPENDIX 3

Species	Status in Hyman (1986)	Status in Shirt (1987)	Revised Status	Page Number
Leiosoma pyrenaeum Brisout,1866	RDB 3	RDB 3	RDB 2	274
Plinthus caliginosus (F.,1775)	Na	-	Na	289
Anchonidium unguiculare (Aubé,1850)	RDB 2	RDB 2	RDB 2	231
Syagrius intrudens Waterhouse,1903	RDB 3	RDB 3	-	
Gronops inaequalis Boheman,1842	List 1	-	RDB K	266
Gronops lunatus (F.,1775)	Nb	-	Nb	267
Pissodes castaneus (Degeer,1775)	Nb	-	-	
Pissodes validirostris (Sahlberg,1834)	RDB 3	RDB 3	RDB 3	289
Magdalis barbicornis (Latreille,1804)	Nb	-	Na	277
Magdalis carbonaria (L.,1758)	Nb	-	Nb	277
Magdalis cerasi (L.,1758)	-	-	Nb	278
Magdalis duplicata Germar,1818	Na	-	Na	278
Magdalis memnonia (Gyllenhal in Faldermann,1837)	RDB 3	RDB 3*	-	
Magdalis phlegmatica (Herbst,1797)	Na	-	Na	278
Anoplus roboris Suffrian,1840	Na	-	Nb	231
Euophryum rufum (Broun,1880)	List 2	-	-	
Pentarthrum huttoni Wollaston,1854	Na	-	-	
Mesites tardii (Curtis,1825)	Nb	-	Nb	280
Cossonus linearis (F.,1775)	Na	-	Na	260
Cossonus parallelepipedus (Herbst,1795)	Nb	-	Nb	260
Rhyncolus ater (L.,1758)	Nb	-	-	
Phloeophagus gracilis (Rosenhauer,1856)	RDB APPENDIX	RDB APPENDIX	EXTINCT	286
Phloeophagus truncorum (Germar,1824)	Na	-	Na	286
Caulotrupodes aeneopiceus (Boheman,1845)	Nb	-	-	
Pselactus spadix (Herbst,1795)	Nb	-	Nb	293
Dryophthorus corticalis (Paykull,1792)	RDB 1	RDB 1	RDB 1	265
Trachodes hispidus (L.,1758)	Na	-	Nb	303
Cryptorhynchus lapathi (L.,1758)	Nb	-	Nb	261
Acalles misellus Boheman,1844	Nb	-	-	
Acalles ptinoides (Marsham,1802)	Nb	-	Nb	230
Acalles roboris Curtis,1835	Nb	-	Nb	230
Stenopelmus rufinasus Gyllenhal,1836	Nb	-	-	
Bagous petro (Herbst,1795)	RDB APPENDIX	RDB APPENDIX	EXTINCT	239
Bagous cylindrus (Paykull,1800)	RDB 2	RDB 2	RDB 2	235
Bagous arduus Sharp,1917	RDB 3	RDB 3	RDB K	234
Bagous argillaceus Gyllenhal,1836	RDB 2	RDB 2	RDB 2	234
Bagous binodulus (Herbst,1795)	RDB 1	RDB 1	EXTINCT	234
Bagous brevis Gyllenhal,1836	RDB 1	RDB 1	RDB 1	235
Bagous collignensis (Herbst,1797)	Nb	-	RDB 3	235
Bagous czwalinai Seidlitz,1891	RDB 1	RDB 1	RDB 1	236
Bagous diglyptus Boheman,1845	RDB 1	RDB 1	RDB 1	236
Bagous frit (Herbst,1795)	RDB 1	RDB 1	RDB 3	236
Bagous limosus (Gyllenhal,1827)	Nb	-	Nb	237
Bagous longitarsis Thomson,1868	RDB 1	RDB 1	RDB 1	237
Bagous lutulosus (Gyllenhal,1827)	Na	-	Na	238
Bagous nodulosus Gyllenhal,1836	RDB 2	RDB 1	RDB 1	239
Bagous subcarinatus Gyllenhal,1836	Na	-	Na	240
Bagous tempestivus (Herbst,1795)	Nb	-	Nb	240
Bagous glabrirostris (Herbst,1795)	Nb	-	Nb	237
Bagous lutosus (Gyllenhal,1813)	Na	RDB 1	RDB 1	238
Bagous lutulentus (Gyllenhal,1813)	Nb	-	Nb	238
Bagous puncticollis Boheman,1854	RDB 1	RDB 1	RDB 1	239
Bagous rudis Sharp,1917	Na	-	-	
Hydronomus alismatis (Marsham,1802)	Nb	-	Nb	270
Dorytomus affinis (Paykull,1800)	RDB 2	RDB 2	RDB 2	262
Dorytomus filirostris (Gyllenhal,1836)	Nb	-	Nb	262
Dorytomus hirtipennis Bedel,1884	Na	-	Na	262
Dorytomus ictor (Herbst,1795)	Nb	-	Nb	263
Dorytomus majalis (Paykull,1792)	Na	-	RDB K	263
Dorytomus salicinus (Gyllenhal,1827)	Na	-	Nb	263

Species	Status in Hyman (1986)	Status in Shirt (1987)	Revised Status	Page Number
Dorytomus salicis Walton,1851	Na	-	Na	264
Dorytomus tremulae (F.,1787)	Na	-	Nb	264
Procas armillatus (F.,1801)	RDB 3	RDB 3	RDB I	292
Procas granulicollis Walton,1848	RDB 3	RDB APPENDIX	RDB I Endemic	292
Notaris aethiops (F.,1792)	Na	-	Na	282
Notaris bimaculatus (F.,1787)	Nb	-	Nb	282
Notaris scirpi (F.,1792)	-	-	Nb	282
Thryogenes scirrhosus (Gyllenhal,1836)	Nb	-	Nb	302
Grypus equiseti (F.,1775)	Nb	-	Nb	267
Pachytychius haematocephalus (Gyllenhal,1836)	RDB 1	RDB 1	RDB 1	285
Pseudostyphlus pillumus (Gyllenhal,1836)	Na	-	Na	293
Orthochaetes insignis (Aubé,1863)	Nb	-	Nb	283
Orthochaetes setiger (Beck,1817)	Nb	-	Nb	283
Smicronyx coecus (Reich,1797)	RDB 3	RDB 3	RDB 3	300
Smicronyx jungermanniae (Reich,1797)	Nb	-	Nb	300
Smicronyx reichi (Gyllenhal,1836)	Nb	-	RDB3	300
Mononychus punctumalbum (Herbst,1784)	Na	-	Na	281
Rutidosoma globulus (Herbst,1795)	Na	-	Na	296
Coeliodes erythroleucos (Gmelin in L.,1790)	Nb	-	Nb	259
Coeliodes nigritarsis Hartmann,1895	Na	-	Na	259
Coeliodes ruber (Marsham,1802)	Nb	-	Nb	259
Zacladus exiguus (Olivier,1807)	Nb	-	Nb	311
Stenocarus umbrinus (Gyllenhal,1837)	Nb	-	Nb	301
Trichosirocalus barnevillei (Brisout,1866)	Nb	-	Nb	305
Trichosirocalus dawsoni (Brisout,1869)	Nb	-	Nb	305
Trichosirocalus horridus (Panzer,1801)	Na	-	Na	306
Trichosirocalus rufulus (Dufour,1851)	Nb	-	Na	306
Ceutorhynchus angulosus Boheman,1845	RDB 3	RDB 3	Na	246
Ceutorhynchus arquatus (Herbst,1795)	RDB 3	RDB 3	RDB I	246
Ceutorhynchus atomus Boheman,1845	Nb	-	Na	246
Ceutorhynchus campestris Gyllenhal,1837	Nb	-	Nb	247
Ceutorhynchus constrictus (Marsham,1802)	-	-	Nb	247
Ceutorhynchus euphorbiae Brisout,1866	Na	-	Na	247
Ceutorhynchus geographicus (Goeze,1777)	Nb	-	Nb	248
Ceutorhynchus hepaticus Gyllenhal,1837	Na	-	RDB K	248
Ceutorhynchus hirtulus Germar,1824	Nb	-	Nb	249
Ceutorhynchus insularis Dieckmann,1971	RDB 1	RDB 1	RDB 1	249
Ceutorhynchus mixtus Mulsant and Rey,1858	Na	-	Nb	249
Ceutorhynchus moelleri Thomson,1868	RDB 3	RDB 3	RDB K	250
Ceutorhynchus parvulus Brisout,1869	RDB 3	RDB 3	RDB 3	250
Ceutorhynchus pectoralis Weise,1895	RDB 3	RDB 3	Na	250
Ceutorhynchus pervicax Weise,1883	Na	-	Na	251
Ceutorhynchus pilosellus Gyllenhal,1837	RDB 2	RDB 2	RDB 2	251
Ceutorhynchus pulvinatus Gyllenhal,1837	Na	-	Na	251
Ceutorhynchus pumilio (Gyllenhal,1827)	Na	-	Na	252
Ceutorhynchus punctiger Sahlberg,1835	-	-	Nb	252
Ceutorhynchus querceti (Gyllenhal,1813)	RDB 2	RDB 2	RDB 2	252
Ceutorhynchus quercicola (Paykull,1792)	Na	-	Na	253
Ceutorhynchus rapae Gyllenhal,1837	Nb	-	Nb	253
Ceutorhynchus resedae (Marsham,1802)	Nb	-	Nb	253
Ceutorhynchus syrites Germar,1824	RDB 3	RDB 3	RDB 1	254
Ceutorhynchus terminatus (Herbst,1795)	Nb	-	Nb	254
Ceutorhynchus thomsoni Kolbe,1900	Nb	-	RDB K	255
Ceutorhynchus triangulum Boheman,1845	Nb	-	Nb	255
Ceutorhynchus trimaculatus (F.,1775)	Nb	-	Nb	255
Ceutorhynchus unguicularis Thomson,1871	Na	-	RDB 3	256
Ceutorhynchus urticae Boheman,1845	Na	-	RDB 3	256
Ceutorhynchus verrucatus Gyllenhal,1837	Na	-	RDB 3	256
Ceutorhynchus viduatus (Gyllenhal,1813)	Na	-	Nb	257
Eubrychius velutus (Beck,1817)	Nb	-	Nb	266

Species	Status in Hyman (1986)	Status in Shirt (1987)	Revised Status	Page Number
Litodactylus leucogaster (Marsham,1802)	Nb	-	Nb	276
Rhinoncus albicinctus Gyllenhal,1837	RDB 1	RDB 1	RDB 1	293
Phytobius canaliculatus Fahraeus,1843	-	-	Nb	287
Phytobius comari (Herbst,1795)	Nb	-	Nb	287
Phytobius muricatus Brisout,1867	Na	-	Na	287
Phytobius olssoni Israelson,1972	RDB 3	RDB 3*	RDB 3	288
Phytobius quadricorniger Colonnelli,1986	Na	-	Na	288
Phytobius quadrinodosus (Gyllenhal,1813)	RDB 3	RDB 3	Na	288
Phytobius waltoni Boheman,1843	Nb	-	Nb	289
Drupenatus nasturtii (Germar,1824)	Nb	-	Nb	265
Tapinotus sellatus (F.,1794)	Na	-	Na	302
Orobitis cyaneus (L.,1758)	Nb	-	-	
Baris analis (Olivier,1790)	RDB 1	RDB 1	RDB 2	240
Baris chlorizans Germar,1824	List 2	-	-	
Baris laticollis (Marsham,1802)	Na	-	Na	241
Baris lepidii Germar,1824	Nb	-	Na	241
Baris picicornis (Marsham,1802)	Nb	-	Nb	242
Baris scolopacea Germar,1818	RDB 3	RDB 3	RDB 3	242
Anthonomus britannus (Desbrochers,1868)	List 2	-	-	
Anthonomus brunnipennis (Curtis,1840)	Nb	-	-	
Anthonomus chevrolati Desbrochers,1868	Na	-	RDB K	231
Anthonomus conspersus Desbrochers,1868	Nb	-	Nb	232
Anthonomus humeralis (Panzer,1795)	Na	-	RDB K	232
Anthonomus piri Kollar,1837	Na	-	RDB K	232
Anthonomus rufus Gyllenhal,1836	Na	-	RDB 3	233
Anthonomus ulmi (Degeer,1775)	Nb	-	Nb	233
Anthonomus varians (Paykull,1792)	Nb	-	Nb	233
Furcipus rectirostris (L.,1758)	Na	-	Nb	266
Brachonyx pineti (Paykull,1792)	Nb	-	-	
Curculio betulae (Stephens,1831)	Nb	-	Nb	261
Curculio rubidus (Gyllenhal,1836)	Nb	-	Nb	261
Curculio villosus F.,1781	-	-	Nb	262
Tychius crassirostris Kirsch,1871	RDB 3	RDB 3*	RDB K	307
Tychius lineatulus Stephens,1831	Nb	-	Na	308
Tychius parallelus (Panzer,1794)	Na	-	Na	308
Tychius polylineatus (Germar,1824)	RDB 3	RDB 3	RDB K	309
Tychius pusillus Germar,1842	Nb	-	Nb	309
Tychius quinquepunctatus (L.,1758)	RDB 2	RDB 2	RDB 2	309
Tychius schneideri (Herbst,1795)	Nb	-	Nb	310
Tychius squamulatus Gyllenhal,1836	Nb	-	Nb	310
Tychius stephensi Gyllenhal,1836	Nb	-	-	
Tychius tibialis Boheman,1843	Nb	-	Na	310
Sibinia arenariae Stephens,1831	Nb	-	Nb	297
Sibinia pellucens (Scopoli,1772)	RDB APPENDIX	RDB APPENDIX	-	
Sibinia potentillae Germar,1824	Nb	-	Nb	297
Sibinia primitus (Herbst,1795)	Nb	-	Nb	297
Sibinia sodalis Germar,1824	Na	-	Na	298
Ellescus bipunctatus (L.,1758)	Nb	-	Nb	265
Ellescus scanicus (Paykull,1792)	List 2	-	-	
Acalyptus carpini (F.,1792)	Nb	-	Nb	230
Miarus distinctus (Boheman,1845)	RDB 3	RDB 3	-	
Miarus graminis (Gyllenhal,1813)	Nb	-	Nb	280
Miarus micros (Germar,1821)	RDB 3	RDB 3	RDB 3	280
Miarus plantarum (Germar,1824)	Na	-	RDB K	281
Mecinus circulatus (Marsham,1802)	Nb	-	Nb	279
Mecinus collaris Germar,1821	Nb	-	Nb	279
Mecinus janthinus Germar,1821	Na	-	Na	279
Gymnetron beccabungae (L.,1761)	Na	-	Na	267
Gymnetron collinum (Gyllenhal,1813)	Na	-	Na	268
Gymnetron labile (Herbst,1795)	Nb	-	-	

Species	Status in Hyman (1986)	Status in Shirt (1987)	Revised Status	Page Number
Gymnetron linariae (Panzer,1795)	Na	-	Na	268
Gymnetron melanarium (Germar,1821)	Na	-	Nb	268
Gymnetron rostellum (Herbst,1795)	Nb	-	Na	269
Gymnetron veronicae (Germar,1821)	Nb	-	Nb	269
Gymnetron villosulum Gyllenhal,1838	Nb	-	Nb	270
Rhynchaenus calceatus Germar,1821	-	-	RDB K	294
Rhynchaenus decoratus (Germar,1821)	RDB 3	RDB 3	RDB I	294
Rhynchaenus foliorum (Müller,1764)	Nb	-	Na	294
Rhynchaenus iota (F.,1787)	Nb	-	Nb	295
Rhynchaenus lonicerae (Herbst,1795)	List 2	-	-	
Rhynchaenus populi (F.,1792)	Na	-	RDB K	295
Rhynchaenus pratensis (Germar,1821)	Nb	-	Nb	295
Rhynchaenus pseudostigma Temperé,1982	-	-	RDB K	296
Rhynchaenus testaceus (Müller,1776)	Na	-	RDB 2	296

Scolytidae

Species	Status in Hyman (1986)	Status in Shirt (1987)	Revised Status	Page Number
Scolytus mali (Bechstein,1805)	-	-	Nb	402
Scolytus ratzeburgi Janson,1856	Na	-	Nb	402
Leperisinus orni (Fuchs,1906)	Nb	-	Nb	400
Xylechinus pilosus (Ratzeburg,1837)	Na	-	-	
Kissophagus hederae (Schmitt,1843)	Na	-	Nb	400
Phloeosinus thujae (Perris,1855)	Nb	-	-	
Hylastes angustatus (Herbst,1793)	Na	-	-	
Hylastes brunneus Erichson,1836	Nb	-	-	
Tomicus minor (Hartig,1834)	RDB 3	RDB 3	RDB 3	403
Polygraphus poligraphus (L.,1758)	Nb	-	-	
Dryocoetinus alni (Georg,1856)	RDB 3	RDB 3	Na	398
Dryocoetes autographus (Ratzeburg,1837)	Nb	-	-	
Triotemnus coryli (Perris,1855)	RDB 3	RDB 3	RDB 1	403
Taphrorychus bicolor (Herbst,1793)	Nb	-	Na	402
Xyloterus lineatus (Olivier,1795)	Nb	-	-	
Xyloterus signatus (F.,1792)	RDB 3	RDB 3	Nb	405
Cryphalus abietis (Ratzeburg,1837)	RDB 3	RDB 3	-	
Cryphalus piceae (Ratzeburg,1837)	Na	-	-	
Cryphalus saltuarius Weise,1891	Na	-	-	
Ernoporus caucasicus Lindemann,1876	RDB 1	RDB 1	RDB 1	399
Ernoporus fagi (F.,1798)	Nb	-	Na	399
Ernoporus tiliae (Panzer,1793)	RDB 3	RDB 3	RDB 1	400
Trypophloeus asperatus (Gyllenhal,1813)	RDB 3	RDB 3	Na	404
Trypophloeus granulatus (Ratzeburg,1837)	RDB APPENDIX	RDB APPENDIX	EXTINCT	404
Xyleborus dispar (F.,1792)	RDB 3	RDB 3	Nb	404
Xyleborus dryographus (Ratzeburg,1837)	Na	-	Nb	405
Pityophthorus lichtensteini (Ratzeburg,1837)	RDB 3	RDB 3	RDB3	401
Pityogenes chalcographus (L.,1761)	RDB 3	RDB 3	-	
Pityogenes quadridens (Hartig,1834)	RDB 3	RDB 3	Na	401
Pityogenes trepanatus (Nördlinger,1848)	RDB 3	RDB 3	Na	401
Ips sexdentatus (Boerner,1767)	Nb	-	-	
Orthotomicus erosus (Wollaston,1857)	Na	-	-	
Orthotomicus suturalis (Gyllenhal,1827)	Nb	-	-	

Platypodidae

Species	Status in Hyman (1986)	Status in Shirt (1987)	Revised Status	Page Number
Platypus cylindrus (F.,1792)	RDB 3	RDB 3	Nb	374
Platypus parallelus (F.,1801)	Na	-	RDB I	374

BIBLIOGRAPHY TO THE INTRODUCTION

ALEXANDER, K.N.A. 1985. The specialist fauna of calcareous grasslands. *Coleopterists Newsletter*, *20*: 2-4.

ALLEN, A.A., HUNTER F.A., JOHNSON, C. AND SKIDMORE, P. *In: Second report to the Nature Conservancy Council on the fauna of the mature timber habitat* by P.T. Harding, 26-32. Unpublished. (Institute of Terrestrial Ecology project No. 405).

ANON., 1981. *Wildlife and Countryside Act 1981*. London, HMSO.

ANON., 1989. *Guidelines for the selection of biological SSSI's*. Peterborough, Nature Conservancy Council.

ANON., 1990. *Environmental Protection Act 1990*. London, HMSO.

BALFOUR-BROWNE, F. 1940. *British water beetles*. Volume 1. London, Ray Society.

BALFOUR-BROWNE, F. 1950. *British water beetles*. Volume 2. London, Ray Society.

BALFOUR-BROWNE, F. 1953. Coleoptera: Hydradephaga. *Handbooks for the Identification of British Insects*, *4*(3): 1-34.

BALFOUR-BROWNE, F. 1958. *British water beetles*. Volume 3. London, Ray Society.

BALL, S.G. 1986. *Terrestrial and freshwater invertebrates with Red Data Book, Notable or Habitat Indicator status*. Invertebrate Site Register report No 66. Nature Conservancy Council, unpublished. (CSD report No. 637).

BECHYNE, J. 1956. *Guide to beetles*. London, Thames and Hudson.

BÍLÝ, S. 1982. *The Buprestidae (Coleoptera) of Fennoscania and Denmark*. Leiden, E.J. Brill. (Fauna Entomologica Scandinavica Volume 10).

BÍLÝ, S. 1990. *A colour guide to beetles*. London, Treasure Press.

BRENDEL, M.J.D. 1975. Coleoptera: Tenebrionidae. *Handbooks for the Identification of British Insects*, *5*(10): 1-22.

BRITTON, E.B. 1956. Coleoptera: Scarabaeoidea. *Handbooks for the Identification of British Insects*, *5*(11): 1-53.

BUCK, F.D. 1954. Coleoptera: Lagriidae to Meloidae. *Handbooks for the Identification of British Insects*, *5*(9): 1-30.

BUCK, F.D. 1955. The British Mycetophagidae and Colydiidae. *Proceedings of the South London Entomological and Natural History Society, 1955*: 53-67.

BUCK, F.D. 1958. The British Anobiidae. *Proceedings of the South London Entomological and Natural History Society, 1958*: 52-64.

BUCKLAND, P.C. 1979. *Thorne Moors: a paleoecological study of a Bronze Age site*. Department of Geography, University of Birmingham. (Occasional publication No 8).

CLARKE, R.O.S. 1973. Coleoptera: Heteroceridae. *Handbooks for the Identification of British Insects*, *5*(2c): 1-15.

COLLINS, N.M. 1987. *Legislation to conserve insects in Europe*, London, Amateur Entomologist's Society.

COOTER, J. 1991. *A Coleopterist's handbook*. 3rd edition. London, Amateur Entomologist's Society.

COOTER, J. & CRIBB, P.W. 1974. *A Coleopterist's handbook*. 2nd edition. Feltham, Amateur Entomologist's Society.

CROWSON, R.A. 1981. *The biology of the Coleoptera*. London, Academic Press.

CROWSON, R.C. 1954. *The natural classification of the families of the Coleoptera*. Hampton, E.W. Classey.

CROWSON, R.C. 1956. Coleoptera. Introduction and keys to families. *Handbooks for the Identification of British Insects*, *4*(1): 1-50.

DANDY, J.E. 1969. *Watsonian vice-counties of Great Britain*. London, Ray Society.

DONISTHORPE, H.ST J.K. 1927. *The guests of British ants. Their habits and life histories*. London, Routledge.

DU CHATENET G. 1990. *Guide des Coleoptreres d'Europe*. Paris, Delachaux and Niestle.

DUFFY, E.A.J. 1952. Coleoptera: Cerambycidae. *Handbooks for the Identification of British Insects*, *5*(12). 1-18.

BIBLIOGRAPHY

DUFFY, E.A.J. 1953. Coleoptera: Scolytidae and Platypodidae. *Handbooks for the Identification of British Insects*, *5*(15): 1-18.

EVERSHAM, B. 1983. *Defining Rare and Notable species - a discussion document.* Invertebrate Site Register report No 49. Nature Conservancy Council, unpublished. (CSD report No. 481).

FALK, S. *A review of the scarce and threatened bees, wasps and ants of Great Britain.* Peterborough, Nature Conservancy Council. (Research and Survey in Nature Conservation No. 35).

FORSYTH, T.G. 1987. *Common ground beetles.* Slough, Richmond Publishing. (Naturalists' Handbooks 8).

FOSTER, G.N. 1981. Atlas of British water beetles. Preliminary edition - part 1. *Balfour-Browne Club Newsletter*, No. 22: 1-18.

FOSTER, G.N. 1983. Atlas of British water beetles. Preliminary edition - part 2. *Balfour-Browne Club Newsletter*, No. 27: 1-23.

FOSTER, G.N. 1984. Atlas of British water beetles. Preliminary edition - part 3. *Balfour-Browne Club Newsletter*, No. 31: 1-22.

FOSTER, G.N. 1985. Atlas of British water beetles. Preliminary edition - part 4. *Balfour-Browne Club Newsletter*, No. 35: 1-22.

FOSTER, G.N. 1986. Nationally Notable species in Britain. *Balfour-Browne Club Newsletter*, No. 37: 19.

FOSTER, G.N. 1987a. Atlas of British water beetles. Preliminary edition - part 5. *Balfour-Browne Club Newsletter*, No. 40: 1-23.

FOSTER, G.N. 1987b. The use of Coleoptera records in assessing the conservation status of wetlands. *In: The use of invertebrates in site assessment for conservation* ed. by M.L. Luff, 8-18. University of Newcastle upon Tyne. Agricultural Environment Research Group.

FOSTER, G.N. 1990. Atlas of British water beetles. Preliminary edition - part 6. *Balfour-Browne Club Newsletter*, No. 48: 1-8.

FOSTER, G.N. & EYRE, M.D. (in press). *Classification and ranking of water beetle communities.* Peterborough, Joint Nature Conservation Committee. (UK Nature No. 1).

FOWLER, W.W. 1887- 1891. *The Coleoptera of the British Islands.* Volumes 1-5 London, Reeve.

FOWLER, W.W. & DONNISTHORPE H. ST.J., 1913. *The Coleoptera of the British Islands.* Volume 6. London, Reeve.

FRIDAY, L.E. 1988. *A key to the adults of the British water beetles.* AIDGAP key. Taunton, Field Studies Council.

FRY, R. & LONSDALE, D. (eds) 1991. *Habitat conservation for insects - a neglected green issue.* London, Amateur Entomologists' Society.

GARLAND, S.P. 1983. Beetles as primary woodland indicators. *Sorby Record*, *21*: 3-38.

GERSTMEIR, R. 1991. *Chequered beetles. Illustrated key to the Cleridae of Europe.* Gimmersheim, Joseph Margraf.

GIRLING, M.A. 1982. Fossil insect faunas from forest sites. *In: Archaeological aspects of woodland ecology ed. by M. Bell and S. Limbrey 129-144.* Symposium of the Association for Environmental Archaeology No 2. (BAR International Series No.146).

HALSTEAD, D.G.H. 1963. Coleoptera: Sphaeritidae & Histeridae. *Handbooks for the Identification of British Insects*, *4*(10): 1-16.

HAMMOND, P.M. 1974. Changes in the British Coleoptera fauna. *In: The changing flora and fauna of Britain* ed. by D.L. Hawksworth, 323-369. Proceedings of a symposium of the Systematics Association at the University of Leicester, April 1973. London, Academic Press.

HANSEN, M. 1987. *The Hydrophiloidea (Coleoptera) of Fennoscandia and Denmark.* Copenhagen, Scandinavian Science Press. (Fauna Entomologica Scandinavica, Volume 18).

HARDE, K.W. & HAMMOND, P.M. 1984. *A field guide in colour to Beetles.* London, Octopus Books.

HARDING, P.T. 1981. Burn wisely - a plea for the conservation of old trees and dead wood. *Nature in Wales*, *17*: 144-148.

HARDING, P.T. & ROSE, F. 1986. *Pasture-woodlands in lowland Britain.* Abbots Ripton, Institute of Terrestrial Ecology.

HICKIN, N.E. 1960. An introduction to the study of the British Lyctidae. *In: Record of the tenth annual* convention of the British Wood Preservation Association. 57-100.

HICKIN, N.E. 1975. *The insect factor in wood decay.* London, Associated Business Programmes, Rentokil Library.

HICKIN, N.E. 1987. *Longhorn beetles of the British Isles.* Aylesbury, Shire Natural History.

HINKS, W.D. 1944. Notes on *Byturus* Latreille, 1796 (Col. Byturidae). *Entomologist's Monthly Magazine, 80*: 178-89.

HINTON, H.E. 1945. *A monograph on the beetles associated with stored products.* Volume 1. London, British Museum (Natural History).

HOLLAND, D.G. 1972. *A key to the larvae, pupae and adults of the British species of Elminthidae.* Windermere, Freshwater Biological Association. (Scientific publication No. 26).

HOLMEN, M. 1987. *The aquatic Adephaga, Haliplidae, Hygrobiidae and Noteridae.* Leiden, E.J. Brill. (Fauna Entomologica Scandinavica, Volume 20).

HUNTER, F.A. 1977. Ecology of pinewood beetles. *In: Native Pinewoods of Scotland* ed. by R.H.J. Bunce and J.R. Jeffers, 42-55. Proceedings of Aviemore Symposium, 1975. Cambridge, Institute of Terrestrial Ecology.

HYMAN, P.S. 1985. *A provisional review of British Coleoptera.* Invertebrate Site Register report No. 60. Peterborough, Nature Conservancy Council, unpublished.

HYMAN, P.S. 1986. *A National Review of British Coleoptera. Ia. A review of the statuses of British Coleoptera (in taxonomic order).* Invertebrate Site Register report No. 64. Nature Conservancy Council, unpublished. (CSD report No. 631).

JCCBI, 1986. *Insect re-establishment - a code of conservation practice.* London. Joint Committee for the Conservation of British Insects.

JESSOP, L. 1986. Coleoptera: Scarabaeoidae. *Handbooks for the Identification of British Insects, 5*(11): 1-53.

JOHNSON, C. 1966. Coleoptera: Clambidae. *Handbooks for the Identification of British Insects, 4*(6a): 1-13.

JOY, N.H. 1932. *A practical handbook of British beetles.* 2 Volumes. London, Witherby. (Reprinted 1976 by Classey, Faringdon).

KEY, R.S. 1989a. NCC's National Review of British Coleoptera. *Coleopterists Newsletter 36*: 18.

KEY, R.S. 1989b. NCC's National Review of Coleoptera. *Coleopterists Newsletter, 37*: 1.

KEY, R.S. 1991. NCC's National Review of British Coleoptera. *Coleopterists Newsletter, 42*: 9-12.

KEY, R.S. 1991. Conservation and the Coleopterist. *In: A Coleopterist's handbook by J. Cooter, 256-274.* London, Amateur Entomologist's Society.

KIRBY, P. (in press) 1991. *Habitat management for invertebrates: a practical handbook.* Sandy, Royal Society for the Protection of Birds.

KLOET, G.S. & HINKS, W.D. 1977. A Check List of British Insects. Part 3: Coleoptera and Strepsiptera. 2nd edition. Revised by R.D. Pope. *Handbooks for the Identification of British Insects, 11*(3): 1-105.

LEVY, B. 1977. Coleoptera: Buprestidae. *Handbooks for the Identification of British Insects, 5*(1b): 1-8.

LINDROTH, C.H. 1974. Coleoptera: Carabidae. *Handbooks for the Identification of British Insects, 4*(2): 1-148.

LINDROTH, C.H. 1985. *The Carabidae (Coleoptera) of Fennoscandia and Denmark.* Leiden, E.J. Brill. (Fauna Entomologica Scandinavica, Volume 15).

LINDROTH, C.H. 1986. *The Carabidae (Coleoptera) of Fennoscandia and Denmark.* Leiden & Copenhagen. E.J. Brill/Scandinavian Science Press. (Fauna Entomologica Scandinavica, Volume 15, part 2).

LINSSEN, E.F. 1959. *Beetles of the British Isles.* 2 Volumes. London, Frederick Warne.

LUFF M.L. ed. 1982. *Preliminary atlas of British Carabidae (Coleoptera).* Abbots Ripton, Biological Records Centre, Institute of Terrestrial Ecology.

MAJERUS, M.E.N., FORGE, H. & WALKER, L. 1990. The geographical distributions of ladybirds in Britain (1984-1989). *British Journal of Entomology and Natural History, 3*: 153-165.

MAJERUS, M. & KEARNS, P. 1989. *Ladybirds.* Slough, Richmond Publishing. (Naturalists' Handbooks 10).

MENDEL, H. 1988. *Provisional Atlas of the click beetles (Coleoptera : Elateroidea) of the British Isles.* Abbots Ripton. Biological Records Centre, Institute of Terrestrial Ecology.

BIBLIOGRAPHY

MOON, A. 1986. *Ladybirds in Dorset*. Dorchester, Dorset Environmental Records Centre.

MORRIS, M.G. 1990. Orthocerus weevils. Coleoptera: Curculionoidea (Nemonychidae, Anthribidae, Urodontidae, Attelabidae and Apionidae). *Handbooks for the Identification of British Insects*, *5*(16): 1-108.

MORRIS, M.G. 1991. *Weevils*. Richmond, Richmond Press. (Naturalists' Handbooks 16).

MUNROE, J.W. 1966. *Pests of stored products*. London, Hutchinson, Rentokil Library

NCC, 1989. *Guidelines for the selection of biological SSSIs*. Peterborough, Nature Conservancy Council.

PEACOCK, E.R. 1977. Coleoptera: Rhizophagidae. *Handbooks for the Identification of British Insects*, *5*(5a): 1-20.

PEARCE, E.J. 1957. Coleoptera: Pselaphidae. *Handbooks for the Identification of British Insects*, *4*(9): 1-32.

POPE, R.D. 1953. Coleoptera: Coccinellidae and Sphindidae. *Handbooks for the Identification of British Insects*, *5*(7): 1-12.

RATCLIFFE, D.A. ed. 1977. *A nature conservation review*. 2 Volumes Cambridge, Cambridge University Press.

RUCKER, W. 1991. *Scavenger beetles. Illustrated key to the Lathridiidae of Europe*. Gimmersheim, Joseph Margraf.

SCHAWALLER, W. 1991. *Carrion beetles. An illustrated key to the Silphidae of Europe*. Gimmersheim, Joseph Margraf.

SHIRT, D.B., ed. 1987. *British Red Data Books: 2. Insects*. Peterborough, Nature Conservancy Council.

SKIDMORE, P. 1991. *Insects of the British cow dung community*. AIDGAP Key. Taunton, Field Studies Council.

SPEIGHT, M.C.D. 1989. *Saproxylic invertebrates and their conservation*. Strasbourg, Council of Europe. (Nature and Environment Series, No. 42).

STORK, N.E. 1990. *The role of ground beetles in ecological and environmental studies*. Andover, Intercept.

STUBBS, A.E. 1972. Wildlife conservation and dead wood. *Journal of the Devon Trust for Nature Conservation*, *4*: 169-182.

THOMAS, J. & COLLINS, M. eds. 1991. Conservation of insects and their habitats. *Procedings of the 15th symposium of the Royal Entomological Society of London*.

THOMPSON, R.T. 1958. Coleoptera: Phalacridae. *Handbooks for the Identification of British Insects*, *5*(5b): 1-17.

TOTTENHAM, C.E. 1954. Coleoptera: Staphylinidae (part). Section (a) Piestinae to Euaesthetinae. *Handbooks for the Identification of British Insects*, *4*(8a): 1-79.

TRAUTNER, J. & GEIGENMüller, 1987. *Tiger beetles, ground beetles. Illustrated key to the Ciccindellidae and Carabidae of Europe*. Gimmersheim, Joseph Margraf.

UNWIN, D.M. 1984. *A field key to the families of the British beetles*. AIDGAP Key. Taunton, Field Studies Council.

UNWIN, D.M. & CORBET, S.A. 1991. *Insects, plants and microclimate*. Richmond. Richmond Press. (Naturalists' Handbooks 15).

WALSH, G.B. & DIBB, J.R. 1954. *Coleopterist's handbook*. Feltham, Amateur Entomologist's Society.

WELLS, S.M., PYLE, R.M. & COLLINS, N.M. 1983. *The IUCN Invertebrate Red Data Book*. Gland, International Union for Conservation of Nature and Natural Resources.

WHITTEN, A.J. 1990. *Recovery: A proposed programme for Britain's protected species*. Nature Conservancy Council, unpublished. (CSD report No. 1089).

WHITTEN, A.J. 1991. Recovery and hope for Britain's rare species. *British Wildlife*, *2*: 219-229.

THE DATA SHEETS

The data sheets are in alphabetic order by family name and then, within a family, by scientific name. The page number on which the data sheets for a given family start is shown in the table of contents and is duplicated overleaf for convenience.

Particular species can be found by consulting Appendix 3 which shows the page number for each species' data sheet and is organised in checklist order, or the generic or specific name (including synonyms) can be looked up in the index.

Page number on which data sheets for families start.

Aderidae	61	Lucanidae	345
Anobiidae	62	Lycidae	345
Anthicidae	67	Lyctidae	347
Anthribidae	69	Lymexylidae	347
Apionidae	72	Melandryidae	348
Attelabidae	83	Meloidae	355
Biphyllidae	87	Melyridae	357
Bostrichidae	87	Mordellidae	361
Bruchidae	87	Mycetophagidae	366
Buprestidae	88	Mycteridae	368
Byrrhidae	92	Oedemeridae	368
Cantharidae	94	Peltidae	371
Carabidae	99	Phalacridae	371
Cerambycidae	155	Phloiophilidae	373
Cerylonidae	169	Platypodidae	374
Chrysomelidae	170	Psephenidae	374
Cisidae	210	Ptinidae	375
Clambidae	213	Pyrochroidae	376
Cleridae	213	Pythidae	377
Coccinellidae	216	Rhizophagidae	377
Colydiidae	220	Salpingidae	380
Corylophidae	226	Scaphidiidae	381
Cucujidae	227	Scarabaeidae	382
Curculionidae	230	Scirtidae	396
Dermestidae	311	Scolytidae	398
Drilidae	313	Scraptiidae	405
Elateridae	314	Silphidae	409
Endomychidae	330	Silvanidae	411
Erotylidae	331	Sphaeritidae	412
Eucnemidae	332	Sphindidae	412
Geotrupidae	334	Tenebrionidae	413
Heteroceridae	335	Tetratomidae	419
Histeridae	335	Throscidae	420
Hypocopridae	344	Trogidae	420
Lampyridae	344	Trogossitidae	421
Limnichidae	344		

ADERUS BREVICORNIS   VULNERABLE

Order COLEOPTERA   Family ADERIDAE

Aderus brevicornis (Perris, 1869). Formerly known as: *Hylophilus brevicornis* (Perris), *Xylophila brevicornis* (Perris), *Xylophilus brevicornis* (Perris), *Xylophilus neglectus* sensu Fowler, 1891 partim not (du Val, 1862).

Distribution Recorded from South Devon, South Hampshire, West Sussex, East Sussex and Berkshire before 1970 and East Sussex and Berkshire from 1970 onwards.

Habitat and ecology Ancient broad-leaved woodland and pasture-woodland. Recorded from oak, beech and elm. Also recorded from under the bark of a dead Scots pine and a single example has been found in the stalk of a fungus thought to be *Hypholoma fasciculare*. Larvae develop in dead wood. Adults have been recorded from July to September.

Status Status revised from RDB 3 (Rare) in Shirt (1987). Very local in Great Britain and on the Continent. Appears to be known from only about 16 individuals in Great Britain. Only recorded from five vice-counties, all in southern England. Recently recorded from just two of these.

Indicator status Grade 1 in Harding & Rose (1986).

Threats Loss of broad-leaved woodland and parkland through, for example, clear-felling and coniferisation. Habitat loss, in particular, through the felling of ancient trees, removal of dead wood from living trees and the destruction or removal of standing and fallen dead wood for reasons such as forest hygiene, aesthetic tidiness, public safety or for use as fire wood.

Management and conservation Ancient trees and both fallen and standing dead timber, especially with the bark attached, should be retained. The removal of dead timber from ancient trees should be avoided. Gaps in the age structure of the tree population should be identified and the continuity of the appropriate dead wood habitat ensured by regeneration, suitable planting and possibly with pollarding.

Published sources Allen, A.A. (1959a), Allen, A.A. (1969a), Allen, A.A. (1985b), Harding, P.T. (1978), Shirt, D.B., ed. (1987).

ADERUS OCULATUS   NOTABLE B

Order COLEOPTERA   Family ADERIDAE

Aderus oculatus (Paykull, 1798). Formerly known as: *Aderus pygmaea* sensu Buck, 1954 not (Degeer, 1774), *Hylophilus pygmaeus* sensu Joy, 1932 not (Degeer), *Xylophila pygmaea* sensu Kloet and Hincks, 1945 not (Degeer), *Xylophilus oculatus* (Paykull), *Xylophilus pygmaeus* sensu Hudson Beare, 1930 not (Degeer).

Distribution England.

Habitat and ecology Broad-leaved woodland and pasture-woodland. Recorded from the stumps and boughs of oak, but possibly with a preference for the tops of stag-horn oaks. Also found on lime, hawthorn, beech, birch and chestnut. Adults have also been recorded from elder blossom. Larvae develop in dead wood and have been found in an oak stump in September. Adults have been noted from June to September.

Status Widespread but local in England. Possibly spreading. Can be common where found.

Indicator status Grade 1 in Garland (1983). Grade 3 in Harding & Rose (1986).

Threats Loss of broad-leaved woodland and parkland through, for example, clear-felling and coniferisation. Habitat loss, in particular, through the felling of ancient trees, removal of dead wood from living trees and the destruction or removal of standing and fallen dead wood for reasons such as forest hygiene, aesthetic tidiness, public safety or for use as fire wood.

Management and conservation Ancient trees and both fallen and standing dead timber, especially with the bark attached, should be retained. The removal of dead timber from ancient trees should be avoided. Gaps in the age structure of the tree population should be identified and the continuity of the appropriate dead wood habitat ensured by regeneration, suitable planting and possibly with pollarding.

ADERUS POPULNEUS   NOTABLE B

Order COLEOPTERA   Family ADERIDAE

Aderus populneus (Creutzer in Panzer, 1796). Formerly known as: *Aderus populnea* (Creutzer in Panzer, 1796), *Hylophilus populneus* (Creutzer in Panzer), *Xylophila populnea* (Creutzer in Panzer), *Xylophilia populnea* (Creutzer in Panzer), *Xylophilus neglectus* sensu Fowler and Donisthorpe, 1913 partim not (du Val, 1862), *Xylophilus populneus* (Creuter in Panzer).

ANOBIIDAE

Distribution South, South East, East Anglia, East Midlands, West Midlands and North East.

Habitat and ecology Broad-leaved woodland, pasture-woodland, scrub, hedges in agricultural land, gardens and also indoors. Recorded from oak, lime, plane, *Salix* and from dry, red-rotten wood in a knot-hole of horse-chestnut. Larvae probably occur in dead wood, although there is a record, which is considered doubtful, of larvae feeding on ash seeds. The species has also been recorded from heaps of grass and manure, and it has been found apparently in association with and feeding on cobwebs. This beetle has been noted at a mercury vapour light trap. Adults have been recorded all year round.

Status Very local and widely distributed in England, recorded from southern England, north to South-east Yorkshire.

Threats Loss of broad-leaved woodland and parkland through, for example, clear-felling and coniferisation. Habitat loss, in particular, through the felling of ancient trees, removal of dead wood from living trees and the destruction or removal of standing and fallen dead wood for reasons such as forest hygiene, aesthetic tidiness, public safety or for use as fire wood.

Management and conservation Ancient trees and both fallen and standing dead timber, especially with the bark attached, should be retained. The removal of dead timber from ancient trees should be avoided. Gaps in the age structure of the tree population should be identified and the continuity of the appropriate dead wood habitat ensured by regeneration, suitable planting and possibly with pollarding.

ANITYS RUBENS   NOTABLE B

Order COLEOPTERA   Family ANOBIIDAE

Anitys rubens (Hoffmann, 1803).

Distribution South, South East, East Anglia, East Midlands, West Midlands and North West.

Habitat and ecology Ancient broad-leaved woodland and pasture-woodland. Associated with decaying oak. Adults have been found in powdery wood mould, very dry fungus-infected dead wood and in red rotten wood. Often found in company with *Mycetophagus piceus* (Mycetophagidae). Adults have been recorded from April to September.

Status Widespread but local in England and recorded as far north as South Lancashire. Possibly under-recorded.

Indicator status Grade 1 in Garland (1983). Grade 1 in Harding & Rose (1986).

Threats Loss of broad-leaved woodland and parkland through, for example, clear-felling and coniferisation. Habitat loss, in particular, through the felling of ancient and fungus-infected trees, removal of dead wood from living trees and the destruction or removal of standing and fallen dead wood for reasons such as forest hygiene, aesthetic tidiness, public safety or for use as fire wood.

Management and conservation Ancient and fungus-infected trees, and both fallen and standing dead timber, especially with the bark attached, should be retained. The removal of dead timber from ancient trees should be avoided. Gaps in the age structure of the tree population should be identified and the continuity of the appropriate dead wood habitat ensured by regeneration, suitable planting and possibly with pollarding.

ANOBIUM INEXSPECTATUM   NOTABLE B

Order COLEOPTERA   Family ANOBIIDAE

Anobium inexspectatum Lohse, 1954.

Distribution South East, South, South West, East Anglia, East Midlands, West Midlands, North West and South Wales.

Habitat and ecology Woodland, pasture-woodland, neglected orchards, quarries and old, ivy covered buildings. Breeds in the stems of old ivy *Hedera helix* growing on trees, rocks and the walls of old buildings. There is also a record of one larva being reared to adult in the stem of a tree lupin *Lupinus arboreus*, as well as a record of this species being found in the dead wood of an old apple tree. Larvae have been found in March and August. Adults have been recorded in January and from May to August.

Status Recently added to the British list from material previously determined as *Anobium punctatum*. Subsequently found to be widespread and local in England, and also recorded in South Wales.

Threats Clear-felling of woodland and parkland and conversion to other land use, and the infilling of quarries. The selective removal of ivy from trees is a further threat.

CAENOCARA AFFINIS   INDETERMINATE

Order COLEOPTERA   Family ANOBIIDAE

Caenocara affinis (Sturm, 1837). Formerly known as: *Caenocara subglossa* sensu auct. Brit. not Mulsant and Rey, 1864, *Coenocara subglossa* sensu auct. Brit. not Mulsant and Rey, 1864.

Distribution Recorded from West Suffolk before 1970.

Habitat and ecology Probably grassland. Occurs in the puff-ball fungus *Lycoperdon perlatum* and possibly other puff-ball fungus species. Larvae have been found in September.

Status Status revised from RDB 1 (Endangered) in Shirt (1987). Known from just eight examples reared from larvae found at Barton Mills, West Suffolk in 1917.

Threats Uncertain, though this species may be threatened by improvement of grassland and conversion to arable agriculture, afforestation and development.

Published sources Donisthorpe, H.St J.K. (1918), Shirt, D.B., ed. (1987).

CAENOCARA BOVISTAE   RARE

Order COLEOPTERA   Family ANOBIIDAE

Caenocara bovistae (Hoffmann, 1803). Formerly known as: *Coenocara bovistae* (Hoffmann, 1803).

Distribution Recorded from South Devon, East Sussex, East Kent, West Kent, Surrey, North Essex, Berkshire, Oxfordshire, East Suffolk, East Norfolk, West Norfolk, Cambridgeshire, Merionethshire, Caernarvonshire, West Lancashire and Cumberland before 1970 and East Sussex and Merionethshire from 1970 onwards.

Habitat and ecology Grassland and coastal shingle. Occurs in the puff-ball fungi *Lycoperdon bovista* and *Bovista plumbea*, and possibly in other puff-ball fungi. Adults have been recorded in April and from June to September.

Status Not listed in the insect Red Data Book (Shirt, 1987). Old records suggest that this species had a widely scattered distribution from southern England, north to Cumberland. Recently recorded from just two vice-counties.

Threats Loss of unimproved grassland through improvement by reseeding or by the application of fertilisers, or by conversion to arable agriculture.

Development, the building of a marina and gravel extraction is destroying the recent East Sussex locality.

Management and conservation Grazing in needed to maintain open conditions. Disturbance of coastal shingle should be avoided.

Published sources Allen, A.A. (1948b), Ashe, G.H. (1922), Fowler, W.W. (1890), Fowler, W.W. & Donisthorpe, H.St J.K. (1913).

DORCATOMA DRESDENSIS   NOTABLE A

Order COLEOPTERA   Family ANOBIIDAE

Dorcatoma dresdensis Herbst, 1792.

Distribution Recorded from East Sussex, West Kent, Surrey, Middlesex, Berkshire, "Suffolk", Cambridgeshire and Huntingdonshire before 1970 and South Hampshire, Berkshire, Oxfordshire, Buckinghamshire, Cambridgeshire and Huntingdonshire from 1970 onwards.

Habitat and ecology Ancient broad-leaved woodland and pasture-woodland. This species has also been recorded indoors. Breeds in hard bracket fungi on trees, such as *Fomes fomentarius* and *Ganoderma*, with records from oak and beech, and also from willow and apple. Also recorded from *Polyporus* fungus. Larvae have been found in April. Adults have been recorded from May to August.

Status Status revised from RDB 1 (Endangered) in Shirt (1987). Only known in southern England and recorded as far north as Huntingdonshire. Recently recorded from six vice-counties. Probably under-recorded.

Indicator status Grade 2 in Harding & Rose (1986).

Threats Loss of broad-leaved woodland and parkland through, for example, clear-felling and coniferisation. Habitat loss, in particular, through the felling of fungus-infected trees, removal of dead wood from living trees and the destruction or removal of fungus-infected standing and fallen dead wood for reasons such as forest hygiene, aesthetic tidiness, public safety or for use as fire wood.

Management and conservation Fungus-infected trees and both fallen and standing dead timber, should be retained. The removal of fungus-infected timber from ancient trees should be avoided. Gaps in the age structure of the tree population should be identified and the continuity of the appropriate dead wood habitat

ANOBIIDAE

ensured by regeneration, suitable planting and possibly with pollarding.

Published sources Donisthorpe, H.St J.K. (1942b), Harding, P.T. (1978), Harding, P.T. & Rose, F. (1986), Hodge, P.J. (1990), Owen, J.A. (1990a), Shirt, D.B., ed. (1987).

DORCATOMA FLAVICORNIS　　　**NOTABLE B**

Order COLEOPTERA　　　Family ANOBIIDAE

Dorcatoma flavicornis (F., 1792).

Distribution South West, South, South East, East Anglia, East Midlands, West Midlands, North East and North Wales.

Habitat and ecology Ancient broad-leaved woodland and pasture-woodland. Adults and larvae live in dead wood, mainly red rotten oak. Adults have been found on the foliage and boughs of oaks, and rotting heartwood of white willow. Larvae have also been found in wych elm, and adults recorded from small-leaved lime and from beneath alders. Larvae have been noted in July. Adults have been recorded from April to August and in October.

Status Very local with a widely scattered distribution from southern England north to Mid-west Yorkshire. There is an old record for Merionethshire in Wales.

Indicator status Grade 1 in Garland (1983). Grade 3 in Harding & Rose (1986).

Threats Loss of broad-leaved woodland and parkland through, for example, clear-felling and coniferisation. Habitat loss, in particular, through the felling of ancient trees, removal of dead wood from living trees and the destruction or removal of standing and fallen dead wood for reasons such as forest hygiene, aesthetic tidiness, public safety or for use as fire wood.

Management and conservation Ancient trees and both fallen and standing dead timber, especially with the bark attached, should be retained. The removal of dead timber from ancient trees should be avoided. Gaps in the age structure of the tree population should be identified and the continuity of the appropriate dead wood habitat ensured by regeneration, suitable planting and possibly with pollarding.

DORCATOMA SERRA　　　**NOTABLE A**

Order COLEOPTERA　　　Family ANOBIIDAE

Dorcatoma serra Panzer, 1795. Formerly known as: *Dorcatoma punctulata* sensu auct. Brit. not Mulsant and Rey, 1864.

Distribution Recorded from South Hampshire, Hertfordshire, Middlesex, Berkshire, East Norfolk, Herefordshire and Worcestershire before 1970 and Berkshire, Oxfordshire, East Suffolk, West Suffolk, East Norfolk, West Norfolk, Herefordshire and Derbyshire from 1970 onwards.

Habitat and ecology Ancient broad-leaved woodland and pasture-woodland. Adults and larvae live in 'soft' bracket fungi on trees, with records from *Inonotus hispus* on ash and *Polyporus squamosus* on beech. Larvae have been recorded in February and August. Adults (including reared material) occur from May to July.

Status Very local and recorded from South Hampshire north to Derbyshire.

Indicator status Grade 2 in Harding & Rose (1986).

Threats Loss of broad-leaved woodland and parkland through, for example, clear-felling and coniferisation. Habitat loss, in particular, through the felling of fungus-infected trees, removal of dead wood from living trees and the destruction or removal of fungus-infected standing and fallen dead wood for reasons such as forest hygiene, aesthetic tidiness, public safety or for use as fire wood.

Management and conservation Fungus-infected trees and both fallen and standing dead timber, should be retained. The removal of fungus-infected timber from ancient trees should be avoided. Gaps in the age structure of the tree population should be identified and the continuity of the appropriate dead wood habitat ensured by regeneration, suitable planting and possibly with pollarding.

Published sources Allen, A.A. (1988c), Buck, F.D. (1957a), Donisthorpe, H.St J.K. (1928), Hallett, H.M. (1951), Harding, P.T. (1978), Harding, P.T. & Rose, F. (1986), Mendel, H. (1979), Mendel, H. (1989), Nash, D.R. (1974a).

DRYOPHILUS ANOBIOIDES — RARE

Order COLEOPTERA   Family ANOBIIDAE

Dryophilus anobioides Chevrolat, 1832.

Distribution Recorded from East Kent, West Kent, Surrey, Berkshire and West Suffolk before 1970 and East Kent and West Suffolk from 1970 onwards.

Habitat and ecology Heathland, scrub, coastal shingle, roadside verges and disturbed ground. Usually in exposed, sandy situations. Larvae develop in the dead stems of broom *Cytisus scoparius* and also, though probably only very rarely, in dead stems of bramble *Rubus fruticosus*. Adults are usually found by beating dead or dying broom, and have been recorded in June and July.

Status Not listed in the insect Red Data Book (Shirt, 1987). Very local. Only known from five vice-counties in south-eastern England, and recently recorded from just two of these.

Threats Loss of habitat through improvement and conversion to arable agriculture, development, afforestation and gravel extraction. Natural succession may be a further threat.

Management and conservation Gaps in the age structure of the broom population should be identified and the continuity of the appropriate dead wood habitat ensured by allowing regeneration. The disturbance of vegetated shingle should be avoided.

Published sources Fowler, W.W. (1890), Fowler, W.W. & Donisthorpe, H.St J.K. (1913), Massee, A.M. (1952).

GASTRALLUS IMMARGINATUS — ENDANGERED

Order COLEOPTERA   Family ANOBIIDAE

Gastrallus immarginatus (Müller, 1821). Formerly known as: *Gastrallus laevigatus* sensu Kloet and Hincks, 1945 not (Olivier, 1790).

Distribution Recorded from Berkshire before 1970 and Berkshire from 1970 onwards.

Habitat and ecology Ancient broad-leaved woodland and pasture-woodland. Breeds in the bark of old field maple. This species has been recorded on a stack of oak, elm and beech logs. On the Continent, this beetle has been found in the wood and bark of old oak and, rarely, in other deciduous trees. In Britain, larvae have been found in April. Adults have been recorded in July and August.

Status Only known from Windsor Forest and Windsor Great Park, Berkshire.

Indicator status Grade 1 in Harding & Rose (1986).

Threats Loss of broad-leaved woodland and parkland through, for example, clear-felling and coniferisation. Habitat loss, in particular, through the felling of ancient maples and the removal of wood from living trees and fallen dead wood for reasons such as forest hygiene, aesthetic tidiness, public safety or for use as fire wood.

Management and conservation Ancient trees should be retained. The removal of timber from ancient trees should be avoided. Gaps in the age structure of the tree population should be identified and the continuity of the appropriate habitat ensured by regeneration, suitable planting and possibly with pollarding. Windsor Forest and Windsor Great Park have been notified as SSSIs.

Published sources Harding, P.T. (1978), Harding, P.T. & Rose, F. (1986), Shirt, D.B., ed. (1987).

HADROBREGMUS DENTICOLLIS — NOTABLE B

Order COLEOPTERA   Family ANOBIIDAE

Hadrobregmus denticollis (Creutzer in Panzer, 1796). Formerly known as: *Anobium denticolle* Creutzer in Panzer, *Dendrobium denticolle* (Creutzer in Panzer).

Distribution South, South East, East Anglia, West Midlands and East Midlands.

Habitat and ecology Broad-leaved woodland, pasture-woodland and isolated trees. Adults and larvae occur in dead wood, the larvae developing and pupating in cells about half an inch below the surface of the rotten wood. The females bore into and deposit eggs in the more solid wood. The whole life-cycle usually takes place within the dead wood. Adults probably only leave this site to disperse and find new breeding sites. The species has been recorded from oak, elm, willow, grey poplar, holly and pear. Also recorded from the rotten stump of a black poplar, under the loose, dry bark of a fallen ash, in a rot hole of an alder and in a dry rotten hawthorn. Larvae have been noted in April. Adults have been recorded in March, April, June, July, September, October and December.

Status Very local. Widely distributed in southern England and recently recorded as far north as West Norfolk. Not known in south-western England. Probably under-recorded.

ANOBIIDAE

Threats Loss of broad-leaved woodland and parkland through, for example, clear-felling and coniferisation. Habitat loss, in particular, through the felling of ancient trees, removal of dead wood from living trees and the destruction or removal of standing and fallen dead wood for reasons such as forest hygiene, aesthetic tidiness, public safety or for use as fire wood.

Management and conservation Ancient trees and both fallen and standing dead timber, especially with the bark attached, should be retained. The removal of dead timber from ancient trees should be avoided. Gaps in the age structure of the tree population should be identified and the continuity of the appropriate dead wood habitat ensured by regeneration, suitable planting and possibly with pollarding.

HEMICOELUS NITIDUS   **INDETERMINATE**

Order COLEOPTERA   Family ANOBIIDAE

Hemicoelus nitidus (Herbst, 1793).

Distribution Recorded from Berkshire and West Suffolk from 1970 onwards.

Habitat and ecology Isolated trees and pasture-woodland. Recorded from a grey poplar and bred from the bough of a dead field maple. The adult has been recorded in July.

Status Not listed in the insect Red Data Book (Shirt, 1987). Added to the British list in 1980 on the strength of a single male beaten from a grey poplar standing amongst a small group of similar trees in the Brecklands at Icklingham, West Suffolk. Subsequently, this species has been bred from field maple at Windsor, Berkshire. These are at present the only known records.

Threats Uncertain, though the clear-felling of the host-trees and conversion of the land to another use is likely to threaten this species. Habitat loss, in particular, maybe through the felling of ancient trees, removal of dead wood from living trees and the destruction or removal of standing and fallen dead wood for reasons such as public safety or for use as fire wood.

Management and conservation Trees and both fallen and standing dead timber, especially with the bark attached, should be retained. The removal of dead timber from trees should be avoided. Gaps in the age structure of the tree population should be identified and the continuity of the appropriate habitat ensured by regeneration or suitable planting.

Published sources Mendel, H. (1982), Mendel, H. (1989), Owen, J.A. (1990a), Owen, J.A. (1990c).

PTINOMORPHUS IMPERIALIS   **NOTABLE B**

Order COLEOPTERA   Family ANOBIIDAE

Ptinomorphus imperialis (L., 1767). Formerly known as: *Hedobia imperialis* (L.).

Distribution England, North Wales, South East Scotland and South West Scotland.

Habitat and ecology Woodland, pasture-woodland, hedgerows and gardens. Once recorded emerging from a piece of walnut furniture. Associated with dead wood. Adults are usually beaten from trees, bushes and hedges, and have also been swept from low growing vegetation and found under bark. Recorded from oak, beech, ash, lime, poplar, hazel, hawthorn, apple, walnut, ivy and southern beech. Cocoons have been found under the bark of crab apple stumps and in dead twigs of *Wistaria*. Adults have been recorded from May to July.

Status Widespread but local in England, also recorded in North Wales and southern Scotland.

Threats Loss of broad-leaved woodland and parkland through practices, for example, clear-felling and coniferisation. Habitat loss, in particular, through the felling of trees, removal of dead wood from living trees and the destruction or removal of standing and fallen dead wood for reasons such as forest hygiene, aesthetic tidiness, public safety or for use as fire wood.

Management and conservation Trees and both fallen and standing dead timber, especially with the bark attached, should be retained. The removal of dead timber from trees should be avoided. Gaps in the age structure of the tree population should be identified and the continuity of the appropriate dead wood habitat ensured by regeneration, suitable planting and possibly with pollarding.

XYLETINUS LONGITARSIS   **VULNERABLE**

Order COLEOPTERA   Family ANOBIIDAE

Xyletinus longitarsis Jansson, 1942. Formerly known as: *Xyletinus ater* sensu auct. Brit. not (Creutzer in Panzer, 1796).

Distribution Recorded from South Devon, South Hampshire, West Sussex, East Kent, West Kent, Surrey, Hertfordshire, Berkshire, East Suffolk, West

Gloucestershire, Herefordshire and Nottinghamshire before 1970 and West Sussex and Herefordshire from 1970 onwards.

Habitat and ecology Ancient broad-leaved woodland. Associated with dead wood. Recorded from an ancient oak. Also found in dead broom *Cytisus scoparius* on a coastal shingle site. Adults have been recorded in May and June.

Status Not listed in the insect Red Data Book (Shirt, 1987). Old records indicate that this species was formerly widely distributed in southern England, with records north to Nottinghamshire. There is an old record for Burton-on-Trent on the Staffordshire and Derbyshire border. There is only one recent record for this beetle.

Indicator status Grade 3 in Garland (1983). Grade 3 in Harding & Rose (1986).

Threats Loss of broad-leaved woodland through, for example, clear-felling and coniferisation. Habitat loss, in particular, through the felling of ancient trees, removal of dead wood from living trees and the destruction or removal of standing and fallen dead wood for reasons such as forest hygiene, aesthetic tidiness, public safety or for use as fire wood. Coastal developments and gravel extraction may be further threats.

Management and conservation Ancient trees and both fallen and standing dead timber, especially with the bark attached, should be retained. The removal of dead timber from ancient trees should be avoided. Gaps in the age structure of the tree population should be identified and the continuity of the appropriate dead wood habitat ensured by regeneration, suitable planting and possibly with pollarding. The disturbance of coastal shingle should be avoided.

Published sources Atty, D.B. (1983), Fowler, W.W. (1890), Fowler, W.W. & Donisthorpe, H.St J.K. (1913), Garland, S.P. (1983), Harding, P.T. (1978), Harding, P.T. & Rose, F. (1986), Johnson, C. (1975a).

ANTHICUS ANGUSTATUS NOTABLE B

Order COLEOPTERA Family ANTHICIDAE

Anthicus angustatus Curtis, 1838.

Distribution South East, South, South West, East Anglia, West Midlands and North West.

Habitat and ecology Sandy shores and probably only occasionally in saltmarshes. Found on and in bare sand

with little or no vegetation. Possibly associated with rotting vegetation in saltmarshes. Adults have been recorded in April.

Status Widespread and very local in England.

Threats Loss of sandy shores and saltmarsh through reclamation, erosion and the construction of sea defences.

Management and conservation Grazing should not be introduced to a saltmarsh where there is no grazing at present.

ANTHICUS BIFASCIATUS NOTABLE B

Order COLEOPTERA Family ANTHICIDAE

Anthicus bifasciatus (Rossi, 1792).

Distribution South East, South, South West, East Anglia, East Midlands and West Midlands.

Habitat and ecology This species has been recorded from farms and in fens. Found in and amongst dung, from decaying compost material, and possibly also amongst rotting vegetation or reed litter. Adults have been recorded in January, April, May and from July to October.

Status Widespread but local in the southern half of England. Very rarely encountered in natural and semi-natural habitats.

Threats Drainage for reasons such as agricultural improvement and development is the primary cause of the loss of fens. Further loss may be through falling water tables because of water abstraction and river engineering schemes. The removal of reed litter is likely to threaten this species.

Management and conservation Cutting or grazing is needed to maintain open conditions in fens.

ANTHICUS BIMACULATUS NOTABLE A

Order COLEOPTERA Family ANTHICIDAE

Anthicus bimaculatus (Illiger, 1801).

Distribution Recorded from North Devon, East Sussex, East Kent, Glamorgan, Cheshire and South Lancashire before 1970 and East Sussex, East Kent, East Norfolk and Anglesey from 1970 onwards.

ANTHICIDAE

Habitat and ecology Sand dunes, sandy areas and the sandy edges of water-filled gravel pits. Probably associated with decaying vegetation. Recorded from under the edge of an algal mat on the edge of a sandy saltmarsh. Adults have been recorded in May, June and August.

Status Widely distributed in England and recorded from North Devon to South Lancashire, there is also an old record from South Wales. Recently recorded from four, widely scattered, vice-counties.

Threats Loss of habitat through coastal developments, the construction of sea defences, erosion, the infilling of gravel pits and natural succession.

Management and conservation Management should aim at maintaining open conditions and encouraging early successional stages.

Published sources Allen, A.A. (1975b), Fowler, W.W. (1891), Fowler, W.W. & Donisthorpe, H.St J.K. (1913), Gilbert, O. (1958), Parry, J.A. (1980a).

ANTHICUS SALINUS	NOTABLE A
Order COLEOPTERA	Family ANTHICIDAE

Anthicus salinus Crotch, 1866. Formerly known as: *Anthicus crotchi* Pic, 1893.

Distribution Recorded from Dorset, South Hampshire, North Hampshire, East Kent and West Kent before 1970 and South Hampshire, East Sussex and East Kent from 1970 onwards.

Habitat and ecology Saltmarshes. Probably associated with rotting vegetation. Adults have been found under vegetation, running in the open on damp or muddy ground, and a single example has been found under a stone. Adults have been recorded in April and May.

Status Only known from southern England. Very local and recently recorded from just three vice-counties.

Threats Loss of saltmarsh through reclamation, erosion and the construction of sea defences.

Management and conservation Grazing should not be introduced to a site where there is no grazing at present.

Published sources Buck, F.D. (1956), Fowler, W.W. (1891).

ANTHICUS SCOTICUS	RARE
Order COLEOPTERA	Family ANTHICIDAE

Anthicus scoticus Rye, 1872.

Distribution Recorded from Caernarvonshire, Cumberland, Ayrshire, Renfrewshire, Midlothian, Fife, Stirlingshire, "Aberdeen", Dunbartonshire and Clyde Islands before 1970 and East Kent from 1970 onwards.

Habitat and ecology Muddy, sandy or silty places such as the margins of lakes and in coastal areas. Probably associated with decaying vegetation. Adults have been recorded in January and August.

Status Not listed in the insect Red Data Book (Shirt, 1987). Old records indicate that this is a northern and western species. However, the only recent records are from East Kent and these may be the result of recent colonisation.

Threats Loss of habitat through reclamation, erosion and the construction of sea defences. Coastal developments are a further threat.

Management and conservation Management should aim at maintaining open conditions.

Published sources Buck, F.D. (1954), Fowler, W.W. (1891), Fowler, W.W. & Donisthorpe, H.St J.K. (1913), Parry, J.A. (1983).

ANTHICUS TRISTIS	ENDANGERED
Order COLEOPTERA	Family ANTHICIDAE

Anthicus tristis Schmidt, 1842.

Distribution Recorded from Dorset, Isle of Wight and South Hampshire before 1970 and Dorset from 1970 onwards.

Habitat and ecology Coastal shingle and sandy expanses at the edge of saltmarshes. Probably associated with decaying vegetation. Adults have been recorded in April.

Status Not listed in the insect Red Data Book (Shirt, 1987). Formerly locally common on the Chesil Beach, Dorset and, until 1989, this species was last recorded in 1926. In 1989 the species was again reported in Dorset.

Threats Shingle beaches are sensitive to the effects of trampling, motorbike access etc., which damages any vegetated sections causing accumulated humus to erode.

Management and conservation Disturbance of the shingle should be avoided. Chesil Beach is in the Chesil and Fleet SSSI.

Published sources Cooter, J. (1989), Fowler, W.W. & Donisthorpe, H.St J.K. (1913).

ANTHRIBUS FASCIATUS　　　　　NOTABLE A
A fungus weevil
Order COLEOPTERA　　　　Family ANTHRIBIDAE

Anthribus fasciatus Forster, 1771. Formerly known as: *Brachytarsus fasciatus* (Forster).

Distribution Recorded from South Devon, Dorset, Isle of Wight, South Hampshire, East Kent, West Kent, Surrey, Hertfordshire, Middlesex, Oxfordshire, East Suffolk, Cambridgeshire, Huntingdonshire, Northamptonshire, West Gloucestershire, Herefordshire, Worcestershire, Staffordshire, Shropshire, Glamorgan, Denbighshire, South Lincolnshire, North Lincolnshire, Leicestershire & Rutland, Nottinghamshire, Derbyshire, Cheshire, South Lancashire and Berwickshire before 1970 and Dorset, Berkshire, East Gloucestershire, Herefordshire, Staffordshire, North Lincolnshire, Leicestershire & Rutland and South-west Yorkshire from 1970 onwards.

Habitat and ecology Woodland, pasture woodland and hedgerows. Associated with a large variety of trees, though possibly with a preference for hawthorn. Adults have also been beaten from oak, field maple and conifers. Larvae are predatory on scale-insects (Hemiptera). Adults have been recorded in May, June and August.

Status Old records suggest that this species was formerly more widespread and recorded from southern England as far north as Berwickshire. Recent records indicate a decline, with a widely scattered distribution from Dorset to North Lincolnshire.

Threats Loss of broad-leaved woodland through, for example, clear-felling and conversion to other land use. This species may be further threatened by the grubbing out and mechanised trimming of hedgerows.

Management and conservation Gaps in the age structure of the tree population should be identified and the continuity of the appropriate habitat ensured by regeneration and suitable planting.

Published sources Atty, D.B. (1983), Cooter, J. (1990d), Fowler, W.W. (1891), Fowler, W.W. & Donisthorpe, H.St J.K. (1913), Morris, M.G. (1990), Skidmore, P., Limbert, M. & Eversham, B. (1985).

ANTHRIBUS NEBULOSUS　　　　　NOTABLE B
A fungus weevil
Order COLEOPTERA　　　　Family ANTHRIBIDAE

Anthribus nebulosus Forster, 1771. Formerly known as: *Anthribus variegatus* Fourcroy, 1785, *Brachytarsus nebulosus* (Forster), *Brachytarsus varius* (F., 1787).

Distribution England, North Wales, South East Scotland and South West Scotland.

Habitat and ecology Broad-leaved woodland, coniferous plantations and hedgerows. Associated with a large variety of trees and shrubs, including oak, lime and willow. Also noted from under sycamore bark, by sweeping under Scots pine, and beating and sweeping under spruce. Larvae are predatory on scale-insects (Hemiptera). Adults have been recorded in February and from April to July.

Status Widespread but local in England, also recorded from southern Scotland and North Wales.

Indicator status Grade 3 in Garland (1983).

Threats Loss of woodland through, for example, clear-felling and conversion to other land use. This species may be further threatened by the grubbing out and mechanised trimming of hedgerows.

Management and conservation Gaps in the age structure of the tree population should be identified and the continuity of the appropriate habitat ensured by regeneration and suitable planting.

CHORAGUS SHEPPARDI　　　　　NOTABLE A
A fungus weevil
Order COLEOPTERA　　　　Family ANTHRIBIDAE

Choragus sheppardi Kirby, 1819.

Distribution Recorded from West Cornwall, South Devon, North Devon, Dorset, Isle of Wight, South Hampshire, West Sussex, East Sussex, East Kent, West Kent, Surrey, South Essex, Hertfordshire, Berkshire, East Suffolk, West Suffolk, East Norfolk, Cambridgeshire, West Gloucestershire, Worcestershire, Warwickshire, Staffordshire, Shropshire, Glamorgan, Leicestershire & Rutland, Derbyshire, Cheshire, South Lancashire, West Lancashire, North-east Yorkshire, South-west Yorkshire, Mid-west Yorkshire and Lanarkshire before 1970 and South Wiltshire, East Sussex, East Suffolk and Cardiganshire from 1970 onwards.

Habitat and ecology Broad-leaved woodland, pasture-woodland and hedgerows. Associated with dead

69

ANTHRIBIDAE

wood, particularly old ivy. Also recorded from hawthorn and hazel. Larvae are found in rotten, fungus-infested or dry, and firm dead wood. Adults have been recorded from April to August and in October.

Status Old records suggest that this species was widely distributed throughout England and recorded in South Wales and South West Scotland. This beetle has been very uncommon during the 20th century and has been recorded recently from only four vice-counties. The beetles hop at the first hint of disturbance and consequently may be under-recorded.

Threats Loss of broad-leaved woodland and parkland through, for example, clear-felling and coniferisation. The grubbing out and mechanised trimming of hedgerows are threats. Habitat loss, in particular, through the felling of trees and the removal of dead wood from living trees and the destruction or removal of standing and fallen dead wood for reasons such as forest hygiene, aesthetic tidiness, public safety or for use as fire wood. This species may be further threatened by the removal of ivy from living, and dead, trees.

Management and conservation Ancient trees and both fallen and standing dead timber, especially with the bark attached, should be retained. The removal of dead timber from ancient trees should be avoided. Gaps in the age structure of the tree population should be identified and the continuity of the appropriate dead wood habitat ensured by regeneration and suitable planting.

Published sources Appleton, D. (1969a), Atty, D.B. (1983), Fowler, W.W. (1891), Fowler, W.W. & Donisthorpe, H.St J.K. (1913), Morris, M.G. (1990), Nash, D.R. (1979a).

PLATYRHINUS RESINOSUS **NOTABLE B**
A fungus weevil
Order COLEOPTERA Family ANTHRIBIDAE

Platyrhinus resinosus (Scopoli, 1763). Formerly known as: *Anthribus resinosus* (Scopoli), *Platyrrhinus latirostris* (F., 1775), *Platyrrhinus resinosus* (F., 1775).

Distribution England, South Wales, North East Scotland and North West Scotland.

Habitat and ecology Broad-leaved woodland, isolated trees and trees in hedgerows. Associated with dead wood, particularly ash, but also on other trees such as birch, beech and sycamore. Also found in areas of burnt birch. Larvae have been noted in June in the fungus *Daldinia concentrica*. Adults have been recorded in January, February, and from April to December, though the main period of adult activity seems to be during April to July.

Status Widespread but local. Old records suggest that this species had a scattered distribution from southern England as far north as Easterness. Recent records indicate that this species is primarily noted from the Midlands, with scattered records as far north as Elgin.

Indicator status Grade 1 in Garland (1983). Grade 3 in Harding & Rose (1986).

Threats Loss of broad-leaved woodland through, for example, clear-felling and coniferisation, the uprooting of hedgerows and felling of isolated trees. Habitat loss, in particular, through the felling of fungus-infected trees, removal of dead wood from living trees and the destruction or removal of standing and fallen dead wood for reasons such as forest hygiene, aesthetic tidiness, public safety or for use as fire wood.

Management and conservation Fungus-infected trees and both fallen and standing dead timber, especially with the bark attached, should be retained. The removal of dead timber from trees should be avoided. Gaps in the age structure of the tree population should be identified and the continuity of the appropriate dead wood habitat ensured by regeneration and suitable planting.

PLATYSTOMOS ALBINUS **NOTABLE B**
A fungus weevil
Order COLEOPTERA Family ANTHRIBIDAE

Platystomos albinus (L., 1758). Formerly known as: *Macrocephalus albinus* (L.), *Platystomus albinus* (L.).

Distribution England and South Wales.

Habitat and ecology Ancient broad-leaved woodland and pasture woodland. Associated with dead wood and dying trees and shrubs, with records from hornbeam, beech, oak, hazel and hawthorn. Larvae are found in dead wood. Adults have been recorded from March to September.

Status Local and widely scattered from southern England as far north as Westmorland.

Indicator status Grade 3 in Harding & Rose (1986).

Threats Loss of broad-leaved woodland and parkland through, for example, clear-felling and coniferisation. Habitat loss, in particular, through the felling of ancient trees, removal of dead wood from living trees and the destruction or removal of standing and fallen dead wood for reasons such as forest hygiene, aesthetic tidiness, public safety or for use as fire wood.

Management and conservation Ancient trees and both fallen and standing dead timber, especially with the bark attached, should be retained. The removal of dead timber from ancient trees should be avoided. Gaps in the age structure of the tree population should be identified and the continuity of the appropriate dead wood habitat ensured by regeneration, suitable planting and possibly with pollarding.

TROPIDERES NIVEIROSTRIS   **VULNERABLE**
A fungus weevil
Order COLEOPTERA   Family ANTHRIBIDAE

Tropideres niveirostris (F., 1798). Formerly known as: *Dissoleucas niveirostris* (F.).

Distribution Recorded from Dorset, South Hampshire, East Kent, West Kent, Surrey, East Suffolk, Huntingdonshire, West Gloucestershire, Worcestershire and Leicestershire & Rutland before 1970 and Dorset and Berkshire from 1970 onwards.

Habitat and ecology Broad-leaved woodland and old, neglected hedges. Associated with dead wood of a variety of trees and shrubs. Particularly in fallen oak boughs of a few inches in diameter, and also in hazel. Larvae are found in dead wood. Adults have been found in the litter at the bottom of hedges. Adults have been recorded in June, July, September and October.

Status Status revised from RDB 3 (Rare) in Shirt (1987). This has been a very uncommon species during the 20th century. Only recorded in southern England and recently recorded from only two examples, each in different vice-counties.

Indicator status Grade 3 in Harding & Rose (1986).

Threats Loss of broad-leaved woodland through, for example, clear-felling and coniferisation, and the grubbing out and mechanised trimming of old hedgerows. Habitat loss, in particular, through the felling of trees, removal of dead wood from living trees and the destruction or removal of standing and fallen dead wood for reasons such as forest hygiene, aesthetic tidiness, public safety or for use as fire wood.

Management and conservation Trees and both fallen and standing dead timber, especially with the bark attached, should be retained. The removal of dead timber from ancient trees should be avoided. Gaps in the age structure of the tree population should be identified and the continuity of the appropriate dead

wood habitat ensured by regeneration, suitable planting and possibly with pollarding.

Published sources Atty, D.B. (1983), Chitty, A.J. (1904), Fowler, W.W. (1891), Fowler, W.W. & Donisthorpe, H.St J.K. (1913), Harding, P.T. (1978), Harding, P.T. & Rose, F. (1986), Morris, M.G. (1986), Morris, M.G. (1990), Shirt, D.B., ed. (1987).

TROPIDERES SEPICOLA   **VULNERABLE**
A fungus weevil
Order COLEOPTERA   Family ANTHRIBIDAE

Tropideres sepicola (F., 1792).

Distribution Recorded from South Hampshire, North Hampshire, West Sussex, East Kent, North Essex, Huntingdonshire, Herefordshire and Leicestershire & Rutland before 1970 and North Hampshire and East Kent from 1970 onwards.

Habitat and ecology Ancient broad-leaved woodland and pasture-woodland. Associated with dead wood, particularly in dead branches of oak, hornbeam and beech, but also on other trees. Larvae are found in dead wood. Adults have been found by sweeping under oaks and have been recorded from May to August.

Status Status revised from RDB 3 (Rare) in Shirt (1987). Very local and only known from southern England. Recently recorded from only two vice-counties.

Indicator status Grade 1 in Harding & Rose (1986).

Threats Loss of broad-leaved woodland and parkland through, for example, clear-felling and coniferisation. Habitat loss, in particular, through the felling of ancient trees, removal of dead wood from living trees and the destruction or removal of standing and fallen dead wood for reasons such as forest hygiene, aesthetic tidiness, public safety or for use as fire wood.

Management and conservation Ancient trees and both fallen and standing dead timber, especially with the bark attached, should be retained. The removal of dead timber from ancient trees should be avoided. Gaps in the age structure of the tree population should be identified and the continuity of the appropriate dead wood habitat ensured by regeneration, suitable planting and possibly with pollarding.

Published sources Appleton, D. (1969a), Fowler, W.W. (1891), Fowler, W.W. & Donisthorpe, H.St J.K. (1913), Harding, P.T. (1978), Harding, P.T. & Rose, F. (1986), Morris, M.G. (1990), Nicholson, G.W. (1931), Shirt, D.B., ed. (1987).

APIONIDAE

APION AFFINE

NOTABLE A

A weevil
Order COLEOPTERA   Family APIONIDAE

Apion affine Kirby, 1808.

Distribution Recorded from West Cornwall, West Sussex, East Sussex, East Kent, West Kent, Surrey, South Essex, North Essex, Hertfordshire, Berkshire, Oxfordshire, Buckinghamshire, "Suffolk", Cambridgeshire, Huntingdonshire, Monmouthshire, Worcestershire, Shropshire, Anglesey, North Lincolnshire, Leicestershire & Rutland, Derbyshire, Cheshire, South Lancashire, West Lancashire, South-east Yorkshire, North-east Yorkshire, South-west Yorkshire, Mid-west Yorkshire, Durham, North Northumberland, Cumberland, Dumfriesshire and Wigtownshire before 1970 and East Sussex, East Kent, Radnorshire and Cumberland from 1970 onwards.

Habitat and ecology Disturbed ground, grassland, sand pits and coastal sites, including stablised shingle. Phytophagous. Associated with sorrel *Rumex acetosa* and possibly also with sheep's sorrel *R. acetosella*. Larvae are reported to occur in galls on the flowers. Adults have been recorded in January, from April to September and in November, though most frequently noted in June.

Status Old records indicate that this species was formerly widespread in England and recorded as far north as Dumfriesshire. Recently recorded from only four, widely scattered, vice-counties.

Threats Improvement of grassland by reseeding or the application of fertilisers, or by conversion to arable agriculture. Gravel extraction, the infilling of sand pits, urban and holiday developments, and natural succession and the invasion of scrub may be further threats.

Management and conservation Grazing, cutting or some other disturbance, such as rotovation, on a rotational basis, is needed to maintain open conditions. The disturbance of vegetated shingle should be avoided.

Published sources Eyre, M.D. (1987), Fowler, W.W. (1891), Fowler, W.W. & Donisthorpe, H.St J.K. (1913), Morris, M.G. (1990).

APION ARMATUM   INSUFFICIENTLY KNOWN

A weevil
Order COLEOPTERA   Family APIONIDAE

Apion armatum Gerstaecker, 1854.

Distribution Recorded from South Hampshire before 1970.

Habitat and ecology On the Continent, this species is found in grassland, field margins in cultivated land and probably ruderal habitats. A species of dry habitats. Phytophagous. Associated with *Centaurea* species.

Status Not listed in the insect Red Data Book (Shirt, 1987). Only known from a single example swept from a damp ride in conifer woodland in the New Forest, South Hampshire. The habitat would appear to be atypical when compared to Continental information.

Published sources Morley, C. (1941), Morris, M.G. (1990).

APION ASTRAGALI

NOTABLE A

A weevil
Order COLEOPTERA   Family APIONIDAE

Apion astragali (Paykull, 1800).

Distribution Recorded from North Somerset, East Sussex, West Kent, Berkshire, Oxfordshire, East Suffolk, Cambridgeshire, Northamptonshire, South Lincolnshire and North Lincolnshire before 1970 and East Kent, Bedfordshire, Northamptonshire and East Gloucestershire from 1970 onwards.

Habitat and ecology Woodland, grassland, roadside verges and a disused quarry. Appears to prefer open situations. Phytophagous. Associated with wild liquorice *Astragalus glycyphyllos*. On the Continent, this species is also associated with other species of *Astragalus*. Larvae occur in flower buds. In Britain, adults have been recorded from May to October.

Status Very local. Old records suggest that this species was widespread in England south of the Humber. Records from northern Britain and Scotland require confirmation. Recently recorded from only four vice-counties. Can be abundant where found.

Threats This species is threatened by clear-felling and conversion to other land use, and improvement of grassland and conversion to arable agriculture. Natural succession and scrub invasion may be a further threat.

Management and conservation Management should aim at encouraging a mosaic of habitats. Open glades and ride margins should be cut on rotation to retain a variety of vegetation structures.

Published sources Atty, D.B. (1983), Fowler, W.W. (1891), Fowler, W.W. & Donisthorpe, H.St J.K. (1913), Morris, M.G. (1990).

APION BRUNNIPES	ENDANGERED
A weevil	
Order COLEOPTERA	Family APIONIDAE

Apion brunnipes Boheman, 1839. Formerly known as: *Apion laevigatum* Kirby, 1808 nec (Paykull, 1792).

Distribution Recorded from North Devon, Dorset, West Kent, Berkshire, East Suffolk and West Suffolk before 1970.

Habitat and ecology Disturbed ground and field-edges. Chiefly in sandy habitats. Phytophagous. Associated with cudweed *Filago* spp. and *Gnaphalium* spp., in particular common cudweed *F. vulgaris*. On the Continent, this species is also associated with narrow-leaved cudweed *F. gallica*. Larvae occur in galls on flower heads and leaves of the growing shoots. In Britain adults have been recorded from July to early October.

Status Status revised from RDB3 (Rare) in Shirt (1987). This has been a very uncommon species in the 20th century. Last recorded in 1937 from Freckenham, West Suffolk and Frostenden, East Suffolk.

Threats This species may be threatened by modern intensive agricultural practices, including the use of pesticides and herbicides. Natural succession and the invasion of scrub may be a further threat.

Published sources Fowler, W.W. (1891), Fowler, W.W. & Donisthorpe, H.St J.K. (1913), Morris, M.G. (1990), Shirt, D.B., ed. (1987).

APION CERDO	NOTABLE B
A weevil	
Order COLEOPTERA	Family APIONIDAE

Apion cerdo Gerstaecker, 1854.

Distribution South, South East, East Anglia, East Midlands, West Midlands, North East, North West, North Wales and South West Scotland.

Habitat and ecology Woodland rides, grassland, hedgerows, floodplain fens and road-side verges. Phytophagous. Associated with *Vicia* spp., particularly tufted vetch *Vicia cracca*. On the Continent, this beetle

is also found on meadow vetchling *Lathyrus pratensis*. Larvae occur in the pods. In Britain, adults have been recorded from May to July and in September.

Status Widespread but local in the Midlands and northern England. Also recorded in southern and eastern England, where it is possibly a recent colonist, and North Wales and South West Scotland.

Threats Improvement of grassland by reseeding or by application of fertilisers, or by conversion to arable agriculture. Lack of, or changes in grazing or cutting regimes, the grubbing out of hedgerows, drainage, and the clear-felling of woodland and conversion to other land use may be further threats.

Management and conservation Grazing or cutting, on a rotational basis, is needed to maintain open conditions. Open glades, ride margins and hedgerows should be managed to retain a variety of vegetation structures.

APION CINERACEUM	NOTABLE A
A weevil	
Order COLEOPTERA	Family APIONIDAE

Apion cineraceum Wencker, 1864. Formerly known as: *Apion annulipes* sensu Fowler, 1891 not Wencker, 1864, *Apion millum* sensu Bach, 1854 not Gyllenhal, 1833.

Distribution Recorded from South Devon, Dorset, Isle of Wight, South Hampshire, East Sussex, East Kent, West Kent, Surrey, Hertfordshire, Berkshire, Oxfordshire, Buckinghamshire, East Suffolk, West Gloucestershire, Cheshire and South Lancashire before 1970 and South Hampshire, East Sussex, East Kent, West Kent and Surrey from 1970 onwards.

Habitat and ecology Grassland, particularly on base rich soils, field margins, old orchards and possibly woodland. Phytophagous. Associated with selfheal *Prunella vulgaris*. Larvae are possibly root-feeders. Adults have been recorded from June to October.

Status Very local. Old records suggest that this species was recorded throughout southern England and with a scattered distribution north to South Lancashire. Recently recorded from only five vice-counties, in southern or south-eastern England.

Threats Improvement of grassland by reseeding or the application of fertilisers, or by conversion to arable agriculture. Lack of, or changes in grazing or cutting regimes, and the use of pesticides and herbicides may be further threats to this species.

APIONIDAE

Management and conservation Grazing, cutting, on a rotational basis, or some other disturbance is needed to maintain open conditions.

Published sources Allen, A.A. (1945b), Atty, D.B. (1983), Easton, A.M. (1946a), Fowler, W.W. (1891), Fowler, W.W. & Donisthorpe, H.St J.K. (1913), Morris, M.G. (1990), Parry, J.A. (1962).

APION CURTISI   NOTABLE A
A weevil
Order COLEOPTERA   Family APIONIDAE

Apion curtisi Stephens, 1831. Formerly known as: *Apion curtisii* Stephens, 1831, *Apion curtulum* Desbrochers,.

Distribution Recorded from West Cornwall, "Devon", North Somerset, Dorset, Isle of Wight, West Sussex, East Sussex, East Kent, West Kent, North Essex and East Suffolk before 1970 and Isle of Wight, South Hampshire, East Sussex and East Kent from 1970 onwards.

Habitat and ecology Maritime cliff, sand dune and possibly other coastal habitats. Very rarely recorded inland. Phytophagous. Recorded from subterranean clover *Trifolium subterraneum*, though possibly also on other species of clover. On the Continent, *A. curtisi* has been found on white clover *T. repens* and strawberry clover *T. fragiferum*. Larvae occur in galls on the rootstocks. In Britain, adults have been recorded throughout the year.

Status Very local. Formerly recorded along the southern coast of England north to Norfolk. Inland records from Cambridgeshire and Leicestershire and perhaps those from Norfolk, West Lancashire, South-east Yorkshire and Cheviot require confirmation. The record for West Kent may refer to Surrey. Recently recorded from only four vice-counties from the Isle of Wight to East Kent. This species is difficult to identify and may be confused with *A. seniculus*. Consequently, the exact status of this species is hard to assess. There is also some confusion over nomenclature; continental literature, almost exclusively, refers to this species as *A. curtulum* and refers to our *A. waltoni* as *A. curtisi*.

Threats Improvement of coastal grassland by reseeding or the application of fertilisers, or by conversion to arable agriculture. Coastal developments, the degradation of suitable habitat through natural succession and scrub invasion or by excessive disturbance of the vegetation are further threats.

Management and conservation Some grazing, cutting or other disturbance may be desirable to maintain the

early successional stages and prevent the invasion of scrub.

Published sources Fowler, W.W. (1891), Fowler, W.W. & Donisthorpe, H.St J.K. (1913), Morris, M.G. (1990).

APION DIFFICILE   NOTABLE A
A weevil
Order COLEOPTERA   Family APIONIDAE

Apion difficile Herbst, 1797. Formerly known as: *Apion kiesenwetteri* Desbrochers, 1870.

Distribution Recorded from Dorset, Isle of Wight, South Hampshire, West Sussex, East Sussex, West Kent, Surrey, East Gloucestershire and Worcestershire before 1970 and Dorset, Isle of Wight, East Sussex, West Kent and East Gloucestershire from 1970 onwards.

Habitat and ecology Grassland, heathland and roadside verges. Phytophagous. Associated with *Genista* spp., possibly exclusively on dyer's greenweed *G. tinctoria*. Larvae occur in the pods, occasionally with up to three larvae in a pod. Adults have been recorded in June, disappear almost entirely in July and August, with the new generation emerging in September and October.

Status Very local and recorded in southern England. Recently recorded from only five vice-counties.

Threats Improvement of grassland by reseeding or the application of fertilisers, or by conversion to arable agriculture. Loss of heathland through conversion to other land use. Lack of, or changes in grazing or cutting regimes may be a threat to this species.

Management and conservation Grazing or cutting, on a rotational basis, is needed to maintain open conditions.

Published sources Allen, A.A. (1952d), Atty, D.B. (1983), Morris, M.G. (1982c), Morris, M.G. (1990).

74

APION DIFFORME — NOTABLE B
A weevil
Order COLEOPTERA — Family APIONIDAE

Apion difforme Ahrens, 1817.

Distribution South East, South, South West, East Anglia, East Midlands, West Midlands and South Wales.

Habitat and ecology Damp grassland, wetland, disturbed ground, hedge-banks and along ditches. Phytophagous. Repeatedly found on knotgrass *Polygonum* species. The true pabulum, however, may prove to be clover *Trifolium* species. Adults have been recorded from late April to September, in litter from September to October and in tussocks from December to April.

Status Widespread but local in southern England as far north as Warwickshire, Leicestershire and Nottinghamshire. Also recorded in South Wales.

Threats Loss of habitat through practices such as improvement and conversion to arable agriculture. Natural succession and scrub invasion may also threaten this species.

Management and conservation Grazing, cutting, on a rotational basis, or some other disturbance, such as rotovation, is needed to maintain open conditions.

APION DISPAR — RARE
A weevil
Order COLEOPTERA — Family APIONIDAE

Apion dispar Germar, 1817.

Distribution Recorded from East Kent before 1970 and East Kent from 1970 onwards.

Habitat and ecology Field margins and disturbed ground. Phytophagous. Chiefly associated with stinking chamomile *Anthemis cotula*, possibly also with mayweeds *Matricaria* and *Tripleurospermum* species. Adults have been recorded in August.

Status Recorded new to Great Britain in 1967 from East Kent. Still only known from a small area around the original site of capture. This species is difficult to identify and may be confused with *A. hookeri*. Consequently, the exact status of this species is hard to assess.

Threats This species may be threatened by modern intensive agricultural practices, including the use of pesticides and herbicides. Natural succession and scrub

invasion, and urban development may be further threats.

Management and conservation Management, such as disturbance through rotovation, should aim at maintaining open conditions and encouraging early successional stages.

Published sources Dolling, W.R. (1974), Morris, M.G. (1990).

APION DISSIMILE — NOTABLE B
A weevil
Order COLEOPTERA — Family APIONIDAE

Apion dissimile Germar, 1817.

Distribution South East, South, South West, East Anglia, East Midlands, West Midlands, North West, South Wales and North Wales.

Habitat and ecology Disturbed ground and sand dunes. The species prefers dry, sandy habitats, on acid or neutral soils, with little vegetation cover. Phytophagous. Associated with hare's-foot clover *Trifolium arvense*. Larvae occur in the flowers. Adults have been recorded throughout the year.

Status Widespread but local in southern England and North West England. Also noted in Wales.

Threats Loss of dune habitat, particularly through afforestation, urban and holiday development. The degradation of remaining habitat by excessive disturbance of the vegetation through activities such as motorbike access, horse-riding and human trampling. This species may be further threatened by natural succession and scrub invasion.

Management and conservation Some grazing and other disturbance, such as rotovation, may be desirable to maintain the early successional stages and prevent the invasion of scrub.

APION FILIROSTRE — NOTABLE B
A weevil
Order COLEOPTERA — Family APIONIDAE

Apion filirostre Kirby, 1808.

Distribution South East, South, South West, East Anglia, East Midlands, West Midlands, North West and Dyfed-Powys.

Habitat and ecology Grassland, field margins, disturbed ground and quarries. The species prefers

APIONIDAE

calcareous soils and can occur on thin soils with sparse vegetation cover. Phytophagous. Associated with black medick *Medicago lupulina*. On the Continent, larvae have been found in the buds of lucerne *M. sativa* and sickle medick *M. sativa* subsp. *falcata*. In Britain, adults have been recorded from May to October.

Status Widespread but local in southern England as far north as East Suffolk and Warwickshire. Also recorded in North West England and mid Wales.

Threats Improvement of grassland by reseeding or the application of fertilisers, or by conversion to arable agriculture. Lack of, or changes in grazing or cutting regimes, the infilling of quarries, urban development, and the use of pesticides and herbicides may also threaten this species.

Management and conservation Grazing, cutting, on a rotational basis, or some other disturbance, such as rotovation, is needed to maintain open conditions.

APION FLAVIMANUM   **NOTABLE A**
A weevil
Order COLEOPTERA   Family APIONIDAE

Apion flavimanum Gyllenhal, 1833.

Distribution Recorded from South Devon, West Sussex, East Sussex, East Kent, West Kent, Surrey, Berkshire, Oxfordshire, Buckinghamshire and "Norfolk" before 1970 and North Hampshire, East Kent, West Kent, Surrey, South Essex, Oxfordshire and Northamptonshire from 1970 onwards.

Habitat and ecology Calcareous grassland and hedgerows. Phytophagous. Associated mainly with marjoram *Origanum vulgare*, also on calamint *Calmintha* spp. and possibly also on wild basil *Clinopodium vulgare*. On the Continent this species is also associated with mint *Mentha* species. Larvae occur in the stems. Adults have been recorded from May to October, though are usually uncommon in the summer months.

Status Very local and formerly widely distributed in southern England. Recently recorded from North Hampshire to Northamptonshire.

Threats Loss of calcareous grassland through development and improvement by reseeding or the application of fertilisers, or by conversion to arable agriculture. Lack of, or changes in grazing or cutting regimes may be a further threat to this species.

Management and conservation Grazing or cutting, on a rotational basis, is needed to maintain open conditions.

Published sources Allen, A.A. (1945b), Allen, A.A. (1960b), Easton, A.M. (1946a), Fowler, W.W. (1891), Morris, M.G. (1990), Morris, M.G. & Rispin, W.E. (1988), Parry, J.A. (1962).

APION GENISTAE   **NOTABLE A**
A weevil
Order COLEOPTERA   Family APIONIDAE

Apion genistae Kirby, 1811.

Distribution Recorded from Dorset, Isle of Wight, South Hampshire, North Hampshire, East Sussex, East Kent, Surrey, South Essex, Berkshire, Buckinghamshire, "Suffolk", East Norfolk, Warwickshire, Durham, Cumberland and Moray before 1970 and Dorset, South Hampshire, Moray and East Inverness & Nairn from 1970 onwards.

Habitat and ecology Grassland and wet heathland. Phytophagous. Associated with *Genista* spp., chiefly, if not exclusively on petty whin *G. anglica*. On the Continent, this species has been recorded from dyer's greenweed *G. tinctoria* and hairy greenweed *G. pilosa*. Larvae occur in the pods. In Britain, adults have been recorded from May to August.

Status Extremely local. Old records suggest that this species was formerly more widespread, with a scattered distribution from southern England to Elgin in Scotland. Recently recorded from just four vice-counties, two in southern England and two in Scotland.

Threats Loss of habitat through development, and improvement by reseeding or by the application of fertilisers, or by conversion to arable agriculture. Lack of, or changes in grazing or cutting regimes, and drainage, may also have contributed to this species decline.

Management and conservation Water tables should be maintained at high levels. Water bodies should be isolated from sources of eutrophication and pollution. Grazing, cutting or some other disturbance, on a rotational basis, is needed to maintain open conditions and encourage early successional stages.

Published sources Eyre, M.D. (1987), Fowler, W.W. (1891), Fowler, W.W. & Donisthorpe, H.St J.K. (1913), Morris, M.G. (1990).

APION GYLLENHALI   NOTABLE B
A weevil
Order COLEOPTERA   Family APIONIDAE

Apion gyllenhali Kirby, 1808. Formerly known as:
Apion gyllenhalii Kirby, 1808.

Distribution South East, South, South West, East
Anglia, East Midlands, North East, North West, Wales,
South East Scotland, South West Scotland and North
East Scotland.

Habitat and ecology Field margins, hedges and
grassland. Phytophagous. Associated with tufted vetch
Vicia cracca and on other *Vicia* species. On the
Continent, this species has also been found on bush
vetch *V. sepium*. Larvae occur in stem galls. In Britain,
adults have been recorded from July to August.

Status Widespread but local in England and Wales.
Recorded in Scotland as far north as Ayr and Fife.

Threats This species is threatened by the grubbing out
of hedgerows, improvement of grassland and
conversion to arable agriculture, and natural succession
and scrub invasion.

Management and conservation Grazing, cutting, on a
rotational basis, or some other disturbance may be
needed to maintain open conditions.

APION INTERMEDIUM   NOTABLE A
A weevil
Order COLEOPTERA   Family APIONIDAE

Apion intermedium Eppelsheim, 1875.

Distribution Recorded from West Sussex, East Sussex,
East Kent and Leicestershire & Rutland from 1970
onwards.

Habitat and ecology Maritime cliff, grassland and
woodland rides on calcareous soils. Phytophagous.
Associated with sainfoin *Onobrychis viciifolia*. Larvae
occur in the stems. Adults have been recorded in June
and August.

Status Recorded new to Great Britain in 1981 from
East Kent and still only known from very few sites.

Threats Improvement of grassland by reseeding or the
application of fertilisers, or by conversion to arable
agriculture. Lack of, or changes in grazing or cutting
regimes, scrub invasion and coastal developments may
be further threats.

Management and conservation Grazing or cutting, on
a rotational basis, may be needed to maintain open
conditions in grassland habitats. Open glades and ride
margins should be managed to retain a variety of
vegetation structures.

Published sources Morris, M.G. (1990), Parry, J.A.
(1982).

APION LAEVICOLLE   NOTABLE A
A weevil
Order COLEOPTERA   Family APIONIDAE

Apion laevicolle Kirby, 1811.

Distribution Recorded from West Cornwall, East
Cornwall, South Devon, North Devon, Dorset, Isle of
Wight, West Sussex, East Sussex, East Kent, West
Kent, Surrey, South Essex, Berkshire and East Suffolk
before 1970 and South Devon, Isle of Wight, East
Sussex and East Kent from 1970 onwards.

Habitat and ecology Maritime cliff, grassland,
stabilised shingle and possibly disturbed ground. The
species is chiefly associated with coastal habitats.
Phytophagous. Recorded from subterraneum clover
Trifolium subterraneum and possibly other species of
clover. On the Continent, this beetle has been found on
white clover *T. repens*. Larvae probably occur in plant
galls. In Britain, adults have been found throughout the
year.

Status Very local and only noted in southern England.
Recently recorded from only four vice-counties, from
South Devon to East Kent.

Threats Improvement by reseeding or the application
of fertilisers, or by conversion to arable agriculture.
Lack of, or changes in grazing or cutting regimes,
scrub invasion and gravel extraction may be further
threats.

Management and conservation Grazing, cutting or
some other disturbance, such as rotovation, on a
rotational basis, is needed to maintain open conditions.

Published sources Fowler, W.W. (1891), Fowler,
W.W. & Donisthorpe, H.St J.K. (1913), Morris, M.G.
(1990), Parry, J.A. (1962).

APIONIDAE

APION LAEVIGATUM   **INDETERMINATE**
A weevil
Order COLEOPTERA   Family APIONIDAE

Apion laevigatum (Paykull, 1792). Formerly known as: *Apion sorbi* (F., 1792).

Distribution Recorded from North Devon, South Somerset, Isle of Wight, South Hampshire, North Hampshire, West Sussex, East Sussex, West Kent, Berkshire, "Suffolk", Cambridgeshire, Monmouthshire, Herefordshire, Merionethshire, Leicestershire & Rutland, Mid-west Yorkshire, Durham, Kirkcudbrightshire and Midlothian before 1970.

Habitat and ecology Disturbed ground and field margins. May prefer base-rich soils. Phytophagous. On the Continent, this species is associated with chamomile *Anthemis*, mayweed *Matricaria* and *Tripleurospermum* species. Larvae occur in flower-heads. In Britain, adults have been recorded in August and September.

Status Not listed in the insect Red Data Book (Shirt, 1987). Old records suggest that this species was formerly widespread and recorded throughout Britain as far north as Midlothian. This species has been very uncommon during the 20th century and it is difficult to know if there has been a genuine decline, there are no recent records. Last recorded in 1948 from Minehead, South Somerset, and Freshwater and Headon Hill, Isle of Wight.

Threats This species may be threatened by development and modern intensive agricultural practices, including the use of pesticides and herbicides. Natural succession and scrub invasion may be further threats.

Published sources Eyre, M.D. (1987), Fowler, W.W. (1891), Fowler, W.W. & Donisthorpe, H.St J.K. (1913), Lloyd, R.W. (1944), Lloyd, R.W. (1948b), Morris, M.G. (1990).

APION LEMOROI   **INSUFFICIENTLY KNOWN**
A weevil
Order COLEOPTERA   Family APIONIDAE

Apion lemoroi Brisout, 1880.

Distribution Recorded from East Kent, Surrey and Cambridgeshire before 1970.

Habitat and ecology Field margins on cultivated land, and disturbed ground. Phytophagous. Associated with knotgrass. Larvae occur in the stems. Adults have been recorded in October.

Status Status revised from RDB 3 (Rare) in Shirt (1987). Only recorded from three sites and first recorded in 1945 from Effingham, Surrey. The date of the last record is uncertain. This species is difficult to identify and may be confused with *A. curtirostre*. Consequently, the exact status of this species is hard to assess.

Threats Uncertain, though this species may be threatened by modern intensive agricultural practices, including the use of pesticides and herbicides, as well as by stubble burning. Natural succession and the invasion of scrub may be a further threat.

Published sources Allen, A.A. (1947d), Easton, A.M. (1946b), Morris, M.G. (1990), Parry, J.A. (1962), Shirt, D.B., ed. (1987).

APION LIMONII   **NOTABLE B**
A weevil
Order COLEOPTERA   Family APIONIDAE

Apion limonii Kirby, 1808.

Distribution South East, South, South West and East Anglia.

Habitat and ecology Saltmarsh and occasionally coastal cliff. Phytophagous. Associated with sea-lavender *Limonium* spp., particularly common sea-lavender *L. vulgare*, but also matted sea-lavender *L. bellidifolium* and rock sea-lavender *L. binervosum*. Larvae probably develop in the roots. Adults have been recorded from late July to September and appear to be more active at dusk.

Status Local. Restricted to the coasts along southern and eastern England as far north as Norfolk. Can be numerous on some sites.

Threats Loss of saltmarsh through reclamation, erosion, and the construction of sea defences. Loss of habitat through cliff stabilisation schemes. Activities that accelerate or reduce the rate of erosion should be avoided. The degradation of suitable habitat through natural succession and the invasion of scrub on stabilised areas, and the overgrazing of saltmarshes are further threats.

Management and conservation In cliff habitats occasional slippages are necessary to maintain habitat continuity. Large areas of unstable cliff are required so that the population does not become isolated and subsequently threatened by individual landslips. Grazing should not be introduced to a saltmarsh where there is no grazing at present.

APION MINIMUM   RARE
A weevil
Order COLEOPTERA   Family APIONIDAE

Apion minimum Herbst, 1797.

Distribution Recorded from West Cornwall, North Somerset, Dorset, East Kent, West Kent, Surrey, Hertfordshire, Middlesex, Oxfordshire, East Suffolk, East Norfolk, Cambridgeshire, Bedfordshire, Glamorgan, Cardiganshire, Merionethshire, North Lincolnshire, Derbyshire, Cheshire, North-east Yorkshire, Dumfriesshire and Kirkcudbrightshire before 1970 and East Norfolk and Northamptonshire from 1970 onwards.

Habitat and ecology Woodland (particularly scrubby areas containing willows), wetland and river margins. Phytophagous. Associated with willows. The larvae are inquilines in the galls of *Pontania* species (Hymenoptera) and probably also on species of gall midge (Diptera). On the Continent, this species has been noted in association with *P. proxima*. In Britain, adults have been recorded in April, June, August and September.

Status Not listed in the insect Red Data Book (Shirt, 1987). Very local. Old records suggest that this species was formerly widespread in southern England with a scattered distribution north to Kirkudbrightshire. Recently recorded from only two vice-counties.

Threats Loss of broad-leaved woodland through, for example, clear-felling and coniferisation. Drainage for reasons such as agricultural improvement and development is the primary cause of the loss of wetlands. Falling water tables because of water abstraction and river engineering schemes may also threaten this species.

Management and conservation Water tables should be maintained at high levels. Water bodies should be isolated from sources of eutrophication and pollution.

Published sources Beare, T.H. (1930a), Fowler, W.W. (1891), Morris, M.G. (1990).

APION PUBESCENS   NOTABLE B
A weevil
Order COLEOPTERA   Family APIONIDAE

Apion pubescens Kirby, 1811.

Distribution England and Wales.

Habitat and ecology Grassland, willow carr and sand dunes. Phytophagous. Recorded from subterranean clover *Trifolium subterraneum* and probably on other species of clover. On the Continent, this species has been found on lesser trefoil *T. dubium*, hop trefoil *T. campestre* and large hop trefoil *T. aureum*. The larva feeds in stem galls. Although not regarded as a foodplant, *A. pubescens* has been found on *Salix* spp. in Britain. Adults have been recorded throughout the year.

Status Widespread but local in England and Wales as far north as North-east Yorkshire and Westmorland. Records for Scotland require confirmation.

Threats Loss of habitat through improvement and conversion to arable agriculture, clear-felling and conversion to other land use, drainage, afforestation, and urban and coastal development. This species may also be threatened by natural succession and scrub invasion.

Management and conservation Open spaces in woodland need to be retained. Open glades and ride margins should be managed to retain a variety of vegetation structures. Grazing or cutting, on a rotational basis, is needed to maintain open conditions.

APION REFLEXUM   NOTABLE A
A weevil
Order COLEOPTERA   Family APIONIDAE

Apion reflexum Gyllenhal, 1833. Formerly known as: *Apion livescerum* Gyllenhal, 1833.

Distribution Recorded from South Somerset, North Somerset, West Sussex, East Sussex, East Kent, West Kent, Surrey, South Essex, Hertfordshire, Berkshire, Oxfordshire, Buckinghamshire, East Suffolk, Cambridgeshire, Bedfordshire, East Gloucestershire, Herefordshire, Warwickshire, South Lancashire and North-east Yorkshire before 1970 and South Wiltshire, East Sussex, East Kent and Bedfordshire from 1970 onwards.

Habitat and ecology Maritime cliff and grassland on calcareous soils. Also noted on chalky roadsides. Phytophagous. Appears to be exclusively associated with sainfoin *Onobrychis viciifolia*. Larvae are possibly found in galls in the flowers. Adults have been recorded throughout the year.

Status Very local. There are old records throughout southern England from South Somerset to East Suffolk, with scattered records as far north as North-east Yorkshire. Recently recorded in only four vice-counties, from South Wiltshire to Bedfordshire.

Threats Improvement of grassland by reseeding or the application of fertilisers, or by conversion to arable

APIONIDAE

agriculture. Lack of, or changes in grazing or cutting regimes, scrub invasion and coastal developments may be further threats.

Management and conservation Grazing or cutting is needed to maintain open conditions on calcareous grassland. Management should be infrequent, or at a low level, to allow development of the foodplant.

Published sources Atty, D.B. (1983), Fowler, W.W. (1891), Fowler, W.W. & Donisthorpe, H.St J.K. (1913), Morris, M.G. (1990).

APION RYEI	ENDEMIC
A weevil	NOTABLE B
Order COLEOPTERA	Family APIONIDAE

Apion ryei Blackburn, 1874.

Distribution North East Scotland.

Habitat and ecology Roadside verges and waste places. Associated with red clover *Trifolium pratense*. Larvae probably feed in the flowerheads. Adults have been recorded in July.

Status Endemic. Only known from the Outer Hebrides, the Orkney and Shetland Isles. Taxonomic status is in doubt and *A.ryei* may prove to be only a form of *A. assimile*.

Threats Uncertain, though development and natural succession may be threats.

Management and conservation Grazing, cutting or some other disturbance, such as rotovation, on a rotational basis, may be needed to maintain open conditions.

Published sources Morris, M.G. (1990), Shirt, D.B., ed. (1987).

APION RUBIGINOSUM	RARE
A weevil	
Order COLEOPTERA	Family APIONIDAE

Apion rubiginosum Grill, 1893. Formerly known as: *Apion sanguineum* (Degeer, 1775).

Distribution Recorded from East Cornwall, North Devon, East Kent, Surrey, South Essex, Berkshire, Oxfordshire, East Suffolk, West Suffolk, East Norfolk, Cambridgeshire, Merionethshire, North-east Yorkshire, South-west Yorkshire, North-west Yorkshire and Westmorland & North Lancashire before 1970 and East

Kent, Berkshire, Cambridgeshire and Cardiganshire from 1970 onwards.

Habitat and ecology Disturbed ground, grassland and coastal shingle. Chiefly on dry, well drained soils. Phytophagous. Associated with sheep's sorrel *Rumex acetosella*. Larvae occur in galls on the roots.

Status Not listed in the insect Red Data Book (Shirt, 1987). Very local and possibly declining. Old records indicate that this species was formerly widespread in England, and also recorded in North Wales. Recently recorded from only four vice-counties, three in southern England and one in mid Wales. This species is difficult to identify and may be confused with *A. haematodes*, *A. cruentatum* and *A. rubens*. Consequently, the exact status of this species is hard to assess.

Threats Improvement of grassland by reseeding or the application of fertilisers, or by conversion to arable agriculture. Lack of, or changes in grazing or cutting regimes, urban development, and gravel extraction, may be further threats this species.

Management and conservation Management, such as disturbance through rotovation, should aim at maintaining open conditions and encouraging early successional stages. The disturbance of coastal shingle should be avoided.

Published sources Boyce, D.C. (1987), Fowler, W.W. (1891), Fowler, W.W. & Donisthorpe, H.St J.K. (1913), Morris, M.G. (1990).

APION SCHOENHERRI	NOTABLE A
A weevil	
Order COLEOPTERA	Family APIONIDAE

Apion schoenherri Boheman, 1839. Formerly known as: *Apion schonherri* Boheman, 1839.

Distribution Recorded from West Cornwall, South Devon, North Devon, North Somerset, Isle of Wight, South Hampshire, North Hampshire, West Sussex, East Sussex, East Kent, West Kent, Surrey, South Essex, Hertfordshire, Berkshire and Oxfordshire before 1970 and North Hampshire, East Sussex, Surrey, South Essex and Berkshire from 1970 onwards.

Habitat and ecology Grassland, stabilised shingle and probably disturbed ground. Primarily coastal. Phytophagous. On the Continent, associated with hare's-foot clover *Trifolium arvense* and lesser trefoil *T. dubium*. In Britain, adults have been recorded from April to August.

Status Very local. This species has been widely recorded in southern England, although the beetle has been very uncommon during the 20th century. Recently recorded from only five vice-counties. An old record for North-east Yorkshire requires confirmation.

Threats Improvement of grassland by reseeding or the application of fertilisers, or by conversion to arable agriculture. Lack of, or changes in grazing or cutting regimes, and gravel extraction may be further threats.

Management and conservation Some grazing, rotational cutting or other disturbance, such as rotovation, is needed to maintain open conditions. Disturbance of stabilised shingle should be avoided.

Published sources Fowler, W.W. (1891), Fowler, W.W. & Donisthorpe, H.St J.K. (1913), Morris, M.G. (1990).

APION SEDI NOTABLE B
A weevil
Order COLEOPTERA Family APIONIDAE

Apion sedi Germar, 1818.

Distribution South East, South, South West, East Anglia, North West, South Wales, North Wales, South East Scotland and South West Scotland.

Habitat and ecology Grassland, sand dunes, stabilised shingle and disturbed ground. Particularly associated with coastal sites. Phytophagous. Found on stonecrop *Sedum*, in particular biting stonecrop *S. acre* and English stonecrop *S. anglicum*, but probably also on other *Sedum* spp. and house-leek *Sempervivum*. On the Continent, *A. sedi* has also been found on orpine *S. telephium*, white stonecrop *S. album* and rock stonecrop *S. forsteranum*. Larvae first mine the leaves and later in the stems. In Britain, adults have been recorded in June, August and November.

Status Local and widely scattered in southern England. More local in the remaining parts of Britain, north to southern Scotland.

Threats Loss of habitat through afforestation, improvement and conversion to arable agriculture, natural succession and the scrub invasion, and urban and holiday development. The degradation of remaining habitat by excessive disturbance of the vegetation through activities such as motorbike access, horse-riding and human trampling.

Management and conservation Some grazing, rotational cutting or other disturbance, such as rotovation, may be needed to maintain open conditions

and to encourage early successional stages. The disturbance of coastal shingle should be avoided.

APION SEMIVITTATUM NOTABLE A
A weevil
Order COLEOPTERA Family APIONIDAE

Apion semivittatum Gyllenhal, 1833.

Distribution Recorded from "Wilts", East Sussex, East Kent, West Kent and South Essex before 1970 and South Wiltshire, East Sussex, East Kent, West Kent, Surrey, South Essex and Middlesex from 1970 onwards.

Habitat and ecology Disturbed ground, field margins on cultivated land, allotments and gardens. Phytophagous. Associated with annual mercury *Mercurialis annua*. This species has also been recorded on dog's mercury *Mercurialis perennis*. Larvae occur at the nodes of the stems. Adults have been found from June to November.

Status Very local. Primarily confined to the London basin and a few sites in South East England, also recorded in South Wiltshire. Often occurs in abundance where found.

Threats This species may be threatened by modern intensive agricultural practices, including the use of pesticides and herbicides. Natural succession and scrub invasion, and urban development may be further threats.

Management and conservation Management, such as disturbance through rotovation, should aim at maintaining open conditions and encouraging early successional stages.

Published sources Allen, A.A. (1961), Allen, A.A. (1979), Allen, A.A. (1988e), Fowler, W.W. (1891), Fowler, W.W. & Donisthorpe, H.St J.K. (1913), Morris, M.G. (1962), Morris, M.G. (1990), Parry, J.A. (1962), Plant, C.W. (1985).

APION SOROR NOTABLE A
A weevil
Order COLEOPTERA Family APIONIDAE

Apion soror Rey, 1895. Formerly known as: *Apion foveatoscutellatum* Wagner, 1906.

Distribution Recorded from East Sussex before 1970 and East Sussex, East Kent and West Kent from 1970 onwards.

APIONIDAE

Habitat and ecology Wetland and river margins. Phytophagous. Associated with marsh-mallow *Althaea officinalis*. Adults have been recorded in June, August and September.

Status Recently recorded as new to Great Britain and at present known from a small number of sites. Previously confused with *A. radiolus*. Probably under-recorded, although the foodplant is itself nationally scarce (Notable B).

Threats Falling water tables because of water abstraction and river engineering schemes may threaten this species. Overgrazing may be a further threat to this species.

Management and conservation Water tables should be maintained at high levels. Management should aim to avoid the invasion of scrub.

Published sources Morris, M.G. (1990).

APION STOLIDUM	NOTABLE B
A weevil	
Order COLEOPTERA	Family APIONIDAE

Apion stolidum Germar, 1817.

Distribution England, Dyfed-Powys and South West Scotland.

Habitat and ecology Field margins, disturbed ground, roadside verges and grassland. Phytophagous. Associated with ox-eye daisy *Leucanthemum vulgare* and less certainly with scentless mayweed *Tripleurospermum inodorum*. On the Continent, this species is also associated with chamomile *Anthemis* spp. Larvae probably occur in the stems or the rootstocks. In Britain, adults have been recorded from May to September.

Status Local and widely distributed in southern England. More local further north to South West Scotland and recorded in mid Wales.

Threats Improvement of grassland by reseeding or the application of fertilisers, or by conversion to arable agriculture. Lack of, or changes in grazing or cutting regimes, and the use of pesticides and herbicides may also threaten this species.

Management and conservation Disturbance, such as rotovation, is needed to maintain open conditions.

APION VARIPES	NOTABLE B
A weevil	
Order COLEOPTERA	Family APIONIDAE

Apion varipes Germar, 1817.

Distribution South East, South, South West, East Anglia, East Midlands, West Midlands, North East, South Wales, North Wales and South East Scotland.

Habitat and ecology Grassland. Also found in woodland rides, clearings, wood margins and disturbed ground. Phytophagous. Associated with red clover *Trifolium pratense* and possibly other *Trifolium* species. On the Continent, this species has been found on hare's-foot clover *T. arvense*. The larva feeds either in flowerheads or in galls. In Britain, adults have been recorded from March to September.

Status Widespread but local in southern England north to South Lincolnshire and Derbyshire. Very local in remaining parts of England, Wales and South East Scotland. Possibly declining.

Threats This species is threatened by improvement of grassland and conversion to arable agriculture, as well as by natural succession. Loss of woodland through, for example, clear-felling and conversion to other land use. Neglect and conversion to high forest may be further threats.

Management and conservation Grazing, cutting, on a rotational basis, or some other disturbance, such as rotovation, may be needed to maintain open conditions. Open spaces in woodland need to be retained. Open glades and ride margins should be managed to retain a variety of vegetation structures.

APION VICINUM	NOTABLE B
A weevil	
Order COLEOPTERA	Family APIONIDAE

Apion vicinum Kirby, 1808.

Distribution England, South Wales and North Wales.

Habitat and ecology Fens and marshland, wet meadows, river margins and grassland. Phytophagous. Associated with mint *Mentha* spp., in particular water mint *M. aquatica*. On the Continent, hosts include corn mint *M. arvensis*, basil-thyme *Acinos arvensis* and cat-mint *Nepeta cataria*. Larvae occur in galls on the stems. In Britain, adults have been recorded mainly in September, though also in August and October.

Status Local and widely distributed in southern England. More local in the rest of England as far north as West Lancashire. Also recorded in Wales.

Threats Drainage and agricultural improvement are the causes of the main losses of fenland and unimproved meadows. Falling water tables due to water abstraction and river improvements make the future of many water meadows and fenland pastures uncertain. Changes in grazing regimes may also adversely affect this species.

Management and conservation Water tables should be maintained at high levels. Water bodies should be isolated from sources of eutrophication and pollution. Some grazing or cutting is needed to maintain open conditions.

NANOPHYES GRACILIS   **RARE**
A weevil
Order COLEOPTERA   Family APIONIDAE

Nanophyes gracilis Redtenbacher, 1849.

Distribution Recorded from South Somerset, Dorset, Isle of Wight, South Hampshire, North Hampshire, East Sussex, Surrey and Berkshire before 1970 and North Hampshire and West Sussex from 1970 onwards.

Habitat and ecology Wetland and wet areas within woodland. Chiefly found in bare, waterlogged situations, often near or under trees. Phytophagous. Associated with water-purslane *Lythrum portula*. Larvae occur in galls on secondary stems. Adults have been recorded in July and September.

Status Not listed in the insect Red Data Book (Shirt, 1987). Very local and only recorded in southern England. Recently recorded in only two vice-counties.

Threats Drainage for reasons such as agricultural improvement and development is the primary cause of the loss of wetlands. Loss of grazing leading to reduced disturbance of margins, natural succession, falling water tables because of water abstraction and river engineering schemes, and clear-felling of woodland and conversion to other land use may also threaten this species.

Management and conservation Water tables should be maintained at high levels. Water bodies should be isolated from sources of eutrophication and pollution. Rotational cutting or some other disturbance is needed to maintain open conditions and encourage early successional stages.

Published sources Drane, A.B. (1990), Fowler, W.W. (1891), Fowler, W.W. & Donisthorpe, H.St J.K. (1913), Morris, M.G. (1990).

BYCTISCUS BETULAE   **NOTABLE B**
Hazel leaf roller
Order COLEOPTERA   Family ATTELABIDAE

Byctiscus betulae (L., 1758). Formerly known as: *Byctiscus betuleti* (F., 1792), *Rhynchites betuleti* (F.).

Distribution England and South Wales.

Habitat and ecology Broad-leaved woodland, in particular hazel coppice. Possibly also associated with scrub. Phytophagous. Polyphagous on a variety of trees and shrubs. Most frequently found on hazel, particularly young coppice, and birch. The larvae are found in leaf rolls, in which the ovipositing female often incorporates several leaves. Adults have been recorded from April to September, the main period of adult activity is in May and June.

Status Widespread but local in England, also recorded in South Wales.

Threats Loss of broad-leaved woodland through, for example, clear-felling and coniferisation. Neglect, in particular the abandonment of coppice management and the conversion to high forest, has led to increased shade and the loss of glades and broad sunny rides.

Management and conservation Existing coppice cycles should be continued and considered for re-introduction in areas of abandoned coppice. Open glades and ride margins should be cut on rotation to retain a variety of vegetation structures.

BYCTISCUS POPULI   **RARE**
Poplar leaf roller
Order COLEOPTERA   Family ATTELABIDAE

Byctiscus populi (L., 1758). Formerly known as: *Rhynchites populi* (L.).

Distribution Recorded from East Cornwall, South Devon, West Sussex, East Sussex, East Kent, West Kent, Surrey, South Essex, North Essex, Hertfordshire, Middlesex, East Suffolk, West Suffolk, East Norfolk, East Gloucestershire and Worcestershire before 1970 and East Sussex, East Kent and Surrey from 1970 onwards.

Habitat and ecology Broad-leaved woodland. Phytophagous. Associated with aspen, white poplar and black poplar. Larvae occur in leaf rolls, usually constructed from a single leaf. Adults have been recorded from May to July and in September, though primarily in June.

Status Not listed in the insect Red Data Book (Shirt, 1987). Old records suggest that this species was widely distributed in southern England. Declining and now very local. Recently recorded from only three vice-counties, all in south-eastern England.

Threats Loss of broad-leaved woodland through, for example, clear-felling and coniferisation. Alteration of the vegetation structure of many remaining woodlands by 20th century management practices, including the selective removal of aspen. Neglect, and conversion to high forest, has led to increased shade and the loss of glades and broad sunny rides.

Management and conservation Open glades and ride margins should be cut on rotation to retain a variety of vegetation structures.

Published sources Allen, A.A. (1955d), Edwards, J.E. (1917), Fincher, F. (1955), Fowler, W.W. (1891), Fowler, W.W. & Donisthorpe, H.St J.K. (1913), Menzies, I.S. (1990), Morris, M.G. (1990), Stretton, G.B. (1943).

RHYNCHITES AURATUS   **EXTINCT**
A leaf-rolling weevil
Order COLEOPTERA   Family ATTELABIDAE

Rhynchites auratus (Scopoii, 1763).

Distribution Recorded from West Kent.

Habitat and ecology Broad-leaved woodland and scrub. Phytophagous. Associated with blackthorn. On the Continent, this species is also associated with wild cherry, dwarf cherry, bird cherry, wild plum and hawthorn. Larvae are found in the kernels.

Status Extinct. Confirmed as recorded from Crayford, West Kent and last noted in 1839. Doubtfully recorded from Cambridgeshire and Dumfriesshire.

Published sources Allen, A.A. (1955d), Fowler, W.W. (1891), Morris, M.G. (1990), Shirt, D.B., ed. (1987).

RHYNCHITES BACCHUS   **EXTINCT**
A leaf-rolling weevil
Order COLEOPTERA   Family ATTELABIDAE

Rhynchites bacchus (L., 1758).

Distribution Recorded from West Kent, East Suffolk and Huntingdonshire.

Habitat and ecology Broad-leaved woodland and scrub. Phytophagous. Associated with rosaceous trees and shrubs, in particular *Prunus* spp., crab apple and hawthorn. Larvae are found in the fruits. Adults have been recorded in June and September.

Status Extinct. Last recorded in 1843 from Birch Wood, near Swanley, West Kent.

Published sources Allen, A.A. (1955d), Allen, A.A. (1962a), Fowler, W.W. (1891), Morris, M.G. (1990), Shirt, D.B., ed. (1987).

RHYNCHITES CAVIFRONS   **NOTABLE B**
A leaf-rolling weevil
Order COLEOPTERA   Family ATTELABIDAE

Rhynchites cavifrons Gyllenhal, 1833. Formerly known as: *Lasiorhynchites cavifrons* (Gyllenhal), *Rhynchites pubescens* sensu Fowler, 1891 not (F., 1775).

Distribution England, South Wales and Dyfed-Powys.

Habitat and ecology Broad-leaved woodland, particularly rides and wood margins. Phytophagous. Associated with oak, hazel and possibly also birch. Larvae are found in one year old twigs. Adults have been recorded from May to early July.

Status Widespread but local in England, also recorded in parts of Wales.

Threats Loss of broad-leaved woodland through, for example, clear-felling and coniferisation. Neglect, and conversion to high forest, has led to increased shade and the loss of glades and broad sunny rides.

Management and conservation Open glades and ride margins should be cut on rotation to retain a variety of vegetation structures.

RHYNCHITES CUPREUS   NOTABLE B
A leaf-rolling weevil
Order COLEOPTERA   Family ATTELABIDAE

Rhynchites cupreus (L., 1758).

Distribution England, Dyfed-Powys, North Wales, South East Scotland, South West Scotland and North West Scotland.

Habitat and ecology Broad-leaved woodland, scrub, roadside verges and isolated trees. Apparently not found in dense woodland. Phytophagous. Associated with rowan and occasionally on other rosaceous plants such as blackthorn. Also recorded from hazel. On the Continent, *R. cupreus* is also associated with wild plum, bird cherry, dwarf cherry and *Sorbus* spp.. Larvae are found in the fruits. In Britain, adults have been recorded from May to August.

Status Widespread but local in England and southern Scotland. Also recorded in mid and North Wales, and North West Scotland.

Threats Loss of broad-leaved woodland through, for example, clear-felling and coniferisation, and the grubbing out of scrub and isolated trees. Neglect, and conversion to high forest may be a further threat.

Management and conservation Open glades and ride margins should be cut on rotation to retain a variety of vegetation structures.

RHYNCHITES INTERPUNCTATUS NOTABLE B
A leaf-rolling weevil
Order COLEOPTERA   Family ATTELABIDAE

Rhynchites interpunctatus Stephens, 1831. Formerly known as: *Caenorhinus interpunctatus* (Stephens).

Distribution South East, South, South West, East Anglia, East Midlands, West Midlands, South Wales and Dyfed-Powys.

Habitat and ecology Broad-leaved woodland. Phytophagous. Associated with oak and hawthorn. Adults have been recorded from mid April to July, and have been found on small twiggy growths on mature trees in April.

Status Widespread but local in the southern half of England and also recorded in South Wales.

Threats Loss of broad-leaved woodland through, for example, clear-felling and coniferisation.

Management and conservation Open glades and ride margins should be cut on rotation to retain a variety of vegetation structures.

RHYNCHITES LONGICEPS   NOTABLE B
A leaf-rolling weevil
Order COLEOPTERA   Family ATTELABIDAE

Rhynchites longiceps Thomson, 1888. Formerly known as: *Caenorhinus longiceps* (Thomson), *Rhynchites harwoodi* Joy, 1911.

Distribution England, South East Scotland and South West Scotland.

Habitat and ecology Broad-leaved woodland and scrub, particularly in damp situations. Also found on lightly wooded heath, carr and fen. Phytophagous. Associated mainly with goat willow, also on birch and possibly other *Salix* species. On the Continent, this species is also associated with osier. Larvae are found in the leaf-buds. In Britain, adults have been recorded mainly in June, though also in July.

Status Widespread but local in England and southern Scotland.

Threats Loss of broad-leaved woodland and scrub through, for example, clear-felling and conversion to other land use. Neglect and conversion to high forest, drainage for reasons such as agricultural improvement, water abstraction and river engineering schemes may be further threats.

Management and conservation Water tables should be maintained at high levels. Open glades and ride margins should be cut on rotation to retain a variety of vegetation structures.

RHYNCHITES OLIVACEUS   NOTABLE A
A leaf-rolling weevil
Order COLEOPTERA   Family ATTELABIDAE

Rhynchites olivaceus Gyllenhal, 1833. Formerly confused under: *Lasiorhynchites ophthalmicus* sensu Kloet and Hincks, 1945 not (Stephens, 1831), *Lasiorhynchites sericeus* sensu Hincks, 1951 not (Herbst, 1797), *Rhynchites ophthalmicus* sensu auct. Brit. not Stephens, *Rhynchites sericeus* sensu Fowler, 1891 not Herbst.

Distribution Recorded from Dorset, East Sussex, West Kent, Surrey, North Essex, Hertfordshire, Berkshire, Oxfordshire, Buckinghamshire, East Suffolk, West Gloucestershire, Monmouthshire, Herefordshire, Worcestershire, Staffordshire, Glamorgan and

ATTELABIDAE

Leicestershire & Rutland before 1970 and West Sussex, West Kent, South Essex, Berkshire, Buckinghamshire and Herefordshire from 1970 onwards.

Habitat and ecology Broad-leaved woodland, particularly recently coppiced oak. Phytophagous. Associated mainly with oak, also found on hazel and occasionally birch. Larvae are found in one year old twigs. Adults have been recorded from May to July.

Status Very local and only recorded in the southern half of England and South Wales. This species has been very uncommon during the 20th century. Recently recorded from only six vice-counties.

Threats Loss of broad-leaved woodland through, for example, clear-felling and coniferisation. Neglect, in particular the abandonment of coppice management and the conversion to high forest, has led to increased shade and the loss of glades and broad sunny rides.

Management and conservation Existing coppice cycles should be continued and considered for re-introduction in areas of abandoned coppice. Open glades and ride margins should be cut on rotation to retain a variety of vegetation structures.

Published sources Allen, A.A. (1946b), Allen, A.A. (1955d), Atty, D.B. (1983), Fincher, F. (1947), Hodge, P.J. (1983), Kirby, P. & Lambert, S.J.J. (1989), Morris, M.G. (1990).

RHYNCHITES PAUXILLUS   **RARE**
A leaf-rolling weevil
Order COLEOPTERA   Family ATTELABIDAE

Rhynchites pauxillus Germar, 1824. Formerly known as: *Caenorhinus pauxillus* (Germar).

Distribution Recorded from "Somerset", Isle of Wight, South Hampshire, East Sussex, West Kent, Surrey, South Essex, Berkshire, Oxfordshire, Buckinghamshire, East Suffolk, Cambridgeshire, Huntingdonshire, East Gloucestershire, Warwickshire, Leicestershire & Rutland, Nottinghamshire, Durham, North Northumberland and Dumfriesshire before 1970 and South Essex, Oxfordshire and Huntingdonshire from 1970 onwards.

Habitat and ecology Broad-leaved woodland, scrub and probably also hedgerows and old orchards. Phytophagous. Associated with rosaceous trees and shrubs, particularly hawthorn and blackthorn, also found on medlar. On the Continent, this species is also recorded from crab apple, wild pear, wild cotoneaster, Midland hawthorn and wild plum. In Britain, larvae are found in the petioles and mid-veins of leaves, these are

partially severed by the ovipositing female and soon fall to the ground, where larvae feed on the dying leaf tissue. Adults have been recorded from April to June.

Status Not listed in the insect Red Data Book (Shirt, 1987). Old records show that this species has been widely recorded in southern England with a scattered distribution north to Dumfriesshire. Recently reported from only three vice-counties, all in southern and eastern England.

Threats Loss of broad-leaved woodland through, for example, clear-felling and coniferisation, the grubbing out of hedgerows, old orchards and scrub, and the mechanised trimming of hedgerows. Neglect, and conversion to high forest may be a further threat.

Management and conservation Open glades and ride margins should be cut on rotation to retain a variety of vegetation structures.

Published sources Allen, A.A. (1955d), Atty, D.B. (1983), Collins, J. (1923), Fowler, W.W. (1891), Fowler, W.W. & Donisthorpe, H.St J.K. (1913), Morris, M.G. (1990).

RHYNCHITES TOMENTOSUS   **NOTABLE B**
A leaf-rolling weevil
Order COLEOPTERA   Family ATTELABIDAE

Rhynchites tomentosus Gyllenhal, 1839. Formerly known as: *Caenorhinus tomentosus* (Gyllenhal), *Rhynchites uncinatus* Thomson, 1865.

Distribution England, Wales and South East Scotland.

Habitat and ecology Woodland, hedgerows and dune slacks. Phytophagous. Associated mainly with aspen, though also with other *Populus* spp. and *Salix* spp., particularly goat willow, osier and creeping willow. Larvae develop in leaf buds. Adults have been recorded from May to August, though primarily in June.

Status Widespread but local in England and Wales, also recorded in South East Scotland.

Threats This species is threatened by the loss of woodland through clear-felling and conversion to other land use, the grubbing out and mechanised trimming of hedgerows, drainage, urban and holiday developments and excessive disturbance of dune vegetation through activities such as horse-riding and motorbike access.

Management and conservation Open glades and ride margins should be cut on rotation to retain a variety of vegetation structures. Water tables should be maintained at high levels. Water bodies should be isolated from sources of eutrophication and pollution.

DIPLOCOELUS FAGI   **NOTABLE B**

Order COLEOPTERA   Family BIPHYLLIDAE

Diplocoelus fagi Guerin-Meneville, 1844.

Distribution South East, South, South West, East Anglia, East Midlands and West Midlands.

Habitat and ecology Ancient broad-leaved woodland and pasture-woodland. Associated with beech, occurring under the bark of dead wood. Adults have been recorded in January and from April to October.

Status Local in southern and central England. This species appears to have spread and increased in recent times.

Indicator status Grade 2 in Harding & Rose (1986).

Threats Loss of broad-leaved woodland and parkland through, for example, clear-felling and coniferisation. Habitat loss, in particular, through the felling of ancient beech trees, removal of dead wood from living trees and the destruction or removal of standing and fallen dead wood for reasons such as forest hygiene, aesthetic tidiness, public safety or for use as fire wood.

Management and conservation Ancient beech trees and both fallen and standing dead timber, especially with the bark attached, should be retained. The removal of dead timber from ancient trees should be avoided. Gaps in the age structure of the beech tree population should be identified and the continuity of the appropriate dead wood habitat ensured by regeneration, suitable planting and possibly with pollarding.

BOSTRICHUS CAPUCINUS   **EXTINCT**
A false powder-post beetle
Order COLEOPTERA   Family BOSTRICHIDAE

Bostrichus capucinus (L., 1758). Formerly known as: *Bostrychus capucinus* (L., 1758).

Distribution Recorded from Dorset, Hertfordshire, Middlesex, East Norfolk, Worcestershire, Nottinghamshire, Derbyshire and Cheshire before 1970.

Habitat and ecology Ancient broad-leaved woodland. Also recorded from a timber yard and a shop floor.

Associated with dead wood, probably oak. The adult has been recorded in July.

Status Status revised from RDB 3 (Rare) in Shirt (1987). Presumed extinct as a native. The beetle is occasionally reported from imported timber. It apparently bred for three years (1906-1908) in a timber yard at Millwall, South Essex. Recently recorded from a shop floor in Northamptonshire. Last recorded in the 'wild' from Burton, the vice-county for this record is unknown for certain though possibly refers to Staffordshire.

Published sources Fowler, W.W. (1890), Fowler, W.W. & Donisthorpe, H.St J.K. (1913), Shirt, D.B., ed. (1987).

BRUCHIDIUS OLIVACEUS   **ENDANGERED**
A seed beetle
Order COLEOPTERA   Family BRUCHIDAE

Bruchidius olivaceus (Germar, 1824). Formerly known as: *Bruchidius canus* sensu auct. Brit. not (Germar, 1824), *Bruchidius unicolor* sensu auct. Brit. not (Olivier, 1795), *Bruchus canus* sensu Fowler, 1890 not Germar, *Laria unicolor* sensu Joy, 1932 not (Olivier).

Distribution Recorded from "Hants", "Sussex", "Kent", Surrey, Berkshire, Oxfordshire and Buckinghamshire before 1970.

Habitat and ecology Calcareous grassland and agricultural land. Adults and larvae are associated with sainfoin *Onobrychis viciifolia*, the larvae developing in the seed pods. The larvae probably overwinter in the seed-pods of the foodplant. Adults have been recorded from June to September.

Status Not listed in the insect Red Data Book (Shirt, 1987). Formerly recorded in several vice-counties in southern England. This species appears to have declined dramatically after around 1920. The last authenticated record for *B. olivaceus* was in 1923 from Cothill, Berkshire.

Threats Loss of unimproved grassland through improvement by reseeding or by the application of fertilisers, or by conversion to arable agriculture. Development, natural succession and possibly the cessation of using sainfoin as a fodder crop may also have contributed to this species' decline.

Published sources Aldridge, R.J.W. & Pope, R.D. (1986).

BRUCHIDAE

BRUCHUS ATOMARIUS   **NOTABLE B**
A seed beetle
Order COLEOPTERA   Family BRUCHIDAE

Bruchus atomarius (L., 1761). Formerly known as:
Bruchus viciae sensu Fowler, 1890 not Olivier, 1795,
Laria atomaria (L.), *Laria viciae* sensu auct. Brit. not
(Olivier).

Distribution South East, South, South West, East
Anglia, East Midlands, West Midlands and
Dyfed-Powys.

Habitat and ecology Rough grassland on neutral or
calcareous soils. Possibly also hedgebanks, roadside
verges and wood margins. Phytophagous. Associated
with vetches *Vicia*, with records from common vetch *V.
sativa*, tufted vetch *V. cracca* and bush vetch *V.
sepium*. Larvae develop in pods of the foodplant.
Adults have been recorded in June and from August to
October.

Status Widespread but local in southern and central
England. Also recorded from mid Wales.

Threats Loss of unimproved grassland through
improvement by reseeding or by the application of
fertilisers, or by conversion to arable agriculture.
Development and natural succession are further threats.

Management and conservation Grazing or cutting, on
a rotational basis, is needed to maintain open
conditions. Woodland margins should be managed to
retain a variety of vegetation structures.

AGRILUS ANGUSTULUS   **NOTABLE B**
A jewel beetle
Order COLEOPTERA   Family BUPRESTIDAE

Agrilus angustulus (Illiger, 1803).

Distribution South East, South, South West, East
Anglia, East Midlands, West Midlands, North East and
South Wales.

Habitat and ecology Broad-leaved woodland and oak
and hazel coppice. Associated with oak and hazel. On
the Continent, this species is also found on beech,
hornbeam and sweet chestnut. In Britain, larvae
develop in the dying branches of oak, particularly those
on which the leaves are still attached. Continental
authors state that the larva develops under the bark of
thin branches and twigs, this stage lasting one year. In
Britain, adults occur from May to mid August. The
main period of adult emergence is probably during
June and July. Adults are very active in hot, sunny
weather.

Status Local and occasionally common where found.
Mainly distributed through southern and central
England. This species is difficult to identify and can be
confused with *A. laticornis*.

Threats Loss of broad-leaved woodland through, for
example, clear-felling and coniferisation. Alteration of
the vegetation structure of many remaining woodlands
by 20th century management practices.

Management and conservation Existing coppice
cycles should be continued and considered for
re-introduction in areas of abandoned coppice. Open
glades and ride margins should be cut on rotation to
retain a variety of vegetation structures.

AGRILUS LATICORNIS   **NOTABLE B**
A jewel beetle
Order COLEOPTERA   Family BUPRESTIDAE

Agrilus laticornis (Illiger, 1803).

Distribution South West, South East, South, East
Anglia, East Midlands, West Midlands and North East.

Habitat and ecology Broad-leaved woodland and
pasture-woodland. Larvae develop in oak and, like *A.
angustulus*, probably occur in dying rather than dead
branches. Continental literature states that the larva
develops under the bark of thin branches and twigs,
this stage lasting one year. In Britain, adults have been
found from late May to early September with the main
period of adult emergence probably occurring in June
and July. Adults are very active in hot, sunny weather.

Status Local, though occassionally common where
found. Distributed through southern and central
England. This species is difficult to identify and can be
confused with *A. angustulus*.

Threats Loss of broad-leaved woodland through, for
example, clear-felling and coniferisation. Alteration of
the vegetation structure of many remaining woodlands
by 20th century management practices leading to an
increase in shade and the loss of glades and broad
sunny rides may also be a threat to this species.

Management and conservation Open glades and ride
margins should be cut on rotation to retain a variety of
vegetation structures.

AGRILUS PANNONICUS   NOTABLE A
Two-spot wood-borer
Order COLEOPTERA   Family BUPRESTIDAE

Agrilus pannonicus (Piller and Mitterpacher, 1783).
Formerly known as: *Agrilus biguttatus* (F., 1777).

Distribution Recorded from, South Hampshire, West Kent, Hertfordshire and Nottinghamshire before 1970 and North Hampshire, West Sussex, Surrey, Middlesex, Berkshire and Leicestershire & Rutland from 1970 onwards.

Habitat and ecology Ancient broad-leaved woodland and pasture-woodland. Associated with oak. Continental authors also list beech and sweet chestnut as host-trees. In Britain, the larvae develop in and under the bark of old, dying and dead trees. Larvae have been recorded in October, November and February. On the Continent, the larval stage takes two years. In Britain, adults leave a characteristic emergence hole in the bark and have been found on the trunks and branches of oak. Adults have been recorded from May to early August, though the main period of emergence is probably during June and early July. Adults are very active in hot, sunny weather.

Status Status revised from RDB 2 in Shirt (1987). Very local in south-eastern England. There are old records from as far north as Nottinghamshire. Possibly increasing, with a number of recent new vice-county records.

Indicator status Grade 2 in Harding & Rose (1986).

Threats Loss of broad-leaved woodland and parkland through, for example, clear-felling and coniferisation. Habitat loss, in particular, through the felling of ancient trees, removal of dead wood from living trees for reasons such as forest hygiene, aesthetic tidiness, public safety or for use as fire wood.

Management and conservation Ancient trees and both fallen and standing dead timber, especially with the bark attached, should be retained. The removal of dead timber from ancient trees should be avoided. Gaps in the age structure of the tree population should be identified and the continuity of the appropriate dead wood habitat ensured by suitable planting and possibly with pollarding.

Published sources Allen, A.A. (1958), Allen, A.A. (1988b), Bílý, S. (1982), Fowler, W.W. & Donisthorpe, H.St J.K. (1913), Garland, S.P. (1983), Harding, P.T. (1978), Harding, P.T. & Rose, F. (1986), Levey, B. (1977), Menzies, I.S. (1990), Owen, J.A. (1990a), Shirt, D.B., ed. (1987), Tozer, D. (1939).

AGRILUS SINUATUS   NOTABLE A
A jewel beetle
Order COLEOPTERA   Family BUPRESTIDAE

Agrilus sinuatus (Olivier, 1790).

Distribution Recorded from South Wiltshire, Dorset, Hertfordshire, Oxfordshire, Bedfordshire, Huntingdonshire and Northamptonshire before 1970 and North Wiltshire, South Hampshire, West Kent, Surrey, Middlesex, Berkshire, Oxfordshire, West Norfolk, East Gloucestershire, West Gloucestershire, Herefordshire and Warwickshire from 1970 onwards.

Habitat and ecology Pasture-woodland, wood edges, hedgerows and downland scrub. Associated with hawthorn. This species appears to favour very old hawthorn bushes. Larvae develop in dying branches and trunks. Adults leave characteristic emergence holes. Adults have been recorded from June to September, with the main period of emergence occurring during July and August. Adults are very active in hot, sunny weather.

Status Status revised from RDB 2 (Vulnerable) in Shirt (1987). Very local, but possibly under-recorded because of the short period of adult activity. Distributed through southern England as far north as Herefordshire.

Threats Loss of broad-leaved woodland and parkland through, for example, clear-felling and coniferisation. The removal of scrub from downland, and the uprooting and mechanised trimming of old hedgerows are further threats to this species. In particular, habitat loss through the removal of dead or dying wood from living shrubs for reasons such as forest hygiene and aesthetic tidiness.

Management and conservation Open spaces in woodland need to be retained. Downland and open glades and ride margins in woodlands should be managed to retain a variety of vegetation structures.

Published sources Alexander, K.N.A. & Grove, S.J. (1990), Allen, A.A. (1947a), Allen, A.A. (1966), Donisthorpe, H.St J.K. (1925b), Levey, B. (1977), Menzies, I.S. (1946), Menzies, I.S. (1954), Menzies, I.S. (1990), Osborne, P.J. (1957), Shirt, D.B., ed. (1987), Speight, M. (1968), Stokes, H.G. (1952), Walker, J.J. (1925).

BUPRESTIDAE

AGRILUS VIRIDIS   NOTABLE A
A jewel beetle
Order COLEOPTERA   Family BUPRESTIDAE

Agrilus viridis (L., 1758).

Distribution Recorded from South Hampshire and East Kent before 1970 and South Wiltshire, South Hampshire, West Sussex, Surrey and Berkshire from 1970 onwards.

Habitat and ecology Broad-leaved woodland and willow beds. Probably confined to areas with old sallows. Associated with goat willow *Salix caprea*, grey willow *S. cinerea* and possibly oak. On the Continent, alder, birch, hornbeam, hazel, beech, lime and *Acer* spp. are also listed as host-trees, the larval stage taking one year. In Britain, larvae develop in old, dying wood. Adults have been recorded from old, decaying grey willow and other *Salix* spp., as well as from stunted oak. Adults occur mainly from June to early August.

Status Status revised from RDB 2 (Vulnerable) in Shirt (1987). Very local in southern and south-eastern England. This species has been confused with *A. angustulus* and *A. laticornis*.

Threats Loss of broad-leaved woodland through, for example, clear-felling and coniferisation. Alteration of the vegetation structure of many remaining woodlands by 20th century management practices. Habitat loss, in particular, through the felling of ancient trees and the removal of dead and dying wood from living trees for reasons such as forest hygiene, aesthetic tidiness, public safety or for use as fire wood.

Management and conservation Old, dying trees should be retained. The removal of dead and dying timber from old trees should be avoided. The continuity of the appropriate habitat may be ensured by suitable pollarding.

Published sources Allen, A.A. (1966), Bílý, S. (1982), Donisthorpe, H.St J.K. (1942a), Fowler, W.W. (1890), Haines, F.H. (1942), Levey, B. (1977), Shirt, D.B., ed. (1987).

ANTHAXIA NITIDULA   ENDANGERED
A jewel beetle
Order COLEOPTERA   Family BUPRESTIDAE

Anthaxia nitidula (L., 1758).

Distribution Recorded from South Hampshire before 1970.

Habitat and ecology Open areas in broad-leaved woodland and wood margins. Not found in dense woodland. Associated with blackthorn and probably other shrub species of the family Rosaceae. Larvae develop beneath the bark of the host-plant. Continental authors state that larval stage lasts one or two years. In Britain, adults are usually found on the flowers of hawthorn, rose and buttercups. Adults have been recorded from mid May to late July.

Status Only known from the Brockenhurst and Lyndhurst areas of the New Forest, South Hampshire. Last recorded in 1954 from Brockenhurst.

Threats Uncertain, though this species may be threatened by high grazing levels. Neglect may lead to an increase of shade and the loss of glades and broad sunny rides.

Management and conservation Open glades and ride margins should be managed on rotation to retain a variety of vegetation structures. The presence of nectar sources such as hawthorn and composite herbs may be particularly important for this species. The removal of dead and dying blackthorn branches should be avoided.

Published sources Allen, A.A. (1966), Bílý, S. (1982), Donisthorpe, H.St J.K. (1925b), Donisthorpe, H.St J.K. (1942a), Levey, B. (1977), Shirt, D.B., ed. (1987).

APHANISTICUS EMARGINATUS ENDANGERED
A jewel beetle
Order COLEOPTERA   Family BUPRESTIDAE

Aphanisticus emarginatus (Olivier, 1790). Formerly known as: *Aphanistiscus emaaginatus* missp.

Distribution Recorded from North Devon, Dorset, Isle of Wight, North Hampshire and Berkshire before 1970.

Habitat and ecology Woodland, wetland and river margins. This species has also been recorded from wet areas in sand dunes. Associated with rushes, particularly jointed rush *Juncus articulatus*, and possibly also on sedges. Recorded from blunt-flowered rush *J. subnodulosus* on the Continent. Larvae probably develop in the stems. In Britain, adults are generally found on the flower heads of the foodplant. Adults have been recorded from late May to late September. The species hibernates in the adult state.

Status Not listed in the insect Red Data Book (Shirt, 1987). Last recorded in 1951 from Longmoor Camp, Liphook, North Hampshire.

Threats Loss of habitat through drainage, felling of woodland and agricultural improvement. Falling water tables due to water abstraction and river engineering make the long term future of many wetland sites look uncertain. Changes in grazing regimes may also adversely affect the invertebrate communities.

Management and conservation Grazing or cutting, on a rotational basis, is needed to maintain open conditions.

Published sources Bílý, S. (1982), Levey, B. (1977), Whicher, L.S. (1952a).

APHANISTICUS PUSILLUS   **NOTABLE B**
A jewel beetle
Order COLEOPTERA   Family BUPRESTIDAE

Aphanisticus pusillus (Olivier, 1790).

Distribution South East, South, South West, East Anglia, East Midlands, North East, North West and South Wales.

Habitat and ecology Often recorded in dry places, particularly downland and heathland. This species has also been noted in wetland. Associated with members of the Juncaceae and Cyperaceae, and possibly restricted to rushes *Juncus* and bog-rushes *Schoenus*. On the Continent, this species may also be associated with sedges *Carex*. Larvae probably develop in the stems. Adults are often swept from black bog-rush *S. nigricans*, and have been found in moss and grass tufts during winter. Adults occur all year round.

Status Widely distributed but local in England and recorded in South Wales.

Threats Uncertain, though agricultural improvement, conversion to arable and urban development are probably the causes of the main loss of habitat that this species frequents. Changes in grazing and cutting regimes may also adversely affect this species.

Management and conservation Grazing or cutting, on a rotational basis, is needed to maintain open conditions.

TRACHYS MINUTA   **VULNERABLE**
A jewel beetle
Order COLEOPTERA   Family BUPRESTIDAE

Trachys minuta (L., 1758). Formerly known as: *Trachys minutus* (L., 1758).

Distribution Recorded from South Wiltshire, Dorset, South Hampshire, North Hampshire, East Sussex, West Kent, Surrey, Hertfordshire, Berkshire, Oxfordshire, Buckinghamshire, East Norfolk, Cambridgeshire, Huntingdonshire, Northamptonshire, Worcestershire, North Lincolnshire and Leicestershire & Rutland before 1970 and South Hampshire, Buckinghamshire, East Norfolk and West Norfolk from 1970 onwards.

Habitat and ecology Broad-leaved woodland. Associated with hazel, hornbeam and sallow. The larvae are leaf miners. On the Continent, this species has also been found mining the leaves of whitebeam and English elm, the larval stage lasting four to six weeks. In Britain, adults occur mainly from mid May to late July. This species hibernates as an adult.

Status Not listed in the insect Red Data Book (Shirt, 1987). Apparently much declined, with modern records from just four vice-counties.

Threats Loss of broad-leaved woodland through, for example, clear-felling and coniferisation. Alteration of the vegetation structure of many remaining woodlands by 20th century management practices. It is possible that the abandonment of coppice management and the conversion to high forest, leading to increased shade and the loss of glades and broad sunny rides has been responsible for a decline in this species.

Management and conservation Open spaces in woodland need to be retained. Existing coppice cycles should be continued and considered for re-introduction in areas of abandoned coppice. Open glades and ride margins should be cut on rotation to retain a variety of vegetation structures.

Published sources Bílý, S. (1982), Fowler, W.W. (1890), Fowler, W.W. & Donisthorpe, H.St J.K. (1913), Levey, B. (1977).

BUPRESTIDAE

TRACHYS SCROBICULATUS   NOTABLE A
A jewel beetle
Order COLEOPTERA   Family BUPRESTIDAE

Trachys scrobiculatus Kiesenwetter, 1857. Formerly known as: *Trachys pumila* sensu auct. not (Illiger, 1803).

Distribution Recorded from South Wiltshire, Dorset, Isle of Wight, South Hampshire, North Hampshire, West Sussex, East Kent, West Kent, Surrey, Berkshire, Oxfordshire, Northamptonshire and Leicestershire & Rutland before 1970 and South Wiltshire, South Hampshire, North Hampshire, West Sussex, Surrey, Northamptonshire, East Gloucestershire and South-east Yorkshire from 1970 onwards.

Habitat and ecology Grassland, woodland and probably quarries. Largely, if not exclusively, confined to chalk and limestone. Associated predominantly with ground ivy *Glechoma hederacea*. Possibly also associated with henbane *Hyoscyamus niger* and white horehound *Marrubium vulgare*. On the Continent, this species has recorded from *Mentha* spp. and lesser calamint *Calamintha nepeta*. The larvae are leaf-miners. In Britain, adults have been found at the roots of the foodplant, by sweeping and from amongst moss. Adults have been recorded in May, June and August.

Status Possibly under-recorded. Very local, distributed through southern England as far north as South-east Yorkshire.

Threats Agricultural improvement, conversion to arable and development are the causes of the main losses of calcareous grassland. There has been a loss of woodland through clear-felling. Changes in grazing and cutting regimes may also adversely affect this species.

Management and conservation In woodland, open glades and ride margins should be managed to retain a variety of vegetation structures. Grazing or cutting, on a rotational basis, is needed to maintain open conditions.

Published sources Bílý, S. (1982), Cooter, J. (1969a), Fowler, W.W. (1890), Fowler, W.W. & Donisthorpe, H.St J.K. (1913), Levey, B. (1977).

BYRRHUS ARIETINUS   NOTABLE B
A pill beetle
Order COLEOPTERA   Family BYRRHIDAE

Byrrhus arietinus Steffahny, 1842. Formerly confused under: *Byrrhus fasciatus* (Forster, 1771).

Distribution West Midlands, North East, North West, North Wales, South East Scotland, North East Scotland and North West Scotland.

Habitat and ecology Moorland, vegetated river shingle, mountain tops and upland areas. Often found beneath stones and at the base of clumps of heather. Also recorded in *Sphagnum* moss in areas almost totally devoid of heather. Adults have been recorded from April to June and also from August to October.

Status Local. A northern and western species. This species is difficult to identify and may be confused with other members of the genus. Consequently, the exact status of this species is hard to assess.

Threats Erosion of montane and moorland habitat may be a problem in areas frequented by large numbers of hill walkers and where livestock densities are too high. Loss of moorland through afforestation. Further threats may be through river engineering, including dredging, level regulation by damming and flood alleviation schemes. In some areas colonisation by Himalayan balsam *Impatiens glandulifera* can reduce the available habitat.

Management and conservation Some grazing may be needed to maintain open conditions. River shingle tends to be mobile and relies on the free flow of river and stream systems. Activities that hinder this flow should be avoided.

CURIMOPSIS NIGRITA   ENDANGERED
Mire Pill Beetle or Bog Hog
Order COLEOPTERA   Family BYRRHIDAE

Curimopsis nigrita (Palm, 1934).

Distribution Recorded from South-west Yorkshire from 1970 onwards.

Habitat and ecology Lowland peat bogs. Adults and larvae feed on moss, occurring just beneath the soil surface, often amongst *Sphagnum* or heather litter. Adults have been found in moss lined tube and have been recorded in April, May and July.

Status Discovered new to Great Britain in 1977 and confined to Thorne Moors and Hatfield Moors, South-west Yorkshire.

Threats Both sites have been damaged by drainage and the commercial stripping of peat. At Hatfield Moors, only a tiny remnant of natural vegetation remains and this is threatened by further peat extraction. Drainage at Thorne Moors has accelerated in the late 1980's and much of the habitat was damaged by fire in 1989. Further threatened by the colonization of the dry cut peat surface by bracken and birch.

Management and conservation The water table should be maintained to keep the surface peat moist. Thorne and Hatfield Moors are notified as SSSIs, part of Thorne Moors is an NNR.

Published sources Buckland, P.C. & Johnson, C. (1983), Johnson, C. (1978), Shirt, D.B., ed. (1987).

CURIMOPSIS SETIGERA NOTABLE A
A pill beetle
Order COLEOPTERA Family BYRRHIDAE

Curimopsis setigera (Illiger, 1798). Formerly known as: *Syncalypta setigera* (Illiger, 1798).

Distribution Recorded from South Devon, North Devon, Dorset, South Hampshire, Oxfordshire, Glamorgan, South Lancashire, South-east Yorkshire, North-east Yorkshire, Westmorland & North Lancashire and Dumfriesshire before 1970 and North Devon, Dorset, South Hampshire and Dumfriesshire from 1970 onwards.

Habitat and ecology Mainly on the coast, with records from undercliffs and brackish dykes. Probably occurs at the roots of plants in sandy places, though also found in grass tussocks. Adults have been recorded in April, May, August and October.

Status Widely distributed and local. Recorded from South Devon, north to Dumfriesshire in Scotland. Recently noted in just four vice-counties.

Threats Cliff stabilisation schemes, the construction of sea defences and coastal developments. Activities that accelerate or reduce the rate of erosion should be avoided. The degradation of suitable habitat through natural succession is a further threat.

Management and conservation In areas of soft-rock cliff, occasional slippages are necessary to maintain habitat continuity. Large areas of unstable cliff are required so that the population does not become isolated and subsequently threatened by individual landslips.

Published sources Johnson, C. (1978).

PORCINOLUS MURINUS NOTABLE B
A pill beetle
Order COLEOPTERA Family BYRRHIDAE

Porcinolus murinus (F., 1794). Formerly known as: *Byrrhus murinus* F.

Distribution South East, South, South West, East Anglia, East Midlands, West Midlands and North East.

Habitat and ecology Heathland, sand dunes and other open habitats on sandy soils. Adults and probably larvae occur at the roots and base of heather and possibly other herbaceous plants. This species has been found at the base of marram grass. Adults have been recorded from May to August.

Status Widespread but local in southern and central England. Also recorded in North East England.

Threats Loss, or fragmentation, of habitat through changes in land use, mainly by conversion to arable agriculture, afforestation and urban development. Coastal development and natural succession may be further threats.

Management and conservation Management should aim for a diversity of successional stages, preferably by grazing or possibly through cutting or scraping.

SIMPLOCARIA MACULOSA INDETERMINATE
A pill beetle
Order COLEOPTERA Family BYRRHIDAE

Simplocaria maculosa Erichson, 1847.

Distribution Recorded from Worcestershire and Mid-west Yorkshire before 1970.

Habitat and ecology Riverbanks. On the Continent, this species has been found amongst moss between stones, on the banks and at the edge of rivers and streams. In Britain, the adult has been recorded in May.

Status Status revised from RDB 3 (Rare) in Shirt (1987). Only known from three examples; two (males) found at Bewdley, Worcestershire last century, and a single female recorded from the River Ouse, Kelfield, South-east Yorkshire in 1956.

Threats Uncertain, though river engineering, including level regulation by damming and flood alleviation schemes may be a threat to this species. In some areas colonisation by Himalayan balsam *Impatiens glandulifera* can reduce the available habitat.

CANTHARIDAE

Published sources Johnson, C. (1966a), Johnson, C. (1978), Shirt, D.B., ed. (1987).

ANCISTRONYCHA ABDOMINALIS NOTABLE B
A soldier beetle
Order COLEOPTERA   Family CANTHARIDAE

Ancistronycha abdominalis (F., 1798). Formerly known as: *Cantharis abdominalis* F.

Distribution West Midlands, Wales, North East, North West and Scotland.

Habitat and ecology Woodland, particularly wood-edges and woodland rides. Also recorded in woodland ravines in moorland. Adults and larvae are probably predatory. The larvae are free-living and active, found on the ground and possibly also on foliage. Adults are noted on flowers and foliage, particularly trees and umbellifers. This species probably overwinters as a full grown larva, with eclosion probably occurring the following April. Adults have been recorded from May to July.

Status Very local. A predominantly northern and western species. Occasionally plentiful at some sites.

Threats Loss of broad-leaved woodland through, for example, clear-felling and conversion to other land use. Neglect, and the conversion to high forest, has led to increased shade and the loss of glades and broad sunny rides.

Management and conservation Management should aim at maintaining open conditions. Open glades and ride margins should be cut on rotation to retain a variety of vegetation structures.

CANTHARIS FUSCA   RARE
A soldier beetle
Order COLEOPTERA   Family CANTHARIDAE

Cantharis fusca L., 1758. Formerly known as: *Telophorus fuscus* (L.).

Distribution Recorded from South Devon, South Wiltshire, Dorset, Isle of Wight, South Hampshire, West Sussex, East Sussex, East Kent, West Kent, Surrey, Hertfordshire, Middlesex, "Suffolk", "Norfolk", Cambridgeshire, Northamptonshire, Cardiganshire, Merionethshire, South-west Yorkshire, Mid-west Yorkshire, Durham and East Lothian before 1970 and North Somerset, South Hampshire, East Sussex and East Kent from 1970 onwards.

Habitat and ecology Fens, wood-edges, coastal shingle, a coastal sea wall and probably also rough grassland, field margins and river margins. The larvae and adults are probably predatory on small insects. The larvae are probably free-living and active, found on the ground and possibly also on foliage. Adults are usually noted on low growing plants, particularly umbellifers. This species probably overwinters as a full grown larva, with eclosion probably occurring the following April. Adults have been recorded from May to July.

Status Not listed in the insect Red Data Book (Shirt, 1987). Old records suggest that this species was formerly widespread, and recorded throughout the southern half of England, with a more scattered distribution as far north as East Lothian. There are recent records for only four vice-counties, all in the south of England.

Threats This species has probably declined because of habitat destruction through drainage of fens, the conversion of rough grassland to arable agriculture, and the clearance of woodlands and conversion to other land use

Management and conservation Water tables should be maintained at high levels. Grazing or cutting, on a rotational basis, may be needed to maintain open conditions. The disturbance of coastal shingle should be avoided.

Published sources Eyre, M.D. & Sheppard, D.A. (1982), Fowler, W.W. (1890), Fowler, W.W. & Donisthorpe, H.St J.K. (1913), Hodge, P.J. (1989).

CANTHARIS OBSCURA   NOTABLE B
A soldier beetle
Order COLEOPTERA   Family CANTHARIDAE

Cantharis obscura L., 1758. Formerly known as: *Telophorus obscurus* (L.).

Distribution East Midlands, West Midlands, North East, North West, Dyfed-Powys, North Wales and Scotland.

Habitat and ecology Broad-leaved and mixed woodland, particularly wood-edges, woodland rides and hedgerows. The larvae and adults are probably predatory. The larvae are probably free-living and active, found on the ground and possibly also on foliage. Adults have been beaten from oak and pine, and probably also occur on other plants. This species probably over-winters as a full grown larva, with eclosion probably occurring the following April. Adults have been recorded in May and June.

Status Local. A predominantly northern and western species.

Threats Loss of broad-leaved woodland through, for example, clear-felling and conversion to other land use. Neglect, and the conversion to high forest, has led to increased shade and the loss of glades and broad sunny rides. The uprooting of hedgerows may also be a further threat.

Management and conservation Open spaces in woodland need to be retained. Open glades and ride margins should be cut on rotation to retain a variety of vegetation structures.

MALTHINUS BALTEATUS   **NOTABLE B**
A soldier beetle
Order COLEOPTERA   Family CANTHARIDAE

Malthinus balteatus Suffrian, 1851.

Distribution South East, South, South West, East Anglia, East Midlands, West Midlands and Dyfed-Powys.

Habitat and ecology Damp broad-leaved woodland, especially carr. Probably predatory. Larvae probably develop in dead twigs and small branches. Adults have been recorded from the foliage of trees and shrubs, such as willow, hazel and lime. Adults have been recorded in June and July, and would seem to be short-lived.

Status Widespread but local in southern England.

Threats Loss of broad-leaved woodland through, for example, clear-felling and coniferisation. This species may be further threatened by drainage and water abstraction schemes, as well as the removal of dead wood from living trees.

Management and conservation Water tables should be maintained at high levels. Gaps in the age structure of the tree population should be identified and the continuity of the appropriate dead wood habitat ensured by regeneration, suitable planting and possibly with pollarding.

MALTHINUS FRONTALIS   **NOTABLE B**
A soldier beetle
Order COLEOPTERA   Family CANTHARIDAE

Malthinus frontalis (Marsham, 1802).

Distribution England, Wales and Scotland.

Habitat and ecology Broad-leaved and, occasionally, coniferous woodland. Also found on more isolated trees. Associated with mature and over-mature trees. Probably predatory. This species has been reared from wood from a large rotting log of white willow. Larvae probably also develop in dead twigs and small branches, and possibly also rot-holes. Adults have been beaten from old oak and from alder, pine, willow, including pollards, and have also been swept beneath oak and spruce. Adults have been recorded from June to August.

Status Widespread but local in England, also recorded in South Wales and North East Scotland. Possibly declining.

Threats Loss of broad-leaved woodland and field edge trees through, for example, clear-felling and conversion to other land use. This species may be further threatened by the removal of dead wood from living trees and the destruction or removal of standing and fallen dead wood for reasons such as forest hygiene, aesthetic tidiness, public safety or for use as fire wood.

Management and conservation Ancient trees and both fallen and standing dead timber, especially with the bark attached, should be retained. The removal of dead timber from ancient trees should be avoided. Gaps in the age structure of the tree population should be identified and the continuity of the appropriate dead wood habitat ensured by regeneration, suitable planting and possibly with pollarding.

MALTHODES BREVICOLLIS   **ENDANGERED**
A soldier beetle
Order COLEOPTERA   Family CANTHARIDAE

Malthodes brevicollis (Paykull, 1798). Formerly known as: *Malthodes nigellus* Kiesenwetter, 1852.

Distribution Recorded from Herefordshire before 1970 and Herefordshire from 1970 onwards.

Habitat and ecology Pasture-woodland. The beetle has been found in all its stages in red-rotten wood of oak. Adults have been found in small cells, sometimes in company with a small collembolan, on which it might prey (A.A. Allen pers. comm.). Adults have been recorded in June. They are probably short-lived and possibly only emerge into the open to disperse.

Status Status revised from RDB 3 (Rare) in Shirt (1987). The only confirmed records are from Moccas Park, Herefordshire. Records of *M. brevicollis* from South Hampshire, South Essex, West Suffolk, Norfolk, Oxfordshire and Mid-west Yorkshire require confirmation as the species has been confused with its close relative *M. crassicornis* and even *M. pumilus*.

Indicator status Grade 1 in Garland (1983). Grade 1 in Harding & Rose (1986).

Threats Loss of broad-leaved woodland and parkland through, for example, clear-felling and coniferisation. Habitat loss, in particular, through the felling of ancient trees, removal of dead wood from living trees and the destruction or removal of standing and fallen dead wood for reasons such as forest hygiene, aesthetic tidiness, public safety or for use as fire wood.

Management and conservation Ancient trees and both fallen and standing dead timber, especially with the bark attached, should be retained. The removal of dead timber from ancient trees should be avoided. Gaps in the age structure of the tree population should be identified and the continuity of the appropriate dead wood habitat ensured by regeneration, suitable planting and possibly with pollarding. Moccas Park is an NNR.

Published sources Allen, A.A. (1942), Black, J.E. (1924), Donisthorpe, H.St J.K. (1926a), Fowler, W.W. (1890), Fowler, W.W. & Donisthorpe, H.St J.K. (1913), Garland, S.P. (1983), Harding, P.T. (1978), Harding, P.T. & Rose, F. (1986), Shirt, D.B., ed. (1987), Tomlin, J.R.le B. (1949).

MALTHODES CRASSICORNIS　　　　　　**RARE**
A soldier beetle
Order COLEOPTERA　　　　　Family CANTHARIDAE

Malthodes crassicornis (Mäklin, 1846).

Distribution Recorded from Surrey, South Essex, North Essex, Hertfordshire, Berkshire, Oxfordshire, East Suffolk, East Gloucestershire and Herefordshire before 1970 and South Essex, Berkshire, East Suffolk and West Gloucestershire from 1970 onwards.

Habitat and ecology Ancient, broad-leaved woodland and pasture-woodland. Also recorded near old willow pollards on a riverbank. The larvae and adults are probably predatory. This species has been reared from red-rotten wood of an old stump. Adults have been found by brushing inside hollow trees, or close by, on tree trunks, rotten logs or canopy foliage. This species

probably overwinters as a full grown larva, with pupation and emergence occurring the following spring. Adults have been recorded in March, May and June, and are probably short-lived.

Status Very local. Only known from southern England, recently recorded from just four vice-counties. Some old records for *M. brevicollis* may refer to this species. Females of this species are difficult to identify.

Indicator status Grade 2 in Harding & Rose (1986).

Threats Loss of broad-leaved woodland and parkland through, for example, clear-felling and coniferisation. Habitat loss, in particular, through the felling of ancient trees, removal of dead wood from living trees and the destruction or removal of standing and fallen dead wood for reasons such as forest hygiene, aesthetic tidiness, public safety or for use as fire wood.

Management and conservation Ancient trees and both fallen and standing dead timber, especially with the bark attached, should be retained. The removal of dead timber from ancient trees should be avoided. Gaps in the age structure of the tree population should be identified and the continuity of the appropriate dead wood habitat ensured by regeneration, suitable planting and possibly with pollarding.

Published sources Allen, A.A. (1937), Allen, A.A. (1942), Allen, A.A. (1951f), Atty, D.B. (1983), Buck, F.D. (1955), Collier, M.J. (1988a), Harding, P.T. (1978), Harding, P.T. & Rose, F. (1986), Last, H. (1943), Shirt, D.B., ed. (1987).

MALTHODES FIBULATUS　　　　　　**NOTABLE B**
A soldier beetle
Order COLEOPTERA　　　　　Family CANTHARIDAE

Malthodes fibulatus Kiesenwetter, 1852.

Distribution South East, South, South West, East Midlands, West Midlands, North East, North West, Dyfed-Powys, South Wales, South East Scotland and South West Scotland.

Habitat and ecology Broad-leaved woodland, hedges, and possibly also gardens and suburban parks. The beetle may have a preference for habitats on calcareous soils. The larvae and adults are probably predatory. Larvae probably develop in dead twigs and boughs. Adults are usually found on the foliage of trees and shrubs, as well as flowers. This species probably overwinters as a full grown larva, with pupation and emergence occurring the following spring. Adults have been recorded from May to June, and seem to be short-lived.

Status Widespread but local throughout England, though not recorded in East Anglia. Also recorded from southern Wales and southern Scotland. Females of this species are difficult to identify.

Threats Loss of broad-leaved woodland through, for example, clear-felling and coniferisation. This species may be further threatened by the uprooting and mechanised trimming of hedgerows, and the removal of dead wood from living trees.

Management and conservation Gaps in the age structure of the tree population should be identified and the continuity of the appropriate dead wood habitat ensured by regeneration, suitable planting and possibly with pollarding. The presence of nectar sources such as hawthorn, umbellifers and composite herbs may also be particularly important for this species.

MALTHODES GUTTIFER	**NOTABLE B**
A soldier beetle	
Order COLEOPTERA	Family CANTHARIDAE

Malthodes guttifer Kiesenwetter, 1852.

Distribution South East, South, South West, East Midlands, West Midlands, North East, North West, Wales and Scotland.

Habitat and ecology Broad-leaved woodland and conifer plantations. The larvae and adults are probably predatory. Larvae have been reared from oak twigs. Adults are usually found on foliage. This species probably overwinters as a full grown larva, with pupation and emergence occurring the following spring. Adults have been recorded in June to early August.

Status Local. Predominantly northern and western Great Britain, with few records in south-eastern England. Females of this species are difficult to identify.

Threats Loss of woodland through, for example, clear-felling and conversion to other land use.

Management and conservation Gaps in the age structure of the tree population should be identified and the continuity of the appropriate habitat ensured by regeneration, suitable planting and possibly with pollarding.

MALTHODES MAURUS	**NOTABLE B**
A soldier beetle	
Order COLEOPTERA	Family CANTHARIDAE

Malthodes maurus (Castelnau, 1840). Formerly known as: *Malthodes misellus* sensu Fowler, 1890 not Kiesenwetter, 1852.

Distribution South, East Anglia, East Midlands, West Midlands, North East, North West, South Wales, South East Scotland and South West Scotland.

Habitat and ecology Broad-leaved woodland, particularly in damp situations. The larvae and adults are probably predatory. Larvae probably develop in dead twigs or small branches. Adults are usually found on the foliage of trees and shrubs. The beetle probably overwinters as a full grown larva, with pupation and emergence occurring the following spring. Adults have been recorded from May to July.

Status Very local. Has been found in the Midlands north to southern Scotland, with very few records in southern England. Possibly declining. Females of this species are dfficult to identify.

Threats Loss of broad-leaved woodland through, for example, clear-felling and coniferisation. This species may be further threatened by the removal of dead wood from living trees.

Management and conservation Gaps in the age structure of the tree population should be identified and the continuity of the appropriate dead wood habitat ensured by regeneration, suitable planting and possibly with pollarding.

RHAGONYCHA ELONGATA	**NOTABLE A**
A soldier beetle	
Order COLEOPTERA	Family CANTHARIDAE

Rhagonycha elongata (Fallén, 1807).

Distribution Recorded from North-west Yorkshire, Kirkcudbrightshire, Wigtownshire, Stirlingshire, East Inverness & Nairn and North Ebudes before 1970 and Kirkcudbrightshire, Roxburghshire, West Perthshire and East Inverness & Nairn from 1970 onwards.

Habitat and ecology Coniferous woodland. Recorded mainly from the canopy foliage of Scots pine. Adults have been recorded from May to July.

Status Very local. Scottish in distribution and especially found in the Highlands. Probably a component of the native pine forest fauna. Occurrences

CANTHARIDAE

outside this range are probably the result of colonisation.

Threats Loss of native pine forest through practices such as clear-felling or conversion to other land use.

Management and conservation Gaps in the age structure of the tree population should be identified and regeneration encouraged to ensure the continuity of suitable habitat.

Published sources Alexander, K.N.A. (1979), Houston, K. & Coulson, J.C. (1972), Welch, R.C. (1975).

RHAGONYCHA LUTEA NOTABLE B
A soldier beetle
Order COLEOPTERA Family CANTHARIDAE

Rhagonycha lutea (Müller, 1764). Formerly known as: *Rhagonycha fuscicornis* (Olivier, 1790).

Distribution South East, South, East Anglia, East Midlands, West Midlands, North East, Dyfed-Powys and South West Scotland.

Habitat and ecology Woodland, parkland, woodland rides, wood-edges and scrubby calcareous grassland. The larvae and adults are probably predatory. The larvae are probably free-living and active, found on the ground and possibly also on foliage. Adults are usually noted on trees and shrubs, and have been recorded from oak, hazel and wild rose. This species probably overwinters as a full grown larva, with eclosion probably occurring the following April. Adults have been recorded from May to July.

Status Widespread but local in England, though not noted in North West England. Also recorded in Wales and South West Scotland.

Threats Loss of broad-leaved woodland through, for example, clear-felling and conversion to other land use. Neglect, and conversion to high forest, has led to increased shade and the loss of glades and broad sunny rides. The loss of calcareous downland through practices such as improvement and conversion to arable agriculture may be a further threat.

Management and conservation Management should aim at maintaining open conditions. Open glades and ride margins should be cut on rotation to retain a variety of vegetation structures.

RHAGONYCHA TRANSLUCIDA NOTABLE B
A soldier beetle
Order COLEOPTERA Family CANTHARIDAE

Rhagonycha translucida (Krynicki, 1832). Formerly known as: *Rhagonycha unicolor* sensu Fowler, 1890 ?not (Curtis, 1840).

Distribution England, Wales, South West Scotland and South East Scotland.

Habitat and ecology Broad-leaved woodland. Particularly woodland rides and clearings, pasture-woodland and bushy places on downland. The larvae and adults are probably predatory. The larvae are probably free-living and active, found on the ground and possibly on foliage. Adults have been swept from low-growing vegetation in woods and beaten from large, old hawthorn bushes, just after flowering. They have also been found on the foliage of a wide range of other trees and shrubs, notably oak and hazel. This species probably overwinters as a full grown larva, with eclosion probably occurring the following April. Adults have been recorded from June to August. This beetle has been recorded at mercury vapour light.

Status Widespread but local south of the Forth-Clyde line. This species may be confused with *Cantharis cryptica* and *C. pallida*. Consequently, the exact status of this species is hard to assess.

Threats Loss of broad-leaved woodland and parkland through, for example, clear-felling and coniferisation. Neglect, and conversion to high forest, has led to increased shade and the loss of glades and broad sunny rides.

Management and conservation Open spaces in woodland need to be retained. Open glades and ride margins should be cut on rotation to retain a variety of vegetation structures.

SILIS RUFICOLLIS NOTABLE B
A soldier beetle
Order COLEOPTERA Family CANTHARIDAE

Silis ruficollis (F., 1775).

Distribution England, South Wales and Dyfed-Powys.

Habitat and ecology River and lake margins, fens, marshes and other wetland habitats. The larvae and adults are probably predatory. The larvae are probably free-living and active, found on the ground and on foliage. Adults are noted on lush, marginal vegetation, and have been swept from common reed. This species probably overwinters as a full grown larva, with eclosion probably occurring the following April. Adults have been recorded from June to August.

Status Formerly a rare and very local species. *S. ruficollis* appears to have increased in abundance over the last 40 years, and is widespread and not uncommon in some areas of southern England and Wales. Extremely local in northern England.

Threats Drainage for reasons such as agricultural improvement and development is the primary cause of the loss of wetland habitats. Falling water tables because of water abstraction and river engineering schemes may also threaten this species.

Management and conservation Water tables should be maintained at high levels. Grazing or cutting, on a rotational basis, may be needed to maintain open conditions.

ACUPALPUS BRUNNIPES   NOTABLE A
A ground beetle
Order COLEOPTERA   Family CARABIDAE

Acupalpus brunnipes (Sturm, 1825). Formerly known as: *Acupalpus brunneipes* (Sturm, 1825).

Distribution Recorded from Dorset, South Hampshire, North Hampshire, Surrey, Hertfordshire, Berkshire and Glamorgan before 1970 and Dorset, West Kent and Surrey from 1970 onwards.

Habitat and ecology Wetland areas, saltmarshes and habitats near water. Found amongst moss and lush vegetation on soft soil or mud. Probably predatory. Adults have been recorded from May to July. This species has been recorded at a mercury vapour light trap.

Status Known only in southern England and South Wales. Recently recorded from only three vice-counties. This species is difficult to identify and may be confused with other members of the genus. Consequently, the exact status of this species is hard to assess.

Threats This species may be threatened by drainage, reclamation, erosion, the construction of coastal defences, improvement and conversion to arable agriculture, and development.

Management and conservation Management should aim at maintaining open conditions and encouraging early successional stages. Water tables should be maintained at high levels. Water bodies should be isolated from sources of eutrophication and pollution.

Published sources Allen, A.A. (1990e), Fowler, W.W. (1887), Fowler, W.W. & Donisthorpe, H.St J.K. (1913), Lindroth, C.H. (1974).

ACUPALPUS CONSPUTUS   NOTABLE B
A ground beetle
Order COLEOPTERA   Family CARABIDAE

Acupalpus consputus (Duftschmid, 1812). Formerly known as: *Anthracus consputus* (Duftschmid).

Distribution England and South Wales.

Habitat and ecology Reservoirs, gravel pits and coastal habitats. Probably predatory. Appears to prefer sparsely vegetated ground on soft soil or mud near water, often occurring under stones. Adults have been recorded from February to August and in December.

Status Widespread but local in England and also recorded in South Wales.

Threats This species is threatened by urban and holiday developments, the infilling of pits and natural succession.

Management and conservation Management should aim at maintaining open conditions and encouraging early successional stages.

ACUPALPUS ELEGANS   EXTINCT
A ground beetle
Order COLEOPTERA   Family CARABIDAE

Acupalpus elegans (Dejean, 1829). Formerly known as: *Stenolophus elegans* (Dejean).

Distribution Recorded from East Kent before 1970.

Habitat and ecology Found in coastal saltmarshes, wet flushes and undercliffs. An exclusively coastal species. Probably predatory. Adults have been recorded in April, June and August.

Status Status revised from RDB 1 (Endangered) in Shirt (1987). A very difficult species to identify, requiring dissection of the male aedeagus for reliable separation from *A. dorsalis*. The single record this century is of an undissected specimen noted in 1952 at Stoke Junction, West Kent and, therefore, requires critical examination. The species was reliably (from dissected material) last recorded in 1874 and 1875 from the Isle of Sheppey, East Kent. There is also a very old record from Deal, East Kent. A pair of male examples from Reigate, Surrey (undated, but from the 19th century) have been dissected and shown to be *A.*

CARABIDAE

elegans. This last locality is an unlikely habitat for the species, being on chalk downland, and the labelling on the specimens is such that that it may not refer to the point of capture. Apart from a single female, which cannot be accurately identified, recorded in 1984 at Aveley, South Essex, other records of *A. elegans* have proved to be erroneous. These include a number of examples from Barton Cliffs, South Hampshire (1981) and a single example from Thorne Moors, South-west Yorkshire (before 1903) which were found to be *A. dorsalis* and *A. luridus*, respectively. A record from Woolmer Bog, North Hampshire in 1954 must be regarded as highly doubtful due to an absence of material and the most unlikely nature of the habitat.

Published sources Allen, S.E. (1953), Lindroth, C.H. (1974), Plant, C.W. & Drane, A.B. (1988), Shirt, D.B., ed. (1987), Walker, J.J. (1900b), Walker, J.J. (1932a).

ACUPALPUS EXIGUUS	**NOTABLE B**
A ground beetle	
Order COLEOPTERA	Family CARABIDAE

Acupalpus exiguus Dejean, 1829.

Distribution South East, South, East Anglia, West Midlands, North West, North East and North Wales.

Habitat and ecology Often found on the coast, on seashores and in saltmarshes. Also inland, on river margins and in grassland on clay soils. Chiefly subterranean. Probably predatory. Found amongst strandline and coastal debris on sand or silt, and amongst debris on mud or silt at the margins of fresh water. Also recorded under the bark of a willow tree and in flood refuse. Adults have been recorded from January to July and in September.

Status Widespread but local in England, possibly more frequent in the Weald of Kent than elsewhere. Also recorded from North Wales. This species is difficult to identify and may be confused with other members of the genus. Consequently, the exact status of this species is hard to assess.

Threats Loss of habitat through reclamation, erosion and the construction of sea defences, as well as improvement and conversion to arable agriculture. River engineering, including dredging, damming and flood alleviation schemes, as well as river pollution may be further threats.

Management and conservation Management should aim at maintaining open conditions and encouraging early successional stages. Activities that hinder the natural flow of river and streams should be avoided.

ACUPALPUS FLAVICOLLIS	**NOTABLE A**
A ground beetle	
Order COLEOPTERA	Family CARABIDAE

Acupalpus flavicollis (Sturm, 1825).

Distribution Recorded from "Devon", Dorset, Isle of Wight, South Hampshire, East Kent, Surrey, Berkshire, East Suffolk, East Norfolk, Cambridgeshire, Huntingdonshire and Warwickshire before 1970 and South Wiltshire, Isle of Wight, South Hampshire, East Sussex, Surrey and South-west Yorkshire from 1970 onwards.

Habitat and ecology On sparsely vegetated sand by rivers and coastal cliffs. Also on acid soils inland in boggy areas with *Sphagnum*. Probably predatory. Adults have been recorded from April, May, August, October and November.

Status Very local and known only from England as far north as South-west Yorkshire. This species is difficult to identify and may be confused with other members of the genus. Consequently, the exact status of this species is hard to assess.

Threats This species is threatened by river engineering, including damming and flood alleviation schemes, urban and holiday developments, cliff stabilisation schemes and the construction of coastal defences, as well as through drainage.

Management and conservation Management should aim at maintaining open conditions and encouraging early successional stages. Activities that affect the natural flow of rivers and streams should be avoided. In areas of cliff occasional slippages are necessary to maintain habitat continuity. Large areas of unstable cliff are required so that the population does not become isolated and subsequently threatened by individual landslips.

Published sources Fowler, W.W. (1891), Fowler, W.W. & Donisthorpe, H.St J.K. (1913), Lindroth, C.H. (1974).

AEPUS MARINUS	**NOTABLE B**
A ground beetle	
Order COLEOPTERA	Family CARABIDAE

Aepus marinus (Ström, 1783).

Distribution South East, South, South West, North East, Wales, South West Scotland, North East Scotland and North West Scotland. A map is given in Luff (1982).

Habitat and ecology Coastal. Found in inter-tidal habitats. Predatory, possibly on springtails (Collembola). Adults usually occur at the mid-water mark, under stones on coarse sand or fine gravel, and probably only rarely in rock crevices. Adults have been recorded from April to July.

Status Widespread but local around the coasts of Great Britain, apparently absent from the East Anglian coastline.

Threats This species may be threatened by marine pollution.

AEPUS ROBINI   **NOTABLE B**
A ground beetle
Order COLEOPTERA   Family CARABIDAE

Aepus robini (Laboulbène, 1849). Formerly known as: *Aepopsis robinii* (Laboulbène), *Aepus robinii* (Laboulbène).

Distribution South East, South West, North East, South Wales, Dyfed-Powys, South East Scotland and South West Scotland. A map is given in Luff (1982).

Habitat and ecology Coastal. Found in inter-tidal habitats with a mixture of rock and sand. Predatory. Adults usually occur in sand-filled rock crevices, often with the springtail *Anurida maritima* (Collembola), on which it probably feeds. Adults have been recorded in June and September.

Status Widespread but local around the coasts of England, Wales and southern Scotland, apparently absent from the East Anglian coastline.

Threats This species may be threatened by marine pollution.

AGONUM ERICETI   **NOTABLE B**
A ground beetle
Order COLEOPTERA   Family CARABIDAE

Agonum ericeti (Panzer, 1809). Formerly known as: *Anchomenus ericeti* (Panzer).

Distribution South, South West, West Midlands, North East, North West, Wales and Scotland. A map is given in Luff (1982).

Habitat and ecology Raised and blanket mires, and occasionally in open moorland. Predatory. A species of wet, acid conditions. Usually found amongst *Sphagnum* and other vegetation. Adults have been recorded in April, June and September.

Status Widespread but local. Primarily a northern species.

Threats The loss of bogs and moorland through drainage, afforestation and conversion to upland pasture.

AGONUM GRACILIPES   **NOTABLE A**
A ground beetle
Order COLEOPTERA   Family CARABIDAE

Agonum gracilipes (Duftschmid, 1812). Formerly known as: *Anchomenus gracilipes* (Duftschmid).

Distribution Recorded from East Suffolk, East Norfolk and South-east Yorkshire before 1970 and East Sussex, East Suffolk, West Norfolk, Cambridgeshire, Cardiganshire and Ayrshire from 1970 onwards.

Habitat and ecology Predominantly in open country on the coast, though recently recorded inland from a wood. Also noted from a fen. Predatory. Mainly found under stones, vegetation and other cover. Has been noted at mercury vapour light. Adults have been recorded in June, July and September.

Status This species may not be resident, records being the result of immigrant individuals. The increasing frequency of recent records could indicate, however, that this species has become established. Most records are from the Brecklands of East Anglia.

Published sources Allen, A.A. (1977d), Flint, J.H. (1984a), Fowler, W.W. (1887), Fowler, W.W. & Donisthorpe, H.St J.K. (1913), Hodge, P.J. (1977a), Lindroth, C.H. (1974), Nash, D.R. (1982b).

AGONUM LIVENS   **NOTABLE B**
A ground beetle
Order COLEOPTERA   Family CARABIDAE

Agonum livens (Gyllenhal, 1810). Formerly known as: *Anchomenus livens* (Gyllenhal).

Distribution South East, South, East Anglia, East Midlands, West Midlands, North East and North West.

Habitat and ecology Alder carr, willow carr, damp woodland, river banks, reservoirs, marshes, wet meadows and fens. Predatory. Found amongst marsh or damp vegetation such as reed litter and the roots of sedges. Also found under bark, in rotten wood of willow and frequently beaten from hornbeam and oak branches. Adults have been recorded from March to August and from October to December.

CARABIDAE

Status Widespread but local throughout England.

Threats The loss of woodland and carr through clear-felling and conversion to other land use. Drainage, water abstraction and river engineering schemes may be further threats to this species.

Management and conservation Water tables should be maintained at high levels. Water bodies should be isolated from sources of eutrophication and pollution.

AGONUM NIGRUM   NOTABLE B
A ground beetle
Order COLEOPTERA   Family CARABIDAE

Agonum nigrum Dejean, 1828. Formerly known as: *Agonum dahli* Preudhomme, 1879, *Anchomenus atratus* sensu auct. not (Duftschmid, 1812), *Anchomenus dahli* (Preudhomme).

Distribution South East, South, South West, East Anglia, East Midlands, North West, Wales and South West Scotland.

Habitat and ecology River banks, estuarine reed-beds, marshes in dune slacks, saltmarshes and the margins of lakes, ponds and gravel pits. Predatory. Found amongst lush vegetation such as sedges, grasses and reeds, on soft soil and mud. Also recorded from under decayng seaweed and driftwood. Adults have been recorded in March, April, June, July and September.

Status Widespread but local in England and Wales, also recorded in South West Scotland.

Threats Drainage for reasons such as agricultural improvement and development is the primary cause of the loss of wetlands. Falling water tables because of water abstraction and river engineering schemes, the infilling of gravel pits and urban and holiday developments may also threaten this species.

Management and conservation Water tables should be maintained at high levels. Water bodies should be isolated from sources of eutrophication and pollution. Management should aim at maintaining open conditions and encouraging early successional stages along pit and lake edges.

AGONUM QUADRIPUNCTATUM ENDANGERED
A ground beetle
Order COLEOPTERA   Family CARABIDAE

Agonum quadripunctatum (Degeer, 1774). Formerly known as *Anchomenus quadripunctatus* (Degeer).

Distribution Recorded from Dorset, West Kent, Surrey, Berkshire, North-east Yorkshire, South Northumberland, Westmorland & North Lancashire, Cumberland, Moray and East Inverness & Nairn before 1970 and Berkshire from 1970 onwards.

Habitat and ecology Usually on sandy or peaty soil and normally found in heathland or woodland. Predatory. The beetle is attracted by forest fires, particularly when conifers are affected. Often found under bark. Adults have been recorded in April, June, July and September.

Status Not listed in the insect Red Data Book (Shirt, 1987). An established immigrant. Reported from Dorset to Easterness, though only recently recorded from Berkshire. By the turn of the 19th century only one example was known in Great Britain.

Published sources Buck, F.D. (1949), Fowler, W.W. (1887), Fowler, W.W. & Donisthorpe, H.St J.K. (1913), Harwood, P. (1922), Johnson, C. (1963a), Lindroth, C.H. (1974).

AGONUM SAHLBERGI   EXTINCT
A ground beetle
Order COLEOPTERA   Family CARABIDAE

Agonum sahlbergi (Chaudoir, 1850). Formerly known as: *Anchomenus sahlbergi* (Chaudoir).

Distribution Recorded from Renfrewshire and Dunbartonshire before 1970.

Habitat and ecology River banks. Predatory. Adults have been found on the edge of a sandy bank and under stones.

Status Status revised from RDB 1+ (Endangered, believed extinct) in Shirt (1987). Probably extinct. A glacial relict species or possibly an accidental introduction. Recorded on only four occasions on the River Clyde near Glasgow. The first was noted around 1864 at Dunglass Castle, Dunbarton, the last in 1914 on the south bank of the Clyde, Renfrewshire. This species is difficult to identify and may be confused with *A. muelleri*.

Published sources Lindroth, C.H. (1960), Lindroth, C.H. (1974), Murphy, J.E. (1918), Shirt, D.B., ed. (1987).

AGONUM SCITULUM   **NOTABLE A**
A ground beetle
Order COLEOPTERA   Family CARABIDAE

Agonum scitulum Dejean, 1828. Formerly known as: *Anchomenus scitulus* (Dejean), *Europhilus scitulus* (Dejean).

Distribution Recorded from "Devon", North Somerset, Dorset, South Hampshire, East Kent, Surrey, Middlesex, Berkshire, Cambridgeshire, West Gloucestershire, Worcestershire, Shropshire, Merionethshire and Derbyshire before 1970 and South Somerset, "Wilts", East Kent, Cambridgeshire, Caernarvonshire, Leicestershire & Rutland, South-east Yorkshire and South-west Yorkshire from 1970 onwards.

Habitat and ecology Wetlands, particularly carr, and riverbanks. Predatory. Found on marshy ground with some vegetation. Adults have been recorded in January, February, April, May and December.

Status This species has a scattered distribution from East Kent north to South-east Yorkshire, with a recent record for Caernarvonshire, Wales. This species is difficult to identify and can be confused with other species of the genus, especially *A. micans*. Consequently, the exact status of this species is hard to assess.

Threats Drainage for reasons such as agricultural improvement and development is the primary cause of the loss of wetlands. Further loss of habitat maybe through water abstraction schemes and river engineering, including dredging, level regulation by damming and flood alleviation schemes. In some areas colonisation by Himalayan balsam *Impatiens glandulifera* can reduce the available habitat. River pollution maybe a further threat.

Management and conservation Activities that hinder the natural flow of rivers should be avoided. Water tables should be maintained at high levels. Water bodies should be isolated from sources of eutrophication and pollution.

Published sources Fowler, W.W. (1887), Fowler, W.W. & Donisthorpe, H.St J.K. (1913), Grensted, L.W. (1931), Lindroth, C.H. (1974).

AGONUM SEXPUNCTATUM   **NOTABLE A**
A ground beetle
Order COLEOPTERA   Family CARABIDAE

Agonum sexpunctatum (L., 1758). Formerly known as: *Anchomenus sexpunctatus* (L.).

Distribution Recorded from Dorset, South Hampshire, North Hampshire, "Kent", Surrey, South Essex, Hertfordshire, Berkshire, Oxfordshire, "Norfolk", Glamorgan and Nottinghamshire before 1970 and Dorset, West Sussex, East Sussex, Surrey and Staffordshire from 1970 onwards. A map is given in Luff (1982).

Habitat and ecology *Sphagnum* bogs, pools, damp ditches and other wet places on sandy heaths. Also found in areas of sparsely vegetated ground on peat. Predatory and gregarious. Found amongst *Sphagnum* and other vegetation. Adults have been recorded from May to August.

Status Very local, recent records are predominantly from southern England, the exception being from Staffordshire. Older records indicate a wider distribution in the southern half of Great Britain.

Threats The loss of bogs through drainage. Much heathland, including the wetter areas, has been lost, or fragmented, through changes in land use, mainly by conversion to arable, forestry and urban development. Further habitat degradation has been through the cessation of traditional heathland management practices.

Management and conservation Management should aim to maintain open conditions and encourage the early successional stages preferably by grazing or by rotational cutting, scraping or burning. Water tables should be maintained at a high level.

Published sources Allen, A.A. (1955c), Anon, (1924), Fowler, W.W. (1887), Fowler, W.W. & Donisthorpe, H.St J.K. (1913), Hodge, P.J. (1977c), Key, R.S. (1990), Lindroth, C.H. (1974), Luff, M.L. (Ed.) (1982).

AGONUM VERSUTUM   **NOTABLE B**
A ground beetle
Order COLEOPTERA   Family CARABIDAE

Agonum versutum Sturm, 1824. Formerly known as: *Anchomenus versutus* (Sturm).

Distribution South East, South, South West, East Anglia, East Midlands, West Midlands and North East.

Habitat and ecology The margins of lakes, ponds and gravel pits. Predatory. Found amongst lush vegetation on mud, in particular bulrushes, sedges and reeds. Adults have been recorded June.

Status Widespread but local in England, known from a single old record in Wales.

Threats This species is threatened by the infilling of ponds and pits, tourist development and natural succession.

Management and conservation Management should aim at maintaining open conditions and encouraging early successional stages along pit and lake edges.

AMARA ALPINA	RARE
A ground beetle	
Order COLEOPTERA	Family CARABIDAE

Amara alpina (Paykull, 1790). Formerly known as: *Cyrtonotus alpinus* (Paykull).

Distribution Recorded from Mid Perthshire, South Aberdeenshire, East Inverness & Nairn and North Ebudes before 1970 and East Inverness & Nairn from 1970 onwards.

Habitat and ecology Restricted to high mountains. Larvae are predatory while adults are phytophagous. Adults are found in grassland areas and have been recorded in June and July.

Status Scottish, with only one recent record. An arctic element of the British fauna.

Threats Erosion of montane habitat may be a problem in areas frequented by large numbers of hill walkers.

Published sources Fowler, W.W. (1887), Lindroth, C.H. (1974), Shirt, D.B., ed. (1987), Walker, J.J. (1900a).

AMARA CONSULARIS	NOTABLE B
A ground beetle	
Order COLEOPTERA	Family CARABIDAE

Amara consularis (Duftschmid, 1812).

Distribution England, South West Scotland and North East Scotland.

Habitat and ecology Heathland, sand dunes, undercliffs, gravel pits, calcareous grassland, arable fields and also in gardens. Larvae are predatory, while adults are phytophagous. A gregarious species. Adults

have been found in a variety of situations including under stones, under driftwood and at the roots of grasses. Adults have been recorded from February to June and in August and December. Possibly more frequent in the early spring or late winter.

Status Widespread but local throughout England and parts of Scotland.

Threats This species is threatened by destruction of habitat and through natural succession.

Management and conservation Management should aim at maintaining open conditions and encouraging early successional stages.

AMARA CURTA	NOTABLE B
A ground beetle	
Order COLEOPTERA	Family CARABIDAE

Amara curta Dejean, 1828.

Distribution South West, South, South East, East Anglia, West Midlands, North West, North East, South Wales and South East Scotland.

Habitat and ecology Heathland, sand dunes, calcareous grassland and also gravel pits. Larvae are predatory, while adults are phytophagous. Found on dry, stony ground, in sparsely vegetated areas on sand or gravel and on calcareous grassland. Also found under a piece of wood in a sand dune. Adults have been recorded in January, from April to September and in November.

Status Local and recently recorded from southern England as far north as North-east Yorkshire. Also noted in South Wales. This species is difficult to identify and may be confused with other members of the genus, especially *A. tibialis* and *A. infima*. Consequently, the exact status of this species is hard to assess.

Threats This species is threatened by destruction of habitat and through natural succession.

Management and conservation Management should aim at maintaining open conditions and encouraging early successional stages.

AMARA EQUESTRIS   NOTABLE B
A ground beetle
Order COLEOPTERA   Family CARABIDAE

Amara equestris (Duftschmid, 1812). Formerly known
as: *Amara patricia* (Duftschmid, 1812).

Distribution South East, South, South West, East
Anglia, West Midlands, Dyfed-Powys, North West,
North Wales and South East Scotland.

Habitat and ecology Heathland, sand dunes, gravel
pits and calcareous grassland, also recorded from a
wood. Larvae are predatory, while adults are
phytophagous. Found on dry, sandy or chalky soil, at
the roots of grass or under dry leaves. Adults have
been recorded from June to October.

Status Widespread but local throughout England and
Wales, also recorded in South East Scotland. This
species is difficult to identify and may be confused
with other members of the genus. Consequently, the
exact status of this species is hard to assess.

Threats This species is threatened by destruction of
habitat and through natural succession.

Management and conservation Management should
aim at maintaining open conditions and encouraging
early successional stages.

AMARA FAMELICA   RARE
A ground beetle
Order COLEOPTERA   Family CARABIDAE

Amara famelica Zimmerman, 1832.

Distribution Recorded from South Hampshire, Surrey,
North Essex, Berkshire, West Suffolk and South-east
Yorkshire before 1970 and East Sussex and North-east
Yorkshire from 1970 onwards.

Habitat and ecology Heathland. Larvae are predatory,
while adults are phytophagous. Found on dry, sandy
heaths. Adults have been recorded in April and May.

Status Not listed in the insect Red Data Book (Shirt,
1987). Recently recorded from only two vice-counties,
old records suggest that this species was more
widespread in England. Recorded as far north as North-
east Yorkshire. This species is difficult to identify and
may be confused with other members of the genus.
Consequently, the exact status of this species is hard to
assess.

Threats Much heathland has been lost, or fragmented,
through changes in land use, mainly by conversion to

arable, forestry and urban development. Further habitat
degradation has been through the cessation of
traditional heathland management practices.

Management and conservation Management should
aim to encourage early successional stages and
maintain open conditions, preferably by grazing or by
rotational cutting, scraping or burning.

Published sources Allen, A.A. (1950f), Fowler, W.W.
& Donisthorpe, H.St J.K. (1913), Hammond, P.M.
(1963), Hodge, P.J. (1977c), Lindroth, C.H. (1974).

AMARA FULVA   NOTABLE B
A ground beetle
Order COLEOPTERA   Family CARABIDAE

Amara fulva (Müller, 1776). Formerly known as:
Cyrtonotus fulvus (Müller).

Distribution South East, South, South West, East
Anglia, East Midlands, North East, North West, North
Wales, Dyfed-Powys and Scotland.

Habitat and ecology Heathland, sand dunes, sand and
gravel pits, riverbanks, river shingle, soft-rock cliffs,
Breckland and woodland rides. Larvae are predatory,
while adults are phytophagous. Found in sparsely
vegetated areas. In Breckland this beetle has been
found under old fence posts and from under pine logs.
Adults have been recorded from April to October.

Status Widespread but local throughout Great Britain.

Threats This species is threatened by destruction of
habitat and through natural succession.

Management and conservation Management should
aim at maintaining open conditions and encouraging
early successional stages.

AMARA FUSCA   ENDANGERED
A ground beetle
Order COLEOPTERA   Family CARABIDAE

Amara fusca Dejean, 1828. Formerly known as: *Amara
complanata var. fusca* Dejean.

Distribution Recorded from West Kent, Glamorgan,
South-west Yorkshire and South Northumberland
before 1970 and Durham from 1970 onwards.

Habitat and ecology Heathland and sand dunes.
Larvae are predatory, while adults are phytophagous.
Found on dry, sandy or gravelly soil with sparse
vegetation.

CARABIDAE

Status Status revised from RDB2 (Vulnerable) in Shirt (1987). Known from just two examples this century; one recorded in 1942 at Swanley, West Kent, the other in 1985 from Co. Durham. This species is difficult to identify and may be confused with other members of the genus.

Published sources Fowler, W.W. (1887), Fowler, W.W. & Donisthorpe, H.St J.K. (1913), Lindroth, C.H. (1974), Shirt, D.B., ed. (1987).

AMARA INFIMA	NOTABLE A
A ground beetle	
Order COLEOPTERA	Family CARABIDAE

Amara infima (Duftschmid, 1812).

Distribution Recorded from Dorset, Isle of Wight, "Hants", East Kent, Surrey, "Norfolk", Bedfordshire and North Lincolnshire before 1970 and Surrey, West Suffolk, West Norfolk and Glamorgan from 1970 onwards.

Habitat and ecology Heathland and sand dunes. Larvae are predatory, while adults are phytophagous. Found in open conditions on dry sandy or gravelly soil, under mats of heather etc.. Adults have been recorded in January and from May to October.

Status Very local. Only known from the southern half of Great Britain.

Threats This species is threatened by destruction of habitat and through natural succession.

Management and conservation Management should aim at maintaining open conditions and encouraging early successional stages.

Published sources Allen, A.A. (1950f), Buck, F.D. (1959a), Fowler, W.W. (1887), Fowler, W.W. & Donisthorpe, H.St J.K. (1913), Key, R.S. (1990), Lindroth, C.H. (1974), Mendel, H. (1980).

AMARA LUCIDA	NOTABLE B
A ground beetle	
Order COLEOPTERA	Family CARABIDAE

Amara lucida (Duftschmid, 1812).

Distribution England, Dyfed-Powys, North Wales and North East Scotland.

Habitat and ecology Predominantly coastal. Sand dunes, sandy areas and coastal shingle. Rarely found on calcareous grassland and in Breckland. Larvae are predatory, while adults are phytophagous. Found under leaves of dune plants etc.. Adults have been recorded from March to September.

Status Widespread but local.

Threats Loss of habitat, particularly through afforestation, urban and holiday development and gravel extraction. The degradation of sand dunes and other sandy areas by excessive disturbance of the vegetation through activities such as motorbike access, horse-riding and human trampling. This species is also threatened by natural succession.

Management and conservation Management should aim at maintaining open conditions and encouraging early successional stages. Disturbance of coastal shingle systems should be avoided.

AMARA NITIDA	NOTABLE A
A ground beetle	
Order COLEOPTERA	Family CARABIDAE

Amara nitida Sturm, 1825.

Distribution Recorded from North Somerset, Dorset, Surrey, Middlesex, Warwickshire, Glamorgan, Montgomeryshire, South Lincolnshire and Cumberland before 1970 and South Hampshire, West Gloucestershire, Merionethshire and Durham from 1970 onwards.

Habitat and ecology Heathland, sand dunes, chalk downland, and also gravel pits. Larvae are predatory, while adults are phytophagous. Recorded both in well vegetated and sparsely vegetated situations. Adults have been recorded in September.

Status Very local with a widely scattered distribution in England and Wales. This species is difficult to identify and may be confused with other members of the genus. Consequently, the exact status of this species is hard to assess.

Threats This species is threatened by destruction of habitat and through natural succession.

Management and conservation Management should aim at maintaining open conditions and encouraging early successional stages.

Published sources Allen, A.A. (1984a), Davidson, W.F. (1960a), Davidson, W.F. (1960b), Donisthorpe, H.St J.K. (1945a), Fowler, W.W. (1887), Fowler, W.W. & Donisthorpe, H.St J.K. (1913), Lindroth, C.H. (1974).

AMARA PRAETERMISSA — NOTABLE B
A ground beetle
Order COLEOPTERA — Family CARABIDAE

Amara praetermissa (Sahlberg, 1827). Formerly known as: *Amara rufocincta* (Sahlberg, 1827).

Distribution England, Wales and Scotland.

Habitat and ecology Chalk grassland and other habitats on chalk, sand dunes, sandy heaths and arable field margins. Also recorded from old railway tracks and in an ornamental deciduous woodland. Larvae are predatory, while adults are phytophagous. Found on chalky and also on gravelly soil, often under dry leaves. Adults have been recorded from April to August.

Status Widespread but local throughout Great Britain.

Threats This species is threatened by destruction of habitat and through natural succession.

Management and conservation Management should aim at maintaining open conditions and encouraging early successional stages.

AMARA QUENSELI — NOTABLE A
A ground beetle
Order COLEOPTERA — Family CARABIDAE

Amara quenseli (Schoenherr, 1806). Formerly known as: *Amara quenselii* (Schoenherr, 1806), *silvicola*.

Distribution Recorded from South Aberdeenshire, Moray and Mid Ebudes before 1970 and Moray, East Inverness & Nairn and North Ebudes from 1970 onwards.

Habitat and ecology Sandy river banks, sand-pits, roadside verges and probably other situations in upland areas. Larvae are predatory, while adults are phytophagous. Found on sandy or gravelly soil with sparse vegetation, often under stones. Adults have been recorded in June and July.

Status Very local and confined to the Highlands of Scotland.

Threats This species is threatened by destruction of habitat and through natural succession.

Management and conservation Management should aim at maintaining open conditions and encouraging early successional stages.

Published sources Cooter, J. & Owen, J.A. (1978), Lindroth, C.H. (1974).

AMARA SPRETA — NOTABLE B
A ground beetle
Order COLEOPTERA — Family CARABIDAE

Amara spreta Dejean, 1831.

Distribution South East, South, South West, East Anglia, East Midlands, West Midlands, North East, South Wales, North Wales and South East Scotland.

Habitat and ecology Sand dunes. Larvae are predatory, while adults are phytophagous. Found on dry, loose sand amongst marram. Adults have been recorded from April to September.

Status Widespread but local in England and Wales, also recorded in South East Scotland. This species is difficult to identify and may be confused with other members of the genus. Consequently, the exact status of this species is hard to assess.

Threats Loss of dune habitat, particularly through afforestation, urban and holiday development. The degradation of remaining habitat by excessive disturbance of the vegetation through activities such as motorbike access, horse-riding and human trampling.

Management and conservation On dunes some grazing and other disturbance may be desirable to maintain early successional stages and prevent the invasion of scrub.

AMARA STRENUA — RARE
A ground beetle
Order COLEOPTERA — Family CARABIDAE

Amara strenua Zimmermann, 1832.

Distribution Recorded from North Somerset, Isle of Wight, East Kent, West Kent and South Essex before 1970 and South Devon, North Somerset and East Kent from 1970 onwards.

Habitat and ecology Saltmarshes, though also recorded inland. Larvae are predatory, while adults are phytophagous. Probably found amongst saltmarsh vegetation. Adults have been recorded in May and June.

Status Not listed in the insect Red Data Book (Shirt, 1987). Very local and recently recorded in only three vice-counties in southern England.

CARABIDAE

Threats Loss of saltmarsh through reclamation, erosion and the construction of sea defences.

Published sources Allen, A.A. (1950f), Fowler, W.W. (1887), Fowler, W.W. & Donisthorpe, H.St J.K. (1913), Lindroth, C.H. (1974).

ANISODACTYLUS NEMORIVAGUS NOTABLE A
A ground beetle
Order COLEOPTERA   Family CARABIDAE

Anisodactylus nemorivagus (Duftschmid, 1812). Formerly known as: *Anisodactylus binotatus var. atricornis* (Stephens, 1835).

Distribution Recorded from South Wiltshire, Dorset, South Hampshire, Surrey, Middlesex, Berkshire, West Suffolk, East Norfolk and Glamorgan before 1970 and South Wiltshire, South Hampshire and Surrey from 1970 onwards.

Habitat and ecology Dry, sandy heaths. Probably phytophagous. Adults probably occur at the roots of plants.

Status Known only from southern England and South Wales. Recently recorded from only three vice-counties.

Threats Much heathland has been lost, or fragmented, through changes in land use, mainly by conversion to arable, forestry and urban development. Further habitat degradation has been through the cessation of traditional heathland management practices.

Management and conservation Management should aim to encourage early successional stages and maintain open conditions, preferably by grazing or by rotational cutting, scraping or burning.

Published sources Fowler, W.W. (1887), Fowler, W.W. & Donisthorpe, H.St J.K. (1913), Lindroth, C.H. (1974).

ANISODACTYLUS POECILOIDES   RARE
A ground beetle
Order COLEOPTERA   Family CARABIDAE

Anisodactylus poeciloides (Stephens, 1828).

Distribution Recorded from West Cornwall, Dorset, Isle of Wight, South Hampshire, West Sussex, East Sussex, East Kent, West Kent, South Essex and North Essex before 1970 and West Sussex, East Kent and West Kent from 1970 onwards.

Habitat and ecology Saltmarshes, salt pans and brackish ditches at the margins of grazing levels. Probably phytophagous. Adults probably occur at the roots of plants. and have been recorded from April to July.

Status Not listed in the insect Red Data Book (Shirt, 1987). Known only from southern England. Very local and recently recorded from only three vice-counties, all in South East England.

Threats Loss of saltmarsh through reclamation, erosion and the construction of sea defences, and the loss of grazing levels through drainage and conversion to arable agriculture. Overgrazing of saltmarshes could be a further threat to this species.

Management and conservation Grazing should not be introduced to a site where there is no grazing at present.

Published sources Fowler, W.W. (1887), Fowler, W.W. & Donisthorpe, H.St J.K. (1913), Lindroth, C.H. (1974), Whicher, L.S. (1952c).

ASAPHIDION PALLIPES   NOTABLE B
A ground beetle
Order COLEOPTERA   Family CARABIDAE

Asaphidion pallipes (Duftschmid, 1812). Formerly known as: *Tachypus pallipes* (Duftschmid).

Distribution South West, South, East Anglia, East Midlands, West Midlands, North East, North West, Wales, South West Scotland, North East Scotland and North West Scotland. A map is given in Luff (1982).

Habitat and ecology Coastal undercliffs, river shingle and the banks of streams. Predatory. Found in sandy soil and in shingle. Adults have been recorded from June to August and in October.

Status Widespread and very local throughout Great Britain.

Threats This species is threatened by cliff stabilisation schemes, the construction of sea defences, river engineering schemes and natural succession. Coastal developments may reduce the amount of available habitat. Activities that accelerate or reduce the rate of erosion should be avoided. The degradation of suitable habitat through natural succession and the invasion of scrub on stabilised areas is a further threat.

Management and conservation Occasional slippages are necessary to maintain habitat continuity. Large areas of unstable cliff are required so that the population does not become isolated and subsequently threatened by individual landslips. Activities that hinder the natural flow of rivers and streams should be avoided.

BADISTER ANOMALUS   **ENDANGERED**
A ground beetle
Order COLEOPTERA   Family CARABIDAE

Badister anomalus (Perris, 1866).

Distribution Recorded from Dorset, East Sussex, "Kent" and East Suffolk before 1970 and East Sussex and West Kent from 1970 onwards.

Habitat and ecology Marshes, boggy areas at the edge of water-filled sand pits and gravel pits, and on shaded margins of fresh, standing water. Probably predatory. Found on mud or in lush vegetation, often amongst bulrushes, sedges and reeds. Adults have been recorded from May to July. This species has been recorded at a mercury vapour light trap.

Status Not listed in the insect Red Data Book (Shirt, 1987). Known only from four vice-counties in southern and eastern England. Recently recorded on a number of occasions from only one site in East Sussex, and a single example at a light trap in West Kent. This species is difficult to identify and may be confused with other members of the genus. Consequently, the exact status of this species is hard to assess. Consequently, it may be under-recorded.

Threats This species is threatened by the infilling of gravel pits and through natural succession.

Management and conservation Management should aim at maintaining open conditions and encouraging early successional stages along pit edges.

Published sources Allen, A.A. (1957b), Allen, A.A. (1990e), Kevan, D.K. (1955a), Kevan, D.K. (1955b), Lindroth, C.H. (1974).

BADISTER DILATATUS   **NOTABLE B**
A ground beetle
Order COLEOPTERA   Family CARABIDAE

Badister dilatatus Chaudoir, 1837.

Distribution South, South West, South East, East Anglia, East Midlands and Wales.

Habitat and ecology Fens, grazing levels, ponds, gravel pits, reed marshes in sand dunes and on shaded margins of fresh, standing water. Probably predatory.

Found on mud or in lush vegetation, often amongst bulrushes, sedges and reeds. Adults have been recorded from March to August and in November.

Status Widespread but local in the southern half of England and in Wales. This species is difficult to identify and may be confused with other members of the genus. Consequently, the exact status of this species is hard to assess.

Threats Drainage for reasons such as agricultural improvement and development is the primary cause of the loss of wetlands. The infilling of gravel pits, natural succession and scrub invasion, and falling water tables because of water abstraction and river engineering schemes may also threaten this species.

Management and conservation Water tables should be maintained at high levels. Water bodies should be isolated from sources of eutrophication and pollution. Management should aim at maintaining open conditions and encouraging early successional stages.

BADISTER MERIDIONALIS   **INDETERMINATE**
A ground beetle
Order COLEOPTERA   Family CARABIDAE

Badister meridionalis Puel, 1925.

Distribution Recorded from Oxfordshire and East Gloucestershire before 1970 and Oxfordshire and East Gloucestershire from 1970 onwards.

Habitat and ecology Bare margins near standing water such as lake shores and in gravel pits. Probably predatory. The adult has been recorded in May.

Status Not listed in the insect Red Data Book (Shirt, 1987). Two recent records. Prior to this, this species was known only from three old records; a pair from the Oxford district recorded by J.J. Walker, and a single male from Tewkesbury, Gloucestershire recorded in 1935 by C.E. Tottenham. These examples were previously considered to be *B. bipustulatus*.

Threats This species may be threatened by the infilling of lakes and pits, and through natural succession.

Published sources Lindroth, C.H. (1971), Lindroth, C.H. (1974).

CARABIDAE

BADISTER PELTATUS　　　　　**NOTABLE A**
A ground beetle
Order COLEOPTERA　　　　Family CARABIDAE

Badister peltatus (Panzer, 1797).

Distribution Recorded from West Cornwall, Dorset, Isle of Wight, South Hampshire, West Sussex, East Sussex, East Kent, Surrey, Middlesex, Berkshire, West Suffolk, West Norfolk, Cambridgeshire and South Lincolnshire before 1970 and East Sussex, East Kent, West Kent, South Essex and West Norfolk from 1970 onwards.

Habitat and ecology Grazing levels, boggy ground at the edge of a water-filled sand pit and on shaded margins of fresh, standing water. Probably predatory. Found on mud or in lush vegetation, often amongst bulrushes, sedges and reeds. Adults have been recorded in May and June.

Status Very local. Known only from the southern half of England. Recently recorded from only five vice-counties, all in southern and eastern England. This species is difficult to identify and may be confused with other members of the genus. Consequently, the exact status of this species is hard to assess.

Threats Drainage for reasons such as agricultural improvement and development is the primary cause of the loss of wetlands. The infilling of pits, natural succession and scrub invasion, and falling water tables because of water abstraction and river engineering schemes may also threaten this species.

Management and conservation Water tables should be maintained at high levels. Water bodies should be isolated from sources of eutrophication and pollution. Management should aim at maintaining open conditions and encouraging early successional stages.

Published sources Fowler, W.W. (1887), Fowler, W.W. & Donisthorpe, H.St J.K. (1913), Kevan, D.K. (1955b), Lindroth, C.H. (1974).

BADISTER UNIPUSTULATUS　　　　**NOTABLE B**
A ground beetle
Order COLEOPTERA　　　　Family CARABIDAE

Badister unipustulatus Bonelli, 1813.

Distribution England.

Habitat and ecology Marshes, fens, willow carr and rides in wet woodland. Probably predatory. Found amongst lush vegetation usually near water, and in reed litter. Adults have been recorded in January, from May to September and in December.

Status Widespread but local throughout England.

Threats Drainage for reasons such as agricultural improvement and development is the primary cause of the loss of marshes and fens. Clear-felling and conversion to other land use, and falling water tables because of water abstraction and river engineering schemes may also threaten this species.

Management and conservation Water tables should be maintained at high levels. Water bodies should be isolated from sources of eutrophication and pollution. Management should aim at maintaining open conditions and encouraging early successional stages.

BEMBIDION ARGENTEOLUM
A ground beetle　　　**INSUFFICIENTLY KNOWN**
Order COLEOPTERA　　　　Family CARABIDAE

Bembidion argenteolum Ahrens, 1812.

Distribution Recorded from East Kent from 1970 onwards.

Habitat and ecology Recorded from a sandy area on the coast. Found in sand by a gravel pit, while in Ireland it was recorded from the shores of a loch. In Ireland, adults have been recorded in June and July, in England, this species has been recorded in August.

Status Not listed in the insect Red Data Book (Shirt, 1987). Formerly known in the British Isles from only three localities on the shores of Loch Neagh, Ireland where, despite an intensive searches, it has not been recorded since 1923. In 1987 a single example of this species was found in a pupal cell on the East Kent coast. Not yet confirmed as an established species in England, and possibly only an immigrant that can establish transitory populations.

Threats Uncertain, though development and natural succession may be threats.

Management and conservation Some disturbance is needed at pit edges to maintain open conditions and encourage early successional stages.

Published sources Johnson, W.F. (1902), Kemp, S.W. (1902).

BEMBIDION BIPUNCTATUM — NOTABLE B
A ground beetle
Order COLEOPTERA   Family CARABIDAE

Bembidion bipunctatum (L., 1761). Formerly known as: *Bembidium bipunctatum* (L., 1761).

Distribution South East, West Midlands, North East, North West, Dyfed-Powys, South Wales and Scotland.

Habitat and ecology Near the margins of water, both inland and on the coast. Predatory. Adults generally occur in sparsely vegetated or bare areas, such as gravel and shingle on the shoreline. Adults have been recorded from May to July.

Status Widespread but local, with a predominantly northern distribution.

Threats This species may be threatened by urban and holiday developments as well as through river engineering schemes, including dredging, level regulation by damming, flood alleviation and the conversion of lakes to reservoirs. In some areas colonisation by Himalayan balsam *Impatiens glandulifera* and natural succession can reduce the available habitat. River pollution maybe a further threat.

Management and conservation An under-valued habitat. Shingle and sand banks tend to be mobile and rely on the free flow of river and stream systems and wave action of lakes etc.. Activities that hinder this flow should be prevented. Disturbance of vegetated coastal shingle should be avoided.

BEMBIDION CLARKI — NOTABLE B
A ground beetle
Order COLEOPTERA   Family CARABIDAE

Bembidion clarki Dawson, 1849. Formerly known as: *Bembidion transparens var. clarki* Dawson, *Bembidium clarki* Dawson.

Distribution England, North Wales and South East Scotland.

Habitat and ecology Most frequently found by the coast. Marshes, willow carr and the margins of standing water. Predatory. Found amongst lush, low vegetation on soft soil or mud at the margins of ponds, lakes, reservoirs, gravel pits and also under sallow bark. Adults probably all year round, and have been recorded from February to December, though mainly noted in May.

Status Widespread but local. Recorded as far north as South East Scotland.

Threats Drainage for reasons such as agricultural improvement and development is the primary cause of the loss of wetlands. Falling water tables because of water abstraction and river engineering schemes, and the infilling of gravel pits etc. and natural succession are further threats.

Management and conservation Water tables should be maintained at high levels. Water bodies should be isolated from sources of eutrophication and pollution.

BEMBIDION EPHIPPIUM — NOTABLE A
A ground beetle
Order COLEOPTERA   Family CARABIDAE

Bembidion ephippium (Marsham, 1802). Formerly known as: *Bembidium ephippium* (Marsham, 1802).

Distribution Recorded from "Cornwall", South Hampshire, East Sussex, East Kent, West Kent, South Essex, North Essex, "Suffolk", East Norfolk and West Norfolk before 1970 and West Sussex, East Sussex, East Kent, South Essex, East Norfolk, West Norfolk and North Lincolnshire from 1970 onwards.

Habitat and ecology Coastal. On the seashore and in saltmarshes. Predatory and gregarious. Found amongst plant and other debris, on fine sandy silt, on open areas of saline mud near pools, in crevices and among rocks wedged against breakwaters. Adults have been recorded in April, May, July and September.

Status Very local. Distributed along the southern and eastern coasts of England, north to North Lincolnshire.

Threats This species is threatened by urban and holiday development, erosion, the construction of coastal defences and land reclamation schemes.

Published sources Collier, M.J. (1988a), Flint, J.H. (1947), Fowler, W.W. (1887), Hodge, P.J. (1977d), Kirby, P. & Lambert, S.J.J. (1989), Lindroth, C.H. (1974), Whicher, L.S. (1953).

CARABIDAE

BEMBIDION FLUVIATILE — NOTABLE B
A ground beetle
Order COLEOPTERA   Family CARABIDAE

Bembidion fluviatile Dejean, 1831. Formerly known as: *Bembidium fluviatile* Dejean, 1831.

Distribution South, East Midlands, West Midlands, North East, North West, South Wales and Dyfed-Powys.

Habitat and ecology River margins and gravel pits. Predatory. Occurs in open areas of clay or sand alongside the waters-edge. Adults are often concealed in cracks and under stones, and frequently burrow in mud. Adults have been recorded in April, May and from August to October.

Status Local in England and Wales. A difficult species to identify and may be confused with other species of the Subgenus *Peryphus* (*Bembidion*).

Threats River engineering, including dredging, level regulation by damming and flood alleviation schemes, and the infilling of gravel pits. In some areas colonisation by Himalayan balsam *Impatiens glandulifera* and natural succession can reduce the available habitat. River pollution is a further threat.

Management and conservation An under-valued habitat. Sand banks tend to be mobile and rely on the free flow of river and stream systems. Activities that hinder this flow should be prevented. Adjacent scrub and rank grassland may be important for the species in winter and to escape late summer floods. Management should aim at maintaining open conditions and encouraging early successional stages along pit edges.

BEMBIDION FUMIGATUM — NOTABLE B
A ground beetle
Order COLEOPTERA   Family CARABIDAE

Bembidion fumigatum (Duftschmid, 1812). Formerly known as: *Bembidium fumigatum* (Duftschmid, 1812).

Distribution South East, South, South West, East Anglia, East Midlands, North West, North East and South Wales.

Habitat and ecology On well vegetated margins of ponds, ditches in fens and other inland situations. Also on the banks of estuaries and on the coast. Predatory. Found amongst wet debris in fens, reed and sedge litter in reed beds, in pond litter and marsh vegetation. Adults have been recorded from February to August and in December.

Status Widespread but local throughout England, also recorded in South Wales.

Threats Drainage for reasons such as agricultural improvement and development is the primary cause of the loss of wetlands and fens. Falling water tables because of water abstraction and river engineering schemes, coastal developments and the conversion of lakes to reservoirs may also threaten this species.

Management and conservation Water tables should be maintained at high levels. Water bodies should be isolated from sources of eutrophication and pollution.

BEMBIDION GILVIPES — NOTABLE B
A ground beetle
Order COLEOPTERA   Family CARABIDAE

Bembidion gilvipes Sturm, 1825. Formerly known as: *Bembidium gilvipes* Sturm, 1825.

Distribution England, South Wales and Dyfed-Powys.

Habitat and ecology River banks, lake margins, gravel pits, lowland peat pools, fens and marshes. Predatory. Found on partially vegetated silt and muddy sand. Also noted in flood debris. Adults probably all year round, but have been recorded in January, February and from April to December.

Status Widespread but local in England and Wales.

Threats Drainage of fens and marshes, the infilling of gravel pits, conversion of lakes to reservoirs, water abstraction schemes and river engineering, including dredging, level regulation by damming and flood alleviation schemes. In some areas colonisation by Himalayan balsam *Impatiens glandulifera* can reduce the available habitat. River pollution maybe a further threat.

Management and conservation In fens and marshes water tables should be maintained at high levels. Water bodies should be isolated from sources of eutrophication and pollution. Activities that hinder the natural flow of rivers and streams should be avoided. Management should aim at maintaining open conditions and encouraging early successional stages along pit edges.

BEMBIDION HUMERALE — ENDANGERED
A ground beetle
Order COLEOPTERA — Family CARABIDAE

Bembidion humerale Sturm, 1825.

Distribution Recorded from South-west Yorkshire from 1970 onwards.

Habitat and ecology Lowland peat bogs. Predatory. Adults are found on peat, particularly where the peat is moist and largely bare of vegetation, usually with a covering of algae and a fringe of overhanging vegetation i.e. damp hollows by peat pools. Adults have been recorded in April and from August to October.

Status Discovered new to Great Britain in 1975 from Thorne Moors near Doncaster, South-west Yorkshire. In 1983 it was found at a further site, Hatfield Moors, South-west Yorkshire, approximately 10 km south of Thorne. The beetle is still extremely localized though may be abundant where found.

Threats Both sites have been very badly damaged by drainage and commercial stripping of peat. This species remains on only a tiny remnant of Hatfield Moors. Drainage at Thorne Moors has accelerated in the late 1980's. Further threatened by recolonisation of the dry, cut peat surface by bracken and birch.

Management and conservation Water tables should be maintained to keep surface peat moist. The species may benefit from bare peat created by small scale turf cutting, provided this is not accompanied by drainage. Thorne and Hatfield Moors are both notified as SSSI's, part of Thorne Moors is an NNR.

Published sources Crossley, R. & Norris, A. (1975), Shirt, D.B., ed. (1987).

BEMBIDION LATERALE — NOTABLE B
A ground beetle
Order COLEOPTERA — Family CARABIDAE

Bembidion laterale (Samouelle, 1819). Formerly known as: *Cillenus lateralis* (Samouelle).

Distribution South East, South, South West, East Anglia, East Midlands, North East, North West, Wales and Scotland.

Habitat and ecology Coastal habitats, particularly saltmarshes and river estuaries. Predatory on sandhoppers, though it has also been found in the burrows of *Bledius tricornis* (Staphylinidae). Occurs in estuarine mud and silt and also found in wet sand on the seashore. Adults have been recorded in April and from June to August.

Status Widespread but local throughout Great Britain.

Threats This species may be threatened by holiday and urban developments as well as land reclamation schemes, erosion and the construction of coastal defences.

BEMBIDION LITORALE — NOTABLE B
A ground beetle
Order COLEOPTERA — Family CARABIDAE

Bembidion litorale (Olivier, 1790). Formerly known as: *Bembidium paludosum* (Panzer, 1794), *Bracteon litorale* (Olivier).

Distribution East Midlands, West Midlands, North East, North West, South Wales, South East Scotland and North East Scotland.

Habitat and ecology The margins of rivers and streams, and rarely standing water. Particularly associated with sand banks and shingle beds. Predatory. Associated with sparsely vegetated or areas of bare sand, silty mud or gravel. Adults have been recorded from April to August.

Status Very local and occasionally common where found. This species has a northern distribution with a few records from the Midlands.

Threats River engineering, including dredging, level regulation by damming and flood alleviation schemes. In some areas colonisation by Himalayan balsam *Impatiens glandulifera* can reduce the available habitat. Livestock access can damage shingle structure and alter vegetation communities. River pollution maybe a further threat.

Management and conservation An under-valued habitat. Shingle and sand banks tend to be mobile and rely on the free flow of river and stream systems. Activities that hinder this flow should be prevented. Adjacent scrub and rank grassland may be important for the species in winter and to escape late summer floods.

CARABIDAE

BEMBIDION LUNATUM   NOTABLE B
A ground beetle
Order COLEOPTERA   Family CARABIDAE

Bembidion lunatum (Duftschmid, 1812). Formerly known as: *Bembidium lunatum* (Duftschmid, 1812).

Distribution South East, South West, East Anglia, East Midlands, West Midlands, North East, North West and South West Scotland.

Habitat and ecology River banks, estuaries and undercliffs. Predatory. Recorded from under stones on bare riverside mud, in tidal refuse, among shingle in inter-tidal areas of estuaries and at the base of soft-rock cliffs. The beetle hibernates as a larva with new generation adults emerging in the late spring. Adults have been recorded in March and from May to September, though mainly in late July and August.

Status Widespread but local in England, recorded in South West Scotland.

Threats This species is threatened by river engineering and cliff stabilisation schemes, and the construction of sea defences. Coastal developments may reduce the amount of available habitat. Activities that accelerate or reduce the rate of erosion should be avoided. The degradation of suitable habitat through natural succession and the invasion of scrub on stabilised areas is a further threat.

Management and conservation In areas of soft rock cliff occasional slippages are necessary to maintain habitat continuity. Large areas of unstable cliff are required so that the population does not become isolated and subsequently threatened by individual landslips. Activities that hinder the natural flow of rivers should be avoided.

BEMBIDION MONTICOLA   NOTABLE B
A ground beetle
Order COLEOPTERA   Family CARABIDAE

Bembidion monticola Sturm, 1825. Formerly known as: *Bembidium monticola* Sturm, 1825.

Distribution South East, South, South West, West Midlands, North East, North West, Dyfed-Powys, North Wales, South East Scotland and South West Scotland.

Habitat and ecology Shaded places, usually near running water. Predatory. Found mainly in cracks and crevices in clay banks on rivers, though also recorded from river shingle and under the bark of tree stumps on river banks. Adults have been recorded from March to October.

Status Widespread but local, not recorded in the far north of Scotland. This species is difficult to identify and may be confused with a other species of the Subgenus *Peryphus* (*Bembidion*). Consequently, the exact status of this species is hard to assess.

Threats River engineering, including dredging, level regulation by damming and flood alleviation schemes. In some areas colonisation by Himalayan balsam *Impatiens glandulifera* can reduce the available habitat. Livestock access can damage shingle structure. River pollution is a further threat.

Management and conservation An under-valued habitat. Activities that hinder the natural flow of rivers and streams should be prevented.

BEMBIDION NIGRICORNE   NOTABLE B
A ground beetle
Order COLEOPTERA   Family CARABIDAE

Bembidion nigricorne Gyllenhal, 1827. Formerly known as: *Bembidium nigricorne* Gyllenhal, 1827.

Distribution England. A map is given in Luff (1982).

Habitat and ecology Heathland and moorland. One example has been found in a clay pit. Predatory. Associated with *Calluna* heathland, particularly poorly vegetated areas of young heather and burnt heathland. Adults occur under heather and stones, and have been recorded in February, from April to August and in November and December.

Status Widespread but local in England.

Threats Much heathland has been lost, or fragmented, through changes in land use, mainly by conversion to arable, forestry and urban development. Further habitat degradation has been through the cessation of traditional heathland management practices. This species is further threatened by the loss of moorland through afforestation.

Management and conservation Management should aim at encouraging the early successional stages, preferably by grazing or by rotational cutting, scraping or burning.

BEMBIDION NIGROPICEUM — NOTABLE A
A ground beetle
Order COLEOPTERA   Family CARABIDAE

Bembidion nigropiceum (Marsham, 1802). Formerly known as: *Lymnaeum nigropiceum* (Marsham).

Distribution Recorded from "Cornwall", South Devon, Dorset, West Kent, South Essex and "Suffolk" before 1970 and Dorset, Isle of Wight, West Sussex, East Sussex, East Kent and West Kent from 1970 onwards.

Habitat and ecology Coastal. Recorded on the sea-shore and from limestone cliffs. Predatory. Found under stones on rocky or sandy sea-shores near the high-water mark. Adults have been recorded from March to June.

Status Very local and recorded along the coast from Dorset to West Kent.

Threats This species is threatened by urban and holiday developments, and possibly marine pollution.

Published sources Allen, A.A. (1989e), Hodge, P.J. (1977d), Lindroth, C.H. (1974).

BEMBIDION OBLIQUUM — NOTABLE B
A ground beetle
Order COLEOPTERA   Family CARABIDAE

Bembidion obliquum Sturm, 1825. Formerly known as: *Bembidium obliquum* Sturm, 1825.

Distribution South East, South, East Anglia, East Midlands, West Midlands, North East and North West.

Habitat and ecology Margins of fresh water ponds, lakes and reservoirs. On acidic soils. Predatory. Appears to prefer open areas of mud that are sparsely vegetated. Adults have been recorded from May to August.

Status Widespread but local in England.

Threats This species is threatened by drainage and natural succession.

Management and conservation Management should aim at maintaining open conditions and encouraging early successional stages.

BEMBIDION OCTOMACULATUM — EXTINCT
A ground beetle
Order COLEOPTERA   Family CARABIDAE

Bembidion octomaculatum (Goeze, 1777). Formerly known as: *Bembidium sturmi* (Panzer, 1804).

Distribution Recorded from Isle of Wight, "Hants", West Sussex, East Kent, Surrey and Middlesex.

Habitat and ecology Margins of fresh water, usually small pools. This species has also been found on the seashore. Predatory.

Status A presumed extinct resident and a possible immigrant. Last recorded from 1875 from Mickleham, Surrey.

Published sources Allen, A.A. (1965d), Lindroth, C.H. (1974), Shirt, D.B., ed. (1987).

BEMBIDION PALLIDIPENNE — NOTABLE B
A ground beetle
Order COLEOPTERA   Family CARABIDAE

Bembidion pallidipenne (Illiger, 1802). Formerly known as: *Bembidium pallidipenne* (Illiger, 1802).

Distribution South East, South, South West, East Anglia, East Midlands, North East, North West, Wales and Scotland. A map is given in Luff (1982).

Habitat and ecology Primarily coastal. On sandy beaches, sand flats backed by dunes, gravel pits and in estuarine areas. Predatory and gregarious. Found on hard inter-tidal sand, next to trickles at the tops of beaches and on damp sand with a coating of algae. Adults have been recorded from April to August.

Status Widespread but local throughout Great Britain.

Threats This species may be threatened at coastal sites by urban and holiday development, and at inland sites by natural succession and the infilling of gravel pits.

Management and conservation Management should aim at maintaining open conditions and encouraging early successional stages along gravel pit edges.

CARABIDAE

BEMBIDION QUADRIPUSTULATUM   NOTABLE B
A ground beetle
Order COLEOPTERA   Family CARABIDAE

Bembidion quadripustulatum Serville, 1821. Formerly known as: *Bembidium quadripustulatum* Serville, 1821.

Distribution South East, South West, East Anglia, East Midlands, West Midlands, North East and North West.

Habitat and ecology Rivers, lakes, pond and gravel pits. Predatory. Found on bare ground and mud at the margins of standing and running water. Adults have been recorded from May to October.

Status Widespread but local in England. Occasionally common where found.

Threats The infilling of gravel pits, the conversion of lakes to reservoirs, and river engineering, including dredging, level regulation by damming and flood alleviation schemes. In some areas colonisation by Himalayan balsam *Impatiens glandulifera* can reduce the available habitat. River pollution maybe a further threat.

Management and conservation Activities that hinder the natural flow of rivers and streams should be avoided. Management should aim at maintaining open conditions and encouraging early successional stages along lake and pit edges.

BEMBIDION SAXATILE   NOTABLE B
A ground beetle
Order COLEOPTERA   Family CARABIDAE

Bembidion saxatile Gyllenhal, 1827. Formerly known as: *Bembidium saxatile* Gyllenhal, 1827.

Distribution England, Dyfed-Powys, South Wales, South East Scotland and South West Scotland.

Habitat and ecology Undercliffs, both coastal and inland. Predatory. Usually occurs among boulder clay debris and under stones at the base of cliffs. Often found near trickles and seepages, particularly on clay and stone mixtures, though may also be found under stones on the seashore. Adults have been recorded from April to November.

Status Widespread but local.

Threats This species is threatened by cliff stabilisation schemes and the construction of sea defences. Coastal developments may reduce the amount of available habitat. Activities that accelerate or reduce the rate of erosion should be avoided. The degradation of suitable habitat through natural succesion and the invasion of scrub on stabilised areas is a further threat.

Management and conservation Occasional slippages are necessary for maintaining habitat continuity. Large areas of unstable cliff are required so that the population does not become isolated and subsequently threatened by individual landslips.

BEMBIDION SCHUEPPELI   NOTABLE A
A ground beetle
Order COLEOPTERA   Family CARABIDAE

Bembidion schueppeli Dejean, 1831. Formerly known as: *Bembidion schuppeli* Dejean, 1831, *Bembidium schuppeli* Dejean, 1831.

Distribution Recorded from Durham, South Northumberland, North Northumberland, Westmorland & North Lancashire, Cumberland, Dumfriesshire, Peeblesshire, Selkirkshire, Roxburghshire, East Lothian, Midlothian, West Lothian and Stirlingshire before 1970 and Durham, South Northumberland, North Northumberland, Selkirkshire, Roxburghshire and Berwickshire from 1970 onwards.

Habitat and ecology A riverbank species. Usually occurs away from the coast in Britain, though also in coastal habitats and in marshy woodland on the Continent. Predatory. Found on damp, fine sand and silt or fine shingle with a 50-100% cover of low herbage. In both open and shady conditions. Adults have been recorded in June.

Status Very local. Found in northern England and southern Scotland.

Threats River engineering, including dredging, level regulation by damming and flood alleviation schemes. In some areas colonisation by Himalayan balsam *Impatiens glandulifera* can reduce the available habitat. Livestock access can damage shingle structure and disturb vegetation communities. River pollution maybe a further threat.

Management and conservation An under-valued habitat. Shingle and sand banks tend to be mobile and rely on the free flow of river and stream systems. Activities that hinder this flow should be avoided. Adjacent scrub and rank grassland may be important for the species in winter and to escape late summer floods.

Published sources Lindroth, C.H. (1974), Lindroth, C.H. (1985), Reid, C.A.M. (1985).

BEMBIDION SEMIPUNCTATUM   NOTABLE A
A ground beetle
Order COLEOPTERA   Family CARABIDAE

Bembidion semipunctatum Donovan, 1806. Formerly known as: *Bembidion adustum* Schaum, 1860, *Bembidium adustum* Schaum.

Distribution Recorded from Cambridgeshire, East Gloucestershire, Herefordshire, Worcestershire, Warwickshire, Glamorgan and Durham before 1970 and East Suffolk, East Gloucestershire and Worcestershire from 1970 onwards.

Habitat and ecology River margins. Predatory. Found on fine sand and muddy banks of rivers. On the Continent, this species is also found at the edge of standing water. In Britain, adults have been recorded in May and June.

Status Scarce, with few recent records. Occasionally plentiful in certain localities.

Threats River engineering, including dredging, level regulation by damming and flood alleviation schemes. In some areas colonisation by Himalayan balsam *Impatiens glandulifera* can reduce the available habitat. River pollution maybe a further threat.

Management and conservation An under-valued habitat. Sand banks tend to be mobile and rely on the free flow of river and stream systems. Activities that hinder this flow should be prevented. Adjacent scrub and rank grassland may be important for the species in winter and to escape late summer floods.

Published sources Atty, D.B. (1983), Lindroth, C.H. (1974), Lindroth, C.H. (1985).

BEMBIDION STOMOIDES   NOTABLE B
A ground beetle
Order COLEOPTERA   Family CARABIDAE

Bembidion stomoides Dejean, 1831. Formerly known as: *Bembidion atroviolaceum* sensu auct. not Dufour, 1820, *Bembidium stomoides* sensu auct. not Dufour, 1820.

Distribution East Anglia, East Midlands, West Midlands, North East, North West, South Wales, Dyfed-Powys and South West Scotland.

Habitat and ecology The banks of rivers and streams, and river shingle. Predatory. Occurs under stones and shingle. Adults have been recorded in April and June.

Status Widespread but local, recorded throughout central Great Britain.

Threats River engineering, including dredging, level regulation by damming and flood alleviation schemes. In some areas colonisation by Himalayan balsam *Impatiens glandulifera* can reduce the available habitat. Livestock access can damage shingle structure. River pollution maybe a further threat.

Management and conservation An under-valued habitat. Shingle and sand banks tend to be mobile and rely on the free flow of river and stream systems. Activities that hinder this flow should be prevented. Adjacent scrub and rank grassland may be important for the species in winter and to escape late summer floods.

BEMBIDION TESTACEUM   NOTABLE B
A ground beetle
Order COLEOPTERA   Family CARABIDAE

Bembidion testaceum (Duftschmid, 1812). Formerly known as: *Bembidium testaceum* (Duftschmid, 1812).

Distribution South East, South West, East Anglia, West Midlands, North East, North West, South Wales, Dyfed-Powys and South West Scotland.

Habitat and ecology Sand and fine gravel at the margins of running water and gravel pits. Predatory. Adults have been recorded from May to August.

Status Widespread but local, recorded as far north as South West Scotland. This species is difficult to identify and may be confused with other species of the Subgenus *Peryphus* (*Bembidion*). Consequently, the exact status of this species is hard to assess.

Threats River engineering, including dredging, level regulation by damming and flood alleviation schemes as well as the infilling of gravel pits. In some areas colonisation by Himalayan balsam *Impatiens glandulifera* can reduce the available habitat. Livestock access can damage shingle structure. River pollution is a further threat.

Management and conservation An under-valued habitat. Shingle and sand banks tend to be mobile and rely on the free flow of river and stream systems. Activities that hinder this flow should be prevented. Adjacent scrub and rank grassland may be important for the species in winter and to escape late summer floods. Management should aim at maintaining open conditions and encouraging early successional stages along pit edges.

CARABIDAE

BEMBIDION VIRENS — RARE
A ground beetle
Order COLEOPTERA   Family CARABIDAE

Bembidion virens Gyllenhal, 1827. Formerly known as: *Bembidium virens* Gyllenhal, 1827.

Distribution Recorded from West Inverness and West Ross before 1970 and West Ross, East Ross and East Sutherland from 1970 onwards.

Habitat and ecology In shingle by lakes and estuaries. Predatory. Adults have been recorded in July and August.

Status Status revised from RDB 1 (Endangered) in Shirt (1987). Extremely local and only known from very few sites, two of which contain sizeable populations of the beetle. Possibly under-recorded.

Threats This species may be threatened by shingle extraction and the conversion of lakes to reservoirs.

Published sources Lindroth, C.H. (1974), Owen, J.A. (1984), Shirt, D.B., ed. (1987).

BLETHISA MULTIPUNCTATA — NOTABLE B
A ground beetle
Order COLEOPTERA   Family CARABIDAE

Blethisa multipunctata (L., 1758). Formerly known as: *Helobium multipunctatum* (L.).

Distribution South East, South West, East Anglia, East Midlands, West Midlands, North East, North West, Wales and Scotland. A map is given in Luff (1982).

Habitat and ecology Fens and lake margins. Predatory. Found in marginal and wetland vegetation, among moss and on soft soils. Adults have been recorded from April to September.

Status Widespread but local throughout Great Britain. Possibly declining in southern and eastern England.

Threats Drainage for reasons such as agricultural improvement and development is the primary cause of the loss of fens. Falling water tables due to water abstraction and river engineering schemes may also threaten this species.

Management and conservation Water tables should be maintained at high levels. Water bodies should be isolated from sources of eutrophication and pollution. Grazing, cutting or some other disturbance, on a rotational basis, may be needed to maintain open conditions.

BRACHINUS CREPITANS — NOTABLE B
Bombardier beetle
Order COLEOPTERA   Family CARABIDAE

Brachinus crepitans (L., 1758). Formerly known as: *Brachynus crepitans* (L., 1758).

Distribution South East, South, South West, East Anglia, East Midlands, West Midlands and South Wales. A map is given in Luff (1982).

Habitat and ecology Grassland and open country, on calcareous soils, chalk and limestone quarries, the margins of arable fields on limestone, clay brick-pits, undercliffs, sea walls and stabilised shingle on the coast. Adults are gregarious in the spring and are probably predatory. Larvae have been reported to be parasitic on pupae of the rove beetle *Ocypus ater* (Staphylinidae). Adults have been found under stones and also under grass mats growing on concrete blocks on the seaward side of sea walls. Adults have been recorded from April to August and in December.

Status Widespread but local in southern and eastern England. More local in western England and South Wales.

Threats Loss of calcareous grassland through improvement by reseeding or by the application of fertilisers, or by conversion to arable agriculture. The infilling of pits and quarries, the construction of coastal defences, cliff stabilisation schemes, and natural succession and scrub invasion may be further threats.

Management and conservation Management should aim at maintaining open conditions and encouraging early successional stages. In cliff situations occasional slippages are necessary to maintain habitat continuity. Large areas of unstable cliff are required so that the population does not become isolated and subsequently threatened by individual landslips.

BRACHINUS SCLOPETA — ENDANGERED
A ground beetle
Order COLEOPTERA   Family CARABIDAE

Brachinus sclopeta (F., 1792).

Distribution Recorded from "Devon", East Sussex, East Kent, Surrey and South Essex before 1970.

Habitat and ecology Recorded from coastal cliffs and possibly only in cracks or crevices on the cliff face. Possibly also associated with calcareous grassland. Probably predatory. Adults have been recorded in October.

Status Not listed in the insect Red Data Book (Shirt, 1987). Only known from a small number of 19th century records and a single 20th century record. This latter record was of an example from Beachy Head, Eastbourne, East Sussex in 1928.

Threats This species may have declined through cliff stabilisation schemes and the construction of sea defences. Coastal developments may have reduced the amount of available habitat. The degradation of suitable habitat through natural succession and the invasion of scrub on stabilised areas may also have contributed to this species decline.

Published sources Allen, A.A. (1985c), Fowler, W.W. (1887), Lindroth, C.H. (1974).

BRADYCELLUS CSIKII　　　**INDETERMINATE**
A ground beetle
Order COLEOPTERA　　　Family CARABIDAE

Bradycellus csikii Laczó, 1912.

Distribution Recorded from Surrey before 1970 and East Suffolk from 1970 onwards.

Habitat and ecology Recorded in Britain from moss on a heavy, chalky soil. On the Continent, this species seems to prefer clay soils in open country. Probably predatory. The adult has been recorded in May.

Status Status revised from RDB 3 (Rare) in Shirt (1987). Only known in Britain from two specimens. One, an old example, detected by C. Lindroth, recorded at Woking, Surrey by G.C. Champion; the other found in moss at Little Blakenham, East Suffolk in 1977 by D. R. Nash.

Published sources Lindroth, C.H. (1974), Nash, D.R. (1979b), Shirt, D.B., ed. (1987).

BRADYCELLUS DISTINCTUS　　　**NOTABLE A**
A ground beetle
Order COLEOPTERA　　　Family CARABIDAE

Bradycellus distinctus (Dejean, 1829).

Distribution Recorded from Dorset, South Hampshire, East Kent, Berkshire, Warwickshire, Staffordshire, Glamorgan, Nottinghamshire, Cheshire and North-east Yorkshire before 1970 and South Devon, Isle of Wight, East Sussex, East Kent, East Suffolk, East Norfolk, Leicestershire & Rutland and South Lancashire from 1970 onwards.

Habitat and ecology Coastal habitats, with a preference for sandy soils. Probably predatory. Probably occurs under stones, at the roots of plants or in coastal debris. Adults have been recorded in January, February, April, June and from August to October.

Status Widespread but local in England, recently recorded from South Devon along the south coast to East Kent and as far north as South Lancashire. Often confused with *B. sharpi*, to which many old records probably refer.

Threats Loss of sandy habitat, particularly through afforestation, urban and holiday development. The degradation of remaining habitat by excessive disturbance of the vegetation through activities such as motorbike access, horse-riding and human trampling.

Management and conservation Management should aim at maintaining open conditions and encouraging early successional stages.

Published sources Allen, A.A. (1959b), Hammond, P.M. (1969), Lindroth, C.H. (1974), Nature Conservancy Council, (1990), Parry, J.A. (1978b), Sharp, D. (1913).

CALATHUS AMBIGUUS　　　**NOTABLE B**
A ground beetle
Order COLEOPTERA　　　Family CARABIDAE

Calathus ambiguus (Paykull, 1790). Formerly known as: *Calathus fuscus* (F., 1792).

Distribution South East, East Anglia, East Midlands, North West, North East, South Wales, North Wales and South West Scotland.

Habitat and ecology Heathland, sand dunes, chalk pits, gravel pits and disused quarries. Predatory. Found on dry, usually sandy ground with sparse vegetation. Also on chalky substrates and amongst marram on sand dunes. Adults have been recorded in April, June and from August to October.

Status Widespread but local, recorded as far north as South West Scotland. This species is difficult to identify and may be confused with *C. erratus*.

Threats The infilling of pits and quarries, urban and holiday developments, and the loss or fragmentation of heathland through changes in land use, mainly by conversion to arable, forestry and urban development. Further habitat degradation has been through the cessation of traditional heathland management practices.

CARABIDAE

Management and conservation Management should aim at maintaining open conditions and encouraging early successional stages.

CALLISTUS LUNATUS **ENDANGERED**
A ground beetle
Order COLEOPTERA Family CARABIDAE

Callistus lunatus (F., 1775).

Distribution Recorded from East Kent, West Kent, Surrey and Berkshire before 1970. A map is given in Luff (1982).

Habitat and ecology Grassland and open country, on calcareous soils. Probably predatory. Adults have been found from March to May and in November.

Status Last recorded with certainty in 1953 from Shoreham, West Kent. There is an unconfirmed record for Box Hill, Surrey in 1983.

Threats Loss of calcareous grassland through improvement by reseeding or by the application of fertilisers, or by conversion to arable agriculture. Lack of, or changes in grazing or cutting regimes may have contributed to this species decline.

Published sources Fowler, W.W. (1887), Fowler, W.W. & Donisthorpe, H.St J.K. (1913), Lindroth, C.H. (1974), Luff, M.L. (Ed.) (1982), Shirt, D.B., ed. (1987).

CALOSOMA INQUISITOR **NOTABLE A**
Caterpillar-hunter
Order COLEOPTERA Family CARABIDAE

Calosoma inquisitor (L., 1758).

Distribution Recorded from South Devon, South Hampshire, West Sussex, East Sussex, West Kent, South Essex, Hertfordshire, Middlesex, Berkshire, East Suffolk, "Norfolk", Cambridgeshire, West Gloucestershire, Warwickshire, Staffordshire, Shropshire, Breconshire, Carmarthenshire, Merionethshire, Caernarvonshire, Leicestershire & Rutland, North-west Yorkshire, Westmorland & North Lancashire, Cumberland and West Inverness before 1970 and South Devon, South Hampshire, Breconshire, Radnorshire, Cardiganshire, Caernarvonshire, North-west Yorkshire and Westmorland & North Lancashire from 1970 onwards. A map is given in Luff (1982).

Habitat and ecology Ancient broad-leaved woodland. Usually found on oak trees. This species is predatory, chiefly on larvae of Lepidoptera (particularly Tortricidae) in canopy foliage. Adults have been recorded under bark, and in cracks and fissures on bark. Adults are primarily found in May and June, after which they aestivate in subterranean situations. Larvae are able to climb in search of prey.

Status Very local and declining. Widely scattered in England and Wales, recently recorded from South Devon and as far north as Westmorland. Population levels of this species may fluctuate with prey availability.

Indicator status Grade 3 in Harding & Rose (1986).

Threats Loss of broad-leaved woodland through, for example, clear-felling and coniferisation.

Management and conservation Gaps in the age structure of the tree population should be identified and continuity ensured through natural regeneration or possibly by suitable planting.

Published sources Atty, D.B. (1983), Boyce, D.C. (1988), Boyce, D.C. (1989), Brown, E.S. (1948a), Eagles, T.R. (1948c), Flint, J.H. (1943b), Fowler, W.W. (1887), Fowler, W.W. & Donisthorpe, H.St J.K. (1913), Garland, S.P. (1983), Harding, P.T. (1978), Harding, P.T. & Rose, F. (1986), Key, R.S. (1983), Lindroth, C.H. (1974), Luff, M.L. (Ed.) (1982), Osborne, P.J. (1953), Roche, P.J.L. (1943).

CARABUS CLATRATUS **NOTABLE A**
A ground beetle
Order COLEOPTERA Family CARABIDAE

Carabus clatratus L., 1761. Formerly known as: *Carabus clathratus* L., 1761.

Distribution Recorded from East Suffolk, South Aberdeenshire, Moray, Argyll Main, Mid Ebudes, North Ebudes, East Ross, West Sutherland and Outer Hebrides before 1970 and West Inverness, Argyll Main, North Ebudes, West Ross and Outer Hebrides from 1970 onwards. A map is given in Luff (1977).

Habitat and ecology Wetland, bogs and lake margins, particularly on moorland and peatland. Predatory. Occurs in wet, heavily vegetated areas, occassionally on drier peaty soils. Adults have been recorded in May and July.

Status Very local and in Great Britain only recorded from Scotland, except for 19th century records from the Halvergate Marshes, East Suffolk.

Threats Loss of peatlands, particularly through drainage and afforestation. Much moorland has been lost through changes in land use, mainly through afforestation.

Published sources Aitken, J.F. (1988), Fowler, W.W. & Donisthorpe, H.St J.K. (1913), Gradwell, G.R. (1953), Lindroth, C.H. (1974), Lloyd, R.W. (1948a), Luff, M.L. (1977).

CARABUS INTRICATUS   **ENDANGERED**
Blue ground beetle
Order COLEOPTERA   Family CARABIDAE

Carabus intricatus L., 1761.

Distribution Recorded from South Devon before 1970 and East Cornwall and South Devon from 1970 onwards. A map is given in Luff (1977).

Habitat and ecology Wet, broad-leaved woodland. Predatory. In stumps and rotten logs, found under the bark where a thick humus layer is present. Also under moss and lichen on old stumps and logs. This species has been noted crawling about on exposed moss-covered logs and tree trunks on a warm, humid day after rain, including two individuals several feet above the ground. Larvae have been noted in August. Adults have been recorded from March to May, and in August and December.

Status Extremely local and restricted to a small area of south-western England. There are unconfirmed reports of the beetle from the Highlands of Scotland during the 1950's. This species possibly also occurs in Somerset, though this has yet to be confirmed.

Indicator status Listed in Speight (1989) Appendix 1.

Threats Loss of broad-leaved woodland through, for example, clear-felling and coniferisation. Habitat loss, in particular, through the felling of ancient trees, removal of dead wood from living trees and the destruction or removal of standing and fallen dead wood for reasons such as forest hygiene, aesthetic tidiness, public safety or for use as fire wood. This species may have suffered from the attention of collectors.

Management and conservation Ancient trees and both fallen and standing dead timber, especially with the bark attached, should be retained. The removal of dead timber from ancient trees should be avoided. Gaps in the age structure of the tree population should be identified and the continuity of the appropriate habitat ensured by natural regeneration or possibly through suitable planting.

Published sources Allen, A.A. (1989f), Fowler, W.W. (1887), Fowler, W.W. & Donisthorpe, H.St J.K. (1913), Lindroth, C.H. (1974), Luff, M.L. (1977), Shirt, D.B., ed. (1987), Speight, M.C.D. (1989).

CARABUS MONILIS   **NOTABLE B**
A ground beetle
Order COLEOPTERA   Family CARABIDAE

Carabus monilis F., 1792.

Distribution England, Dyfed-Powys, South East Scotland, South West Scotland and North West Scotland. A map is given in Luff (1982).

Habitat and ecology Not habitat specific. This beetle has been found on sand in heathy clearings in woods, in willow carr, old woodland, thick scrub and secondary woodland. Also noted on limestone dales, from cultivated land along field edges, in hay meadows, in gardens and on river shingle beds. Predatory, this species forages on the ground. Adults have been recorded from April to June, and in August and September.

Status Local and declining. Widely distributed in Great Britain and very local in Scotland.

Threats Uncertain, though this species is probably threatened by loss of habitat through practices such as agricultural improvement, conversion to arable, afforestation and river engineering schemes.

CARABUS NITENS   **NOTABLE B**
A ground beetle
Order COLEOPTERA   Family CARABIDAE

Carabus nitens L., 1758.

Distribution England, South East Scotland and South West Scotland. A map is given in Luff (1982).

Habitat and ecology Heathland and moorland, particularly at the edges of *Sphagnum* bogs and in wet heathy areas with *Sphagnum* moss. It has also been recorded from sea-cliffs. Predatory, this beetle forages on the ground. Adults have been recorded from March to May, and in July and August.

Status Local and with a disjunct distribution. This species is primarily found in southern England and northern England, with a scattered distribution through the rest of England and southern Scotland.

Threats Much heathland has been lost, or fragmented, through changes in land use, mainly by conversion to

CARABIDAE

arable agriculture, forestry and urban development. Further habitat degradation has been through the cessation of traditional heathland management practices. This species is further threatened by the afforestation of its moorland habitats.

Management and conservation Management of heathland should aim for a diversity of successional stages, preferably by grazing or by rotational cutting, scraping or burning.

CHLAENIUS NIGRICORNIS	**NOTABLE B**
A ground beetle	
Order COLEOPTERA	Family CARABIDAE

Chlaenius nigricornis (F., 1787).

Distribution England and Wales.

Habitat and ecology Marshes, grazing levels, the margins of lakes and rivers, poor fen at the edges of raised mires, dried-up ponds, moorland, canal banks, parkland and in coastal habitats. Probably predatory. Found on mud, under stones or in lush vegetation, often amongst bulrushes, sedges and reeds. Adults have been recorded from March to August, and in October and November.

Status Widespread but local in England and Wales.

Threats This species is probably threatened by drainage for reasons such as agricultural improvement and development. Afforestation, urban and holiday developments, falling water tables because of water abstraction and river engineering schemes, and natural succession and scrub invasion may be further threats.

Management and conservation Water tables should be maintained at high levels. Water bodies should be isolated from sources of eutrophication and pollution. Grazing, cutting or some other disturbance, on a rotational basis, may be needed to encourage early successional stages and maintain open conditions.

CHLAENIUS NITIDULUS	**ENDANGERED**
A ground beetle	
Order COLEOPTERA	Family CARABIDAE

Chlaenius nitidulus (Schrank, 1781). Formerly known as: *Chlaenius schrankii* (Duftschmid, 1812).

Distribution Recorded from Dorset, Isle of Wight and East Sussex before 1970.

Habitat and ecology Amongst vegetation on silt or clay and in damp places by the coast. Possibly also associated with cliff seepages. Probably predatory. Adults have been recorded from April to June.

Status Possibly extinct. Last recorded in 1930 from Charmouth, Dorset.

Threats This species may have declined because of urban and holiday developments.

Published sources Allen, J.W. & Nicholson, G.W. (1924c), Fowler, W.W. (1887), Lindroth, C.H. (1974), Shirt, D.B., ed. (1987).

CHLAENIUS TRISTIS	**ENDANGERED**
A ground beetle	
Order COLEOPTERA	Family CARABIDAE

Chlaenius tristis (Schaller, 1783). Formerly known as: *Chlaenius holosericeus* (F., 1787).

Distribution Recorded from Berkshire, "Norfolk", Cambridgeshire, Huntingdonshire and South-east Yorkshire before 1970 and Caernarvonshire from 1970 onwards.

Habitat and ecology Lush vegetation on soft soil or mud at the margins of standing water. Also recorded amongst grass tussocks in an area of *Sphagnum* and bog myrtle. Probably predatory. Adults have been recorded in June.

Status Only one recent record, in 1976 from Cors Geirch, Caernarvonshire.

Threats This species may be threatened by drainage, and natural succession and scrub invasion.

Management and conservation Water tables should be maintained at high levels. Water bodies should be isolated from sources of eutrophication and pollution. Management should aim at maintaining open conditions and encouraging early successional stages. Cors Geirch is an NNR.

Published sources Edwards, J. (1893), Fowler, W.W. (1887), Lindroth, C.H. (1974), Morley, C. (1904), Shirt, D.B., ed. (1987).

CICINDELA GERMANICA — RARE
A tiger beetle
Order COLEOPTERA   Family CARABIDAE

Cicindela germanica L., 1758.

Distribution Recorded from South Devon, Dorset, Isle of Wight, South Hampshire and Carmarthenshire before 1970 and Dorset and Isle of Wight from 1970 onwards. A map is given in Luff (1982).

Habitat and ecology Damp sandy or silty areas almost devoid of vegetation, such as sandy undercliffs, landslips and seepage sites. Predatory. Larvae are found in burrows in damp sand. Adults have been found running rapidly over sand in bare or sparsely vegetated areas, and are apparently reluctant to take flight. Adults have been recorded from June to August.

Status Extremely local and restricted in distribution with modern records for the Isle of Wight and Dorset only.

Threats This species is threatened by cliff stabilisation schemes and the construction of sea defences. Coastal developments may reduce the amount of available habitat. Activities that accelerate or reduce the rate of erosion should be avoided. The degradation of suitable habitat through natural succession and the invasion of scrub on stabilised areas is a further threat.

Management and conservation Occasional slippages are necessary for maintaining habitat continuity. Large areas of unstable cliff are required so that the population does not become isolated and subsequently threatened by individual landslips.

Published sources Fowler, W.W. & Donisthorpe, H.St J.K. (1913), Fraser, F.C. (1949), Lindroth, C.H. (1974), Luff, M.L. (Ed.) (1982), Shirt, D.B., ed. (1987).

CICINDELA HYBRIDA — VULNERABLE
A tiger beetle
Order COLEOPTERA   Family CARABIDAE

Cicindela hybrida L., 1758.

Distribution Recorded from West Cornwall, North Devon, Isle of Wight, "Suffolk", East Norfolk, Glamorgan, Merionethshire, Cheshire, South Lancashire, Cumberland and Fife before 1970 and South Lancashire and Cumberland from 1970 onwards. A map is given in Luff (1978).

Habitat and ecology On open sandy areas on or near the coast and for short distances inland. Typically on sand dunes. Predatory. Larvae probably develop in burrows in the sand. Adults are active hunters, flying and running rapidly in search of prey, and have been recorded from April to August.

Status Status revised from RDB 3 (Rare) in Shirt (1987). Formerly widespread and local around the coasts of England and Wales. Now restricted to the north-western coast of England.

Threats Loss of dune habitat, particularly through afforestation, urban and holiday development. The degradation of remaining habitat by excessive disturbance of the vegetation through activities such as motorbike access, horse-riding and human trampling.

Management and conservation On dunes some grazing and other disturbance may be desirable to maintain the early successional stages and prevent the invasion of scrub.

Published sources Appleton, D. (1971), Fowler, W.W. (1887), Lindroth, C.H. (1974), Luff, M.L. (1978), Morley, C. (1904), Sharpe, J.S. (1946), Shirt, D.B., ed. (1987), Skidmore, P. & Johnson, C. (1969).

CICINDELA MARITIMA — NOTABLE B
A tiger beetle
Order COLEOPTERA   Family CARABIDAE

Cicindela maritima Latreille et Dejean, 1822. Formerly known as: *Cicindela hybrida ssp. maritima* Latreille and Dejean, *Cicindela hybrida var. maritima* Latreille and Dejean.

Distribution South East, South, South West, East Anglia, East Midlands, West Midlands and Wales. A map is given in Luff (1978).

Habitat and ecology On sand along the driftline and in inter-tidal areas of beaches. Also in sand dunes. The beetle is almost exclusively coastal. Predatory. Larvae live in burrows in open areas of hard packed sand. Adults are active hunters, flying and running rapidly in search of prey, and have been recorded from May to August.

Status Widespread but local in the southern half of England and Wales.

Threats Loss of dunes and sandy beaches, particularly through afforestation, urban and holiday development. The degradation of remaining habitat by excessive disturbance of the vegetation through activities such as motorbike access, horse-riding and human trampling.

CARABIDAE

Management and conservation On dunes some grazing and other disturbance may be desirable to maintain the early successional stages and prevent the invasion of scrub.

CICINDELA SYLVATICA	NOTABLE A
Wood tiger beetle	
Order COLEOPTERA	Family CARABIDAE

Cicindela sylvatica L., 1758.

Distribution Recorded from South Hampshire, North Hampshire, East Kent, Surrey, Cambridgeshire, West Gloucestershire and North Lincolnshire before 1970 and Dorset, South Hampshire, North Hampshire, West Sussex and Surrey from 1970 onwards. A map is given in Luff (1982).

Habitat and ecology Heathland and heathy areas in open coniferous woodland. Predatory. Occurs on dry, sandy soils, favouring sunny situations. Larvae live in burrows in the soil. Adults are found amongst heather, where they run about rapidly in search of prey, and have been recorded from March to July.

Status Very local and only found in southern and south-eastern England, from Dorset to Surrey.

Threats Much heathland has been lost, or fragmented, through changes in land use, mainly by conversion to arable agriculture, forestry and urban development. Further habitat degradation has been through the cessation of traditional heathland management practices.

Management and conservation Management should aim for a diversity of successional stages from bare ground to mature heath, preferably by grazing or by rotational cutting, scraping or burning.

Published sources Atty, D.B. (1983), Fowler, W.W. (1887), Fowler, W.W. & Donisthorpe, H.St J.K. (1913), Lindroth, C.H. (1974), Luff, M.L. (Ed.) (1982).

CYMINDIS AXILLARIS	NOTABLE A
A ground beetle	
Order COLEOPTERA	Family CARABIDAE

Cymindis axillaris (F., 1794).

Distribution Recorded from Dorset, Isle of Wight, South Hampshire, West Sussex, East Sussex, East Kent, West Kent, Surrey, East Suffolk, West Suffolk, West Norfolk, West Gloucestershire, Glamorgan and North Lincolnshire before 1970 and East Sussex, East Suffolk, West Suffolk, East Norfolk, West Norfolk and North Lincolnshire from 1970 onwards.

Habitat and ecology Grassland and open country, on calcareous soils, Breckland, heathland, coastal sandy grassland and stabilised shingle. Probably predatory. Probably occurs under stones or at the roots of plants. Adults have been recorded in April, August and September.

Status Very local and only recorded south of the Humber in England and in South Wales. Recently recorded from only six vice-counties from East Sussex through East Anglia to North Lincolnshire.

Threats This species is threatened by gravel extraction, urban and holiday developments, afforestation, and improvement and conversion to arable agriculture. Natural succession and scrub invasion may be a further threat.

Management and conservation Disturbance of coastal shingle should be avoided. Management should aim at maintaining open conditions and encouraging early successional stages.

Published sources Atty, D.B. (1983), Fowler, W.W. (1887), Hammond, P.M. (1982), Key, R.S. (1990), Lindroth, C.H. (1974), Nature Conservancy Council, (1988), Nature Conservancy Council, (1990).

CYMINDIS MACULARIS	ENDANGERED
A ground beetle	
Order COLEOPTERA	Family CARABIDAE

Cymindis macularis (Fischer von Waldheim, 1824).

Distribution Recorded from West Suffolk before 1970 and West Suffolk from 1970 onwards.

Habitat and ecology Probably grassland on sandy soils. Probably predatory. Adults have been recorded in June.

Status Not listed in the insect Red Data Book (Shirt, 1987). Recorded new to Great Britain in 1982 from a head-capsule found in a stone curlew pellet collected in September 1980 near Icklingham, West Suffolk. Subsequently, three examples of *C. macularis* were discovered in material from Barton Mills, West Suffolk in 1966. Also recorded from near Thetford in West Suffolk.

Threats Uncertain, though this species is probably threatened by improvement and conversion to arable agriculture, and natural succession and scrub invasion.

Management and conservation Management should aim at maintaining open conditions and encouraging early successional stages.

Published sources Hammond, P.M. (1982), Williams, S.A. (1984).

CYMINDIS VAPORARIORUM   NOTABLE B
A ground beetle
Order COLEOPTERA   Family CARABIDAE

Cymindis vaporariorum (L., 1758).

Distribution West Midlands, North East, North West, North Wales and Scotland.

Habitat and ecology Mountain tops, peat and heather moorland, *Sphagnum* bogs. Frequently on sandy soils. Probably predatory. Found under stones. Adults have been recorded in April and from June to September.

Status Local. A northern species.

Threats Erosion of montane and moorland habitat may be a problem in areas frequented by large numbers of hill walkers and where livestock densities are too high. Loss of moorland through afforestation and conversion to upland pasture may be further threats.

Management and conservation Management should aim at maintaining open conditions and encouraging early successional stages.

DEMETRIAS IMPERIALIS   NOTABLE B
A ground beetle
Order COLEOPTERA   Family CARABIDAE

Demetrias imperialis (Germar, 1824). Formerly known as: *Aetophorus imperialis* (Germar), *Risophilus imperialis* (Germar).

Distribution South East, South, South West, East Anglia and East Midlands. A map is given in Luff (1982).

Habitat and ecology Fens, broads, marshes, ponds, gravel pits, brackish marshes and tidal rivers. Probably predatory. Found in reed beds and amongst bulrushes, also in flood litter. Often in association with *Odacantha melanura* (Carabidae). Adults have been recorded from January to October. Found hibernating in dead bulrush stems.

Status Formerly not uncommon in the East Anglian fens and Thames marshes. Around the turn of the century this species declined dramatically. Since then this species has increased substantially, and is now locally common in some areas of southern England. Possibly still spreading.

Threats Drainage for reasons such as agricultural improvement and development is the primary cause of the loss of wetlands. The infilling of gravel pits, and falling water tables because of water abstraction and river engineering schemes may also threaten this species.

Management and conservation Water tables should be maintained at high levels. Water bodies should be isolated from sources of eutrophication and pollution. Cutting, on a rotational basis, may be needed to maintain open conditions.

DEMETRIAS MONOSTIGMA   NOTABLE B
A ground beetle
Order COLEOPTERA   Family CARABIDAE

Demetrias monostigma Samouelle, 1819. Formerly known as: *Demetrias unipunctatus* (Germar, 1824), *Risophilus monostigma* (Samouelle).

Distribution South East, South, East Anglia, East Midlands, North East, South Wales and North Wales.

Habitat and ecology Sand dunes and coastal habitats. Also in fens and marshes. Probably predatory. Found mainly at the roots of marram on sand dunes and lyme grass on sandy seashores. Also found under driftwood, decaying seaweed and strandline debris, in tussocks, and amongst bulrushes, sedges and in reed beds in fens. Adults have been recorded in January and from April to September.

Status Widespread but local in southern and eastern England, also recorded from parts of Wales.

Threats Loss of dune habitat, particularly through afforestation, urban and holiday development. Drainage for reasons such as agricultural improvement and development is the primary cause of the loss of fens and marshes. Excessive disturbance of dunes through horse-riding and motorbike access, as well as falling water tables because of water abstraction and river engineering schemes may also threaten this species.

Management and conservation Water tables should be maintained at high levels in fen and marshes. Water bodies should be isolated from sources of eutrophication and pollution.

CARABIDAE

DIACHROMUS GERMANUS — EXTINCT

A ground beetle
Order COLEOPTERA   Family CARABIDAE

Diachromus germanus (L., 1758).

Distribution Recorded from "Cornwall", South Devon, Isle of Wight, East Sussex and East Kent.

Habitat and ecology Open, dry grassland. Probably phytophagous. Adults probably occur under stones or at the roots of plants.

Status Presumed extinct. Last recorded in about 1839 from the Isle of Wight.

Published sources Fowler, W.W. (1887), Lindroth, C.H. (1974), Shirt, D.B., ed. (1987).

DICHEIROTRICHUS OBSOLETUS — NOTABLE B

A ground beetle
Order COLEOPTERA   Family CARABIDAE

Dicheirotrichus obsoletus (Dejean, 1829). Formerly known as: *Dichirotrichus obsoletus* (Dejean, 1829).

Distribution South East, South, South West, East Anglia, East Midlands, North East and South West Scotland.

Habitat and ecology Saltmarshes, wet sand on the coast and the sea-shore, particularly at the strandline. Found on sand and clay. Probably predatory. Noted from under seaweed and driftwood, often in company with sandhoppers (Amphipoda). Adults have been recorded from April to November.

Status Widespread but local around the coasts of southern and eastern England, also recorded in South West Scotland.

Threats Loss of habitat through reclamation, erosion and the construction of sea defences.

Management and conservation On saltmarshes grazing should not be introduced where there is no grazing at present.

DROMIUS LONGICEPS — NOTABLE A

A ground beetle
Order COLEOPTERA   Family CARABIDAE

Dromius longiceps Dejean, 1826.

Distribution Recorded from West Suffolk, East Norfolk, West Norfolk, Cambridgeshire, Huntingdonshire, North Lincolnshire, South-east Yorkshire and South-west Yorkshire before 1970 and South Essex, East Norfolk, West Norfolk, Cambridgeshire, North Lincolnshire, South-east Yorkshire, South-west Yorkshire and Mid-west Yorkshire from 1970 onwards.

Habitat and ecology Fens, broads, marshes and brackish situations. Probably predatory. Found mainly in reed beds, though occasionally in other situations. Adults and larvae have been found in reed stems and it has been recorded as emerged from the stem of reed galled by the fly *Lipara lucens* (Diptera). Adults have been found all year round.

Status Status revised from RDB 2 (Vulnerable) in Shirt (1987). Recently recorded from East Anglia through North Lincolnshire as far north as Mid-west Yorkshire.

Threats Drainage for reasons such as agricultural improvement and development is the primary cause of the loss of fens. Reclamation, and falling water tables because of water abstraction and river engineering schemes may also threaten this species.

Management and conservation Water tables should be maintained at high levels. Water bodies should be isolated from sources of eutrophication and pollution. Cutting, on a rotational basis, may be needed to maintain open conditions.

Published sources Edwards, J. (1893), Flint, J.H. (1984b), Foster, A.P. (1989), Fowler, W.W. (1887), Fowler, W.W. & Donisthorpe, H.St J.K. (1913), Key, R.S. (1990), Lindroth, C.H. (1974), Shirt, D.B., ed. (1987).

DROMIUS QUADRISIGNATUS — ENDANGERED

A ground beetle
Order COLEOPTERA   Family CARABIDAE

Dromius quadrisignatus Dejean, 1825.

Distribution Recorded from South Devon, Dorset, South Hampshire, West Sussex, East Kent, Surrey, Berkshire, Oxfordshire, "Norfolk", Cambridgeshire, West Gloucestershire, Glamorgan, Nottinghamshire and Fife before 1970 and Middlesex from 1970 onwards.

Habitat and ecology Broad-leaved woodland and pasture-woodland. Probably predatory. Among dead branches and twigs and under bark of broad-leaved trees. Also noted from an old hawthorn. Adults have probably been recorded all year round.

Status Status revised from RDB 3 (Rare) in Shirt (1987). Only one recent record. Old records suggest that this species had a widely scattered distribution over southern England and South Wales, with records from Nottinghamshire and from Fife in Scotland.

Threats Loss of broad-leaved woodland and parkland through, for example, clear-felling and coniferisation. Habitat loss, in particular, through the felling of ancient trees, removal of dead wood from living trees and the destruction or removal of standing and fallen dead wood for reasons such as forest hygiene, aesthetic tidiness, public safety or for use as fire wood.

Management and conservation Ancient trees and both fallen and standing dead timber, especially with the bark attached, should be retained. The removal of dead timber from ancient trees should be avoided. Gaps in the age structure of the tree population should be identified and the continuity of the appropriate dead wood habitat ensured by suitable planting and possibly with pollarding.

Published sources Allen, A.A. (1957a), Atty, D.B. (1983), Edwards, J. (1893), Fowler, W.W. (1887), Fowler, W.W. & Donisthorpe, H.St J.K. (1913), Gilmour, E.F. (1946), Lindroth, C.H. (1974), Shirt, D.B., ed. (1987).

DROMIUS SIGMA   NOTABLE A
A ground beetle
Order COLEOPTERA   Family CARABIDAE

Dromius sigma (Rossi, 1790).

Distribution Recorded from West Sussex, West Kent, Surrey, South Essex, Middlesex, East Suffolk, West Norfolk, Cambridgeshire, Huntingdonshire, North-east Yorkshire, South-west Yorkshire, Mid-west Yorkshire and Cumberland before 1970 and West Norfolk, Nottinghamshire, South-west Yorkshire and Mid-west Yorkshire from 1970 onwards.

Habitat and ecology Fens, marshes, old gravel pits, poor fen and moorland. Probably predatory. Found on soft soil or mud at the margins of standing water, amongst lush vegetation, and often in large tussocks of hair-grass. Adults are probably found all year round.

Status Status revised from RDB 2 (Vulnerable) in Shirt (1987). Very local, old records suggest that this species

had a scattered distribution in England being recorded as far north as Cumberland. There are recent records for only five vice-counties.

Threats Drainage for reasons such as agricultural improvement and development is the primary cause of the loss of wetlands. The infilling of gravel pits, afforestation of moorland, and falling water tables because of water abstraction and river engineering schemes may also threaten this species.

Management and conservation Water tables should be maintained at high levels. Water bodies should be isolated from sources of eutrophication and pollution. Management should aim at maintaining open conditions and encouraging early successional stages.

Published sources Britten, H. (1943), Flint, J.H. (1984b), Fowler, W.W. (1887), Fowler, W.W. & Donisthorpe, H.St J.K. (1913), Hammond, P.M. (1963), Key, R.S. (1990), Lindroth, C.H. (1974), Morley, C. (1904), Shirt, D.B., ed. (1987), Walsh, G.B. & Rimington, F.C. (1953).

DROMIUS VECTENSIS   RARE
A ground beetle
Order COLEOPTERA   Family CARABIDAE

Dromius vectensis Rye, 1872. Formerly known as: *Dromius insignis* sensu auct. not Lucas, 1846.

Distribution Recorded from South Devon, Dorset, Isle of Wight, West Sussex, East Sussex, East Kent and West Kent before 1970 and Dorset, East Sussex and East Kent from 1970 onwards.

Habitat and ecology Dry sand or shingle with moderately dense vegetation chiefly on the coast. The drier parts of saltmarshes, in gravel pits and possibly also sand dunes and sandy river banks. Probably predatory. Found under refuse on wet saline mud and under old broom bushes. Adults have been recorded from May to October.

Status Not listed in the insect Red Data Book (Shirt, 1987). Very local and formerly recorded from South Devon to West Kent. Recently recorded from only three vice-counties.

Threats This species is threatened by gravel extraction, reclamation, the construction of coastal defences, and urban and holiday developments.

Management and conservation Disturbance of coastal shingle sites should be avoided. On saltmarshes grazing should not be introduced where there is no grazing at present.

CARABIDAE

Published sources Allen, A.A. (1953), Fowler, W.W. (1887), Fowler, W.W. & Donisthorpe, H.St J.K. (1913), Lindroth, C.H. (1974), Nature Conservancy Council, (1990).

DRYPTA DENTATA	**ENDANGERED**
A ground beetle	
Order COLEOPTERA	Family CARABIDAE

Drypta dentata (Rossi, 1790).

Distribution Recorded from Dorset, Isle of Wight, South Hampshire, East Sussex, East Kent and West Kent before 1970 and Dorset and Isle of Wight from 1970 onwards.

Habitat and ecology Stable, well vegetated coastal landslips near seepages. Also on silt or clay soils by rivers. Probably predatory. Adults have been found in small depressions in the ground, under soft, dense grass matting on damp clay. These were in the vicinity of shrubs, such as gorse, on a well vegetated landslip area. Adults were completely concealed and could easily have missed detection. Adults have been recorded in April, May and July.

Status Extremely local. Recently discovered at a site in the Isle of Wight and a site in Dorset. Possibly overlooked and hence under-recorded. Formerly recorded along the south coast of England from Dorset to Kent. A record of this species from Brownsea Island, Dorset in 1977 must be treated as unconfirmed.

Threats This species may be threatened by cliff stabilisation schemes and the construction of sea defences. Coastal developments may reduce the amount of available habitat. Activities that accelerate or reduce the rate of erosion should be avoided. The degradation of suitable habitat through natural succession and the invasion of scrub on stabilised areas is a further threat.

Management and conservation Occasional slippages are necessary to maintain habitat continuity. Large areas of unstable cliff are required so that the population does not become isolated and subsequently threatened by individual landslips.

Published sources Drane, A.B. (1990), Fowler, W.W. (1887), Fowler, W.W. & Donisthorpe, H.St J.K. (1913), Lindroth, C.H. (1974), Mendel, H. (1988b), Shirt, D.B., ed. (1987).

DYSCHIRIUS ANGUSTATUS	**RARE**
A ground beetle	
Order COLEOPTERA	Family CARABIDAE

Dyschirius angustatus (Ahrens, 1830).

Distribution Recorded from South Hampshire, East Kent, South-east Yorkshire, Cumberland and Moray before 1970 and West Sussex and East Sussex from 1970 onwards.

Habitat and ecology Usually in bare sand near water. Often near the coast. It has been found on the upper shore under a stone, from dunes among marram and on the banks of a pond. Predatory, probably on one or more species of *Bledius* (Staphylinidae). It has been found in association with, or in the vicinity of *B. subterraneus*, *B. opacus*, *B. occidentalis* and possibly also *B. arcticus*. Adults usually occur in burrows in the sand and have been found from May to July and in October.

Status Very local and recently recorded from only East and West Sussex. There are scattered, but older, records for this species as far north as Elgin, Scotland.

Threats Loss of habitat, particularly through urban and holiday development and, in certain areas, natural succession.

Management and conservation Management should aim at maintaining open conditions and encouraging the early successional stages.

Published sources Fowler, W.W. (1887), Fowler, W.W. & Donisthorpe, H.St J.K. (1913), Lindroth, C.H. (1974), Parry, J.A. (1975), Shirt, D.B., ed. (1987).

DYSCHIRIUS EXTENSUS	**ENDANGERED**
A ground beetle	
Order COLEOPTERA	Family CARABIDAE

Dyschirius extensus Putzeys, 1845.

Distribution Recorded from West Sussex, East Kent and North Essex before 1970.

Habitat and ecology Sandy places by the coast. Predatory, probably on one or more species of *Bledius* (Staphylinidae). Adults probably occur in burrows in the sand.

Status Status revised from RDB 3 (Rare) in Shirt (1987). No recent records. Last recorded in 1940 from Deal, East Kent. A published record of this species from Newborough Warren, Anglesey (Gilbert, 1958) has since proved to be of *D. politus*.

Threats This species may have declined because of the loss of habitat through urban and holiday development.

Published sources Fowler, W.W. (1887), Gilbert, O. (1958), Lindroth, C.H. (1974), Shirt, D.B., ed. (1987), Walker, J.J. (1900b).

DYSCHIRIUS IMPUNCTIPENNIS   NOTABLE B
A ground beetle
Order COLEOPTERA   Family CARABIDAE

Dyschirius impunctipennis Dawson, 1854.

Distribution South East, South West, East Anglia, East Midlands, West Midlands, Dyfed-Powys, North West, North Wales, South West Scotland and North East Scotland.

Habitat and ecology Sandy places by the coast and saltmarshes. Predatory, probably on one or more species of *Bledius* (Staphylinidae) and usually found in association with *B. fergussoni*. Adults usually occur in burrows in hard sand and also in the sides of saltmarsh creeks, and have been recorded from June.

Status Widespread but local throughout Great Britain.

Threats Loss of habitat, particularly through urban and holiday development, erosion, land reclamation and the construction of coastal defences.

Management and conservation Management should aim at maintaining open conditions and encouraging early successional stages.

DYSCHIRIUS NITIDUS   NOTABLE A
A ground beetle
Order COLEOPTERA   Family CARABIDAE

Dyschirius nitidus (Dejean, 1825).

Distribution Recorded from Dorset, West Sussex, East Kent, North Essex, East Suffolk, "Norfolk", Merionethshire, "Lincs", Cheshire, South Lancashire, West Lancashire, South-east Yorkshire, North-east Yorkshire, Cumberland, Dumfriesshire and Moray before 1970 and East Sussex, East Kent, South Essex, North Essex, Glamorgan, Cumberland and Dumfriesshire from 1970 onwards.

Habitat and ecology Primarily coastal. This species frequents sandy places and the edges of saltmarshes. Predatory, probably on one or more species of *Bledius* (Staphylinidae). Adults probably occur in burrows in the sand and have been recorded in May, June, August and September.

Status Widespread and very local. Recently recorded in south-eastern England, South Wales, North West England and from South West Scotland.

Threats Loss of habitat through urban and holiday development, land reclamation and the construction of sea defences.

Management and conservation Management should aim at maintaining open conditions and encouraging early successional stages.

Published sources Fowler, W.W. (1887), Fowler, W.W. & Donisthorpe, H.St J.K. (1913), Key, R.S. (1990), Lindroth, C.H. (1974), Owen, J.A. (1988a).

DYSCHIRIUS OBSCURUS   VULNERABLE
A ground beetle
Order COLEOPTERA   Family CARABIDAE

Dyschirius obscurus (Gyllenhal, 1827).

Distribution Recorded from East Sussex, East Kent, East Norfolk and "Scotland VC" before 1970 and East Kent, East Suffolk and East Norfolk from 1970 onwards.

Habitat and ecology Gravel pits. In bare sand bordering water. Predatory, probably on one or more species of *Bledius* (Staphylinidae). Adults probably occur in burrows in the sand and have been recorded from April to September.

Status Status revised from RDB 1 (Endangered) in Shirt (1987). This species has been found in a small number of new sites during recent years and is currently known from three vice-counties in southern and eastern England.

Threats This species is threatened by the infilling of gravel pits, tourist development and natural succession.

Management and conservation Management should aim at maintaining open conditions and encouraging early successional stages along pit edges.

Published sources Collier, M.J. (1988a), Collier, M.J. (1988b), Lindroth, C.H. (1974), Nature Conservancy Council, (1990), Shirt, D.B., ed. (1987).

CARABIDAE

ELAPHRUS LAPPONICUS   NOTABLE A
A ground beetle
Order COLEOPTERA   Family CARABIDAE

Elaphrus lapponicus Gyllenhal, 1810.

Distribution Recorded from Mid-west Yorkshire, Cumberland, Stirlingshire, West Perthshire, Mid Perthshire, Angus, South Aberdeenshire, North Aberdeenshire, Moray, West Inverness, Outer Hebrides and Shetland before 1970 and West Perthshire, Mid Perthshire and East Ross from 1970 onwards. A map is given in Luff (1982).

Habitat and ecology Wetland, bogs and stream margins, especially swift mountain streams. A high altitude species. Predatory. Often in moss.

Status Very local in Scotland, formerly recorded in northern England.

Threats This species may be threatened by drainage and afforestation.

Published sources Fowler, W.W. (1887), Lindroth, C.H. (1974), Waterston, A.R. (1981).

ELAPHRUS ULIGINOSUS   NOTABLE B
A ground beetle
Order COLEOPTERA   Family CARABIDAE

Elaphrus uliginosus F., 1792.

Distribution South West, South, South East, East Midlands, West Midlands, Wales and Scotland.

Habitat and ecology The margins of lakes and ponds, and in bogs, fens and reedbeds. Continental literature states that this beetle is found in rich fens as well as oligotrophic bogs on acid soils. Predatory. Adults have been recorded from May to July.

Status Widely distributed and local, recently noted in southern England, much of Wales and from one vice-county in Scotland.

Threats Drainage for reasons such as agricultural improvement, afforestation and development is the primary cause of the loss of fens and bogs. Falling water tables through river engineering schemes may also threaten this species.

Management and conservation Water tables should be maintained at high levels. Water bodies should be isolated from sources of eutrophication and pollution.

HARPALUS ARDOSIACUS   NOTABLE B
A ground beetle
Order COLEOPTERA   Family CARABIDAE

Harpalus ardosiacus Lutschnik, 1922. Formerly known as: *Harpalus rotundicollis* sensu auct. not Kolenati, 1845, *Ophonus diffinis* (Dejean, 1829), *Ophonus rotundicollis* sensu auct. not (Kolenati).

Distribution South East, South, South West, East Anglia, West Midlands, North East and South Wales.

Habitat and ecology On chalk or, more rarely, sand and clay soils. Found on cultivated land, undercliffs, cliff tops, sea walls, upper levels of beaches, saltmarshes and infill sites. Phytophagous, feeding mainly on seeds. Adults have been found under stones and have been recorded in January, from May to August and in October.

Status Widespread but local, not known in Scotland. This species is difficult to identify and may be confused with other members of the genus. Consequently, the exact status of this species is hard to assess.

Threats This species is threatened by habitat destruction and through natural succession. The tidying up of field margins and possibly the use of pesticides are additional threats.

Management and conservation Management should aim at maintaining open conditions and encouraging early successional stages.

HARPALUS AZUREUS   NOTABLE B
A ground beetle
Order COLEOPTERA   Family CARABIDAE

Harpalus azureus (F., 1775). Formerly known as: *Harpalus subquadratus* sensu auct. not Dejean, 1829, *Ophonus azureus* (F.).

Distribution South East, South, South West, East Anglia, East Midlands, West Midlands, North West and South Wales.

Habitat and ecology Predominantly on calcareous soils, though also recorded from saline habitats on clay soils. Found on chalk grassland, in quarries, on cliff tops, undercliffs, sea-walls and on sand. Phytophagous, feeding mainly on seeds. Probably occurs under stones. Adults have been recorded from April to August.

Status Widespread but local, not known in Scotland.

Threats This species is threatened by habitat destruction and through natural succession.

Management and conservation Management should aim at maintaining open conditions and encouraging early successional stages.

HARPALUS CORDATUS   **RARE**
A ground beetle
Order COLEOPTERA   Family CARABIDAE

Harpalus cordatus (Duftschmid, 1812). Formerly known as: *Ophonus cordatus* (Duftschmid).

Distribution Recorded from Dorset, Isle of Wight, East Sussex, East Kent, Surrey, East Gloucestershire and South-west Yorkshire before 1970 and East Sussex and East Kent from 1970 onwards.

Habitat and ecology Sand dunes, though occasionally recorded inland on chalk grassland. Phytophagous, feeding mainly on seeds. Adults are probably found under stones and at the roots of plants. Adults have been recorded in April and July.

Status Not listed in the insect Red Data Book (Shirt, 1987). Formerly recorded from Dorset to South-west Yorkshire, the only recent records are for two vice-counties in the extreme south-east of the country. This species is difficult to identify and may be confused with other members of the genus. Consequently, the exact status of this species is hard to assess.

Threats Loss of dune habitat, particularly through afforestation, urban and holiday development. The degradation of remaining habitat by excessive disturbance of the vegetation through activities such as motorbike access, horse-riding and human trampling. Improvement, conversion to arable agriculture and development cause the main losses of calcareous grassland.

Management and conservation Some grazing and other disturbance may be desirable to maintain the early successional stages and prevent the invasion of scrub.

Published sources Fowler, W.W. (1887), Fowler, W.W. & Donisthorpe, H.St J.K. (1913), Lindroth, C.H. (1974), Sharp, D. (1912b).

HARPALUS CUPREUS   **ENDANGERED**
A ground beetle
Order COLEOPTERA   Family CARABIDAE

Harpalus cupreus Dejean, 1829.

Distribution Recorded from Isle of Wight before 1970.

Habitat and ecology Cultivated land and field margins on chalk, sandy or possibly clay soils. Phytophagous, feeding mainly on seeds. Adults probably occur under stones or at the roots of plants. Adults have been recorded in April, July, October and November.

Status Only known from the Isle of Wight where it has been recorded from Sandown, Ryde, Cowes, Bembridge and Alverstone. Last recorded in 1914 from Sandown. Possibly accidentally introduced in the 19th century. This species is difficult to identify and may be confused with other members of the genus.

Threats Loss of suitable habitat through improvement by reseeding or by the application of fertilisers, or by conversion to arable agriculture. The tidying up of field edges and the use of pesticides may have contributed to the decline of this species.

Published sources Fowler, W.W. (1887), Fowler, W.W. & Donisthorpe, H.St J.K. (1913), Lindroth, C.H. (1974), Shirt, D.B., ed. (1987).

HARPALUS DIMIDIATUS   **NOTABLE A**
A ground beetle
Order COLEOPTERA   Family CARABIDAE

Harpalus dimidiatus (Rossi, 1790). Formerly known as: *Harpalus caspius* (von Steven, 1806).

Distribution Recorded from South Devon, Dorset, Isle of Wight, South Hampshire, East Sussex, East Kent, West Kent, Surrey, South Essex, Berkshire and Oxfordshire before 1970 and North Somerset, Dorset, Surrey and Worcestershire from 1970 onwards.

Habitat and ecology Grassland and open ground on calcareous soils. Phytophagous, feeding mainly on seeds. Adults probably occur under stones or at the roots of plants. Adults have been recorded from March to May and in August.

Status Only known from southern England, recently recorded from only four vice-counties. This species is difficult to identify and may be confused with other members of the genus. Consequently, the exact status of this species is hard to assess.

CARABIDAE

Threats Loss of calcareous grassland through improvement by reseeding or by the application of fertilisers, or by conversion to arable agriculture. Lack of, or changes in grazing or cutting regimes may be a further threat.

Management and conservation Management should aim at maintaining open conditions and encouraging early successional stages.

Published sources Fowler, W.W. (1887), Greenslade, P.J.M. (1963a), Lindroth, C.H. (1974), Whitehead, P.F. (1989b).

HARPALUS FROELICHI   **VULNERABLE**
A ground beetle
Order COLEOPTERA   Family CARABIDAE

Harpalus froelichi Sturm, 1818. Formerly known as: *Harpalus froelichii* Sturm, 1818, *Harpalus frolichii* Sturm, 1818, *Harpalus tardus* sensu Fowler, 1887 not (Panzer, 1796).

Distribution Recorded from Dorset, North Essex, East Suffolk, West Norfolk and "Yorks" before 1970 and West Suffolk from 1970 onwards.

Habitat and ecology Heathland and open ground on sandy soils. Also on calcareous grassland, open ground on chalk and cultivated land. Phytophagous, feeding mainly on seeds. Adults probably occur under stones or at the roots of plants. This species has been noted at mercury vapour light. Adults have been recorded from April to August.

Status Not listed in the insect Red Data Book (Shirt, 1987). Extremely local and recently recorded from only one vice-county. Formerly recorded over a wide area of southern and eastern England and reported from as far north as Yorkshire. This species is difficult to identify and may be confused with other members of the genus. Consequently, the exact status of this species is hard to assess.

Threats Much heathland and Breckland has been lost, or fragmented, through changes in land use, mainly by conversion to arable, forestry and urban development. Further habitat degradation has been through the cessation of traditional heathland management practices. The loss of calcareous grassland through improvement and conversion to arable agriculture may also have contributed to the decline of this species.

Management and conservation Management should aim at maintaining open conditions and encouraging early successional stages.

Published sources Foster, A.P. (1988), Hammond, P.M. (1968), Lindroth, C.H. (1974).

HARPALUS HONESTUS   **ENDANGERED**
A ground beetle
Order COLEOPTERA   Family CARABIDAE

Harpalus honestus (Duftschmid, 1812). Formerly known as: *Harpalus ignavus* (Duftschmid, 1812), *Harpalus ignavus* sensu Fowler, 1887 partim.

Distribution Recorded from West Kent and Berkshire before 1970.

Habitat and ecology Grassland, open ground and cultivated land on calcareous soils. Possibly also sandy soils. Phytophagous, feeding mainly on seeds. Adults probably occur under stones or at the roots of plants.

Status Reliably known only from a very old record from Charlton, West Kent, noted in about 1795 and from more than one example found at Streatley, Berkshire in 1905. A record by from Box Hill, Surrey by W. West requires confirmation. A further record from Foxhall, Suffolk by C. Morley is more doubtful, probably being based on a bluish *H. rufitarsis* (A.A. Allen pers. comm.).

Threats Loss of calcareous grassland through improvement by reseeding or by the application of fertilisers, or by conversion to intensive arable agriculture may have contributed to this species decline.

Published sources Allen, A.A. (1964c), Holland, W. (1905), Lindroth, C.H. (1974), Shirt, D.B., ed. (1987).

HARPALUS MELANCHOLICUS   **ENDANGERED**
A ground beetle
Order COLEOPTERA   Family CARABIDAE

Harpalus melancholicus Dejean, 1829.

Distribution Recorded from West Cornwall, East Cornwall, South Devon, North Somerset, Dorset, Isle of Wight, East Kent, West Kent, South Essex, Oxfordshire, "Norfolk", Glamorgan, Pembrokeshire and Caernarvonshire before 1970.

Habitat and ecology Sand dunes, though also recorded inland. Phytophagous, feeding mainly on seeds. Adults probably occur under stones or at the roots of plants.

Status Not listed in the insect Red Data Book (Shirt, 1987). No recent records. Last recorded in 1964 from West Cornwall. This species is difficult to identify and may be confused with other members of the genus. Consequently, the exact status of this species is hard to assess.

Threats Loss of dune habitat, particularly through afforestation, urban and holiday development. The degradation of remaining habitat by excessive disturbance of the vegetation through activities such as horse-riding and human trampling may also have contributed to the decline of this species.

Published sources Allen, A.A. (1952c), Fowler, W.W. (1887), Fowler, W.W. & Donisthorpe, H.St J.K. (1913), Lindroth, C.H. (1974).

HARPALUS MELLETI   NOTABLE A
A ground beetle
Order COLEOPTERA   Family CARABIDAE

Harpalus melleti Heer, 1837. Formerly known as: *Harpalus brevicollis* sensu Jeannel, 1942 ?Serville, 1821, *Harpalus parallelus* sensu Fowler, 1887 partim not Dejean, 1829, *Ophonus championi* (Sharp, 1912), *Ophonus rupicoloides* (Sharp, 1912).

Distribution Recorded from West Cornwall, East Cornwall, Dorset, Isle of Wight, East Kent, West Kent, Surrey, South Essex, Cambridgeshire, Huntingdonshire, East Gloucestershire and South Lincolnshire before 1970 and East Cornwall, Dorset, Surrey and Northamptonshire from 1970 onwards.

Habitat and ecology Calcareous grassland, open ground on chalky or alluvial soils. Phytophagous, feeding mainly on seeds. Adults probably occur under stones and at the roots of vegetation. Adults have been recorded in April, May and August.

Status Only known from the southern half of England, recently recorded from only four vice-counties. This species is difficult to identify and may be confused with other members of the genus. Consequently, the exact status of this species is hard to assess.

Threats Loss of calcareous grassland through improvement by reseeding or by the application of fertilisers, or by conversion to arable agriculture. Lack of, or changes in grazing or cutting regimes may threaten this species.

Management and conservation Grazing or cutting, on a rotational basis, is needed to maintain open conditions.

Published sources Atty, D.B. (1983), Hammond, P.M. (1963), Lindroth, C.H. (1974), Sharp, D. (1912b).

HARPALUS OBSCURUS   ENDANGERED
A ground beetle
Order COLEOPTERA   Family CARABIDAE

Harpalus obscurus (F., 1792). Formerly known as: *Ophonus obscurus* (F.).

Distribution Recorded from Dorset, North Essex, Oxfordshire, Cambridgeshire, Northamptonshire and East Gloucestershire before 1970.

Habitat and ecology Calcareous grassland, an oolitic limestone quarry and open ground on chalky soils. Phytophagous, feeding mainly on seeds. Adults have been found at the roots of grass and probably also occur under stones. Adults have been recorded in June and August. This species is difficult to identify and may be confused with other members of the genus. Consequently, this species may be under-recorded.

Status Not listed in the insect Red Data Book (Shirt, 1987). No recent records. Last recorded in 1926 from Dorset.

Threats Loss of calcareous grassland through improvement by reseeding or by the application of fertilisers, or by conversion to arable agriculture have probably contributed to the decline of this species. Lack of, or changes in grazing or cutting regimes may also have been contributory factors.

Published sources Atty, D.B. (1983), Fowler, W.W. (1887), Fowler, W.W. & Donisthorpe, H.St J.K. (1913), Lindroth, C.H. (1974), Twinn, D.C. (1952).

HARPALUS PARALLELUS   RARE
A ground beetle
Order COLEOPTERA   Family CARABIDAE

Harpalus parallelus Dejean, 1829. Formerly known as: *Harpalus parallelus* sensu Fowler, 1887 partim, *Harpalus zigzag* sensu auct. not Costa, 1882.

Distribution Recorded from Dorset, Isle of Wight, South Hampshire, East Sussex, East Kent, West Kent, Surrey, Oxfordshire, East Suffolk, East Norfolk and Bedfordshire before 1970 and East Sussex, East Kent and Northamptonshire from 1970 onwards.

Habitat and ecology A predominantly coastal species of calcareous grassland, chalk cliffs and open ground on chalky soils. Phytophagous, feeding mainly on seeds. Adults have been found under large stones amongst short-turf and have been recorded in June and August.

CARABIDAE

Status Not listed in the insect Red Data Book (Shirt, 1987). Only known from the southern half of England, recently recorded from only three vice-counties. This species is difficult to identify and may be confused with other members of the genus. Consequently, the exact status of this species is hard to assess.

Threats Loss of calcareous grassland through improvement by reseeding or by the application of fertilisers, or by conversion to arable agriculture, and through urban and holiday developments. Lack of, or changes in grazing or cutting regimes may be a further threat. Cliff stabilisation schemes and the construction of sea defences may also have contributed to this speceis decline.

Management and conservation Management should aim at maintaining open conditions and encouraging early successional stages.

Published sources Fowler, W.W. (1887), Fowler, W.W. & Donisthorpe, H.St J.K. (1913), Hodge, P.J. (1990), Key, R.S. (1990), Lindroth, C.H. (1974).

HARPALUS PUNCTATULUS	NOTABLE A
A ground beetle	
Order COLEOPTERA	Family CARABIDAE

Harpalus punctatulus (Duftschmid, 1812). Formerly known as: *Ophonus punctatulus* (Duftschmid).

Distribution Recorded from North Somerset, Isle of Wight, South Hampshire, East Sussex, East Kent, West Kent, Surrey, North Essex, Berkshire, East Suffolk, East Norfolk, West Norfolk, Cambridgeshire, Bedfordshire, Glamorgan and Nottinghamshire before 1970 and South Hampshire, Oxfordshire, East Suffolk, West Suffolk and West Norfolk from 1970 onwards.

Habitat and ecology Calcareous grassland and open ground on chalky or sandy soils. Also recorded from *Calluna* heath. Phytophagous, feeding mainly on seeds. Adults probably occur under stones and at the roots of plants. Adults have been recorded in June and August.

Status Only known from the southern half of England, recently recorded from only five vice-counties. This species is difficult to identify and may be confused with other members of the genus. Consequently, the exact status of this species is hard to assess.

Threats Loss of calcareous grassland through improvement by reseeding or by the application of fertilisers, or by conversion to arable agriculture. Much heathland has been lost, or fragmented, through changes in land use, mainly by conversion to arable, forestry and urban development. Lack of, or changes in, grazing or cutting regimes may be a further threat.

Management and conservation Management should aim at maintaining open conditions and encouraging early successional stages.

Published sources Fowler, W.W. (1887), Fowler, W.W. & Donisthorpe, H.St J.K. (1913), Lindroth, C.H. (1974).

HARPALUS PUNCTICOLLIS	RARE
A ground beetle	
Order COLEOPTERA	Family CARABIDAE

Harpalus puncticollis (Paykull, 1798). Formerly known as: *Harpalus puncticollis* sensu Fowler, 1887 partim, *Ophonus puncticollis* (Paykull).

Distribution Recorded from West Cornwall, "Somerset", East Kent, Surrey, North Essex, Middlesex, Cambridgeshire, South Lincolnshire, Nottinghamshire, Cheshire and South-west Yorkshire before 1970 and Buckinghamshire and West Gloucestershire from 1970 onwards.

Habitat and ecology Grassland and open ground on calcareous soils. Phytophagous, feeding mainly on seeds. Adults probably occur under stones or at the roots of plants. Adults have been recorded in May and August.

Status Not listed in the insect Red Data Book (Shirt, 1987). Recorded in England north to South-west Yorkshire. Now very local and only recorded from three vice-counties, all in southern England. This species is difficult to identify and may be confused with other members of the genus. Consequently, the exact status of this species is hard to assess.

Threats Loss of calcareous grassland through improvement by reseeding or by the application of fertilisers, or by conversion to arable agriculture. Lack of, or changes in grazing or cutting regimes may be a further threat.

Management and conservation Management should aim at maintaining open conditions and encouraging early successional stages.

Published sources Fowler, W.W. & Donisthorpe, H.St J.K. (1913), Lindroth, C.H. (1974), Philp, E.G. (1965).

HARPALUS QUADRIPUNCTATUS   NOTABLE A
A ground beetle
Order COLEOPTERA   Family CARABIDAE

Harpalus quadripunctatus Dejean, 1829.

Distribution Recorded from North Somerset, South Hampshire, South Aberdeenshire, Banffshire, Moray, East Inverness & Nairn, Outer Hebrides and Shetland before 1970 and Derbyshire, Mid-west Yorkshire, Angus, South Aberdeenshire, Moray, East Inverness & Nairn and Shetland from 1970 onwards.

Habitat and ecology Mainly on gravelly soils and moraine. Phytophagous, feeding mainly on seeds. Recorded from moss and under leaves amongst trees and bushes. Adults have been recorded from May to July.

Status Widespread and very local. Recent records indicate that this species has a northern distribution, though there are old records from southern England. This species is difficult to identify and may be confused with other members of the genus. Consequently, the exact status of this species is hard to assess.

Threats This species may be threatened by afforestation, agricultural improvement and conversion to other land use.

Management and conservation Management should aim at maintaining open conditions and encouraging early successional stages.

Published sources Fowler, W.W. (1887), Horsfield, D. (1981), Lindroth, C.H. (1974).

HARPALUS RUPICOLA   NOTABLE B
A ground beetle
Order COLEOPTERA   Family CARABIDAE

Harpalus rupicola Sturm, 1818. Formerly known as: *Ophonus rupicola* (Sturm).

Distribution South East, South, South West, East Anglia, East Midlands, West Midlands and North East.

Habitat and ecology Calcareous grassland, open ground on chalky soils, coastal shingle and the drier parts of saltmarshes. Phytophagous, feeding mainly on seeds. Adults probably occur under stones or at the roots of vegetation. Adults have been recorded from June to August.

Status Widespread but local in England. This species is difficult to identify and may be confused with other

members of the genus. Consequently, the exact status of this species is hard to assess.

Threats Loss of calcareous grassland through improvement by reseeding or by the application of fertilisers, or by conversion to arable agriculture. Lack of, or changes in grazing or cutting regimes may be a further threat. Loss of coastal shingle through gravel extraction and development and the loss of saltmarsh through reclamation, erosion and the construction of coastal defences.

Management and conservation Management should aim at maintaining open conditions and encouraging early successional stages.

HARPALUS SABULICOLA   RARE
A ground beetle
Order COLEOPTERA   Family CARABIDAE

Harpalus sabulicola (Panzer, 1796). Formerly known as: *Ophonus sabulicola* (Panzer).

Distribution Recorded from East Cornwall, Dorset, North Hampshire, East Sussex, East Kent, West Kent, Surrey, North Essex, Hertfordshire, East Suffolk, Cambridgeshire, East Gloucestershire and Glamorgan before 1970 and Dorset, North Hampshire and East Sussex from 1970 onwards.

Habitat and ecology On dry, usually chalky soils, occasionally in sandy areas. Phytophagous, feeding mainly on seeds. Adults probably occur under stones or at the roots of vegetation. Adults have been recorded in October.

Status Not listed in the insect Red Data Book (Shirt, 1987). Only known from southern England being recorded as far north as Cambridgeshire. Recently recorded from only three vice-counties. This species is difficult to identify and may be confused with other members of the genus. Consequently, the exact status of this species is hard to assess.

Threats Loss of calcareous grassland through improvement by reseeding or by the application of fertilisers, or by conversion to arable agriculture. Lack of, or changes in grazing or cutting regimes may be a further threat.

Management and conservation Management should aim at maintaining open conditions and encouraging early successional stages.

Published sources Atty, D.B. (1983), Fowler, W.W. (1887), Fowler, W.W. & Donisthorpe, H.St J.K. (1913), Lindroth, C.H. (1974).

CARABIDAE

HARPALUS SCHAUBERGERIANUS NOTABLE B
A ground beetle
Order COLEOPTERA   Family CARABIDAE

Harpalus schaubergerianus Puel, 1937. Formerly confused under: *Harpalus rufibarbis* sensu Fowler, 1887 partim not (F., 1792), *Ophonus rufibarbis* sensu auct. not (F.).

Distribution South East, South, South West, East Anglia, East Midlands and North East.

Habitat and ecology Calcareous grassland, chalk pits, coastal cliffs and occasionally on sand. Phytophagous, feeding mainly on seeds. Adults have been found under stones, under piles of bark and in ditch litter. Adults have been recorded from April to August, though mainly noted in May.

Status Widespread but local throughout much of England. Possibly under-recorded as this species is difficult to identify and can be confused with *H. rufibarbis*.

Threats Loss of calcareous grassland through improvement by reseeding or by the application of fertilisers, or by conversion to arable agriculture. Lack of, or changes in grazing or cutting regimes may be a further threat. Cliff stabilisation schemes, the construction of sea defences and the infilling of chalk pits may also affect this species.

Management and conservation Management should aim at maintaining open conditions and encouraging early successional stages.

Published sources Atty, D.B. (1983), Hammond, P.M. (1963), Johnson, C. (1963b), Lindroth, C.H. (1974), Nash, D.R. (1974b), Nash, D.R. (1979b).

HARPALUS SERRIPES   NOTABLE B
A ground beetle
Order COLEOPTERA   Family CARABIDAE

Harpalus serripes (Quensel in Schoenherr, 1806).

Distribution South East, South, South West, East Anglia, West Midlands and South Wales.

Habitat and ecology Primarily coastal. Sand dunes, coastal shingle, a cliff top, a sea wall and recorded from the Woolwich Beds overlaid on chalk. Phytophagous, feeding mainly on seeds. Adults probably occur under stones or at the roots of plants and have been recorded in April, May, July September and October.

Status Widespread but local in the southern half of England and South Wales.

Threats Loss of coastal habitat through urban and holiday developments, cliff stabilisation schemes, the construction of coastal defences, afforestation, and improvement and conversion to arable agriculture. Lack of, or changes in, grazing and cutting regimes may be a further threat to this species.

Management and conservation Management should aim at maintaining open conditions and encouraging early successional stages.

HARPALUS SERVUS   NOTABLE B
A ground beetle
Order COLEOPTERA   Family CARABIDAE

Harpalus servus (Duftschmid, 1812).

Distribution South East, South, South West, East Anglia, North East and South Wales.

Habitat and ecology Sand dunes and Breckland. Phytophagous, feeding mainly on seeds. Recorded amongst hawkweeds on sand. Adults have been recorded in May, June, September and October.

Status Widespread but local in southern England and South Wales. Also recorded in North East England. This species is difficult to identify and may be confused with other members of the genus. Consequently, the exact status of this species is hard to assess.

Threats Loss of habitat, particularly through afforestation, urban and holiday development, and natural succession and scrub invasion. The degradation of remaining habitat by excessive disturbance of the vegetation through activities such as motorbike access, horse-riding and human trampling.

Management and conservation On dunes some grazing and other disturbance may be desirable to maintain the early successional stages and prevent the invasion of scrub.

HARPALUS SMARAGDINUS   **NOTABLE B**
A ground beetle
Order COLEOPTERA   Family CARABIDAE

Harpalus smaragdinus (Duftschmid, 1812). Formerly known as: *Harpus discoideus* Erichson, 1837.

Distribution South East, South, South West, East Anglia, East Midlands and North Wales.

Habitat and ecology Sand pits, heathland on sandy soils and in sandy fields. Phytophagous, feeding mainly on seeds. Adults have been recorded under leaf-litter and probably occur under stones or at the roots of plants. Adults have been recorded from June to September.

Status Widespread but local in southern England, also recorded in North Wales. This species is difficult to identify and may be confused with other members of the genus. Consequently, the exact status of this species is hard to assess.

Threats Much heathland has been lost, or fragmented, through changes in land use, mainly by conversion to arable, forestry and urban development. Further habitat degradation has been through the cessation of traditional heathland management practices. The infilling of pits may be a further threat.

Management and conservation Management should aim to encourage early successional stages and maintain open conditions, preferably by grazing or by rotational cutting, scraping or burning.

HARPALUS TENEBROSUS   **NOTABLE A**
A ground beetle
Order COLEOPTERA   Family CARABIDAE

Harpalus tenebrosus Dejean, 1829. Formerly known as: *Harpalus tenebrosus ssp. centralis* Schauberger, 1929.

Distribution Recorded from West Cornwall, East Cornwall, South Devon, North Devon, Isle of Wight, East Sussex, East Kent, West Kent, Surrey, East Suffolk, East Norfolk, West Gloucestershire and Glamorgan before 1970 and West Cornwall, East Cornwall, South Devon, Dorset, East Norfolk and West Norfolk from 1970 onwards.

Habitat and ecology Primarily coastal. On dry, sandy or chalky ground. Typically recorded from coastal cliffs. Phytophagous, feeding mainly on seeds. Adults probably occur under stones or at the roots of plants and have been recorded in June, August, September and November.

Status Known only from southern England and South Wales. Now very local and only recorded in south-western England and East Anglia. This species is difficult to identify and may be confused with other members of the genus. Consequently, the exact status of this species is hard to assess.

Threats This species is threatened by cliff stabilisation schemes and the construction of sea defences. Coastal developments may reduce the amount of available habitat. Activities that accelerate or reduce the rate of erosion should be avoided. The degradation of suitable habitat through natural succession and the invasion of scrub on stabilised areas is a further threat.

Management and conservation Occasional slippages are necessary to maintain habitat continuity. Large areas of unstable cliff are required so that the population does not become isolated and subsequently threatened by individual landslips.

Published sources Atty, D.B. (1983), Fowler, W.W. (1887), Fowler, W.W. & Donisthorpe, H.St J.K. (1913), Lindroth, C.H. (1974).

HARPALUS VERNALIS   **NOTABLE A**
A ground beetle
Order COLEOPTERA   Family CARABIDAE

Harpalus vernalis (Duftschmid, 1812). Formerly known as: *Harpalus picipennis* sensu auct. not (Duftschmid, 1812).

Distribution Recorded from South Devon, Dorset, South Hampshire, Middlesex, West Suffolk, East Norfolk, West Norfolk and Glamorgan before 1970 and West Suffolk, West Norfolk and Northamptonshire from 1970 onwards.

Habitat and ecology Found on open or disturbed ground on sandy or gravelly soils. Phytophagous, feeding mainly on seeds. Adults occur under stones in bare or sparsely vegetated ground and have been recorded in May, June, August and September.

Status Known only from southern England and South Wales. Now very local and recently recorded only from the Brecklands and Northamptonshire.

Threats Much sandy habitat has been lost, or fragmented, through changes in land use, mainly by conversion to arable, forestry and urban development. Further habitat degradation has been through the cessation of traditional management practices.

Management and conservation Management should aim to encourage early successional stages and maintain open conditions, preferably by grazing or by rotational cutting or scraping.

Published sources Collier, M.J. (1990), Fowler, W.W. (1887), Fowler, W.W. & Donisthorpe, H.St J.K. (1913), Lindroth, C.H. (1974), Morris, M.G. & Rispin, W.E. (1988).

LEBIA CHLOROCEPHALA　　　　　**NOTABLE B**
A ground beetle
Order COLEOPTERA　　　　Family CARABIDAE

Lebia chlorocephala (Hoffmannsegg, 1803).

Distribution England, Wales, South East Scotland and South West Scotland. A map is given in Luff (1982).

Habitat and ecology Grassland and open country, on calcareous soils. Also recorded from woodland, boulder clay cliffs and river shingle. Adults are probably predatory. The larvae are parasitic on the pupae of various species of Chrysomelidae, including *Chrysolina polita* and *C. varians*, though probably also *C. staphylea* and other species of *Chrysolina*. Adults, and doubtless larvae, are found at the roots of plants, particularly those used by the host species of *Chrysolina*. Adults have often been found in grass tussocks in winter. There is also a record from under bark. Adults have been recorded from January to June and from October to December.

Status Widespread but local throughout the southern half of England, more local in the rest of the country though not reported in northern Scotland.

Threats Loss of calcareous grassland through improvement by reseeding or by the application of fertilisers, or by conversion to arable agriculture. Clear-felling of woodland and conversion to other land use, cliff stabilisation and coastal defence schemes, river engineering schemes, and a lack of, or changes in grazing or cutting regimes may be further threats.

Management and conservation Management should aim at maintaining open conditions. In woodlands, open glades and ride margins should be cut on rotation to retain a variety of vegetation structures.

LEBIA CRUXMINOR　　　　　**ENDANGERED**
A ground beetle
Order COLEOPTERA　　　　Family CARABIDAE

Lebia cruxminor (L., 1758).

Distribution Recorded from East Cornwall, South Hampshire, North Hampshire, East Sussex, West Kent, Surrey, Berkshire, West Gloucestershire, Shropshire and Cumberland before 1970 and East Cornwall and East Sussex from 1970 onwards.

Habitat and ecology Dry meadows, downland and woodland. Adults are probably predatory. The larvae are parasitic, probably specifically on *Galeruca tanaceti* (Chrysomelidae). Adults, and doubtless larvae, occur at the roots of plants, particularly those utilised by *G. tanaceti*. Adults have been recorded in April, May and August.

Status Old records suggest that this species had a scattered distribution throughout England as far north as Cumberland. There are recent records for two vice-counties, though this species is now only regularly recorded at Ditchling Common in East Sussex.

Threats This species probably declined through the loss of unimproved grassland because of improvement by reseeding or by the application of fertilisers, or by conversion to arable agriculture. Lack of, or changes in grazing or cutting regimes may also have contributed to this species decline.

Management and conservation Grazing or cutting, on a rotational basis, is needed to maintain open conditions. Ditchling Common is notified as an SSSI.

Published sources Day, F.H. (1930), Fowler, W.W. (1887), Fowler, W.W. & Donisthorpe, H.St J.K. (1913), Lindroth, C.H. (1974), Shirt, D.B., ed. (1987).

LEBIA CYANOCEPHALA　　　　　**ENDANGERED**
A ground beetle
Order COLEOPTERA　　　　Family CARABIDAE

Lebia cyanocephala (L., 1758).

Distribution Recorded from "Devon", Dorset, South Hampshire, East Kent, West Kent, Surrey, Berkshire and North-west Yorkshire before 1970.

Habitat and ecology Grassland and open country, on calcareous soils. Adults are probably predatory. Larvae are parasitic on one or more species of the Chrysomelidae including *Chrysolina hyperici*.

Status Not listed in the insect Red Data Book (Shirt, 1987). Last recorded in 1951 from Chipstead, Surrey.

Threats This species has probably declined through the loss of calacareous grassland because of improvement by reseeding or by the application of fertilisers, or by conversion to arable agriculture. Lack of, or changes in grazing or cutting regimes may also have contributed to this species decine.

Published sources Fowler, W.W. (1887), Fowler, W.W. & Donisthorpe, H.St J.K. (1913), Lindroth, C.H. (1974).

LEBIA MARGINATA	EXTINCT
A ground beetle	
Order COLEOPTERA	Family CARABIDAE

Lebia marginata (Fourcroy, 1785). Formerly known as: *Lebia haemorrhoidalis* (F., 1787).

Distribution Recorded from "Wilts" and Shropshire.

Habitat and ecology Probably grassland. Adults are probably predatory. Larvae are parasitic, probably on one or more species of the Chrysomelidae.

Status Presumed extinct. Last recorded in the 19th century from near Devizes, Wiltshire.

Published sources Fowler, W.W. (1887), Lindroth, C.H. (1974), Shirt, D.B., ed. (1987).

LEBIA SCAPULARIS	EXTINCT
A ground beetle	
Order COLEOPTERA	Family CARABIDAE

Lebia scapularis (Fourcroy, 1785). Formerly known as: *Lebia turcica* (F., 1787).

Distribution Recorded from East Sussex.

Habitat and ecology Probably grassland. Adults are probably predatory, while the larvae are almost certainly parasitic. Known in Italy as a larval parasite on the pupa of *Galerucella luteola* (Chrysomelidae). *L. scapularis* may have been associated with one or more species of *Galerucella* in Britain.

Status Presumed extinct. Last recorded in 1883 from Hastings, East Sussex.

Published sources Fowler, W.W. (1887), Lindroth, C.H. (1974), Shirt, D.B., ed. (1987).

LEISTUS MONTANUS	NOTABLE A
A ground beetle	
Order COLEOPTERA	Family CARABIDAE

Leistus montanus Stephens, 1827.

Distribution Recorded from Merionethshire, Caemarvonshire, North Northumberland, Cumberland, Clyde Islands, North Ebudes, West Ross and Outer Hebrides before 1970 and Caemarvonshire, Westmorland & North Lancashire, North Ebudes, West Ross and Outer Hebrides from 1970 onwards.

Habitat and ecology Montane habitats. The few records at or near sea level are probably due to individuals being washed down by flooding. Predatory, this beetle forages on the ground. Usually found in fairly dry habitats and has been noted under stones. Adults have been recorded from May to September.

Status Status revised from RDB 3 (Rare) in Shirt (1987). Very local and with a northern and western distribution. Possibly under-recorded because of the inaccessability of its habitat.

Threats This species may be threatened in parts of its range by erosion of its habitat through fell-walking.

Published sources Davidson, W.F. (1954), Davidson, W.F. (1961), Day, F.H. (1943), Fergusson, A. (1922), Fowler, W.W. (1887), Fowler, W.W. & Donisthorpe, H.St J.K. (1913), Lindroth, C.H. (1974), Shirt, D.B., ed. (1987), Skidmore, P. & Johnson, C. (1969), Thomas, J. (1972).

LICINUS DEPRESSUS	NOTABLE B
A ground beetle	
Order COLEOPTERA	Family CARABIDAE

Licinus depressus (Paykull, 1790).

Distribution South East, South, South West, East Anglia, East Midlands, West Midlands and North East. A map is given in Luff (1982).

Habitat and ecology Grassland on limestone dales and chalk downland, chalk quarries and coastal downland. Also recorded from gravel pits and on sandy or gravelly soils. Predatory. Continental literature states that the larvae, and probably also the adults feed on snails. Probably occurs under stones and possibly has a preference for more sparsely vegetated areas. In Britain, it has been found in company with *L. punctatulus* on coastal downland. Adults have been recorded from April to December.

CARABIDAE

Status Widespread but local in England, though not recorded in North West England.

Threats Loss of calcareous grassland through improvement by reseeding or by the application of fertilisers, or by conversion to arable agriculture. Urban and holiday developments, the infilling of gravel pits and quarries, and a lack of, or changes in grazing or cutting regimes may be further threats.

Management and conservation Management should aim to encourage early successional stages and maintain open conditions.

LICINUS PUNCTATULUS	**NOTABLE A**
A ground beetle	
Order COLEOPTERA	Family CARABIDAE

Licinus punctatulus (F., 1792). Formerly known as: *Licinus silphoides* sensu (F., 1792) not (Rossi, 1790).

Distribution Recorded from North Devon, South Somerset, North Wiltshire, Dorset, Isle of Wight, West Sussex, East Sussex, East Kent, West Kent, Surrey, Middlesex, Berkshire, Northamptonshire and South Lincolnshire before 1970 and North Somerset, Dorset, East Sussex, East Kent, Oxfordshire and Glamorgan from 1970 onwards.

Habitat and ecology Usually near the coast, frequenting cliff tops, coastal shingle, downland and other habitats on chalky soils. Also inland in limestone quarries and chalk pits. Predatory, the larvae possibly feed on snails. Found under stones. Adults have been recorded in February and from May to December.

Status Very local. Known only from the southern half of England and South Wales. Recently recorded from only six vice-counties from Glamorganshire to East Kent.

Threats Loss of calcareous grassland through improvement by reseeding or by the application of fertilisers, or by conversion to arable agriculture. Further loss of habitat through gravel extraction, urban and holiday developments, cliff stabilisation schemes, and the infilling of pits and quarries. Lack of, or changes in grazing or cutting regimes may be a further threat.

Management and conservation Management should aim at maintaining open conditions and encouraging early successional stages.

Published sources Alexander, K.N.A. & Clements, D.K. (1988), Anon, (1939), Foster, A.P. (1988), Fowler, W.W. (1887), Fowler, W.W. & Donisthorpe,

H.St J.K. (1913), Hodge, P.J. (1990), Key, R.S. (1990), Lindroth, C.H. (1974), Nature Conservancy Council, (1990), Tozer, D. (1953).

LIONYCHUS QUADRILLUM	**RARE**
A ground beetle	
Order COLEOPTERA	Family CARABIDAE

Lionychus quadrillum (Duftschmid, 1812).

Distribution Recorded from West Cornwall, South Devon, South Hampshire, East Kent, South Essex, East Suffolk, South Lincolnshire and North-east Yorkshire before 1970 and North Essex, East Suffolk, Carmarthenshire and Cardiganshire from 1970 onwards.

Habitat and ecology River shingle or sand. Also on the coast above the high-water mark, and from a saltmarsh. Probably predatory. Adults are very active in spring sunshine, but are found under stones in cool weather. Adults have been recorded from March to September, though adults are less frequent in July and August.

Status Very local, old records suggest that this species formerly had a scattered distribution in England as far north as North-east Yorkshire. Recently recorded from only four vice-counties, including two from central Wales.

Threats River engineering, including dredging, level regulation by damming and flood alleviation schemes. In some areas colonisation by Himalayan balsam *Impatiens glandulifera* can reduce the available habitat. Livestock access can damage the shingle structure. Urban and holiday developments, and river pollution may be further threats.

Management and conservation An under-valued habitat. Shingle and sand banks tend to be mobile and rely on the free flow of river and stream systems. Activities that hinder this flow should be avoided. Adjacent scrub and rank grassland is thought to be important for the species in winter and to escape late summer floods.

Published sources Ashe, G.H. (1944a), Boyce, D.C. (1989), Boyce, D.C. (1990), Boyce, D.C. & Fowles, A.P. (1988), Fowler, W.W. (1887), Fowler, W.W. & Donisthorpe, H.St J.K. (1913), Fowles, A.P. (1989), Fowles, A.P. & Morgan, I.K. (1987), Key, R.S. (1990), Keys, J.H. (1922), Lindroth, C.H. (1974), Morgan, I.K. (1988), Morgan, I.K. (1990), Morley, C. (1904), Shirt, D.B., ed. (1987), Walker, J.J. (1932a), Walsh, G.B. & Rimington, F.C. (1953).

MASOREUS WETTERHALLI   NOTABLE A
A ground beetle
Order COLEOPTERA   Family CARABIDAE

Masoreus wetterhalli (Gyllenhal, 1813). Formerly known as: *Masoreus wetterhali* (Gyllenhal, 1813), *Masoreus wetterhalii* (Gyllenhal, 1813), *Masoreus wetterhallii* (Gyllenhal, 1813).

Distribution Recorded from Dorset, East Sussex, East Kent, South Essex, North Essex, East Norfolk and West Norfolk before 1970 and West Cornwall, Dorset, East Sussex, East Kent, West Suffolk, East Norfolk and West Norfolk from 1970 onwards.

Habitat and ecology Sand dunes, coastal cliffs, coastal shingle, sea walls and on sparsely vegetated sand or gravel, often near the coast. Probably predatory. Adults usually hide under mats of vegetation such as heather and thyme, and have also been found under pieces of old wood. Adults have been recorded in January, February, from April to September and in November.

Status Very local. Recorded from coastal vice-counties from West Cornwall to West Norfolk.

Threats Loss of dune habitat, particularly through afforestation, urban and holiday development. Cliff stabilisation schemes, the construction of sea defences, gravel extraction, and the degradation of remaining habitat by natural succession and the invasion of scrub on stabilised areas are further threats.

Management and conservation Disturbance of coastal shingle should be avoided. In areas of cliff occasional slippages are necessary to maintain habitat continuity. Large areas of unstable cliff are required so that the population does not become isolated and subsequently threatened by individual landslips. Some grazing and other disturbance may be desirable to maintain the early successional stages and prevent the invasion of scrub.

Published sources Fowler, W.W. (1887), Lindroth, C.H. (1974), Mendel, H. (1980).

MISCODERA ARCTICA   NOTABLE B
A ground beetle
Order COLEOPTERA   Family CARABIDAE

Miscodera arctica (Paykull, 1798).

Distribution East Midlands, West Midlands, North East, North West, Dyfed-Powys, North Wales and Scotland. A map is given in Luff (1982).

Habitat and ecology Predominantly upland habitats on dry soils. Mainly moorland, though occasionally from lowland heathland, in wooded areas and also from the silty banks of a river. Predatory, probably on adults and larvae of *Byrrhus* (Byrrhidae). Adults have been found under stones and in moss and have been recorded from April to August.

Status Local. A northern and western species.

Threats Loss of habitat through afforestation, improvement and conversion to arable agriculture. Further habitat degradation has been through the cessation of traditional management practices.

Management and conservation Management should aim for a diversity of successional stages preferably through grazing.

NEBRIA COMPLANATA   NOTABLE A
A ground beetle
Order COLEOPTERA   Family CARABIDAE

Nebria complanata (L., 1767). Formerly known as: *Eurynebria complanata* (L.).

Distribution Recorded from North Devon, North Somerset, Glamorgan, Carmarthenshire and Pembrokeshire before 1970 and North Devon, Glamorgan and Carmarthenshire from 1970 onwards. A map is given in Luff (1982).

Habitat and ecology Beaches, particularly along the strand line. Gregarious and predatory, probably feeding on sandhoppers. Occurs in or near the tidal zone, on bare sand or sandy clay. This beetle is found under beach debris, in particular driftwood, seaweed and flat stones. Adults have been found from May to October, and through the winter when adults move inland into dune areas.

Status Extremely local, but may be abundant where found. Restricted to coastal sites along the Bristol Channel.

Threats The removal of driftwood and other tidal debris through beach tidying schemes and fires at beach parties. Pollution may be a further threat. Because the juxtaposition of beach and dune is important, the loss of dunes through holiday and afforestation may further threaten this species.

Published sources Fowler, W.W. (1887), Greenslade, P.J.M. (1963b), Henderson, M.K. (1988), Lindroth, C.H. (1974), Luff, M.L. (Ed.) (1982), Morgan, I.K. (1988).

CARABIDAE

NEBRIA LIVIDA — NOTABLE A

NEBRIA LIVIDA   **NOTABLE A**
A ground beetle
Order COLEOPTERA   Family CARABIDAE

Nebria livida (L., 1758).

Distribution Recorded from East Suffolk, East Norfolk, Staffordshire, South-east Yorkshire and North-east Yorkshire before 1970 and East Suffolk, East Norfolk, South-east Yorkshire and North-east Yorkshire from 1970 onwards. A map is given in Luff (1982).

Habitat and ecology On boulder-clay cliffs, and sandy shores often mixed with clay. Almost entirely coastal. Predatory and nocturnal, adults hiding during the day in cracks in clay, under lumps of fallen boulder clay at the base of cliffs and under refuse, seaweed or driftwood. This species tends to be gregarious. Adults have been recorded from May to September.

Status Very local and with a disjunct distribution from East Suffolk and East Norfolk to the coasts of South East and North-east Yorkshire. The only recent inland site is a sandpit in East Suffolk.

Threats This species is threatened by cliff stabilisation schemes and the construction of sea defences. Coastal developments may reduce the amount of available habitat. Activities that accelerate or reduce the rate of erosion should be avoided. Beach tidying schemes and the burning of driftwood at beach parties may be further threats.

Management and conservation Occasional slippages are necessary to maintain habitat continuity. Large areas of unstable cliff are required so that the population does not become isolated and subsequently threatened by individual landslips.

Published sources Collier, M.J. (1988a), Collier, M.J. (1988b), Cox, L.G. (1921), Flint, J.H. (1988), Fowler, W.W. (1887), Lindroth, C.H. (1974), Luff, M.L. (Ed.) (1982), Whicher, L.S. (1953).

NEBRIA NIVALIS — NOTABLE A

NEBRIA NIVALIS   **NOTABLE A**
A ground beetle
Order COLEOPTERA   Family CARABIDAE

Nebria nivalis (Paykull, 1790).

Distribution Recorded from South Aberdeenshire, Moray, East Inverness & Nairn, West Inverness, East Ross and Shetland before 1970 and Caernarvonshire, Cumberland, East Inverness & Nairn, South Ebudes, Mid Ebudes, North Ebudes, West Ross and West Sutherland from 1970 onwards. A map is given in Luff (1982).

Habitat and ecology Found mainly at very high altitudes, frequenting frost shattered summits. This species has been noted at a lower altitude on Skye in Scotland. Predatory and probably nocturnal. Occurs under stones among *Rhacomitrium* moss and lichen heath. Usually found in association with *N. gyllenhali*. This species is reputed to be associated with permanent snowfields but it is not restricted to this habitat in Britain. Adults have been recorded mainly in May and June and also in July and August.

Status Status revised from RDB 3 (Rare) in Shirt (1987). Very local and recorded from Caernarvonshire, North Wales through the Lake District of England to Scotland. Possibly under-recorded because of the inaccessability of its habitat.

Threats Erosion of montane habitat may be a problem in areas frequented by large numbers of hill walkers.

Published sources Benson, R.B. (1967), Blair, K.G. (1950), Garland, S.P. & Lee, J. (1981), Horsfield, D. (1988), Johnson, C. (1967), Key, R.S. (1981), Lindroth, C.H. (1974), Luff, M.L. (Ed.) (1982), Owen, J.A. (1988a), Shirt, D.B., ed. (1987), Welch, R.C. (1980), Welch, R.C. (1983a).

NOTIOPHILUS AESTHUANS — NOTABLE B

NOTIOPHILUS AESTHUANS   **NOTABLE B**
A ground beetle
Order COLEOPTERA   Family CARABIDAE

Notiophilus aesthuans (Motschulsky, 1864). Formerly known as: *Notiophilus pusillus* Waterhouse, 1833.

Distribution South, East Anglia, East Midlands, West Midlands, North East, North West, South West Scotland, North East Scotland and North West Scotland. A map is given in Luff (1982).

Habitat and ecology Frequents dry areas including montane lichen heath, spoil heaps on or near lead mine workings, quarries and also stream and river margins. Noted on one occasion from a sand dune. This species has also been found in areas of sugar limestone. Predatory. Appears to prefer bare, gravelly ground with a short-turf and sparse vegetation. Adults have been found under stones, among sand and shingle, and have been recorded from April to September.

Status A local northern and western species, very local in the remaining parts of Great Britain. This species is difficult to identify and has been confused with *N. aquaticus*.

Threats This species may be threatened by river engineering schemes, dredging, level regulation by damming and quarry reclamation projects.

Management and conservation Bare and sparsely vegetated ground and areas of shingle are under-valued habitats. In river and stream systems, shingle banks tend to be mobile and rely on the free flow of water. Activities that hinder this flow should be prevented. Adjacent scrub and rank grassland may be important for the species in winter and to escape late summer floods.

NOTIOPHILUS QUADRIPUNCTATUS
A ground beetle   **NOTABLE B**
Order COLEOPTERA   Family CARABIDAE

Notiophilus quadripunctatus Dejean, 1826. Formerly known as: *Notiophilus quadriguttatus* Fowler, 1887.

Distribution South East, South, East Anglia, East Midlands, North East, North West, South Wales and Dyfed-Powys.

Habitat and ecology Sparsely vegetated and bare ground such as recently dry drainage ditches, undercliffs, heathland and peatland. Predatory. Adults have been recorded in April and from June to September.

Status Widely distributed and local throughout England and Wales. This species is difficult to identify and has been confused with *N. biguttatus*.

Threats The loss of bare and sparsely vegetated land through natural succession. Loss of suitable habitat through changes in land use by conversion to arable agriculture, forestry, urban development and cliff stabilisation schemes and the construction of sea defences.

Management and conservation Some grazing and other disturbance may be desirable to maintain the early successional stages and prevent the invasion of scrub. In areas of soft rock cliff occasional slippages are necessary to maintain habitat continuity. Large areas of unstable cliff are required so that the population does not become isolated and subsequently threatened by individual landslips.

ODACANTHA MELANURA   **NOTABLE B**
A ground beetle
Order COLEOPTERA   Family CARABIDAE

Odacantha melanura (L., 1767). Formerly known as: *Colliuris melanura* (L.).

Distribution South East, South, South West, East Anglia, East Midlands and South Wales. A map is given in Luff (1982).

Habitat and ecology Fens, marshes, broads, grazing levels and the margins of standing water. Probably predatory. Predominantly associated with reed in reed beds, where it can be found under the outer, dead sheathing leaves of the flowering stems. Also amongst bulrushes, particularly in the dead stems. Adults have been recorded from April to June, in August, September and November.

Status Widespread but local in southern and eastern England, more local in south-western England and South Wales.

Threats Drainage for reasons such as agricultural improvement and development is the primary cause of the loss of wetlands. Falling water tables because of water abstraction and river engineering schemes may also threaten this species.

Management and conservation Water tables should be maintained at high levels. Water bodies should be isolated from sources of eutrophication and pollution. Cutting may be needed to maintain open conditions.

OMOPHRON LIMBATUM   **ENDANGERED**
A ground beetle
Order COLEOPTERA   Family CARABIDAE

Omophron limbatum (F., 1777).

Distribution Recorded from East Sussex before 1970 and East Sussex and East Kent from 1970 onwards.

Habitat and ecology On and in sand bordering flooded gravel pits. Predatory. Adults live in burrows in sand and have been recorded mainly in May though also from July to October.

Status Known from very few sites in a limited area on the south-east coast of England. Possibly indigenous in the 19th century, reliably first recorded in Great Britain in 1969. This species presence in Great Britain is probably the result of recent (re)colonisation.

Threats This species is threatened by the infilling of gravel pits, tourist development and natural succession.

CARABIDAE

Management and conservation Management should aim at maintaining open conditions and encouraging early successional stages along the pit edges.

Published sources Hammond, P.M., Smith, K.G.V., Else, G.R. & Allen, G.W. (1989), Lindroth, C.H. (1974), Nature Conservancy Council, (1990), Shirt, D.B., ed. (1987).

OODES HELOPIOIDES　　　　　　**NOTABLE B**
A ground beetle
Order COLEOPTERA　　　　Family CARABIDAE

Oodes helopioides (F., 1792).

Distribution England, Dyfed-Powys and South Wales.

Habitat and ecology Fens, grazing levels, marshes, wet heaths, water meadows and the margins of standing water. Probably predatory. Semi-aquatic, adults can forage under water. Found on soft soil or mud, amongst lush vegetation, and under loose bark at the base of willows in water meadows. Adults have been recorded in February, from May to September and in November.

Status Widespread but local in England and Wales.

Threats Drainage for reasons such as agricultural improvement and development is the primary cause of the loss of fens. Falling water tables because of water abstraction and river engineering schemes may also threaten this species.

Management and conservation Water tables should be maintained at high levels. Water bodies should be isolated from sources of eutrophication and pollution. Grazing, cutting or some other disturbance, on a rotational basis, may be needed to encourage early successional stages and maintain open conditions.

PANAGAEUS BIPUSTULATUS　　　**NOTABLE B**
A ground beetle
Order COLEOPTERA　　　　Family CARABIDAE

Panagaeus bipustulatus (F., 1775). Formerly known as: *Panagaeus quadripustulatus* Sturm, 1815.

Distribution South East, South, South West, East Anglia, East Midlands, West Midlands, North East and South Wales. A map is given in Luff (1982).

Habitat and ecology Calcareous grassland, dry meadows, sand dunes, coastal shingle, heathland, sand pits and coastal sandstone undercliffs. Probably predatory. Recorded under stones and concrete rubble, amongst sparse vegetation at the back of a sandy beach under red sandstone cliffs, under cow pats in a dry meadow and at the edge of a water-filled sand pit. Adults have been recorded in January, February and from May to September.

Status Widespread but local in England, also recorded in South Wales.

Threats This species is threatened by loss of habitat through improvement and conversion to arable agriculture, gravel extraction, afforestation, the infilling of pits, cliff stabilisation schemes, and urban and holiday devlopments. Natural succession and scrub invasion may be a further threat.

Management and conservation Management should aim at maintaining open conditions and encouraging early successional stages.

PANAGAEUS CRUXMAJOR　　　**ENDANGERED**
A ground beetle
Order COLEOPTERA　　　　Family CARABIDAE

Panagaeus cruxmajor (L., 1758).

Distribution Recorded from North Devon, North Somerset, Isle of Wight, East Sussex, East Kent, West Kent, South Essex, Berkshire, East Suffolk, East Norfolk, Cambridgeshire, Huntingdonshire, Northamptonshire, West Gloucestershire, Glamorgan, Carmarthenshire, North Lincolnshire and South-west Yorkshire before 1970 and Carmarthenshire from 1970 onwards.

Habitat and ecology Amongst lush vegetation on soft soil or mud at the margins of standing or slowly running water, in fens and, in South Wales, amongst reeds in slightly brackish fens and dune slacks. Probably predatory. Adults have been found in moss and under driftwood. Adults have been recorded in March and April, and probably occur all year round.

Status Status revised from RDB 2 (Vulnerable) in Shirt (1987). Now extremely local. Formerly widespread through central and southern England. The beetle used to be regularly recorded in the Cambridgeshire and East Anglian fens. At present, the only known population occurs on the coast of Carmarthenshire. There is an unconfirmed record for Mid-west Yorkshire during the 1970's.

Threats Drainage for reasons such as agricultural improvement and development is the primary cause of the loss of fens. Falling water tables because of water abstraction and river engineering schemes may also have contributed to this species decline. Coastal developments may be a further threat to this species.

Management and conservation Water tables should be maintained at high levels. Water bodies should be isolated from sources of eutrophication and pollution. Cutting or some other disturbance, on a rotational basis, may be needed to encourage early successional stages and maintain open conditions.

Published sources Atty, D.B. (1983), Buck, F.D. (1962), Eagles, T.R. (1955b), Edwards, J. (1893), Flint, J.H. (1947), Fowler, W.W. (1887), Fowler, W.W. & Donisthorpe, H.St J.K. (1913), Fowles, A.P. & Morgan, I.K. (1987), Lindroth, C.H. (1974), Shirt, D.B., ed. (1987), Tomlin, J.R.le B. (1921), Wilson, W.A. (1958).

PATROBUS SEPTENTRIONIS   **NOTABLE B**
A ground beetle
Order COLEOPTERA   Family CARABIDAE

Patrobus septentrionis Dejean, 1828.

Distribution North East, North West, North Wales and Scotland.

Habitat and ecology Predominantly montane habitats. Also in grassland and heathlands. Generally a species of high altitudes (above 760m) though in the Shetlands it occurs below 300m. Predatory. Usually found in damp gullies, under stones and in moss by streams. Adults have been recorded in June and July.

Status Widespread but local in the Scottish Highlands, with a few scattered records in northern England and North Wales.

Threats Uncertain, though this species may be threatened by erosion of montane habitat in areas frequented by large numbers of hill walkers and where livestock densities are to high, as well as through the loss of moorland through afforestation.

PELOPHILA BOREALIS   **RARE**
A ground beetle
Order COLEOPTERA   Family CARABIDAE

Pelophila borealis (Paykull, 1790).

Distribution Recorded from Orkney and Shetland before 1970 and East Inverness & Nairn, Orkney and Shetland from 1970 onwards.

Habitat and ecology At the margins of fresh water, both lakes and slow-running rivers. Found on silty or muddy substrates with some vegetation. This species

has been reported in wet muddy flushes on moorland. Predatory.

Status Not listed in the insect Red Data Book (Shirt, 1987). Very local and recently recorded only from Easterness, Orkney and Shetland. Formerly recorded from as far south as South Devon, though this record requires confirmation. The species' occurence in Perthshire also needs to be confirmed. Records for Merionethshire and Derbyshire are almost certainly erroneous.

Threats This species may be threatened by the loss of its habitat through river engineering schemes, including dredging, level regulation by damming and flood alleviation. Drainage, and scrub invasion on lake edges as well as pollution may be further threats.

Management and conservation An under-valued habitat. This habitat tends to be mobile and relies on the free flow of river and stream systems, and wave action on lakes. Activities that hinder this flow should be prevented. Adjacent scrub and rank grassland may be important for the species in winter and to escape late summer floods.

Published sources Fowler, W.W. (1887), Lindroth, C.H. (1974).

PERILEPTUS AREOLATUS   **NOTABLE A**
A ground beetle
Order COLEOPTERA   Family CARABIDAE

Perileptus areolatus (Creutzer, 1799).

Distribution Recorded from "Cornwall", South Devon, Herefordshire, Worcestershire, Shropshire, Breconshire, Caernarvonshire, Westmorland & North Lancashire and Dumfriesshire before 1970 and Herefordshire, Radnorshire, Cardiganshire, Montgomeryshire and Caernarvonshire from 1970 onwards.

Habitat and ecology River shingle. Predatory. Adults and probably larvae live in open areas of shingle near the waters-edge, amongst fine sand and under larger pebbles. Adults have been recorded from April to July.

Status Very local. A western species.

Threats River engineering, including dredging, level regulation by damming and flood alleviation schemes. In some areas colonisation by Himalayan balsam *Impatiens glandulifera* and natural succession can reduce the available habitat. Livestock access can damage shingle structure. River pollution maybe a further threat.

CARABIDAE

Management and conservation An under-valued habitat. Shingle and sand banks tend to be mobile and rely on the free flow of river and stream systems. Activities that hinder this flow should be prevented. Adjacent scrub and rank grassland may be important for the species in winter and to escape late summer floods.

Published sources Allen, A.A. (1954e), Boyce, D.C. (1989), Boyce, D.C. (1990), Boyce, D.C. & Fowles, A.P. (1988), Fowler, W.W. (1887), Fowler, W.W. & Donisthorpe, H.St J.K. (1913), Lindroth, C.H. (1974), Parry, J.A. (1978a).

PLATYDERUS RUFICOLLIS	**NOTABLE B**

A ground beetle
Order COLEOPTERA   Family CARABIDAE

Platyderus ruficollis (Marsam, 1802).

Distribution England and North Wales. A map is given in Luff (1982).

Habitat and ecology On dry, usually sandy or chalky ground. Recorded from heathland, sand dunes, clay cliffs, gravel pits, chalk and limestone quarries, a brick works and grassland. Also in urban areas such as gardens and churchyards. Predatory. Found amongst leaves, moss and other vegetation, and under stones. Adults have been recorded from March to November.

Status Widespread but local in England and North Wales, very local outside southern and eastern England.

Threats This species is threatened by loss of habitat through urban and holiday developments, improvement and conversion to arable agriculture, and the infilling of pits and quarries.

Management and conservation Management should aim at maintaining open conditions and encouraging early successional stages.

POGONUS LITTORALIS	**NOTABLE B**

A ground beetle
Order COLEOPTERA   Family CARABIDAE

Pogonus littoralis (Duftschmid, 1812). Formerly known as: *Pogonus litoralis* (Duftschmid, 1812).

Distribution South East, South, South West, East Anglia, East Midlands and South Wales.

Habitat and ecology Coastal habitats, particularly saltmarshes. Predatory. On clay and mud, and also under seaweed. Adults have been recorded from May to July and in September and October.

Status Local and recorded from the southern and eastern coasts of England. Also noted in South Wales.

Threats Loss of saltmarsh through reclamation, erosion and the construction of sea defences.

POGONUS LURIDIPENNIS	**RARE**

A ground beetle
Order COLEOPTERA   Family CARABIDAE

Pogonus luridipennis (Germar, 1822).

Distribution Recorded from Isle of Wight, South Hampshire, West Sussex, East Sussex, East Kent, East Suffolk, East Norfolk, West Gloucestershire and North Lincolnshire before 1970 and Dorset, East Norfolk and North Lincolnshire from 1970 onwards.

Habitat and ecology Predominantly coastal habitats, particularly saltmarshes. Also inland, up tidal rivers. Predatory. Found under stones on muddy sand and under seaweed. Adults have been recorded in June and September.

Status Not listed in the insect Red Data Book (Shirt, 1987). Very local and only recently recorded from three vice-counties. Southern and eastern coasts of England.

Threats Loss of habitat through urban and holiday development, reclamation, erosion and the construction of sea defences.

Management and conservation Effort should initially ensure the protection of intact saltmarsh and estuarine systems.

Published sources Collier, M.J. (1988a), Drane, A.B. (1990), Fowler, W.W. (1887), Fowler, W.W. & Donisthorpe, H.St J.K. (1913), Lindroth, C.H. (1974).

POLISTICHUS CONNEXUS	**VULNERABLE**

A ground beetle
Order COLEOPTERA   Family CARABIDAE

Polistichus connexus (Fourcroy, 1785). Formerly known as: *Polystichus connexus* (Fourcroy, 1785), *Polystichus vittatus* Brulle, 1834.

Distribution Recorded from Dorset, East Sussex, East Kent, South Essex, North Essex, Berkshire, East

Suffolk and East Norfolk before 1970 and East Sussex, East Kent, West Kent and Surrey from 1970 onwards.

Habitat and ecology Saltmarshes, coastal undercliffs, the sea-shore, and on silt or clay soils by rivers, as well as from damp heathland. This species was also recorded from the roots of trees or stumps in Windsor Great Park, Berkshire in the 1930s and 1940s (A.A. Allen pers. comm.). Probably predatory. Adults have been found in cracks and crevices in undercliffs, the beetle probably also occurs under stones. Adults have been recorded from April to June.

Status Very local, formerly recorded between Dorset and East Norfolk. Recently recorded fom only four vice-counties, all in south-eastern England.

Threats This species is threatened by cliff stabilisation schemes, the construction of sea defences and reclamation. Coastal developments may reduce the amount of available habitat. Activities that accelerate or reduce the rate of erosion should be avoided. Drainage, and the degradation of suitable habitat through natural succession and the invasion of scrub on stabilised areas is a further threat.

Management and conservation In cliff situations, occasional slippages are necessary to maintain habitat continuity. Large areas of unstable cliff are required so that the population does not become isolated and subsequently threatened by individual landslips. Water tables should be maintained at high levels.

Published sources Anon, (1932), Buck, F.D. (1955), Edwards, J. (1893), Flint, J.H. (1946), Fowler, W.W. (1887), Fowler, W.W. & Donisthorpe, H.St J.K. (1913), Lindroth, C.H. (1974), Shirt, D.B., ed. (1987), Walker, J.J. (1932a).

PTEROSTICHUS AETHIOPS   **NOTABLE B**
A ground beetle
Order COLEOPTERA   Family CARABIDAE

Pterostichus aethiops (Panzer, 1796). Formerly known as: *Feronia aethiops* (Panzer).

Distribution South West, North East, North West, Dyfed-Powys, North Wales, South East Scotland, South West Scotland and North West Scotland.

Habitat and ecology Montane habitats, upland areas of grass, moorland, bogs and woodland. This species has also been found at lower altitudes. Predatory. Often found under stones. Adults have been recorded from January to October. This species may migrate from moorland (summer) to woodland (winter).

Status Widespread but local. A northern and western species.

Threats Loss of moorland and bogs through conversion to other land use and by drainage.

Management and conservation The juxtaposition of open moorland and woodland may be important for this species.

PTEROSTICHUS ANGUSTATUS   **NOTABLE B**
A ground beetle
Order COLEOPTERA   Family CARABIDAE

Pterostichus angustatus (Duftschmid, 1812). Formerly known as: *Feronia angustata* (Duftschmid), *Lyperosomus angustatus* (Duftschmid).

Distribution England, South Wales and South East Scotland. A map is given in Luff (1982).

Habitat and ecology On sandy or peaty soils. Heathland, also in woodland and wetlands. Predatory. Mainly found on dry heaths, often on burnt ground. Adults have been recorded in January and from April to October.

Status An established immigrant in Great Britain from around 1900. Widespread but local. Still spreading northwards.

Threats Loss of habitat through changes in land use, mainly by conversion to arable agriculture, clear-felling and afforestation, drainage and urban development.

Management and conservation Management should aim to encourage the early successional stages and maintain open conditions.

PTEROSTICHUS ANTHRACINUS   **NOTABLE B**
A ground beetle
Order COLEOPTERA   Family CARABIDAE

Pterostichus anthracinus (Panzer, 1795). Formerly known as: *Feronia anthracina* (Panzer), *Feronia antracina* (Panzer).

Distribution England, South Wales and South West Scotland.

Habitat and ecology The margins of lakes, ponds, reservoirs, gravel pits, rivers, ditches and in fenland and carr. Predatory. Found among damp, lush vegetation on soft soil or mud. Adults have been recorded in January, March, May, and from August to October.

CARABIDAE

Status Widespread but local in England, recorded from South Wales and South West Scotland.

Threats The infilling of gravel pits, drainage, water abstraction schemes and river engineering, including dredging, level regulation by damming and flood alleviation schemes. In some areas colonisation by Himalayan balsam *Impatiens glandulifera* can reduce the available habitat. River pollution maybe a further threat.

Management and conservation Water tables should be maintained at high levels. Water bodies should be isolated from sources of eutrophication and pollution. Activities that hinder the natural flow of rivers and streams should be avoided. Management should aim at maintaining open conditions and encouraging early successional stages along pit and lake edges.

PTEROSTICHUS ATERRIMUS   ENDANGERED
A ground beetle
Order COLEOPTERA   Family CARABIDAE

Pterostichus aterrimus (Herbst, 1784). Formerly known as: *Feronia aterrima* (Herbst), *Lyperosomus aterrimus* (Herbst).

Distribution Recorded from South Hampshire, East Norfolk and Cambridgeshire before 1970 and South Hampshire from 1970 onwards.

Habitat and ecology *Sphagnum* bogs, fens and acid pools. Predatory. Found at the edge of water on muddy or peaty soils. Adults have been recorded in June and September.

Status Originally found in the fens of East Anglia but not recorded there since 1910. Found from 1969 to 1973 in a formerly marshy area south of Denny Wood, New Forest, South Hampshire. Not noted since 1973.

Threats This species has probably declined because of drainage for reasons such as agricultural improvement and development. Falling water tables because of river engineering schemes may also have contributed to this species decline.

Published sources Appleton, D. (1969b), Fowler, W.W. (1887), Lindroth, C.H. (1974), Shirt, D.B., ed. (1987).

PTEROSTICHUS CRISTATUS   NOTABLE B
A ground beetle
Order COLEOPTERA   Family CARABIDAE

Pterostichus cristatus (Dufour, 1820). Formerly known as: *Feronia cristata* (Dufour), *Lyperosomus cristatus* (Dufour), *Pterostichus cristatus ssp. parumpunctatus* Germar, 1824.

Distribution North East, North West, South East Scotland, South West Scotland and North East Scotland. A map is given in Luff (1982).

Habitat and ecology Broad-leaved woodland. Also recorded from damp grassland, field margins, roadside verges and on the seashore. Predatory and gregarious. This beetle appears to prefer stoney flushes in woods being found on sand and under stones. Also found in leaf litter. Adults have been recorded from April to October.

Status Local. A species of northern England, with a few scattered records in Scotland.

Threats This species may be threatened by the clear-felling of woodland for conversion to other land use. Drainage for reasons such as agricultural improvement may be a further threat.

Management and conservation Water tables should be maintained at high levels.

PTEROSTICHUS GRACILIS   NOTABLE B
A ground beetle
Order COLEOPTERA   Family CARABIDAE

Pterostichus gracilis (Dejean, 1828). Formerly known as: *Feronia gracilis* (Dejean).

Distribution South East, South West, East Anglia, East Midlands, North East, North West, South Wales, Dyfed-Powys, South East Scotland, South West Scotland and North East Scotland.

Habitat and ecology The margins of lakes, ponds and reservoirs, in fens, on riverbanks and in wet grassland. Predatory. Found among damp, lush vegetation on soft soil or mud, and under stones. Adults have been recorded from March to July and in October, and have been found hibernating under bark.

Status Widespread but local throughout Great Britain.

Threats The infilling of lakes and ponds, and drainage for reasons such as agricultural improvement and development. Falling water tables because of water abstraction and river engineering schemes may also threaten this species.

Management and conservation Water tables should be maintained at high levels. Water bodies should be isolated from sources of eutrophication and pollution. Management should aim at maintaining open conditions and encouraging early successional stages along lake and pond edges. Activities that hinder the natural flow of rivers and streams should be avoided.

PTEROSTICHUS KUGELANNI   ENDANGERED
A ground beetle
Order COLEOPTERA   Family CARABIDAE

Pterostichus kugelanni (Panzer, 1797). Formerly known as: *Feronia kugelanni* (Panzer), *Poecilus dimidiatus* (Olivier, 1795), *Pterostichus dimidiatus* (Olivier).

Distribution Recorded from West Cornwall, East Cornwall, South Devon, North Devon, Dorset, Isle of Wight, South Hampshire, North Hampshire, West Sussex, East Sussex, East Kent, West Kent, Surrey, Middlesex, Oxfordshire, East Norfolk and Glamorgan before 1970 and South Hampshire from 1970 onwards.

Habitat and ecology Heathland and sand pits. Predatory. Found on sandy or gravelly soils, probably under stones and amongst herbage. Adults have been recorded primarily in April and May, though also in July.

Status Not listed in the insect Red Data Book (Shirt, 1987). Only one recent record, from South Hampshire in 1970.

Threats This species has probably declined because of the loss and fragmentation of heathland through changes in land use, mainly by conversion to arable, forestry and urban development. Further habitat degradation has been through the cessation of traditional heathland management practices.

Management and conservation Management should aim to encourage the early successional stages and open conditions, preferably by grazing or by rotational cutting, scraping or burning.

Published sources Brown, S.C.S. (1954), Fowler, W.W. (1887), Fowler, W.W. & Donisthorpe, H.St J.K. (1913), Lindroth, C.H. (1974), Side, K.C. (1957).

PTEROSTICHUS LEPIDUS   NOTABLE B
A ground beetle
Order COLEOPTERA   Family CARABIDAE

Pterostichus lepidus (Leske, 1785). Formerly known as: *Feronia lepida* (Leske), *Poecilus lepidus* (Leske).

Distribution South East, South, South West, East Anglia, West Midlands, North East, North West, Dyfed-Powys, South East Scotland, South West Scotland and North East Scotland. A map is given in Luff (1982).

Habitat and ecology On sandy or gravelly soils. Heathland, particularly dry heaths, though probably also grassland. Predatory. Found on open, sun exposed ground among sparse vegetation. Adults have been recorded from May to August.

Status Widespread but local throughout Great Britain.

Threats Much heathland has been lost, or fragmented, through changes in land use, mainly by conversion to arable agriculture, forestry and urban development. Further habitat degradation has been through the cessation of traditional heathland management practices.

Management and conservation Management should aim to encourage early successional stages and open conditions, preferably by grazing or by rotational cutting, scraping or burning.

PTEROSTICHUS LONGICOLLIS   NOTABLE B
A ground beetle
Order COLEOPTERA   Family CARABIDAE

Pterostichus longicollis (Duftschmid, 1812). Formerly known as: *Feronia longicollis* (Duftschmid), *Pterostichus inaequalis* (Marsham, 1802).

Distribution South East, South, South West, East Anglia, East Midlands, West Midlands, North East, South Wales and North Wales.

Habitat and ecology Bare margins of lakes and ponds, also river banks, gravel and clay pits. This species may have a preference for calcareous substrates. Predatory. Appears to be subterranean, found in the soil, under stones and in flood litter. It has also been found in an old stump on a riverbank. Adults have been recorded from February to June, and in August and September.

Status Widespread but local in England and Wales.

Threats The infilling of ponds and pits and natural succession.

CARABIDAE

Management and conservation Management should aim at maintaining open conditions and encouraging early successional stages along pond, lake and pit edges.

PTEROSTICHUS OBLONGOPUNCTATUS
A ground beetle   **NOTABLE B**
Order COLEOPTERA   Family CARABIDAE

Pterostichus oblongopunctatus (F., 1787). Formerly known as: *Feronia oblongopunctata* F., 1787.

Distribution South East, South, South West, East Midlands, West Midlands, North East, North West, Dyfed-Powys, North Wales and Scotland. A map is given in Luff (1982).

Habitat and ecology Broad-leaved woodland. On all kinds of soil. Predatory. Nearly always found under bark. Adults have been recorded throughout the year.

Status A predominantly northern species with very few records in the southern the south-eastern England

Threats The loss of broad-leaved woodland through, for example, clear-felling and afforestation.

Management and conservation Dead timber, especially with the bark attached, should be retained. The removal of dead timber from ancient trees should be avoided. Gaps in the age structure of the tree population should be identified and the continuity of the appropriate habitat ensured by natural regeneration, suitable planting and possibly with pollarding.

STENOLOPHUS SKRIMSHIRANUS   NOTABLE A
A ground beetle
Order COLEOPTERA   Family CARABIDAE

Stenolophus skrimshiranus Stephens, 1828.

Distribution Recorded from Isle of Wight, South Hampshire, East Sussex, East Kent, West Kent, South Essex, North Essex, Middlesex, East Suffolk, West Norfolk, Cambridgeshire, Huntingdonshire and South Lincolnshire before 1970 and North Somerset, Isle of Wight, South Hampshire, East Sussex, East Kent, West Kent, South Essex and Huntingdonshire from 1970 onwards.

Habitat and ecology Marshes, fens and the margins of standing water. Frequently found near the coast. Probably predatory. Noted amongst lush vegetation, on mud, under timber in a fen and under clods of loose peat in a damp gully. Adults have been recorded in May, June and November.

Status Widespread and very local in southern England.

Threats Drainage for reasons such as agricultural improvement and development is the primary cause of the loss of fens. Falling water tables because of water abstraction and river engineering schemes may also threaten this species.

Management and conservation Water tables should be maintained at high levels. Water bodies should be isolated from sources of eutrophication and pollution. Grazing, cutting or some other disturbance, on a rotational basis, may be needed to maintain open conditions.

Published sources Fowler, W.W. (1887), Fowler, W.W. & Donisthorpe, H.St J.K. (1913), Hodge, P.J. (1989), Key, R.S. (1990), Lindroth, C.H. (1974).

STENOLOPHUS TEUTONUS   NOTABLE B
A ground beetle
Order COLEOPTERA   Family CARABIDAE

Stenolophus teutonus (Schrank, 1781). Formerly known as: *Stenolophus abdominalis* Gene, 1836.

Distribution South East, South, East Anglia, East Midlands and West Midlands.

Habitat and ecology Bare ground at the margins of standing water such as in gravel pits and ornamental lakes. Also recorded from the edges of a pond in woodland, coastal chines and clay undercliffs. Probably predatory. Found in grass tussocks and wet vehicle tracks but probably occurs in a variety of situations including under stones and at the roots of plants. Adults have been recorded in May, June and August.

Status Widespread but local in the southern half of England

Threats This species is threatened by the infilling of lakes and pits, falling water tables because of water abstraction and river engineering schemes, pollution, as well as through cliff stabilisation schemes and the construction of sea defences. Coastal developments may reduce the amount of available habitat. Activities that accelerate or reduce the rate of erosion should be avoided. The degradation of suitable habitat through natural succession and the invasion of scrub is a further threat.

Management and conservation Occasional slippages are necessary to maintain habitat continuity. Large areas of unstable cliff are required so that the population does not become isolated and subsequently threatened by individual landslips. Management should aim at maintaining open conditions and encouraging early successional stages along pit and lake edges.

TACHYS BISTRIATUS   **NOTABLE B**
A ground beetle
Order COLEOPTERA   Family CARABIDAE

Tachys bistriatus (Duftschmid, 1812).

Distribution South East, South, South West, East Midlands, East Anglia, North East and South Wales.

Habitat and ecology River banks, saltmarshes, part-flooded gravel pits and coastal habitats. Predatory. Found on damp sand or clay by running water and wet sand on the coast. Adults have been recorded from April to July.

Status Widespread but local.

Threats Urban and holiday developments, the infilling of gravel pits, and river engineering, including dredging, level regulation by damming and flood alleviation schemes. In some areas colonisation by Himalayan balsam *Impatiens glandulifera* can reduce the available habitat. River pollution is a further threat.

Management and conservation Sand banks tend to be mobile and rely on the free flow of river and stream systems. Activities that hinder this flow should be avoided. Adjacent scrub and rank grassland may be important for the species in winter and to escape late summer floods. Management should aim at maintaining open conditions and encouraging early successional stages along pit edges.

TACHYS EDMONDSI   **ENDANGERED**
A ground beetle   **ENDEMIC**
Order COLEOPTERA   Family CARABIDAE

Tachys edmondsi Moore, 1956. Formerly known as: *Tachys piceus* Edmonds, 1934.

Distribution Recorded from South Hampshire before 1970.

Habitat and ecology Predatory. Found in wet *Sphagnum* on bogs.

Status Endemic. Status revised from RDB 5 (Category 3) (Endemic; Rare) in Shirt (1987). Only known from *Sphagnum* bogs in the New Forest, South Hampshire. Not recorded in recent years.

Threats This species may have declined as a result of drainage and through fires, particularly during drought.

Management and conservation Much of the New Forest is notified as an SSSI.

Published sources Edmonds, T.H. (1934), Lindroth, C.H. (1974), Shirt, D.B., ed. (1987).

TACHYS MICROS   **NOTABLE A**
A ground beetle
Order COLEOPTERA   Family CARABIDAE

Tachys micros (Fischer von Waldheim, 1828).

Distribution Recorded from Dorset, South Hampshire and East Sussex before 1970 and Dorset, South Hampshire, East Sussex and Caernarvonshire from 1970 onwards. A map is published in Luff (1982).

Habitat and ecology Coastal habitats. Particularly undercliffs, landslips and the seashore. Predatory. Found on damp sand at the base of coastal cliffs and on the seashore. Also noted under small sandstone blocks lying on damp sand. Often in areas of seepages. Adults have been recorded from April to June and in September.

Status Status revised from RDB 3 (Rare) in Shirt (1987). Very local and recorded from very few sites between Dorset and East Sussex. Also noted in Caernarvonshire.

Threats This species is threatened by cliff stabilisation schemes and the construction of sea defences. Coastal developments may reduce the amount of available habitat. Activities that accelerate or reduce the rate of erosion should be avoided. The degradation of suitable habitat through natural succession and the invasion of scrub on stabilised areas is a further threat.

Management and conservation Occasional slippages are necessary to maintain habitat continuity. Large areas of unstable cliff are required so that the population does not become isolated and subsequently threatened by individual landslips.

Published sources Allen, J.W. & Nicholson, G.W. (1924a), Bedwell, E.C. (1943a), Lindroth, C.H. (1974), Luff, M.L. (Ed.) (1982), Shirt, D.B., ed. (1987).

CARABIDAE

TACHYS PARVULUS   NOTABLE B
A ground beetle
Order COLEOPTERA   Family CARABIDAE

Tachys parvulus (Dejean, 1831).

Distribution South West, South East, South, East Midlands, West Midlands, Dyfed-Powys and North West.

Habitat and ecology Coastal sites, river shingle, the margins of lakes and gravel pits, gardens, rubbish tips, old walls and paths. Predatory. Found near tiny trickles and noted from shingle with little or no vegetation cover, and probably also on bare sand in coastal situations and bare mud or soil by lake margins. Recorded under pearlwort growing in cracks in a concrete path and behind a loose piece of brick in an old, partly decayed red brick wall. Adults have been found from March to September.

Status Very local and found over a widely scattered area in southern England, also recorded in South Wales. Possibly increasing.

Threats Urban and coastal development, the infilling of gravel pits, and river engineering, including dredging, level regulation by damming and flood alleviation schemes. In some areas colonisation by Himalayan balsam *Impatiens glandulifera* and natural succession can reduce the available habitat. River pollution maybe a further threat.

Management and conservation An under-valued habitat. Shingle and sand banks tend to be mobile and rely on the free flow of river and stream systems. Activities that hinder this flow should be avoided. Adjacent scrub and rank grassland may be important for the species in winter and to escape late summer floods. Management should aim at maintaining open conditions and encouraging early successional stages along lake and pit edges.

TACHYS SCUTELLARIS   NOTABLE A
A ground beetle
Order COLEOPTERA   Family CARABIDAE

Tachys scutellaris Stephens, 1828.

Distribution Recorded from North Devon, Isle of Wight, South Hampshire, East Sussex, East Kent, East Norfolk, West Norfolk and North Lincolnshire before 1970 and North Devon, Dorset, East Kent, South Essex, North Essex, East Norfolk and West Norfolk from 1970 onwards.

Habitat and ecology Saltmarshes, also wetland near the coast. A species associated with saline or brackish water. Predatory. Occurs on sandy silts and mud. Adults have been recorded from May to September.

Status Status revised from RDB 3 (Rare) in Shirt (1987). Very local and restricted to the southern and eastern coasts of England.

Threats Loss of saltmarsh through reclamation, erosion and the construction of sea defences.

Published sources Collier, M.J. (1988a), Eagles, T.R. (1952), Fowler, W.W. (1887), Fowler, W.W. & Donisthorpe, H.St J.K. (1913), Lindroth, C.H. (1974), Shirt, D.B., ed. (1987), Walker, J.J. (1927), Walker, J.J. (1932a), Whicher, L.S. (1953).

TACHYS WALKERIANUS   ENDANGERED
A ground beetle
Order COLEOPTERA   Family CARABIDAE

Tachys walkerianus Sharp, 1913.

Distribution Recorded from South Hampshire and Surrey before 1970 and Surrey from 1970 onwards.

Habitat and ecology Predatory. Found in wet *Sphagnum* on bogs.

Status Not listed in the insect Red Data Book (Shirt, 1987). Only one recent record.

Threats This species may have declined as a result of drainage and through fires, particularly during drought.

Published sources Lindroth, C.H. (1974).

THALASSOPHILUS LONGICORNIS NOTABLE A
A ground beetle
Order COLEOPTERA   Family CARABIDAE

Thalassophilus longicornis (Sturm, 1825). Formerly known as: *Trechus longicornis* (Sturm).

Distribution Recorded from East Kent, Shropshire, Cardiganshire, Merionethshire, Caernarvonshire, South Northumberland, Westmorland & North Lancashire, Cumberland, Dumfriesshire, Roxburghshire and Argyll Main before 1970 and Radnorshire, Caernarvonshire and Argyll Main from 1970 onwards.

Habitat and ecology River shingle. Predatory. Adults, and probably larvae, live in open areas of shingle near the water's edge, nearly always occurring at a depth of many inches amongst the finer shingle. Adults have been recorded in June and July.

Status Very local with a northern and western distribution, though also recorded in East Kent. Probably under-recorded.

Threats River engineering, including dredging, level regulation by damming and flood alleviation schemes. In some areas colonisation by Himalayan balsam *Impatiens glandulifera* and natural succession can reduce the available habitat. Livestock access can damage shingle structure. River pollution maybe a further threat.

Management and conservation An under-valued habitat. Shingle and sand banks tend to be mobile and rely on the free flow of river and stream systems. Activities that hinder this flow should be prevented. Adjacent scrub and rank grassland may be important for the species in winter and to escape late summer floods.

Published sources Allen, A.A. (1976a), Fowler, W.W. (1887), Lindroth, C.H. (1974), Parry, J.A. (1978a).

TRECHUS DISCUS   **NOTABLE B**
A ground beetle
Order COLEOPTERA   Family CARABIDAE

Trechus discus (F., 1792). Formerly known as: *Lasiotrechus discus* (F.).

Distribution South East, South, East Anglia, East Midlands, West Midlands, North East, North West, South Wales and Dyfed-Powys.

Habitat and ecology The margins of rivers, ditches and gravel pits. Predatory. Adults have been found amongst vegetation and moss on fine silt and mud, also in cracks and crevices particularly in clay ditch banks and in the burrows of small mammals. This species has been noted at light. Adults have been recorded from June to October.

Status Widespread but local in England and Wales.

Threats River engineering, including dredging, level regulation by damming and flood alleviation schemes. In some areas colonisation by Himalayan balsam *Impatiens glandulifera* and natural succession can reduce the available habitat. River pollution maybe a further threat.

Management and conservation Silty and muddy areas rely on the free flow of river and stream systems. Activities that hinder this flow should be prevented.

Adjacent scrub and rank grassland may be important for the species in winter and to escape late summer floods. Management should aim at maintaining open conditions and encouraging early successional stages along pit edges.

TRECHUS FULVUS   **NOTABLE B**
A ground beetle
Order COLEOPTERA   Family CARABIDAE

Trechus fulvus Dejean, 1831. Formerly known as: *Trechus lapidosus* Dawson, 1849.

Distribution South East, South, South West, East Anglia, West Midlands, North East, Wales, South East Scotland, South West Scotland and North East Scotland.

Habitat and ecology Coastal. Under stones on rocky or sandy sea-shores near the high-water mark, especially close to fresh water springs or trickles. It has also been found at the base of sandstone cliffs a few inches down in damp shingle, on dunes and from coastal shingle. Predatory. Adults have been recorded in January, February, April, June, September and November.

Status Widespread but local. Recorded throughout Great Britain.

Threats This species may be threatened by urban and holiday development, gravel extraction and possibly marine pollution.

Management and conservation On shingle sites disturbance should be avoided.

TRECHUS RIVULARIS   **RARE**
A ground beetle
Order COLEOPTERA   Family CARABIDAE

Trechus rivularis (Gyllenhal, 1810).

Distribution Recorded from East Suffolk, East Norfolk, Cambridgeshire and Huntingdonshire before 1970 and East Norfolk, Cambridgeshire, Montgomeryshire, Caernarvonshire, Mid-west Yorkshire and South Northumberland from 1970 onwards.

Habitat and ecology Fens, and upland mires and mosses. Predatory. Adults have been found in litter and grass tussocks while larvae have been found in vole runs. Adults have been recorded from June to August.

CARABIDAE

Status Status revised from RDB 1 (Endangered) in Shirt (1987). Very local and currently known from a very few vice-counties in England and Wales. May be under-recorded because of the remoteness of upland sites.

Threats Loss of moorland through afforestation and conversion to upland pasture. Drainage for reasons such as agricultural improvement and development is the primary cause of the loss of fens. Falling water tables due to river engineering schemes may also threaten this species.

Management and conservation Water tables should be maintained at high levels.

Published sources Lindroth, C.H. (1974), Shirt, D.B., ed. (1987).

TRECHUS RUBENS	NOTABLE B
A ground beetle	
Order COLEOPTERA	Family CARABIDAE

Trechus rubens (F., 1792).

Distribution South East, South West, East Anglia, East Midlands, West Midlands, North East, North West, North Wales and Scotland.

Habitat and ecology Wetland and the margins of ponds and rivers, often near coniferous woodland. Predatory. Adults occur under stones and amongst vegetation debris. This species has been noted at light. Adults have been recorded in April, May, July and August.

Status Widespread but local throughout Great Britain.

Threats Drainage for reasons such as agricultural improvement and development is the primary cause of the loss of wetland habitats. Water abstraction and river engineering schemes including dredging, level regulation by damming and flood alleviation, and pollution may also threaten this species.

Management and conservation Water tables should be maintained at high levels. Water bodies should be isolated from sources of eutrophication and pollution.

TRECHUS SUBNOTATUS	ENDANGERED
A ground beetle	
Order COLEOPTERA	Family CARABIDAE

Trechus subnotatus Dejean, 1831.

Distribution Recorded from South Devon and South-west Yorkshire before 1970 and South Devon and South-west Yorkshire from 1970 onwards.

Habitat and ecology Uncertain. This species has been recorded from a moor, a wood, a grammar school playing field and a builders tip in a wood. Predatory. Adults have been found in and amongst plant debris, soil and rubble. Adults have been recorded from February to May, and in August and November.

Status Introduced, probably on more than one occasion, and established but not spreading. A record for this species on Dartmoor in 1932 does suggest that it may also be a native species.

Published sources Allen, A.A. (1950a), Aubrook, E.W. (1972), Flint, J.H. (1984b), Flint, J.H. (1988), Lindroth, C.H. (1974), Shirt, D.B., ed. (1987).

ZABRUS TENEBRIOIDES	NOTABLE A
A ground beetle	
Order COLEOPTERA	Family CARABIDAE

Zabrus tenebrioides (Goeze, 1777). Formerly known as: *Zabrus gibbus* (F., 1794), *Zabrus tenebroides* (F., 1794).

Distribution Recorded from "Wilts", Dorset, Isle of Wight, South Hampshire, North Hampshire, West Sussex, East Sussex, East Kent, Surrey, South Essex, North Essex, East Norfolk, Cambridgeshire, Glamorgan and Leicestershire & Rutland before 1970 and South Hampshire, West Sussex, East Sussex, East Kent, Berkshire and Cambridgeshire from 1970 onwards. A map is given in Luff (1982).

Habitat and ecology On chalky or sandy soils. Coastal habitats and cultivated land. Phytophagous, the larvae feeding on the young shoots of grasses and cereals. Adults have been found in numbers under stones and dead vegetation at the margins of wheat fields. Adults have been recorded in August and September.

Status Only known from the southern half of England and Wales. Declining, recently recorded in south-eastern England from South Hampshire to Cambridgeshire. Can be found in large numbers.

Threats Loss of habitat through improvement by reseeding or by the application of fertilisers, or by conversion to intensive arable agriculture. The tidying up of field margins and the application of pesticides have probably contributed to the decline of this species. Lack of, or changes in grazing or cutting regimes may be a further threat.

Management and conservation Management should aim at maintaining open conditions and encouraging early successional stages.

Published sources Fowler, W.W. (1887), Fowler, W.W. & Donisthorpe, H.St J.K. (1913), Hodge, P.J. (1977b), Lindroth, C.H. (1974), Luff, M.L. (Ed.) (1982).

ACANTHOCINUS AEDILIS   **NOTABLE B**
The Timberman
Order COLEOPTERA   Family CERAMBYCIDAE

Acanthocinus aedilis (L., 1758).

Distribution Scotland.

Habitat and ecology Coniferous woodland. Associated with Scots pine. On the Continent, this species has also been noted on spruce. In Britain, the beetle is found in recently dead pines in which the needles are still present but are brown and brittle. Females have been noted ovipositing in cut logs utilising holes made by beetles of the family Scolytidae. Larvae feed entirely within the bark of trunks and branches and appear to take two years to develop. Depending upon the thickness of the bark pupation occurs in a cell either entirely within the bark, usually lower down on the trunk, or in a cell between the bark and the outer sapwood or in a cell beneath the bark and entirely within the outer sapwood. Adult emergence normally begins in July and continues throughout August and September. The beetle may also overwinter in the adult state within the pupal cell as there are records of adult emergence in March. Continental literature states that the life cycle lasts one year.

Status Exclusively Scottish and local. The beetle appears to be a component of the native pine forest fauna. All other records are probably either adventive or established introductions.

Threats Loss of native pine forest through, for example, clear-felling. Habitat loss in the remaining areas through the felling of ancient trees and the destruction or removal of standing and fallen dead wood for reasons such as forest hygeine, aesthetic tidiness, public safety or for use as fire wood.

Management and conservation Ancient trees and both standing and fallen dead timber, especially with the

bark attached, should be retained. Gaps in the age structure of the tree population should be identified and regeneration encouraged to ensure the continuity of the dead wood habitats.

ACMAEOPS COLLARIS   **ENDANGERED**
A longhorn beetle
Order COLEOPTERA   Family CERAMBYCIDAE

Acmaeops collaris (L., 1758). Formerly known as: *Pachyta collaris* (L.).

Distribution Recorded from East Cornwall, South Devon, South Hampshire, East Sussex, East Kent, West Kent, Surrey, Hertfordshire, Cambridgeshire, East Gloucestershire, Herefordshire, Worcestershire, North Lincolnshire, Leicestershire & Rutland, Cheshire and South Lancashire before 1970 and Cheshire from 1970 onwards.

Habitat and ecology Broad-leaved woodland, especially on steep slopes on sandy soil. Chiefly associated with oak and probably on other broad-leaved trees. Not associated with sweet chestnut hop poles as was sometimes claimed. This species may have been associated with ash hop poles as, in 19th century Kent, this was the main tree used for this purpose. Larvae occur under loose dry bark of dead exposed rotten roots. They do not construct galleries but crawl actively on the surface of the roots. Pupation occurs in the soil. Adults have been recorded from April to July. Adults visit flowers such as the blossom of hawthorn, apple and *Viburnum* spp. as well as cow parsley and meadowsweet. Continental literature states that the life cycle lasts two years.

Status Formerly widespread but very local in central and southern England. This species appears to have declined dramatically with the only modern record from Cheshire. Colonies may still exist in the Wyre Forest district of Worcestershire and Shropshire.

Threats Loss of broad-leaved woodland through, for example, clear-felling and coniferisation. The loss of old oaks in hedgerows and field boundaries is a further threat to this species. Habitat loss, in particular, through the felling of ancient trees and the destruction or removal of standing dead wood and stumps for reasons such as forest hygiene, aesthetic tidiness, public safety or for use as fire wood.

Management and conservation Ancient trees and standing dead timber, especially with the bark attached, should be retained. The removal of dead timber from ancient trees should be avoided. Gaps in the age structure of the tree population should be identified and the continuity of the appropriate dead wood habitat

CERAMBYCIDAE

ensured by suitable planting and possibly with pollarding. The presence of nectar sources such as hawthorn and umbellifers may also be particularly important for this species.

Published sources Bílý, S. & Mehl, O. (1989), Shirt, D.B., ed. (1987).

ANAGLYPTUS MYSTICUS NOTABLE B
A longhorn beetle
Order COLEOPTERA Family CERAMBYCIDAE

Anaglyptus mysticus (L., 1758). Formerly known as: *Anaclyptus mysticus* (L.), *Clytus mysticus* (L.).

Distribution England and Wales.

Habitat and ecology Woodland, scrub and hedgerows. Larvae develop in dead wood. Continental literature states that oviposition takes place in crevices and emergence holes of other insects, the larvae feeding on very dry dead branches and boles. Trees that have been scorched by fire seem particularly prone to attack. In Britain, new generation adults have been recorded from ash logs and an old maple stump. Also recorded from aspen, hawthorn, holly, oak, wild pear, rowan, elm, beech, lime and cultivated apple. Probably associated with a number of other trees and shrubs. Adults visit flowers such as ox-eye daisy, hogweed and hawthorn blossom. Adults have been recorded from April to July. Continental literature states that the life cycle takes two or three years.

Status Widespread, but local in England and Wales.

Indicator status Grade 3 in Garland (1983).

Threats Loss of broad-leaved woodland through, for example, clear-felling and coniferisation. Alteration of the vegetation structure of many remaining woodlands by 20th century management practices, including the clearing of under-storey shrubs at ride edges. The removal of dead wood for reasons such as forest hygiene, aesthetic tidiness or for use as fire wood.

Management and conservation Both fallen and standing dead timber, especially with the bark attached, should be retained. The removal of dead timber from ancient trees should be avoided. Gaps in the age structure of the tree and shrub population should be identified and the continuity of the appropriate dead wood habitat ensured by suitable planting. Open glades and ride margins should be cut on rotation to retain a variety of vegetation structures. Nectar sources such as hawthorn, umbellifers and composite herbs may be particularly important for this species.

AROMIA MOSCHATA NOTABLE B
Musk beetle
Order COLEOPTERA Family CERAMBYCIDAE

Aromia moschata (L., 1758).

Distribution England, South Wales, North Wales, South West Scotland, South East Scotland and North East Scotland.

Habitat and ecology Wetland and wet woodland, in particular fens and willow carr. Also along river margins. The host-trees are common sallow, white willow and osier, probably also on a number of other *Salix* species. The beetle appears to have a preference for healthy, younger or mature trees, rather than overmature ones, though there are records from the latter case. Continental authors state that oviposition takes place either in a trunk or at the base of stout branches of healthy trees. The larva bores into the sapwood and on reaching the heartwood tunnels in a vertical direction for a distance of 30cm to 40cm. The larva pupates just below the surface of the bark. In Britain, fully grown larvae have been recorded in January. Adults occur from May to September with the main period of emergence probably occurring from about mid June to early August. Adults have been recorded from alder, birch, lime, Lombardy poplar and flowers such as those of angelica, cow parsley and chervil. Continental literature states that the life cycle lasts three years.

Status Formerly more widespread. Much declined with modern records widely distributed from southern England through to northern England, also recorded from the Hebrides.

Threats Loss of habitat through drainage, clear-felling of wet woodland and riverside trees. Cessation of pollarding may also threaten this species.

Management and conservation Gaps in the age structure of the tree population should be identified and the continuity of the appropriate habitat ensured by suitable planting and by pollarding in rotation. The presence of nectar sources such as umbellifers may also be important for this species.

CERAMBYCIDAE

GRACILIA MINUTA VULNERABLE
A longhorn beetle
Order COLEOPTERA Family CERAMBYCIDAE

Gracilia minuta (F., 1781).

Distribution Recorded from East Cornwall, South Devon, North Somerset, North Wiltshire, Dorset, Isle of Wight, South Hampshire, East Sussex, East Kent, West Kent, Surrey, South Essex, Hertfordshire, Middlesex, Berkshire, Oxfordshire, Buckinghamshire, East Suffolk, West Suffolk, East Norfolk, West Norfolk, Cambridgeshire, East Gloucestershire, West Gloucestershire, Herefordshire, Worcestershire, Warwickshire, Staffordshire, Glamorgan, Leicestershire & Rutland, Nottinghamshire, Derbyshire, Cheshire, South Lancashire, South-east Yorkshire, South-west Yorkshire, Mid-west Yorkshire, Durham and Cumberland before 1970 and South Hampshire, East Kent and West Kent from 1970 onwards.

Habitat and ecology Woodland and scrub. Associated with blackthorn, elm, hazel, lime and osier. This species has also been found breeding in bramble and loganberry canes and in the stems of dog rose. The larvae probably develop in twigs and small branches. On the Continent, the larvae excavate longitudinal galleries under bark. In Britain, the beetle has also been frequently recorded from wickerwork. In the wild adults have been recorded from May to August. Continental literature states that the life cycle lasts one year.

Status Not listed in the insect Red Data Book (Shirt, 1987). Formerly widespread occurring throughout England north to Cumberland, also recorded in south Wales. Apparently much declined with very few modern records.

Threats Uncertain, though this species is probably threatened by the loss of broad-leaved woodland through, for example, clear-felling and coniferisation. The loss of under-storey and shrub species through alteration of the vegetation structure of many remaining woodlands by 20th century management practices.

Management and conservation Open glades and ride margins should be cut on rotation to retain a variety of vegetation structures.

Published sources Atty, D.B. (1983), Bílý, S. & Mehl, O. (1989), Hardy, J.R. & Standen, R. (1917).

GRAMMOPTERA USTULATA RARE
A longhorn beetle
Order COLEOPTERA Family CERAMBYCIDAE

Grammoptera ustulata (Schaller, 1783). Formerly known as: *Grammoptera praeusta* (F., 1787).

Distribution Recorded from South Hampshire, North Hampshire, East Sussex, Middlesex, Berkshire, Buckinghamshire, East Gloucestershire and West Gloucestershire before 1970 and South Hampshire, North Hampshire, Berkshire, East Gloucestershire and West Gloucestershire from 1970 onwards.

Habitat and ecology Ancient broad-leaved woodland and pasture woodland. The larvae are probably associated with dead wood. On the Continent, oak, walnut, sweet chestnut and *Acer* spp. are listed as host-trees, the larvae developing in decaying twigs and slender branches. In Britain, adults frequently visit flowers, in particular hawthorn blossom, also that of *Viburnum* species. Adults have been recorded from April to August. Continental literature states that the life cycle lasts one year.

Status Very local and possibly declining. This species appears to be confined to central southern England.

Indicator status Grade 1 in Harding & Rose (1986).

Threats Loss of broad-leaved woodland and parkland through, for example, clear-felling and coniferisation. Habitat loss, in particular, through the felling of ancient trees, removal of dead wood from living trees and the destruction or removal of standing and fallen dead wood for reasons such as forest hygiene, aesthetic tidiness, public safety or for use as fire wood.

Management and conservation Ancient trees and both fallen and standing dead timber, especially with the bark attached, should be retained. The removal of dead timber from ancient trees should be avoided. Gaps in the age structure of the tree population should be identified and the continuity of the appropriate dead wood habitat ensured by suitable planting and possibly with pollarding. The presence of nectar sources such as hawthorn may also be particularly important for this species.

Published sources Atty, D.B. (1983), Bílý, S. & Mehl, O. (1989), Harding, P.T. (1978), Harding, P.T. & Rose, F. (1986), Hunter, F.A. (1953), Shirt, D.B., ed. (1987).

157

CERAMBYCIDAE

GRAMMOPTERA VARIEGATA NOTABLE A
A longhorn beetle
Order COLEOPTERA Family CERAMBYCIDAE

Grammoptera variegata (Germar, 1824). Formerly known as: *Grammoptera analis* (Herrich-Schaeffer in Panzer, 1832).

Distribution Recorded from North Somerset, Isle of Wight, South Hampshire, West Sussex, East Kent, West Kent, Surrey, South Essex, North Essex, Middlesex, Berkshire, Oxfordshire, Buckinghamshire, East Suffolk, West Norfolk, Cambridgeshire, Huntingdonshire, East Gloucestershire, West Gloucestershire, Staffordshire, North Lincolnshire, Leicestershire & Rutland, Nottinghamshire, Derbyshire, Cheshire, South-east Yorkshire, North-east Yorkshire, Mid-west Yorkshire and North-west Yorkshire before 1970 and North Hampshire, West Sussex, Surrey, Hertfordshire, Oxfordshire, Northamptonshire and West Gloucestershire from 1970 onwards.

Habitat and ecology Broad-leaved woodland. The larvae are probably associated with dead wood. The species has been recorded mainly from oak. Also noted from sweet chestnut, pear, gorse and *Viburnum* species. On the Continent, the larval habitat is in recently dead and decaying twigs and slender branches, pupation occurring in the sapwood. In Britain, adults frequently visit flowers, in particular hawthorn blossom. Adults occur from May to July. Continental literature states that the life cycle lasts one year.

Status Very local and possibly declining. Modern records are distributed from southern England as far north as Northamptonshire.

Indicator status Grade 3 in Harding & Rose (1986).

Threats Uncertain, though this species is probably threatened by loss of broad-leaved woodland through, for example, clear-felling and coniferisation. Habitat loss, in particular, through the felling of ancient trees and shrubs, removal of dead wood from living trees and the destruction or removal of standing and fallen dead wood for reasons such as forest hygiene, aesthetic tidiness, public safety or for use as fire wood.

Management and conservation Ancient trees, shrubs and both fallen and standing dead timber, especially with the bark attached, should be retained. The removal of dead timber from ancient trees should be avoided. Gaps in the age structure of the tree population should be identified and the continuity of the appropriate dead wood habitat ensured by suitable planting and possibly with pollarding. The presence of nectar sources such as hawthorn may also be particularly important for this species.

Published sources Atty, D.B. (1983), Bílý, S. & Mehl, O. (1989), Flint, J.H. (1943a), Garland, S.P. (1983), Harding, P.T. (1978), Harding, P.T. & Rose, F. (1986), Hunter, F.A. (1953).

JUDOLIA SEXMACULATA NOTABLE A
A longhorn beetle
Order COLEOPTERA Family CERAMBYCIDAE

Judolia sexmaculata (L., 1758). Formerly known as: *Leptura sexmaculata* (L.), *Pachyta sexmaculata* (L.).

Distribution Recorded from Mid Perthshire, Moray and East Inverness & Nairn before 1970 and South Aberdeenshire, East Inverness & Nairn and West Ross from 1970 onwards.

Habitat and ecology Coniferous woodland. Associated mainly with Scots pine, though also recorded from spruce. Larvae develop in decaying roots. Adults occur in June and July and rarely in August. Adults visit flowers, in particular hogweed and rowan blossom. Continental literature states that the life cycle probably lasts two years.

Status Exclusively Scottish and very local. The beetle appears to be a component of the native pine forest fauna.

Threats Loss of native pine forest through, for example, clear-felling and conversion to plantation forest. Habitat loss in the remaining areas through the felling of ancient trees and the destruction or removal of standing and fallen dead wood for reasons such as forest hygiene, aesthetic tidiness, public safety or for use as fire wood.

Management and conservation Ancient trees and both standing and fallen dead timber, especially with the bark attached, should be retained. Gaps in the age structure of the tree population should be identified and regeneration encouraged to ensure the continuity of the dead wood habitats.

Published sources Bílý, S. & Mehl, O. (1989), Carter, A.E.J. (1924).

LAMIA TEXTOR   ENDANGERED
Pine sawyer
Order COLEOPTERA   Family CERAMBYCIDAE

Lamia textor (L., 1758).

Distribution Recorded from North Somerset, South Hampshire, East Sussex, Cambridgeshire, West Gloucestershire, Warwickshire, Merionethshire, Mid Perthshire and Argyll Main before 1970.

Habitat and ecology Wet woodland, willow carr, osier beds and willows along river margins. Associated with old willow, sallow, osier and doubtfully also with aspen. On the Continent, this species has also been recorded from poplar and birch. Oviposition takes place in the bark of healthy slender trunks and branches. In Britain, the larvae develop in healthy roots or boles and often leave scant evidence above ground of their presence. Continental authors state that the larvae pupate in the trunk, usually near the base or in a root. In Britain, adults have been recorded from June to September. Continental literature states that the life cycle takes two to four years.

Status Status revised from RDB 2 (Vulnerable) in Shirt (1987). No recent records. This species may be under-recorded as it is both crepuscular in habit and also well camouflaged. A record for East Kent requires confirmation. Last recorded in 1953 from Rannoch, Mid Perthshire.

Indicator status Listed in Speight (1989) Appendix 1.

Threats Drainage and clear-felling for reasons such as agricultural improvement. The felling of riverside trees may also threaten this species.

Management and conservation Gaps in the age structure of the tree population should be identified and the continuity of the appropriate habitat ensured by suitable planting.

Published sources Atty, D.B. (1983), Bílý, S. & Mehl, O. (1989), Shirt, D.B., ed. (1987), Speight, M.C.D. (1989), Uhthoff-Kaufmann, R.R. (1991b).

LEPTURA FULVA   RARE
A longhorn beetle
Order COLEOPTERA   Family CERAMBYCIDAE

Leptura fulva Degeer, 1775.

Distribution Recorded from East Cornwall, South Devon, South Wiltshire, Isle of Wight, South Hampshire, North Hampshire, Berkshire, Herefordshire, Glamorgan and Derbyshire before 1970 and Dorset,

South Hampshire and North Hampshire from 1970 onwards.

Habitat and ecology Probably broad-leaved woodland. Larvae probably develop in dead wood and are mainly associated with aspen, beech and possibly willow. Adults have been noted at flowers on calcareous downland adjacent to a beech woodland. The beetle has also been observed emerging from sleepers on a disused railway track. Adults frequently visit flowers, in particular umbellifers such as wild angelica and hogweed, though it has also been noted on thistles and the blossom of dog rose. Adults have been recorded from June to August.

Status Not listed in the insect Red Data Book (Shirt, 1987). Very local and much declined. There are modern records for three vice-counties, all in the south of England.

Threats Uncertain, though this species is probably threatened by the loss of broad-leaved woodland through, for example, clear-felling and coniferisation. Habitat loss, in particular, through the felling of ancient trees, removal of dead wood from living trees and the destruction or removal of standing and fallen dead wood for reasons such as forest hygiene, aesthetic tidiness, public safety or for use as fire wood.

Management and conservation Ancient trees and both fallen and standing dead timber, especially with the bark attached, should be retained. The removal of dead timber from ancient trees should be avoided. Gaps in the age structure of the tree population should be identified and the continuity of the appropriate dead wood habitat ensured by suitable planting and possibly with pollarding. Open glades and ride margins should be cut on rotation to retain a variety of vegetation structures. The presence of nectar sources such as umbellifers and composite herbs may also be particularly important for this species.

LEPTURA SANGUINOLENTA   RARE
A longhorn beetle
Order COLEOPTERA   Family CERAMBYCIDAE

Leptura sanguinolenta L., 1761.

Distribution Recorded from South Devon, South Hampshire, Surrey, North Essex, East Suffolk, East Norfolk, Huntingdonshire, South-west Yorkshire, Moray and East Inverness & Nairn before 1970 and Moray and East Inverness & Nairn from 1970 onwards.

Habitat and ecology Coniferous woodland and small pines in quaking peat bogs. Associated with Scots pine, though also recorded from rowan and silver fir. The

CERAMBYCIDAE

larvae probably develop in dead or dying wood. Adults occur predominantly in June and July. Adults frequently visit flowers, in particular umbellifers such as hogweed, and thistles. Continental literature states that the life cycle lasts two years.

Status Not listed in the insect Red Data Book (Shirt, 1987). Appears to have two main centres of distribution in Great Britain, the Scottish Highlands and southern England, particularly East Anglia. Recent records only from the former area where the species is very local. In southern and eastern England this species may have been resident in the 19th century, though some records may refer to chance importations. There are no recent records from this part of Great Britain.

Threats Loss of native pine forest through, for example, clear-felling or conversion to plantation forest. Habitat loss in the remaining areas through the felling of ancient trees and the destruction or removal of standing and fallen dead wood for reasons such as forest hygiene, aesthetic tidiness, public safety or for use as fire wood.

Management and conservation Ancient trees and both standing and fallen dead timber, especially with the bark attached, should be retained. Gaps in the age structure of the tree population should be identified and regeneration encouraged to ensure the continuity of the dead wood habitats. The presence of nectar sources such as umbellifers may also be particularly important for this species.

Published sources Allen, A.A. (1972b), Bílý, S. & Mehl, O. (1989), Else, G.R. (1970).

LEPTURA SCUTELLATA   **NOTABLE A**
A longhorn beetle
Order COLEOPTERA   Family CERAMBYCIDAE

Leptura scutellata F., 1781.

Distribution Recorded from South Hampshire, East Kent, West Kent, South Essex, Hertfordshire, Berkshire, Buckinghamshire, Herefordshire and Nottinghamshire before 1970 and South Hampshire, East Kent, West Kent, Surrey, South Essex, North Essex and Berkshire from 1970 onwards.

Habitat and ecology Ancient broad-leaved woodland and pasture-woodland. Associated mainly with beech. Also recorded from oak, birch, hornbeam, sycamore and lime. On the Continent, this species has also been noted from alder and hazel. In Britain, larvae develop in dead wood, continental authors stating that the larvae feed deep in the wood of sun-exposed dead and decaying stumps and branches, pupation occurring in

the sapwood. In Britain, adults occur from March to August though most frequently recorded in July. Adults occasionally visit flowers such as umbellifers and the blossom of bramble and rose. Continental literature states that the life cycles lasts two or three years.

Status Very local. Restricted to southern and south-eastern England.

Indicator status Grade 1 in Harding & Rose (1986). Listed in Speight (1989) Appendix 1.

Threats Loss of broad-leaved woodland and parkland through, for example, clear-felling and coniferisation. Habitat loss, in particular, through the felling of ancient trees, removal of dead wood from living trees and the destruction or removal of standing and fallen dead wood for reasons such as forest hygiene, aesthetic tidiness, public safety or for use as fire wood.

Management and conservation Ancient trees and both fallen and standing dead timber, especially with the bark attached, should be retained. The removal of dead timber from ancient trees should be avoided. Gaps in the age structure of the tree population should be identified and the continuity of the appropriate dead wood habitat ensured by suitable planting and possibly with pollarding. The presence of nectar sources such as bramble and umbellifers may also be particularly important for this species.

Published sources Allen, A.A. (1952b), Bílý, S. & Mehl, O. (1989), Cooter, J. (1977c), Eagles, T.R. (1953b), Garland, S.P. (1983), Harding, P.T. (1978), Harding, P.T. & Rose, F. (1986), Hunter, F.A. (1951), Hunter, F.A. (1953), Speight, M.C.D. (1989).

LEPTURA SEXGUTTATA   **RARE**
A longhorn beetle
Order COLEOPTERA   Family CERAMBYCIDAE

Leptura sexguttata F., 1775. Formerly known as: *Anoplodera sexguttata* (F.).

Distribution Recorded from North Devon, North Wiltshire, South Wiltshire, South Hampshire, West Kent, Surrey, Merionethshire, North Lincolnshire and North-east Yorkshire before 1970 and North Devon, South Hampshire, Huntingdonshire and North-east Yorkshire from 1970 onwards.

Habitat and ecology Probably associated with broad-leaved woodland. The larvae have been found in the dead wood of oak. On the Continent, this species has also been reported from hornbeam and beech. The eggs are laid in decaying trunks and branches, the larvae feeding in dry and hard wood. Pupation takes place in the outer sapwood. This is probably also the case in Britain, A.A.Allen (pers.comm.) suggests that the species does not seem to be associated with old timber, but more likely with smaller boughs and possibly stumps. Adults emerge in June and July. There is a record for January. Adults frequently visit flowers, in particular umbellifers, and the blossom of bramble, holly, dog rose and *Viburnum* species. Continental literature states that the life cycle lasts two or three years.

Status Declining and very local, with few recent records. Fairly regularly noted in the New Forest, South Hampshire and North-east Yorkshire.

Indicator status Grade 1 in Garland (1983).

Threats Loss of broad-leaved woodland through, for example, clear-felling and coniferisation. Habitat loss, in particular, through the felling of ancient trees, removal of dead wood from living trees and the destruction or removal of standing and fallen dead wood for reasons such as forest hygiene, aesthetic tidiness, public safety or for use as fire wood.

Management and conservation Ancient trees and both fallen and standing dead timber, especially with the bark attached, should be retained. The removal of dead timber from ancient trees should be avoided. Gaps in the age structure of the tree population should be identified and the continuity of the appropriate dead wood habitat ensured by suitable planting and possibly with pollarding. The presence of nectar sources such as bramble may also be particularly important for this species. Much of the New Forest is notified as an SSSI.

Published sources Bílý, S. & Mehl, O. (1989), Garland, S.P. (1983), Hunter, F.A. (1953), Hunter, F.A. (1959), Shirt, D.B., ed. (1987), Wollaston, V. (1843).

LEPTURA VIRENS   **EXTINCT**
A longhorn beetle
Order COLEOPTERA   Family CERAMBYCIDAE

Leptura virens L., 1758.

Distribution Recorded from West Gloucestershire before 1970.

Habitat and ecology Woodland. The larvae are probably associated with dead wood. On the Continent, this beetle is a northern forest/sub alpine species, the larvae being associated with decaying fallen trunks and stumps of conifers.

Status Not listed in the insect Red Data Book (Shirt, 1987). Presumed extinct. Only known from two 19th

century records, both from the Forest of Dean, West Gloucestershire.

Published sources Atty, D.B. (1983), Bílý, S. & Mehl, O. (1989).

MESOSA NEBULOSA   **RARE**
A longhorn beetle
Order COLEOPTERA   Family CERAMBYCIDAE

Mesosa nebulosa (F., 1781). Formerly known as: *Mesosa nubila* (Gmelin in L., 1790).

Distribution Recorded from South Hampshire, East Kent, West Kent, Surrey, South Essex, North Essex, Berkshire, East Suffolk, Cambridgeshire, Huntingdonshire, West Gloucestershire, Monmouthshire, Herefordshire and Worcestershire before 1970 and South Hampshire, East Kent, South Essex, Berkshire, East Suffolk, East Gloucestershire and West Gloucestershire from 1970 onwards.

Habitat and ecology Broad-leaved woodland and pasture-woodland. Associated mainly with oak, though also recorded from aspen, beech, hawthorn, hazel and willow. Larvae develop in dead wood and may prefer the topmost branches of oak. Pupation takes place in July or August. Adults have been recorded in March, May and June, the adults being fully developed in August or September and usually remaining in the pupal cell until the following May or June. Continental authors state that the life cycle lasts two or three years.

Status Very local, recent records emanating from southern England only. Possibly under-recorded due to the inaccessible habits of this species.

Indicator status Grade 2 in Harding & Rose (1986).

Threats Loss of broad-leaved woodland and parkland through, for example, clear-felling and coniferisation. Habitat loss, in particular, through the felling of ancient trees, removal of dead wood from living trees and the destruction or removal of standing and fallen dead wood for reasons such as forest hygiene, aesthetic tidiness, public safety or for use as fire wood.

Management and conservation Ancient trees and both fallen and standing dead timber, especially with the bark attached, should be retained. The removal of dead timber from ancient trees should be avoided. Gaps in the age structure of the tree population should be identified and the continuity of the appropriate dead wood habitat ensured by suitable planting.

Published sources Allen, A.A. (1955e), Atty, D.B. (1983), Bílý, S. & Mehl, O. (1989), Buck, F.D.

CERAMBYCIDAE

(1957b), Buck, F.D. (1960a), Harding, P.T. (1978), Hunter, F.A. (1959), Massee, A.M. (1958), Shirt, D.B., ed. (1987), Uhthoff-Kaufmann, R.R. (1991b).

MOLORCHUS UMBELLATARUM   NOTABLE A
A longhorn beetle
Order COLEOPTERA   Family CERAMBYCIDAE

Molorchus umbellatarum (von Schreber, 1759). Formerly known as: *Caenoptera umbellatarum* (von Schreber).

Distribution Recorded from Dorset, South Hampshire, North Hampshire, West Sussex, East Sussex, East Kent, West Kent, Surrey, North Essex, Middlesex, Berkshire, Buckinghamshire, West Norfolk, Cambridgeshire, Huntingdonshire, Northamptonshire, East Gloucestershire, West Gloucestershire, Monmouthshire, Herefordshire, Worcestershire, Leicestershire & Rutland, Nottinghamshire and Mid-west Yorkshire before 1970 and West Sussex, Surrey, Huntingdonshire, Herefordshire and Worcestershire from 1970 onwards.

Habitat and ecology Broad-leaved woodland, scrub and hedgerows. The larvae probably develop in dead wood, possibly that of old, wild rose bushes. On the Continent, apple, dogwood and guelder-rose are all listed as host-plants. In Britain, adults have been recorded from crab apple, dog rose, dogwood and hawthorn, and on the flowers of hawthorn, privet and hogweed. Adults have been recorded from May to July. Continental literature states that the life cycle lasts two years.

Status Very local and much declined. Modern records are distributed from southern England to Herefordshire and Worcestershire.

Threats Loss of broad-leaved woodland through, for example, clear-felling and coniferisation. Alteration of the vegetation structure of many remaining woodlands by 20th century management practices. The removal of dead and dying shrubs for reasons such as forest hygiene and aesthetic tidiness.

Management and conservation Open glades and ride margins should be cut on rotation to retain a variety of vegetation structures. Gaps in the age structure of the shrub population should be identified and the continuity of the appropriate dead wood habitat ensured by suitable planting. The presence of nectar sources such as hawthorn and umbellifers may be particularly important for this species.

Published sources Allen, A.A. (1955e), Atty, D.B. (1983), Bílý, S. & Mehl, O. (1989), Jones, R.A.

(1977a), Lloyd, R.W. (1943), Lloyd, R.W. (1945), Uhthoff-Kaufmann, R.R. (1990).

OBEREA OCULATA   ENDANGERED
A longhorn beetle
Order COLEOPTERA   Family CERAMBYCIDAE

Oberea oculata (L., 1758).

Distribution Recorded from East Kent, Oxfordshire, West Suffolk, West Norfolk, Cambridgeshire and Cumberland before 1970 and Cambridgeshire from 1970 onwards.

Habitat and ecology Fens and marshes. Eggs are laid on the smooth bark of twigs and slender stems of living healthy sallow bushes, and possibly willows. The larva bores a straight gallery, 30cm or more in length, in the pith channel or sapwood of wider stems. An accumulation of ejected frass clinging to the twigs is the only external indication that larvae are present. Adults occur from June to September, though there is a record for mid-winter, and have been recorded from buckthorn, sea buckthorn, alder and umbellifers, as well as from sallows and willows. Adults tend to remain motionless on the upper branches of sallows etc., but sometimes fly very actively in sunny conditions and in sultry, thundery weather. Continental literature states that the life cycle takes one or two years.

Status Formerly not uncommon in the fens of East Anglia and Cambridgeshire, with records from a number of other localities. This century *O. oculata* has only been recorded from Wicken Fen, Cambridgeshire.

Threats Uncertain, though this beetle may be threatened by drainage for reasons such as agricultural improvement.

Management and conservation Wicken Fen is managed as a nature reserve by the National Trust.

Published sources Bílý, S. & Mehl, O. (1989), Shirt, D.B., ed. (1987).

OBRIUM CANTHARINUM   EXTINCT
A longhorn beetle
Order COLEOPTERA   Family CERAMBYCIDAE

Obrium cantharinum (L., 1767).

Distribution Recorded from South Devon, East Sussex, East Kent, South Essex, North Essex and Hertfordshire.

Habitat and ecology Broad-leaved woodland and orchards. Associated with aspen and possibly oak, crab apple and other fruit trees. Larvae develop in dead wood beneath the bark. On the Continent, the larval habitat is in recently dead and sun-exposed branches and trunks. In Britain, adults have been recorded from June to August. Continental literature states that the life cycle lasts two years.

Status Presumed extinct. There is a single 20th century record for this beetle though this is atypical in distribution when compared to other records. Primarily recorded from south-eastern England. The majority of the records for *O. cantharinum* are from South Essex and chiefly within the old limits of Epping Forest i.e., Wanstead, Leytonstone and Epping. Substantial areas of ancient woodland and old orchards still remain in this part of South Essex and it is possible that *O. cantharinum* may still occur in the area. Last recorded in 1929 from Bovey Tracy, South Devon.

Published sources Bílý, S. & Mehl, O. (1989), Kaufmann, R.R.U. (1947), Kaufmann, R.R.U. (1985).

PHYMATODES ALNI NOTABLE B
A longhorn beetle
Order COLEOPTERA Family CERAMBYCIDAE

Phymatodes alni (L., 1767). Formerly known as: *Callidium alni* (L.), *Poecilium alni* (L.).

Distribution England and North Wales.

Habitat and ecology Woodland, scrub and hedgerows. The larvae probably develop in dead wood of small boughs and possibly also in twigs. It has frequently been found on dead hedgerow shrubs. Recorded from alder, aspen, elm, hazel, oak, hawthorn and from under willows. Adults have been found from April to July. Continental literature states that the life cycle lasts one year.

Status Widespread but local in England and North Wales.

Threats Loss of broad-leaved woodland through, for example, clear-felling and coniferisation. Alteration of the vegetation structure of many remaining woodlands by 20th century management practices, including the clearance of under-storey shrubs along ride edges. The removal of dead wood for reasons such as forest hygiene, aesthetic hygiene or for use as fire wood.

Management and conservation Both fallen and standing dead timber, especially with the bark attached, should be retained. The removal of dead timber from ancient trees should be avoided. Gaps in the age structure of the tree and shrub population should be identified and the continuity of the appropriate dead

wood habitat ensured by suitable planting and pollarding. Open glades and ride margins should be cut on rotation to retain a variety of vegetation structures.

PHYTOECIA CYLINDRICA NOTABLE B
A longhorn beetle
Order COLEOPTERA Family CERAMBYCIDAE

Phytoecia cylindrica (L., 1758).

Distribution South East, South, South West, East Anglia, East Midlands, West Midlands and South Wales.

Habitat and ecology Hedgerows, field margins, roadside verges and probably wood edges. On the Continent, the larvae develop in the stems of cow parsley, hogweed, wild carrot and rough chervil. The eggs are laid singly in the walls of the stem just below the flowerhead. The larva tunnels down to the base of the stem. This action kills the stem, though the larva continues to feed on the inner walls until mature. The pupal cell is found at the base of a hollowed stem. The larval stage does not exceed three months. In Britain, adults occur from March until July, with the main period of adult emergence period probably occurring from late June to early July. Adults visit flowers with records from cow parsley and hawthorn blossom. Continental literature states that the life cycle takes one year.

Status Widely distributed, though infrequently recorded.

Threats This species may be threatened by agricultural improvement, and the 'tidying up' of field margins and roadside verges. Changes in grazing or cutting regimes may also adversely affect this species.

Management and conservation The importance of field margins, wood edges etc. for their invertebrate communities should be taken into account for any management. Mowing regimes adjusted to accomodate this beetles life cycle.

PLAGIONOTUS ARCUATUS EXTINCT
A longhorn beetle
Order COLEOPTERA Family CERAMBYCIDAE

Plagionotus arcuatus (L., 1758). Formerly known as: *Clytus arcuatus* (L.).

Distribution Recorded from West Kent, South Essex, Hertfordshire, Middlesex, "Suffolk", East Norfolk, South Northumberland and Cumberland.

CERAMBYCIDAE

Habitat and ecology Woodland. The larvae probably develop in dead wood, possibly with a preference for cherry and other old fruit trees. On the Continent, this species has been recorded from oak, beech and hornbeam and *Salix* spp., oviposition taking place in crevices in the bark of branches and recently felled or dead trees.

Status Presumed extinct. Twentieth century records are regarded as importations. Many old records are probably also the result of importation. The only convincing records suggesting this species to be a former resident emanate from Epping and Hainault Forests, South Essex and Highgate Woods, Middlesex (A.A. Allen pers. comm.).

Published sources Bílý, S. & Mehl, O. (1989), Fowler, W.W. (1890).

POGONOCHERUS FASCICULATUS NOTABLE B
A longhorn beetle
Order COLEOPTERA   Family CERAMBYCIDAE

Pogonocherus fasciculatus (Degeer, 1775). Formerly known as: *Pogonochaerus fasciculatus* (Degeer, 1775).

Distribution Scotland and East Anglia.

Habitat and ecology Coniferous woodland. Associated with Scots pine. On the Continent, this species has been found on spruce. Larvae develop in dead wood. Continental literature states that the larva feeds under the bark of 2cm to 4cm thick, recently dead branches, seldom occurring in the trunks. The larva pupates in a short pupal cell in the sapwood. In Britain, adults have been recorded from April to September. Continental literature states that the life cycle lasts two years.

Status Exclusively Scottish and local. The beetle appears to be a component of the native pine forest fauna. All other records are probably either adventives, or as in East Anglia, the result of successful establishments.

Threats Loss of native pine forest through, for example, clear-felling and conversion to other land use. Habitat loss in the remaining areas through the felling of ancient trees and the destruction or removal of standing and fallen dead wood for reasons such as forest hygeine, aesthetic tidiness, public safety or for use as fire wood.

Management and conservation Ancient trees and both standing and fallen dead timber, especially with the bark attached, should be retained. Gaps in the age structure of the tree population should be identified and regeneration encouraged to ensure the continuity of the dead wood habitats.

PRIONUS CORIARIUS   NOTABLE A
Sawyer beetle
Order COLEOPTERA   Family CERAMBYCIDAE

Prionus coriarius (L., 1758).

Distribution Recorded from West Cornwall, East Cornwall, South Devon, North Devon, South Somerset, North Wiltshire, Dorset, South Hampshire, North Hampshire, West Sussex, East Kent, West Kent, Surrey, South Essex, North Essex, Hertfordshire, Middlesex, Berkshire, Buckinghamshire, East Suffolk, West Suffolk, East Norfolk, West Norfolk, Cambridgeshire, East Gloucestershire, West Gloucestershire, Herefordshire, Worcestershire, Warwickshire, Staffordshire, Shropshire, Glamorgan, Denbighshire, Nottinghamshire, Cheshire, South Lancashire, West Lancashire and Westmorland & North Lancashire before 1970 and West Sussex, East Sussex, East Kent, West Kent, Surrey, South Essex, Berkshire, East Suffolk and Flintshire from 1970 onwards.

Habitat and ecology Broad-leaved woodland and pasture woodland. Chiefly associated with oak, beech and birch. Also recorded from apple, ash, cherry, plum, elm, holly, hornbeam, horse chestnut, pine and willow. On the Continent, this species has also been noted from sweet chestnut, alder and spruce. The eggs are inserted into crevices in the bark, usually at the base of trunks or in exposed roots. In Britain, the larvae feed on the roots of the host trees and on dead wood in stumps. Continental authors state that when the food supply is exhausted the larva is able to travel through the soil to nearby roots. In Britain pupation occurs in earthen cocoons among the roots of the host-tree. The main period of emergence is from mid July until late September, though individuals have been recorded in March and November. Continental literature states that the life cycles takes three to four years.

Status Formerly widespread. This robust beetle appears to have declined. The majority of modern records are from south-east England. A record for South Wiltshire requires confirmation.

Indicator status Grade 3 in Harding & Rose (1986).

Threats Loss of broad-leaved woodland and parkland through, for example, clear-felling and coniferisation. Habitat loss, in particular, through the felling of ancient trees and the uprooting of stumps for reasons such as forest hygiene, aesthetic tidiness, public safety or for use as fire wood.

Management and conservation Ancient trees and stumps, especially with the bark attached, should be retained. Gaps in the age structure of the tree population should be identified and the continuity of appropriate dead wood habitat ensured by suitable planting.

Published sources Atty, D.B. (1983), Bílý, S. & Mehl, O. (1989), Champion, G.C. (1926), Ellis, E.A. (1943), Ferry, R.S. (1952), Garland, S.P. (1983), Harding, P.T. (1978), Harding, P.T. & Rose, F. (1986), Hardy, J.R. & Standen, R. (1917), Hunter, F.A. (1959), Imms, A.D. (1947), Lloyd, R.W. (1938), Roberts, M. (1959), Symes, H. (1952), Uhthoff-Kaufmann, R.R. (1991a), Verdcourt, B. (1944), Weal, R.D. (1953).

PYRRHIDIUM SANGUINEUM   VULNERABLE
A longhorn beetle
Order COLEOPTERA   Family CERAMBYCIDAE

Pyrrhidium sanguineum (L., 1758). Formerly known as: *Callidium sanguineum* (L.).

Distribution Recorded from South Devon, Dorset, Herefordshire and Anglesey before 1970 and Monmouthshire, Herefordshire, Shropshire, Breconshire and Radnorshire from 1970 onwards.

Habitat and ecology Ancient broad-leaved woodland and pasture-woodland. Associated with oak. Continental literature states that this species has also been recorded on beech, hornbeam, elm and apple. In Britain, oviposition has been recorded in May on a dead oak branch. Larvae develop in dead wood at the bark and sapwood interface. Larvae have been recorded in June. Pupae have been recorded in March, May and June and newly emerging adults in April, May and June. Cooter (1982) suggests that *P. sanguineum* is found "during the spring and early summer" on "oak that has fallen the previous late summer - autumn, and takes two years normally, three rarely, to attain maturity". Continental literature states that the life cycle takes one year.

Status Formerly considered possibly extinct in Britain. *P. sanguineum* was found to be well established and breeding at Moccas Park, Herefordshire in the late 1940's. This beetle has since been discovered in a small area of the Welsh Marches, including records from Radnorshire, Monmouthshire, Breconshire and Shropshire.

Indicator status Grade 1 in Harding & Rose (1986).

Threats Loss of broad-leaved woodland and parkland through, for example, clear-felling and coniferisation. Habitat loss, in particular, through the felling of ancient trees, removal of dead wood from living trees and the destruction or removal of standing and fallen dead wood for reasons such as forest hygiene, aesthetic tidiness, public safety or for use as fire wood.

Management and conservation Ancient trees and both fallen and standing dead timber, especially with the bark attached, should be retained. The removal of dead timber from ancient trees should be prevented. Gaps in the age structure of the tree population should be identified and the continuity of the appropriate dead wood habitat ensured by suitable planting and possibly with pollarding. Moccas Park is an NNR.

Published sources Allen, A.A. & Lloyd, R.W. (1951), Bílý, S. & Mehl, O. (1989), Cooter, J. (1977c), Cooter, J. (1980a), Cooter, J. (1982), Garland, S.P. (1983), Green, M.E. (1972), Harding, P.T. (1978), Harding, P.T. & Rose, F. (1986), Horton, G.A.N. (1980), Hunter, F.A. (1959), Morris, M.G. (1973b), Shirt, D.B., ed. (1987), Zatloukal-Williams, R.G.Z. (1973).

RHAGIUM INQUISITOR   NOTABLE B
A longhorn beetle
Order COLEOPTERA   Family CERAMBYCIDAE

Rhagium inquisitor (L., 1758). Formerly known as: *Rhagium indagator* F., 1787.

Distribution Scotland, East Midlands, East Anglia, North East and North West.

Habitat and ecology Woodland. Associated mainly with dead Scots pine, though possibly also on birch. Larvae are wood-borers under bark. On the Continent, this species has also been found on larch, oak and spruce, the larvae preferring moist and decaying wood. In Britain, pupae have been noted in August. Adults have been recorded in June and August. Continental literature states that the life cycle lasts two years, occasionally three years.

Status As a native species this beetle is exclusively Scottish and very local. Probably a component of the native pine forest fauna. Records from other parts of Britain are considered to introductions or adventives.

Threats Loss of native pine forest through, for example, clear-felling. Habitat loss in the remaining areas through the felling of ancient trees and the destruction or removal of standing and fallen dead wood for reasons such as forest hygiene, aesthetic tidiness, public safety or for use as fire wood.

Management and conservation Ancient trees and both standing and fallen dead timber, especially with the bark attached, should be retained. Gaps in the age

CERAMBYCIDAE

structure of the tree population should be identified and regeneration encouraged to ensure the continuity of the dead wood habitats.

SAPERDA CARCHARIAS   NOTABLE A
Poplar borer
Order COLEOPTERA   Family CERAMBYCIDAE

Saperda carcharias (L., 1758).

Distribution Recorded from East Kent, South Essex, North Essex, Hertfordshire, Middlesex, Berkshire, East Suffolk, West Suffolk, East Norfolk, Cambridgeshire, Huntingdonshire, Northamptonshire, West Gloucestershire, South Lincolnshire, North Lincolnshire, Nottinghamshire, Derbyshire, South Lancashire, North-east Yorkshire, South-west Yorkshire, Mid-west Yorkshire, South Aberdeenshire, Moray and East Inverness & Nairn before 1970 and East Suffolk, Huntingdonshire, Northamptonshire, South Lincolnshire and East Inverness & Nairn from 1970 onwards.

Habitat and ecology Predominantly wet woodland, fens and carr. Also found in other types of woodland, along river margins and in gardens. Primarily associated with aspen, though also on black poplar, Lombardy poplar, willow and oak. Continental literature states that oviposition takes place in the bark of trunks and larger branches of healthy trees, isolated trees or small groups of trees being particularly favoured. The young larva bores into the sapwood. Here it tunnels deeper until it reaches the heartwood. The larva bores a gallery of about 20cm to 30cm in length. The larva pupates in the gallery. The adults are active at dusk and feed on the leaves of the larval host-tree. In Britain, adults have been recorded from June to September. Continental literature states that the life cycle lasts two to four years.

Status Formerly widespread but now very local, though possibly under-recorded. Can be common in some localities.

Threats Drainage and clear-felling for reasons such as agricultural improvement. The felling of riverside trees and windbelt trees in fenland areas may also be a threat to this beetle.

Management and conservation Gaps in the age structure of the tree population should be identified and the continuity of the appropriate habitat ensured by suitable planting and by pollarding in rotation.

Published sources Atty, D.B. (1983), Bílý, S. & Mehl, O. (1989), Eagles, T.R. (1946), Hunter, F.A. (1953), Massee, A.M. (1956).

SAPERDA SCALARIS   NOTABLE A
A longhorn beetle
Order COLEOPTERA   Family CERAMBYCIDAE

Saperda scalaris (L., 1758).

Distribution Recorded from South Hampshire, Surrey, Hertfordshire, Cambridgeshire, Staffordshire, Cardiganshire, Merionethshire, Caernarvonshire, Nottinghamshire, Derbyshire, Cheshire, South Lancashire, North-east Yorkshire, South-west Yorkshire, Durham, South Northumberland, Cumberland, Dumfriesshire, Kirkcudbrightshire, Lanarkshire, Stirlingshire, Mid Perthshire, South Aberdeenshire, Moray, East Inverness & Nairn, West Inverness, Argyll Main, East Sutherland and West Sutherland before 1970 and Staffordshire, Cardiganshire, Caernarvonshire, Nottinghamshire, Derbyshire, North-east Yorkshire, Durham, South Northumberland, Mid Perthshire, East Inverness & Nairn, West Inverness and East Ross from 1970 onwards.

Habitat and ecology Broad-leaved woodland. Larvae develop in dead wood and have been recorded from birch logs, under and within oak bark and in chestnut. Pupae have also been recorded from beech and alder. On the Continent, hazel, poplar and elm, amongst others, have also been noted as host-trees. In Britain, adults oviposit in freshly dead wood and slightly decayed trunks and bigger branches. Larvae take two years to develop with the final and penultimate instars being recorded in April. Pupation occurs in a cell in the sapwood or if thick enough within the bark. Pupation starts from about late March to early April and takes about five weeks. Adults have been recorded from May to July. Adults feed on leaves and have been found on beech, alder, aspen, birch, oak, poplar, rowan, sallow, willow and hawthorn. This species has also been noted visiting flowers. Continental literature states that the life cycle lasts one or two years.

Status Widespread but very local. In recent years this species has been noted from north Midlands, through northern England to Scotland, with a few records from Wales.

Indicator status Grade 2 in Garland (1983). Grade 3 in Harding & Rose (1986).

Threats Loss of broad-leaved woodland through, for example, clear-felling and coniferisation. Habitat loss, in particular, through the felling of ancient trees, removal of dead wood from living trees and the destruction or removal of standing and fallen dead wood for reasons such as forest hygiene, aesthetic tidiness, public safety or for use as fire wood.

Management and conservation Ancient trees and both fallen and standing dead timber, especially with the bark attached, should be retained. The removal of dead timber from ancient trees should be avoided. Gaps in the age structure of the tree population should be identified and the continuity of the appropriate dead wood habitat ensured by suitable planting and possibly with pollarding. The presence of nectar sources may also be particularly important for this species.

Published sources Bílý, S. & Mehl, O. (1989), Davies, R.D. & Burrows, L.W. (1952), Fraser, M.G. (1948), Garland, S.P. (1983), Harding, P.T. (1978), Harding, P.T. & Rose, F. (1986), Morgan, M.J. (1980).

STENOSTOLA DUBIA   **NOTABLE B**
A longhorn beetle
Order COLEOPTERA   Family CERAMBYCIDAE

Stenostola dubia (Laicharting, 1784). Formerly known as: *Stenostola ferrea* sensu auct. not (Schrank, 1776).

Distribution England, Dyfed-Powys and North Wales.

Habitat and ecology Broad-leaved woodland, parkland and probably also on isolated trees. Associated with small-leaved lime and large-leaved lime. Adults are known to eat the leaves of the host-tree. Adults have also been recorded from alder, elm, hazel, oak, rowan, sallow, willow, and on hawthorn blossom. Continental literature states that the eggs are laid in recently dead or recently cut branches, especiallly those with a diameter from 2cm to 5cm. In Britain, larvae develop and pupate in dead twigs and branches. Larvae and pupae are found in May with adults occurring from May to July, though the main period of adult emergence probably occurs in June. Continental sources state that the life cycle takes two years to complete.

Status Widely distributed, though very local. Possibly increasing in parts of the country, can be common in some areas.

Indicator status Grade 2 in Garland (1983).

Threats Loss of broad-leaved woodland and parkland through, for example, clear-felling and coniferisation. Habitat loss, in particular,through the felling of ancient trees, removal of dead wood from living trees and the destruction or removal of standing and fallen dead wood for reasons such as forest hygiene, aesthetic tidiness, public safety or for use as fire wood.

Management and conservation Both fallen and standing dead timber, especially with the bark attached, should be retained. The removal of dead timber from living trees should be avoided. Gaps in the age

structure of the tree population should be identified and the continuity of the appropriate dead wood habitat ensured by suitable planting and pollarding. The presence of nectar sources such as hawthorn may also be important for this species.

STRANGALIA ATTENUATA   **EXTINCT**
A longhorn beetle
Order COLEOPTERA   Family CERAMBYCIDAE

Strangalia attenuata (L., 1758).

Distribution Recorded from South Wiltshire, South Essex and Berkshire.

Habitat and ecology Probably woodland. The host-tree may be oak or possibly horse chestnut. The larvae probably develop in dead wood. Adults probably visit flowers. On the Continent, this species has been recorded from oak, birch, alder and horse chestnut. Continental literature states that the life cycle is two years, occasionally more.

Status Presumed extinct. Last recorded in the 19th century (probably about 1845) from Windsor Forest, Berkshire.

Published sources Allen, A.A. (1957d), Bílý, S. & Mehl, O. (1989), Fowler, W.W. (1890).

STRANGALIA AURULENTA   **NOTABLE A**
A longhorn beetle
Order COLEOPTERA   Family CERAMBYCIDAE

Strangalia aurulenta (F., 1792). Formerly known as: *Leptura aurulenta* F.

Distribution Recorded from West Cornwall, East Cornwall, South Devon, North Devon, South Somerset, North Somerset, South Hampshire, West Sussex, Hertfordshire, Middlesex, West Gloucestershire, Glamorgan and Radnorshire before 1970 and West Cornwall, East Cornwall, South Devon, South Somerset, North Somerset, South Hampshire and West Sussex from 1970 onwards.

Habitat and ecology Broad-leaved woodland, particularly in the more open parts and pasture-woodland. Chiefly associated with oak. Also recorded from aspen, ash, alder, birch, cherry, pear, elm, horse chestnut, sweet chestnut, walnut, willow and beech. Adults have been noted flying over oak logs. Larvae develop in dead wood and stumps. Pupae have been found in June. Adults have been recorded mainly in June and July, though also in August and September.

CERAMBYCIDAE

Adults frequently visit flowers such as cow parsley, and the blossom of bramble and broom.

Status A local, predominantly southern and western species. Unthoff-Kaufmann (1988) states that this species is "apparently spreading slowly in Wales". Some records may represent adventitious examples.

Indicator status Grade 3 in Harding & Rose (1986). Listed in Speight (1989) Appendix 1.

Threats Loss of broad-leaved woodland and parkland through, for example, clear-felling and coniferisation. Habitat loss, in particular, through the felling of ancient trees, removal of dead wood from living trees and the destruction or removal of standing and fallen dead wood for reasons such as forest hygiene, aesthetic tidiness, public safety or for use as fire wood. Neglect, in particular the abandonment of coppice management and the conversion to high forest, has led to increased shade and the loss of glades and broad sunny rides.

Management and conservation Ancient trees and both fallen and standing dead timber, especially with the bark attached, should be retained. The removal of dead timber from ancient trees should be avoided. Gaps in the age structure of the tree population should be identified and the continuity of the appropriate dead wood habitat ensured by suitable planting and pollarding. Open glades and ride margins should be cut on rotation to retain a variety of vegetation structures. The presence of nectar sources such as bramble and umbellifers may also be particularly important for this species.

Published sources Allen, A.A. (1957d), Allen, A.A. (1967b), Cooter, J. (1966), Ferry, R.S. (1953), Hare, D. & Jeffrey, P. (1953), Huggins, H.C. (1953), Kaufmann, R.R.U. (1988), Mitchell, A.V. (1927), Perkins, R.C.L. (1926), Speight, M.C.D. (1989).

STRANGALIA NIGRA	NOTABLE A

A longhorn beetle
Order COLEOPTERA   Family CERAMBYCIDAE

Strangalia nigra (L., 1758). Formerly known as: *Leptura nigra* L.

Distribution Recorded from West Cornwall, East Cornwall, South Devon, North Somerset, South Wiltshire, Dorset, South Hampshire, North Hampshire, West Sussex, East Sussex, East Kent, West Kent, Surrey, South Essex, Berkshire, Oxfordshire, East Suffolk, West Suffolk, East Norfolk, West Norfolk, Northamptonshire, West Gloucestershire, Worcestershire, Staffordshire, Shropshire, Glamorgan, Leicestershire & Rutland and Cheshire before 1970 and

West Sussex, East Sussex, Surrey, Northamptonshire and Shropshire from 1970 onwards.

Habitat and ecology Probably broad-leaved woodland. The larvae are probably associated with dead wood of deciduous trees. Continental literature states that this species is recorded from birch and hazel. In Britain, adults occur from May to July and frequently visit flowers such as umbellifers, the blossom of hawthorn and dog rose, spurge, buttercup and ox-eye daisy. It has also been recorded from European larch *Larix decidua*.

Status Very local and declining, modern records emanating from a few sites in south-eastern England, Northamptonshire and Shropshire.

Threats Uncertain, though this species will probably be threatened by the loss of broad-leaved woodland through, for example, clear-felling and coniferisation. Habitat loss, in particular, through the felling of ancient trees, removal of dead wood from living trees and the destruction or removal of standing and fallen dead wood for reasons such as forest hygiene, aesthetic tidiness, public safety or for use as fire wood.

Management and conservation Ancient trees and both fallen and standing dead timber, especially with the bark attached, should be retained. The removal of dead timber from ancient trees should be avoided. Gaps in the age structure of the tree population should be identified and the continuity of the appropriate dead wood habitat ensured by suitable planting and possibly with pollarding. The presence of nectar sources such as hawthorn, umbellifers and composite herbs may also be particularly important for this species.

Published sources Bílý, S. & Mehl, O. (1989), Hunter, F.A. (1951), Hunter, F.A. (1953).

STRANGALIA REVESTITA	ENDANGERED

A longhorn beetle
Order COLEOPTERA   Family CERAMBYCIDAE

Strangalia revestita (L., 1767).

Distribution Recorded from South Hampshire, North Hampshire, West Kent, South Essex, Middlesex, Berkshire, Cambridgeshire, Bedfordshire and Warwickshire before 1970 and Surrey from 1970 onwards.

Habitat and ecology Woodland. This species may be associated with wild fruit trees or oak. On the Continent, this species has been recorded from oak, beech, elm and pine, the larval habitat being being dead trunks, stumps and decaying branches. In Britain, adults have been recorded in June only, and probably visit wild flowers. A single adult has been beaten from oak. Continental literature states that the life cycle lasts two or three years.

Status Status revised from RDB 3 (Rare) in Shirt (1987). Only one modern record.

Indicator status Grade 2 in Harding & Rose (1986).

Threats Uncertain, though the loss of broad-leaved woodland and parkland trees through, for example, clear-felling and coniferisation is likely to threaten this species. Habitat loss, in particular, through the felling of ancient trees, removal of dead wood from living trees and the destruction or removal of standing and fallen dead wood for reasons such as forest hygiene, aesthetic tidiness, public safety or for use as fire wood.

Management and conservation Ancient trees and both fallen and standing dead timber, especially with the bark attached, should be retained. The removal of dead timber from ancient trees should be avoided. Gaps in the age structure of the tree population should be identified and the continuity of the appropriate dead wood habitat ensured by suitable planting and possibly with pollarding. The presence of nectar sources such as hawthorn, umbellifers and composite herbs may also be particularly important for this species.

Published sources Allen, A.A. (1972a), Bílý, S. & Mehl, O. (1989), Harding, P.T. (1978), Shirt, D.B., ed. (1987).

ANOMMATUS DIECKI
INSUFFICIENTLY KNOWN
Order COLEOPTERA   Family CERYLONIDAE

Anommatus diecki Reitter, 1875.

Distribution Recorded from Cheshire from 1970 onwards.

Habitat and ecology Wooded escarpments in river valleys. Recorded from fissures and the underside of elm wood, lying on or buried in clayey soil. Like its relative *A. duodecimstriatus* it appears to be partly subterranean. Adults have been recorded in May, July and November.

Status Not listed in the insect Red Data Book (Shirt, 1987). Described new to Great Britain from examples found in 1985 and 1986 in Cheshire.

Threats Uncertain, though clear-felling and conversion to other land use may be a threat. Habitat loss, in particular, maybe through the removal of fallen dead wood.

Management and conservation Trees and both fallen and standing dead timber, especially with the bark attached, should be retained. The removal of dead timber from trees should be avoided. Gaps in the age structure of the tree population should be identified and the continuity of the appropriate dead wood habitat ensured by regeneration or suitable planting.

Published sources Eccles, T.M. & Bowestead, S. (1987).

ANOMMATUS DUODECIMSTRIATUS
NOTABLE A
Order COLEOPTERA   Family CERYLONIDAE

Anommatus duodecimstriatus (Müller, 1821). Formerly known as: *Anommatus duodecemstriatus* (Müller, 1821).

Distribution Recorded from South Hampshire, East Gloucestershire and West Gloucestershire before 1970 and East Kent, Worcestershire, Cheshire and South Lancashire from 1970 onwards.

Habitat and ecology Woodland and probably gardens. Partly subterranean, and most likely to be encountered in rotten or old potatoes. Also recorded under elm wood, occasionally under bark and from ivy. Adults have been recorded in April, May, July and from September to November.

Status Probably under-recorded. Recorded from South Hampshire to South Lancashire. Recently recorded from only four vice-counties.

Threats Uncertain, though the clear-feeling of woodland and conversion to other land use, and development may be threats. In woodland, habitat loss may be through the removal of fallen dead wood.

Management and conservation In woodland, Trees and both fallen and standing dead timber, especially with the bark attached, should be retained. The removal of dead timber from trees should be avoided. Gaps in the age structure of the tree population should be identified and the continuity of the appropriate dead wood habitat ensured by regeneration or suitable planting.

Published sources Atty, D.B. (1983), Eccles, T.M. & Bowestead, S. (1987).

CERYLONIDAE

CERYLON FAGI   **NOTABLE B**

Order COLEOPTERA   Family CERYLONIDAE

Cerylon fagi Brisout, 1867.

Distribution England, South Wales and South West Scotland.

Habitat and ecology Ancient broad-leaved woodland and pasture-woodland. Under bark and in fungus-infected wood. Recorded from beech and oak. Occasionally also recorded from ash. Adults have been recorded from March to October.

Status Widespread but local in England, also recorded in South Wales and in South West Scotland. Most frequently encountered in southern and south-eastern England.

Indicator status Grade 3 in Harding & Rose (1986).

Threats Loss of broad-leaved woodland and parkland through, for example, clear-felling and coniferisation. Habitat loss, in particular, through the felling of fungus-infected trees, removal of dead wood from living trees and the destruction or removal of standing and fallen dead wood for reasons such as forest hygiene, aesthetic tidiness, public safety or for use as fire wood.

Management and conservation Fungus-infected trees and both fallen and standing dead timber, especially with the bark attached, should be retained. The removal of dead timber from fungus-infected trees should be avoided. Gaps in the age structure of the tree population should be identified and the continuity of the appropriate dead wood habitat ensured by regeneration, suitable planting and possibly with pollarding.

MURMIDIUS OVALIS   **EXTINCT**

Order COLEOPTERA   Family CERYLONIDAE

Murmidius ovalis (Beck, 1817).

Distribution Recorded from Surrey and Cambridgeshire before 1970.

Habitat and ecology Recorded from a wood. This species can also occur indoors in food stores etc. Found in plant refuse, haystack refuse and also in stored food products. The adult has been recorded in December.

Status Not listed in the insect Red Data Book (Shirt, 1987). Presumed extinct. The last record was in 1831 from Madingley Wood, Cambridgeshire.

Published sources Fowler, W.W. (1891).

AGELASTICA ALNI   INSUFFICIENTLY KNOWN
A leaf beetle
Order COLEOPTERA   Family CHRYSOMELIDAE

Agelastica alni (L., 1758).

Distribution Recorded from West Cornwall, South Devon, South Hampshire, East Kent, South Essex and West Gloucestershire before 1970.

Habitat and ecology Wetland, particularly alder carr. Also along river margins and in wet flushes in woodland. Phytophagous. Associated with alder and possibly hazel. Larvae feed on foliage of the host-tree in May and June, and then drop to the ground and pupate in the soil. New generation adults emerge between July and August. Adults aestivate then become active for a short period during autumn. They also hibernate during winter and re-emerge the following April.

Status Status revised from RDB Appendix (Extinct) in Shirt (1987). Status uncertain and possibly only an immigrant to the British shores. Only known from a small number of records, the most recent being of several examples in the New Forest, South Hampshire in 1946

Published sources Bedwell, E.C. (1936b), Britten, H. (1937), Donisthorpe, H.St J.K. (1936b), Fowler, W.W. (1890), Fowler, W.W. & Donisthorpe, H.St J.K. (1913), May, A.H. (1933), Shirt, D.B., ed. (1987).

ALTICA BREVICOLLIS   **NOTABLE A**
A flea beetle
Order COLEOPTERA   Family CHRYSOMELIDAE

Altica brevicollis Foudras, 1860. Formerly known as: *Haltica brevicollis* Foudras.

Distribution Recorded from Dorset, Isle of Wight, South Hampshire, East Sussex, East Kent, West Kent, Surrey, North Essex, Berkshire, Oxfordshire, Buckinghamshire, East Suffolk, West Suffolk, East Norfolk, Huntingdonshire, Northamptonshire, East Gloucestershire, West Gloucestershire, Warwickshire, Derbyshire, North-east Yorkshire and South-west Yorkshire before 1970 and South Somerset, Dorset, Hertfordshire, West Suffolk, Northamptonshire and West Gloucestershire from 1970 onwards.

Habitat and ecology Broad-leaved woodland and possibly also scrub. Phytophagous. Associated with hazel. Larvae probably occur between June and August and are free-living on the foliage of the host-plant, pupation probably occurring in the soil between mid July and mid September. Adults have been recorded from April to October.

Status Possibly declining. Old records indicate that this species was formerly widely distributed in southern England, with scattered records north to North-east Yorkshire. Recently recorded from six vice-counties from Dorset to Northamptonshire. This species is difficult to identify and may be confused with other members of the genus. Consequently, the exact status of this species is hard to assess.

Threats Loss of broad-leaved woodland through, for example, clear-felling and coniferisation. Neglect, possibly including the abandonment of coppice management, and conversion to high forest may be a further threat.

Management and conservation Existing coppice cycles should be continued and considered for re-introduction in areas of abandoned coppice. Open glades and ride margins should be cut on rotation to retain a variety of vegetation structures.

Published sources Atty, D.B. (1983), Fowler, W.W. (1890), Fowler, W.W. & Donisthorpe, H.St J.K. (1913).

ALTICA ERICETI   **NOTABLE B**
A flea beetle
Order COLEOPTERA   Family CHRYSOMELIDAE

Altica ericeti (Allard, 1859). Formerly known as: *Haltica ericeti* (Allard).

Distribution South East, South, South West, East Anglia, East Midlands, West Midlands and Wales.

Habitat and ecology Heathland and raised mire. Phytophagous. Associated with cross-leaved heath *Erica tetralix* and bell heather *E. cinerea*. Larvae are free-living on the foliage of the foodplant during April and May, with pupation probably occurring in June and July. Adults have been recorded from April to September.

Status Widespread but local in southern Great Britain. Possibly only a race of *A. britteni* (although *ericeti* is the older name). This species is difficult to identify and may be confused with other members of the genus. Consequently, the exact status of this species is hard to assess.

Threats Much heathland has been lost, or fragmented, through changes in land use, mainly by conversion to arable, forestry and urban development. Further habitat degradation has been through the cessation of traditional heathland management practices. Drainage may be a further threat.

Management and conservation Management should aim for a diversity of successional stages from bare ground to mature heath, preferably by grazing, rotational cutting or possibly by scraping or burning. Water tables should be maintained at high levels.

APHTHONA NIGRICEPS   **NOTABLE A**
A flea beetle
Order COLEOPTERA   Family CHRYSOMELIDAE

Aphthona nigriceps (Redtenbacher, 1842). Formerly known as: *Aphthona pallida* sensu auct. Brit. not Bach, 1856.

Distribution Recorded from North Devon, South Wiltshire, Isle of Wight, West Sussex, East Sussex, South Essex, Oxfordshire, Huntingdonshire, Worcestershire, Staffordshire, Glamorgan, Leicestershire & Rutland, South-east Yorkshire, North-east Yorkshire, Mid-west Yorkshire, Durham, Cumberland and Fife before 1970 and North Wiltshire, Dorset, Leicestershire & Rutland, Derbyshire, Mid-west Yorkshire, Durham, Westmorland & North Lancashire and Cumberland from 1970 onwards.

Habitat and ecology Grassland, wetland, fens, river margins and parkland. Phytophagous. Associated with meadow crane's-bill *Geranium pratense*. Adults have been recorded from June to October.

Status Very local and with a widely scattered distribution in England, from Dorset to Cumberland. There are old records for Wales and Scotland. This species is difficult to identify and may be confused with other members of the genus. Consequently, the exact status of this species is hard to assess.

Threats Loss of habitat through improvement and conversion to arable agriculture, afforestation and development. Natural succession is a further threat.

Management and conservation Limited grazing or cutting may be needed to maintain open conditions.

Published sources Eyre, M.D. & Cox, M.L. (1987), Fowler, W.W. (1890), Fowler, W.W. & Donisthorpe, H.St J.K. (1913).

CHRYSOMELIDAE

APTEROPEDA GLOBOSA — NOTABLE B
A flea beetle
Order COLEOPTERA   Family CHRYSOMELIDAE

Apteropeda globosa (Illiger, 1794).

Distribution South West, South, South East, East Anglia, West Midlands, North East, North West, Dyfed-Powys and North Wales.

Habitat and ecology Particularly on calcareous soils. Grassland, woodland and probably also wetland. Phytophagous. Found on bramble *Rubus fruticosus* agg. and also recorded in leaf-litter under bramble in woodlands. On the Continent, this species is probably associated with a variety of foodplants such as selfheal *Prunella*, woundwort *Stachys*, germander *Teucrium*, speedwell *Veronica*, bugle *Ajuga* and dead-nettle *Lamium* and possibly also with plantain *Plantago* and ground-ivy *Glechoma hederacea*. The beetle may be nocturnal. In Britain, adults have been recorded from April to July and in September and October.

Status Very local and with a widely scattered distribution in Engalnd and Wales, and recorded as far north as Westmorland. This species is difficult to identify and may be confused with other members of the genus. Consequently, the exact status of this species is hard to assess.

Threats Loss of habitat through improvement and conversion to arable agriculture, development and clear-felling and conversion to other land use. Drainage and natural succession may be further threats.

Management and conservation In woodland, open glades and ride margins should be managed to retain a variety of vegetation structures. Grazing or cutting, on a rotational basis, may be needed to maintain open conditions.

APTEROPEDA SPLENDIDA — ENDANGERED
A flea beetle
Order COLEOPTERA   Family CHRYSOMELIDAE

Apteropeda splendida Allard, 1860.

Distribution Recorded from Dorset, East Sussex, Berkshire, West Norfolk, Herefordshire, Glamorgan and "Lancs" before 1970.

Habitat and ecology Possibly associated with wetland, grassland and sandhills. Phytophagous. This species mines the leaves of greater plantain *Plantago major* and hoary plantain *P. media*. Possibly also associated with bugle *Ajuga* and speedwell *Veronica*. Adults have been recorded in August.

Status Not listed in the insect Red Data Book (Shirt, 1987). Only recorded on a very few occasions, the last record being in 1931 from Ashdown Forest, East Sussex.

Threats Uncertain, but this species may be threatened by loss of habitat through improvement and conversion to arable agriculture, development, afforestation, drainage and natural succession.

Published sources Day, F.H. (1932), Fowler, W.W. (1890), Fowler, W.W. & Donisthorpe, H.St J.K. (1913), Harde, K.W. (1966).

BROMIUS OBSCURUS — ENDANGERED
A leaf beetle
Order COLEOPTERA   Family CHRYSOMELIDAE

Bromius obscurus (L., 1758). Formerly known as: *Adoxus obscurus* (L.).

Distribution Recorded from "Lincs" and Cheshire before 1970 and Cheshire from 1970 onwards.

Habitat and ecology Recorded from a disused railway embankment near a river. On light, sandy soils. Phytophagous. Associated with willow-herbs, in particular rose-bay willowherb *Chamaenerion angustifolium*. The ova are laid in batches either in the soil or at the base of the stem. The larvae feed in groups on the roots of the foodplant during the summer and early autumn, then penetrate deep into the ground to overwinter. Pupation occurs in an earthen cocoon during March or early April. Adults have been recorded from May to October. The species may be parthenogenetic and flightless.

Status Known only from one site near Bosley, Cheshire, where it was first discovered in 1979 and last recorded in 1985. There is also a 19th century record for Lincolnshire.

Threats Uncertain, though this species may be threatened by change in land use for reasons such as development. Natural succession may be a further threat.

Management and conservation Some disturbance, on a rotational basis, may be needed to maintain open conditions and prevent scrub invasion.

Published sources Kendall, P. (1981), Shirt, D.B., ed. (1987), Stephens, J.F. (1839).

CALOMICRUS CIRCUMFUSUS NOTABLE A
A leaf beetle
Order COLEOPTERA   Family CHRYSOMELIDAE

Calomicrus circumfusus (Marsham, 1802). Formerly known as: *Luperus circumfuscus* (Marsham, 1802), *Luperus circumfusus* (Marsham), *Luperus nigrofasciatus* sensu Fowler, 1890 not (Goeze, 1777).

Distribution Recorded from East Cornwall, South Devon, North Devon, Dorset, Isle of Wight, South Hampshire, East Sussex, West Kent, Surrey, North Essex, East Suffolk, East Norfolk, East Gloucestershire, Glamorgan, North Northumberland and Wigtownshire before 1970 and West Cornwall, East Cornwall, South Devon, Dorset, South Hampshire, East Sussex, South Lincolnshire and South-west Yorkshire from 1970 onwards.

Habitat and ecology Heathland, grassland, maritime cliff, scrub and possibly also disturbed ground. Phytophagous. Associated with gorse *Ulex europaeus*, western gorse *U. gallii*, dwarf gorse *U. minor*, dyer's greenweed *Genista tinctoria* and possibly other *Genista* spp. and broom *Cytisus scoparius*. Adults have been recorded from June to September.

Status Widespread and very local in southern England, with scattered records north to Wigtownshire in Scotland. Recently recorded from eight vice-counties.

Threats Loss, or fragmentation, of heathland and grassland through changes in land use, mainly by conversion to arable agriculture, forestry and urban development. Natural succession is a further threat.

Management and conservation Grazing, cutting, or possibly scraping, on a rotational basis, may be needed to maintain open conditions.

Published sources Alexander, K.N.A. & Grove, S.J. (1990), Allen, A.A. (1959c), Allen, A.A. (1959d), Atty, D.B. (1983), Fowler, W.W. (1890), Fowler, W.W. & Donisthorpe, H.St J.K. (1913).

CASSIDA DENTICOLLIS ENDANGERED
A tortoise beetle
Order COLEOPTERA   Family CHRYSOMELIDAE

Cassida denticollis Suffrian, 1844. Formerly known as: *Cassida chloris* sensu auct. Brit. not Suffrian, 1844.

Distribution Recorded from South Hampshire, West Sussex, Glamorgan, Dumfriesshire and Kirkcudbrightshire before 1970.

Habitat and ecology Possibly associated with disturbed ground and grassland. Phytophagous. Probably associated with thistles, with a record from creeping thistle *Cirsium arvense* in Wales. On the Continent, this species has been found on yarrow *Achillea millefolium*, tansy *Tanacetum vulgare* and field wormwood *Artemisia campestris*.

Status Status revised from RDB 3 (Rare) in Shirt (1987). There are old records for five vice-counties from southern England to South West Scotland. Also noted in South Wales. Last recorded before 1963 in Kircudbrightshire.

Threats Uncertain, though this species may have declined through loss of habitat by change in land use.

Published sources Kevan, D.K. (1963b), Shirt, D.B., ed. (1987).

CASSIDA HEMISPHAERICA NOTABLE A
A tortoise beetle
Order COLEOPTERA   Family CHRYSOMELIDAE

Cassida hemisphaerica Herbst, 1799.

Distribution Recorded from East Cornwall, South Devon, North Devon, Dorset, Isle of Wight, West Sussex, East Sussex, East Kent, West Kent, Surrey, South Essex, Hertfordshire, Berkshire, "Suffolk", East Norfolk, Cambridgeshire, Staffordshire, Pembrokeshire, Merionethshire, Denbighshire, South-east Yorkshire, North-east Yorkshire and Cumberland before 1970 and West Cornwall, East Kent, South Essex, Monmouthshire, Radnorshire, Pembrokeshire, Montgomeryshire, Anglesey and Ayrshire from 1970 onwards.

Habitat and ecology Coastal shingle and grassland. Phytophagous. Associated with berry catchfly *Cucubalus baccifer*, ragged-robin *Lychnis flos-cuculi* and campion *Silene*, and possibly other related species. Larvae are probably free-living on the foodplant during the summer, with new generation adults emerging in late summer and early autumn. Adults have been recorded in January, from March to September and in November.

Status Old records show that this species has been widely recorded in southern England with a scattered distribution as far north as Cumberland. Recently reported from West Cornwall to Ayrshire in Scotland.

Threats Loss of grassland through improvement and conversion to arable agriculture, development and natural succession. Gravel extraction may be a further threat.

CHRYSOMELIDAE

Management and conservation Grazing or cutting, on a rotational basis, is needed to maintain open conditions. Disturbance of coastal shingle should be avoided.

Published sources Anon, (1933), Crowson, R.A. (1976), Fowler, W.W. (1890), Fowler, W.W. & Donisthorpe, H.St J.K. (1913), Marriner, T.F. (1938), Morgan, M.J. (1984).

CASSIDA NEBULOSA   **INDETERMINATE**
A tortoise beetle
Order COLEOPTERA   Family CHRYSOMELIDAE

Cassida nebulosa L., 1758.

Distribution Recorded from East Cornwall, West Kent, Surrey, North Essex, Berkshire, West Suffolk, East Norfolk, Cambridgeshire, Huntingdonshire and South Lincolnshire before 1970 and Dorset from 1970 onwards.

Habitat and ecology Disturbed ground, field margins, and probably grassland and scrub. Phytophagous. Associated with goosefoot *Chenopodium*. Larvae are probably free-living on the foodplant during the summer, with new generation adults emerging in late summer and early autumn. Adults have been recorded from May to September.

Status Not listed in the insect Red Data Book (Shirt, 1987). Old records suggest that this species was formerly more widespread and noted from East Cornwall to South Lincolnshire. Only one recent record, that from Dorset.

Threats Loss of habitat through improvement and conversion to arable agriculture, and development. Natural succession, and the use of herbicides and pesticides may be a further threat.

Management and conservation Disturbance, such as rotovation, is needed to maintain open conditions and encourage early successional stages.

Published sources Allen, A.A. (1950b), Dinnage, H. (1945), Fowler, W.W. (1890), Fowler, W.W. & Donisthorpe, H.St J.K. (1913), Morris, M.G. (1989), Turk, F.A. (1942).

CASSIDA NOBILIS   **NOTABLE B**
A tortoise beetle
Order COLEOPTERA   Family CHRYSOMELIDAE

Cassida nobilis L., 1758.

Distribution England, South Wales, Dyfed-Powys, South West Scotland and North West Scotland.

Habitat and ecology Sandy and chalky soils. Phytophagous. Associated with corn spurrey *Spergula arvensis* and possibly other Caryophyllaceae. On the Continent, this species has also be found on a species of the goosefoot family (Chenopodiaceae) and sea sandwort *Honkenya peploides*. In Britain, larvae are probably free-living on the foodplant during the summer and early autumn. Adults have been recorded from March to September.

Status Widespread but local in England, and recorded in parts of Wales and western Scotland. This species is difficult to identify and can be confused with *C. vittata*. Consequently, the exact status of this species is hard to assess.

Threats Loss of habitat through improvement and conversion to arable agriculture, and development. Natural succession may be a further threat.

Management and conservation Some grazing, rotational cutting or other disturbance, such as rotovation, is needed to maintain open conditions.

CASSIDA PRASINA   **NOTABLE B**
A tortoise beetle
Order COLEOPTERA   Family CHRYSOMELIDAE

Cassida prasina Illiger, 1798. Formerly known as: *Cassida sanguinolenta* sensu auct. Brit. not Müller, 1776.

Distribution England, Dyfed-Powys, North Wales, South West Scotland and North East Scotland.

Habitat and ecology Grassland, disturbed ground and probably scrub. Phytophagous. Associated with yarrow *Achillea millefolium* and possibly also thistles *Carduus*. Larvae are probably free-living on the foodplant during the summer. Adults have been recorded in January, February and from May to October.

Status Widespread but local in England and recorded in parts of Wales and Scotland.

Threats Loss of habitat through improvement and conversion to arable agriculture, and development. Natural succession may be a further threat.

Management and conservation Grazing, cutting or some other disturbance, on a rotational basis, is needed to maintain open conditions.

CHAETOCNEMA AEROSA
A flea beetle   **INSUFFICIENTLY KNOWN**
Order COLEOPTERA   Family CHRYSOMELIDAE

Chaetocnema aerosa (Letzner, 1847).

Distribution Recorded from South Hampshire and Surrey before 1970.

Habitat and ecology Possibly associated with wetland and water margins. Phytophagous. On the Continent, this species is associated with club-rushes. Possibly also associated with rushes and sedges. Adults have been recorded from June to September.

Status Not listed in the insect Red Data Book (Shirt, 1987). Found on just five occasions and from just two sites. Only recorded this century on Bookham Common, Surrey where it was found once in 1915 and on three occasions in 1961. Possibly under-recorded. This species is difficult to identify and may be confused with *C. hortensis*.

Threats Uncertain, though this species may be threatened by drainage and water abstraction schemes.

Published sources Henderson, J.L. (1961), MacKechnie Jarvis, C. (1967).

CHAETOCNEMA CONDUCTA
A flea beetle   **INDETERMINATE**
Order COLEOPTERA   Family CHRYSOMELIDAE

Chaetocnema conducta (Motschulsky, 1838).

Distribution Recorded from North-east Yorkshire before 1970.

Habitat and ecology The only records in Britain are from a wooded river valley. Phytophagous. Possibly associated with members of the Polygonaceae. On the Continent, this species is probably associated with rushes and common spiked-rush *Eleocharis palustris*. In Britain, the adult has been recorded in May.

Status Status revised from RDB 3 (Rare) in Shirt (1987). Recorded on just two occasions in 1911 and 1936 in Great Britain, from the Forge Valley, near Scarborough, North-east Yorkshire.

Published sources Fowler, W.W. & Donisthorpe, H.St J.K. (1913), Shirt, D.B., ed. (1987), Walsh, G.B. & Rimington, F.C. (1953).

CHAETOCNEMA SAHLBERGI   NOTABLE A
A flea beetle
Order COLEOPTERA   Family CHRYSOMELIDAE

Chaetocnema sahlbergi (Gyllenhal, 1827). Formerly known as: *Chaetocnema sahlbergii* (Gyllenhal, 1827).

Distribution Recorded from South Devon, North Devon, Dorset, Isle of Wight, South Hampshire, West Sussex, West Kent, East Suffolk, East Norfolk, Cambridgeshire, Caernarvonshire, South Lancashire, North-east Yorkshire and Cumberland before 1970 and South Essex, West Gloucestershire, Glamorgan, Denbighshire, Westmorland & North Lancashire and Cumberland from 1970 onwards.

Habitat and ecology Coastal habitats, particularly saltmarshes, though with one confirmed inland record from Quy Fen, Cambridgeshire. There are a number of other inland records, though these are not confirmed and at present must be regarded as highly doubtful. Phytophagous. Probably associated with sedges, and possibly also with grasses, glasswort *Salicornia* and sea-milkwort *Glaux maritima*. Adults have been recorded from March to June and in August and September.

Status Possibly declining. Old records suggest that this species had a widely scattered distribution in southern England, with records north to Cumberland. Recently recorded from six, widely scattered, vice-counties.

Threats Coastal developments. Loss of saltmarsh through reclamation, erosion and the construction of sea defences. Overgrazing of saltmarshes could be a further threat to this species.

Management and conservation Grazing should not be introduced to a saltmarsh where there is no grazing at present.

Published sources Atty, D.B. (1983), Collier, M.J. (1988a), Fowler, W.W. (1890), Fowler, W.W. & Donisthorpe, H.St J.K. (1913), Marriner, T.F. (1938), Morgan, M.J. (1984), Read, R.W.J. (1988).

CHRYSOMELIDAE

CHAETOCNEMA SUBCOERULEA   NOTABLE B
A flea beetle
Order COLEOPTERA   Family CHRYSOMELIDAE

Chaetocnema subcoerulea (Kutschera, 1864).

Distribution South East, South, South West, East Anglia, North East and South Wales.

Habitat and ecology Wetland, wet heathland and damp grassland. Phytophagous. Probably associated with rushes and possibly also with sedges. Adults have been recorded from February to November.

Status Widespread but local in southern England and also recorded in North East England and South Wales.

Threats Drainage for reasons such as agricultural improvement and development. Falling water tables because of water abstraction and river engineering schemes and natural succession may also threaten this species.

Management and conservation Grazing or cutting, on a rotational basis, is needed to maintain open conditions.

CHALCOIDES NITIDULA   NOTABLE B
A flea beetle
Order COLEOPTERA   Family CHRYSOMELIDAE

Chalcoides nitidula (L., 1758). Formerly known as: *Crepidodera nitidula* (L.).

Distribution South West, South, South East, East Anglia, East Midlands and West Midlands.

Habitat and ecology Broad-leaved woodland. Phytophagous. Associated with aspen and grey poplar, possibly preferring young trees. Larvae probably feed on the roots of the host-trees during June, with pupation occurring at the end of June. Adults have been recorded in February, March, and from May to September.

Status Widespread and very local in southern England, with records north to Nottinghamshire. Can be common where found.

Threats Loss of broad-leaved woodland through, for example, clear-felling and coniferisation. Neglect and conversion to high forest, and the selective removal of aspen may be further threats.

Management and conservation Open glades and ride margins should be cut on rotation to retain a variety of vegetation structures.

CHRYSOLINA CEREALIS   ENDANGERED
Rainbow leaf beetle
Order COLEOPTERA   Family CHRYSOMELIDAE

Chrysolina cerealis (L., 1767). Formerly known as: *Chrysomela cerealis* L.

Distribution Recorded from Caernarvonshire before 1970 and Caernarvonshire from 1970 onwards.

Habitat and ecology Montane grassland. Phytophagous. Associated with wild thyme *Thymus drucei*. Larvae feed on the leaves of wild thyme in the spring (at lower altitude) and summer (at higher altitude) and have been found on plants growing in deep, narrow crevices between large boulders. In captivity, larvae have been reared on dead thyme. Adults have been recorded on the foodplant between June and October.

Status Only known from the Snowdon area of Wales. Can be numerous where found. This species is listed on Schedule 5 of the Wildlife and Countryside Act 1981.

Threats Uncontrolled burning may be a threat to this species. Erosion of montane habitat may be a problem in areas frequented by large numbers of hill walkers and where livestock densities are too high.

Management and conservation Y Wyddfa - Snowdon is an NNR. Part of the area is owned by the National Trust.

Published sources Fowler, W.W. (1890), Shirt, D.B., ed. (1987).

CHRYSOLINA CRASSICORNIS   VULNERABLE
A leaf beetle
Order COLEOPTERA   Family CHRYSOMELIDAE

Chrysolina crassicornis (Helliesin, 1911). Formerly known as: *Chrysolina latecincta* (Demaison, 1896), *Chrysolina sanguinolenta* sensu Joy, 1932 not (L., 1758), *Chrysomela crassicornis* Helliesin, *Chrysomela sanguinolenta* sensu Fowler, 1890 not L.

Distribution Recorded from Renfrewshire, Argyll Main, "Sutherland", Caithness, Orkney and Shetland before 1970 and Argyll Main and Orkney from 1970 onwards.

Habitat and ecology Cliff tops, frequenting dry grasslands and sandy hills in maritime situations. Phytophagous. Associated with plantain *Plantago*, and doubtfully with toadflax *Linaria*. Oviposition has been observed on *P. maritima*. Larvae feed externally on the foodplant during the summer, and have been noted feeding on the normal and hairy forms of sea plantain *P. maritima* growing in crevices at the top of sea cliffs. In captivity, this species has been reared on ribwort plantain *P. lanceolata*, greater plantain *P. major*, buck's-horn plantain *P. coronopus* and also yellow toadflax *Linaria vulgaris*. Adults have been recorded from January to September, and probably overwinter in this stage.

Status Status revised from RDB 1 (Endangered) in Shirt (1987). Only known from Scotland, and recently recorded from just two vice-counties. Can be numerous where found.

Threats Coastal developments may reduce the amount of available habitat. Activities that accelerate or reduce the rate of erosion should be avoided. The degradation of suitable habitat through natural succession may be a further threat.

Management and conservation Disturbance, on a rotational basis, may be needed to maintain open conditions.

Published sources Drummond, D.C. (1956), Fowler, W.W. (1890), Owen, J.A. (1990a), Shirt, D.B., ed. (1987).

CHRYSOLINA GRAMINIS   **NOTABLE A**
A leaf beetle
Order COLEOPTERA   Family CHRYSOMELIDAE

Chrysolina graminis (L., 1758). Formerly known as: *Chrysomela graminis* L.

Distribution Recorded from South Devon, South Wiltshire, East Kent, Oxfordshire, West Suffolk, Cambridgeshire, Huntingdonshire, West Gloucestershire, South Lincolnshire, North Lincolnshire, Derbyshire, South-east Yorkshire, South-west Yorkshire, Mid-west Yorkshire and Cumberland before 1970 and West Cornwall, East Cornwall, East Norfolk, Cambridgeshire, North-east Yorkshire and Mid-west Yorkshire from 1970 onwards.

Habitat and ecology Wetland, flood meadows and dry grassland. Phytophagous. Associated with tansy *Tanacetum vulgare* and also water mint *Mentha aquatica*. On the Continent, this species is also associated with gipsywort *Lycopus europaeus*, marsh woundwort *Stachys palustris* and sneezewort *Achillea ptarmica*. Larvae feed on the foodplant in June and July. New generation adults emerge from late July to late August. Adults enter hibernation in September and re-appear the following June.

Status Old records show that this species has been widely recorded in England, from South Devon to Cumberland. Recently reported from only six, widely scattered, vice-counties. Can be numerous where found.

Threats Improvement aand conversion to arable agriculture, development, drainage, and falling water tables because of water abstraction schemes. Natural succession may also threaten this species.

Management and conservation Water tables should be maintained at high levels. Grazing or cutting, on a rotational basis, may be needed to maintain open conditions.

Published sources Atty, D.B. (1983), Buck, F.D. (1959b), Drummond, D.C. (1952), Fowler, W.W. (1890), Marriner, T.F. (1938), Turner, H.J. (1921).

CHRYSOLINA HAEMOPTERA   **NOTABLE B**
A leaf beetle
Order COLEOPTERA   Family CHRYSOMELIDAE

Chrysolina haemoptera (L., 1758). Formerly known as: *Chrysomela haemoptera* L.

Distribution England, North Wales, South East Scotland and South West Scotland.

Habitat and ecology Dry sandy coasts and sand dunes with a sward of herbaceous plant species but little grass. The species has been recorded inland and has occurred in a fen. Phytophagous. Associated with buck's-horn plantain *Plantago coronopus*, adults have also been found feeding on ribwort plantain *P. lanceolata* and hoary plantain *P. media*. In captivity, this species is also known to feed on sea plantain *P. maritima*, greater plantain *P. major* and common toadflax *Linaria vulgaris*. Larvae probably feed on the foodplants during the summer. Adults have been recorded from July to September, and probably overwinter in this stage.

Status Widespread but local in England, also recorded in North Wales and southern Scotland.

Threats Loss of habitat, particularly through afforestation, urban and holiday development. The degradation of dune habitat by excessive disturbance of the vegetation through activities such as motorbike access, horse-riding and human trampling.

Management and conservation Some grazing, rotational cutting or some other disturbance, such as rotovation, is needed to maintain open conditions.

CHRYSOMELIDAE

CHRYSOLINA MARGINATA   NOTABLE A
A leaf beetle
Order COLEOPTERA   Family CHRYSOMELIDAE

Chrysolina marginata (L., 1758). Formerly known as: *Chrysomela marginata* L.

Distribution Recorded from West Sussex, East Kent, Surrey, South Essex, West Suffolk, West Gloucestershire, Glamorgan, Merionethshire, North Lincolnshire, South-east Yorkshire, North-east Yorkshire, South-west Yorkshire, North-west Yorkshire, South Northumberland, North Northumberland, Westmorland & North Lancashire, Cumberland, Dumfriesshire, Midlothian, South Aberdeenshire, North Aberdeenshire, East Inverness & Nairn and Orkney before 1970 and West Suffolk, North-west Yorkshire, Durham, South Northumberland and Moray from 1970 onwards.

Habitat and ecology Well grazed, short-turf grassland, river margins and dry, sandy habitats. Phytophagous. Associated with yarrow *Achillea millefolium*, also recorded from sea plantain *Plantago maritima* and wild mignonette *Reseda lutea*. On the Continent, this species has also been associated with scentless mayweed *Tripleurospermum inodorum*, *Artemisia* and *Chrysanthemum*. Larvae feed on yarrow. Both larvae and adults are nocturnal, sheltering at the base of the foodplants during the day. In Britain, this beetle probably has two egg-laying periods, the main one being in autumn (September to mid-November), with a second period in spring (mid April to mid July). Larvae have been recorded from March to late June. Adults have been noted in July, August and October.

Status Possibly declining, old records suggest that this species formerly had a widely scattered distribution throughout Britain. Recently recorded from five, widely scattered, vice-counties.

Threats Loss of habitat through improvement and conversion to arable agriculture, development and afforestation. Natural succession may be a further threat.

Management and conservation Grazing, cutting or some other disturbance, such as rotovation, on a rotational basis, is needed to maintain open conditions.

Published sources Atty, D.B. (1983), Eyre, M.D. & Cox, M.L. (1987), Fowler, W.W. (1890), Fowler, W.W. & Donisthorpe, H.St J.K. (1913), Fryer, J.C.F. & Fryer, H.F. (1923c), Marriner, T.F. (1938), Morgan, M.J. (1984), Nelson, J.M. (1978), Shaw, H.K.A. (1949).

CHRYSOLINA ORICALCIA   NOTABLE B
A leaf beetle
Order COLEOPTERA   Family CHRYSOMELIDAE

Chrysolina oricalcia (Müller, 1776). Formerly known as: *Chrysolina orichalcia* (Müller, 1776), *Chrysomela orichalcia* (Müller, 1776).

Distribution England, North Wales, South East Scotland and South West Scotland.

Habitat and ecology River banks, downland and hedges. Also found in woodland rides, on a roadside verge and possibly also disturbed ground and wetland. Phytophagous. Associated with members of the parsley family Umbelliferae, with records from upright hedge-parsley *Torilis japonica*, cow parsley *Anthriscus sylvestris*, cowbane *Cicuta virosa* and hemlock *Conium maculatum*. This beetle has also been recorded from black horehound *Ballota nigra*. On the Continent, this species has also been noted on ground-elder *Aegopodium podagraria* and apparently also black poplar *Populus*. Larvae have fed on cow parsley in captivity and probably feed on foodplants during the summer. Adults have been recorded in June.

Status Widespread but local in England, also recorded in North Wales and southern Scotland.

Threats Loss of habitat through change in land use because of improvement and conversion to arable agriculture, development and afforestation. Drainage, water abstraction schemes and natural succession may be further threats.

Management and conservation Grazing or cutting may be needed to maintain open conditions.

CHRYSOLINA SANGUINOLENTA   NOTABLE A
A leaf beetle
Order COLEOPTERA   Family CHRYSOMELIDAE

Chrysolina sanguinolenta (L., 1758). Formerly known as: *Chrysolina marginalis* sensu Joy, 1932 not (Duftschmid, 1825), *Chrysomela marginalis* sensu Fowler, 1890 not Duftschmid.

Distribution Recorded from South Devon, North Devon, South Wiltshire, Dorset, South Hampshire, West Sussex, East Kent, West Kent, Surrey, North Essex, Hertfordshire, Berkshire, Oxfordshire, Buckinghamshire, East Suffolk, West Suffolk, East Norfolk, West Norfolk, Cambridgeshire, Bedfordshire, Huntingdonshire, Northamptonshire, East Gloucestershire, West Gloucestershire, Worcestershire, Glamorgan, Cheshire, South Lancashire, South-east Yorkshire, South-west Yorkshire and Mid-west

Yorkshire before 1970 and East Sussex, Surrey, South Essex, Hertfordshire, East Suffolk, West Suffolk, East Norfolk, West Norfolk and South-west Yorkshire from 1970 onwards.

Habitat and ecology Open ground, grassland, hedgebanks and field margins. Has been recorded in woodland and possibly associated with disturbed ground. Phytophagous. Associated with common toadflax *Linaria vulgaris*. The beetle has also been found feeding on snapdragon *Antirrhinum majus*. Larvae occur on the foodplants during the summer and are probably nocturnal, remaining at the base of the plants during the day. In captivity, this species is also known to feed on purple toadflax *L. purpurea*. The beetle has been recorded from common toadflax growing in tall, dense, grassy rides. Adults have been recorded from April to October.

Status Possibly declining. Very local and widespread in southern England, with scattered records north to Mid-west Yorkshire. Also recorded in South Wales.

Threats Loss of unimproved grassland through improvement by reseeding or by the application of fertilisers, or by conversion to arable agriculture. Development, afforestation, natural succession and the use of herbicides and pesticides may be further threats.

Management and conservation Disturbance, on a rotational basis, is needed to maintain open conditions.

Published sources Atty, D.B. (1983), Fowler, W.W. (1890), Fowler, W.W. & Donisthorpe, H.St J.K. (1913).

CHRYSOLINA VIOLACEA　　　　**NOTABLE B**
A leaf beetle
Order COLEOPTERA　　　Family CHRYSOMELIDAE

Chrysolina violacea (Müller, 1776). Formerly known as: *Chrysolina goettingensis* sensu (L., 1761) not (L., 1758), *Chrysomela goettingensis* sensu (L., 1761) not (L., 1758).

Distribution England and South Wales.

Habitat and ecology Calcareous grassland. Possibly also scrub and disturbed ground on base-rich soils. Phytophagous. Associated with ground-ivy *Glechoma hederacea*, possibly also on bedstraws *Galium* and other plants. Adults have been recorded from April to September.

Status Widespread but local in England, becoming more local in central and northern England. Also recorded in South Wales.

Threats Loss of unimproved grassland through improvement by reseeding or by the application of fertilisers, or by conversion to arable agriculture. Natural succession may be a further threat.

Management and conservation Grazing or cutting, on a rotational basis, is needed to maintain open conditions. Some areas of scrub should be retained.

CHRYSOMELA TREMULA　　　　**ENDANGERED**
A leaf beetle
Order COLEOPTERA　　　Family CHRYSOMELIDAE

Chrysomela tremula F., 1787. Formerly known as: *Chrysomela tremulae* F., 1787, *Melasoma longicolle* Suffrian, 1851, *Melasoma tremulae* Suffrian, 1851.

Distribution Recorded from South Hampshire, East Sussex, East Kent, West Kent, Surrey, South Essex, North Essex, Hertfordshire, Middlesex, Berkshire, Oxfordshire, East Suffolk, "Norfolk", Cambridgeshire, Bedfordshire, Huntingdonshire, West Gloucestershire, Herefordshire, Worcestershire, Warwickshire, Glamorgan and North Lincolnshire before 1970.

Habitat and ecology Broad-leaved woodland. Phytophagous. Associated with poplars, particularly young aspen. Also recorded from grey willow. The ova are laid in batches on the underside of leaves of the host-tree. Larvae feed on the leaves of the host-tree from May to July, with pupation occurring between June and August. Adults have been recorded from May to August.

Status Probably extinct. Formerly widespread in southern Great Britain and recorded as far north as North Lincolnshire. This species declined dramatically from around the late 1940's, with the last record being in 1957 from Bookham Common, Surrey.

Threats This species may have declined because of the loss of woodland and conversion to other land use. The selective removal of aspen, and neglect and conversion to high forest may also have contributed to this species' decline.

Published sources Allen, A.A. (1948a), Atty, D.B. (1983), Bromley, P.J. (1947), Brown, E.S. (1948b), Fowler, W.W. (1890), Fowler, W.W. & Donisthorpe, H.St J.K. (1913), Last, H. (1946), Shirt, D.B., ed. (1987).

CHRYSOMELIDAE

CLYTRA LAEVIUSCULA   EXTINCT
A leaf beetle
Order COLEOPTERA   Family CHRYSOMELIDAE

Clytra laeviuscula Ratzeburg, 1837. Formerly known as: *Clythra laeviuscula* Ratzeburg, 1837.

Distribution Recorded from Surrey, Berkshire and Mid Perthshire.

Habitat and ecology Recorded from woodland and grassland. Phytophagous. Recorded from a Caledonian pine and birch wood and a southern downland site. On the Continent, this species is associated with ash, *Prunus* and *Salix*. Larvae are myrmecophilous, requiring the nests of certain species of ant (Hymenoptera) to complete their development i.e., *Formica sanguinea* (known to occur at the Caledonian site), *Lasius niger* and *L. alienus*. Development probably takes one year. In Britain, adults have been recorded in October.

Status Presumed extinct. Last recorded in 1895 from Streatley, Berkshire.

Published sources Allen, A.A. (1976c), Shirt, D.B., ed. (1987).

CREPIDODERA IMPRESSA   NOTABLE A
A flea beetle
Order COLEOPTERA   Family CHRYSOMELIDAE

Crepidodera impressa (F., 1801).

Distribution Recorded from "Devon", Dorset, Isle of Wight, South Hampshire, West Sussex, East Kent, West Kent, North Essex and Glamorgan before 1970 and Dorset, Isle of Wight, South Hampshire, East Kent, West Kent, North Essex, East Suffolk and West Norfolk from 1970 onwards.

Habitat and ecology Saltmarshes. Phytophagous. Associated with common sea-lavender *Limonium vulgare*. Adults have been recorded from May to September.

Status Very local. Recorded along the coasts of southern Great Britain, from Glamorganshire to West Norfolk.

Threats Loss of saltmarsh through reclamation, erosion and the construction of sea defences. Overgrazing of saltmarshes could be a further threat to this species.

Management and conservation Grazing should not be introduced to a saltmarsh where there is no grazing at present.

Published sources Booth, R.G. (1978), Fowler, W.W. & Donisthorpe, H.St J.K. (1913), Hammond, P.M. (1959), Nash, D.R. (1977).

CRYPTOCEPHALUS AUREOLUS   NOTABLE B
A leaf beetle
Order COLEOPTERA   Family CHRYSOMELIDAE

Cryptocephalus aureolus Suffrian, 1847.

Distribution England, Wales and North East Scotland.

Habitat and ecology Primarily found on lightly grazed grassland, particularly on base-rich soils. Also recorded from sand dunes. Phytophagous. Associated with herbs, particularly hawkweeds *Hieracium*. and also on rock-roses *Helianthemum*. Adults may feed on pollen. Larvae are cased and free living on foliage, sometimes occurring in leaf-litter and possibly ant associated. Development probably takes one year. Adults have been recorded from May to July.

Status Widespread but local in England and Wales. Also recorded in North East Scotland. This species is difficult to identify and can be confused with *C. hypochaeridis*. Consequently, the exact status of this species is hard to assess.

Threats Loss of unimproved grassland through improvement by reseeding or by the application of fertilisers, or by conversion to arable agriculture. Loss of dune habitat, particularly through afforestation, and urban and holiday development. Natural succession, overgrazing and the degradation of dune habitat by excessive disturbance of the vegetation through activities such as motorbike access, horse-riding and human trampling may be further threats.

Management and conservation Grazing, cutting or possibly some other disturbance is needed to maintain open conditions.

CRYPTOCEPHALUS BIGUTTATUS
A leaf beetle   **VULNERABLE**
Order COLEOPTERA   Family CHRYSOMELIDAE

Cryptocephalus biguttatus (Scopoli, 1763).

Distribution Recorded from South Devon, Dorset, South Hampshire, North Hampshire, East Kent, Surrey, Berkshire, Staffordshire and South Lancashire before 1970 and Dorset, West Sussex, Surrey and North-east Yorkshire from 1970 onwards.

Habitat and ecology Wet heathland and bog. Phytophagous. Associated with cross-leaved heath

Erica tetralix in wet and boggy heathland areas. Larvae are cased and free-living on foliage, sometimes occurring in leaf-litter and possibly ant associated. Development probably takes one year. Adults have been recorded from May to July.

Status Old records show that this species has been widely recorded in southern England, with a scattered distribution north to South Lancashire. Recently reported from just four vice-counties.

Threats Loss of habitat through improvement and conversion to arable agriculture, drainage, afforestation, and development. Further habitat degradation has been through the cessation of traditional heathland management practices.

Management and conservation Water tables should be maintained at high levels. Management should aim for a diversity of successional stages from bare ground to mature heath, preferably by grazing or by rotational cutting, scraping or possibly burning.

Published sources Allen, A.A. (1970a), Fowler, W.W. (1890), Jones, R.A. (1988), Nicholson, G.W. (1921), Shirt, D.B., ed. (1987).

CRYPTOCEPHALUS BILINEATUS NOTABLE B
A leaf beetle
Order COLEOPTERA Family CHRYSOMELIDAE

Cryptocephalus bilineatus (L., 1767).

Distribution South East, South, East Anglia, West Midlands and East Midlands.

Habitat and ecology Calcareous grassland. Phytophagous. Associated with kidney vetch *Anthyllis vulneraria*. On the Continent, this species has been recorded from ox-eye daisy *Leucanthemum vulgare*. Larvae are probably cased and free-living on foliage, and may be ant associated. Development probably takes one year. In Britain, adults have been recorded in June and July.

Status Local in southern and south-eastern England, and recorded north to the Midlands.

Threats Loss of unimproved grassland through improvement by reseeding or by the application of fertilisers, or by conversion to arable agriculture. Natural succession and overgrazing are further threats.

Management and conservation Grazing or cutting, on a rotational basis, is needed to maintain open conditions.

CRYPTOCEPHALUS BIPUNCTATUS
A leaf beetle **NOTABLE B**
Order COLEOPTERA Family CHRYSOMELIDAE

Cryptocephalus bipunctatus (L., 1758).

Distribution England, South Wales, North Wales, South East Scotland, South West Scotland and North East Scotland.

Habitat and ecology Broad-leaved woodland, scrubby areas on downland, and open heathland. Phytophagous. Adults have been recorded from a variety of broad-leaved trees and also from rock-rose *Helianthemum*. Hazel, grey willow and birch may be the preferred hosts. Larvae are cased and free-living on foliage, sometimes occurring in leaf-litter and possibly ant associated. Development probably takes one year. Adults have been recorded from late April to July.

Status Widespread but local in Great Britain.

Threats Loss of broad-leaved woodland through, for example, clear-felling and coniferisation. The grubbing out of scrub, and the loss of heathland through afforestation, development, and improvement and conversion to arable agricuture are further threats.

Management and conservation In woodland and areas of scrub, open glades and ride margins should be cut on rotation to retain a variety of vegetation structures. On heathland, management should aim for a diversity of successional stages from bare ground to mature heath, preferably by grazing or by rotational cutting, scraping or possibly burning.

CRYPTOCEPHALUS CORYLI ENDANGERED
A leaf beetle
Order COLEOPTERA Family CHRYSOMELIDAE

Cryptocephalus coryli (L., 1758).

Distribution Recorded from West Kent, Surrey, Berkshire, East Norfolk, Bedfordshire, Northamptonshire, Staffordshire, North Lincolnshire, Nottinghamshire and "Inverness" before 1970 and Surrey and Berkshire from 1970 onwards.

Habitat and ecology Broad-leaved woodland. Phytophagous. Adults have been recorded from birch, oak, hazel and hawthorn blossom. On the Continent, this species is also associated with alder and *Salix*. Larvae are cased and free-living on foliage, sometimes occurring in leaf-litter and possibly ant associated. Development probably takes one year. In Britain, adults have been recorded from April to June.

CHRYSOMELIDAE

Status Old records suggest that this species had a widely scattered distribution from southern England to Inverness in Scotland. This species appears to have declined from about the middle of this century and is now known from just two sites.

Threats This species has probably declined through the loss of broad-leaved woodland through, for example, clear-felling and coniferisation. Neglect and conversion to high forest may also have contributed to this species' decline.

Management and conservation Open glades and ride margins should be cut on rotation to retain a variety of vegetation structures.

Published sources Fowler, W.W. (1890), Fowler, W.W. & Donisthorpe, H.St J.K. (1913), Harwood, P. (1947), Shirt, D.B., ed. (1987).

CRYPTOCEPHALUS DECEMMACULATUS
A leaf beetle   **VULNERABLE**
Order COLEOPTERA   Family CHRYSOMELIDAE

Cryptocephalus decemmaculatus (L., 1758).

Distribution Recorded from East Sussex, Staffordshire, Mid Perthshire and South Aberdeenshire before 1970 and Staffordshire, Cheshire and Mid Perthshire from 1970 onwards.

Habitat and ecology Broad-leaved woodland. Phytophagous. Adults have been recorded from dwarf sallows and birch. On the Continent, this species has also been noted on alder. Larvae are cased and free-living on foliage, sometimes occurring in leaf-litter and possibly ant associated. Development probably takes one year. In Britain, adults have been recorded from June to August.

Status Only known from five vice-counties, and recently recorded from just three.

Threats Loss of broad-leaved woodland through, for example, clear-felling and coniferisation. Neglect and conversion to high forest may be a further threat.

Management and conservation Open glades and ride margins should be cut on rotation to retain a variety of vegetation structures.

Published sources Allen, A.A. (1960d), Allen, A.A. (1970a), Fowler, W.W. (1891), Shirt, D.B., ed. (1987), Stott, C.E. (1929).

CRYPTOCEPHALUS EXIGUUS   ENDANGERED
A leaf beetle
Order COLEOPTERA   Family CHRYSOMELIDAE

Cryptocephalus exiguus Schneider, 1792.

Distribution Recorded from East Suffolk, West Suffolk, East Norfolk and North Lincolnshire before 1970 and West Suffolk from 1970 onwards.

Habitat and ecology Bogs, fens and broads. Recently found in open grassland. Phytophagous. Although adults have been recorded from birch and grey willow, the larval foodplant is almost certainly a species of wetland thistle. Adults have been recorded on flowers of melancholy thistle *Cirsium helenioides* in shady places. Larvae are cased and free-living on foliage, sometimes occurring in leaf-litter and possibly ant associated. Development probably takes one year. Adults have been recorded in June and July.

Status Until recently the last known records of *C. exiguus* were in 1910 from Freshney Bog, North Lincolnshire. However, in 1980 a single example was found at Pashford Poors' Fen, West Suffolk. A second example was noted at the same site in 1986.

Threats This species is threatened by loss of habitat through a falling water table because of abstraction bore holes adjacent to the only known site for the beetle. Drainage, water abstraction schemes, and improvement and conversion to arable agriculture have probably contributed to this species' decline.

Management and conservation Water tables should be maintained at high levels. Grazing, on a rotational basis, may be needed to maintain open conditions. Pashford Poors' Fen has been notified as an SSSI and is a Suffolk Wildlife Trust reserve.

Published sources Fowler, W.W. (1890), Fowler, W.W. & Donisthorpe, H.St J.K. (1913), Shirt, D.B., ed. (1987).

CRYPTOCEPHALUS FRONTALIS   NOTABLE A
A leaf beetle
Order COLEOPTERA   Family CHRYSOMELIDAE

Cryptocephalus frontalis Marsham, 1802.

Distribution Recorded from West Sussex, East Sussex, East Kent, West Kent, North Essex, Middlesex, Oxfordshire, "Suffolk", East Norfolk, West Norfolk, Cambridgeshire, Bedfordshire, East Gloucestershire, West Gloucestershire, South Lincolnshire and Leicestershire & Rutland before 1970 and East Sussex,

East Kent, Surrey, South Essex and Oxfordshire from 1970 onwards.

Habitat and ecology Hedgerows in cultivated land and along roadside verges. Probably also associated with broad-leaved woodland and scrub. Phytophagous. Most records are from hawthorn. On the Continent, this species has also been recorded on silver birch and aspen. Larvae are cased and free-living on foliage, sometimes occurring in leaf-litter and possibly ant associated. Development probably takes one year. In Britain, adults have been recorded in June and July.

Status Old records show that this species has been widely recorded in southern England, with scattered records north to South Lincolnshire. Recently noted from just five vice-counties, all in south-eastern England.

Threats The grubbing out of hedgerows and scrub, and the loss of broad-leaved woodland through, for example, clear-felling and coniferisation. The use of pesticides and mechanised hedge trimming may be further threats.

Management and conservation Cutting may be needed, this should be undertaken on a rotational basis. In woodlands, open glades and ride margins should be managed to retain a variety of vegetation structures.

Published sources Atty, D.B. (1983), Fowler, W.W. (1890), Fowler, W.W. & Donisthorpe, H.St J.K. (1913), Henderson, C.W. (1944).

CRYPTOCEPHALUS NITIDULUS ENDANGERED
A leaf beetle
Order COLEOPTERA   Family CHRYSOMELIDAE

Cryptocephalus nitidulus F., 1787. Formerly known as: *Cryptocephalus ochrostoma* Harold, 1872.

Distribution Recorded from Dorset, South Hampshire, West Kent, Surrey, Oxfordshire, East Gloucestershire, West Gloucestershire and Nottinghamshire before 1970 and Surrey from 1970 onwards.

Habitat and ecology Broad-leaved woodland and downland scrub. Phytophagous. Associated with young birch and hazel. Larvae are cased and free-living on foliage, sometimes occurring in leaf-litter and possibly ant associated. Development probably takes one year. Adults have been recorded in May and June.

Status Old records suggest that this species had a widely scattered distribution in southern England and was recorded as far north as Nottinghamshire. This species appears to have declined since around the

middle of this century and has recently been recorded from just one vice-county.

Threats This species may have declined because of the loss of broad-leaved woodland through, for example, clear-felling and coniferisation. Neglect and conversion to high forest may also have contributed to this species' decline.

Management and conservation Open glades and ride margins should be cut on rotation to retain a variety of vegetation structures.

Published sources Atty, D.B. (1983), Drane, A.B. (1990), Fowler, W.W. (1890), Fowler, W.W. & Donisthorpe, H.St J.K. (1913), Halstead, A.J. (1988), Shirt, D.B., ed. (1987).

CRYPTOCEPHALUS PARVULUS   NOTABLE B
A leaf beetle
Order COLEOPTERA   Family CHRYSOMELIDAE

Cryptocephalus parvulus Müller, 1776.

Distribution South West, South, South East, East Anglia, East Midlands, West Midlands, North West, North Wales, South East Scotland and North West Scotland.

Habitat and ecology Broad-leaved woodland and scrub. Phytophagous. Associated with birch. On the Continent, this species has also been recorded from oak. In Britain, larvae feed on birch leaves, particularly those that are brown and have a fungal infection. Larvae probably require two years to achieve development to adult. Adults have been recorded from April to September.

Status Widespread and very local, recorded from southern England through to northern Scotland.

Threats Loss of broad-leaved woodland through, for example, clear-felling and coniferisation. Neglect and conversion to high forest, and the grubbing out of scrub are further threats.

Management and conservation Open glades and ride margins should be cut on rotation to retain a variety of vegetation structures.

CHRYSOMELIDAE

CRYPTOCEPHALUS PRIMARIUS
A leaf beetle   **ENDANGERED**
Order COLEOPTERA   Family CHRYSOMELIDAE

Cryptocephalus primarius Harold, 1872.

Distribution Recorded from Berkshire, Cambridgeshire, West Gloucestershire and Mid Perthshire before 1970.

Habitat and ecology Calcareous grassland, particularly on dry hillsides. Phytophagous. Associated with common rock-rose *Helianthemum nummularium*. On the Continent, this species is associated with hazel and white willow. Larvae are cased and probably free-living on foliage, although they have been found at the roots of common rock-rose. They may also occur in leaf-litter and possibly be ant associated. Development probably takes at least one year. In Britain, adults have been recorded in May and June.

Status Only known from four, widely scattered, vice-counties and none recently. Last recorded in 1955 from Cholsey, Berkshire.

Threats This species has probably declined because of the loss of calcareous grassland through improvement by reseeding or by the application of fertilisers, or by conversion to arable agriculture. Natural succession may also have contributed to this species decline.

Management and conservation Grazing or cutting, on a rotational basis, is needed to maintain open conditions.

Published sources Atty, D.B. (1983), Fowler, W.W. (1890), Hobby, B.M. (1955), Richards, O.W. (1927), Shirt, D.B., ed. (1987), Skidmore, P. (1966).

CRYPTOCEPHALUS PUNCTIGER   NOTABLE A
A leaf beetle
Order COLEOPTERA   Family CHRYSOMELIDAE

Cryptocephalus punctiger Paykull, 1799.

Distribution Recorded from East Sussex, West Kent, Surrey, Staffordshire, Denbighshire, South Lincolnshire, Nottinghamshire, South-west Yorkshire, Moray and East Inverness & Nairn before 1970 and Dorset, West Sussex, Surrey and Staffordshire from 1970 onwards.

Habitat and ecology Broad-leaved woodland. Phytophagous. Associated with birch and also recorded from goat willow. On the Continent, this species has also been noted on oak. Larvae are cased and free-living on foliage, sometimes occurring in leaf-litter and possibly ant associated. Development probably takes one year. In Britain, adults have been recorded in June and July.

Status Old records show that this species has been widely recorded in England, and was noted as far north as Easterness in Scotland. Recently reported from just four vice-counties.

Threats Loss of broad-leaved woodland through, for example, clear-felling and coniferisation. Neglect and conversion to high forest may be a further threat.

Management and conservation Open glades and ride margins should be cut on rotation to retain a variety of vegetation structures.

Published sources Fowler, W.W. (1890), Harwood, P. (1947), Morgan, M.J. (1984).

CRYPTOCEPHALUS QUERCETI   VULNERABLE
A leaf beetle
Order COLEOPTERA   Family CHRYSOMELIDAE

Cryptocephalus querceti Suffrian, 1848.

Distribution Recorded from South Hampshire, North Essex, Berkshire, Nottinghamshire and South Lancashire before 1970 and Berkshire and Nottinghamshire from 1970 onwards.

Habitat and ecology Ancient broad-leaved woodland. Phytophagous. Associated with ancient oaks and also recorded from hawthorn. On the Continent, this species has also been noted on birch. Larvae are cased and free-living on foliage, sometimes occurring in leaf-litter and possibly ant associated. Development probably takes one year. In Britain, adults have been recorded from June to August.

Status Formerly only known from four sites. Since around the middle of this century this species was only recorded from Windsor Forest and Windsor Great Park, Berkshire, where it is well established, until 1965 when it was re-discovered in Sherwood Forest, Nottinghamshire.

Indicator status Grade 1 in Harding & Rose (1986).

Threats Loss of broad-leaved woodland and parkland through, for example, clear-felling and coniferisation. Habitat loss, in particular, through the felling of ancient trees for reasons such as forest hygiene, aesthetic tidiness, public safety or for use as fire wood.

Management and conservation Ancient trees should be retained. Gaps in the age structure of the tree population should be identified and the continuity of the appropriate habitat ensured by regeneration, suitable planting and possibly with pollarding. Windsor Forest and parts of Windsor Great Park are notified as SSSIs. A large part of Sherwood Forest is notified as the Birklands and Bilhaugh SSSI.

Published sources Bedwell, E.C. (1926b), Fowler, W.W. (1890), Garland, S.P. (1983), Harding, P.T. (1978), Johnson, C. (1965), Shirt, D.B., ed. (1987).

CRYPTOCEPHALUS SEXPUNCTATUS
A leaf beetle   **VULNERABLE**
Order COLEOPTERA   Family CHRYSOMELIDAE

Cryptocephalus sexpunctatus (L., 1758).

Distribution Recorded from South Wiltshire, Dorset, South Hampshire, West Sussex, East Sussex, East Kent, West Kent, Surrey, South Essex, North Essex, East Suffolk, East Gloucestershire, West Gloucestershire, Worcestershire, Warwickshire, North Lincolnshire, Dumfriesshire, Ayrshire and Midlothian before 1970 and West Sussex and South Essex from 1970 onwards.

Habitat and ecology Broad-leaved woodland. Phytophagous. Recorded from hazel, birch, aspen, crack willow and young oak. Possibly also on other tree species. Adults have also been noted on wood spurge blossom. On the Continent, this species is also associated with willow and hawthorn. Larvae are cased and free-living on foliage, sometimes occurring in leaf-litter and possibly ant associated. Development probably takes one year. In Britain, adults have been recorded from May to early July.

Status Old records indicate that this species was widely distributed in southern England, with scattered records north to Ayrshire in Scotland. Formerly common in some areas. Recently recorded from two vice-counties in southern and eastern England.

Threats Loss of broad-leaved woodland through, for example, clear-felling and coniferisation. Neglect and conversion to high forest is a further threat.

Management and conservation Open glades and ride margins should be cut on rotation to retain a variety of vegetation structures.

Published sources Atty, D.B. (1983), Cox, D. (1947), Cox, D. (1948), Fowler, W.W. (1890), Fowler, W.W. & Donisthorpe, H.St J.K. (1913), Massee, A.M. (1947), Shirt, D.B., ed. (1987).

CRYPTOCEPHALUS VIOLACEUS   **EXTINCT**
A leaf beetle
Order COLEOPTERA   Family CHRYSOMELIDAE

Cryptocephalus violaceus Laicharting, 1781.

Distribution Recorded from East Kent and Cambridgeshire.

Habitat and ecology Phytophagous. On the Continent, this species is associated with goat willow. In Britain, the adult has been recorded in June.

Status Extinct. Last recorded in 1864 at Folkestone, East Kent.

Published sources Allen, A.A. (1976d), Shirt, D.B., ed. (1987).

DIBOLIA CYNOGLOSSI   **ENDANGERED**
A flea beetle
Order COLEOPTERA   Family CHRYSOMELIDAE

Dibolia cynoglossi (Koch, 1803).

Distribution Recorded from South Devon, East Sussex, Cambridgeshire, West Gloucestershire and North Lincolnshire before 1970 and East Sussex and East Kent from 1970 onwards.

Habitat and ecology Coastal shingle, broad-leaved woodland rides, clearings and wood margins, and possibly grassland. Phytophagous. Associated with red hemp-nettle *Galeopsis angustifolia*, *G. ladanum* v. *canescens* and hound's tongue *Cynoglossum*. Possibly also on mint *Mentha*, clary *Salvia*, woundwort *Stachys* and black horehound *Ballota nigra*. On the Continent, this species is also associated with horehound *Marrubium*. Larvae mine the leaves of the foodplant during the summer. In Britain, adults have been recorded from April to September and probably overwinter in this stage.

Status Only known from five vice-counties. Recently recorded from only two sites; Rye Harbour, East Sussex and Dungeness, East Kent. Only recorded on coastal shingle since around the mid 1940's. Possibly under-recorded as the adult beetle is extremely active, jumping at the slightest disturbance.

Threats This species may be threatened by loss of habitat through gravel extraction and possibly natural succession. The clear-felling of woodland and conversion to other land use may be a further threat.

Management and conservation Disturbance of coastal shingle should be avoided. In woodland, open glades

CHRYSOMELIDAE

and ride margins should be cut on rotation to retain a variety of vegetation structures. Rye Harbour has been notified as a SSSI and is also a Local Nature Reserve, Dungeness has been notified as an SSSI.

Published sources Fowler, W.W. & Donisthorpe, H.St J.K. (1913), Fryer, J.C.F. & Fryer, H.F. (1923b), Shirt, D.B., ed. (1987).

DONACIA AQUATICA **RARE**
A reed beetle
Order COLEOPTERA Family CHRYSOMELIDAE

Donacia aquatica (L., 1758). Formerly known as: *Donacia dentipes* F., 1792.

Distribution Recorded from "Devon", South Wiltshire, Dorset, South Hampshire, West Sussex, East Sussex, Surrey, South Essex, East Suffolk, West Suffolk, East Norfolk, Cambridgeshire, Worcestershire, Merionethshire, North Lincolnshire, Derbyshire, Cheshire, South-east Yorkshire, North-east Yorkshire, South-west Yorkshire, Mid-west Yorkshire, North Northumberland, Cumberland, East Inverness & Nairn and Argyll Main before 1970 and West Sussex, Westmorland & North Lancashire, Roxburghshire and East Inverness & Nairn from 1970 onwards.

Habitat and ecology Aquatic and semi-aquatic habitats, such as freshwater lakes, ponds and possibly ditches. Phytophagous. Associated with sedges, sweet-grass and bur-reeds. On the Continent, this species has also been recorded from greater spearwort *Ranunculus lingua*. In Britain, floating sweet-grass *G. fluitans* may be the preferred foodplant. Larvae are aquatic on foodplants. Adults are found on emergent vegetation and have been recorded from May to July.

Status Not listed in the insect Red Data Book (Shirt, 1987). Old records indicate that this species was formerly widely distributed in southern England, with a scattered distribution north to central Scotland. Recently recorded from just four vice-counties; one in southern England, one in northern England and two in Scotland. Populations of this species can be very restricted in larger areas of apparently suitable habitat.

Threats Loss of habitat through falling water tables because of water abstraction schemes, and the infilling of lakes and ponds. Water pollution and natural succession may be further threats.

Management and conservation Water tables should be maintained at high levels. Water bodies should be isolated from sources of pollution and eutrophication. Clearance of the emergent vegetation may be needed to maintain open conditions, this should be undertaken on

a rotational basis and be aimed at maintaining the plant populations.

Published sources Alexander, K.N.A. & Clements, D.K. (1988), Allen, A.A. (1954c), Cribb, J. (1954), Fowler, W.W. (1890), Fowler, W.W. & Donisthorpe, H.St J.K. (1913), Stainforth, T. (1944).

DONACIA BICOLORA **VULNERABLE**
A reed beetle
Order COLEOPTERA Family CHRYSOMELIDAE

Donacia bicolora Zschach, 1788.

Distribution Recorded from North Somerset, Dorset, Isle of Wight, South Hampshire, East Sussex, East Kent, Surrey, South Essex, "Suffolk", East Norfolk, Cambridgeshire, Worcestershire, Warwickshire, Staffordshire, Glamorgan, Leicestershire & Rutland, Cheshire, South Lancashire and North Northumberland before 1970 and Dorset and South Hampshire from 1970 onwards.

Habitat and ecology Aquatic and semi-aquatic habitats, such as freshwater lakes, ponds and ditches. Phytophagous. Associated with branched bur-reed *Sparganium erectum*, and possibly also arrowhead *Sagittaria*, sweet-grass *Glyceria* and sedges *Carex*. Larvae are probably aquatic on the foodplants. Adults are found on emergent and marginal vegetation, and have been recorded in June.

Status Not listed in the insect Red Data Book (Shirt, 1987). Old records indicate that this species was widely distributed in southern England, with scattered records north to Northumberland. Recently recorded from just two vice-counties.

Threats Loss of habitat through falling water tables because of water abstraction schemes, and the infilling of lakes and ponds. Water pollution and natural succession may be further threats.

Management and conservation Water tables should be maintained at high levels. Water bodies should be isolated from sources of pollution and eutrophication. Clearance of emergent vegetation may be needed to maintain open conditions, this should be undertaken on a rotational basis and be aimed at maintaining the plant populations.

Published sources Fowler, W.W. (1890), Fowler, W.W. & Donisthorpe, H.St J.K. (1913).

DONACIA CINEREA   NOTABLE B
A reed beetle
Order COLEOPTERA   Family CHRYSOMELIDAE

Donacia cinerea Herbst, 1784.

Distribution South East, South, East Anglia, East Midlands, West Midlands, North East, North West, South Wales and Dyfed-Powys.

Habitat and ecology Aquatic and semi-aquatic habitats, such as freshwater lakes, ponds and ditches. Phytophagous. Associated with reed *Phragmites australis*, and also reedmace *Typha* and bur-reed *Sparganium*. On the Continent, this species is also associated with sedges. Larvae occur on the roots of the foodplants. In Britain, adults are found on emergent and marginal vegetation, and have been recorded in May and June.

Status Widespread but local in England and parts of Wales.

Threats Loss of habitat through falling water tables because of water abstraction schemes, and the infilling of lakes and ponds. Water pollution and natural succession may be further threats.

Management and conservation Water tables should be maintained at high levels. Water bodies should be isolated from sources of pollution and eutrophication. Clearance of the edges of ditches, ponds, etc. should be undertaken on a rotational basis and be aimed at maintaining aquatic and emergent plant populations.

DONACIA CLAVIPES   NOTABLE B
A reed beetle
Order COLEOPTERA   Family CHRYSOMELIDAE

Donacia clavipes Fabricius, 1792. Formerly known as: *Donacia menyanthidis* Gyllenhal, 1813, *Donacia menyanthis* Fabricius, 1801.

Distribution England, Wales and Scotland.

Habitat and ecology Aquatic and semi-aquatic habitats, such as freshwater lakes, ponds and ditches. Phytophagous. Associated with reed *Phragmites australis* and bur-reed *Sparganium*. Adults are found on emergent and marginal vegetation, and have been recorded in June.

Status Widespread but local in Great Britain.

Threats Loss of habitat through falling water tables because of water abstraction schemes, and the infilling

of lakes and ponds. Water pollution and natural succession may be further threats.

Management and conservation Water tables should be maintained at high levels. Water bodies should be isolated from sources of pollution and eutrophication. Clearance of the edges of ponds, lakes, dykes, etc. should be undertaken on a rotational basis and be aimed at maintaining the aquatic and emergent plant populations.

DONACIA CRASSIPES   NOTABLE B
A reed beetle
Order COLEOPTERA   Family CHRYSOMELIDAE

Donacia crassipes F., 1775.

Distribution England, Dyfed-Powys, South East Scotland, South West Scotland and North West Scotland.

Habitat and ecology Aquatic habitats, such as freshwater lakes, ponds and possibly ditches. Phytophagous. Associated with white water-lily *Nymphaea alba* and yellow water-lily *Nuphar lutea*. Larvae are aquatic on the foodplants. Adults have been recorded from April to August.

Status Possibly under-recorded. Widespread and very local in England, becoming more local in Scotland and recorded as far north as West Ross.

Threats Loss of habitat through falling water tables because of water abstraction schemes, and the infilling of lakes and ponds. Water pollution and natural succession may be further threats.

Management and conservation Water tables should be maintained at high levels. Water bodies should be isolated from sources of pollution and eutrophication. Clearance of emergent vegetation may be needed to maintain open conditions, this should be undertaken on a rotational basis and be aimed at maintaining the plant populations.

DONACIA DENTATA   NOTABLE A
A reed beetle
Order COLEOPTERA   Family CHRYSOMELIDAE

Donacia dentata Hoppe, 1795.

Distribution Recorded from South Somerset, North Somerset, South Wiltshire, West Sussex, East Sussex, East Kent, Surrey, South Essex, Hertfordshire, Middlesex, Berkshire, Oxfordshire, East Suffolk, East Norfolk, West Norfolk, Cambridgeshire, East

CHRYSOMELIDAE

Gloucestershire, Warwickshire, Glamorgan, South-east Yorkshire and Cumberland before 1970 and South Somerset, North Somerset, West Sussex, East Sussex, East Kent, East Suffolk and Leicestershire & Rutland from 1970 onwards.

Habitat and ecology Aquatic habitats, such as canals, slow flowing rivers and ditches. Phytophagous. Associated with arrowhead *Sagittaria sagittifolia*, and occasionally recorded from pondweed *Potamogeton*. The beetle appears to prefer large stands of arrowhead rather than isolated plants. On the Continent, this species is also associated with water-plantain *Alisma*. In Britain, adults have been recorded in March and from July to September.

Status Old records show that this species has been widely recorded in southern England, with scattered records north to Cumberland. Recently noted from only six vice-counties, all in southern England.

Threats River engineering, including dredging, level regulation by damming and flood alleviation schemes. Water pollution, the use of motor-boats and natural succession may be further threats.

Management and conservation Water tables should be maintained at high levels. Water bodies should be isolated from sources of pollution and eutrophication. Clearance of emergent vegetation may be needed to maintain open conditions, this should be undertaken on a rotational basis and be aimed at maintaining the plant populations.

Published sources Atty, D.B. (1983), Fowler, W.W. (1890), Fowler, W.W. & Donisthorpe, H.St J.K. (1913), Marriner, T.F. (1938), Parry, J.A. (1979), Parry, J.A. (1980b), Stainforth, T. (1944), Walker, J.J. (1932b).

DONACIA IMPRESSA   **NOTABLE A**
A reed beetle
Order COLEOPTERA   Family CHRYSOMELIDAE

Donacia impressa Paykull, 1799.

Distribution Recorded from South Somerset, Dorset, South Hampshire, West Sussex, East Sussex, East Kent, Surrey, South Essex, Berkshire, Oxfordshire, East Suffolk, East Norfolk, Huntingdonshire, West Gloucestershire, Worcestershire, Staffordshire, Pembrokeshire, Merionethshire, Caernarvonshire, Derbyshire, South-east Yorkshire, South-west Yorkshire, Westmorland & North Lancashire, Cumberland and Ayrshire before 1970 and West Sussex, East Sussex, Oxfordshire, Worcestershire, Warwickshire, Radnorshire, Caernarvonshire,

Westmorland & North Lancashire and Ayrshire from 1970 onwards.

Habitat and ecology Aquatic and semi-aquatic habitats, such as freshwater lakes, ponds, meres, ditches and canals. Phytophagous. Associated with bulrush *Scirpus lacustris*, possibly also with sedges. Larvae are aquatic on the roots of the foodplant. Adults are found on emergent vegetation and have been recorded from April to September and in November.

Status Old records show that this species has been widely recorded in southern England, with scattered records north to Ayrshire in Scotland. Also found in parts of Wales. Recently recorded from nine vice-counties.

Threats Loss of habitat through falling water tables because of water abstraction schemes, and the infilling of lakes and ponds. Water pollution, the use of motor-boats and natural succession may be further threats.

Management and conservation Water tables should be maintained at high levels. Water bodies should be isolated from sources of pollution and eutrophication. Clearance of emergent vegetation may be needed to maintain open conditions, this should be undertaken on a rotational basis and be aimed at maintaining the plant populations.

Published sources Allen, A.A. (1954b), Atty, D.B. (1983), Cribb, J. (1954), Fowler, W.W. (1890), Fowler, W.W. & Donisthorpe, H.St J.K. (1913), Lane, S.A. (1990), Marriner, T.F. (1938), Stainforth, T. (1944), Walker, J.J. (1932b).

DONACIA OBSCURA   **NOTABLE A**
A reed beetle
Order COLEOPTERA   Family CHRYSOMELIDAE

Donacia obscura Gyllenhal, 1813.

Distribution Recorded from Dorset, West Sussex, Berkshire, East Suffolk, East Norfolk, West Gloucestershire, Caernarvonshire, Cheshire, Cumberland, Dumfriesshire, Kirkcudbrightshire, Lanarkshire, Moray, East Inverness & Nairn, Argyll Main, Mid Ebudes and West Ross before 1970 and Warwickshire, Staffordshire, Radnorshire, Carmarthenshire, Caernarvonshire, Denbighshire, Westmorland & North Lancashire, South Aberdeenshire, East Inverness & Nairn, Argyll Main and South Ebudes from 1970 onwards.

Habitat and ecology Aquatic and semi-aquatic habitats in uplands, fens and woodlands, such as freshwater lakes and ponds. Phytophagous. Associated with club-rushes, sedges, especially bottle sedge *Carex rostrata*, and possibly water-lilies (Nymphaeaceae). Larvae probably develop at the roots of the foodplants. Adults are found on emergent vegetation and have been recorded from April to July.

Status Status revised from RDB 2 (Vulnerable) in Shirt (1987). Old records indicate that this species had a scattered distribution through southern England, parts of Wales and north to West Ross in Scotland. Recently recorded from Carmarthenshire, north to South Ebudes in Scotland.

Threats Drainage for reasons such as agricultural improvement and development is the primary cause of the loss of fens. Falling water tables because of water abstraction, the infilling of lakes and ponds, water pollution and natural succession may also threaten this species.

Management and conservation Water tables should be maintained at high levels. Water bodies should be isolated from sources of pollution and eutrophication. Clearance may be needed to maintain open conditions, this should be undertaken on a rotational basis and be aimed at maintaining aquatic plant populations.

Published sources Allen, A.A. (1954c), Atty, D.B. (1983), Fowler, W.W. (1890), Fowler, W.W. & Donisthorpe, H.St J.K. (1913), Hodge, P.J. (1990), Lane, S.A. (1990), Marriner, T.F. (1938), Morgan, I.K. (1988), Shirt, D.B., ed. (1987).

DONACIA SPARGANII   **NOTABLE A**
A reed beetle
Order COLEOPTERA   Family CHRYSOMELIDAE

Donacia sparganii Ahrens, 1810.

Distribution Recorded from South Devon, Dorset, South Hampshire, East Sussex, East Kent, West Kent, Surrey, South Essex, North Essex, Hertfordshire, Middlesex, Berkshire, Oxfordshire, East Suffolk, East Norfolk, Cambridgeshire, East Gloucestershire, Staffordshire, North Lincolnshire, Cheshire, South Lancashire, West Lancashire, South-east Yorkshire, South-west Yorkshire and Mid-west Yorkshire before 1970 and East Cornwall, East Sussex, Cambridgeshire, West Gloucestershire, Carmarthenshire and Caernarvonshire from 1970 onwards.

Habitat and ecology Aquatic and semi-aquatic habitats, such as freshwater lakes, ponds, ditches and slow-flowing rivers. Possibly prefers large, well oxygenated water-bodies. Phytophagous. Associated with unbranched bur-reed *Sparganium emersum* and

flowering rush *Butomus umbellatus*, and also recorded in the flower-heads of yellow water-lily *Nuphar lutea*. Adults are found on emergent vegetation and have been recorded from July to September.

Status Old records show that this species has been widely recorded in southern England, with a scattered distribution north to Mid-west Yorkshire. Recently noted from only six vice-counties, including two in Wales.

Threats River engineering, including dredging, level regulation by damming and flood alleviation schemes. The infilling of lakes and ponds, water pollution and natural succession may be further threats.

Management and conservation Water tables should be maintained at high levels. Water bodies should be isolated from sources of pollution and eutrophication. Clearance of emergent vegetation may be needed to maintain open conditions, this should be undertaken on a rotational basis and be aimed at maintaining the plant populations.

Published sources Atty, D.B. (1983), Cribb, J. (1954), Fowler, W.W. (1890), Fowler, W.W. & Donisthorpe, H.St J.K. (1913), Parry, J.A. (1979), Stainforth, T. (1944), Walker, J.J. (1932b).

DONACIA THALASSINA   **NOTABLE B**
A reed beetle
Order COLEOPTERA   Family CHRYSOMELIDAE

Donacia thalassina Germar, 1811.

Distribution England, Dyfed-Powys, North Wales and Scotland.

Habitat and ecology Aquatic and semi-aquatic habitats, such as freshwater lakes, ponds and ditches. Phytophagous. Associated with club-rushes and sedges. On the Continent, this species has been recorded from common spike-rush *Eleocharis palustris*. Larvae develop at the roots of the foodplants. In Britain, adults are found on emergent and marginal vegetation, and have been recorded in May and June.

Status Widespread and very local. Recorded from southern England, through northern England to the Outer Hebrides in Scotland.

Threats Loss of habitat through falling water tables because of water abstraction schemes, and the infilling of lakes and ponds. Water pollution and natural succession may be further threats.

CHRYSOMELIDAE

Management and conservation Water tables should be maintained at high levels. Water bodies should be isolated from pollution and eutrophication. Clearance of emergent vegetation may be needed to maintain open conditions, this should be undertaken on a rotational basis and be aimed at maintaining the plant populations.

EPITRIX ATROPAE   **NOTABLE B**
A flea beetle
Order COLEOPTERA   Family CHRYSOMELIDAE

Epitrix atropae Foudras, 1860.

Distribution South East, South, East Midlands, West Midlands and North East.

Habitat and ecology Woodland rides, clearings and wood margins, scrub, grassland, and disturbed ground. Primarily on base-rich soils. Phytophagous. Associated with deadly nightshade *Atropa bella-donna*, possibly also with henbane *Hyoscyamus niger* and, more doubtfully, with thorn-apple *Datura stramonium*. Larvae feed on the roots of the host-plant between mid June and mid August, with pupation occurring between mid July and late August. New generation adults emerge during August and feed on the foliage of the host-plant until late September after which they hibernate, reappearing the following April, and have also been recorded in June and July.

Status Widespread but local in southern England, though not known in the South West and East Anglia. Also recorded in North East England. Can be common where found.

Threats Loss of woodland through, for example, clear-felling and conversion to other land use. Loss of grassland through improvement and conversion to arable agriculture and development. Neglect and conversion to high forest and the use of herbicides may be further threats.

Management and conservation Grazing or some other disturbance, on a rotational basis, may be need to maintain open conditions. In woodland, open glades and ride margins should be cut on rotation to retain a variety of vegetation structures.

GALERUCA INTERRUPTA   **ENDANGERED**
A leaf beetle
Order COLEOPTERA   Family CHRYSOMELIDAE

Galeruca interrupta Illiger, 1802. Formerly known as: *Adimonia oelandica* (Boheman, 1849), *Galeruca circumdata* sensu auct. ?not (Duftschmid).

Distribution Recorded from North Devon, Dorset and Cambridgeshire before 1970.

Habitat and ecology Phytophagous. On the Continent, this species may be associated with a member of the Cruciferae or possibly field wormwood *Artemisia campestris*. In Britain, this beetle has been recorded on sallows in marshy habitats, and has also been recorded on creeping willow. Adults have been recorded from June to August.

Status Probably extinct. Last recorded in 1919 from Sherborne, Dorset.

Published sources Shirt, D.B., ed. (1987).

GYNANDROPHTHALMA AFFINIS
A leaf beetle   **ENDANGERED**
Order COLEOPTERA   Family CHRYSOMELIDAE

Gynandrophthalma affinis (Illiger, 1794).

Distribution Recorded from Oxfordshire and East Gloucestershire before 1970.

Habitat and ecology Broad-leaved woodland. Phytophagous. Recorded mainly from hazel and birch. On the Continent, this species is possibly also associated with oak, elder, elm and *Salix*. Larvae are probably myrmecophilous, living in ants' nests, development probably taking one year. Adults may be short-lived. In Britain, adults have been recorded in June.

Status Possibly under-recorded. Only found in numbers from Wychwood Forest, Oxfordshire between 1898 and 1910, and again in the early 1950's. Last recorded (a single individual) in 1965 from Brassey Nature Reserve, East Gloucestershire.

Threats This species may have declined because of the loss of broad-leaved woodland through, for example, clear-felling and coniferisation. Neglect and conversion to high forest may also have contributed to this species' decline.

Published sources Atty, D.B. (1983), Lloyd, R.W. (1951), Shirt, D.B., ed. (1987).

HYDROTHASSA HANNOVERIANA   RARE
A leaf beetle
Order COLEOPTERA   Family CHRYSOMELIDAE

Hydrothassa hannoveriana (F., 1775). Formerly known as: *Hydrothassa hannoverana* (F., 1775).

Distribution Recorded from South Hampshire, South-east Yorkshire, South-west Yorkshire, Mid-west Yorkshire, Durham, Cumberland and Orkney before 1970 and South Hampshire, Mid-west Yorkshire and North-west Yorkshire from 1970 onwards.

Habitat and ecology Marshes and peat bogs. Phytophagous. Associated with marsh-marigold *Caltha palustris*, usually growing on deep moss. Adults tend to occur at the roots rather than on the flowers. Adults tend to be nocturnal and have been recorded from April to July.

Status Possibly under-recorded. Very local and with a disjunct distribution. Recorded in South Hampshire, northern England and Orkney in Scotland. Recently reported from just three vice-counties.

Threats Drainage for reasons such as agricultural improvement and development is the primary cause of the loss of wetland. Falling water tables because of water abstraction schemes and erosion in areas where livestock densities are too high may also threaten this species.

Management and conservation Water tables should be maintained at high levels.

Published sources Allen, A.A. (1972c), Anon, (1960), Drummond, D.C. (1956), Flint, J.H. (1973), Fowler, W.W. (1890), Fowler, W.W. & Donisthorpe, H.St J.K. (1913), Marriner, T.F. (1938), Shirt, D.B., ed. (1987).

HYPOCASSIDA SUBFERRUGINEA   EXTINCT
A tortoise beetle
Order COLEOPTERA   Family CHRYSOMELIDAE

Hypocassida subferruginea (Schrank, 1776). Formerly known as: *Cassida subferruginea* Schrank.

Distribution Recorded from "Devon" and Glamorgan.

Habitat and ecology Probably associated with field margins, disturbed ground and wetland. Possibly also coastal habitats. Phytophagous. On the Continent, this species is associated with bindweed *Convolvulus.* and possibly also associated with yarrow *Achillea millefolium*.

Status Extinct. Last recorded in the 19th century from Swansea, Glamorganshire.

Published sources Fowler, W.W. (1890), Shirt, D.B., ed. (1987).

LABIDOSTOMIS TRIDENTATA   ENDANGERED
A leaf beetle
Order COLEOPTERA   Family CHRYSOMELIDAE

Labidostomis tridentata (L., 1758).

Distribution Recorded from North Hampshire, East Sussex, East Kent, West Kent, Surrey, Berkshire, Worcestershire and North-east Yorkshire before 1970.

Habitat and ecology Open ground in woodland. Phytophagous. Associated with young (about 5 years old) birch. On the Continent, this species has also been associated within hazel and *Salix*. Larvae are myrmecophilous, living in ants' nests, and take one year to complete development. In Britain, adults occur on the host-plant between May and July.

Status Very infrequently recorded and probably under-recorded. Old records suggest that this species had a scattered distribution from southern England through to North-east Yorkshire. Last recorded in 1951 from Ham Street Woods, East Kent and in the mid-1950s from Abbot's Wood, East Sussex. This species has occurred in profusion, such as at Oaken Wood, East Malling, West Kent, in 1945.

Threats This species probably declined because of the loss of broad-leaved woodland through, for example, clear-felling and coniferisation. Neglect, in particular the abandonment of coppice management and the conversion to high forest, has led to increased shade and the loss of glades and broad sunny rides and may also have contributed to this species' decline.

Published sources Fowler, W.W. (1890), Fowler, W.W. & Donisthorpe, H.St J.K. (1913), Massee, A.M. (1945a), Shirt, D.B., ed. (1987).

LONGITARSUS ABSYNTHII   NOTABLE A
A flea beetle
Order COLEOPTERA   Family CHRYSOMELIDAE

Longitarsus absynthii Kutschera, 1862. Formerly known as: *Longitarsus absinthii* Kutschera, 1862.

Distribution Recorded from South Hampshire, East Kent, West Kent, South Essex and North Essex before 1970 and East Kent, West Kent and East Suffolk from 1970 onwards.

CHRYSOMELIDAE

Habitat and ecology Saltmarsh, coastal cliff, and rough ground near the sea and sea walls. Phytophagous. Associated with sea wormwood *Artemisia maritima*. Adults have been recorded in February, from May to July and from September to November.

Status Only known in southern and south-eastern England. Recently recorded from just three vice-counties. Can be common where found. This species is difficult to identify and may be confused with other members of the genus. Consequently, the exact status of this species is hard to assess.

Threats Coastal developments and the loss of saltmarsh through reclamation and erosion. Overgrazing and sea wall maintenance may be a further threat to this species.

Management and conservation Grazing should not be introduced to a saltmarsh where there is no grazing at present.

Published sources Fowler, W.W. (1890), Owen, J.A. (1990b).

LONGITARSUS AERUGINOSUS   ENDANGERED
A flea beetle
Order COLEOPTERA   Family CHRYSOMELIDAE

Longitarsus aeruginosus (Foudras, 1860).

Distribution Recorded from Dorset, Isle of Wight, South Hampshire and Surrey before 1970.

Habitat and ecology Coastal habitats and river banks. Phytophagous. Associated with hemp-agrimony *Eupatorium cannabinum*. Adults have been recorded in July.

Status Not listed in the insect Red Data Book (Shirt, 1987). Only known from southern England and last recorded in 1925 from Charmouth, Dorset. This species is difficult to identify and may be confused with other members of the genus.

Threats Loss of habitat because of coastal developments, river engineering schemes and improvement and conversion to other land use.

Published sources Donisthorpe, H.St J.K. (1944).

LONGITARSUS AGILIS   NOTABLE A
A flea beetle
Order COLEOPTERA   Family CHRYSOMELIDAE

Longitarsus agilis (Rye, 1868).

Distribution Recorded from South Devon, East Sussex, East Kent, West Kent, Surrey, Hertfordshire, Berkshire, Oxfordshire, East Suffolk, West Suffolk, East Norfolk and Cambridgeshire before 1970 and "Somerset", East Sussex, East Kent, West Kent and Surrey from 1970 onwards.

Habitat and ecology Downland, heathland, hedgebanks, grass verges, open areas in woodland and lake margins. Phytophagous. Associated with figwort *Scrophularia*. Adults have been recorded from May to October and in December.

Status Very local and only known in southern and south-eastern England. Recently recorded from four vice-counties. This species is difficult to identify and may be confused with other members of the genus. Consequently, the exact status of this species is hard to assess.

Threats Loss of habitat through changes in land use, such as improvement and conversion to arable agriculture, development and afforestation. Natural succession is a further threat.

Management and conservation Grazing, cutting or some other distrubance is needed to maintain open conditions.

LONGITARSUS ANCHUSAE   NOTABLE B
A flea beetle
Order COLEOPTERA   Family CHRYSOMELIDAE

Longitarsus anchusae (Paykull, 1799).

Distribution England, South West Scotland and South East Scotland.

Habitat and ecology Open areas such as cliffs and downs, probably also rough grassland, disturbed ground and coastal habitats. Phytophagous. Associated with members of the borage family Boraginaceae, particularly viper's bugloss *Echium vulgare*, comfrey *Symphytum* and hound's-tongue *Cynoglossum officinale*. Adults have been recorded from April to June and in August.

Status Old records suggest that this species has been widely recorded in southern England, with a scattered distribution north to Peebleshire in Scotland. Recently reported in southern England only. This species is difficult to identify and may be confused with other members of the genus. Consequently, the exact status of this species is hard to assess.

Threats Loss of unimproved grassland through improvement by reseeding or by the application of fertilisers, or by conversion to arable agriculture. Coastal developments and natural succession may be further threats.

Management and conservation Disturbance, such as rotovation, is needed to maintain open conditions.

LONGITARSUS BALLOTAE   **NOTABLE B**
A flea beetle
Order COLEOPTERA   Family CHRYSOMELIDAE

Longitarsus ballotae (Marsham, 1802). Formerly known as: *Longitarsus cerinus* sensu Fowler, 1890 partim not (Foudras, 1860).

Distribution South East, South, East Anglia, East Midlands and South Wales.

Habitat and ecology Grassland, especially chalk grassland, hedgerows, disturbed ground and field margins. Phytophagous. Associated with black horehound *Ballota nigra* and white horehound *Marrubium vulgare*. Adults have been recorded from May to November.

Status Widespread but local in southern England, also recorded in South Wales. This species is difficult to identify and may be confused with other members of the genus. Consequently, the exact status of this species is hard to assess.

Threats Loss of habitat through improvement by reseeding or by the application of fertilisers, or by conversion to arable agriculture. Natural succession, and the use of herbicides and pesticides may be further threats.

Management and conservation Disturbance, such as rotovation, is needed to maintain open conditions.

LONGITARSUS BEAREI
A flea beetle   **INSUFFICIENTLY KNOWN**
Order COLEOPTERA   Family CHRYSOMELIDAE

Longitarsus bearei Kevan, 1967.

Distribution Recorded from Isle of Wight before 1970.

Habitat and ecology Possibly coastal. Phytophagous. The adult has been recorded in March.

Status Not listed in the insect Red Data Book (Shirt, 1987). Described new to science in 1967. This species is still only known from the unique type specimen collected in 1905 from Sandown, Isle of Wight.

Published sources Kevan, D.K. (1967).

LONGITARSUS BRUNNEUS   **NOTABLE B**
A flea beetle
Order COLEOPTERA   Family CHRYSOMELIDAE

Longitarsus brunneus (Duftschmid, 1825). Formerly known as: *Longitarsus castaneus* sensu auct. Brit. not (Duftschmid, 1825).

Distribution England, South Wales and South West Scotland.

Habitat and ecology Fens, though probably in other wetland habitats and on the coast. Phytophagous. Associated with meadow-rue *Thalictrum*, sea aster *Aster tripolium* and probably other members of the Compositae. Adults have been recorded in January, from March to June and from August to December.

Status Widespread but local in southern England, with scattered records north to the Clyde Isles in Scotland. Recently recorded from South Somerset, north to Mid-west Yorkshire. This species is difficult to identify and may be confused with other members of the genus. Consequently, the exact status of this species is hard to assess.

Threats Drainage for reasons such as agricultural improvement and development is the primary cause of the loss of fens. Falling water tables because of water abstraction and river engineering schemes, and coastal developments may also threaten this species.

Management and conservation Water tables should be maintained at high levels. Cutting, on a rotational basis, may be needed to maintain open conditions.

LONGITARSUS CURTUS   **NOTABLE A**
A flea beetle
Order COLEOPTERA   Family CHRYSOMELIDAE

Longitarsus curtus (Allard, 1860).

Distribution Recorded from North Devon, South Hampshire, West Sussex, West Kent, Berkshire, Oxfordshire, Buckinghamshire, Huntingdonshire, Caernarvonshire, Midlothian and Fife before 1970 and South Hampshire, West Sussex, East Sussex, Surrey, Oxfordshire, Huntingdonshire, Northamptonshire, East Gloucestershire and Durham from 1970 onwards.

CHRYSOMELIDAE

Habitat and ecology Calcareous hillsides, also other grassland habitats and woodland. Phytophagous. Associated with comfrey *Symphytum*, lungwort *Pulmonaria*, viper's-bugloss *Echium vulgare* and possibly also field forget-me-not *Myosotis arvensis*. Adults have been recorded from March to November.

Status Very local and widely distributed in Great Britain, with old records from southern England, Caernarvonshire in Wales and north to Fife in Scotland. Recently recorded from nine vice-counties, from South Hampshire to Durham. This species is difficult to identify and may be confused with other members of the genus. Consequently, the exact status of this species is hard to assess.

Threats Loss of unimproved grassland through improvement by reseeding or by the application of fertilisers, or by conversion to arable agriculture. Loss of woodland through clear-felling and conversion to other land use, and natural succession may be further threats.

Management and conservation Some disturbance, such as rotovation, is needed to maintain open conditions. In woodlands, open glades and ride margins should be cut on rotation to retain a variety of vegetation structures.

Published sources Atty, D.B. (1983), Fowler, W.W. & Donisthorpe, H.St J.K. (1913), Morgan, M.J. (1984).

LONGITARSUS DORSALIS   **NOTABLE B**
A flea beetle
Order COLEOPTERA   Family CHRYSOMELIDAE

Longitarsus dorsalis (F., 1781).

Distribution South East, South, South West, East Anglia and East Midlands.

Habitat and ecology Primarily found on calcareous or sandy soils. Grassland, maritime cliff, limestone quarries, and woodland rides and clearings. Phytophagous. Associated with ragwort *Senecio*. Adults have been recorded from March to June, and in September and December.

Status Widespread but local in southern England.

Threats Loss of unimproved grassland through improvement by reseeding or by the application of fertilisers, or by conversion to arable agriculture. The infilling of quarries, loss of woodland through clear-felling and conversion to other land use, natural succession and the use of herbicides are further threats.

Management and conservation Grazing, cutting or some other disturbance, such as rotovation, on a rotational basis, may be needed to maintain open conditions. In woodlands, open glades and ride margins should be cut on rotation to retain a variety of vegetation structures.

LONGITARSUS FERRUGINEUS   **ENDANGERED**
Mint flea beetle
Order COLEOPTERA   Family CHRYSOMELIDAE

Longitarsus ferrugineus (Foudras, 1860). Formerly known as: *Longitarsus rubiginosus var. ferrugineus* (Foudras), *Longitarsus waterhousei* Kutschera, 1864.

Distribution Recorded from West Cornwall, North Devon, Isle of Wight, North Hampshire, East Kent, West Kent, Surrey, South Essex, Hertfordshire, Middlesex, Berkshire, Oxfordshire, East Suffolk, West Suffolk, Cambridgeshire, West Gloucestershire, South Lincolnshire and Leicestershire & Rutland before 1970.

Habitat and ecology Probably wetland, woodland rides, clearings and wood margins, marginal vegetation, gardens and possibly grassland and disturbed ground. Phytophagous. Associated with mint *Mentha*, particularly water mint *M. aquatica*, but also on other species such as spearmint *M. spicata* and peppermint *M. x piperata*. The beetle might also occur on cat-mint *Nepeta cataria*. Adults have been recorded from July to December.

Status Not listed in the insect Red Data Book (Shirt, 1987). Old records indicate that this species was formerly widely distributed in southern England and recorded as far north as South Lincolnshire. Last recorded from Buckingham Palace Gardens in 1963, where it may have been introduced. This species is difficult to identify and may be confused with other members of the genus. Consequently, the exact status of this species is hard to assess.

Threats Uncertain, though loss of habitat and conversion to other land use may have contributed to this species' decline.

Published sources Allen, A.A. (1960a), Atty, D.B. (1983), Fowler, W.W. (1890), Fowler, W.W. & Donisthorpe, H.St J.K. (1913).

LONGITARSUS FOWLERI   NOTABLE A
A flea beetle
Order COLEOPTERA   Family CHRYSOMELIDAE

Longitarsus fowleri Allen, 1967. Formerly confused under: *Longitarsus abdominalis* sensu Fowler, 1890 not (Duftschmid, 1825), *Longitarsus lycopi?* sensu auct. partim not (Foudras, 1860).

Distribution Recorded from Dorset, East Sussex, West Kent, Surrey and Middlesex before 1970 and Dorset, Isle of Wight, West Sussex, East Sussex, East Kent, West Kent, South Essex and Northamptonshire from 1970 onwards.

Habitat and ecology Short-turf calcareous grassland, disturbed ground and coastal land-slips. Favours warm, south-facing slopes. Phytophagous. Associated with teasel *Dipsacus*, also recorded from thyme *Thymus* and possibly also associated with ground-ivy *Glechoma hederacea*. Predominantly associated with young teasel plants in the early spring, dispersing as the plants grow larger. Adults have been recorded from April to July, and in September and October.

Status Described new to science in 1967. Only recorded from a few localities in southern England, and recently recorded from eight vice-counties. This species is difficult to identify and may be confused with other members of the genus. Consequently, the exact status of this species is hard to assess.

Threats Loss of unimproved grassland through improvement by reseeding or by the application of fertilisers, or by conversion to arable agriculture. Cliff stabilisation schemes, the construction of sea defences and coastal developments may reduce the amount of available habitat. Activities that accelerate or reduce the rate of erosion should be avoided. The degradation of suitable habitat through natural succession is a further threat.

Management and conservation In areas of coastal cliffs, occasional slippages are necessary to maintain habitat continuity. Large areas of unstable cliff are required so that the population does not become isolated and subsequently threatened by individual landslips. In areas of grassland, grazing, cutting or some other disturbance, on a rotational basis, may be needed to maintain open conditions.

Published sources Allen, A.A. (1967a), Williams, S.A. (1979).

LONGITARSUS GANGLBAUERI   NOTABLE A
A flea beetle
Order COLEOPTERA   Family CHRYSOMELIDAE

Longitarsus ganglbaueri Heikertinger, 1912. Formerly known as: *Longitarsus picipes* sensu Fowler, 1890 not (Stephens, 1831), *Longitarsus senecionis* Brisout, 1873.

Distribution Recorded from West Sussex, East Kent, West Kent, Surrey, Hertfordshire, Berkshire, Oxfordshire, East Suffolk, Cambridgeshire, Derbyshire, South-east Yorkshire, North-east Yorkshire, North-west Yorkshire, South Northumberland, North Northumberland and Cumberland before 1970 and East Sussex, East Kent, Oxfordshire, Northamptonshire, Warwickshire, Durham and West Lothian from 1970 onwards.

Habitat and ecology Coastal shingle, disturbed ground, disused quarries and other sites where the foodplant may grow. Phytophagous. Associated with ragwort *Senecio*. Adults have been recorded from April to October and in December.

Status Widespread and very local. Old records indicate that this species has been widely recorded from southern England north to Cumberland. Recently reported from six vice-counties from southern England to West Lothian in Scotland. Not known in Wales. This species is difficult to identify and may be confused with other members of the genus. Consequently, the exact status of this species is hard to assess.

Threats Loss of habitat and conversion to other land use. Gravel extraction, the infilling of quarries, the use of herbicides and natural succession may be further threats.

Management and conservation Some grazing, rotational cutting or other disturbance, such as rotovation, is needed to maintain open conditions. The disturbance of coastal shingle should be avoided.

Published sources Crowson, R.A. (1987), Easton, A.M. (1947), Eyre, M.D. & Cox, M.L. (1987), Lane, S.A. (1990), Pickard-Cambridge, A.W. (1946).

LONGITARSUS LONGISETA
A flea beetle   **INSUFFICIENTLY KNOWN**
Order COLEOPTERA   Family CHRYSOMELIDAE

Longitarsus longiseta Weise, 1889. Formerly known as: *Longitarsus clarus* Allen, 1967.

Distribution Recorded from East Kent before 1970.

CHRYSOMELIDAE

Habitat and ecology Grassland, possibly preferring shaded areas. Phytophagous. Although recorded from black nightshade *Solanum nigrum*, the true foodplant is probably hoary plantain *Plantago media*, as on the Continent. In Britain, the adult has been recorded in October.

Status Not listed in the insect Red Data Book (Shirt, 1987). Only known in Great Britain from a single example recorded in 1951 from outside Church Wood, Blean, East Kent. This species is difficult to identify and may be confused with other members of the genus.

Published sources Allen, A.A. (1967a).

LONGITARSUS LYCOPI   **NOTABLE B**
A flea beetle
Order COLEOPTERA   Family CHRYSOMELIDAE

Longitarsus lycopi (Foudras, 1860).

Distribution South West, South, South East, East Anglia, East Midlands and West Midlands.

Habitat and ecology Calcareous grassland, and occasionally in open woodland and wetland. Phytophagous. Associated with common calamint *Calamintha ascendens* in grassland, selfheal *Prunella* in open woodland and gipsywort *Lycopus europaeus* in wetland habitats. Very doubtfully associated with mint *Mentha* and cat-mint *Nepeta cataria*. Adults have been recorded in every month of the year.

Status Widespread and very local in southern England, with scattered, records north to Derbyshire. Recently recorded as far north as Cambridgeshire. This species is difficult to identify and may be confused with other members of the genus. Consequently, the exact status of this species is hard to assess.

Threats Loss of unimproved grassland through improvement by reseeding or by the application of fertilisers, or by conversion to arable agriculture. Loss of wetland through drainage and water abstraction schemes, and the loss of woodland through clear-felling and conversion to other land use. Natural succession and overgrazing may be further threats.

Management and conservation Grazing or cutting, on a rotational basis, is needed to maintain open conditions. In areas of wetland, water tables should be maintained at high levels. In woodlands, open glades and ride margins should be cut on rotation to retain a variety of vegetation structures.

LONGITARSUS NASTURTII   **NOTABLE B**
A flea beetle
Order COLEOPTERA   Family CHRYSOMELIDAE

Longitarsus nasturtii (F., 1792).

Distribution South West, South, South East, East Anglia, East Midlands, West Midlands and North East.

Habitat and ecology Field margins, and probably also grassland, maritime cliff, woodland margins, rides and clearings, and disturbed ground. Phytophagous. Associated with comfrey *Symphytum*, viper's bugloss *Echium vulgare* and possibly other members of the borage family. Adults have been recorded from January to September.

Status Widespread and very local. Possibly declining. Old records indicate that this species has been widely recorded in southern England and noted north to North-east Yorkshire. Recently reported as far north as North Lincolnshire. Can be common where found. This species is difficult to identify and may be confused with other members of the genus, particularly *L. suturellus* and *L. atricillus*. Consequently, the exact status of this species is hard to assess.

Threats Loss of grassland through improvement and conversion to arable agriculture and development. Loss of woodland through clear-felling and conversion to other land use. Neglect and conversion to high forest, natural succession, and the use of herbicides and pesticides may be further threats.

Management and conservation Disturbance, on a rotational basis, is needed to maintain open conditions. In woodlands, open glades and ride margins should be cut on rotation to retain a variety of vegetation structures.

LONGITARSUS NIGERRIMUS   **ENDANGERED**
A flea beetle
Order COLEOPTERA   Family CHRYSOMELIDAE

Longitarsus nigerrimus (Gyllenhal, 1827).

Distribution Recorded from Dorset and South Hampshire before 1970.

Habitat and ecology Bogs and ponds. Phytophagous. Probably associated with greater bladderwort *Utricularia vulgaris* as on the Continent, though recorded from rushes and *Sphagnum* moss in ponds and peaty bogs. Larvae probably develop during the summer at the roots of the foodplant, with new generation adults probably emerging in late September and October. Adults probably overwinter and have been recorded during February, May, June, September and October.

Status Only known from Dorset and South Hampshire. Last recorded in 1933 from Hurn, South Hampshire. Records for South and North Lincolnshire and Middlesbrough, North-east Yorkshire require confirmation. This species is difficult to identify and may be confused with other members of the genus.

Threats Uncertain, though this species may be threatened by drainage for reasons such as agricultural improvement and development. Falling water tables because of water abstraction and river engineering schemes, pollution and eutrophication, the infilling of ponds and natural succession may also be a threat.

Published sources Harwood, P. (1928), Shirt, D.B., ed. (1987), Thornley, A. & Wallace, W. (1907).

LONGITARSUS NIGROFASCIATUS NOTABLE A
A flea beetle
Order COLEOPTERA   Family CHRYSOMELIDAE

Longitarsus nigrofasciatus (Goeze, 1777). Formerly known as: *Longitarsus distinguendus* (Rye, 1872), *Longitarsus patruelis* (Allard, 1866).

Distribution Recorded from North Devon, Isle of Wight, West Sussex, East Sussex, East Kent, West Kent, Surrey, Hertfordshire, Oxfordshire, East Suffolk, Cambridgeshire, North Lincolnshire, South Lancashire, South-east Yorkshire, South-west Yorkshire and Cumberland before 1970 and North Devon, West Sussex, East Kent, West Kent and Surrey from 1970 onwards.

Habitat and ecology Calcareous grassland and maritime cliff. Phytophagous. Associated with figwort *Scrophularia* and mullein *Verbascum*. Adults have been recorded from January to October and in December.

Status Old records show that this species has been widely recorded in southern England, with a scattered distribution north to Cumberland. Recently reported from only five vice-counties, all in southern England. This species is difficult to identify and may be confused with other members of the genus. Consequently, the exact status of this species is hard to assess.

Threats Loss of unimproved grassland through improvement by reseeding or by the application of fertilisers, or by conversion to arable agriculture. Natural succession may be a further threat.

Management and conservation Grazing, cutting or some other disturbance, such as rotovation, on a rotational basis, is needed to maintain open conditions.

Published sources Fowler, W.W. (1890), Fowler, W.W. & Donisthorpe, H.St J.K. (1913), Marriner, T.F. (1938).

LONGITARSUS OCHROLEUCUS   NOTABLE B
A flea beetle
Order COLEOPTERA   Family CHRYSOMELIDAE

Longitarsus ochroleucus (Marsham, 1802).

Distribution England, South Wales, South East Scotland and South West Scotland.

Habitat and ecology Probably grassland and disturbed ground. Recorded from a chalk pit. Phytophagous. Recorded from Oxford ragwort *Senecio squalidus*, though probably also on other species of ragwort. Adults have been recorded from May to September.

Status Widespread but local in England, also recorded South East Scotland. There are old records for Main Argyll and one old record for South Wales. Can be common where found.

Threats Loss of grassland through improvement and conversion to arable agriculture. The infilling of chalk pits, natural succession and the use of herbicides may be further threats.

Management and conservation Grazing, cutting or some other disturbance, such as rotovation, on a rotational basis, is needed to maintain open conditions.

LONGITARSUS PARVULUS   NOTABLE A
A flea beetle
Order COLEOPTERA   Family CHRYSOMELIDAE

Longitarsus parvulus (Paykull, 1799). Formerly known as: *Longitarsus ater* sensu Fowler, 1890 not (F., 1775).

Distribution Recorded from South Devon, South Hampshire, West Sussex, East Sussex, East Kent, West Kent, Surrey, Hertfordshire, Berkshire, West Suffolk, East Norfolk, West Norfolk, Cambridgeshire, East Gloucestershire, Leicestershire & Rutland, South-east Yorkshire and Cumberland before 1970 and West Sussex, East Kent, West Kent and Cambridgeshire from 1970 onwards.

Habitat and ecology Chalk grassland and probably field margins and disturbed ground. Phytophagous. Has been recorded from perennial flax *Linum perenne* ssp.

CHRYSOMELIDAE

anglicum, but in times of abundance probably occurs on other species of flax. On the Continent, this species has been noted on flax *L. usitatissimum*. In Britain, adults have been recorded in all months of the year.

Status Possibly declining. Old records indicate that this species was formerly widespread in southern England, with scattered records north to Cumberland. Recently recorded from only four vice-counties, all in south-eastern England. Flax is now popular as a crop and this could lead to an increase in the distribution of the species. This beetle is difficult to identify and may be confused with other members of the genus. Consequently, it may be under-recorded.

Threats Loss of grassland through improvement and conversion to arable agriculture and development. Natural succession and the use of herbicides and pesticides may be be further threats.

Management and conservation Grazing, cutting or some other disturbance, such as rotovation, on a rotational basis, may be needed to maintain open conditions.

Published sources Allen, A.A. (1945b), Allen, A.A. (1950c), Atty, D.B. (1983), Dinnage, H. (1952), Fowler, W.W. (1890).

LONGITARSUS PLANTAGOMARITIMUS
A flea beetle   **NOTABLE B**
Order COLEOPTERA   Family CHRYSOMELIDAE

Longitarsus plantagomaritimus Dollman, 1912.

Distribution England, South Wales, South East Scotland, South West Scotland and North West Scotland.

Habitat and ecology Predominantly saltmarshes and possibly also maritime cliff and coastal shingle. Phytophagous. Associated with sea plantain *Plantago maritima*. Adults have been recorded from April to October.

Status Widespread and very local, with scattered records from Dorset north to Midlothian in Scotland. There are records of *L. nigerrimus* from Middlesbrough, North-east Yorkshire and Grantham, South Lincolnshire which are published as possibly referring to this species. This species is difficult to identify and may be confused with other members of the genus, particularly *L. melanocephalus* and *L. kutscherae*. Consequently, the exact status of this species is hard to assess.

Threats Loss of saltmarsh through reclamation, erosion and the construction of sea defences. Coastal developments, gravel extraction and overgrazing of saltmarshes could be a further threat to this species.

Management and conservation Grazing should not be introduced to a saltmarsh where there is no grazing at present. Disturbance of vegetated shingle should be avoided.

LONGITARSUS QUADRIGUTTATUS
A flea beetle   **NOTABLE A**
Order COLEOPTERA   Family CHRYSOMELIDAE

Longitarsus quadriguttatus (Pontoppidan, 1763).

Distribution Recorded from North Hampshire, West Sussex, East Kent, South Essex, Hertfordshire, Oxfordshire, East Suffolk, West Suffolk and Leicestershire & Rutland before 1970 and West Sussex, East Sussex, East Suffolk, West Suffolk and West Norfolk from 1970 onwards.

Habitat and ecology Calcareous grassland, also disturbed ground and lane margins. Phytophagous. Associated with hound's-tongue *Cynoglossum officinale*, and also recorded from viper's-bugloss *Echium vulgare*. Adults have been recorded from May to September.

Status Status revised from RDB 3 (Rare) in Shirt (1987). Very local and with a scattered distribution in southern and south-eastern England. Recorded from North Hampshire through East Anglia to West Norfolk. Recently recorded from only four vice-counties.

Threats Loss of calcareous grassland through improvement and conversion to arable agriculture, development and afforestation. Natural succession may be a further threat.

Management and conservation Disturbance, such as rotovation, may be needed to maintain open conditions.

Published sources Allen, A.A. (1956c), Fowler, W.W. (1890), Fowler, W.W. & Donisthorpe, H.St J.K. (1913), Massee, A.M. (1955), Shirt, D.B., ed. (1987).

LONGITARSUS RUTILUS NOTABLE A
A flea beetle
Order COLEOPTERA   Family CHRYSOMELIDAE

Longitarsus rutilus (Illiger, 1807).

Distribution Recorded from West Cornwall, East Cornwall, South Devon, South Hampshire, East Sussex, East Kent, West Kent, Surrey and East Norfolk before 1970 and West Cornwall, Dorset, Isle of Wight, East Sussex, East Kent, West Kent and Berkshire from 1970 onwards.

Habitat and ecology Wetland and water margins, such as pond and stream margins. Also recorded from an old clay pit and an east facing chalky slope. Possibly preferring shady conditions. Phytophagous. Associated with water figwort *Scrophularia aquatica* and balm-leaved figwort *S. scorodonia*. Larvae probably occur at the roots of the foodplant. Adults have been recorded from March to December.

Status Status revised from RDB 2 (Vulnerable) in Shirt (1987). Very local and widely distributed in southern England. Recently recorded from seven vice-counties. This species is difficult to identify and may be confused with other members of the genus. Consequently, the exact status of this species is hard to assess.

Threats Drainage for reasons such as agricultural improvement and development. Falling water tables because of water abstraction and river engineering schemes, the infilling of pits and natural succession may also threaten this species.

Management and conservation Water tables should be maintained at high levels. Grazing or cutting, on a rotational basis, may be needed to maintain open conditions.

Published sources Allen, A.A. (1978a), Fowler, W.W. (1890), Fowler, W.W. & Donisthorpe, H.St J.K. (1913), MacKechnie-Jarvis, C. (1969), Shirt, D.B., ed. (1987).

LONGITARSUS SUTURALIS NOTABLE B
A flea beetle
Order COLEOPTERA   Family CHRYSOMELIDAE

Longitarsus suturalis (Marsham, 1802).

Distribution England, South Wales, North Wales, South West Scotland and North East Scotland.

Habitat and ecology Recorded from quarries. Probably occurs in a variety of habitats including hedgebanks, field margins and disturbed ground. Phytophagous. On the Continent, this species is associated with gromwell *Lithospermum*. In Britain, adults have been recorded in February, April, May, August, September and December.

Status Widespread and very local in England and Wales, also recorded north to Main Argyll in Scotland. This species is difficult to identify and may be confused with other members of the genus, particularly *L. atricillus* and *L. nasturtii*. Consequently, the exact status of this species is hard to assess.

Threats Loss of habitat through improvement and conversion to arable agriculture and development. The infilling of quarries, natural succession and the use of herbicides and pesticides may be further threats.

Management and conservation Some grazing, cutting or some other disturbance may be necessary to maintain open conditions.

LONGITARSUS TABIDUS NOTABLE B
A flea beetle
Order COLEOPTERA   Family CHRYSOMELIDAE

Longitarsus tabidus (F., 1775). Formerly confused under: *Longitarsus jacobaeae* sensu Fowler, 1890 partim not (Waterhouse, 1858).

Distribution South East, South, South West, East Anglia, East Midlands, West Midlands, North East and South Wales.

Habitat and ecology Calcareous grassland and possibly also coastal dunes. Phytophagous. Associated with mullein *Verbascum*. Adults have been recorded from April to December.

Status Widespread but local in England, also recorded in South Wales. Only known from one old record in North East England. Can be common where found. This species is difficult to identify and may be confused with other members of the genus.

Threats Loss of grassland through improvement and conversion to arable agriculture and development. Loss of dunes particularly through coastal developments and afforestation. Natural succession and degradation of habitat by excessive disturbance of the vegetation through activities such as motorbike access, horse-riding and human trampling.

Management and conservation Disturbance, such as rotovation, is needed to maintain open conditions.

CHRYSOMELIDAE

LUPERUS FLAVIPES   **NOTABLE B**
A leaf beetle
Order COLEOPTERA   Family CHRYSOMELIDAE

Luperus flavipes (L., 1767).

Distribution England, North Wales, South East Scotland, South West Scotland and North East Scotland.

Habitat and ecology Broad-leaved woodland, parkland, scrub, heathland and disused railway lines. Phytophagous. Associated with birch and willow, possibly also on oak. The larvae probably feed at the roots of grasses. The beetle probably overwinters in the larval or pupal stage. Adults have been recorded from May to August.

Status Widespread but local in England, also recorded in North Wales and much of Scotland.

Threats Loss of broad-leaved woodland through, for example, clear-felling and coniferisation. Neglect and conversion to high forest may be a further threat. Much heathland has been lost, or fragmented, through changes in land use, mainly by conversion to arable, forestry and urban development. Further habitat degradation has been through the cessation of traditional heathland management practices.

Management and conservation In woodland, open glades and ride margins should be cut on rotation to retain a variety of vegetation structures. On heathland, management should aim for a diversity of successional stages from bare ground to scrub, preferably by grazing, rotational cutting, or possibly scraping or burning.

LYTHRARIA SALICARIAE   **NOTABLE B**
A flea beetle
Order COLEOPTERA   Family CHRYSOMELIDAE

Lythraria salicariae (Paykull, 1800). Formerly known as: *Ochrosis salicariae* (Paykull).

Distribution England and North Wales.

Habitat and ecology Wetland and wet woodland. Phytophagous. Associated with purple-loosestrife *Lythrum salicaria*, yellow pimpernel *Lysimachia nemorum* and yellow loostrife *L. vulgaris*. Adults have been recorded from April to October.

Status Widespread but local in England, also recorded from North Wales.

Threats Drainage for reasons such as agricultural improvement and development. Falling water tables because of water abstraction and river engineering schemes, and the clear felling of woodland and conversion to other land use may also threaten this species.

Management and conservation Water tables should be maintained at high levels. Cutting, on a rotational basis, may be needed to maintain open conditions.

MACROPLEA APPENDICULATA   **RARE**
A reed beetle
Order COLEOPTERA   Family CHRYSOMELIDAE

Macroplea appendiculata (Panzer, 1794). Formerly known as: *Haemonia appendiculata* (Panzer).

Distribution Recorded from West Cornwall, East Kent, Berkshire, Oxfordshire, Buckinghamshire, Cambridgeshire, Staffordshire, Nottinghamshire, South-east Yorkshire, North-east Yorkshire, Mid-west Yorkshire, North-west Yorkshire, Cumberland, Renfrewshire, Fife, East Perthshire, West Inverness and Argyll Main before 1970 and West Cornwall, Mid-west Yorkshire, Cumberland and South Ebudes from 1970 onwards.

Habitat and ecology Aquatic habitats. Found in rivers, and freshwater lakes and ponds. Possibly also found in ditches. Predominantly associated with running water. Phytophagous. Associated with pondweeds *Potamogeton* and *Myriophyllum*, particularly shining pondweed *P. lucens*, spiked water-milfoil *M. spicatum*, and also with *P. pectinatus* x *filiformis*. Very doubtfully associated with bur-reed *Sparganium erectum* and sea club-rush *Scirpus maritimus*. Larvae are aquatic, feeding and pupating on the roots of the foodplant. The adults rarely, if ever, emerge from the water, and generally occur from June to August, though they have also been recorded from September to November. The species overwinters in the adult stage, either in cocoons on the foodplant or in marginal vegetation.

Status Possibly under-recorded. Old records suggest that this species was formerly widespread in southern England with scattered records north to central Scotland. This species appears to have declined and is now extremely local in southern England. Recently recorded from just four, widely scattered, vice-counties. This beetle is difficult to identify and may be confused with *M. mutica*.

Threats River engineering, including dredging, level regulation by damming and flood alleviation schemes. The infilling of ponds and lakes, falling water tables through water abstraction schemes, water pollution, the use of motor-boats and natural succession may be further threats.

Management and conservation Water tables should be maintained at high levels. Water bodies should be isolated from sources of pollution and eutrophication. Clearance of emergent vegetation may be needed, this should be undertaken on a rotational basis and be aimed at maintaining the aquatic plant populations.

Published sources Aubrook, E.W. (1963), Fowler, W.W. & Donisthorpe, H.St J.K. (1913), Russell, H.M. (1951), Shirt, D.B., ed. (1987), Stainforth, T. (1944), Walker, J.J. (1932b).

MACROPLEA MUTICA   **NOTABLE A**
A reed beetle
Order COLEOPTERA   Family CHRYSOMELIDAE

Macroplea mutica (F., 1792). Formerly known as: *Haemonia curtisi* (Lacordaire, 1845), *Haemonia mutica* (F.).

Distribution Recorded from East Sussex, East Kent, West Kent, North Essex, East Suffolk, East Norfolk, West Norfolk, South-east Yorkshire, North-east Yorkshire, South-west Yorkshire and Durham before 1970 and East Sussex, East Kent, West Kent, East Norfolk, North Lincolnshire, South-west Yorkshire, Mid-west Yorkshire and Cumberland from 1970 onwards.

Habitat and ecology Primarily coastal, and associated with brackish lakes, ponds and ditches. This species has also been recorded from a saline lagoon inland. Phytophagous. Associated with fennel-leaved pondweed *Potamogeton pectinatus* and possibly also eel-grass *Zostera marina*. Larvae probably feed and pupate on the foodplant under water. Adults rarely, if ever, emerge from the water, and have been recorded from May to August.

Status Status revised from RDB 3 (Rare) in Shirt (1987). Probably under-recorded. Widespread and very local in the eastern half of England, being recorded from East Sussex to Mid-west Yorkshire. Recently recorded from eight vice-counties. This species is difficult to identify and may be confused with *M. appendiculata*.

Threats Loss of habitat through reclamation, sea defence improvements and the infilling of pond and

lakes. Water pollution and natural succession may be further threats.

Management and conservation Water bodies should be isolated from sources of pollution and eutrophication. Water tables should be maintained at high levels. Clearance of emergent vegetation may be needed, this should be undertaken on a rotational basis and be aimed at maintaining the aquatic plant populations.

Published sources Anon, (1935), Eyre, M.D. & Cox, M.L. (1987), Fowler, W.W. (1890), Fowler, W.W. & Donisthorpe, H.St J.K. (1913), Marriner, T.F. (1938), Parry, J.A. (1979), Shirt, D.B., ed. (1987), Stainforth, T. (1944).

MANTURA CHRYSANTHEMI   **NOTABLE A**
A flea beetle
Order COLEOPTERA   Family CHRYSOMELIDAE

Mantura chrysanthemi (Koch, 1803).

Distribution Recorded from South Devon, North Devon, East Suffolk, West Suffolk, East Norfolk, West Norfolk, Cambridgeshire, East Gloucestershire, West Gloucestershire, Cardiganshire, Caernarvonshire, Anglesey, Cheshire and South Lancashire before 1970 and East Cornwall, North Devon, East Suffolk, West Suffolk, West Norfolk, Caernarvonshire and Anglesey from 1970 onwards.

Habitat and ecology Heathland, wood margins and disturbed ground. Phytophagous. Associated with sheep's-sorrel *Rumex acetosella*. Larvae are probably leaf miners on the foodplant from late June to late August, with pupation occurring between mid July and early September. Adults have been recorded fro May to October.

Status Only known in Great Britain as far north as South Lancashire. Recently recorded from only seven vice-counties.

Threats Loss and fragmentation of heathland through development, improvement and conversion to arable agriculture and afforestation. Degradation of heathland through the cessation of traditional heathland management practices and natural succession are further threats.

Management and conservation Grazing, cutting or some other disturbance, such as scraping or rotovation, are needed to maintain open conditions. Wood margins should be managed to retain a variety of vegetation structures.

CHRYSOMELIDAE

Published sources Atty, D.B. (1983), Collier, M.J. (1990), Fowler, W.W. (1890), Fowler, W.W. & Donisthorpe, H.St J.K. (1913), Hodge, P.J. (1990).

MANTURA OBTUSATA   NOTABLE B
A flea beetle
Order COLEOPTERA   Family CHRYSOMELIDAE

Mantura obtusata (Gyllenhal, 1813).

Distribution England, Dyfed-Powys and North Wales.

Habitat and ecology Predominantly damp meadows, wet areas in woodland and other wetland areas. Phytophagous. Associated with small-leaved docks *Rumex* and possibly exclusive to sorrel *R. acetosa*. Larvae are probably leaf-miners on the foodplant from late June to late August, with pupation occurring between mid July and early September. Adults have been recorded from January to November.

Status Widespread but local, recorded from southern England, north to Cumberland.

Threats Drainage for reasons such as agricultural improvement and development. Falling water tables because of water abstraction and river engineering schemes, neglect and conversion to high forest, and natural succession may also threaten this species.

Management and conservation Water tables should be maintained at high levels. Grazing, cutting or some other disturbance, on a rotational basis, is needed to maintain open conditions. In woodland, open glades and ride margins should be managed to retain a variety of vegetation structures.

MANTURA RUSTICA   NOTABLE B
A flea beetle
Order COLEOPTERA   Family CHRYSOMELIDAE

Mantura rustica (L., 1767).

Distribution England, Dyfed-Powys, North Wales, South East Scotland, South West Scotland and North East Scotland.

Habitat and ecology A species of sandy soils. Dry grassland, woodland rides, clearings, wood margins, hedgebanks, field margins, disturbed ground, and possibly coastal habitats. Phytophagous. Associated with broad-leaved docks *Rumex*, particularly broad-leaved dock *R. obtusifolius*. On the Continent, this species is also associated with *Polygonum*. Larvae are leaf-miners on the foodplant occurring between late June and late August, with pupation occurring between mid July and early September. Adults have been recorded from all months of the year.

Status Widespread but local in Great Britain.

Threats Loss of habitat through a change in land use such as conversion to arable agriculture, development, afforestation and the clear-felling of woodland. Natural succession and the use of herbicides and pesticides may be further threats.

Management and conservation Some grazing, rotational cutting or some other disturbance, such as rotovation, is needed to maintain open conditions.

MNIOPHILA MUSCORUM   NOTABLE B
A flea beetle
Order COLEOPTERA   Family CHRYSOMELIDAE

Mniophila muscorum (Koch, 1803).

Distribution South East, South, South West, East Midlands, West Midlands, North East, North West, Dyfed-Powys, North Wales, South East Scotland, South West Scotland and North East Scotland.

Habitat and ecology Woodland and lightly wooded areas. Phytophagous. Found in moss on trees, on tree stumps, on rocks, on walls and on chalky slopes. On the Continent, this species is apparently associated with foxglove *Digitalis*, plantain *Plantago* and possibly germander *Teucrium*. In Britain, the larvae live and develop in moss, pupating in late September and early October. Adults have been recorded from March to June and from August to December.

Status Widespread but local in Britain. Possibly under-recorded.

Threats Loss of habitat through conversion to other land use.

OCHROSIS VENTRALIS   RARE
A leaf beetle
Order COLEOPTERA   Family CHRYSOMELIDAE

Ochrosis ventralis (Illiger, 1807). Formerly known as: *Crepidodera ventralis* (Illiger).

Distribution Recorded from South Devon, Isle of Wight, West Sussex, East Sussex, East Kent, West Kent, Surrey, North Essex, East Norfolk, East Gloucestershire, Herefordshire, Denbighshire, Derbyshire, West Lancashire and Cumberland before 1970 and East Kent, Bedfordshire and South-west Yorkshire from 1970 onwards.

Habitat and ecology Disturbed ground and other habitats in which the foodplants may grow, particularly on chalky or sandy substrates. Phytophagous. Associated with bittersweet *Solanum dulcamara*, and also recorded from *Matricaria*. On the Continent, this species has been noted on scarlet pimpernel *Anagalis arvensis*. Oviposition probably takes place in spring, with new generation adults emerging in late summer. Adults have been recorded from April to June and in August and September.

Status Not listed in the insect Red Data Book (Shirt, 1987). Possibly declining. Old records indicate that this species was formerly widely distributed in southern England, with scattered records north to Cumberland. Also noted in North Wales. Recently recorded from only three vice-counties.

Threats Loss of habitat through improvement and conversion to arable agriculture and development. Natural succession may be a further threat.

Management and conservation Disturbance, such as rotovation, is needed to maintain open conditions.

Published sources Atty, D.B. (1983), Fowler, W.W. (1890), Fowler, W.W. & Donisthorpe, H.St J.K. (1913), Marriner, T.F. (1938), Morgan, M.J. (1984).

ORSODACNE LINEOLA **NOTABLE B**
A leaf beetle
Order COLEOPTERA Family CHRYSOMELIDAE

Orsodacne lineola (Panzer, 1795).

Distribution South East, South, East Anglia, East Midlands, West Midlands, North East and South West Scotland.

Habitat and ecology Broad-leaved woodland, parkland, scrub and heathland. Phytophagous. Associated with hawthorn, and probably also on *Prunus*, *Sorbus* and other rosaceous plants. Continental literature also cites *Rosa*, *Mespilus* and *Pyrus* as foodplants. The larvae are probably either external or internal root-feeders. This species probably overwinters as a first instar larva, feeding up in the spring with adults emerging during summer. In Britain, the adults are often recorded on the flowers of hawthorn. Adults have been recorded in May, June and August.

Status Widespread but local in England and also recorded from South West Scotland. Not as gregarious as *O. cerasi*.

Threats Loss of broad-leaved woodland through, for example, clear-felling and coniferisation. Neglect and

conversion to high forest, the loss of heathland through improvement and conversion to arable agriculture, development and afforestation may be further threats.

Management and conservation Open glades and ride margins should be cut on rotation to retain a variety of vegetation structures. On heathland, management should aim for a diversity of successional stages and include areas of scrub. The presence of nectar sources, such as hawthorn, may be particularly important for this species.

OULEMA ERICHSONI **ENDANGERED**
A leaf beetle
Order COLEOPTERA Family CHRYSOMELIDAE

Oulema erichsoni (Suffrian, 1841). Formerly known as: *Lema erichsoni* Suffrian, *Lema erichsonii* Suffrian.

Distribution Recorded from South Devon, East Sussex and East Kent before 1970 and North Somerset from 1970 onwards.

Habitat and ecology Known from an area under commercial peat extraction. Phytophagous. Found in association with young plants of sweet-grass, probably floating sweet-grass *Glyceria fluitans*, chiefly in the damp bottoms of recently cut peat trenches where there is little other vegetation. Possibly also associated with a single (unspecified) species of grass. Adults have been found from late May to July.

Status Status revised from RDB 3 (Rare) in Shirt (1987). There are old records from three vice-counties all in southern England. Recently recorded from a single vice-county, and fourth in total, where it has been recorded in three 1km squares.

Threats This species is threatened by drainage and the drying out of the cut peat surface, as well as through natural succession.

Management and conservation Water tables should be maintained to keep surface peat moist. The species may benefit from the early successional stages created by small scale turf cutting, provided that this is not accompanied by drainage.

Published sources Allen, A.A. (1976c), Champion, G.C. (1897), Hodge, P.J. (1989), Shirt, D.B., ed. (1987).

CHRYSOMELIDAE

PHAEDON CONCINNUS   NOTABLE B
A leaf beetle
Order COLEOPTERA   Family CHRYSOMELIDAE

Phaedon concinnus Stephens, 1831. Formerly known as: *Phaedon armoraciae var. concinnus* Stephens.

Distribution South West, England, Wales, South West Scotland and North West Scotland.

Habitat and ecology Saltmarshes, often at the edges of pools frequented by cattle. Phytophagous. Chiefly associated with sea arrowgrass *Triglochin maritima*, also recorded from sea plantain *Plantago maritima* and possibly associated with scurvygrass *Cochlearia*. Adults appear to be most frequent in April, and have also been recorded in March and July.

Status Old records indicate that this species has been widely reported in southern England, with scattered records throughout Great Britain as far north as North Ebudes in Scotland. Possibly more local than formerly, and recently recorded from East Norfolk to Westerness in Scotland.

Threats Loss of saltmarsh through reclamation, erosion and the construction of sea defences. Overgrazing of saltmarshes could be a further threat to this species.

Management and conservation Grazing should not be introduced to a saltmarsh where there is no grazing at present.

PHYLLODECTA POLARIS   RARE
A leaf beetle
Order COLEOPTERA   Family CHRYSOMELIDAE

Phyllodecta polaris Schneider, 1886.

Distribution Recorded from East Inverness & Nairn, West Ross and East Ross before 1970 and West Ross, East Ross and West Sutherland from 1970 onwards.

Habitat and ecology Montane habitats, this species has been recorded at 950m above sea level. Phytophagous. Associated with dwarf willow *Salix herbacea* and possibly other montane *Salix* species. Ova are laid in July and larvae have been found in August. Adults have been recorded May to September.

Status Possibly under-recorded. A high altitude species that has only been recorded from four vice-counties, and just three recently.

Threats Erosion of montane habitat may be a problem in areas frequented by large numbers of hill walkers and where livestock densities are too high.

Published sources Horsfield, D. (1987), Morris, M.G. (1970), Owen, J.A. (1983a), Owen, J.A. (1988a), Owen, J.A. (1988c), Owen, J.A. (1990a), Shirt, D.B., ed. (1987).

PHYLLOTRETA AEREA   NOTABLE B
A flea beetle
Order COLEOPTERA   Family CHRYSOMELIDAE

Phyllotreta aerea Allard, 1859. Formerly known as: *Phyllotreta punctulata* sensu Fowler, 1890 ?not (Marsham, 1802).

Distribution England and South Wales.

Habitat and ecology Disturbed ground and gardens. Phytophagous. Recorded from radish *Raphanus sativus*, horse-radish *Armoracia rusticana* and turnip *Brassica rapa*, but probably also on other species of the Cruciferae. Larvae probably feed at the roots of the foodplant during May and early June, with pupation probably occurring in June or July. Adults have been recorded from May to October.

Status Widespread but local in southern and central England, also recorded in North West England and South Wales. There is are old records for North East England. This species is difficult to identify and may be confused with other members of the genus. Consequently, the exact status of this species is hard to assess.

Threats Natural succession, and the use of herbicides and pesticides may threaten this species.

Management and conservation Cutting, or some other disturbance, on a rotational basis, may be necessary to maintain open conditions.

PHYLLOTRETA CRUCIFERAE   NOTABLE B
Turnip flea beetle
Order COLEOPTERA   Family CHRYSOMELIDAE

Phyllotreta cruciferae (Goeze, 1777). Formerly known as: *Phyllotreta atra var. cruciferae* (Goeze).

Distribution South East, South, South West, East Anglia, East Midlands, West Midlands, North East, Dyfed-Powys, North Wales and South West Scotland.

Habitat and ecology Coastal habitats, disturbed ground and probably also in gardens. Phytophagous. Recorded from pepperwort *Lepidium*, sea rocket *Cakile maritima* and turnip *Brassica rapa*, probably also on other species of the Cruciferae. Larvae feed on the roots of the foodplant between May and July, with pupation occurring towards the end of July and in August. Adults have been recorded in January, from March to October and in December.

Status Widespread but local in England, also recorded in parts of Wales and South West Scotland. This species is difficult to identify and may be confused with other members of the genus. Consequently, the exact status of this species is hard to assess.

Threats This species is probably threatened by coastal developments, natural succession, and the use of herbicides and pesticides.

Management and conservation Grazing, cutting or some other disturbance, such as rotovation, on a rotational basis, may be needed to maintain open conditions.

PHYLLOTRETA VITTATA NOTABLE A
A flea beetle
Order COLEOPTERA Family CHRYSOMELIDAE

Phyllotreta vittata (F., 1801). Formerly known as: *Phyllotreta sinuata* sensu auct. Brit. not (Stephens, 1831).

Distribution Recorded from West Cornwall, North Devon, South Hampshire, East Sussex, East Kent, West Kent, Berkshire, Oxfordshire, East Suffolk, West Norfolk, Cambridgeshire, Huntingdonshire, Herefordshire, Warwickshire, Glamorgan, Merionethshire, Caernarvonshire, Derbyshire, South Lancashire, North-east Yorkshire, Westmorland & North Lancashire, Cumberland, Roxburghshire, East Lothian, Stirlingshire and Dunbartonshire before 1970 and West Cornwall, East Kent, West Suffolk, West Norfolk, Huntingdonshire, Cardiganshire and Mid-west Yorkshire from 1970 onwards.

Habitat and ecology Disturbed ground and probably in a variety of other habitats, such as river shingle, roadside verges, hedgebanks and gardens. Phytophagous. Recorded from turnip *Brassica rapa*, water cress *Nasturtium officinale* and probably on other species of the Cruciferae. Larvae probably feed at the roots of the foodplants during May and June, with pupation occurring in the soil during June or July. Adults have been recorded in January, from March to June, August, and in November and December.

Status Old records show that this species has been widely recorded in England, and was also noted in Wales and as far north as Dunbartonshire in Scotland. Recently reported from seven, widely scattered, vice-counties, six of these in England. This species is difficult to identify and may be confused with other

members of the genus. Consequently, the exact status of this species is hard to assess.

Threats Natural succession, and the use of herbicides and pesticides may threaten this species. River engineering schemes may be a further threat.

Management and conservation Cutting, grazing or some other disturbance is needed to maintain open conditions. River shingle tends to be mobile and relies on the free flow of river and stream systems. Activities that hinder this flow should be avoided.

Published sources Allen, A.A. (1977c), Fowler, W.W. (1890), Fowler, W.W. & Donisthorpe, H.St J.K. (1913).

PHYTODECTA DECEMNOTATA NOTABLE B
A leaf beetle
Order COLEOPTERA Family CHRYSOMELIDAE

Phytodecta decemnotata (Marsham, 1802). Formerly known as: Phytodecta *rufipes* (De Geer, 1775).

Distribution South East, South, East Anglia, East Midlands, West Midlands, North West, South East Scotland, South West Scotland and North East Scotland.

Habitat and ecology Broad-leaved woodland. Phytophagous. Associated with aspen and, very doubtfully, with goat willow. Larvae feed on the foliage of the host-tree between early May and early June. Pupation occurs in the soil between the end of May and the end of June. The species overwinters in the adult stage, reappearing the following spring. Adults have been recorded from May to August and in November.

Status Widespread but local in England, and recorded throughout much of Scotland.

Threats Loss of broad-leaved woodland through, for example, clear-felling and coniferisation. Alteration of the vegetation structure of many remaining woodlands by 20th century management practices, especially the selective removal of aspen. Neglect and conversion to high forest may be a further threat.

Management and conservation Open glades and ride margins should be cut on rotation to retain a variety of vegetation structures.

CHRYSOMELIDAE

PILEMOSTOMA FASTUOSA   NOTABLE A
A leaf beetle
Order COLEOPTERA   Family CHRYSOMELIDAE

Pilemostoma fastuosa (Schaller, 1783). Formerly known as: *Cassida fastuosa* Schaller.

Distribution Recorded from North Devon, Dorset, Isle of Wight, South Hampshire, West Sussex, East Sussex, East Kent, West Kent, Surrey, Berkshire, Oxfordshire, Glamorgan and South Lancashire before 1970 and North Devon, Dorset, South Hampshire, East Sussex, West Kent, West Suffolk and Carmarthenshire from 1970 onwards.

Habitat and ecology Calcareous soils. Preferring slopes in open or lightly shaded situations, also grassland and disturbed ground. Sometimes found besides streams. Phytophagous. Associated with ploughman's-spikenard *Inula conyza* and, rarely, common fleabane *Pulicaria dysenterica*. Possibly also associated with other *Inula* species and species of ragwort *Senecio*. Larvae are probably free-living on the foodplant during July and August, with pupation probably occurring from mid July until mid September. Adults have been recorded from February to July and in September.

Status Very local. Old records show that this species has been widely recorded in southern England and was also found in South Lancashire and South Wales. Recently reported from seven vice-counties.

Threats Loss of unimproved grassland through improvement by reseeding or by the application of fertilisers, or by conversion to arable agriculture. Development and natural succession are further threats.

Management and conservation Grazing, cutting or some other distrubance, on a rotational basis, is needed to maintain open conditions.

Published sources Eagles, T.R. (1955a), Fowler, W.W. (1890), Fowler, W.W. & Donisthorpe, H.St J.K. (1913).

PLATEUMARIS AFFINIS   NOTABLE B
A reed beetle
Order COLEOPTERA   Family CHRYSOMELIDAE

Plateumaris affinis (Kunze, 1818).

Distribution England, Wales, South West Scotland and North East Scotland.

Habitat and ecology Aquatic and semi-aquatic habitats, such as freshwater lakes, ponds, ditches, fens, acid bogs, and slow-flowing rivers. Phytophagous.

Associated with sedges, and has also been found on the leaves of cotton grass. Adults are found on emergent vegetation and have been recorded from May to July.

Status Widespread but local in England and Wales, and also recorded in parts of Scotland.

Threats Drainage for reasons such as agricultural improvement and development. River engineering, including dredging, level regulation by damming and flood alleviation schemes. The infilling of lakes and ponds, water pollution, the use of motor-boats and natural succession may be further threats.

Management and conservation Water tables should be maintained at high levels. Water bodies should be isolated from sources of pollution and eutrophication. Clearance of emergent vegetation may be needed to maintain open conditions, this should be undertaken on a rotational basis and be aimed at maintaining aquatic plant populations.

PLATEUMARIS BRACCATA   NOTABLE A
A reed beetle
Order COLEOPTERA   Family CHRYSOMELIDAE

Plateumaris braccata (Scopoli, 1772).

Distribution Recorded from North Somerset, West Sussex, East Sussex, East Kent, West Kent, Surrey, South Essex, North Essex, Oxfordshire, East Suffolk, West Suffolk, East Norfolk, West Norfolk, Cambridgeshire, Bedfordshire, West Gloucestershire, Glamorgan, North Lincolnshire, Leicestershire & Rutland and South-east Yorkshire before 1970 and North Somerset, West Sussex, Berkshire, East Norfolk, West Norfolk, Huntingdonshire, Worcestershire, Glamorgan and Carmarthenshire from 1970 onwards.

Habitat and ecology Aquatic and semi-aquatic habitats, such as brackish ditches, slow-flowing rivers and estuaries. Primarily found near the coast. Phytophagous. Associated with reed *Phragmites australis*. Larvae occur at the roots of the foodplant, and pupate in cocoons. Adults are found on emergent vegetation and have been recorded in April, from June to August and in October.

Status Old records show that this species was widely recorded in southern and eastern England, with a scattered distribution north to South-east Yorkshire and also noted in southern Wales. Recently reported from nine vice-counties.

Threats River engineering, including dredging, level regulation by damming and flood alleviation schemes. Coastal defence improvements, water abstraction schemes, water pollution and natural succession may be further threats.

Management and conservation Water tables should be maintained at high levels. Water bodies should be isolated from sources of pollution and eutrophication. Clearance of marginal vegetation may be needed to maintain open conditions, this should be undertaken on a rotational basis and be aimed at maintaining the emergent plant populations.

Published sources Anon, (1925), Atty, D.B. (1983), Cribb, J. (1954), Fowler, W.W. (1890), Fowler, W.W. & Donisthorpe, H.St J.K. (1913), Parry, J.A. (1979), Pavett, P.M. (1988), Stainforth, T. (1944).

PODAGRICA FUSCICORNIS	NOTABLE B
A flea beetle	
Order COLEOPTERA	Family CHRYSOMELIDAE

Podagrica fuscicornis (L., 1767).

Distribution South East, South, South West, East Anglia, East Midlands, West Midlands, North East and South Wales.

Habitat and ecology Grassland, scrub and disturbed ground. Probably also in coastals habitats. Phytophagous. Associated with musk mallow *Malva moschata*, probably also on other species of *Malva*. On the Continent, this species is also associated with marsh-mallow *Althaea*. There is probably only one generation a year in Britain. Overwintered larvae feed on the roots of the foodplant in May and early June, with pupation occurring between early and late June. Adults have been recorded from June to September and hibernate in the stems of the foodplant or in leaf-debris.

Status Widespread but local in England, though not known in North West England. Also recorded in South Wales.

Threats Loss of unimproved grassland through improvement by reseeding or by the application of fertilisers, or by conversion to arable agriculture. Coastal developments, overgrazing and natural succession are further threats.

Management and conservation Some grazing, cutting or other disturbance, on a rotational basis, may be needed to maintain open conditons.

PODAGRICA FUSCIPES	NOTABLE A
A flea beetle	
Order COLEOPTERA	Family CHRYSOMELIDAE

Podagrica fuscipes (F., 1775).

Distribution Recorded from West Cornwall, South Devon, North Devon, North Somerset, Dorset, Isle of Wight, South Hampshire, East Sussex, East Kent, West Kent, Surrey, South Essex, Cambridgeshire, Huntingdonshire and East Gloucestershire before 1970 and East Sussex, East Kent, West Kent and South Essex from 1970 onwards.

Habitat and ecology Grassland, scrub, wood margins, disturbed ground and coastal sites. Phytophagous. Associated with mallow *Malva* and probably marsh-mallow *Althaea*. There is probably only one generation a year in Britain, with larval development occurring in spring and early summer and adult emergence in late summer and early autumn. Larvae probably feed on the roots of the foodplant. Adults have been recorded from May to September.

Status Possibly declining. Old records indicate that this species was formerly widely distributed in southern England, with records north to Huntingdonshire. Recently recorded from only four vice-counties, all in south-eastern England.

Threats Loss of grassland through improvement and conversion to arable agriculture. Coastal and urban developments, overgrazing and natural succession are further threats.

Management and conservation Some grazing, cutting or other disturbance, on a rotational basis, may be needed to maintain open conditions.

Published sources Allen, A.A. (1954g), Atty, D.B. (1983), Fowler, W.W. (1890).

PSYLLIODES ATTENUATA	ENDANGERED
Hop flea beetle	
Order COLEOPTERA	Family CHRYSOMELIDAE

Psylliodes attenuata (Koch, 1803).

Distribution Recorded from West Sussex, East Sussex, East Kent, West Kent, Surrey, Berkshire, Oxfordshire, East Suffolk, Huntingdonshire, Worcestershire, Anglesey, South-west Yorkshire, Mid-west Yorkshire and "Perth" before 1970 and East Kent and South-west Yorkshire from 1970 onwards.

Habitat and ecology Cultivated land, particularly hop-fields and field margins. Phytophagous. Associated

CHRYSOMELIDAE

with hop *Humulus lupulus*. On the Continent, this species has also been recorded from nettle *Urtica*. Larvae are probably root-feeders during May and June. Adults have been recorded from April to June.

Status Not listed in the insect Red Data Book (Shirt, 1987). Old records indicate that this species was widely distributed in England, with records north to Mid-west Yorkshire, and recorded from Anglesey in Wales and Perthshire in Scotland. This species was not uncommon at the turn of the century, particularly in the hop fields of south-eastern England. This species has declined and has recently been recorded from just two vice-counties in southern England. This species is difficult to identify and may be confused with other members of the genus. Consequently, the exact status of this species is hard to assess.

Threats The grubbing out and mechanised trimming of hedgerows, improvement and change in land use and the use of herbicides and pesticides.

Management and conservation Hedgerows should be managed, on a rotational basis, to retain a variety of vegetation structures.

Published sources Fowler, W.W. (1890).

PSYLLIODES CHALCOMERA   **NOTABLE B**
A flea beetle
Order COLEOPTERA   Family CHRYSOMELIDAE

Psylliodes chalcomera (Illiger, 1807).

Distribution South West, South, South East, East Anglia, East Midlands, West Midlands, North East, North Wales and South East Scotland.

Habitat and ecology Grassland, disturbed ground, coastal sites and riverside vegetation. Phytophagous. Associated with thistles *Carduus* and possibly also with species of *Cirsium*. Larvae are probably root-feeders during May and June, with new generation adults emerging between mid July and late August. Adults have been recorded from March to September.

Status Old records indicate that this species has been widely recorded in southern England, with a scattered distribution north to East Lothian in Scotland. Recently reported in southern England, north to West Norfolk and also noted in Caernarvonshire, Wales. This species is difficult to identify and may be confused with other members of the genus. Consequently, the exact status of this species is hard to assess.

Threats Loss of habitat through improvement and conversion to arable agriculture and development. Natural succession is a further threat.

Management and conservation Disturbance, such as rotovation, is needed to maintain open conditions.

PSYLLIODES HYOSCYAMI   **ENDANGERED**
Henbane flea beetle
Order COLEOPTERA   Family CHRYSOMELIDAE

Psylliodes hyoscyami (L., 1758).

Distribution Recorded from West Cornwall, Surrey, Hertfordshire, Berkshire, Oxfordshire, Cambridgeshire, Huntingdonshire, Leicestershire & Rutland, South Lancashire, Mid-west Yorkshire, Durham and West Lothian before 1970.

Habitat and ecology Sandy soils, especially near the coast. Disturbed ground. Phytophagous. Associated with henbane *Hyoscyamus niger*. Oviposition commences in early May and continues during June. Larvae mine the leaf-stalks, leaf-blades, stems and tap-roots of the foodplant in late May and June, with pupation occurring in late June and July. The species overwinters in long grass in the adult stage. Adults have been recorded from April to August.

Status Old records indicate a widely scattered distribution in England from West Cornwall to Durham, also recorded in West Lothian in Scotland. Last recorded in 1930 from Oxfordshire. This species is difficult to identify and may be confused with other members of the genus.

Threats The foodplant is very local and sporadic, though with a widely scattered distribution. Coastal developments, improvement and conversion to arable agriculture, the use of herbicides and natural succession have probably all contributed to this species' decline.

Published sources Bagnall, R.S. (1905), Fowler, W.W. (1890), Fowler, W.W. & Donisthorpe, H.St J.K. (1913), Shirt, D.B., ed. (1987).

PSYLLIODES LURIDIPENNIS   **VULNERABLE**
A flea beetle   **ENDEMIC**
Order COLEOPTERA   Family CHRYSOMELIDAE

Psylliodes luridipennis Kutschera, 1864. Formerly known as: *Psylliodes chrysocephala* var. *luridipennis* Kutschera, *Psylliodes hospes* sensu auct. not Wollaston, 1854.

Distribution Recorded from North Devon before 1970 and North Devon from 1970 onwards.

Habitat and ecology Maritime cliff and coastal shingle. Phytophagous. Associated with Lundy cabbage *Rhynchosinapis wrightii*. Larvae probably occur during the winter at the roots or mining the roots of the foodplant. Adults have been recorded in April and from June to August.

Status Status revised from RDB 1 (Endangered) in Shirt (1987). The species is apparently endemic to the Isle of Lundy, North Devon. Can be numerous where found. This species is difficult to identify and may be confused with other members of the genus. Consequently, the exact status of this species is hard to assess. The foodplant is listed as RDB 3 (Rare) in Perring and Farrel (1983).

Threats The foodplant is threatened through grazing by goats, sheep and deer, as well as through increasing tourist pressure.

Published sources Perring, F.H. & Farrell, L. (1983), Shirt, D.B., ed. (1987).

PSYLLIODES LUTEOLA
A flea beetle   **INSUFFICIENTLY KNOWN**
Order COLEOPTERA   Family CHRYSOMELIDAE

Psylliodes luteola (Müller, 1776).

Distribution Recorded from Dorset, East Sussex, South Essex, Oxfordshire, Buckinghamshire, Huntingdonshire, Worcestershire and Nottinghamshire before 1970 and South Wiltshire and Oxfordshire from 1970 onwards.

Habitat and ecology Probably hedgerows, woodland and disturbed ground. Phytophagous. Usually associated with bittersweet *Solanum dulcamara*, possibly on a number of other hosts. Adults have been recorded from July to November.

Status Not listed in the insect Red Data Book (Shirt, 1987). There are old records from Dorset to Nottinghamshire. Recently recorded from two vice-counties. This species is difficult to identify and may be confused with other members of the genus. Consequently, the exact status of this species is hard to assess.

Threats Uncertain, but possibly loss of habitat through clear-felling of woodland and conversion to other land use, improvement and conversion to arable agriculture, afforestation, development and natural succession.

Management and conservation Some grazing, cutting or other disturbance, on a rotational basis, may be needed to maintain open conditions. In woodland, open glades and ride margins should be managed to retain a variety of vegetation structures.

Published sources Fowler, W.W. (1890).

PSYLLIODES SOPHIAE   **RARE**
A flea beetle
Order COLEOPTERA   Family CHRYSOMELIDAE

Psylliodes sophiae Heikertinger, 1914. Formerly known as: *Psylliodes cyanoptera* sensu auct. not (Illiger, 1807).

Distribution Recorded from West Suffolk, West Norfolk, Cambridgeshire and Huntingdonshire before 1970 and West Suffolk from 1970 onwards.

Habitat and ecology Particularly on sandy soils. Disturbed ground and probably also grassland. Phytophagous. Associated with flixweed *Descurainia sophia*. Larvae probably mine the roots and stems of the foodplant in late May and June, with pupation probably occurring in late June and July. Adults have been recorded from June to August.

Status Primarily a Breckland species, recently recorded from only one vice-county. Can be common where found. There is an old record from Bristol that requires confirmation. This species is difficult to identify and may be confused with other members of the genus. Consequently, the exact status of this species is hard to assess.

Threats Loss of unimproved grassland through improvement and conversion to arable agriculture, development and afforestation. Natural succession is a further threat.

Management and conservation Disturbance, such as scraping or rotovation, is needed to maintain open conditions.

Published sources Fowler, W.W. (1890), Fowler, W.W. & Donisthorpe, H.St J.K. (1913), Shirt, D.B., ed. (1987).

CHRYSOMELIDAE

ZEUGOPHORA FLAVICOLLIS   VULNERABLE
A leaf beetle
Order COLEOPTERA   Family CHRYSOMELIDAE

Zeugophora flavicollis (Marsham, 1802).

Distribution Recorded from South Hampshire, North Hampshire, East Sussex, East Kent, West Kent, Surrey, South Essex, North Essex, Hertfordshire, "Suffolk", Huntingdonshire, Worcestershire, Warwickshire, Leicestershire & Rutland, South Lancashire and Westmorland & North Lancashire before 1970 and Surrey, Bedfordshire and Warwickshire from 1970 onwards.

Habitat and ecology Broad-leaved woodland. Phytophagous. Associated with aspen. On the Continent, this species has also been recorded on Italian poplar and goat willow. Larvae mine the leaves of the host-tree during the summer. In Britain, adults occur on the host-tree between May and July and again in September and October, and probably overwinter in this stage.

Status Status revised from RDB 1 (Endangered) in Shirt (1987). Old records indicate that this species was formerly widespread in southern England with scattered records north to Westmorland. Recently recorded from just three vice-counties.

Threats Loss of broad-leaved woodland through, for example, clear-felling and coniferisation. Neglect and conversion to high forest, and the selective removal of aspen are further threats.

Management and conservation Open glades and ride margins should be cut on rotation to retain a variety of vegetation structures. Gaps in the age structure of the tree population should be identified and the continuity of the appropriate habitat ensured by regeneration or possibly through suitable planting.

Published sources Alexander, K.N.A. & Clements, D.K. (1988), Buck, F.D. (1955), Cox, D. (1947), Cox, D. (1948), Fowler, W.W. (1890), Fowler, W.W. & Donisthorpe, H.St J.K. (1913), Lane, S.A. (1990), Menzies, I.S. (1990), Shirt, D.B., ed. (1987).

ZEUGOPHORA TURNERI   NOTABLE A
A leaf beetle
Order COLEOPTERA   Family CHRYSOMELIDAE

Zeugophora turneri Power, 1863.

Distribution Recorded from Moray, East Inverness & Nairn and West Ross before 1970 and Mid Perthshire, South Aberdeenshire, Moray, East Inverness & Nairn, West Sutherland and Caithness from 1970 onwards.

Habitat and ecology Broad-leaved woodland. Phytophagous. Associated with aspen. On the Continent, this species is associated with *Populus*. Larvae are leaf-miners. In Britain, adults have been recorded from June to August and probably overwinter in this stage.

Status Very local and only known from central and northern Scotland.

Threats Loss of woodland through, for example, clear-felling and afforestation. Neglect and conversion to high forest, and the selective removal of aspen are further threats.

Management and conservation Open glades and ride margins should be cut on rotation to retain a variety of vegetation structures. Gaps in the age structure of the tree population should be identified and regeneration or suitable planting encouraged to ensure the continuity of suitable habitat.

Published sources Ashe, G.H. (1952).

CIS COLUBER   RARE
Order COLEOPTERA   Family CISIDAE

Cis coluber Abeille, 1874. Formerly known as: *Cis latifrons* Pool, 1917.

Distribution Recorded from South Hampshire, East Sussex, East Kent, Berkshire and East Inverness & Nairn before 1970 and North Somerset, West Sussex and Berkshire from 1970 onwards.

Habitat and ecology Ancient broad-leaved woodland. Probably in fungi on trees or under bark. Recorded from the bough of an oak, from under oak, from a sallow bush and tapped from dead alder twigs. Adults have been noted in March, July and August.

Status Recorded from southern England and East Inverness and Nairn in Scotland. Known from very few records and recently noted from just three vice-counties. This species is difficult to identify and may be confused with other members of the genus. Consequently, the exact status of this species is hard to assess.

Indicator status Grade 2 in Harding & Rose (1986).

Threats Loss of broad-leaved woodland through, for example, clear-felling and coniferisation. Habitat loss, in particular, may be through the felling of fungus-infected trees, removal of dead wood from living trees and the destruction or removal of standing and fallen dead wood for reasons such as forest hygiene, aesthetic tidiness, public safety or for use as fire wood.

Management and conservation Fungus-infected trees and both fallen and standing dead timber, especially with the bark attached, should be retained. The removal of dead timber from fungus-infected trees should be avoided. Gaps in the age structure of the tree population should be identified and the continuity of the appropriate habitat ensured by regeneration, suitable planting and possibly with pollarding.

Published sources Allen, A.A. (1951c), Harding, P.T. (1978), Harding, P.T. & Rose, F. (1986), Shirt, D.B., ed. (1987).

CIS DENTATUS   RARE

Order COLEOPTERA   Family CISIDAE

Cis dentatus Mellié, 1848.

Distribution Recorded from East Inverness & Nairn before 1970 and East Inverness & Nairn from 1970 onwards.

Habitat and ecology Coniferous woodland. Found under the bark of a dead standing pine and from a dead birch supporting *Piptoporus betulinus*. Adults have been recorded in July and September.

Status Not listed in the insect Red Data Book (Shirt, 1987). Extremely local and only known from one vice-county in Scotland. A record of this species from the Isle of Wight has been found to be an aberrant example of *C. alni*.

Threats Loss of native pine forest through practices such as clear-felling or conversion to plantation forest. Habitat loss in the remaining areas through the felling of ancient and fungus-infected trees and the destruction or removal of standing and fallen dead wood for reasons such as forest hygiene, aesthetic tidiness, public safety or for use as fire wood.

Management and conservation Ancient and fungus-infected, trees and both standing and fallen dead timber, especially with the bark attached, should be retained. Gaps in the age structure of the tree population should be identified and regeneration encouraged to ensure the continuity of the appropriate habitat.

Published sources Allen, A.A. (1990c), Aubrook, E.W. (1970), Fowler, W.W. & Donisthorpe, H.St J.K. (1913), Owen, J.A. & Carter, I.S. (1988).

CIS FESTIVUS   NOTABLE B

Order COLEOPTERA   Family CISIDAE

Cis festivus (Panzer, 1793).

Distribution England, Dyfed-Powys, North Wales, South East Scotland, North East Scotland and North West Scotland.

Habitat and ecology Woodland. In fungi on trees and dead wood. Recorded from old, dry *Piptoporus betulinus* on birch, in *Stereum* on hazel and probably on other trees, in fungi on hornbeam, willow and on an oak fence post. Also under fungus-infected bark of a cypress. Adults have been recorded in May and June.

Status Apparently widespread and local in Great Britain. However, this species has been confused with *C. vestitus* to which many records may refer. Consequently, this species may be more local than the records imply.

Threats Loss of woodland through, for example, clear-felling and conversion to other land use. Habitat loss, in particular, through the felling of fungus-infected trees, removal of dead wood from living trees and the destruction or removal of standing and fallen dead wood for reasons such as forest hygiene, aesthetic tidiness, public safety or for use as fire wood.

Management and conservation Fungus-infected trees and both fallen and standing dead timber, especially with the bark attached, should be retained. The removal of dead timber from fungus-infected trees should be avoided. Gaps in the age structure of the tree population should be identified and the continuity of the appropriate habitat ensured by regeneration, suitable planting and possibly with pollarding.

CISIDAE

CIS JACQUEMARTI   NOTABLE B

Order COLEOPTERA   Family CISIDAE

Cis jacquemarti Mellié, 1848.

Distribution South East Scotland, North East Scotland and North West Scotland.

Habitat and ecology Birch woodland. Found in the fungus *Fomes fomentarius* on dead birch trees. Adults have been recorded in August.

Status Only known from Scotland and very local. This species is difficult to identify and may be confused with other members of the genus. Consequently, the exact status of this species is hard to assess. An old record for Windsor, Berkshire is in error.

Threats Loss of birch woodland through, for example, clear-felling and conversion to plantation forest. Habitat loss in the remaining areas through the felling of fungus-infected trees and the destruction or removal of standing and fallen dead wood for reasons such as forest hygiene, aesthetic tidiness, public safety or for use as fire wood.

Management and conservation Fungus-infected trees and both standing and fallen dead timber, especially with the bark attached, should be retained. Gaps in the age structure of the tree population should be identified and regeneration encouraged to ensure the continuity of the appropriate habitat.

CIS LINEATOCRIBRATUS   NOTABLE B

Order COLEOPTERA   Family CISIDAE

Cis lineatocribratus Mellié, 1848.

Distribution South, North West, South East Scotland and North East Scotland.

Habitat and ecology Woodland. In *Polyporus nigrinus* and probably other polypore fungi on trees. Adults have been recorded in June.

Status Apart from the New Forest, South Hampshire, where this species became established for a time, this beetle is only known from Scotland and northern England. This species is difficult to identify and may be confused with other members of the genus. Consequently, the exact status of this species is hard to assess.

Threats Loss of woodland through, for example, clear-felling or conversion to other land use. Habitat loss in the remaining areas through the felling of fungus-infected trees and the destruction or removal of standing and fallen dead wood for reasons such as forest hygiene, aesthetic tidiness, public safety or for use as fire wood.

Management and conservation Fungus-infected trees and both standing and fallen dead timber, especially with the bark attached, should be retained. Gaps in the age structure of the tree population should be identified and regeneration encouraged to ensure the continuity of the appropriate habitat.

RHOPALODONTUS PERFORATUS   RARE

Order COLEOPTERA   Family CISIDAE

Rhopalodontus perforatus (Gyllenhal, 1813).

Distribution Recorded from Mid Perthshire and South Aberdeenshire before 1970 and Mid Perthshire, East Inverness & Nairn and West Inverness from 1970 onwards.

Habitat and ecology Birch woodland. In polypore fungi on birch. Usually associated with *Fomes fomentarius*, and often found with *Bolitophagus reticulatus* (Tenebrionidae). Also recorded from *Piptoporus betulinus*. Adults have been noted in June and August.

Status Not listed in the insect Red Data Book (Shirt, 1987). Only known from Scotland and very local. Recently recorded from just three vice-counties. Can be common where found.

Threats Loss of birch woodland through, for example, clear-felling and conversion to plantation forest. Habitat loss in the remaining areas through the felling of fungus-infected trees and the destruction or removal of standing and fallen dead wood for reasons such as forest hygiene, aesthetic tidiness, public safety or for use as fire wood.

Management and conservation Fungus-infected trees and both standing and fallen dead timber, especially with the bark attached, should be retained. Gaps in the age structure of the tree population should be identified and regeneration encouraged to ensure the continuity of the appropriate habitat.

Published sources Fowler, W.W. (1890), Massee, A.M. (1967).

SULCACIS BICORNIS   NOTABLE B

Order COLEOPTERA   Family CISIDAE

Sulcacis bicornis (Mellié, 1848). Formerly known as: *Rhopalodontus fronticornis* sensu auct. Brit. not (Panzer,1809).

Distribution South East, South, South West, East Anglia, East Midlands and West Midlands.

Habitat and ecology Broad-leaved woodland. In bracket-fungi on trees, with records from beech, alder, willow and from *Piptoporus betulinus* on birch. Adults have been recorded in April.

Status Widespread but local in the southern half of England.

Threats Loss of broad-leaved woodland through, for example, clear-felling and coniferisation. Habitat loss, in particular, through the felling of fungus-infected trees, removal of dead wood from living trees and the destruction or removal of standing and fallen dead wood for reasons such as forest hygiene, aesthetic tidiness, public safety or for use as fire wood.

Management and conservation Fungus-infected trees and both fallen and standing dead timber, especially with the bark attached, should be retained. The removal of dead timber from fungus-infected trees should be avoided. Gaps in the age structure of the tree population should be identified and the continuity of the appropriate habitat ensured by regeneration, suitable planting and possibly with pollarding.

CLAMBUS NIGRELLUS   NOTABLE A

Order COLEOPTERA   Family CLAMBIDAE

Clambus nigrellus Reitter, 1914. Formerly confused under: *Clambus minutus* (Sturm, 1807).

Distribution Recorded from Surrey, Worcestershire and Mid-west Yorkshire before 1970 and Northamptonshire, Worcestershire, Staffordshire, Cardiganshire, Derbyshire, Roxburghshire and Berwickshire from 1970 onwards.

Habitat and ecology River banks. Also recorded from garden compost. Found in decaying vegetation and probably also in moss. Recorded from grass cuttings, though almost certainly in a variety of other decaying vegetation. Also recorded from partially submerged logs in a river. Adults have been noted in May, July and August.

Status Very local and widely scattered in England and Scotland. Recorded between Surrey and Berwickshire in Scotland and recently found in only seven vice-counties.

Threats River engineering, level regulation by damming and flood alleviation schemes. In some areas colonisation by Himalayan balsam *Impatiens glandulifera* can reduce the available habitat.

Management and conservation Grazing or cutting may be needed to maintain open conditions.

Published sources Johnson, C. (1966b).

CLAMBUS PALLIDULUS
INSUFFICIENTLY KNOWN
Order COLEOPTERA   Family CLAMBIDAE

Clambus pallidulus Reitter, 1911. Formerly confused under: *Clambus minutus* (Sturm, 1807).

Distribution Recorded from East Kent, West Kent, Berkshire and Oxfordshire before 1970 and Worcestershire from 1970 onwards.

Habitat and ecology Found in decaying vegetation and probably also moss. A single example has been found in a hollow apple tree.

Status Not listed in the insect Red Data Book (Shirt, 1987). Recorded on only five occasions.

Published sources Johnson, C. (1966b).

KORYNETES CAERULEUS   NOTABLE B
A checkered beetle
Order COLEOPTERA   Family CLERIDAE

Korynetes caeruleus (Degeer, 1775). Formerly known as: *Corynetes coeruleus* (Degeer, 1775).

Distribution England.

Habitat and ecology Ancient broad-leaved woodland, pasture-woodland, structural timber and in other habitats where the species is associated with old bones and other carrion. Found in dead wood where the larvae prey on wood-boring beetles such as *Anobium* and *Xestobium rufovillosum* (Anobiidae) and possibly bark beetles (Scolytidae). On the Continent, this species has been reported preying on the scolytid *Ips typographus*. The beetle also occurs in synanthropic situations where it probably preys on beetles, such as dermestids (Dermestidae), and other invertebrates. In Britain, adults have been recorded from April to July.

CLERIDAE

Status Widespread but local in England.

Indicator status Grade 3 in Garland (1983). Grade 3 in Harding & Rose (1986).

Threats Loss of broad-leaved woodland and parkland through, for example, clear-felling and coniferisation. Habitat loss, in particular, through the felling of ancient trees, removal of dead wood from living trees and the destruction or removal of standing and fallen dead wood for reasons such as forest hygiene, aesthetic tidiness, public safety or for use as fire wood. This is also an opportunistic species which will require a continuity of carrion availability.

Management and conservation Ancient trees and both fallen and standing dead timber, especially with the bark attached, should be retained. The removal of dead timber from ancient trees should be avoided. Gaps in the age structure of the tree population should be identified and the continuity of the appropriate dead wood habitat ensured by regeneration, suitable planting and possibly with pollarding.

OPILO MOLLIS   **NOTABLE B**
A checkered beetle
Order COLEOPTERA   Family CLERIDAE

Opilo mollis (L., 1758). Formerly known as: *Opilio mollis* (L., 1758).

Distribution South East, South, East Anglia, East Midlands, West Midlands, North East and North Wales.

Habitat and ecology Ancient broad-leaved woodland, pasture-woodland and on isolated trees. Predatory on anobiid beetles (Anobiidae). The larvae live in dead wood where they feed on wood-boring beetles, and also the dead wood and contents of the host's burrows. The species has been recorded from oak with the beetle *Xestobium rufovillosum*, and has been reared in midland hawthorn containing the beetle *Anobium fulvicorne*. A larva was recorded in June which, after pupation for four weeks, emerged as an adult in early August. A larva found in March produced an adult in June. Larvae have also been recorded in a fallen elm branch, in poplar and bred from a bracket fungus on willow containing the larvae of *Dorcatoma dresdensis*. Adults have been noted under the bark of elm and Norway spruce, and also from willow, poplar, beech and sycamore. Adults have been recorded in February and from June to September.

Status Widespread but local in southern and central England, also recorded in North East England and North Wales.

Indicator status Grade 1 in Garland (1983). Grade 3 in Harding & Rose (1986).

Threats Loss of broad-leaved woodland and parkland through, for example, clear-felling and conversion to other land use. Habitat loss, in particular, through the felling of ancient trees, removal of dead wood from living trees and the destruction or removal of standing and fallen dead wood for reasons such as forest hygiene, aesthetic tidiness, public safety or for use as fire wood.

Management and conservation Ancient trees and both fallen and standing dead timber, especially with the bark attached, should be retained. The removal of dead timber from ancient trees should be avoided. Gaps in the age structure of the tree population should be identified and the continuity of the appropriate dead wood habitat ensured by regeneration, suitable planting and possibly with pollarding.

TARSOSTENUS UNIVITTATUS   **EXTINCT**
A checkered beetle
Order COLEOPTERA   Family CLERIDAE

Tarsostenus univittatus (Rossi, 1792).

Distribution Recorded from Middlesex and "Gloucs".

Habitat and ecology Woodland. The species is predatory on the immature stages of the beetle *Lyctus linearis* and possibly other species of *Lyctus* (Lyctidae).

Status Presumed extinct. Examples are occasionally found in timber yards and other similar situations. There are records for South Essex, North Essex, Buckinghamshire and Yorkshire, these are all considered to be accidental introductions.

Published sources Fowler, W.W. (1890), Fowler, W.W. & Donisthorpe, H.St J.K. (1913), Shirt, D.B., ed. (1987).

THANASIMUS RUFIPES   **RARE**
An ant beetle
Order COLEOPTERA   Family CLERIDAE

Thanasimus rufipes (Brahm, 1797).

Distribution Recorded from Moray before 1970 and East Inverness & Nairn and East Ross from 1970 onwards.

Habitat and ecology Coniferous woodland. Probably predatory. Adults have been beaten from the tops of felled Scots pine in July and September.

Status Not listed in the insect Red Data Book (Shirt, 1987). Only known from three Scottish vice-counties, and only two recently.

Threats Loss of native pine forest through practices such as clear-felling or conversion to plantation forest. Habitat loss in the remaining areas through the felling of ancient trees and the destruction or removal of standing and fallen dead wood for reasons such as forest hygiene, aesthetic tidiness, public safety or for use as fire wood.

Management and conservation Ancient trees and both standing and fallen dead timber, especially with the bark attached, should be retained. Gaps in the age structure of the tree population should be identified and regeneration encouraged to ensure the continuity of the dead wood habitats.

Published sources Beare, T.H. (1912), Fowler, W.W. & Donisthorpe, H.St J.K. (1913).

TILLOIDEA UNIFASCIATUS   **EXTINCT**
A checkered beetle
Order COLEOPTERA   Family CLERIDAE

Tilloidea unifasciatus (F., 1787). Formerly known as: *Tillus unifasciatus* (F.).

Distribution Recorded from "Hants", Surrey, Hertfordshire and Berkshire.

Habitat and ecology Woodland and probably pasture-woodland. The species is predatory on the immature stages of the beetles *Lyctus brunneus* and *L. linearis* (Lyctidae), which are associated with the dead wood of oak, beech and ash. The majority of the records for *T. unifasciata* are from fresh oak palings. The adult has been recorded in July.

Status Presumed extinct, the last record being in 1877 from Upper Norwood, Surrey.

Published sources Fowler, W.W. (1890), Shirt, D.B., ed. (1987).

TILLUS ELONGATUS   **NOTABLE B**
A checkered beetle
Order COLEOPTERA   Family CLERIDAE

Tillus elongatus (L., 1758).

Distribution England and South Wales.

Habitat and ecology Ancient broad-leaved woodland and pasture-woodland. Also found in hardwood

structural timbers. Predatory on anobiid beetles (Anobiidae) in dead wood, in particular *Ptilinus pectinicornis* in dead beech. Also on *Anobium* species. This beetle has been recorded from oak, hazel, black poplar, holly and ivy, and occasionally noted on the blossom of hawthorn and elder. Adults occur under bark and have been recorded from April to September.

Status Widespread but local in England, also recorded in South Wales.

Indicator status Grade 3 in Harding & Rose (1986).

Threats Loss of broad-leaved woodland and parkland through, for example, clear-felling and coniferisation. Habitat loss, in particular, through the felling of ancient trees, removal of dead wood from living trees and the destruction or removal of standing and fallen dead wood for reasons such as forest hygiene, aesthetic tidiness, public safety or for use as fire wood.

Management and conservation Ancient trees and both fallen and standing dead timber, especially with the bark attached, should be retained. The removal of dead timber from ancient trees should be avoided. Gaps in the age structure of the tree population should be identified and the continuity of the appropriate dead wood habitat ensured by regeneration, suitable planting and possibly with pollarding. The presence of nectar sources such as hawthorn and elder may also be particularly important for this species.

TRICHODES ALVEARIUS   **EXTINCT**
Bee-eating beetle
Order COLEOPTERA   Family CLERIDAE

Trichodes alvearius (F., 1792).

Distribution Recorded from Surrey and South Lancashire.

Habitat and ecology Parasitic, or possibly scavenging, in the nests and hives of the bees *Osmia*, *Megachile* (Hymenoptera) and probably *Apis* (Hymenoptera). On the Continent, adults have been noted on flowers, particularly umbellifers. In Britain, adults have been recorded in June and July.

Status Presumed extinct as a native. An example imported with peaches from Spain was recorded in July 1953 at Wyre Piddle, Worcestershire. Another example was noted by the River Lea near Tottenham, Middlesex in June 1950 and is regarded as a probable import.

Published sources Allen, A.A. (1967d), Collingwood, C.A. (1953), Shirt, D.B., ed. (1987).

TRICHODES APIARIUS   EXTINCT
Bee-eating beetle
Order COLEOPTERA   Family CLERIDAE

Trichodes apiarius (L., 1758).

Distribution Recorded from South Hampshire, East Kent, "Norfolk" and South Lancashire.

Habitat and ecology Parasitic, or possibly scavenging, in the nests and hives of bees, probably *Osmia*, *Megachile* (Hymenoptera) and *Apis* (Hymenoptera). An adult has been noted from hawthorn. On the Continent, adults have been recorded on flowers, particularly umbellifers.

Status Presumed extinct. Last recorded in around 1830 from Coombe Wood, near Dover, East Kent.

Published sources Allen, A.A. (1967d), Shirt, D.B., ed. (1987).

ADONIA VARIEGATA   NOTABLE B
Adonis' Ladybird
Order COLEOPTERA   Family COCCINELLIDAE

Adonia variegata (Goeze, 1777).

Distribution England and South Wales. A map is given in Harding *et al.* (1986) and Majerus *et al.* (1990).

Habitat and ecology Heathland, grassland, parkland, sand dunes, riverbanks and wasteground. A mainly coastal species. Predatory. Found by general sweeping and also noted on thistles, knapweed, broom, gorse and bramble. Adults have been recorded in February and from June to September. This species probably overwinters in plant litter in dry situations.

Status Widespread but local in southern and eastern England, very local in the rest of England and also recorded in South Wales.

Threats Loss of habitat through change in land use. Coastal developments may be a further threat to this species.

CLITOSTETHUS ARCUATUS   ENDANGERED
A ladybird
Order COLEOPTERA   Family COCCINELLIDAE

Clitostethus arcuatus (Rossi, 1794). Formerly known as: *Scymnus arcuatus* (Rossi).

Distribution Recorded from Surrey, Berkshire, Oxfordshire, East Suffolk and Leicestershire & Rutland before 1970 and Oxfordshire and East Suffolk from 1970 onwards.

Habitat and ecology Possibly associated with deciduous and coniferous woodland. Predatory. Most records are from beating ivy, particularly that growing on buildings and walls and possibly also on quarry and rock faces. The beetle has been found by beating *Viburnum tinus* infested with whitefly. Adults have been recorded in July and August. This species probably overwinters as an adult.

Status Extremely local, populations are probably very small. Possibly over-looked. Recently recorded from only two vice-counties.

Threats This species may be threatened by the removal of ivy from trees, buildings and walls etc.

Published sources Fowler, W.W. (1889), Fowler, W.W. & Donisthorpe, H.St J.K. (1913), Majerus, M. & Kearns, P. (1989), Shirt, D.B., ed. (1987).

COCCINELLA MAGNIFICA   NOTABLE A
Scarce Seven-spot ladybird
Order COLEOPTERA   Family COCCINELLIDAE

Coccinella magnifica Redtenbacher, 1843. Formerly known as: *Coccinella distincta* Faldermann, 1837 nec Thunberg, 1781, *Coccinella divaricata* sensu auct. not Olivier, 1808.

Distribution Recorded from North Hampshire, East Sussex, East Kent, West Kent, Surrey, South Essex and Worcestershire before 1970 and South Devon, South Wiltshire, Dorset, North Hampshire, West Sussex, East Kent, Surrey, South Essex, West Suffolk, Cambridgeshire, Bedfordshire, West Gloucestershire, Herefordshire, Mid-west Yorkshire, Durham and East Lothian from 1970 onwards. A map is given in Harding *et al.* (1986), Majerus (1989) and Majerus *et al.* (1990).

Habitat and ecology Woodland and heathland. Predatory. Associated with the wood ant *Formica rufa* (Hymenoptera), though in captivity this species has been reared without ant contact. It seems probable that the ants provide *C. magnifica* with habitat that is free from competing aphid predators and reduced numbers of ladybird predators. In Scandinavia, this beetle has been observed with *F. lugubris* (R.S. Key pers. comm.). The early stages are passed near the nests of the host ant. Eggs are laid on foliage in May and June. The larvae feed on aphids for about 25 to 30 days. They then pupate on foliage and emerge as adults approximately 10 days later. Usually, there is only one generation a year, occasionally there may be two. Adults have been recorded all year round. This species overwinters as an adult in gorse bushes, plant litter etc.

Status Status revised from RDB 3 (Rare) in Shirt (1987). Very local. Recorded in southern England as far north as Herefordshire, also noted in northern England and from two records in Scotland. The vice-county is of the northern-most record in Scotland is unknown. Possibly overlooked because of this species superficial similarity to *C. septempunctata*.

Threats Loss of woodland through, for example, clear-felling and conversion to other land use. Much heathland has been lost, or fragmented, through changes in land use, mainly by conversion to arable, forestry and urban development. *Formica rufa*, and hence *C. magnifica*, has probably suffered through lack of coppicing in woodlands in southern England.

Management and conservation Management of heathland should aim for a diversity of successional stages, preferably by animal grazing or by rotational cutting, scraping or burning. Open glades and margins in woodlands should be managed to retain a variety of vegetation structures.

Published sources Atty, D.B. (1983), Fowler, W.W. (1889), Fowler, W.W. & Donisthorpe, H.St J.K. (1913), Harding, P.T., Greene, D.M., Preston, C.D., Arnold, H.R. & Eversham, B.C. (1986), Majerus, M. & Kearns, P. (1989), Majerus, M.E.N. (1989), Majerus, M.E.N., Forge, H. & Walker, L. (1990), Nash, D.R. (1981), Owen, J.A. (1990a), Shirt, D.B., ed. (1987).

COCCINELLA QUINQUEPUNCTATA   **RARE**
Five-spot ladybird
Order COLEOPTERA   Family COCCINELLIDAE

Coccinella quinquepunctata L., 1758.

Distribution Recorded from South Devon, Dorset, North-east Yorkshire, Durham, North Northumberland, Cumberland, East Lothian, Moray and East Inverness & Nairn before 1970 and Carmarthenshire, Cardiganshire, Westmorland & North Lancashire, Moray and East Inverness & Nairn from 1970 onwards. A map is given in Harding *et al.* (1986) and Majerus *et al.* (1990).

Habitat and ecology River shingle and stream margins. Predatory. Recorded feeding on aphids found amongst leaf-axils of young broom plants. Also found on prostrate sallow bushes, knapweed, creeping thistle, angelica and from moss on rocks in streams. Larvae have been found from May to August, adults from April to October. This species overwinters as an adult in litter and under stones.

Status Very local and with a disjunct distribution. Recently noted in West Wales, North West England and North East Scotland. The species is capable of high altitude dispersal, this may account for some of the records listed.

Threats River engineering, including dredging, level regulation by damming, and flood alleviation schemes. In some areas colonisation by Himalayan balsam *Impatiens glandulifera* can reduce the available habitat. Livestock access can damage shingle structure and disturb vegetation communities. River pollution maybe a further threat.

Management and conservation An under-valued habitat. Shingle and sand banks tend to be mobile and rely on the free-flow of river and stream systems. Activities that hinder this flow should be prevented. Adjacent scrub and rank grassland may be important for the species in winter and to escape late summer floods.

Published sources Asmole, N.P., Nelson, J.M., Shaw, M.R. & Garside, A. (1983), Boyce, D.C. (1989), Boyce, D.C. (1990), Boyce, D.C. & Fowles, A.P. (1988), Day, F.H. (1909), Fowler, W.W. (1889), Fowler, W.W. & Donisthorpe, H.St J.K. (1913), Fowles, A.P. (1989), Fowles, A.P. & Morgan, I.K. (1987), Harding, P.T., Greene, D.M., Preston, C.D., Arnold, H.R. & Eversham, B.C. (1986), Majerus, M. & Kearns, P. (1989), Majerus, M.E.N. & Fowles, A.P. (1989), Majerus, M.E.N., Forge, H. & Walker, L. (1990), Morgan, I.K. (1988), Morgan, I.K. (1989), Morgan, I.K. (1990), Shirt, D.B., ed. (1987).

EXOCHOMUS NIGROMACULATUS
A ladybird   **INDETERMINATE**
Order COLEOPTERA   Family COCCINELLIDAE

Exochomus nigromaculatus (Goeze, 1777).

Distribution Recorded from "Bristol district", Berkshire and South-west Yorkshire before 1970.

Habitat and ecology Recorded on heathland. Predatory. This beetle has been beaten from willow. Adults have been recorded in June and September. On the Continent, this species has been recorded from May

COCCINELLIDAE

to September, the optimum flight period being given as August and September.

Status Not listed in the insect Red Data Book (Shirt, 1987). Formerly known from just three specimens; one recorded near Bristol, undated but before 1831, and two noted at Windsor, Berkshire in 1816. Subsequently, the beetle was omitted from the main checklists this century, and has only recently been reinstated as a British species on the basis of a single specimen found in 1967 near Doncaster, South-west Yorkshire.

Threats Uncertain, though the site near Doncaster has been threatened with housing development.

Published sources Majerus, M. & Kearns, P. (1989), Skidmore, P. (1985a).

HIPPODAMIA TREDECIMPUNCTATA
13-spot ladybird   **INSUFFICIENTLY KNOWN**
Order COLEOPTERA   Family COCCINELLIDAE

Hippodamia tredecimpunctata (L., 1758). Formerly known as: *Hippodamia tredecempunctata* (L., 1758).

Distribution Recorded from West Sussex, East Sussex, East Kent, West Kent, Surrey, South Essex, North Essex, Oxfordshire, "Suffolk", East Norfolk, Huntingdonshire, West Lancashire, South-east Yorkshire, North-east Yorkshire, South-west Yorkshire, Durham, South Northumberland and Cumberland before 1970. A map is given in Harding *et al.* (1986).

Habitat and ecology This species has been recorded in wetland and marshes. Recorded from common reed, yarrow and nettle. Adults have been recorded in May, June and from August to October.

Status Status revised from RDB3 (Rare) in Shirt (1987). No recent records. This species has been recorded throughout England, though with a bias to eastern England, and has been recorded as far north as Cumberland. Last noted in 1952 from the Castle Museum grounds, Hastings, East Sussex. Possibly only an immigrant to these shores.

Threats Uncertain, though if resident this species may be threatened by loss of wetland sites through practices such as drainage.

Published sources Baines, J.M. (1952), Buck, F.D. (1953), Day, F.H. (1909), Fowler, W.W. (1889), Fowler, W.W. & Donisthorpe, H.St J.K. (1913), Harding, P.T., Greene, D.M., Preston, C.D., Arnold, H.R. & Eversham, B.C. (1986), Majerus, M. & Kearns, P. (1989), Shirt, D.B., ed. (1987), Wakely, S. (1953).

HYPERASPIS PSEUDOPUSTULATA
A ladybird   **NOTABLE B**
Order COLEOPTERA   Family COCCINELLIDAE

Hyperaspis pseudopustulata Mulsant, 1853. Formerly known as: *Hyperaspis reppensis* sensu auct. Brit. not (Herbst, 1783).

Distribution England, Wales, South East Scotland, South West Scotland and North East Scotland. A map is given in Harding *et al.* (1986).

Habitat and ecology Deciduous woodland, heathland, grassland, coastal marshes and saltings. Predatory. Usually found amongst herbage or in moss on or below trees. Also beaten from the flowers of dwarf gorse. Adults have been recorded from April to October.

Status Widespread but local throughout Great Britain, becoming extremely local and infrequently recorded in northern England and Scotland.

Threats Loss of habitat through change in land use. Coastal developments may be a further threat to this species.

NEPHUS BISIGNATUS
A ladybird   **EXTINCT**
Order COLEOPTERA   Family COCCINELLIDAE

Nephus bisignatus (Boheman, 1850).

Distribution Recorded from East Sussex and East Kent. A map is given in Harding *et al.* (1986).

Habitat and ecology Uncertain, though recorded from the coast. Probably predatory. Found on low vegetation. On the Continent, adults have been recorded in May and November.

Status Presumed extinct. Only known from three examples recorded in the 19th century.

Published sources Harding, P.T., Greene, D.M., Preston, C.D., Arnold, H.R. & Eversham, B.C. (1986), Majerus, M. & Kearns, P. (1989), Pope, R.D. (1973), Shirt, D.B., ed. (1987).

NEPHUS QUADRIMACULATUS   VULNERABLE
A ladybird
Order COLEOPTERA   Family COCCINELLIDAE

Nephus quadrimaculatus (Herbst, 1783). Formerly known as: *Scymnus pulchellus* (Herbst, 1797), *Scymnus quadrimaculatus* (Herbst).

Distribution Recorded from "Kent", East Suffolk, "Norfolk" and Cambridgeshire before 1970 and Berkshire, Buckinghamshire and East Suffolk from 1970 onwards. A map is given in Harding *et al.* (1986).

Habitat and ecology Recorded from coniferous woodland and an old church. Predatory, and has been noted feeding on the coccid *Phenacoccus aceris* (Hemiptera). Found on pine, oak and ivy. Adults have been recorded from May to October.

Status Known from very few records and recently noted from just two vice-counties. There is also one old record from "near Manchester". Possibly over-looked because of its small size.

Threats This species may be threatened by the removal of ivy from trees, buildings and walls etc.

Published sources Harding, P.T., Greene, D.M., Preston, C.D., Arnold, H.R. & Eversham, B.C. (1986), Majerus, M. & Kearns, P. (1989), Pope, R.D. (1973), Shirt, D.B., ed. (1987).

PLATYNASPIS LUTEORUBRA   NOTABLE A
A ladybird
Order COLEOPTERA   Family COCCINELLIDAE

Platynaspis luteorubra (Goeze, 1777).

Distribution Recorded from "Cornwall", Dorset, East Sussex, East Kent, Surrey, Hertfordshire, Oxfordshire, East Suffolk, Glamorgan and Nottinghamshire before 1970 and North Devon, West Sussex, East Sussex, East Kent, Berkshire, West Norfolk and Northamptonshire from 1970 onwards. A map is given in Harding *et al.* (1986).

Habitat and ecology Woodland, grassland, hedgerows and coastal shingle. Probably predatory. Recorded at the roots of grass, by beating dead hedgerow shrubs and hawthorn blossom, from under broom bushes and, particularly during winter, from under the bark of firs and willows. Possibly lives in association with ants such as *Lasius niger* (Hymenoptera). A larva has been found in the underground galleries of *L. niger* preying on *Aphis scaliae* (Hemiptera). Adults have been recorded in May, June and August.

Status Very local in southern England, recorded as far north as Nottinghamshire. Also found in South Wales.

Threats Loss of habitat through conversion to other land use. Gravel extraction and coastal developments may be further threats.

Management and conservation The disturbance of coastal shingle should be avoided.

Published sources Collier, M.J. (1990), Fowler, W.W. (1889), Fowler, W.W. & Donisthorpe, H.St J.K. (1913), Harding, P.T., Greene, D.M., Preston, C.D., Arnold, H.R. & Eversham, B.C. (1986), Majerus, M. & Kearns, P. (1989), Pontin, A.J. (1959).

SCYMNUS FEMORALIS   NOTABLE B
A ladybird
Order COLEOPTERA   Family COCCINELLIDAE

Scymnus femoralis (Gyllenhal, 1827). Formerly confused under: *Scymnus pygmaeus* sensu Fowler, 1889 partim not (Fourcroy, 1785), *Scymnus rubromaculatus* sensu auct. Brit. not (Goeze, 1777).

Distribution South East, South, South West, East Anglia, East Midlands, West Midlands, North West, South Wales, South East Scotland and North East Scotland. A map is given in Harding *et al.* (1986).

Habitat and ecology Grassland and woodland rides, particularly on chalky or sandy soils. Predatory. Occurs at the roots of grasses and low-growing plants. Adults have been recorded from March, April, from June to August and in October and November.

Status Widespread but local in southern and eastern England. Very local outside this area, recorded as far north as North East Scotland. This species is difficult to identify and may be confused with *S. auritus*. Consequently, the exact status of this species is hard to assess.

Threats Loss of unimproved grassland through agricultural improvement by reseeding or by the application of fertiliser, or by conversion to arable. Changes in woodland management leading to a loss of open conditions. Lack of, or a change in grazing or cutting regimes may threaten this species.

Management and conservation Grazing or cutting, on a rotational basis, is needed to maintain open conditions.

COCCINELLIDAE

SCYMNUS LIMBATUS   **NOTABLE B**
A ladybird
Order COLEOPTERA   Family COCCINELLIDAE

Scymnus limbatus Stephens, 1832. Formerly known as: *Pullus limbatus* ab. limbatus Stephens, *Scymnus suturalis ab. limbatus* Stephens, *Scymnus suturalis var. limbatus*.

Distribution South East, South, South West, East Anglia, East Midlands, West Midlands, North West, North East, South East Scotland, South West Scotland and North East Scotland. A map is given in Harding *et al.* (1986).

Habitat and ecology Wetlands, including fens and wet woodland. Probably predatory. Associated with willows, sallows and poplars. Also recorded from leaf litter. Adults have been recorded from May to October.

Status Widespread but local in Great Britain, not known from Wales. This species is difficult to identify and may be confused with other species of *Scymnus*. Consequently, the exact status of this species is hard to assess.

Threats Drainage for reasons such as agricultural improvement is the primary cause of the loss of wetland and carr.

Management and conservation Water tables should be maintained at high levels.

SCYMNUS SCHMIDTI   **NOTABLE B**
A ladybird
Order COLEOPTERA   Family COCCINELLIDAE

Scymnus schmidti Fuersch, 1958. Formerly confused under: *Scymnus frontalis* (F., 1787), *Scymnus frontalis var. immaculatus* sensu Joy not Suffrian, 1843, *Scymnus pygmaeus* sensu Fowler, 1889 partim not (Fourcroy, 1785).

Distribution South East, South, South West, East Anglia, North West, North Wales, South West Scotland, South East Scotland and North East Scotland. A map is given in Harding *et al.* (1986).

Habitat and ecology Woodland, sand dunes, downland, coastal limestone grassland and heathland. This species appears to prefer sites on well drained soils. This species has a coastal distribution north of the Bristol - Wash line. Predatory. Adults are found all year round.

Status This species has a disjunct distribution. Local in south-eastern and southern England, also recorded from North Wales to North East Scotland. This species is difficult to identify and can be confused with other species of *Scymnus*. Consequently, the exact status of this species is hard to assess.

Threats Loss of habitat through change in land use.

VIBIDIA DUODECIMGUTTATA
12-spot ladybird   **INSUFFICIENTLY KNOWN**
Order COLEOPTERA   Family COCCINELLIDAE

Vibidia duodecimguttata (Poda, 1761). Formerly known as: *Halyzia duodecimguttata* (Poda).

Distribution Recorded from Dorset, Berkshire, Oxfordshire, West Gloucestershire and Dumfriesshire before 1970. A map is given Harding *et al.* (1986).

Habitat and ecology Possibly associated with woodland. Mycophagous. Beaten from the foliage of trees and shrubs. Adults have been recorded in April and June.

Status Status revised from RDB Appendix (Extinct) in Shirt (1987). Only known in Great Britain from very few records. Last recorded in 1905 from Wychwood, Oxfordshire. Possibly only an immigrant to these shores.

Published sources Atty, D.B. (1983), Campbell, J.M. (1985), Fowler, W.W. (1889), Harding, P.T., Greene, D.M., Preston, C.D., Arnold, H.R. & Eversham, B.C. (1986), Majerus, M. & Kearns, P. (1989), Shirt, D.B., ed. (1987).

AULONIUM TRISULCUM   **NOTABLE A**

Order COLEOPTERA   Family COLYDIIDAE

Aulonium trisulcum (Fourcroy, 1785). Formerly known as: *Aulonium trisulcatum* (Fourcroy, 1785).

Distribution Recorded from Middlesex, Berkshire, East Gloucestershire and South Lincolnshire before 1970 and East Sussex, West Kent, Surrey, South Essex, Middlesex, Buckinghamshire, Bedfordshire, Huntingdonshire, Northamptonshire, East Gloucestershire, West Gloucestershire, Herefordshire, Worcestershire, Staffordshire and Leicestershire & Rutland from 1970 onwards.

Habitat and ecology Broad-leaved woodland and pasture-woodland. Predatory on the larvae and pupae of elm bark beetles in the genus *Scolytus* (Scolytidae). Adults are usually found under bark and have been recorded from June to September. This species has been noted at mercury vapour light traps.

Status Formerly extremely local. The beetle appeared to increase in distribution with the spread of Dutch elm disease. This species may now have declined because of the loss of elms.

Threats Loss of broad-leaved woodland and parkland through, for example, clear-felling and coniferisation. Habitat loss, in particular, through the effects of Dutch elm disease, the felling of trees, removal of dead wood from living trees and the destruction or removal of standing and fallen dead wood for reasons such as forest hygiene, aesthetic tidiness, public safety or for use as fire wood.

Management and conservation Where posible, trees and both fallen and standing dead timber, especially with the bark attached, should be retained. The removal of dead timber from trees should be avoided. Gaps in the age structure of the tree population should be identified and the continuity of the appropriate dead wood habitat ensured by regeneration or suitable planting.

Published sources Atty, D.B. (1983), Fowler, W.W. & Donisthorpe, H.St J.K. (1913), Hodge, P.J. (1977e).

CICONES UNDATUS   **ENDANGERED**

Order COLEOPTERA   Family COLYDIIDAE

Cicones undatus Guerin Mcneville, 1829. Formerly known as: *Cicones undata* Guerin-Meneville, 1829.

Distribution Recorded from Berkshire from 1970 onwards.

Habitat and ecology Pasture-woodland. Associated with fungus-infected bark or wood. So far, only recorded under bark and in dry, dead wood of sycamore in company with *Synchita separanda* (Colydiidae). Adults have been recorded from April to June.

Status Not listed in the insect Red Data Book (Shirt, 1987). Only known from a few dead sycamore trees in Windsor Great Park, Berkshire. First recorded in 1984.

Threats Habitat loss, in particular, through the felling of trees, removal of dead wood from living trees and the destruction or removal of standing and fallen dead wood for reasons such as forest hygiene, aesthetic tidiness, public safety or for use as fire wood.

Management and conservation Trees and both fallen and standing dead timber, especially with the bark attached, should be retained. The removal of dead timber from trees should be avoided. Gaps in the age structure of the tree population should be identified and the continuity of the appropriate dead wood habitat ensured through some regeneration. Parts of Windsor Great Park are notified as an SSSI.

CICONES VARIEGATUS   **NOTABLE A**

Order COLEOPTERA   Family COLYDIIDAE

Cicones variegatus (Hellwig, 1792). Formerly known as: *Cicones variegata* (Hellwig, 1792).

Distribution Recorded from South Hampshire, North Hampshire, West Sussex, East Kent, West Kent, Surrey, South Essex, North Essex, Hertfordshire, Berkshire and East Gloucestershire before 1970 and South Hampshire, West Sussex, Surrey, Hertfordshire, Berkshire, Oxfordshire and Buckinghamshire from 1970 onwards.

Habitat and ecology Ancient broad-leaved woodland and pasture-woodland. Associated with dry, decayed bark or wood, particularly if infected with the fungus *Ustulina vulgaris*. The beetle is associated with beech and hornbeam, a single example has been found under the bark of a dead sycamore in company with *C. undatus*. Adults have been recorded from March to July, and in September and October.

Status Very local and only known in southern England, north to Oxfordshire. Recently recorded from seven vice-counties. Can be numerous where found.

Indicator status Grade 2 in Harding & Rose (1986).

Threats Loss of broad-leaved woodland and parkland through, for example, clear-felling and coniferisation. Habitat loss, in particular, through the felling of ancient and fungus-infected trees, removal of dead wood from living trees and the destruction or removal of standing and fallen dead wood for reasons such as forest hygiene, aesthetic tidiness, public safety or for use as fire wood.

Management and conservation Ancient and fungus-infected trees, and both fallen and standing dead timber, especially with the bark attached, should be retained. The removal of dead timber from ancient and fungus-infected trees should be avoided. Gaps in the age structure of the tree population should be identified and the continuity of the appropriate dead wood habitat ensured by regeneration, suitable planting and possibly with pollarding.

COLYDIIDAE

Published sources Atty, D.B. (1983), Buck, F.D. (1955), Cooter, J. (1972), Donisthorpe, H.St J.K. (1939a), Eagles, T.R. (1953a), Fowler, W.W. (1889), Fowler, W.W. & Donisthorpe, H.St J.K. (1913), Halstead, A.J. (1988), Hammond, P.M. (1979), Harding, P.T. (1978).

COLYDIUM ELONGATUM   RARE

Order COLEOPTERA   Family COLYDIIDAE

Colydium elongatum (F., 1787).

Distribution Recorded from North Wiltshire, Dorset and South Hampshire before 1970 and North Wiltshire, South Wiltshire, South Hampshire, North Hampshire, Surrey and Berkshire from 1970 onwards.

Habitat and ecology Ancient broad-leaved woodland and pasture-woodland. Lives in the burrows of, and possibly predatory on, wood-boring beetles in the family Scolytidae, in particular *Platypus cylindrus*, *Xyloterus signatus* and *X. domesticus*, mainly in beech and oak, though also in birch and on sycamore. This species has also been found under the bark of burnt trunks. Adults have been recorded from March to September.

Status Very local and only known from southern England. Recently recorded from six vice-counties. This species is unlikely to have been overlooked in the past and its recent appearance in some historically well-worked sites may represent a spread in distribution.

Indicator status Grade 1 in Harding & Rose (1986).

Threats Loss of broad-leaved woodland and parkland through, for example, clear-felling and coniferisation. Habitat loss, in particular, through the felling of ancient trees, removal of dead wood from living trees and the destruction or removal of standing and fallen dead wood for reasons such as forest hygiene, aesthetic tidiness, public safety or for use as fire wood.

Management and conservation Ancient trees and both fallen and standing dead timber, especially with the bark attached, should be retained. The removal of dead timber from ancient trees should be avoided. Gaps in the age structure of the tree population should be identified and the continuity of the appropriate dead wood habitat ensured by regeneration, suitable planting and possibly with pollarding.

Published sources Halstead, A. (1990), Harding, P.T. (1978), Harding, P.T. & Rose, F. (1986), Harwood, P. (1956), MacKechnie-Jarvis, C. (1976), Menzies, I.S. (1990), Menzies, I.S., Nash, D.R. & Owen, J.A. (1991), Porter, D.A. (1989), Shirt, D.B., ed. (1987).

ENDOPHLOEUS MARKOVICHIANUS
ENDANGERED

Order COLEOPTERA   Family COLYDIIDAE

Endophloeus markovichianus (Piller and Mitterpacher, 1783). Formerly known as: *Endophloeus spinulosus* (Latreille, 1807).

Distribution Recorded from South Hampshire before 1970.

Habitat and ecology Ancient broad-leaved woodland. Possibly associated with wood-boring beetles such as *Leptura scutellata* (Cerambycidae). Found under bark and in dead wood of beech. Adults have been recorded in March and August.

Status Not listed in the insect Red Data Book (Shirt, 1987). Only known from the New Forest, South Hampshire, with only one record this century, in 1927.

Threats Loss of broad-leaved woodland through, for example, clear-felling and coniferisation. Habitat loss, in particular, through the felling of ancient trees, removal of dead wood from living trees and the destruction or removal of standing and fallen dead wood for reasons such as forest hygiene, aesthetic tidiness, public safety or for use as fire wood.

Management and conservation Ancient trees and both fallen and standing dead timber, especially with the bark attached, should be retained. The removal of dead timber from ancient trees should be avoided. Gaps in the age structure of the tree population should be identified and the continuity of the appropriate dead wood habitat ensured by regeneration, suitable planting and possibly with pollarding. Much of the New Forest is notified as an SSSI.

Published sources Fowler, W.W. (1889).

LANGELANDIA ANOPHTHALMA   RARE

Order COLEOPTERA   Family COLYDIIDAE

Langelandia anophthalma Aubé, 1843.

Distribution Recorded from South Devon, East Kent, West Kent and Berkshire before 1970 and "Cornwall" and East Sussex from 1970 onwards.

Habitat and ecology Subterranean. Probably associated with decaying underground vegetation. Early records are from seed potatoes, while later records include examples from wood mould at the roots of a blown elm, under a log and under a railway sleeper at the base of a coastal cliff. Adults have been recorded in February, from May to August and in October.

Status Not listed in the insect Red Data Book (Shirt, 1987). Very local and only recorded in southern England as far north as Berkshire. Recently recorded from just two vice-counties.

Published sources Allen, A.A. (1954a), Booth, R.G. (1977), Donisthorpe, H.St J.K. (1936a), Wood, T. (1886).

MYRMECHIXENUS SUBTERRANEUS
INDETERMINATE
Order COLEOPTERA   Family COLYDIIDAE

Myrmechixenus subterraneus Chevrolat, 1835. Formerly known as: *Myrmecoxenus subterraneus* (Chevrolat).

Distribution Recorded from South Hampshire and Derbyshire before 1970.

Habitat and ecology Woodland and possibly heathland. Myrmecophilous. Associated with the wood ants *Formica rufa* and *F. lugubris* (Hymenoptera). Adults have been recorded in May, June and October.

Status Not listed in the insect Red Data Book (Shirt, 1987). Only recorded from two localities and on a total of just three occasions; in 1920 and 1930 from Hurn, near Ringwood, South Hampshire, and in 1956 from near Cromford, Derbyshire.

Threats Uncertain, though the loss of habitat through the clear-felling of woodland and conversion to other land use may threaten this species. The loss and fragmentation of heathland through improvement and conversion to arable agriculture, development and afforestation. Natural succession may be a further threat.

Management and conservation Open glades and ride margins should be managed to retain a variety of vegetation structures. Grazing or cutting, on a rotational basis, may be needed to maintain open conditions.

Published sources Collingwood, C.A. (1957), Donisthorpe, H.St J.K. (1930a), Harwood, P. (1930).

MYRMECHIXENUS VAPORARIORUM   RARE
Order COLEOPTERA   Family COLYDIIDAE

Myrmechixenus vaporariorum Guerin-Meneville, 1843. Formerly known as: *Myrmecoxenus vaporariorum* (Guerin-Meneville).

Distribution Recorded from North Somerset, West Kent, Surrey, Middlesex, Berkshire, Oxfordshire, Warwickshire, Cheshire and Mid-west Yorkshire before 1970 and Northamptonshire and Worcestershire from 1970 onwards.

Habitat and ecology Probably associated with dung heaps and vegetable refuse. Recorded from an ants nest. Adults have been noted in July and August.

Status Not listed in the insect Red Data Book (Shirt, 1987). Widely scattered distribution in England, recorded from West Kent to Mid-west Yorkshire. Recently recorded from just two vice-counties.

Published sources Drane, A.B. (1990), Fowler, W.W. (1889), Fowler, W.W. & Donisthorpe, H.St J.K. (1913).

ORTHOCERUS CLAVICORNIS   NOTABLE B
Order COLEOPTERA   Family COLYDIIDAE

Orthocerus clavicornis (L., 1758). Formerly known as: *Orthocerus muticus* (L., 1767).

Distribution South East, South, South West, East Anglia, East Midlands, North East, North West, South Wales, South East Scotland, South West Scotland and North West Scotland.

Habitat and ecology Sandy areas, in particular sand dunes. Also found on heathland. Associated with *Peltigera* lichens, perhaps with a preference for *P. canina*. Adults have been recorded from April to June and in August and October.

Status Widely distributed and local in Great Britain.

Threats Loss of dune and heathland, particularly through afforestation, urban and holiday development. Improvement and conversion to arable agriculture, and natural succession may be further threats. The degradation of remaining habitat by excessive disturbance of the vegetation through activities such as motorbike access, horse-riding and human trampling.

Management and conservation Rabbit grazing, or possibly some other disturbance, may be desirable to

223

COLYDIIDAE

maintain the early successional stages and prevent the invasion of scrub.

OXYLAEMUS CYLINDRICUS   EXTINCT

Order COLEOPTERA   Family COLYDIIDAE

Oxylaemus cylindricus (Panzer, 1796).

Distribution Recorded from South Hampshire and Nottinghamshire.

Habitat and ecology Ancient broad-leaved woodland. Probably predatory on other beetles in dead wood.

Status Presumed extinct. Last recorded in the 19th century from the New Forest, South Hampshire.

Indicator status Grade 2 in Harding & Rose (1986).

Published sources Carr, J.W. (1916), Garland, S.P. (1983), Harding, P.T. (1978), Harding, P.T. & Rose, F. (1986), Shirt, D.B., ed. (1987).

OXYLAEMUS VARIOLOSUS   RARE

Order COLEOPTERA   Family COLYDIIDAE

Oxylaemus variolosus (Dufour, 1843).

Distribution Recorded from East Sussex, West Kent, Surrey, Middlesex and Berkshire before 1970 and South Hampshire, Huntingdonshire and Worcestershire from 1970 onwards.

Habitat and ecology Ancient broad-leaved woodland. Probably predatory on other beetles in dead wood. The beetle has been found in litter at the base of a tree stump and also recorded in the fungus *Collybia fusipes* growing at the base of a large American red oak. Adults have been noted in May and June.

Status Only known from South Hampshire to West Kent and as far north as Huntingdonshire. Recently recorded from just three vice-counties.

Indicator status Grade 2 in Harding & Rose (1986).

Threats Loss of broad-leaved woodland through, for example, clear-felling and coniferisation. Habitat loss, in particular, through the felling of ancient trees, removal of dead wood from living trees and the destruction or removal of standing and fallen dead wood for reasons such as forest hygiene, aesthetic tidiness, public safety or for use as fire wood.

Management and conservation Ancient trees and both fallen and standing dead timber, especially with the bark attached, should be retained. The removal of dead timber from ancient trees should be avoided. Gaps in the age structure of the tree population should be identified and the continuity of the appropriate dead wood habitat ensured by regeneration, suitable planting and possibly with pollarding.

Published sources Anon, (1972), Fowler, W.W. (1889), Fowler, W.W. & Donisthorpe, H.St J.K. (1913), Harding, P.T. (1978), Harding, P.T. & Rose, F. (1986), Shirt, D.B., ed. (1987).

SYNCHITA HUMERALIS   NOTABLE B

Order COLEOPTERA   Family COLYDIIDAE

Synchita humeralis (F., 1792). Formerly known as: *Synchita juglandis* (Hellwig, 1792).

Distribution South, South East, East Anglia, East Midlands, West Midlands, North West, North East, South East Scotland and North East Scotland.

Habitat and ecology Broad-leaved woodland and pasture-woodland. Also recorded from a wooded fen and a sand dune. Under and in fungus-infected bark or wood. It has been found in the fungus *Daldinia concentrica* growing on birch and, occasionally, on beech. Recorded from a variety of other trees including hawthorn, hazel, alder and also found in fallen pine cones. Adults have been noted from March to September.

Status Widely distributed and very local. Recorded from South Hampshire, north to East Inverness and Nairn in Scotland.

Indicator status Grade 2 in Garland (1983). Grade 3 in Harding & Rose (1986).

Threats Loss of broad-leaved woodland and parkland through, for example, clear-felling and conversion to other land use. The loss of dunes through coastal developments and afforestation may be a further threats to this species. Habitat loss, in particular, through the felling of fungus-infected trees, removal of dead wood from living trees and the destruction or removal of standing and fallen dead wood for reasons such as forest hygiene, aesthetic tidiness, public safety or for use as fire wood.

Management and conservation Fungus-infected trees and both fallen and standing dead timber, especially with the bark attached, should be retained. The removal of dead timber from fungus-infected trees should be avoided. Gaps in the age structure of the tree population should be identified and the continuity of the appropriate habitat ensured by regeneration, suitable planting and possibly with pollarding.

SYNCHITA SEPARANDA   RARE

Order COLEOPTERA   Family COLYDIIDAE

Synchita separanda (Reitter, 1882).

Distribution Recorded from West Kent and Berkshire before 1970 and West Kent, Surrey and Berkshire from 1970 onwards.

Habitat and ecology Ancient broad-leaved woodland and pasture-woodland, and also found in a garden. Associated with fungus-infected bark and wood. Old records are mainly from beech. Recently recorded under sycamore bark in company with *Cicones undatus* (Colydiidae). On the Continent, this species has also been recorded from lime. In Britain, adults have been noted from May to September.

Status Only known from three sites: Windsor Forest and Great Park, Berkshire; Knole Park, near Sevenoaks, West Kent; and a garden in Peckham, West Kent. First added to the British list in 1964. Formerly confused with *Synchita humeralis*, some examples of which have been found to refer to this species. Can be common where found.

Indicator status Grade 1 in Harding & Rose (1986).

Threats Loss of broad-leaved woodland and parkland through, for example, clear-felling and coniferisation. Habitat loss, in particular, through the felling of fungus-infected trees, removal of dead wood from living trees and the destruction or removal of standing and fallen dead wood for reasons such as forest hygiene, aesthetic tidiness, public safety or for use as fire wood.

Management and conservation Fungus-infected trees and both fallen and standing dead timber, especially with the bark attached, should be retained. The removal of dead timber from fungus-infected trees should be avoided. Gaps in the age structure of the tree population should be identified and the continuity of the appropriate habitat ensured by regeneration, suitable planting and possibly with pollarding. Knole Park, Windsor Forest and parts of Windsor Great Park are notified as SSSIs.

Published sources Allen, A.A. (1964a), Harding, P.T. (1978), Harding, P.T. & Rose, F. (1986), Jones, R.A. (1987), Shirt, D.B., ed. (1987).

TEREDUS CYLINDRICUS   ENDANGERED

Order COLEOPTERA   Family COLYDIIDAE

Teredus cylindricus (Olivier, 1790). Formerly known as: *Teredus nitidus* (F., 1792).

Distribution Recorded from Berkshire and Nottinghamshire before 1970 and Berkshire from 1970 onwards.

Habitat and ecology Ancient broad-leaved woodland and pasture-woodland. Probably predatory on other beetles. Recorded in and under the bark of old oaks infested with beetles of the genera *Xestobium*, *Anobium*, *Ptilinus* (Anobiidae) or *Dryocoetes* (Scolytidae), or the ant *Lasius brunneus* (Hymenoptera). Also recorded from the fungus *Polyporus sulphureus* growing on oak, under the bark of sycamore and from sweet chestnut. On the Continent, this species has been recorded from beech Adults have been noted in April, June, July, August and October.

Status Only known from two areas: Sherwood Forest, Nottinghamshire, where it was last recorded in 1964, and from Windsor Forest and Windsor Great Park, and also Silwood Park, in Berkshire.

Indicator status Grade 1 in Harding & Rose (1986).

Threats Loss of broad-leaved woodland and parkland through, for example, clear-felling and coniferisation. Habitat loss, in particular, through the felling of ancient trees, removal of dead wood from living trees and the destruction or removal of standing and fallen dead wood for reasons such as forest hygiene, aesthetic tidiness, public safety or for use as fire wood.

Management and conservation Ancient trees and both fallen and standing dead timber, especially with the bark attached, should be retained. The removal of dead timber from ancient trees should be avoided. Gaps in the age structure of the tree population should be identified and the continuity of the appropriate dead wood habitat ensured by regeneration, suitable planting and possibly with pollarding. Windsor Forest and parts of Windsor Great Park are notified as SSSIs. Much of Sherwood Forest is notified as the Birklands and Bilhaugh SSSI.

Published sources Bedwell, E.C. (1926b), Copestake, D.R. (1990b), Harding, P.T. (1978), Harding, P.T. & Rose, F. (1986), Shirt, D.B., ed. (1987).

CORYLOPHIDAE

ORTHOPERUS AEQUALIS
INSUFFICIENTLY KNOWN
Order COLEOPTERA   Family CORYLOPHIDAE

Orthoperus aequalis Sharp, 1885. Formerly known as: *Orthoperus nitidulus* Allen, 1942.

Distribution Recorded from West Sussex and Berkshire before 1970 and East Sussex, Surrey and South Lancashire from 1970 onwards.

Habitat and ecology Ancient broad-leaved woodland and pasture-woodland. Recorded under the bark of a freshly sawn beech, on beech logs, on the cut ends of felled oak trees and from loose bark of fire damaged sycamore. The adults have been recorded in June, August and November.

Status Not listed in the insect Red Data Book (Shirt, 1987). Only recorded from four vice-counties, all in England. Recently recorded from just three vice-counties. This species is difficult to identify and may be confused with other members of the genus. Consequently, the exact status of this species is hard to assess.

Threats Uncertain, though this species may be threatened by loss of broad-leaved woodland and parkland through, for example, clear-felling and coniferisation.

Published sources Allen, A.A. (1970d), Cooter, J. (1977a).

ORTHOPERUS BRUNNIPES   RARE

Order COLEOPTERA   Family CORYLOPHIDAE

Orthoperus brunnipes (Gyllenhal, 1808). Formerly confused under: *Orthoperus brunnipes* sensu auct. partim, *Orthoperus kluki* Wankowicz, 1865.

Distribution Recorded from South Devon, South Hampshire, East Kent, West Kent, Surrey, South Essex, Oxfordshire, East Suffolk, West Suffolk, "Norfolk", Cambridgeshire, Huntingdonshire, Nottinghamshire, South-east Yorkshire and Roxburghshire before 1970 and West Kent, Berkshire and South-east Yorkshire from 1970 onwards.

Habitat and ecology Wetlands, fens, marshy or damp places and has been found in pasture-woodland. Recorded from reed beds and by sweeping grass under oak trees. Probably also occurs amongst vegetable debris, moss etc. Adults have been recorded in June and August.

Status Widespread and very local. Recorded from southern England with scattered records north to Roxburghshire in Scotland. Recently found in just three vice-counties. This species is difficult to identify and may be confused with other members of the genus. Consequently, the exact status of this species is hard to assess.

Threats Drainage for reasons such as agricultural improvement and development is the primary cause of the loss of fens. Falling water tables because of water abstraction and river engineering schemes may also threaten this species.

Management and conservation Water tables should be maintained at high levels. Reed cutting should be undertaken on rotation.

Published sources Allen, A.A. (1970d), Carr, J.W. (1916), Flint, J.H. (1988), Fowler, W.W. (1889), Fowler, W.W. & Donisthorpe, H.St J.K. (1913), Shirt, D.B., ed. (1987).

ORTHOPERUS NIGRESCENS   NOTABLE B

Order COLEOPTERA   Family CORYLOPHIDAE

Orthoperus nigrescens Stephens, 1829. Formerly known as: *Orthoperus coriaceus* sensu auct. Brit. not Mulsant and Rey, 1861.

Distribution South, South West, East Anglia and Dyfed-Powys.

Habitat and ecology Broad-leaved woodland and also recorded from a garden. Found in damp situations from under oaks and other deciduous trees, recorded from moss on oaks, from traveller's-joy, in the fungus *Piptoporus betulinus*, in cut grass, from red rot in a coppiced ash tree and from spruce. Adults have been recorded in March, from June to September and in December.

Status Local in parts of southern England and recently recorded in mid Wales. This species is difficult to identify and may be confused with other members of the genus. Consequently, the exact status of this species is hard to assess.

Threats Uncertain, though this species may be threatened by the loss of broad-leaved woodland through, for example, clear-felling and conversion to other land use.

RYPOBIUS RUFICOLLIS
INSUFFICIENTLY KNOWN
Order COLEOPTERA   Family CORYLOPHIDAE

Rypobius ruficollis (du Val, 1854).

Distribution Recorded from Dorset and West Sussex before 1970.

Habitat and ecology Probably associated with damp or marshy habitats near the coast. Probably occurs amongst low growing plants or in decaying vegetable debris. Adults have been recorded in August and September.

Status Status revised from RDB 3 (Rare) in Shirt (1987). Only known from three examples in Great Britain; a singleton found in 1919 from Chichester, West Sussex, and two noted in 1931 from Studland, Dorset. This species is difficult to identify and may be confused with other members of the family. Consequently, the exact status of this species is hard to assess.

Threats Uncertain, though this species may be threatened by drainage for reasons such as agricultural improvement and development. Coastal developments may be a further threat.

Published sources Hammond, P.M. (1971), Shirt, D.B., ed. (1987).

CRYPTOLESTES SPARTII   NOTABLE A
A flat bark beetle
Order COLEOPTERA   Family CUCUJIDAE

Cryptolestes spartii (Curtis, 1834). Formerly known as: *Laemophloeus ater* (Olivier, 1795).

Distribution Recorded from "Wilts", Isle of Wight, South Hampshire, East Kent, West Kent, Surrey, Berkshire, West Suffolk and West Gloucestershire before 1970 and East Kent, East Suffolk, West Suffolk and West Norfolk from 1970 onwards.

Habitat and ecology Coastal shingle and probably heathland, scrub, roadside verges and hedgebanks. Found in the dead stems of broom *Cytisus scoparius*. This species probably preys on the beetle *Hylastinus obscurus* (Scolytidae). Also recorded under the bark of elm, beech, gorse, from burnt pine and in a granary. Adults have been recorded from April to July and in October.

Status Only known from southern England and recorded as far north as West Norfolk. Recently recorded from only four vice-counties.

Threats Loss of coastal shingle through coastal developments and gravel extraction. This species may also be threatened by the loss of unimproved grassland and heathland through improvement by reseeding or by the application of fertilisers, or by conversion to arable agriculture, afforestation, development and natural succession.

Management and conservation Disturbance of coastal shingle should be avoided. On other sites, grazing, cutting or some other disturbance may be needed to maintain open conditions. Gaps in the age structure of the broom population should be identified and the continuity of the appropriate habitat ensured by regeneration.

Published sources Atty, D.B. (1983), Fowler, W.W. (1889), Fowler, W.W. & Donisthorpe, H.St J.K. (1913).

DENDROPHAGUS CRENATUS   NOTABLE B
A flat bark beetle
Order COLEOPTERA   Family CUCUJIDAE

Dendrophagus crenatus (Paykull, 1799).

Distribution Scotland.

Habitat and ecology Ancient pine forests, birch woodland, oak woods, spruce and probably pine plantations. Larvae and adults are found under loose bark of pine, oak, beech, birch, alder and spruce. The larvae are fungivorous and have been recorded feeding on a fungus identified as *Ceratocystis radicicola*. Larvae probably overwinter. Larvae found in June produced adults in July. Adults have been recorded from June to August.

Status Local and only known in Scotland.

Threats Loss of woodland through, for example, clear-felling and conversion to other land use. Habitat loss in the remaining areas through the felling of trees and the destruction or removal of fungus-infected standing and fallen dead wood for reasons such as forest hygiene, aesthetic tidiness, public safety or for use as fire wood.

Management and conservation Fungus-infected trees and both standing and fallen dead timber, especially with the bark attached, should be retained. Gaps in the age structure of the tree population should be identified and regeneration encouraged to ensure the continuity of the dead wood habitats.

CUCUJIDAE

LAEMOPHLOEUS MONILIS   ENDANGERED
A flat bark beetle
Order COLEOPTERA   Family CUCUJIDAE

Laemophloeus monilis (F., 1787).

Distribution Recorded from West Sussex and Berkshire before 1970 and West Sussex from 1970 onwards.

Habitat and ecology Ancient broad-leaved woodland and pasture-woodland. Recorded from under the bark and cut ends of beech. On the Continent, has also been found under the bark of plane, from the cones of conifers, in the burrows of the bark beetle *Taphrorychus bicolor* (Scolytidae), on lime trees and under the bark of dead lime. In Britain, adults have been recorded in May and from July to October.

Status Known from just two localities and recorded on very few occasions.

Indicator status Grade 1 in Harding & Rose (1986).

Threats Loss of broad-leaved woodland and parkland through, for example, clear-felling and coniferisation. Habitat loss, in particular, through the felling of ancient beech trees, removal of dead wood from living trees and the destruction or removal of standing and fallen dead wood for reasons such as forest hygiene, aesthetic tidiness, public safety or for use as fire wood.

Management and conservation Ancient beech trees and both fallen and standing dead timber, especially with the bark attached, should be retained. The removal of dead timber from ancient trees should be avoided. Gaps in the age structure of the beech tree population should be identified and the continuity of the appropriate dead wood habitat ensured by regeneration, suitable planting and possibly with pollarding.

Published sources Harding, P.T. (1978), Harding, P.T. & Rose, F. (1986), Shirt, D.B., ed. (1987).

LEPTOPHLOEUS CLEMATIDIS   ENDANGERED
A flat bark beetle
Order COLEOPTERA   Family CUCUJIDAE

Leptophloeus clematidis (Erichson, 1846). Formerly known as: *Laemophloeus clematidis* Erichson.

Distribution Recorded from West Kent and Oxfordshire before 1970 and East Suffolk from 1970 onwards.

Habitat and ecology Woodland and hedgerows. Associated with small, dead stems of traveller's joy

Clematis vitalba and a predator on the bark beetle *Xylocleptes bispinus* (Scolytidae). Adults are usually found under the loose bark of the foodplant or occasionally in the burrows of the bark beetle within the stem, and appear to be confined to one or two stems only, per plant. Adults have been recorded in April.

Status Status revised from RDB 2 (Vulnerable) in Shirt (1987). Formerly known from three localities in North Kent and one site in Oxfordshire. Recently discovered at a single site in East Suffolk.

Threats Loss of habitat for reasons such as development, afforestation and improvement and conversion to arable agriculture. The selective removal of dead traveller's joy may be a further threat.

Management and conservation Gaps in the age structure of the foodplant population should be identified and regeneration encouraged to ensure the continuity of the appropriate habitats.

Published sources Shirt, D.B., ed. (1987).

NOTOLAEMUS UNIFASCIATUS   NOTABLE A
A flat bark beetle
Order COLEOPTERA   Family CUCUJIDAE

Notolaemus unifasciatus (Latreille, 1804). Formerly known as: *Laemophloeus bimaculatus* (Paykull, 1801).

Distribution Recorded from South Hampshire, East Kent, West Kent, Surrey, South Essex, Middlesex, Berkshire, East Suffolk and Worcestershire before 1970 and South Hampshire, Surrey, Middlesex, Berkshire, East Suffolk, Northamptonshire, Herefordshire and Leicestershire & Rutland from 1970 onwards.

Habitat and ecology Ancient broad-leaved woodland and pasture-woodland. Found under the bark of dead beech, and also associated with oak and hornbeam. This species has also been found on turkey oak. Adults have been recorded from May to July.

Status Status revised from RDB 3 (Rare) in Shirt (1987). Very local in southern England and recorded as far north as Leicestershire. Recently recorded from seven vice-counties.

Indicator status Grade 2 in Harding & Rose (1986).

Threats Loss of broad-leaved woodland and parkland through, for example, clear-felling and coniferisation. Habitat loss, in particular, through the felling of ancient trees, removal of dead wood from living trees and the destruction or removal of standing and fallen dead

wood for reasons such as forest hygiene, aesthetic
tidiness, public safety or for use as fire wood.

Management and conservation Ancient trees and both
fallen and standing dead timber, especially with the
bark attached, should be retained. The removal of dead
timber from ancient trees should be avoided. Gaps in
the age structure of the tree population should be
identified and the continuity of the appropriate dead
wood habitat ensured by regeneration, suitable planting
and possibly with pollarding.

Published sources Fowler, W.W. (1889), Fowler,
W.W. & Donisthorpe, H.St J.K. (1913), Harding, P.T.
(1978), Harding, P.T. & Rose, F. (1986), Shirt, D.B.,
ed. (1987), Skidmore, P. (1972).

PEDIACUS DEPRESSUS	NOTABLE A

A flat bark beetle
Order COLEOPTERA　　　Family CUCUJIDAE

Pediacus depressus (Herbst, 1797).

Distribution Recorded from North Somerset, South
Hampshire, North Hampshire, East Sussex, West Kent,
Surrey, North Essex, Middlesex, Berkshire,
Buckinghamshire, West Suffolk, East Norfolk,
Cambridgeshire, Huntingdonshire, Cheshire, South
Lancashire, North-east Yorkshire and Stirlingshire
before 1970 and East Sussex, Surrey, West Suffolk,
South Lancashire, North-east Yorkshire and
Stirlingshire from 1970 onwards.

Habitat and ecology Ancient broad-leaved woodland,
pasture-woodland and also recorded from conifers.
Found under oak bark, at sap and under sappy bark of
pine and from cut larch boards. The beetle has also
been recorded from burrows of the goat moth *Cossus
cossus* (Lepidoptera) in oak. Adults and larvae are
probably fungivorous having been found on a number
of occasions in the presence of a yellow fungus. Adults
have been recorded from June to August and in
October.

Status Very local and widely scattered in England and
also recorded from Stirlingshire in Scotland. Recently
recorded from six, widely scattered, vice-counties. This
species is difficult to identify and may be confused
with *P. dermestoides*. Consequently, the exact status of
this species is hard to assess.

Indicator status Grade 2 in Harding & Rose (1986).

Threats Loss of woodland through, for example,
clear-felling and conversion to other land use. Habitat
loss, in particular, through the felling of ancient trees,
removal of dead wood from living trees and the

destruction or removal of standing and fallen dead
wood for reasons such as forest hygiene, aesthetic
tidiness, public safety or for use as fire wood.

Management and conservation Ancient trees and both
fallen and standing dead timber, especially with the
bark attached, should be retained. The removal of dead
timber from ancient trees should be avoided. Gaps in
the age structure of the tree population should be
identified and the continuity of the appropriate dead
wood habitat ensured by regeneration, suitable planting
and possibly with pollarding.

Published sources Allen, A.A. (1956b), Allen, A.A.
(1957e), Anon, (1935), Cooter, J. (1977d), Fowler,
W.W. (1889), Fowler, W.W. & Donisthorpe, H.St J.K.
(1913), Harding, P.T. (1978), Harding, P.T. & Rose, F.
(1986), Johnson, C. (1963c), Moore, B.P. (1958).

ULEIOTA PLANATA	NOTABLE A

A flat bark beetle
Order COLEOPTERA　　　Family CUCUJIDAE

Uleiota planata (L., 1761). Formerly known as:
Brontes planatus (L.).

Distribution Recorded from South Hampshire, North
Hampshire, West Kent, Berkshire and Cumberland
before 1970 and North Hampshire, East Sussex, West
Kent, Surrey, South Essex, Berkshire, East
Gloucestershire, Carmarthenshire and South Lancashire
from 1970 onwards.

Habitat and ecology Ancient broad-leaved woodland.
Occurs under the bark of beech. Also recorded from
wych elm, oak, silver birch, sweet chestnut, sycamore
and on one occasion from walnut. The larvae also
occur under bark and are probably fungivorous. Larvae
have been found from August to October, pupae have
been noted in October. Adults have been recorded in
January, February, from May to October and in
December.

Status Status revised from RDB 2 (Vulnerable) in Shirt
(1987). Recently recorded from nine vice-counties from
West Kent to South Lancashire. There is an old record
from South Aberdeen which is considered doubtful and
probably refers to *Dendrophagus crenatus*.

Indicator status Grade 1 in Harding & Rose (1986).

Threats Loss of broad-leaved woodland through, for
example, clear-felling and coniferisation. Habitat loss,
in particular, through the felling of ancient trees,
removal of dead wood from living trees and the
destruction or removal of standing and fallen dead

CURCULIONIDAE

wood for reasons such as forest hygiene, aesthetic tidiness, public safety or for use as fire wood.

Management and conservation Ancient trees and both fallen and standing dead timber, especially with the bark attached, should be retained. The removal of dead timber from ancient trees should be avoided. Gaps in the age structure of the tree population should be identified and the continuity of the appropriate dead wood habitat ensured by regeneration, suitable planting and possibly with pollarding.

Published sources Alexander, K.N.A. & Clements, D.K. (1988), Harding, P.T. (1978), Harding, P.T. & Rose, F. (1986), Hodge, P.J. (1990), Pavett, P.M. (1987), Pavett, P.M. (1988), Shirt, D.B., ed. (1987).

ACALLES PTINOIDES   NOTABLE B
A Weevel
Order COLEOPTERA   Family CURCULIONIDAE

Acalles ptinoides (Marsham, 1802).

Distribution England, Wales, South East Scotland, South West Scotland and North West Scotland.

Habitat and ecology Broad-leaved and mixed woodland, also recorded on heathland. Phytophagous. Often recorded from leaf-litter and from under heather. Also from old hawthorn and hazel, in rotten beech and under beech bark. On the Continent, this species has also been associated with oak. In Britain, adults have been recorded from February to November.

Status Widespread but local in Great Britain, not recorded in North East Scotland.

Threats Loss, or fragmentation, of heathland through changes in land use, mainly by conversion to arable, forestry and urban development. Loss of woodland through clear-felling and conversion to other land use. The removal of dead wood from living trees and the destruction or removal of standing and fallen dead wood for reasons such as forest hygiene, aesthetic tidiness, public safety or for use as fire wood may be a further threat.

Management and conservation Heathland should be managed to retain the range of successional stages. In woodland, open glades and ride margins should be cut on rotation to retain a variety of vegetation structures.

ACALLES ROBORIS   NOTABLE B
A weevil
Order COLEOPTERA   Family CURCULIONIDAE

Acalles roboris Curtis, 1835.

Distribution England, Dyfed-Powys, South Wales, South East Scotland and South West Scotland.

Habitat and ecology Broad-leaved woodland. Phytophagous. Often recorded from oak leaf-litter. Also noted from gnarled oak boughs. On the Continent, this species is also associated with beech. In Britain, adults have been recorded from March to July and in September.

Status Widespread but local in England, also recorded in parts of Wales and southern Scotland. Possibly under-recorded.

Threats Loss of broad-leaved woodland through, for example, clear-felling and coniferisation.

ACALYPTUS CARPINI   NOTABLE B
A weevil
Order COLEOPTERA   Family CURCULIONIDAE

Acalyptus carpini (F., 1792).

Distribution South East, South, South West, East Anglia and East Midlands.

Habitat and ecology Wetland, particularly fens and bogs. Phytophagous. Associated with species of *Salix*. Adults have been recorded from April to August.

Status Widespread but local in southern England.

Threats Drainage for reasons such as agricultural improvement and development is the primary cause of the loss of wetland. Falling water tables because of water abstraction and river engineering schemes may also threaten this species.

Management and conservation Water tables should be maintained at high levels. Gaps in the age structure of the tree population should be identified and the continuity of the appropriate habitat ensured by regeneration.

ANCHONIDIUM UNGUICULARE VULNERABLE
A weevil
Order COLEOPTERA   Family CURCULIONIDAE

Anchonidium unguiculare (Aubé, 1850).

Distribution Recorded from West Cornwall and South Devon before 1970 and West Cornwall and South Devon from 1970 onwards.

Habitat and ecology Sessile oak woodland and sea cliffs. On acidic soils. Phytophagous. Found in leaf litter and mosses. Adults have been recorded in from March to September.

Status Very local and recorded from only two vice-counties.

Threats Loss of broad-leaved woodland through, for example, clear-felling and conversion to other land use. This species may also be threatened by cliff stabilisation schemes and the construction of sea defences.

Management and conservation Large areas of cliff are required so that the population does not become isolated and subsequently threatened by individual landslips.

Published sources Alexander, K.N.A. (1986), Bedwell, E.C. (1936a), Keys, J.H. (1916), Shirt, D.B., ed. (1987).

ANOPLUS ROBORIS   NOTABLE B
A weevil
Order COLEOPTERA   Family CURCULIONIDAE

Anoplus roboris Suffrian, 1840.

Distribution South West, South, South East, East Anglia, West Midlands, Dyfed-Powys, North Wales, North East, North West, South East Scotland, South West Scotland and North East Scotland.

Habitat and ecology Wetland, particularly alder carr and wet areas of broad-leaved woodland. Also noted from river margins. Phytophagous. Associated with alder. Adults have been recorded in January and from April to August.

Status Widespread but local. Recently recorded from southern England, with a scattered distribution as far north as Mid Perthshire in Scotland.

Threats Drainage, and falling water tables because of water abstraction and river engineering schemes may threaten this species.

Management and conservation Water tables should be maintained at high levels. Gaps in the age structure of the alder population should be identified and the continuity of the appropriate habitat ensured by regeneration or suitable planting.

ANTHONOMUS CHEVROLATI
A weevil   **INSUFFICIENTLY KNOWN**
Order COLEOPTERA   Family CURCULIONIDAE

Anthonomus chevrolati Desbrochers, 1868.

Distribution Recorded from Dorset, South Hampshire, East Sussex, East Kent, West Kent, Surrey, South Essex, Oxfordshire and Buckinghamshire before 1970.

Habitat and ecology Woodland, scrub and hedges in cultivated land. Phytophagous. Associated with hawthorn, adults feeding on the bark and leaf buds. On the Continent, this species may also be associated with *Sorbus* species. In Britain, eggs are laid in the vegetative buds, probably as early as January or February. By the middle of April pupae as well as larvae may be found in deformed leaf-trusses. The adult weevils probably emerge in May and June and normally aestivate. They become active in autumn and remain on the host-plant, either continuously or intermittently, throughout the winter and early spring. Adults have been recorded from January to March, May to July, and in November and December, though mainly from December to February.

Status Not listed in the insect Red Data Book (Shirt, 1987). Old records suggest that this species was widely reported in southern England as far north as Oxfordshire. Last recorded in 1966 from East Malling, West Kent. This species has been recorded in numbers on some sites. Possibly under-recorded because of the time of year that this species appears to be most active. This species is difficult to identify and may be confused with other members of the genus. Consequently, the exact status of this species is hard to assess.

Threats Neglect and conversion to high forest, the grubbing out of scrub and hedgerows, the mechanical trimming of hedgerows, and clear-felling of woodland and conversion to other land use may be threats to this species.

Management and conservation Open glades, ride margins and hedgerows should be cut on rotation to retain a variety of vegetation structures.

Published sources Fowler, W.W. (1891), Morris, M.G. (1964), Morris, M.G. (1976a), Nicholson, G.W. (1930).

CURCULIONIDAE

ANTHONOMUS CONSPERSUS   NOTABLE B
A weevil
Order COLEOPTERA   Family CURCULIONIDAE

Anthonomus conspersus Desbrochers, 1868. Formerly known as: *Anthonomus pedicularius var. conspersus* Desbrochers.

Distribution West Midlands, North East, North West, South West Scotland, North East Scotland and North West Scotland.

Habitat and ecology Woodland and scrub. Phytophagous. Associated with rowan. Eggs are laid in the flower buds of the host, and the larvae develop in "capped" blossoms. On the Continent, larvae have been found in early June, with adult emergence occurring in about mid June. In Britain, adults have been recorded from May to August.

Status Widespread but local in Scotland and northern England. This species is difficult to identify and may be confused with other members of the genus. Consequently, the exact status of this species is hard to assess.

Threats Loss of woodland through, for example, clear-felling and conversion to other land use. Neglect and conversion to high forest, and the grubbing out of scrub may be further threats.

Management and conservation Woodland should be managed to be kept open and retain a variety of vegetation structures.

ANTHONOMUS HUMERALIS
A weevil   INSUFFICIENTLY KNOWN
Order COLEOPTERA   Family CURCULIONIDAE

Anthonomus humeralis (Panzer, 1795).

Distribution Recorded from South Hampshire, West Kent, Surrey and Cumberland before 1970 and Cardiganshire from 1970 onwards.

Habitat and ecology Woodland, orchards and probably scrub. Phytophagous. Associated with apple and crab-apple. On the Continent, this species has also been recorded from dwarf cherry, wild cherry and bird cherry. The larvae feed in "capped" blossoms of the host-plant. Adults and pupae have been found in mid May with new generation adults emerging in late May. In Britain, adults have been recorded in February, May, July and August.

Status Not listed in the insect Red Data Book (Shirt, 1987). There are old records for only four vice-counties and a recent record of three individuals in Wales. This species is difficult to identify and may be confused with other members of the genus. Consequently, the exact status of this species is hard to assess.

Threats Probably threatened by the loss of broad-leaved woodland through, for example, clear-felling and conversion to other land use. Neglect and conversion to high forest, the grubbing out of orchards and scrub, and the use of herbicides and pesticides may be further threats.

Management and conservation Open glades and ride margins should be cut on rotation to retain a variety of vegetation structures.

Published sources Champion, G.C. (1924), Morris, M.G. (1976a).

ANTHONOMUS PIRI
Apple bud weevil   INSUFFICIENTLY KNOWN
Order COLEOPTERA   Family CURCULIONIDAE

Anthonomus piri Kollar, 1837. Formerly known as: *Anthonomus cinctus* Redtenbacher, 1858, *Anthomus pyri* Redtenbacher, 1858.

Distribution Recorded from Isle of Wight, East Kent, West Kent, Surrey, West Norfolk and Cambridgeshire before 1970 and West Norfolk from 1970 onwards.

Habitat and ecology Orchards and probably also woodland and scrub. Phytophagous. Associated with apple. On the Continent, the weevil has also been recorded on pear. In Britain, eggs have been observed in early February but were probably laid earlier. Larvae have been found as late as mid May. Fully grown larvae have been found in hollowed-out fruit buds of apple. These pupated by late April, with new generation adults emerging in early June. Adults aestivate after emerging from buds in June. The weevil may occasionally be found in the shoot buds of apple. Adults have been recorded in January, March, May, June, August, September and November.

Status Not listed in the insect Red Data Book (Shirt, 1987). Only recorded in southern and south-eastern England. Last recorded in 1970 from West Norfolk. This species is difficult to identify and may be confused with other members of the genus. Consequently, the exact status of this species is hard to assess.

Threats This species is threatened by the grubbing out of orchards, and the use of herbicides and possibly also pesticides. The clear-felling of woodland, neglect and conversion to high forest, and grubbing out of scrub may be further threats.

Management and conservation Open glades and ride margins should be cut on rotation to retain a variety of vegetation structures.

Published sources Donisthorpe, H.St J.K. (1924), Duffield, C.A.W. (1922), Harwood, P. (1921), Massee, A.M. (1961), Morris, M.G. (1976a).

ANTHONOMUS RUFUS   **RARE**
A weevil
Order COLEOPTERA   Family CURCULIONIDAE

Anthonomus rufus Gyllenhal, 1836.

Distribution Recorded from West Cornwall, East Sussex, South Essex, Cardiganshire and Westmorland & North Lancashire before 1970 and West Cornwall, Dorset and Cardiganshire from 1970 onwards.

Habitat and ecology Woodland, scrub and hedges, near the coast. Phytophagous. Associated with blackthorn. Larvae probably occur in the flower buds. Adults have been found in March and also recorded from April to July.

Status Not listed in the insect Red Data Book (Shirt, 1987). Extremely local and recently recorded from just three vice-counties. Can be numerous where found. This species is difficult to identify and may be confused with other members of the genus. Consequently, the exact status of this species is hard to assess.

Threats Loss of woodland through, for example, clear-felling and conversion to other land use. Neglect and conversion to high forest, the grubbing out of scrub and hedgerows, and mechanised trimming of hedgerows are also threats to this species.

Management and conservation Open glades and ride margins should be cut on rotation to retain a variety of vegetation structures.

Published sources Donisthorpe, H.St J.K. (1935), Hammond, P.M. (1979), Morris, M.G. (1966), Morris, M.G. (1976a).

ANTHONOMUS ULMI   **NOTABLE B**
A weevil
Order COLEOPTERA   Family CURCULIONIDAE

Anthonomus ulmi (Degeer, 1775). Formerly known as: *Anthonomus inversus* Bedel, 1884, *Anthonomus inversus* sensu auct. partim.

Distribution England, South Wales, North Wales, South East Scotland, South West Scotland and North East Scotland.

Habitat and ecology Broad-leaved woodland, parkland and hedgerows. Phytophagous. Associated chiefly with English elm and also on wych elm. On the Continent, eggs and first instar larvae have been found in flower buds in mid January, with fully grown larvae being found from early April until early May. The larvae pupate in hollowed-out buds, and emerge between early May and early July. Adult weevils then aestivate during the summer after which they re-emerge in around September. In Britain, adults have been recorded in January, from May to August and in October, though predominantly in May and June.

Status Widespread but local in Britain, though declining because of Dutch elm disease. This species is difficult to identify and may be confused with other members of the genus.

Threats Dutch elm disease is the main cause of this species' decline. Further threats are through the loss of broad-leaved woodland through, for example, clear-felling and coniferisation, and the grubbing out of hedgerows.

Management and conservation Open glades and ride margins should be cut on rotation to retain a variety of vegetation structures. This species requires elms that are mature enough to produce reproductive (flowering) buds.

ANTHONOMUS VARIANS   **NOTABLE B**
A weevil
Order COLEOPTERA   Family CURCULIONIDAE

Anthonomus varians (Paykull, 1792). Formerly known as: *Anthonomus phyllocola* Herbst, 1795.

Distribution South East Scotland, North East Scotland and North West Scotland.

Habitat and ecology Native pine forest. Phytophagous. Associated with Scots pine. On the Continent, this species has been recorded on Norway spruce. Eggs and larvae develop in male catkins of the host-tree. Larvae

CURCULIONIDAE

feed on pollen. In Britain, adults have been recorded in March and from May to August.

Status Only known from the Highlands of Scotland. This species does not appear to have spread to conifer plantations. This species is difficult to identify and may be confused with other members of the genus. Consequently, the exact status of this species is hard to assess.

Threats Loss of native pine forest through practices such as clear-felling or conversion to plantation forest. Overgrazing, preventing regeneration, may be a further threat.

Management and conservation Gaps in the age structure of the tree population should be identified and regeneration encouraged to ensure the continuity of suitable habitat.

BAGOUS ARDUUS   INSUFFICIENTLY KNOWN
A weevil
Order COLEOPTERA   Family CURCULIONIDAE

Bagous arduus Sharp, 1917.

Distribution Recorded from "London district" before 1970.

Habitat and ecology Uncertain, but probably associated with wetland. Phytophagous. Biology unknown.

Status Status revised from RDB 3 (Rare) in Shirt (1987). Reliably known only from Sharp's type specimen from the "London district". Five examples recorded by Champion from Woking, Surrey are not this species but are referable to *B. longitarsis* (A.A. Allen pers. comm.). Similarly, two examples noted by Walker from the Isle of Sheppey, East Kent are also unlikely to be *B. arduus*, though examination of this material is still required. There is also an unconfirmed record of this species from Oxford. This species is not known on the Continent.

Published sources Sharp, D. (1917b), Shirt, D.B., ed. (1987), Walker, J.J. (1932a).

BAGOUS ARGILLACEUS   VULNERABLE
A weevil
Order COLEOPTERA   Family CURCULIONIDAE

Bagous argillaceus Gyllenhal, 1836.

Distribution Recorded from South Hampshire, East Kent, West Kent, South Essex and West Norfolk before 1970 and East Kent from 1970 onwards.

Habitat and ecology Brackish ditches and ponds. The species is coastal in Britain. On the Continent, this beetle has also been recorded inland on saline soils. Phytophagous. Possibly associated with grasses. Adults have been recorded in June and November.

Status Extremely local, though may be found in numbers where it occurs. Only known from southern and eastern England. Recently recorded from just one vice-county. This species is difficult to identify and may be confused with other members of the genus. Consequently, the exact status of this species is hard to assess.

Threats This species may be threatened by coastal and urban development, as well as through drainage for reasons such as agricultural improvement and changes to oteher land use. Falling water tables because of water abstraction schemes, the use of herbicides and pesticides on adjacent areas of land, the infilling of ponds, and natural succession may be further threats.

Management and conservation Water tables should be maintained at high levels. Water bodies should be isolated from sources of eutrophication and pollution. Ditch and pond clearance should be undertaken on a rotational basis, and should aim at maintaining open conditions and a variety of early successional stages.

Published sources Allen, A.A. (1950e), Anon, (1935), Fowler, W.W. (1891), Sharp, D. (1917b), Shirt, D.B., ed. (1987).

BAGOUS BINODULUS   EXTINCT
A weevil
Order COLEOPTERA   Family CURCULIONIDAE

Bagous binodulus (Herbst, 1795).

Distribution Recorded from East Norfolk before 1970.

Habitat and ecology Aquatic habitats, including, broads, ditches and ponds. Phytophagous. Associated with water soldier *Stratiotes aloides*, the larvae being found on the fleshy leaves.

234

Status Presumed extinct. Status revised from RDB 1 (Endangered) in Shirt (1987). Only known from the Norfolk Broads. Last recorded in 1861 from Horning Fen, East Norfolk.

Published sources Fowler, W.W. (1891), Newbery, E.A. (1902), Shirt, D.B., ed. (1987).

BAGOUS BREVIS   **ENDANGERED**
A weevil
Order COLEOPTERA   Family CURCULIONIDAE

Bagous brevis Gyllenhal, 1836.

Distribution Recorded from South Hampshire and Surrey before 1970 and South Hampshire from 1970 onwards.

Habitat and ecology Recorded in and on the banks of ponds and floating bogs. Phytophagous. On the Continent, this species is associated with lesser spearwort *Ranunculus flammula*. In Britain, adults have been recorded from March to June and in August.

Status Extremely local and only known from two ponds in the New Forest, South Hampshire this century. There is also a doubtful record from East Kent. This species is difficult to identify and may be confused with other members of the genus. Consequently, the exact status of this species is hard to assess.

Threats This species is threatened by drainage and a change in land use. Falling water table because of water abstraction schemes, the infilling of ponds and natural succession may be further threats.

Management and conservation Management should aim at encouraging open conditions and ensuring the presence of the foodplant. Water tables should be maintained at high levels.

Published sources Fowler, W.W. (1891), Fowler, W.W. & Donisthorpe, H.St J.K. (1913), Newbery, E.A. (1902), Sharp, D. (1917b), Shirt, D.B., ed. (1987).

BAGOUS COLLIGNENSIS   **RARE**
A weevil
Order COLEOPTERA   Family CURCULIONIDAE

Bagous collignensis (Herbst, 1797). Formerly known as: *Bagous claudicans* Boheman, 1845, and confused under: *Bagous frit* sensu auct. partim not (Herbst, 1795).

Distribution Recorded from West Cornwall, South Devon, Isle of Wight, South Hampshire, East Sussex, East Kent, West Kent, Surrey, Hertfordshire, Oxfordshire, Buckinghamshire, East Norfolk, West Norfolk, Herefordshire, Cumberland and Kintyre before 1970 and South Hampshire from 1970 onwards.

Habitat and ecology Aquatic and wetland habitats such as ponds and riverbanks. Phytophagous. On the Continent, this species is associated with water horsetail *Equisetum fluviatile*. The larvae develop and pupate in the stalks of the foodplant. In Britain, adults have been recorded from May to July, and in October and November.

Status Not listed in the insect Red Data Book (Shirt, 1987). Old records suggest that this beetle had a widely scattered distribution from southern England, north to Kintyre in Scotland. Recently recorded from only one vice-county. This species is difficult to identify and may be confused with other members of the genus. Consequently, the exact status of this species is hard to assess.

Threats This species has probably declined because of loss of habitat through drainange for reasons such as agricultural improvement and development. Falling water tables because of water abstraction schemes, the use of herbicides and pesticides, the increase in the use of motor-boats, natural succession and the infilling of ponds may have further contributed to this species' decline.

Management and conservation Water tables should be maintained at high levels. Water bodies should be isolated from sources of eutrophication and pollution. Grazing or cutting, on a rotational basis, is needed to maintain open conditions and encourage early successional stages.

Published sources Fowler, W.W. & Donisthorpe, H.St J.K. (1913), Newbery, E.A. (1902), Sharp, D. (1917b).

BAGOUS CYLINDRUS   **VULNERABLE**
A weevil
Order COLEOPTERA   Family CURCULIONIDAE

Bagous cylindrus (Paykull, 1800).

Distribution Recorded from East Sussex, East Kent, West Kent, South Essex, North Essex, Middlesex and Bedfordshire before 1970 and East Sussex, East Kent, West Kent and South Essex from 1970 onwards.

Habitat and ecology Aquatic habitats such as dykes, ditches and ponds. Phytophagous. Probably associated with grasses. On the Continent, this species has been

CURCULIONIDAE

associated with floating sweet-grass *Glyceria fluitans*, plicate sweet-grass *G. plicata* and orange foxtail *Alopecurus aequalis*. In Britain, adults have been recorded from March to July and in October.

Status Restricted to eastern and south-eastern England. Recently recorded from only four vice-counties. Can be numerous where found.

Threats This species is threatened by loss of habitat through drainage for reasons such as agricutural improvement and development. Falling water tables because of water abstraction schemes, the use of herbicides and pesticides, the infilling of ponds, and natural succession may also affect this species.

Management and conservation Water tables should be maintained at high levels. Water bodies should be isolated from sources of eutrophication and pollution. Ditch and pond clearance should be undertaken on a rotational basis, and aim at maintaining open conditions and a variety of early successional stages.

Published sources Allen, A.A. (1948c), Fowler, W.W. (1891), Fowler, W.W. & Donisthorpe, H.St J.K. (1913), Newbery, E.A. (1902), Shirt, D.B., ed. (1987).

BAGOUS CZWALINAI   **ENDANGERED**
A weevil
Order COLEOPTERA   Family CURCULIONIDAE

Bagous czwalinai Seidlitz, 1891. Formerly known as: *Bagous heasleri* Newbery, 1902, *Bagous tempestivus var. heasleri* Newbery.

Distribution Recorded from South Hampshire before 1970 and South Hampshire from 1970 onwards.

Habitat and ecology *Sphagnum* bogs. Phytophagous. Adults have been recorded in June and August.

Status Extremely local. Only known from a very few small *Sphagnum* bogs in the New Forest, South Hampshire.

Threats This species may be threatened by drainage for reasons such as a change in land use, as well as through natural succession.

Management and conservation Water tables should be maintained at high levels. Grazing or cutting, on a rotational basis, may be needed to maintain open conditions and encourage early successional stages.

Published sources Cooter, J. (1977b), Drane, A.B. (1990), Shirt, D.B., ed. (1987).

BAGOUS DIGLYPTUS   **ENDANGERED**
A weevil
Order COLEOPTERA   Family CURCULIONIDAE

Bagous diglyptus Boheman, 1845.

Distribution Recorded from East Suffolk, West Suffolk, East Norfolk and Derbyshire before 1970.

Habitat and ecology Uncertain, but probably a species of wetland habitats, particularly river banks and ditches. On the Continent, it has been found in dry habitats. Phytophagous. On the Continent, this beetle has been associated with species of lichen in the genus *Peltigera*.

Status Last recorded in 1906 from Stalham Broad, East Norfolk. This species is difficult to identify and may be confused with other members of the genus.

Threats Uncertain, though this species may have declined because of drainage for reasons such as agricultural improvement and development. Falling water tables because of water abstraction and river engineering schemes, the use of herbicides and pesticides, and the increase in the use of motor-boats may also have contributed to this species' decline.

Published sources Fowler, W.W. (1891), Fowler, W.W. & Donisthorpe, H.St J.K. (1913), Sharp, D. (1917b), Shirt, D.B., ed. (1987).

BAGOUS FRIT   **RARE**
A weevil
Order COLEOPTERA   Family CURCULIONIDAE

Bagous frit (Herbst, 1795).

Distribution Recorded from Dorset, South Hampshire and East Norfolk before 1970 and South Hampshire, Carmarthenshire and Caernarvonshire from 1970 onwards.

Habitat and ecology *Sphagnum* bogs. Phytophagous. Pupae have been found in July, and adults have been recorded in April, May, July and August.

Status Status revised from RDB 1 (Endangered) in Shirt (1987). Very local. Recently discovered as new to Wales. The only other modern records are from South Hampshire. An old record for Herefordshire requires confirmation. This species is difficult to identify and may be confused with other members of the genus. Consequently, the exact status of this species is hard to assess.

Threats This species is threatened by inadvertent interference during forestry operations, natural succession and drainage.

Management and conservation Management should aim at maintaining open conditions and preventing the sites from drying out. Water tables should be maintained at high levels.

Published sources Shirt, D.B., ed. (1987), Tomlin, J.R.le B. (1949).

BAGOUS GLABRIROSTRIS   **NOTABLE B**
A weevil
Order COLEOPTERA   Family CURCULIONIDAE

Bagous glabrirostris (Herbst, 1795).

Distribution South East, South, South West, East Anglia and East Midlands.

Habitat and ecology Wetland and aquatic habitats. Phytophagous and possibly polyphagous. Recorded from sedge refuse. Adults have been recorded in March, May, June and August. This species has been recorded at a mercury vapour light trap.

Status Widespread but local in southern and eastern England. This species is difficult to identify and may be confused with other members of the genus. Consequently, the exact status of this species is hard to assess.

Threats This species is threatened by drainage for reasons such as agricultural improvement and development. falling water tables because of water abstraction schemes, the use of herbicides and pesticides, and natural succession may be further threats.

Management and conservation Water tables should be maintained at high levels. Water bodies should be isolated from sources of eutrophication and pollution. Grazing or cutting, on a rotational basis, is needed to maintain open conditions and encourage early successional stages.

BAGOUS LIMOSUS   **NOTABLE B**
A weevil
Order COLEOPTERA   Family CURCULIONIDAE

Bagous limosus (Gyllenhal, 1827).

Distribution South East, South, South West, East Anglia, East Midlands, North East and North West.

Habitat and ecology *Sphagnum* bogs, ponds, extremely slow moving water and brackish ditches. Recorded inland and near the coast. Phytophagous. Associated with pondweed *Potamogeton*. Adults have been recorded from March to November.

Status Widespread but local in England. Can be numerous where found. This species is difficult to identify and may be confused with other members of the genus. Consequently, the exact status of this species is hard to assess.

Threats This species is threatened by drainage, for reasons such as agricultural improvement and development. Falling water tables because of water abstraction and river engineering schemes, the use of herbicides and pesticides, the increase in the use of motor-boats, the infilling of ponds, and natural succession may also threaten this species.

Management and conservation Water tables should be maintained at high levels. Water bodies should be isolated from sources of eutrophication and pollution. Ditch and pond clearance should be undertaken on a rotational basis, and should aim at maintaining the aquatic plant populations.

BAGOUS LONGITARSIS   **ENDANGERED**
A weevil
Order COLEOPTERA   Family CURCULIONIDAE

Bagous longitarsis Thomson, 1868. Formerly known as: *Bagous tomlini* Sharp, 1917.

Distribution Recorded from South Hampshire, East Sussex, East Kent, West Kent and Surrey before 1970 and East Kent from 1970 onwards.

Habitat and ecology Ponds, ditches and dykes. Phytophagous. Probably associated with water-milfoil *Myriophyllum*. Adults have been recorded in April, June and September.

Status Only recorded from southern and eastern England, with just one recent site. Can be numerous where found. This species is difficult to identify and may be confused with other members of the genus.

CURCULIONIDAE

Consequently, the exact status of this species is hard to assess.

Threats This species may be threatened by drainage for reasons such as agricultural improvement and development. Falling water tables because of water abstraction and river engineering schemes, the use of herbicides and pesticides, and the infilling of ponds may also threaten this species.

Management and conservation Water tables should be maintained at high levels. Water bodies should be isolated from sources of eutrophication and pollution. Ditch and pond clearance should be undertaken on a rotational basis, and should aim at maintaining the aquatic plant populations.

Published sources Sharp, D. (1917b), Shirt, D.B., ed. (1987).

BAGOUS LUTOSUS   **ENDANGERED**
A weevil
Order COLEOPTERA   Family CURCULIONIDAE

Bagous lutosus (Gyllenhal, 1813).

Distribution Recorded from South Hampshire, North Hampshire, Surrey, West Norfolk and Leicestershire & Rutland before 1970.

Habitat and ecology Aquatic and semi-aquatic habitats such as ditches, ponds, lakes and reservoirs. Phytophagous. Associated with branched bur-reed *Sparganium erectum*. Adults have been recorded in May and June.

Status Not listed in the insect Red Data Book (Shirt, 1987). Very few records this century. Last recorded in 1964 from Fleet Pond, North Hampshire. This species is difficult to identify and may be confused with other members of the genus. Consequently, the exact status of this species is hard to assess.

Threats This species has probably declined because of drainage for reasons such as agricultural improvement and development. falling water tables because of water abstraction schemes, the use of herbicides and pesticides, natural succession, and the infilling of ponds and lakes may have contributed to the species' decline.

Management and conservation Water tables should be maintained at high levels. Water bodies should be isolated from sources of eutrophication and pollution. Ditch and pond clearance should be undertaken on a rotational basis, and should aim at encouraging open conditions and ensuring the presence of the foodplant.

Published sources Beare, T.H. (1934b), Fowler, W.W. (1891), Fowler, W.W. & Donisthorpe, H.St J.K. (1913), Henderson, C.W. (1944), Sharp, D. (1917a).

BAGOUS LUTULENTUS   **NOTABLE B**
A weevil
Order COLEOPTERA   Family CURCULIONIDAE

Bagous lutulentus (Gyllenhal, 1813). Formerly known as: *Bagous glabrirostris var. nigritarsis* Thomson, 1865, *Bagous nigritarsis* Thomson.

Distribution South East, South, South West, East Anglia, North West and Wales.

Habitat and ecology Aquatic and wetland habitats. Phytophagous. Associated with water horsetail *Equisetum fluviatile*. The larvae develop and pupate in the stalks of the foodplant. Adults have been recorded in February, May, June and September.

Status Widespread but local in parts of England and Wales. This species is difficult to identify and may be confused with other members of the genus. Consequently, the exact status of this species is hard to assess.

Threats This species is threatened by drainage for reasons such as agricultural improvement and development. Falling water tables because of water abstraction schemes, the use of herbicides and pesticides, and natural succession may be further threats.

Management and conservation Water tables should be maintained at high levels. Water bodies should be isolated from sources of eutrophication and pollution. Clearance in aquatic habitats should be undertaken on a rotational basis, and should aim at encouraging open conditions and ensuring the presence of the foodplant.

BAGOUS LUTULOSUS   **NOTABLE A**
A weevil
Order COLEOPTERA   Family CURCULIONIDAE

Bagous lutulosus (Gyllenhal, 1827).

Distribution Recorded from Isle of Wight, South Hampshire, North Hampshire, East Sussex, East Kent, West Kent, Surrey, Berkshire, West Norfolk, Glamorgan and South Lancashire before 1970 and South Hampshire, West Sussex and South-east Yorkshire from 1970 onwards.

Habitat and ecology Damp, sandy areas. Occasionally in very wet situations. Phytophagous. On the Continent, this species is associated with rushes. In Britain, adults have been recorded from April to September and in November.

Status Old records suggest that this species was formerly widespread in England as far north as South Lancashire, and also recorded in South Wales. Recently recorded from just three vice-counties. This species is difficult to identify and may be confused with other members of the genus. Consequently, the exact status of this species is hard to assess.

Threats This species may be threatened by drainage for reasons such as agricultural improvement and development. Natural succession is a further threat.

Management and conservation Water tables should be maintained at high levels. Water bodies should be isolated from sources of eutrophication and pollution. Grazing or cutting, on a rotational basis, is needed to maintain open conditions and encourage the early successional stages.

Published sources Flint, J.H. (1988), Fowler, W.W. (1891), Newbery, E.A. (1902).

BAGOUS NODULOSUS ENDANGERED
A weevil
Order COLEOPTERA Family CURCULIONIDAE

Bagous nodulosus Gyllenhal, 1836.

Distribution Recorded from South Hampshire, West Sussex, East Sussex, East Kent, West Kent, Surrey, Middlesex, East Suffolk and Huntingdonshire before 1970 and South Somerset from 1970 onwards.

Habitat and ecology Ditches, dykes and ponds. Phytophagous. Associated with flowering rush *Butomus umbellatus*. The larvae develop and pupate in the leaf and flowering stems. Adults have been recorded in May and June.

Status Recently recorded in only one site. Formerly recorded in southern and south-eastern England as far north as Huntingdonshire. This species is difficult to identify and may be confused with other members of the genus. Consequently, the exact status of this species is hard to assess.

Threats This species may be threatened by drainage for reasons such as agricultural improvement. Falling water tables because of water abstraction schemes, the use of herbicides and pesticides, the infilling of ponds, and natural succession may be further threats.

Management and conservation Water tables should be maintained at high levels. Water bodies should be isolated from sources of eutrophication and pollution. Ditch and pond clearance should be undertaken on a rotational basis, and should aim at maintaining populations of the foodplant.

Published sources Fowler, W.W. (1891), Fowler, W.W. & Donisthorpe, H.St J.K. (1913), Newbery, E.A. (1902), Shirt, D.B., ed. (1987).

BAGOUS PETRO EXTINCT
A weevil
Order COLEOPTERA Family CURCULIONIDAE

Bagous petro (Herbst, 1795). Formerly known as: *Ephimeropus petro* (Herbst), *Helmidomorphus petro* (Herbst).

Distribution Recorded from Mid-west Yorkshire.

Habitat and ecology Bogs and fens. Phytophagous. Associated with greater bladderwort *Utricularia vulgaris*. Adults have been recorded in August.

Status Presumed extinct. Only known from Askham Bog, Mid-west Yorkshire. Last recorded in 1895.

Published sources Fowler, W.W. & Donisthorpe, H.St J.K. (1913), Newbery, E.A. (1902), Shirt, D.B., ed. (1987).

BAGOUS PUNCTICOLLIS ENDANGERED
A weevil
Order COLEOPTERA Family CURCULIONIDAE

Bagous puncticollis Boheman, 1854.

Distribution Recorded from Isle of Wight, North Hampshire, West Kent and Surrey before 1970 and East Sussex from 1970 onwards.

Habitat and ecology Ponds, ditches and dykes. Phytophagous. On the Continent, this species is associated with a variety of water-plants such as pondweed *Elodea canadensis*, frogbit *Hydrocharis morsus-ranae* and water-soldier *Stratiotes aloides*.

Status Only recorded in southern England. Recently recorded from just one site. This species is difficult to identify and may be confused with other members of the genus. Consequently, the exact status of this species is hard to assess.

Threats This species has probably declined because of drainage for reasons such as agricultural improvement

CURCULIONIDAE

and development. Falling water tables because of water abstraction schemes, the use of herbicides and pesticides, natural succession and the infilling of ponds may have also contributed to this species' decline.

Management and conservation Water tables should be maintained at high levels. Water bodies should be isolated from sources of eutrophication and pollution. Ditch and pond clearance should be undertaken on a rotational basis, and should aim at maintaining aquatic plant populations.

Published sources Sharp, D. (1917a), Shirt, D.B., ed. (1987).

BAGOUS SUBCARINATUS **NOTABLE A**
A weevil
Order COLEOPTERA   Family CURCULIONIDAE

Bagous subcarinatus Gyllenhal, 1836. Formerly confused under: *Bagous frit* sensu Newbery, 1902 not (Herbst, 1795).

Distribution Recorded from North Somerset, East Sussex, East Kent, West Kent, Middlesex and Cambridgeshire before 1970 and North Somerset, East Sussex, East Kent, West Kent and South Essex from 1970 onwards.

Habitat and ecology Fresh-water and brackish habitats such as ditches and dykes. Phytophagous. Associated with soft hornwort *Ceratophyllum submersum*. Adults have been recorded from April to June.

Status Very local and only recorded in southern England as far north as Cambridgeshire. Recently recorded from just five vice-counties. This species is difficult to identify and may be confused with other members of the genus. Consequently, the exact status of this species is hard to assess.

Threats This species is threatened by drainage for reasons such as agricultural improvement and development. Falling water tables because of water abstraction schemes, the use of herbicides and pesticides, and natural succession may be further threats.

Management and conservation Water tables should be maintained at high levels. Water bodies should be isolated from sources of eutrophication and pollution. Ditch clearance should be undertaken on a rotational basis, and should aim at maintaining aquatic plant populations.

Published sources Allen, A.A. (1948c), Newbery, E.A. (1902), Sharp, D. (1917a).

BAGOUS TEMPESTIVUS **NOTABLE B**
A weevil
Order COLEOPTERA   Family CURCULIONIDAE

Bagous tempestivus (Herbst, 1795). Formerly known as: *Bagous cnemerythrus* (Marsham, 1802).

Distribution England, South Wales and Dyfed-Powys.

Habitat and ecology Wetland and aquatic habitats, including brackish ditches. Phytophagous. Associated with a variety of wetland and aquatic plants, in particular sedges, pondweeds and arrowhead *Sagittaria sagittifolia*. Adults have been recorded from April to June.

Status Widespread but local in England, more local in northern England. Also recorded in parts of Wales. This species is difficult to identify and may be confused with other members of the genus. Consequently, the exact status of this species is hard to assess.

Threats This species is threatened by drainage for reasons such as agricultural improvement and development. Falling water tables because of water abstraction schemes, the use of herbicides and pesticides, and natural succession may be further threats.

Management and conservation Water tables should be maintained at high levels. Water bodies should be isolated from sources of eutrophication and pollution. Ditch clearance should be undertaken on a rotational basis, and should aim at maintaining aquatic plant populations.

BARIS ANALIS **VULNERABLE**
A weevil
Order COLEOPTERA   Family CURCULIONIDAE

Baris analis (Olivier, 1790).

Distribution Recorded from Isle of Wight before 1970 and Dorset and Isle of Wight from 1970 onwards.

Habitat and ecology Coastal cliff, grassland and possibly disturbed ground. Phytophagous. Associated with common fleabane *Pulicaria dysenterica*. The weevil probably overwinters in the adult stage. Adults have been recorded from March to June.

Status Status revised from RDB 1 (Endangered) in Shirt (1987). Extremely local and known from just two vice-counties. Formerly known only from Sandown, Isle of Wight and recorded there up until 1887. This species was not recorded for nearly 100 years, until 1984 when it was found at East Wight, Isle of Wight. Also recently recorded from Dorset.

Threats This species is threatened by cliff stabilisation schemes and the construction of sea defences. Coastal developments may reduce the amount of available habitat. Activities that accelerate or reduce the rate of erosion should be avoided. The degradation of suitable habitat through natural succession and the invasion of scrub on stabilised areas is a further threat.

Management and conservation Occasional slippages are necessary to maintain habitat continuity. Large areas of unstable cliff are required so that the population does not become isolated and subsequently threatened by individual landslips.

Published sources Appleton, D. (1986), Cooter, J. (1990a), Drane, A.B. (1990), Fowler, W.W. (1891), Shirt, D.B., ed. (1987).

BARIS LATICOLLIS	**NOTABLE A**
A weevil	
Order COLEOPTERA	Family CURCULIONIDAE

Baris laticollis (Marsham, 1802).

Distribution Recorded from "Cornwall", South Devon, Dorset, Isle of Wight, South Hampshire, East Sussex, East Kent, West Kent, Surrey, South Essex, West Suffolk, Cambridgeshire, East Gloucestershire, Leicestershire & Rutland and North-east Yorkshire before 1970 and Dorset, East Sussex, Surrey and South Essex from 1970 onwards.

Habitat and ecology Coastal cliff and undercliff, coastal shingle, disturbed ground, gardens and allotments. This species may prefer coastal habitats. Phytophagous. Associated predominantly with hedge mustard *Sisymbrium officinale*, also on wild and cultivated cabbages *Brassica* and probably on other Cruciferae. On the Continent, this weevil has also been recorded from garden radish *Raphanus sativus*, wallflower cabbage *Rhynchosinapis cheiranthos*, treacle mustard *Erysimum cheiranthoides*, sea rocket *Cakile maritima*, hedge mustard *S. officinale* and tall rocket *S. altissimum*. In Britain, eggs are laid from late April until early June in punctures made in the root and basal stem of the foodplant. Larvae tunnel the root and basal stem between April and September, with pupation occurring between mid July and September. The main period of pupation is in August and September, with adult emergence starting in August. The weevil overwinters in the adult stage, although some larvae and pupae may overwinter to complete development

the following year. The weevil overwinters in the soil, re-emerging the following year about April.

Status Very local and recently recorded from only four vice-counties, all in south-eastern England. Old records suggest that this species was formerly more widespread in southern England and had a scattered distribution as far north as North-east Yorkshire. Can be numerous where found.

Threats This species is threatened by cliff stabilisation schemes and the construction of sea defences. Coastal developments may reduce the amount of available habitat. Activities that accelerate or reduce the rate of erosion should be avoided. Improvement and conversion to arable agriculture, gravel extraction, and natural succession are further threats.

Management and conservation In areas of unstable cliff, occasional slippages are necessary to maintain habitat continuity. Large areas of unstable cliff are required so that the population does not become isolated and subsequently threatened by individual landslips. Except in areas of coastal shingle, which should be left unmanaged, grazing, cutting or possibly some other disturbance is needed to maintain open conditions.

Published sources Atty, D.B. (1983), Fowler, W.W. (1891), Fowler, W.W. & Donisthorpe, H.St J.K. (1913), Lloyd, R.W. (1945), Owen, J.A. (1989c).

BARIS LEPIDII	**NOTABLE A**
A weevil	
Order COLEOPTERA	Family CURCULIONIDAE

Baris lepidii Germar, 1824.

Distribution Recorded from South Devon, Dorset, Isle of Wight, South Hampshire, West Sussex, East Sussex, East Kent, West Kent, Surrey, South Essex, Hertfordshire, Middlesex, Berkshire, Oxfordshire, West Norfolk, Cambridgeshire, Huntingdonshire, East Gloucestershire, Worcestershire, Warwickshire, Shropshire, North Lincolnshire, Leicestershire & Rutland, Nottinghamshire, North-east Yorkshire and Mid-west Yorkshire before 1970 and East Sussex, East Kent, Surrey, West Suffolk, Worcestershire and Leicestershire & Rutland from 1970 onwards.

Habitat and ecology Wetland, especially riverbanks and ditches. Phytophagous. Associated with water-cress *Nasturtium officinale*, creeping yellow-cress *Rorippa sylvestris*, winter-cress *Barbarea vulgaris*, dittander *Lepidium latifolium*, and probably with other Cruciferae. Adults have been recorded in January and from May to September.

CURCULIONIDAE

Status Very local and recently recorded from only six vice-counties. Old records suggest that this species was formerly more widespread in England and recorded as far north as North-east Yorkshire.

Threats Drainage for reasons such as agricultural improvement and development. Falling water tables because of water abstraction, river engineering schemes, natural succession and the use of motor boats may also threaten this species.

Management and conservation Water tables should be maintained at high levels. Water bodies should be isolated from sources of eutrophication and pollution. Ditch clearance should be undertaken on a rotational basis and should aim at maintaining the aquatic plant populations.

Published sources Atty, D.B. (1983), Fowler, W.W. (1891), Fowler, W.W. & Donisthorpe, H.St J.K. (1913), Whitehead, P.F. (1990a).

BARIS PICICORNIS   **NOTABLE B**
A weevil
Order COLEOPTERA   Family CURCULIONIDAE

Baris picicornis (Marsham, 1802).

Distribution South East, South, South West, East Anglia, East Midlands and West Midlands.

Habitat and ecology Disturbed ground and grassland. Particularly on calcareous soils. Phytophagous. Associated with wild mignonette *Reseda lutea*. On the Continent, this species has also been recorded from weld *R. luteola*. Larvae feed in the stems. In Britain, adults have been recorded from April to October.

Status Widespread but local in England, though not recorded in North East and North West England.

Threats This species is threatened by loss of habitat through improvement and conversion to arable agriculture, development, and natural succession.

Management and conservation Disturbance, such as rotovation, is necessary to maintain open conditions.

BARIS SCOLOPACEA   **RARE**
A weevil
Order COLEOPTERA   Family CURCULIONIDAE

Baris scolopacea Germar, 1818.

Distribution Recorded from West Sussex, East Kent, West Kent and South Essex before 1970 and East Kent, West Kent, South Essex and North Essex from 1970 onwards.

Habitat and ecology Saltmarsh. Phytophagous. Predominantly associated with sea-purslane *Halimione portulacoides*, though reared from grass-leaved orache *Atriplex littoralis*. On the Continent, the weevil has also been recorded from *Chenopodium*, glasswort *Salicornia* and annual sea-blite *Suaeda maritima*. In Britain, adults have been recorded from June to September.

Status Only known from the coasts of south eastern England. Can be numerous where found.

Threats Loss of saltmarsh through reclamation, erosion and the construction of sea defences. Overgrazing of saltmarshes could be a further threat to this species.

Management and conservation Grazing should not be introduced to a site where there is no grazing at present.

Published sources Allen, A.A. (1949a), Donisthorpe, H.St J.K. (1921), Fowler, W.W. (1891), Massee, A.M. (1949), Shirt, D.B., ed. (1987), Wakely, S. (1951).

BARYNOTUS SQUAMOSUS   **NOTABLE B**
A weevil
Order COLEOPTERA   Family CURCULIONIDAE

Barynotus squamosus Germar, 1824. Formerly known as: *Barynotus schoenherri* (Zetterstedt, 1838), *Barynotus schonherri* (Zetterstedt, 1838).

Distribution South West, East Midlands, West Midlands, North East, North West, North Wales and Scotland.

Habitat and ecology Predominantly open ground and primarily found in upland situations. Phytophagous and probably polyphagous. Parthenogenetic. Adults have been recorded from May to September

Status Widespread but local, predominantly a northern species.

Threats Uncertain, though this species may be threatened by erosion through high densities of livestock. Afforestation may be a further threat.

BARYPEITHES PYRENAEUS   INDETERMINATE
A weevil
Order COLEOPTERA   Family CURCULIONIDAE

Barypeithes pyrenaeus Seidlitz, 1868. Formerly known as: *Barypithes pyrenaeus* Seidlitz, 1868.

Distribution Recorded from East Cornwall and South Devon before 1970 and East Cornwall and South-west Yorkshire from 1970 onwards.

Habitat and ecology Uncertain, but probably associated with grassland, scrub and woodland. Phytophagous and probably polyphagous at the roots of plants. Adults have been found resting on the leaves of primrose growing at the side of a track in oak woodland. Adults have been recorded in May and September.

Status Not listed in the insect Red Data Book (Shirt, 1987). Extremely local. Formerly restricted to a few areas around Plymouth and recorded until 1911. There were no further records until in 1985 when a single example was found in South-west Yorkshire. Two further individuals were recorded in East Cornwall in 1989.

Threats Uncertain, though possibly threatened by the loss of woodland through clear-felling and conversion to other land use. Neglect, and the conversion to high forest, has led to increased shade and the loss of glades and broad sunny rides.

Published sources Flint, J.H. (1988), Fowler, W.W. (1891), Keys, J.H. (1911).

BARYPEITHES SULCIFRONS   NOTABLE B
A weevil
Order COLEOPTERA   Family CURCULIONIDAE

Barypeithes sulcifrons (Boheman, 1843). Formerly known as: *Barypithes sulcifrons* (Boheman, 1843).

Distribution South West, East Anglia, East Midlands, North East, North West, Dyfed-Powys, North Wales, South East Scotland and South West Scotland.

Habitat and ecology Grassland, coastal cliffs and possibly also disturbed ground. Phytophagous and polyphagous at the roots of plants. Adults have been recorded from March to June and in November.

Status Widespread but local in Great Britain from South West England to southern Scotland. This species has a restricted range in western Europe.

Threats Loss of unimproved grassland through improvement by reseeding or by the application of fertilisers, or by conversion to arable agriculture. Lack of, or changes in grazing or cutting regimes, natural succession and scrub invasion may threaten this species.

Management and conservation Grazing, cutting or some other disturbance, such as rotovation, on a rotational basis, is needed to maintain open conditions.

BRACHYSOMUS ECHINATUS   NOTABLE B
A weevil
Order COLEOPTERA   Family CURCULIONIDAE

Brachysomus echinatus (Bonsdorff, 1785).

Distribution England, South Wales, North Wales, South East Scotland, South West Scotland and North West Scotland.

Habitat and ecology Grassland and coastal cliffs. On calcareous soils. Phytophagous and polyphagous. Probably occurs at the roots of plants. Parthenogenetic. Adults have been found in moss and have been recorded from April to August and in November.

Status Widespread but local in England, parts of Wales and recorded as far north as North West Scotland.

Threats Loss of calcareous grassland through improvement by reseeding or by the application of fertilisers, or by conversion to arable agriculture. Lack of, or changes in grazing or cutting regimes may threaten this species.

Management and conservation Grazing or cutting, on a rotational basis, is needed to maintain open conditions.

BRACHYSOMUS HIRTUS   RARE
A weevil
Order COLEOPTERA   Family CURCULIONIDAE

Brachysomus hirtus (Boheman, 1845).

Distribution Recorded from South Wiltshire, Dorset, South Hampshire, West Sussex, East Kent, West Kent, Surrey, Oxfordshire, Buckinghamshire, West Gloucestershire and Leicestershire & Rutland before 1970 and East Kent and Surrey from 1970 onwards.

CURCULIONIDAE

Habitat and ecology Chalky hillsides and other grassland situations. Not restricted to calcareous soils. Phytophagous and probably polyphagous. Found amongst moss in shady situations, also noted under a piece of rotting wood. This species may be associated with primroses. Adults have been recorded in February, April, May and October.

Status Extremely local. Only known from southern England and recently recorded from only two vice-counties.

Threats This species has probably declined because of the loss of unimproved grassland through improvement by reseeding or by the application of fertilisers, or by conversion to arable agriculture. Lack of, or changes in grazing or cutting regimes may have also contributed to this species' decline.

Management and conservation Grazing or cutting, on a rotational basis, is needed to maintain open conditions.

Published sources Atty, D.B. (1983), Fowler, W.W. (1891), Fowler, W.W. & Donisthorpe, H.St J.K. (1913), Shirt, D.B., ed. (1987), Woodroffe, G.E. (1966).

CAENOPSIS FISSIROSTRIS   NOTABLE B
A weevil
Order COLEOPTERA   Family CURCULIONIDAE

Caenopsis fissirostris (Walton,J., 1847).

Distribution England and Wales.

Habitat and ecology Woodland, heathland, sand pits and disturbed ground. Phytophagous and polyphagous. The larvae almost certainly feed on the roots of plants, while the adults tend to be found in leaf-litter and at the base of plants. Adults have been recorded in January, from March to July, and in September and November.

Status Widespread but local in England and Wales.

Threats The loss of habitat through conversion to other land use such as arable agriculture and development, as well as the the infilling of sand pits. Neglect, natural succession and scrub invasion may be further threats.

Management and conservation Management should aim at maintaining open conditions.

CATHORMIOCERUS ATTAPHILUS
A weevil   **ENDANGERED**
Order COLEOPTERA   Family CURCULIONIDAE

Cathormiocerus attaphilus Brisout, 1880. Formerly known as: *Trachyphloeus attaphilus* (Brisout).

Distribution Recorded from West Cornwall and South Devon before 1970 and West Cornwall and South Devon from 1970 onwards.

Habitat and ecology Coastal cliffs. Phytophagous and probably polyphagous. The larvae feed on the roots of plants, while the adults occur at the base of plants, particularly buck's horn plantain *Plantago coronopus*. Also found in litter under heather. Parthenogenetic. Adults have been recorded in June, July, September and October.

Status Exceedingly local, with very few recent records. First discovered in 1917 on the Lizard, West Cornwall and recorded as recently as 1986. In addition to the Cornish records, there is also a recent record from Wembury near Plymouth, South Devon. This species is also extremely localised in its restricted range in western Europe.

Threats This species may be threatened by erosion in areas frequented by large numbers of walkers, as well as through coastal developments and natural succession and scrub invasion.

Management and conservation Some grazing, cutting or other disturbance is needed to maintain open conditions and encourage the early successional stages.

Published sources Beare, T.H. (1934a), Keys, J.H. (1921), Keys, J.H. (1923), Shirt, D.B., ed. (1987).

CATHORMIOCERUS BRITANNICUS
A weevil   **ENDANGERED, ENDEMIC**
Order COLEOPTERA   Family CURCULIONIDAE

Cathormiocerus britannicus Blair, 1934.

Distribution Recorded from West Cornwall and East Cornwall before 1970 and West Cornwall from 1970 onwards.

Habitat and ecology Coastal cliffs. Phytophagous and probably polyphagous. The larvae feed on the roots of plants while the adults occur at the base of plants, particularly ribwort plantain *Plantago lanceolata*. Also recorded in litter under heather and under thyme. Parthenogenetic. Adults have been recorded from March to June and from September to December.

Status Endemic. Extremely localised with few recent records. Apart from its original discovery in 1908 at Tintagel, East Cornwall, all other records are from the Lizard, West Cornwall.

Threats This species may be threatened by erosion through large numbers of walkers. Coastal developments, natural succession and scrub invasion may be further threats.

Management and conservation Some grazing, cutting or other disturbance is needed to maintain open conditions and encourage early successional stages.

Published sources Beare, T.H. (1934a), Shirt, D.B., ed. (1987).

CATHORMIOCERUS MARITIMUS   RARE
A weevil
Order COLEOPTERA   Family CURCULIONIDAE

Cathormiocerus maritimus Rye, 1874.

Distribution Recorded from West Cornwall, East Cornwall, North Devon, Dorset and South Hampshire before 1970 and West Cornwall, North Devon and South Hampshire from 1970 onwards.

Habitat and ecology Coastal cliffs and rough open ground near the coast. Phytophagous and probably polyphagous. The larvae probably feed on the roots of plants, while the adults occur at the base of plants, particularly buck's-horn plantain *Plantago coronopus*. Parthenogentic. Adults have been recorded from March to June, September and in November.

Status Very local and recently recorded from only three vice-counties. Only known on the coast from South Hampshire to North Devon. Can be numerous where found. This species is also extremely localised in its restricted range in western Europe.

Threats This species may be threatened by erosion in areas frequented by large numbers of walkers. Coastal developments, natural succession and scrub invasion may be further threats.

Management and conservation Some grazing, cutting or other disturbance, such as rotovation, may be needed to maintain open conditions and encourage early successional stages.

Published sources Fowler, W.W. (1891), Fowler, W.W. & Donisthorpe, H.St J.K. (1913), Shirt, D.B., ed. (1987).

CATHORMIOCERUS MYRMECOPHILUS RARE
A weevil
Order COLEOPTERA   Family CURCULIONIDAE

Cathormiocerus myrmecophilus (Seidlitz, 1868). Formerly known as: *Trachyphloeus myrmecophilus* Seidlitz.

Distribution Recorded from West Cornwall, East Cornwall, South Devon, South Hampshire and East Sussex before 1970 and West Cornwall, East Cornwall, South Devon, South Hampshire and East Sussex from 1970 onwards.

Habitat and ecology Coastal cliffs. Phytophagous and probably polyphagous. The larvae probably feed on the roots of plants, while the adults occur at the base of plants. Parthenogenetic. Adults have been found in March, April, August and September. Probably overwinters as an adult.

Status Very local and recorded along the coast from East Sussex to West Cornwall. Recently recorded from only five vice-counties. Can be numerous where found. This species is also extremely localised in its restricted range in western Europe.

Threats This species may be threatened by erosion in areas frequented by large numbers of walkers. Coastal developments, cliff stabilisation schemes, and natural succession and scrub invasion may be further threats.

Management and conservation In areas of unstable cliff occasional slippages are necessary to maintain habitat continuity. Large areas of unstable cliff are required so that the population does not become isolated and subsequently threatened by individual landslips. In more stable areas grazing or cutting may be needed to maintain open conditions and encourage early successional stages.

Published sources Fowler, W.W. (1891), Fowler, W.W. & Donisthorpe, H.St J.K. (1913), Shirt, D.B., ed. (1987).

CATHORMIOCERUS SOCIUS   VULNERABLE
A weevil
Order COLEOPTERA   Family CURCULIONIDAE

Cathormiocerus socius Boheman, 1843.

Distribution Recorded from Isle of Wight before 1970 and Isle of Wight from 1970 onwards.

Habitat and ecology Coastal cliffs. Phytophagous and probably polyphagous. The larvae feed on the roots of plants, while adults occur at the base of plants.

CURCULIONIDAE

Possibly associated with plantain *Plantago*. Mainly found in chalky areas, often by the side of coastal paths. Parthenogenetic. Adults have been recorded from March to September.

Status Only known from a very small number of sites on the south and east coasts of the Isle of Wight. This species has a restricted distribution on the Continent, being only known from France and Spain.

Threats This species may be threatened by erosion in areas frequented by large numbers of walkers. Coastal developments and natural succession and scrub invasion may be further threats.

Management and conservation Some grazing, cutting or other disturbance is needed to maintain open conditions and encourage early successional stages.

Published sources Fowler, W.W. (1891), Fowler, W.W. & Donisthorpe, H.St J.K. (1913), Shirt, D.B., ed. (1987).

CEUTORHYNCHUS ANGULOSUS   NOTABLE A
A weevil
Order COLEOPTERA   Family CURCULIONIDAE

Ceutorhynchus angulosus Boheman, 1845. Formerly known as: *Ceuthorhynchus angulosus* Boheman, 1845, *Ceuthorrhynchus angulosus* Boheman, 1845.

Distribution Recorded from West Norfolk, Cambridgeshire, Huntingdonshire, Caernarvonshire, South Lancashire and South-west Yorkshire before 1970 and North Somerset, West Suffolk, Cardiganshire and South-west Yorkshire from 1970 onwards.

Habitat and ecology Peatland, particularly fenland, and riverbanks. Phytophagous. Associated with hemp-nettle *Galeopsis* spp., in particular common hemp-nettle *G. tetrahit*, and to a lesser extent large-flowered hemp-nettle *G. speciosa*. Also recorded on marsh woundwort *Stachys palustris* and hedge woundwort *S. sylvatica*. The larvae are stem borers. Adults occur on the foliage of the foodplant and have been found in May, June and September.

Status Very local and recently recorded from only four, widely scattered vice-counties. Old records show that this species has been recorded, with a scattered distribution, in England as far north as South-west Yorkshire and was also recorded in Wales. A record for Hertfordshire requires confirmation.

Threats This species is threatened by commercial peat cutting, drainage and natural succession.

Management and conservation Water tables should be maintained at levels that keep the surface peat moist. This species may thrive in areas of small scale turf cutting, providing this is not accompanied by drainage. It is possible that lower populations may be maintained where the foodplant occurs in relatively undisturbed situations.

Published sources Fordham, W.J. (1942), Fowler, W.W. & Donisthorpe, H.St J.K. (1913), Fryer, J.C.F. (1929b), Hodge, P.J. (1989), de Worms, C.G.M. & Payne, R.M. (1952).

CEUTORHYNCHUS ARQUATUS
A weevil   INDETERMINATE
Order COLEOPTERA   Family CURCULIONIDAE

Ceutorhynchus arquatus (Herbst, 1795). Formerly known as: *Ceuthorhynchus arquatus* (Herbst, 1795), *Ceuthorrhynchus arcuatus* (Herbst, 1795), *Ceuthorrhynchus arquatus* (Herbst, 1795).

Distribution Recorded from Surrey, North Essex, East Norfolk, North Lincolnshire, Nottinghamshire and South Lancashire before 1970.

Habitat and ecology Wetland. Phytophagous. Associated with gipsywort *Lycopus europaeus*, possibly also on *Mentha* species.

Status Status revised from RDB 3 (Rare) in Shirt (1987). This century, there are only two unconfirmed records for this species, the last being in 1940 from East Anglia.

Published sources Fowler, W.W. (1891), Shirt, D.B., ed. (1987).

CEUTORHYNCHUS ATOMUS   NOTABLE A
A weevil
Order COLEOPTERA   Family CURCULIONIDAE

Ceutorhynchus atomus Boheman, 1845. Formerly known as: *Ceuthorhynchus atomus* Boheman, 1845, *Ceuthorrhynchus atomus* Boheman, 1845, *Ceuthorrhynchus setosus* Boheman, 1845.

Distribution Recorded from East Cornwall, South Devon, East Sussex, West Kent, Surrey, Hertfordshire, Berkshire, West Suffolk, East Norfolk, Cheshire, South-east Yorkshire, Mid-west Yorkshire, Cumberland and Kirkcudbrightshire before 1970 and North Devon, East Kent, Surrey, West Suffolk, Glamorgan, Cardiganshire and Anglesey from 1970 onwards.

Habitat and ecology Disturbed ground, particularly on chalky or sandy substrates. Also recorded from coastal shingle, saltmarsh and on waste ground. Phytophagous. Associated with thale cress *Arabidopsis thaliana*, though occasionally on shepherd's cress *Teesdalia nudicaulis*. Also recorded from scurvygrass *Cochlearia* and hairy bitter cress *Arabis hirsuta*. Doubtfully associated with wild candy tuft *Iberis amara* and watercress *Nasturtium officinale*. On the Continent, this species has also been recorded on whitlowgrass *Draba* species. In Britain, adults have been recorded from April to August.

Status Very local and recently recorded from seven vice-counties, all in the southern half of Britain. Old records suggest that this species was formerly more widespread in Britain and recorded as far north as Kircudbrightshire in Scotland.

Threats This species is threatened by change in land use, such as through conversion to arable agriculture, and urban and coastal development. Natural succession may be a further threat.

Management and conservation Except on coastal shingle, which should be left unmanaged, disturbance, such as rotovation, is necessary to maintain open conditions and encourage early successional stages.

Published sources Cooter, J. (1990c), Fowler, W.W. (1891), Fowler, W.W. & Donisthorpe, H.St J.K. (1913), Shaw, S. (1954).

CEUTORHYNCHUS CAMPESTRIS   NOTABLE B
A weevil
Order COLEOPTERA   Family CURCULIONIDAE

Ceutorhynchus campestris Gyllenhal, 1837. Formerly known as: *Ceuthorhynchus chrysanthemi* sensu auct. Brit. not (Germar, 1824), *Ceuthorrhynchus chrysanthemi* sensu auct. Brit. not (Germar, 1824).

Distribution England, Dyfed-Powys and North Wales.

Habitat and ecology Disturbed ground, roadside verges and field margins. Phytophagous. Associated with ox-eye daisy *Leucanthemum vulgare*. Adults have been recorded from April to August.

Status Widespread but local in England, also recorded in parts of Wales.

Threats This species is threatened by changes in land use, such as conversion to arable, and urban and tourist

development. Natural succession may be a further threat.

Management and conservation Some grazing, rotational cutting or other disturbance, such as rotovation, is necessary to maintain open conditions and encourage early successional stages.

CEUTORHYNCHUS CONSTRICTUS NOTABLE B
A weevil
Order COLEOPTERA   Family CURCULIONIDAE

Ceutorhynchus constrictus (Marsham, 1802). Formerly known as: *Ceuthorhynchus constrictus* (Marsham, 1802), *Ceuthorrhynchus constrictus* (Marsham, 1802).

Distribution South East, South, South West, East Anglia, East Midlands, West Midlands, North East, South Wales, South East Scotland and North East Scotland.

Habitat and ecology Disturbed ground, roadside verges and field margins. Phytophagous. Associated with garlic mustard *Alliaria petiolata*. The larvae feed in the pods of the foodplant. Adults have been recorded from March to June.

Status Widespread but local in England, and recorded in parts of Wales and Scotland. This species is difficult to identify and may be confused with other members of the genus. Consequently, the exact status of this species is hard to assess.

Threats This species is threatened by natural succession, the use of herbicides and pesticides, and possibly a change in land use.

Management and conservation Disturbance, such as rotovation, is needed to maintain open conditions.

CEUTORHYNCHUS EUPHORBIAE   NOTABLE A
A weevil
Order COLEOPTERA   Family CURCULIONIDAE

Ceutorhynchus euphorbiae Brisout, 1866. Formerly known as: *Ceuthorhynchus euphorbiae* Brisout, 1866, *Ceuthorrhynchus euphorbiae* Brisout, 1866.

Distribution Recorded from South Devon, North Devon, East Kent, West Kent, Surrey, Hertfordshire, Berkshire, Oxfordshire, Buckinghamshire, West Suffolk, Huntingdonshire, East Gloucestershire, Warwickshire, South Lancashire, Cumberland, Dumfriesshire and Dunbartonshire before 1970 and

CURCULIONIDAE

East Kent, Surrey, Oxfordshire, Radnorshire, Westmorland & North Lancashire, Cumberland and Lanarkshire from 1970 onwards.

Habitat and ecology Downland, chalky hillsides, field margins, roadside verges and possibly wetland. Phytophagous. Recorded from ground-ivy *Glechoma hederacea*, viper's-bugloss *Echium vulgare*, common ragwort *Senecio jacobaea* and speedwell *Veronica*. The beetle is probably also associated with forget-me-not *Myosotis*. On the Continent, this species has been associated with field forget-me-not *M. arvensis* and water forget-me-not *M. scorpioides*. In Britain, adults have been recorded in January, from March to June, and in September and November.

Status Very local and recently recorded from seven, widely scattered, vice-counties. Old records show that this species has been widely recorded from South Devon to Dunbarton in Scotland.

Threats Loss of calcareous grassland through improvement by reseeding or by the application of fertilisers, or by conversion to arable agriculture. Development, the lack of, or changes in grazing or cutting regimes, and the use of herbicides and pesticides may also threaten this species.

Management and conservation Grazing, cutting or some other disturbance, such as rotovation, on a rotational basis, is needed to maintain open conditions and encourage early successional stages.

Published sources Atty, D.B. (1983), Fowler, W.W. (1891), Fowler, W.W. & Donisthorpe, H.St J.K. (1913), Owen, J.A. (1989a), Read, R.W.J. (1989c).

CEUTORHYNCHUS GEOGRAPHICUS
A weevil   **NOTABLE B**
Order COLEOPTERA   Family CURCULIONIDAE

Ceutorhynchus geographicus (Goeze, 1777). Formerly known as: *Ceuthorhynchus geographicus* (Goeze, 1777), *Ceuthorrhynchus geographicus* (Goeze, 1777).

Distribution South East, South, South West, East Anglia, East Midlands, West Midlands, North East, South Wales and South East Scotland.

Habitat and ecology Disturbed ground, grassland and coastal habitats, including coastal shingle, on sandy and chalky soils. Phytophagous. Associated with viper's-bugloss *Echium vulgare*. Larvae develop on the roots of the foodplant and pupate in a cocoon in the soil. Adults have been recorded from April to October.

Status Widespread but local in England, also recorded in South Wales and South East Scotland.

Threats Loss of habitat through improvement and conversion to arable agriculture, urban and tourist development, gravel extraction and natural succession.

Management and conservation Except on coastal shingle, which can be left unmanaged, disturbance, such as rotovation, is necessary to maintain open conditions and encourage early successional stages.

CEUTORHYNCHUS HEPATICUS
A weevil   **INSUFFICIENTLY KNOWN**
Order COLEOPTERA   Family CURCULIONIDAE

Ceutorhynchus hepaticus Gyllenhal, 1837. Formerly known as: *Ceuthorhynchus hepaticus* Gyllenhal, 1837, *Ceuthorrhynchidius hepaticus* (Gyllenhal), *Ceuthorrhynchus hepaticus* (Gyllenhal).

Distribution Recorded from East Sussex, East Kent, Surrey, East Suffolk, East Norfolk, Cambridgeshire, Huntingdonshire and Derbyshire before 1970.

Habitat and ecology Disturbed ground and field margins. Phytophagous. Almost certainly associated with a variety of species of Cruciferae, this beetle has been recorded from the introduced wallflower cabbage *Rhynchosinapis cheiranthos*. On the Continent, this species has also been found on hedge mustard *Sisymbrium officinale*. In Britain, adults have been recorded in August.

Status Not listed in the insect Red Data Book (Shirt, 1987). No recent records, and possibly not recorded this century. This species is difficult to identify and may be confused with other members of the genus. Consequently, the exact status of this species is hard to assess.

Threats Uncertain, though this species may have declined through changes in land use such as conversion to arable agriculture.

Published sources Fowler, W.W. (1891).

CEUTORHYNCHUS HIRTULUS   NOTABLE B
A weevil
Order COLEOPTERA   Family CURCULIONIDAE

Ceutorhynchus hirtulus Germar, 1824. Formerly known as: *Ceuthorhynchus hirtulus* Germar, 1824, *Ceuthorrhynchus hirtulus* Germar, 1824.

Distribution South East, South, South West, East Anglia, West Midlands, North East, North West, South Wales, Dyfed-Powys, South East Scotland, South West Scotland and North West Scotland.

Habitat and ecology Sand dunes, though also recorded from disturbed ground and rarely in wetland and woodland. A predominantly maritime species of dry, sandy open ground. Phytophagous. Associated with common whitlowgrass *Erophila verna*, though probably also on flixweed *Descurainia sophia*, hedge mustard *Sisymbrium officinale*, water-cress *Nasturtium officinale*, and possibly other Cruciferae. On the Continent, this species has also been recorded from thale cress *Arabidopsis thaliana*. The larva has been recorded in galls on the stems of *E. verna*, pupation occurring in the soil. In Britain, adults have been recorded from May to September.

Status Widespread but local in England, also recorded in parts of Wales and Scotland.

Threats Loss of dune and sandy habitats, particularly through afforestation, urban and holiday development. The degradation of remaining habitat by excessive disturbance of the vegetation through activities such as motorbike access, horse-riding and human trampling. Natural succession may be a further threat.

Management and conservation On dunes and sandy soils, some grazing or other disturbance, such as rotovation, may be desirable to maintain the early successional stages and prevent the invasion of scrub.

CEUTORHYNCHUS INSULARIS   ENDANGERED
A weevil
Order COLEOPTERA   Family CURCULIONIDAE

Ceutorhynchus insularis Dieckmann, 1971. Formerly known as: *Ceuthorhynchus contractus* sensu auct. partim not (Marsham, 1802), *Ceuthorrhynchus contractus var. pallipes* Crotch, 1865.

Distribution Recorded from Outer Hebrides from 1970 onwards.

Habitat and ecology Coastal sites, particularly grassland. Phytophagous. Associated with common scurvygrass *Cochlearia officinalis*.

Status Extremely local. Described new to science in 1971 and known in Great Britain only from a small number of examples from the island of St Kilda, Outer Hebrides. A published record for Lundy may be in error. Outside Great Britain, the weevil is known only from a few islands off the southern coast of Iceland.

Management and conservation St Kilda is an NNR.

Published sources Shirt, D.B., ed. (1987).

CEUTORHYNCHUS MIXTUS   NOTABLE B
A weevil
Order COLEOPTERA   Family CURCULIONIDAE

Ceutorhynchus mixtus Mulsant and Rey, 1858. Formerly known as: *Ceuthorhynchus mixtus* Mulsant and Rey, 1858, *Ceuthorrhynchidius mixtus* (Mulsant and Rey), *Ceuthorrhynchus mixtus* (Mulsant and Rey).

Distribution South West, South, South East, East Anglia, East Midlands, West Midlands, Dyfed-Powys and North West.

Habitat and ecology Disturbed ground and lightly wooded heathland. Phytophagous. Associated with climbing corydalis *Corydalis claviculata*, common fumitory *Fumaria officinalis* and other species of fumitory. On the Continent, this species has also been recorded from fine-leaved fumitory *F. parviflora*. The larvae feed in stem galls. In Britain, adults have been recorded from April to August; this beetle has also been found in December from oak leaf-litter and humus.

Status Local. Recently recorded, with a scattered distribution, from North Devon and Dorset to Cumberland in England, and also recorded in parts of Wales. This species is difficult to identify and may be confused with other members of the genus. Consequently, the exact status of this species is hard to assess.

Threats Loss of habitat through improvement and conversion to arable, clear-felling and conversion to other land use, development and natural succession. The use of herbicides and pesticides may be a further threat.

Management and conservation Grazing, cutting or some other disturbance, such as rotovation, may be necessary to maintain open conditions and encourage early successional stages.

CURCULIONIDAE

CEUTORHYNCHUS MOELLERI
A weevil   **INSUFFICIENTLY KNOWN**
Order COLEOPTERA   Family CURCULIONIDAE

Ceutorhynchus moelleri Thomson, 1868. Formerly known as: *Ceuthorhynchus moelleri* Thomson, 1868, *Ceuthorhynchus molleri* Thomson, 1868, *Ceuthorrhynchus molleri* Thomson, 1868.

Distribution Recorded from North Hampshire, West Kent, Surrey, Berkshire and Oxfordshire before 1970 and Oxfordshire from 1970 onwards.

Habitat and ecology Downland. Phytophagous. Possibly associated with hawkweed *Hieracium* and hawkbit *Leontodon*. Adults have been recorded from May to July.

Status Status revised from RDB 3 (Rare) in Shirt (1987). Extremely local and only known in southern England. Recently recorded from just one site, an NNR, in Oxfordshire. This species is difficult to identify and may be confused with other members of the genus. Consequently, the exact status of this species is hard to assess.

Threats This species may have declined through loss of downland because of improvement and conversion to arable agriculture, development and natural succession.

Management and conservation Grazing, or possibly cutting, is needed to maintain open conditions.

Published sources Fowler, W.W. (1891), Massee, A.M. (1963), Shirt, D.B., ed. (1987).

CEUTORHYNCHUS PARVULUS   **RARE**
A weevil
Order COLEOPTERA   Family CURCULIONIDAE

Ceutorhynchus parvulus Brisout, 1869. Formerly known as: *Ceuthorhynchus parvulus* Brisout, 1869, *Ceuthorrhynchus parvulus* Brisout, 1869.

Distribution Recorded from West Cornwall, East Cornwall, South Devon and North Devon before 1970 and East Cornwall and North Devon from 1970 onwards.

Habitat and ecology Grassland and disturbed ground on or near the coast. Probably also sand-dunes. Phytophagous. Associated with Smith's pepperwort *Lepidium heterophyllum*. On the Continent, this species has also been recorded from field pepperwort *L. campestre*. In Britain, adults have been recorded in May and June.

Status Very local and only known in South West England. Recently recorded from just two vice-counties. This species is difficult to identify and may be confused with other members of the genus. Consequently, the exact status of this species is hard to assess.

Threats This species is threatened by improvement and conversion to arable agriculture, urban and coastal development, and natural succession.

Management and conservation Disturbance, such as rotovation, is needed to maintain open conditions and encourage early successional stages.

Published sources Fowler, W.W. (1891), Henderson, J.L. (1951), Shirt, D.B., ed. (1987), Walker, J.J. (1931).

CEUTORHYNCHUS PECTORALIS   **NOTABLE A**
A weevil
Order COLEOPTERA   Family CURCULIONIDAE

Ceutorhynchus pectoralis Weise, 1895. Formerly known as: *Ceuthorhynchus chalybaeus* sensu auct. Brit. ?Germar, 1824, *Ceuthorrhynchus chalybaeus* sensu auct. Brit. ?Germar, 1824.

Distribution Recorded from West Kent, Surrey, East Norfolk, Dumfriesshire, Peebleshire, Berwickshire and Midlothian before 1970 and West Sussex, East Sussex, South Essex, Westmorland & North Lancashire and Berwickshire from 1970 onwards.

Habitat and ecology Wetland and woodland. Phytophagous. Associated with cuckooflower *Cardamine pratensis* and probably large bitter-cress *C. amara*. Possibly also associated with other related species. On the Continent, this species has been associated with field penny-cress *Thlaspi arvense*, marsh yellow-cress *Rorippa islandica* and winter-cress *Barbarea*. In Britain, adults have been recorded in February, April, July, August, October and November.

Status Status revised from RDB 3 (Rare) in Shirt (1987). Very local and recently recorded from five, widely scattered, vice-counties, from West Sussex to Berwickshire in Scotland. Usually recorded as single individuals. Dumfriesshire is listed in the distribution section, though this record may refer to Kircudbrightshire.

Threats This species is threatened by loss of habitat and conversion to other land use, such as conversion to arable agriculture and development. Drainage, and natural succession are further threats.

Management and conservation Grazing or cutting, on a rotational basis, is needed to maintain open conditions.

Published sources Allen, A.A. (1945a), Collier, M.J. (1988a), Hodge, P.J. (1980), Kevan, D.K. (1959), Shirt, D.B., ed. (1987).

CEUTORHYNCHUS PERVICAX   NOTABLE A
A weevil
Order COLEOPTERA   Family CURCULIONIDAE

Ceutorhynchus pervicax Weise, 1883. Formerly known as: *Ceuthorhynchus suturelius* sensu auct. not Gyllenhal, 1837, *Ceuthorrhynchus suturellus* sensu auct. not Gyllenhal, 1837.

Distribution Recorded from East Sussex, East Kent and West Kent before 1970 and West Sussex, East Sussex, East Kent and Surrey from 1970 onwards.

Habitat and ecology Wetland, damp pastures and roadside verges. Phytophagous. Associated with cuckooflower *Cardamine pratensis* and large bitter-cress *C. amara*. On the Continent, this species is associated with *Cardamine* spp. and *Dentaria* species. In Britain, adults have been recorded from March to May.

Status Very local, though possibly increasing in distribution. Only known from South East England.

Threats This species is threatened by loss of habitat through improvement and conversion to arable agriculture, development, drainage and natural succession.

Management and conservation Grazing or cutting, on a rotational basis, is needed to maintain open conditions.

Published sources Allen, A.A. (1957c), Fowler, W.W. (1891), Hodge, P.J. (1980).

CEUTORHYNCHUS PILOSELLUS
A weevil   **VULNERABLE**
Order COLEOPTERA   Family CURCULIONIDAE

Ceutorhynchus pilosellus Gyllenhal, 1837. Formerly known as: *Ceuthorhynchus pilosellus* Gyllenhal, 1837, *Ceuthorrhynchus pilosellus* Gyllenhal, 1837.

Distribution Recorded from West Cornwall, South Devon, East Kent, West Kent, Surrey and Berkshire before 1970 and West Cornwall and Glamorgan from 1970 onwards.

Habitat and ecology Grassland on sand dunes, downland, sandy habitats and probably disturbed ground. Phytophagous. Associated with lesser dandelion *Taraxacum laevigatum*. The larvae probably feed in the capitula. Adults have been found under the leaves of the foodplant and have been recorded in January, April, May and September.

Status Very local and recently recorded from only two sites, one being Merthyr Mawr Warren in Glamorganshire. Only recorded in southern England and South Wales.

Threats Loss of dune habitat, particularly through afforestation, urban and holiday development. The degradation of remaining habitat by excessive disturbance of the vegetation through activities such as motorbike access, horse-riding and human trampling. Improvement and conversion to arable agriculture, and natural succession are further threats.

Management and conservation Grazing, cutting or possibly some other disturbance may be desirable to maintain the early successional stages and prevent the invasion of scrub. Merthyr Mawr Warren has been designated as an SSSI.

Published sources Cooter, J. (1990c), Fowler, W.W. (1891), Fowler, W.W. & Donisthorpe, H.St J.K. (1913), Shirt, D.B., ed. (1987).

CEUTORHYNCHUS PULVINATUS   NOTABLE A
A weevil
Order COLEOPTERA   Family CURCULIONIDAE

Ceutorhynchus pulvinatus Gyllenhal, 1837. Formerly known as: *Ceuthorhynchus pulvinatus* Gyllenhal, 1837, *Ceuthorrhynchidius pulvinatus* (Gyllenhal), *Ceuthorrhynchus pulvinatus* (Gyllenhal).

Distribution Recorded from East Suffolk, West Suffolk, East Norfolk, West Norfolk and Cambridgeshire before 1970 and East Suffolk, West Suffolk and West Norfolk from 1970 onwards.

Habitat and ecology Disturbed ground. Phytophagous. Associated with flixweed *Descurainia sophia*. Adults have been recorded from June to October.

Status Very local but can be numerous where it occurs. Primarily found in the Breckland of East Anglia. This species is difficult to identify and may be confused with other members of the genus. Consequently, the exact status of this species is hard to assess.

CURCULIONIDAE

Threats Loss of habitat through improvement and conversion to arable agriculture, afforestation, development and natural succession.

Management and conservation Disturbance, such as rotovation, is needed to maintain open conditions and encourage early successional stages.

Published sources Allen, A.A. (1981d), Fowler, W.W. (1891), Fryer, J.C.F. (1929a).

CEUTORHYNCHUS PUMILIO   NOTABLE A
A weevil
Order COLEOPTERA   Family CURCULIONIDAE

Ceutorhynchus pumilio (Gyllenhal, 1827). Formerly known as: *Ceuthorhynchus posthumus* sensu auct. Brit. not Germar, 1824, *Ceuthorrhynchidius posthumus* sensu auct. Brit. not (Germar), *Coeliodes posthumus* sensu auct. Brit. not (Germar).

Distribution Recorded from East Sussex, East Kent, Surrey, Middlesex, East Suffolk, West Suffolk, Worcestershire, Nottinghamshire, West Lancashire and Cumberland before 1970 and East Kent, West Suffolk, West Norfolk and South-east Yorkshire from 1970 onwards.

Habitat and ecology Disturbed ground and coastal habitats, including coastal shingle, particularly in sandy places. Phytophagous. Associated with shepherd's cress *Teesdalia nudicaulis*. The larvae feed on unripe seeds in the pods. Adults have been recorded from April to June and in September.

Status Very local and recently recorded from only four vice-counties. Old records show that this species has been widely recorded throughout England from East Kent to Cumberland. This beetle can be numerous where found.

Threats This species is threatened by urban and tourist development, improvement and conversion to arable agriculture, and gravel extraction. Natural succession may be a further threat.

Management and conservation Except on coastal shingle, which should be left unmanaged, disturbance, such as rotovation, is needed to maintain open conditions and encourage early successional stages.

Published sources Collier, M.J. (1988a), Fowler, W.W. (1891), Fowler, W.W. & Donisthorpe, H.St J.K. (1913).

CEUTORHYNCHUS PUNCTIGER   NOTABLE B
A weevil
Order COLEOPTERA   Family CURCULIONIDAE

Ceutorhynchus punctiger Sahlberg, 1835. Formerly known as: *Ceuthorhynchus punctiger* Sahlberg, 1835, *Ceuthorrhynchus punctiger* Sahlberg, 1835.

Distribution England, South Wales and Dyfed-Powys.

Habitat and ecology Particularly on sandy soils. Found in grassland on coastal sand dunes, though also on other grasslands. Phytophagous. Associated with dandelion *Taraxacum officinale*. Adults have been recorded from May to August.

Status Widespread but local in England, also recorded in South Wales. There is an old record from Cardiganshire, Wales. A record from North West Scotland requires confirmation.

Threats Loss of dune habitat, particularly through afforestation, and urban and holiday development. The degradation of remaining habitat by excessive disturbance of the vegetation through activities such as motorbike access, horse-riding and human trampling. Improvement and conversion to arable agriculture, and natural succession are further threats.

Management and conservation Some grazing or other disturbance may be desirable to maintain the early successional stages and prevent the invasion of scrub.

CEUTORHYNCHUS QUERCETI   VULNERABLE
A weevil
Order COLEOPTERA   Family CURCULIONIDAE

Ceutorhynchus querceti (Gyllenhal, 1813). Formerly known as: *Ceuthorhynchus querceti* (Gyllenhal, 1813), *Ceuthorrhynchus querceti* (Gyllenhal, 1813), *Coeliodes querceti* (Gyllenhal).

Distribution Recorded from East Norfolk before 1970 and East Norfolk and West Norfolk from 1970 onwards.

Habitat and ecology Wetland, including a riverbank. Phytophagous. Associated with marsh yellow-cress *Rorippa islandica*, and possibly also with great yellow-cress *R. amphibia*. The larvae feed in the stems of the foodplant. Adults have been recorded in June and September.

Status Known in Great Britain only from a few sites in the Norfolk Broads and recently found at a single locality in West Norfolk.

Threats Loss of habitat through drainage, water abstraction schemes, river engineering schemes, development, and improvement and conversion to arable agriculture. The use of motorboats may be a further threat.

Management and conservation Water tables should be maintained at high levels. Water bodies should be isolated from sources of eutrophication and pollution. Grazing or cutting, on a rotational basis, is needed to maintain open conditions.

Published sources Fowler, W.W. & Donisthorpe, H.St J.K. (1913), Morris, M.G. & Collier, M.J. (1989), Shirt, D.B., ed. (1987).

CEUTORHYNCHUS QUERCICOLA NOTABLE A
A weevil
Order COLEOPTERA   Family CURCULIONIDAE

Ceutorhynchus quercicola (Paykull, 1792). Formerly known as: *Ceuthorhynchus quercicola* (Paykull, 1792), *Ceuthorrhynchidius quercicola* (Paykull), *Ceuthorrhynchus quercicola* (Paykull).

Distribution Recorded from East Cornwall, South Devon, North Devon, Isle of Wight, East Sussex, East Kent, West Kent, Surrey, Hertfordshire, Middlesex, Berkshire, Oxfordshire, Buckinghamshire, Cambridgeshire, East Gloucestershire, Shropshire, Glamorgan, Caernarvonshire, Derbyshire, South Lancashire, South-west Yorkshire, South Northumberland, North Northumberland, Cumberland, Dumfriesshire, Peeblesshire, Roxburghshire, Berwickshire, East Lothian, Midlothian and Fife before 1970 and Dorset, East Sussex, West Suffolk, North Northumberland and Cumberland from 1970 onwards.

Habitat and ecology Disturbed ground. Phytophagous. Associated with common fumitory *Fumaria officinalis*, and possibly other species of fumitory. The larvae feed in galls on the lower stems of the food plant. Adults have been recorded from May to August.

Status Very local and recently recorded from only five, widely scattered, vice-counties. Old records show that this species has been widely recorded from East Cornwall throughout England, north to Fife in Scotland. There are also old records from Wales. This species is difficult to identify and may be confused with other members of the genus. Consequently, the exact status of this species is hard to assess.

Threats This species is threatened by improvement and conversion to arable agriculture, development, and natural succession.

Management and conservation Disturbance, such as rotovation, is needed to maintain open conditions and encourage early successional stages.

Published sources Allen, A.A. (1947e), Atty, D.B. (1983), Eyre, M.D. (1987), Fowler, W.W. (1891), Fowler, W.W. & Donisthorpe, H.St J.K. (1913).

CEUTORHYNCHUS RAPAE   NOTABLE B
A weevil
Order COLEOPTERA   Family CURCULIONIDAE

Ceutorhynchus rapae Gyllenhal, 1837. Formerly known as: *Ceuthorhynchus rapae* Gyllenhal, 1837, *Ceuthorrhynchus rapae* Gyllenhal, 1837.

Distribution South East, South, South West, East Anglia, East Midlands, North West and South Wales.

Habitat and ecology Disturbed ground, gardens, hedgebanks, ditches and roadside verges. Phytophagous. Associated with members of the Cruciferae, with records from hedge mustard *Sisymbrium officinale*, flixweed *Descurainia sophia*, garlic mustard *Alliaria petiolata*, field pepperwort *Lepidium campestre*, horse-radish *Armoracia rusticana* and wallflower *Cheiranthus cheiri*. On the Continent, this species has also been reported from wild cabbage *Brassica oleracea*. The larvae feed in stems. In Britain, adults have been recorded from April to August and in October.

Status Widespread but local in England, and recorded in South Wales.

Threats This species is probably threatened by development and natural succession.

Management and conservation Disturbance, such as rotovation, may be necessary to maintain open conditions and encourage early successional stages.

CEUTORHYNCHUS RESEDAE   NOTABLE B
A weevil
Order COLEOPTERA   Family CURCULIONIDAE

Ceutorhynchus resedae (Marsham, 1802). Formerly known as: *Ceuthorhynchus resedae* (Marsham, 1802), *Ceuthorrhynchus resedae* (Marsham, 1802).

CURCULIONIDAE

Distribution South East, South, South West, East Anglia, East Midlands, West Midlands, North East, South Wales and South East Scotland.

Habitat and ecology Disturbed ground. This species has been recorded from chalk downland and Breckland. Phytophagous. Associated with weld *Reseda luteola* and possibly also on wild mignonette *R. lutea*, white mignonette *R. alba* and corn mignonette *R. phyteuma*. Adults have been recorded in January and from May to August.

Status Widespread but local in England, and also recorded in South Wales and South East Scotland. This species is difficult to identify and may be confused with other members of the genus. Consequently, the exact status of this species is hard to assess.

Threats Loss of habitat through development, afforestation, improvement and conversion to arable agriculture, and natural succession.

Management and conservation Disturbance, such as rotovation, is needed to maintain open conditions and encourage early successional stages.

CEUTORHYNCHUS SYRITES   ENDANGERED
A weevil
Order COLEOPTERA   Family CURCULIONIDAE

Ceutorhynchus syrites Germar, 1824. Formerly known as: *Ceuthorhynchus syrites* Germar, 1824, *Ceuthorrhynchus syrites* Germar, 1824.

Distribution Recorded from West Cornwall, South Devon, West Kent, Middlesex, Oxfordshire, East Gloucestershire, Worcestershire and Warwickshire before 1970.

Habitat and ecology Probably found on disturbed ground and in grassland, perhaps with a preference for calcareous soils. Phytophagous. Possibly associated with species of Cruciferae. On the Continent, this species has been associated with hedge mustard *Sisymbrium officinale*, tall rocket *S. altissimum*, flixweed *Descurainia sophia*, hoary cress *Cardaria draba*, charlock *Sinapis arvensis*, wild radish *Raphanus raphanistrum*, garden radish *R. sativus*, sea-kale *Crambe maritima*, common scurvygrass *Cochlearia officinalis*, wild cabbage *Brassica oleracea*, rape *B. napus* and garlic mustard *Alliaria petiolata*. In Britain, adults have been recorded from May to July and in October.

Status Status revised from RDB 3 (Rare) in Shirt (1987). No recent records. Last recorded in 1966 from Aston Rowant, Oxfordshire.

Threats This species may have declined through the loss of habitat because of improvement and conversion to arable agriculture and development. Natural succession may have also contributed to this beetle's decline.

Published sources Fowler, W.W. (1891), Shirt, D.B., ed. (1987).

CEUTORHYNCHUS TERMINATUS NOTABLE B
A weevil
Order COLEOPTERA   Family CURCULIONIDAE

Ceutorhynchus terminatus (Herbst, 1795). Formerly known as: *Ceuthorhynchus terminatus* (Herbst, 1795), *Ceuthorrhynchidius terminatus* (Herbst), *Ceuthorrhynchus terminatus* (Herbst).

Distribution England, South Wales, North Wales, South East Scotland and South West Scotland.

Habitat and ecology Grassland, coastal cliffs, undercliffs and disturbed ground. Predominantly coastal but also occurring inland. Phytophagous. Associated with wild carrot *Daucus carota*. On the Continent, this species has been recorded from a variety of umbellifers including wild celery *Apium graveolens*, hemlock *Conium maculatum* and garden parsley *Petroselinum crispum*. Larvae feed in the stems and petioles of the foodplant. In Britain, adults have been recorded from March to June and in August.

Status Widespread but local in England, also recorded in parts of Wales and Scotland.

Threats This species is threatened by cliff stabilisation schemes and the construction of sea defences. Coastal developments may reduce the amount of available habitat. Activities that accelerate or reduce the rate of erosion should be avoided. Improvement and conversion to arable agriculture, and natural succession are further threats.

Management and conservation In areas of unstable cliff, occasional slippages are necessary to maintain habitat continuity. Large areas of unstable cliff are required so that the population does not become isolated and subsequently threatened by individual landslips. In other habitats grazing, cutting or possibly some other disturbance is needed to maintain open conditions.

CEUTORHYNCHUS THOMSONI
A weevil INSUFFICIENTLY KNOWN
Order COLEOPTERA Family CURCULIONIDAE

Ceutorhynchus thomsoni Kolbe, 1900. Formerly known as: *Ceuthorhynchus moguntiacus* sensu auct. not Schultze, 1895, *Ceuthorrhynchus chalybaeus var. viridipennis* sensu Fowler, 1891 ?Brisout, *Ceuthorrhynchus moguntiacus* sensu Fowler, 1891 ?Brisout.

Distribution Recorded from North Devon, East Kent, West Kent, Surrey, Hertfordshire, Berkshire, Oxfordshire, Cambridgeshire, East Gloucestershire and Denbighshire before 1970 and Berkshire from 1970 onwards.

Habitat and ecology Hedgerows, field margins and wood-edges, especially on chalky soils. Phytophagous. Associated with garlic mustard *Alliaria petiolata* and probably on other species of Cruciferae. On the Continent, this species has been recorded from perennial wall-rocket *Diplotaxis tenuifolia*. In Britain, adults have been recorded from March to August.

Status Not listed in the insect Red Data Book (Shirt, 1987). Only known from southern England and an old record in North Wales. Recently recorded from just one vice-county. This species is very difficult to identify and may be confused with other members of the genus. Consequently, the exact status of this species is hard to assess.

Threats Uncertain, although this species is probably threatened by loss of habitat through change in land use. The use of herbicides and pesticides may be a further threat.

Management and conservation Some grazing or cutting may be needed to maintain open conditions.

Published sources Atty, D.B. (1983), Kevan, D.K. (1961).

CEUTORHYNCHUS TRIANGULUM NOTABLE B
A weevil
Order COLEOPTERA Family CURCULIONIDAE

Ceutorhynchus triangulum Boheman, 1845. Formerly known as: *Ceuthorhynchus triangulum* Boheman, 1845, *Ceuthorrhynchus triangulum* Boheman, 1845.

Distribution South East, South, South West, East Anglia, East Midlands, North West and South West Scotland.

Habitat and ecology Roadside verges, field margins, grassland and disturbed ground. Phytophagous. Associated with yarrow *Achillea millefolium*. The larvae feed in the upper stems of the foodplant. Adults have been recorded from May to September.

Status Widespread but local in southern England, also recorded from Cumberland to Dumfriesshire in Scotland.

Threats Loss of habitat through improvement and conversion to arable agriculture, afforestation, and development. Natural succession, and the use of herbicides and pesticides may be further threats.

Management and conservation Grazing, cutting or possibly some other disturbance may be necessary to maintain open conditions.

CEUTORHYNCHUS TRIMACULATUS
A weevil NOTABLE B
Order COLEOPTERA Family CURCULIONIDAE

Ceutorhynchus trimaculatus (F., 1775). Formerly known as: *Ceuthorhynchus trimaculatus* (F., 1775), *Cuethorrhynchus trimaculatus* (F., 1775).

Distribution South East, South, South West, East Anglia, East Midlands, West Midlands, North East and South Wales.

Habitat and ecology Grassland, field margins and possibly other disturbed ground. Phytophagous. Associated with musk thistle *Carduus nutans*. On the Continent, this beetle is also associated with *Cirsium* species. In Britain, adults have been noted from late April to September.

Status Widespread but local in England, also recorded in South Wales.

Threats Loss of grassland through improvement by reseeding or by the application of fertilisers, or by conversion to arable agriculture. Development, the lack of, or changes in grazing or cutting regimes, and the use of herbicides and pesticides also threaten this species.

Management and conservation Grazing, cutting or some other disturbance, such as rotovation, on a rotational basis, is needed to maintain open conditions.

CURCULIONIDAE

CEUTORHYNCHUS UNGUICULARIS   RARE
A weevil
Order COLEOPTERA   Family CURCULIONIDAE

Ceutorhynchus unguicularis Thomson, 1871.

Distribution Recorded from North Wiltshire, Buckinghamshire and West Suffolk before 1970 and East Sussex, Buckinghamshire and West Suffolk from 1970 onwards.

Habitat and ecology Grassland and disturbed ground on base-rich or dry sandy soils. Phytophagous. Associated with hairy rock-cress *Arabis hirsuta*. On the Continent, this species is also recorded from tower mustard *Turritis glabra*. In Britain, adults have been recorded from May to July.

Status Not listed in the insect Red Data Book (Shirt, 1987). Recorded new to Britain from near Mildenhall, West Suffolk in 1962. Subsequently found in a further three vice-counties. All records are from southern England. This species is difficult to identify and may be confused with other members of the genus. Consequently, the exact status of this species is hard to assess.

Threats Loss of habitat through improvement and conversion to arable agriculture, afforestation, and development. Natural succession is a further threat.

Management and conservation Grazing, cutting or some other disturbance, such as rotovation, on a rotational basis, is needed to maintain open conditions.

Published sources Jones, R.A. (1977c), Morris, M.G. (1965b), Morris, M.G. (1968).

CEUTORHYNCHUS URTICAE   RARE
A weevil
Order COLEOPTERA   Family CURCULIONIDAE

Ceutorhynchus urticae Boheman, 1845.

Distribution Recorded from Dorset, South Hampshire, West Sussex, East Kent, West Kent, Surrey, Hertfordshire, Berkshire, Buckinghamshire, Cambridgeshire, Denbighshire and Leicestershire & Rutland before 1970 and South Hampshire and East Gloucestershire from 1970 onwards.

Habitat and ecology Wetland, woodland and field margins. Phytophagous. Associated with woundwort *Stachys*, the main host is probably hedge woundwort *S. sylvatica*. On the Continent, this species has been recorded from marsh woundwort *S. palustris*. In Britain, adults have been recorded from May to July.

Status Not listed in the insect Red Data Book (Shirt, 1987). Very local and recently recorded from just two vice-counties. Old records suggest that this species was formerly more widespread and found throughout southern England as far north as Leicestershire, also recorded in North Wales.

Threats Loss of habitat through clear-felling and conversion to other land use, improvement and conversion to arable agriculture, and development. Drainage, water abstraction schemes, natural succession and the use of herbicides and pesticides may also have contributed to this species' decline.

Management and conservation Water tables should be maintained at high levels. Grazing or cutting, on a rotational basis, is needed to maintain open conditions.

Published sources Fowler, W.W. (1891), Fowler, W.W. & Donisthorpe, H.St J.K. (1913), Owen, J.A. (1989a).

CEUTORHYNCHUS VERRUCATUS   RARE
A weevil
Order COLEOPTERA   Family CURCULIONIDAE

Ceutorhynchus verrucatus Gyllenhal, 1837. Formerly known as: *Ceuthorhynchus verrucatus* Gyllenhal, 1837, *Ceuthorrhynchus verrucatus* Gyllenhal, 1837.

Distribution Recorded from "Cornwall", South Devon, South Hampshire, West Sussex, East Sussex, East Kent, South Essex and East Suffolk before 1970 and East Sussex and East Kent from 1970 onwards.

Habitat and ecology Coastal shingle. Phytophagous. Associated with yellow-horned-poppy *Glaucium flavum*. Adults, and probably larvae, occur inside a cavity in the main tap-root and can also be found at the base of the plant. Adults have been recorded in May, July and September.

Status Not listed in the insect Red Data Book (Shirt, 1987). Very local, though possibly under-recorded because of the secretive nature of this species. Restricted to the southern and south-eastern coasts, and recently recorded from only two vice-counties. This species is difficult to identify and may be confused with other members of the genus.

Threats Activities likely to result in the disturbance or destruction of the foodplant, in particular land reclamation, gravel and aggregate extraction, and development.

Management and conservation Disturbance of coastal shingle should be avoided.

Published sources Fowler, W.W. (1891), Fowler, W.W. & Donisthorpe, H.St J.K. (1913).

CEUTORHYNCHUS VIDUATUS   NOTABLE B
A weevil
Order COLEOPTERA   Family CURCULIONIDAE

Ceutorhynchus viduatus (Gyllenhal, 1813). Formerly known as: *Ceuthorhynchus viduatus* (Gyllenhal, 1813), *Ceuthorrhynchus viduatus* (Gyllenhal, 1813).

Distribution England, Dyfed-Powys, South East Scotland and South West Scotland.

Habitat and ecology Wetland, damp meadows and field margins. Phytophagous. Associated with marsh woundwort *Stachys palustris* and field woundwort *S. arvensis*. The larvae feed in the stems. Adults have been recorded in March and from May to September.

Status Widespread but local in England and southern Scotland, also recorded in Wales.

Threats This species is threatened by improvement and conversion to arable agriculture, development, drainage and water abstraction schemes. Natural succession and the use of herbicides and pesticides may be a further threat.

Management and conservation Water tables should be maintained at high levels. Grazing or cutting, on a rotational basis, is needed to maintain open conditions.

CHROMODERUS AFFINIS   EXTINCT
A weevil
Order COLEOPTERA   Family CURCULIONIDAE

Chromoderus affinis (Schrank, 1781). Formerly known as: *Chromoderus fasciatus* (Müller, 1776), *Cleonus albidus* (Schrank).

Distribution Recorded from North Essex, "Suffolk", West Norfolk and Midlothian before 1970.

Habitat and ecology Uncertain, but possibly associated with sand dunes and disturbed ground. Phytophagous. On the Continent, this species has been associated with fat hen *Chenopodium album*, oak-leaved goosefoot *C. glaucum*, maple-leaved goosefoot *C. hybridum*, grass-leaved orache *Atriplex littoralis*, common orache *A. patula*, spear-leaved orache *A. hastata*, beet *Beta vulgaris* and prickly saltwort *Salsola kali*. Adults and larvae probably occur at the roots of the foodplant.

Status Presumed extinct. Status revised from RDB 3 (Rare) in Shirt (1987). Only doubtfully resident, all records are possibly the result of casual immigration. Last recorded in 1883 from near Colchester, North Essex.

Published sources Fowler, W.W. (1891), Shirt, D.B., ed. (1987).

CIONUS LONGICOLLIS   NOTABLE A
A weevil
Order COLEOPTERA   Family CURCULIONIDAE

Cionus longicollis Brisout, 1863.

Distribution Recorded from West Norfolk before 1970 and South Hampshire, North Hampshire, West Suffolk, West Norfolk and Cambridgeshire from 1970 onwards.

Habitat and ecology Grassland, roadside verges and disturbed ground. Particularly found in sandy areas. Phytophagous. Associated with great mullein *Verbascum thapsus*. Adults have been recorded from June to September.

Status Extremely local and only known in southern England. This species is well established in the Brecklands of East Anglia.

Threats Loss of sandy grassland through improvement by reseeding or by the application of fertilisers, or by conversion to arable agriculture. Lack of, or changes in grazing or cutting regimes, and the use of herbicides and pesticides may also threaten this species.

Management and conservation Occasional disturbance, followed by a period of non-management, is needed to maintain open conditions, encourage early successional stages and ensure continuity of populations of the food-plant.

Published sources Allen, A.A. (1974), Fowler, W.W. & Donisthorpe, H.St J.K. (1913).

CIONUS NIGRITARSIS   NOTABLE A
A weevil
Order COLEOPTERA   Family CURCULIONIDAE

Cionus nigritarsis Reitter, 1904. Formerly known as: *Cionus thapsi* Reitter, 1904, *Cionus thapsus* sensu auct. Brit. not (F., 1792).

Distribution Recorded from Dorset, South Hampshire, East Sussex, Surrey, Hertfordshire, Berkshire, Oxfordshire, Buckinghamshire, West Suffolk, East Norfolk, Huntingdonshire, East Gloucestershire, Glamorgan and Denbighshire before 1970 and South

CURCULIONIDAE

Hampshire, Surrey, West Suffolk, Monmouthshire and Herefordshire from 1970 onwards.

Habitat and ecology Grassland and downland, also roadside verges, hedgebanks, and possibly field margins and disturbed ground. Particularly associated with calcareous soils. Phytophagous. Associated with great mullein *Verbascum thapsus*, dark mullein *V. nigrum* and common figwort *Scrophularia nodosa*. On the Continent, this species has also been found on white mullein *V. lychnitis*. Larvae are probably free-living on the foodplant. Adults have been recorded from May to September.

Status Very local and recently recorded from only five vice-counties. Old records suggest that this species was formerly more widespread in the southern half of England and Wales.

Threats Loss of grassland through improvement by reseeding or by the application of fertilisers, or by conversion to arable agriculture. Afforestation, the lack of, or changes in grazing or cutting regimes, and the use of herbicides and pesticides may threaten this species. An intensive cutting regime on roadside verges may also be a threat.

Management and conservation Disturbance, followed by a period of non-management, is needed to maintain open conditions, encourage early successional stages and maintain populations of the food-plant(s).

Published sources Allen, A.A. (1974), Atty, D.B. (1983), Cooter, J. & Welch, R.C. (1981), Fowler, W.W. (1891), Fowler, W.W. & Donisthorpe, H.St J.K. (1913).

CLEONUS PIGER NOTABLE B
A weevil
Order COLEOPTERA Family CURCULIONIDAE

Cleonus piger (Scopoli, 1763). Formerly known as: *Cleonus sulcirostris* (L., 1767).

Distribution England, Wales and South East Scotland.

Habitat and ecology Sand dunes, unstable cliffs and sandy areas. Primarily coastal, though also recorded inland. Phytophagous. The larvae feed in the stems of creeping thistle *Cirsium arvense*. On the Continent, this species has been recorded on cabbage thistle *Cirsium oleraceum*, musk thistle *Carduus nutans*, welted thistle *C. acanthoides*, cotton thistle *Onopordum acanthium* and greater burdock *Arctium lappa*. Larvae form swellings in the stems of the foodplant and, in Britain, have been recorded in late May, July and early September. Pupae have been found in early September,

while adults have been noted in January, from April to September and in December.

Status Widespread but local in England. In Wales, this species has only recently been recorded in Pembrokeshire. There are no recent records for Scotland.

Threats Loss of sandy habitat, particularly through afforestation, urban and holiday development, and improvement and conversion to arable agriculture. The degradation of remaining habitat by excessive disturbance of the vegetation through activities such as motorbike access, horse-riding and human trampling. This species may be further threatened by cliff stabilisation schemes.

Management and conservation Grazing or some other disturbance may be necessary to maintain open conditions and encourage early successional stages. In areas of unstable cliff, occasional slippages are necessary to maintain habitat continuity. Large areas of unstable cliff are required so that the population does not become isolated and subsequently threatened by individual landslips.

CNEORHINUS PLUMBEUS NOTABLE B
A weevil
Order COLEOPTERA Family CURCULIONIDAE

Cneorhinus plumbeus (Marsham, 1802). Formerly known as: *Atactogenus exaratus* (Marsham, 1802), *Cneorrhinus exaratus* (Marsham), *Cneorrhinus plumbeus* (Marsham).

Distribution England, South Wales and South West Scotland.

Habitat and ecology Grassland, sandpits, hedgebanks, wet meadows and short turf by the coast. Phytophagous and polyphagous, larvae feed on the roots of plants. Often found on hedgerow shrubs. Adults have been recorded from April to August.

Status Widespread but local in England, also recorded in South Wales and South West Scotland.

Threats Uncertain, though this species may be threatened by loss of habitat through improvement and conversion to arable agriculture, development, the infilling of pits, and the grubbing out and mechanised trimming of hedgerows.

COELIODES ERYTHROLEUCOS — NOTABLE B
A weevil
Order COLEOPTERA   Family CURCULIONIDAE

Coeliodes erythroleucos (Gmelin in L., 1790). Formerly known as: *Coeliodes erythroleucus* (Gmelin in L., 1790).

Distribution England, South Wales, Dyfed-Powys, South East Scotland, South West Scotland and North East Scotland.

Habitat and ecology Broad-leaved woodland, oak scrub and wooded heathland. Phytophagous. Associated with oak. Eggs are laid in the buds and the larvae feed on the female flowers. Adults have been recorded from April to August.

Status Widespread but local. Recorded throughout England, parts of Wales and as far north as North East Scotland.

Threats Loss of habitat through, for example, clear-felling and coniferisation.

Management and conservation Gaps in the age structure of the tree population should be identified and the continuity of the appropriate habitat ensured by regeneration, suitable planting and possibly with pollarding.

COELIODES NIGRITARSIS — NOTABLE A
A weevil
Order COLEOPTERA   Family CURCULIONIDAE

Coeliodes nigritarsis Hartmann, 1895.

Distribution Recorded from Moray before 1970 and Cumberland, Dumfriesshire, East Inverness & Nairn, West Inverness and West Sutherland from 1970 onwards.

Habitat and ecology Birch woodland. Phytophagous. Associated with birch. Adults have been recorded from April to early August.

Status Predominantly Scottish and recently discovered at a site in Cumberland. Possibly more widespread in Scotland than records suggest. An old record for Cambridgeshire requires confirmation.

Threats Loss of birch woodland through practices such as clear-felling or conversion to plantation forest.

Management and conservation Gaps in the age structure of the tree population should be identified and

regeneration encouraged to ensure the continuity of habitat.

Published sources Allen, A.A. (1950d), Drane, A.B. (1990), Morris, M.G. (1976b), Sinclair, M. (1985).

COELIODES RUBER — NOTABLE B
A weevil
Order COLEOPTERA   Family CURCULIONIDAE

Coeliodes ruber (Marsham, 1802).

Distribution England, Wales, South East Scotland, South West Scotland and North East Scotland.

Habitat and ecology Broad-leaved woodland. Phytophagous. Associated with oak and hazel. Adults have been recorded from March to May and in July.

Status Widespread but local. Recorded throughout Great Britain, though not known from North West Scotland.

Threats Loss of broad-leaved woodland through, for example, clear-felling and coniferisation. Neglect and conversion to high forest, has led to increased shade and the loss of glades and broad sunny rides.

Management and conservation Open glades and ride margins should be cut on rotation to retain a variety of vegetation structures. Gaps in the age structure of the tree population should be identified and the continuity of the appropriate habitat ensured by regeneration, suitable planting and possibly with pollarding.

CONIOCLEONUS HOLLBERGI — EXTINCT
A weevil
Order COLEOPTERA   Family CURCULIONIDAE

Coniocleonus hollbergi (Fahraeus, 1842). Formerly known as: *Cleonus glaucus* (F., 1787), *Coniocleonus glaucus* (F.).

Distribution Recorded from Surrey.

Habitat and ecology Heathland. Phytophagous. Adults and probably larvae occur at the roots of heather *Calluna vulgaris*. Adults have been recorded in June.

Status Presumed extinct. Last recorded in 1815 near Chobham Common, Surrey. This species is very similar to *C. nebulosus* and may be confused with it.

Published sources Fowler, W.W. (1891), Shirt, D.B., ed. (1987).

CURCULIONIDAE

CONIOCLEONUS NEBULOSUS   RARE
A weevil
Order COLEOPTERA   Family CURCULIONIDAE

Coniocleonus nebulosus (L., 1758). Formerly known as: *Cleonus nebulosus* (L.).

Distribution Recorded from Dorset, South Hampshire, North Hampshire, Surrey, Berkshire, East Suffolk, East Norfolk, Cambridgeshire, Staffordshire, North Lincolnshire and Derbyshire before 1970 and West Cornwall and Dorset from 1970 onwards.

Habitat and ecology Heathland. Phytophagous. The larvae feed in the stems of mature heather *Calluna vulgaris*. Adults are probably found at the roots of the foodplant and have been recorded in April and May.

Status Not listed in the insect Red Data Book (Shirt, 1987). Only known in southern England, and found as far north as Derbyshire. Recently recorded from only two vice-counties, both in South West England.

Threats Loss, or fragmentation of heathland through changes in land use, by conversion to arable, afforestation and urban development. Further habitat degradation has been through the cessation of traditional heathland management practices. Fires are likely to prevent the development of mature heather.

Management and conservation Management should aim for a diversity of successional stages from bare ground to mature heath, preferably by grazing, rotational cutting or scraping.

Published sources Eagles, T.R. (1951), Fowler, W.W. (1891), Fowler, W.W. & Donisthorpe, H.St J.K. (1913).

COSSONUS LINEARIS   NOTABLE A
A weevil
Order COLEOPTERA   Family CURCULIONIDAE

Cossonus linearis (F., 1775).

Distribution Recorded from East Kent, Buckinghamshire, East Suffolk, West Suffolk, East Norfolk, Huntingdonshire and North Lincolnshire before 1970 and East Sussex, West Kent, Surrey, West Suffolk, Cambridgeshire and Huntingdonshire from 1970 onwards.

Habitat and ecology Broad-leaved woodland, pasture-woodland and wetland, particularly in broads and fens. Often recorded near the coast. Phytophagous. Associated with poplar and also recorded from pine. On the Continent, this species has been recorded from willow. Larvae and adults occur in dead wood. In

Britain, adults have been recorded from March to June, September and in October. This species has been recorded at a mercury vapour light trap.

Status Very local and only recorded in England south of the Humber. Recently recorded from just six vice-counties.

Threats Loss of broad-leaved woodland and parkland through, for example, clear-felling and conversion to other land use. Habitat loss, in particular, through the felling of trees, removal of dead wood from living trees and the destruction or removal of standing and fallen dead wood for reasons such as forest hygiene, aesthetic tidiness, public safety or for use as fire wood.

Management and conservation Both fallen and standing dead timber, especially with the bark attached, should be retained. The removal of dead timber from trees should be avoided. Gaps in the age structure of the tree population should be identified and the continuity of the appropriate dead wood habitat ensured by regeneration, suitable planting and possibly with pollarding.

Published sources Allen, A.A. (1990d), Donisthorpe, H.St J.K. (1939b), Drane, A.B. (1978), Eagles, T.R. (1955c), Hodge, P.J. (1990), Nash, D.R. (1978).

COSSONUS PARALLELEPIPEDUS   NOTABLE B
A weevil
Order COLEOPTERA   Family CURCULIONIDAE

Cossonus parallelepipedus (Herbst, 1795). Formerly known as: *Cossonus ferrugineus* sensu Fowler, 1891 ?Clairville, 1798, *Cossonus parallelopipedus* sensu Fowler, 1891 ?Clairville, 1798.

Distribution South East, South, South West, East Anglia, East Midlands, West Midlands, South Wales and North West.

Habitat and ecology Broad-leaved woodland, pasture-woodland and also isolated trees. Phytophagous. Associated with oak, poplar, elm and willow. Larvae and adults occur in dead wood. Larvae have been recorded in May with adult emergence occurring the following April. Adults probably occur all year round and have been recorded in January, from March to July, and from October to December.

Status Widespread but local in southern and central Great Britain.

Indicator status Grade 3 in Harding & Rose (1986).

260

Threats Loss of broad-leaved woodland, parkland and isolated trees through clear-felling and conversion to other land use. Habitat loss, in particular, through the felling of trees, removal of dead wood from living trees and the destruction or removal of standing and fallen dead wood for reasons such as forest hygiene, aesthetic tidiness, public safety or for use as fire wood.

Management and conservation Both fallen and standing dead timber, especially with the bark attached, should be retained. The removal of dead timber from trees should be avoided. Gaps in the age structure of the tree population should be identified and the continuity of the appropriate dead wood habitat ensured by regeneration, suitable planting and possibly with pollarding.

CRYPTORHYNCHUS LAPATHI   **NOTABLE B**
Willow beetle
Order COLEOPTERA   Family CURCULIONIDAE

Cryptorhynchus lapathi (L., 1758). Formerly known as: *Cryptorhynchidius lapathi* (L.).

Distribution England, South Wales, North Wales and Scotland.

Habitat and ecology Wetland. Particularly fens, willow carr and osier beds. Also recorded on willows along lake and river margins. Phytophagous. Associated mainly with willows, especially osier, though also recorded from poplar suckers. On the Continent, this species has been associated with crack willow, grey willow, goat willow, white willow, purple willow, Italian poplar, balsam poplar, aspen, white poplar, alder, and grey alder. The larvae form galls in the twigs, and pupate around October. In Britain, adults have been recorded from May to July and in the gall in February.

Status Widespread but local in Britain. Can be numerous where found. This species has been noted as a pest of basket osiers.

Threats This species is threatened by clear-felling and conversion to other land use. Drainage, and falling water tables because of water abstraction and river engineering schemes may also threaten this species.

Management and conservation Water tables should be maintained at high levels. Gaps in the age structure of the willow population should be identified and the continuity of the appropriate habitat ensured by encouraging regeneration, suitable planting and possibly pollarding.

CURCULIO BETULAE   **NOTABLE B**
A weevil
Order COLEOPTERA   Family CURCULIONIDAE

Curculio betulae (Stephens, 1831). Formerly known as: *Balaninus betulae* Stephens, *Balaninus cerasorum* sensu (Paykull, 1792) not (F., 1775).

Distribution England, South Wales and North Wales.

Habitat and ecology Broad-leaved woodland and moorland. Phytophagous. Associated with birch. On the Continent, this species has also been recorded from alder, oak, blackthorn, dwarf cherry and grey willow. The larvae develop in fruits or catkins of the host-plants. In Britain, adults have been recorded from June to September.

Status Widespread but local in England and parts of Wales.

Threats Loss of woodland through, for example, clear-felling and conversion to other land use. Neglect and conversion to high forest may be a further threat.

Management and conservation Open glades and ride margins should be cut on rotation to retain a variety of vegetation structures.

CURCULIO RUBIDUS   **NOTABLE B**
A weevil
Order COLEOPTERA   Family CURCULIONIDAE

Curculio rubidus (Gyllenhal, 1836). Formerly known as: *Balaninus rubidus* Gyllenhal.

Distribution South East, South, South West, East Anglia, West Midlands, East Midlands and North West.

Habitat and ecology Broad-leaved woodland. Phytophagous. Associated with birch. On the Continent, this species is also recorded from grey willow. This latter record lends support to a possible association with *Populus* spp. in Britain. The larvae have been recorded from birch catkins. In Britain, adults have been recorded from July to September.

Status Widespread but local in England, though not recorded in North East England.

Threats Loss of woodland through, for example, clear-felling and conversion to other land use. Neglect and conversion to high forest is a further threat.

Management and conservation Open glades and ride margins should be cut on rotation to retain a variety of vegetation structures.

CURCULIONIDAE

CURCULIO VILLOSUS   NOTABLE B
A weevil
Order COLEOPTERA   Family CURCULIONIDAE

Curculio villosus F., 1781. Formerly known as:
Balaninus villosus (F.).

Distribution England, South Wales, North Wales,
South East Scotland and South West Scotland.

Habitat and ecology Broad-leaved woodland.
Phytophagous. Associated with oak. The larvae develop
as inquilines in 'oak-apple' galls induced by *Biorhiza
pallida* (Hymenoptera), and very occasionally as
inquilines of bedeguar galls on rose, induced by
Diplolepis rosae (Hymenoptera). Adults have been
recorded from April to June and in September.

Status Widespread but local in England, parts of Wales
and southern Scotland.

Threats Loss of broad-leaved woodland through, for
example, clear-felling and coniferisation. Neglect and
conversion to high forest may be a further threat.

Management and conservation Open glades and ride
margins should be cut on rotation to retain a variety of
vegetation structures.

DORYTOMUS AFFINIS   VULNERABLE
A weevil
Order COLEOPTERA   Family CURCULIONIDAE

Dorytomus affinis (Paykull, 1800).

Distribution Recorded from East Kent, West Kent,
Cambridgeshire and Huntingdonshire before 1970 and
Huntingdonshire from 1970 onwards.

Habitat and ecology Broad-leaved woodland.
Phytophagous. Associated with aspen. Larvae feed in
the catkins, predominantly female catkins. Adults have
been recorded from March to August.

Status Recently recorded only from Monks Wood,
Huntingdonshire, where it can occur in numbers.
Records for Leicestershire, Glamorganshire and Dorset
all require confirmation. This species is difficult to
identify and has been confused with other members of
the genus. Consequently, the exact status of this species
is hard to assess.

Threats Loss of broad-leaved woodland through, for
example, clear-felling and coniferisation. The selective
removal of aspen is a further threat.

Management and conservation Open glades and ride
margins should be cut on rotation to retain a variety of
vegetation structures. Gaps in the age structure of the
aspen population should be identified and the
continuity of the appropriate habitat ensured by
regeneration or suitable planting. Monks Wood is an
NNR.

Published sources Allen, A.A. (1967c), Shirt, D.B., ed.
(1987).

DORYTOMUS FILIROSTRIS   NOTABLE B
A weevil
Order COLEOPTERA   Family CURCULIONIDAE

Dorytomus filirostris (Gyllenhal, 1836).

Distribution East Anglia, East Midlands, West
Midlands, North East and North West.

Habitat and ecology Wetland, fens, and trees in
cultivated land. Phytophagous. Associated with black
poplar and Italian poplar. Doubtfully associated with
willows. The larvae develop in the catkins. Adults have
been recorded in May, August and September.

Status Widespread but local in central eastern England.
Probably a late arrival to the British Isles, possibly
spreading. This species is difficult to identify and may
be confused with other members of the genus.
Consequently, the exact status of this species is hard to
assess.

Threats This species is threatened by the felling of its
host-trees.

Management and conservation Gaps in the age
structure of the tree population should be identified and
the continuity of the appropriate habitat ensured by
suitable planting and possibly with pollarding.

DORYTOMUS HIRTIPENNIS   NOTABLE A
A weevil
Order COLEOPTERA   Family CURCULIONIDAE

Dorytomus hirtipennis Bedel, 1884.

Distribution Recorded from Dorset, Surrey, North
Essex, Berkshire, Oxfordshire, East Suffolk, West
Suffolk, East Norfolk, Cambridgeshire,
Huntingdonshire, Northamptonshire, Cheshire, South
Lancashire and Durham before 1970 and East Norfolk
and Northamptonshire from 1970 onwards.

Habitat and ecology Wetland, willow carr, and
woodland along the margins of rivers and lakes.

Phytophagous. Associated with white willow, and probably also on goat willow and crack willow. Eggs are laid in the catkins about mid April and the larvae develop to maturity by late April to early May, depending upon whether they develop in male or female catkins (female catkins mature later than male catkins). Larvae probably pupate in the soil during May. Adults have been recorded from February to October and overwinter on the host-tree, often under bark.

Status Old records suggest that this species formerly had a scattered distribution from southern England as far north as Durham. Recently recorded from only two vice-counties. This species is difficult to identify and may be confused with other members of the genus. Consequently, the exact status of this species is hard to assess.

Threats This species is threatened by drainage and clear-felling, for reasons such as agricultural improvement and development. Falling water tables because of water abstraction schemes may also threaten this species.

Management and conservation Water tables should be maintained at high levels. Gaps in the age structure of the willow population should be identified and the continuity of the appropriate habitat ensured by regeneration, suitable planting and possibly with pollarding.

Published sources Beare, T.H. (1927), Fowler, W.W. (1891), Fowler, W.W. & Donisthorpe, H.St J.K. (1913), Hodge, P.J. (1990), Morris, M.G. (1969).

DORYTOMUS ICTOR   **NOTABLE B**
A weevil
Order COLEOPTERA   Family CURCULIONIDAE

Dorytomus ictor (Herbst, 1795). Formerly known as: *Dorytomus validirostris* (Gyllenhal, 1836).

Distribution South East, South, South West, East Anglia, West Midlands, North East and Dyfed-Powys.

Habitat and ecology Trees in cultivated land, in fields and on riverbanks. Phytophagous. Associated with Italian poplar, and probably black poplar. Doubtfully associated with aspen. The larvae probably develop in the catkins. Adults have been recorded in June, July and November.

Status Widespread but local in central and southern Great Britain. This species is difficult to identify and may be confused with other members of the genus.

Consequently, the exact status of this species is hard to assess.

Threats This species is threatened by the felling of its host-trees.

Management and conservation Gaps in the age structure of the poplar population should be identified and the continuity of the appropriate habitat ensured by suitable planting.

DORYTOMUS MAJALIS
A weevil   **INSUFFICIENTLY KNOWN**
Order COLEOPTERA   Family CURCULIONIDAE

Dorytomus majalis (Paykull, 1792).

Distribution Recorded from Cheshire, Durham, North Northumberland, Cumberland, Dumfriesshire, Kirkcudbrightshire and Stirlingshire before 1970.

Habitat and ecology Trees in cultivated land. Probably also found in wetland and doubtfully in broad-leaved woodland. Phytophagous. Associated with grey willow, eared willow and goat willow. The larvae probably develop in the catkins.

Status Not listed in the insect Red Data Book (Shirt, 1987). No recent records. Last recorded from Langwathby and Honeypot, near Penrith, Cumberland, the date is not recorded but is pre-1960. This species is difficult to identify and may be confused with other members of the genus. Consequently, the exact status of this species is hard to assess.

Threats Uncertain, though this species may have declined because of loss of habitat through felling for reasons such as agricultural improvement and development.

Published sources Davidson, W.F. (1961), Eyre, M.D. (1987), Fowler, W.W. (1891).

DORYTOMUS SALICINUS   **NOTABLE B**
A weevil
Order COLEOPTERA   Family CURCULIONIDAE

Dorytomus salicinus (Gyllenhal, 1827).

Distribution East Anglia, East Midlands, West Midlands, North East, North West and South West Scotland.

Habitat and ecology Wetland, broad-leaved woodland and probably also trees in cultivated land. Phytophagous. Associated with grey willow and

CURCULIONIDAE

probably other *Salix* species. On the Continent, this species has also been noted from eared willow and goat willow. The larvae probably develop in the catkins. In Britain, adults have been recorded in March, April and early August.

Status Widespread but local in central and northern England, also recorded in South West Scotland. Can be numerous where found. A record for North Wiltshire requires confirmation. This species is difficult to identify and may be confused with other members of the genus. Consequently, the exact status of this species is hard to assess.

Threats Drainage and clear-felling for reasons such as agricultural improvement and conversion to other land use. Falling water tables because of water abstraction schemes may be a further threat.

Management and conservation Water tables should be maintained at high levels. Gaps in the age structure of the tree population should be identified and the continuity of the appropriate habitat ensured by regeneration, suitable planting and possibly with pollarding.

DORYTOMUS SALICIS　　　　　　**NOTABLE A**
A weevil
Order COLEOPTERA　　　　Family CURCULIONIDAE

Dorytomus salicis Walton, 1851.

Distribution Recorded from North Hampshire, West Kent, Surrey, Hertfordshire, South-east Yorkshire, Mid-west Yorkshire and North Northumberland before 1970 and South-east Yorkshire, North-east Yorkshire and South-west Yorkshire from 1970 onwards.

Habitat and ecology Probably associated with broad-leaved woodland, wetland and possibly with trees in cultivated land. Phytophagous. Associated with creeping willow, and probably grey willow and other *Salix*. The larvae probably develop in the catkins. Adults have been recorded from March to June.

Status Old records show that this species has been recorded from North Hampshire, north to North Northumberland. Recently recorded from only three vice-counties. A record from North Lincolnshire requires confirmation. This species is difficult to identify and may be confused with other members of the genus. Consequently, the exact status of this species is hard to assess.

Threats Uncertain, though this species is probably threatened by loss of habitat and conversion to other land use. Drainage, falling water tables because of

water abstraction schemes, lack of grazing and natural succession may be further threats.

Management and conservation Water tables should be maintained at high levels. Some grazing may be needed to maintain creeping willow population.

Published sources Eyre, M.D. (1987), Fowler, W.W. (1891), Skidmore, P., Limbert, M. & Eversham, B. (1985).

DORYTOMUS TREMULAE　　　　　**NOTABLE B**
A weevil
Order COLEOPTERA　　　　Family CURCULIONIDAE

Dorytomus tremulae (F., 1787).

Distribution England, South Wales and North East Scotland.

Habitat and ecology Broad-leaved woodland and isolated trees. Probably also trees in cultivated land. Phytophagous. Associated with aspen, grey poplar and white poplar. Larvae develop in the vegetative buds and shoots. Adults have also been found under bark. Adults have been recorded from February to October and in December. Adults emerged in May from larvae collected in April.

Status Widespread but local with a widely scattered distribution from Surrey, north to Elgin in Scotland. Old records suggest that this species may have been more widespread and recorded from southern England, north to Easterness in Scotland. This species is difficult to identify and may be confused with other members of the genus. Consequently, the exact status of this species is hard to assess.

Threats This species may be threatened by clear-felling and conversion to other land use.

Management and conservation Gaps in the age structure of the tree population should be identified and the continuity of the appropriate habitat ensured by regeneration or suitable planting.

Published sources Fowler, W.W. (1891), Fowler, W.W. & Donisthorpe, H.St J.K. (1913).

DRUPENATUS NASTURTII   NOTABLE B
A weevil
Order COLEOPTERA   Family CURCULIONIDAE

Drupenatus nasturtii (Germar, 1824). Formerly known as: *Ceuthorrhynchus nasturtii* (Germar), *Drusenatus nasturtii* (Germar), *Poophagus nasturtii* (Germar).

Distribution England and Wales.

Habitat and ecology Wetland, particularly base-rich streams and drainage ditches. Phytophagous. Associated with water-cress *Nasturtium officinale*. Adults have been recorded in January and from April to October.

Status Widespread but local in England and Wales.

Threats Drainage for reasons such as agricultural improvement and development. Falling water tables because of water abstraction, river engineering schemes, and natural succession may also threaten this species.

Management and conservation Water tables should be maintained at high levels. Water bodies should be isolated from sources of eutrophication and pollution. Grazing or cutting is needed to maintain open conditions. Ditch clearance should be undertaken on a rotational basis and should aim at maintaining the aquatic plant populations.

DRYOPHTHORUS CORTICALIS ENDANGERED
A weevil
Order COLEOPTERA   Family CURCULIONIDAE

Dryophthorus corticalis (Paykull, 1792).

Distribution Recorded from Berkshire before 1970 and Berkshire from 1970 onwards.

Habitat and ecology Ancient broad-leaved woodland and pasture-woodland. Phytophagous. In red-rotten wood, tough stringy wood and under the bark of deciduous trees. Found in oak, often in association with the ant *Lasius brunneus* (Hymenoptera). On the Continent, this species has also been found on pine, willow and holly. The larvae are woodfeeders. In Britain, adults have been recorded from June to August, October and in December.

Status Only known from Windsor Forest and Windsor Great Park, Berkshire. This species can be found in some numbers on individual trees.

Indicator status Grade 1 in Harding & Rose (1986).

Threats Loss of broad-leaved woodland and parkland through, for example, clear-felling and coniferisation. Habitat loss, in particular, through the felling of ancient trees, removal of dead wood from living trees and the destruction or removal of standing and fallen dead wood for reasons such as forest hygiene, aesthetic tidiness, public safety or for use as fire wood.

Management and conservation Ancient trees and both fallen and standing dead timber, especially with the bark attached, should be retained. The removal of dead timber from ancient trees should be avoided. Gaps in the age structure of the tree population should be identified and the continuity of the appropriate dead wood habitat ensured by regeneration, suitable planting and possibly with pollarding. Windsor Forest and parts of Windsor Great Park are notified as SSSIs.

Published sources Copestake, D.R. (1990a), Donisthorpe, H.St J.K. (1925a), Harding, P.T. (1978), Harding, P.T. & Rose, F. (1986), Owen, J.A. (1983b), Shirt, D.B., ed. (1987).

ELLESCUS BIPUNCTATUS   NOTABLE B
A weevil
Order COLEOPTERA   Family CURCULIONIDAE

Ellescus bipunctatus (L., 1758). Formerly known as: *Elleschus bipunctatus* (L., 1758).

Distribution England, Dyfed-Powys, South East Scotland and South West Scotland.

Habitat and ecology Woodland, particularly in wetter areas. Phytophagous. Associated with grey willow. On the Continent, this species has also been recorded from goat willow and osier. Larvae feed in the buds. In Britain, larvae have been found in July. Adults have been recorded from April to June and August.

Status Widespread but local in England and southern Scotland. Can be numerous where found. There is an old record for mid Wales.

Threats Loss of woodland through, for example, clear-felling and conversion to other land use. Neglect and conversion to high forest, drainage and water abstraction schemes may be further threats.

Management and conservation Open glades and ride margins should be cut on rotation to retain a variety of vegetation structures. Water tables should be maintained at high levels.

CURCULIONIDAE

EUBRYCHIUS VELUTUS   NOTABLE B
A weevil
Order COLEOPTERA   Family CURCULIONIDAE

Eubrychius velutus (Beck, 1817). Formerly known as: *Eubrychius velatus* (Beck, 1817), *Litodactylus velatus* (Germar, 1818), *Phytobius velatus* (Germar).

Distribution England, South Wales, North Wales and South East Scotland.

Habitat and ecology Slow moving and stagnant water, including brackish ditches. Phytophagous and aquatic. Associated with water-milfoil *Myriophyllum* species. On the Continent, this species has been recorded from spiked water-milfoil *M. spicatum* and whorled water-milfoil *M. verticillatum*. The larva forms a cocoon on the foodplant. In Britain, adults have been recorded from April to June and from August to October.

Status Widespread but local in England, also recorded from parts of Wales and in South East Scotland.

Threats Loss of habitat through drainage for reasons such as agricultural improvement and development. Falling water tables because of water abstraction schemes, use of herbicides and pesticides, and natural succession are further threats.

Management and conservation Water tables should be maintained at high levels. Water bodies should be isolated from sources of eutrophication and pollution. Ditch clearance should be undertaken on a rotational basis and should aim to maintain the aquatic plant populations and well oxygenated water.

FURCIPUS RECTIROSTRIS   NOTABLE B
A weevil
Order COLEOPTERA   Family CURCULIONIDAE

Furcipus rectirostris (L., 1758). Formerly known as: *Anthonomus druparum*.

Distribution North East, North West, Dyfed-Powys and North West Scotland.

Habitat and ecology River margins, hedgerows and isolated trees in meadows. Phytophagous. Associated with bird cherry. On the Continent, this species is also associated with wild cherry, dwarf cherry, cherry plum, bullace, and blackthorn. The larvae develop and pupate in the fruits of the foodplant in August and September. In Britain, adults have been recorded in May, June, August and September.

Status Recorded new to Britain in 1981 from examples found in 1979 from Cumberland. It has since been found to be widely distributed in the Lake District, North Yorkshire, central Wales, and also recorded from North West Scotland.

Threats Uncertain, though this species is likely to be threatened by the felling of the host-tree and the grubbing out of hedgerows.

Management and conservation Gaps in the age structure of the tree population should be identified and regeneration encouraged to ensure the continuity of suitable habitat.

GRONOPS INAEQUALIS
A weevil   INSUFFICIENTLY KNOWN
Order COLEOPTERA   Family CURCULIONIDAE

Gronops inaequalis Boheman, 1842.

Distribution Recorded from East Kent from 1970 onwards.

Habitat and ecology Disturbed ground and derelict areas. In Britain, only known from an old land-fill site, which dates back to Victorian times. Found at the roots and under the basal leaves of hastate orache *Atriplex prostrata*. On the Continent, it is associated with fat-hen *Chenopodium album* and possibly other members of the Chenopodiaceae. The beetle is predominantly nocturnal, hiding by day under the foodplant. Adults have been recorded in August.

Status Not listed in the insect Red Data Book (Shirt, 1987). Possibly introduced. An eastern European species which has been spreading westwards. Recorded as new to Britain in 1982, and now well established at one site in East Kent.

Threats Uncertain, though this species may be threatened by a change in land use of its only site. Natural succession may be a further threat.

Management and conservation Uncertain, though grazing, cutting or some other disturbance may be needed to maintain open conditions.

Published sources Clemons, L. (1983).

GRONOPS LUNATUS — NOTABLE B
A weevil
Order COLEOPTERA   Family CURCULIONIDAE

Gronops lunatus (F., 1775).

Distribution South East, South, South West, East Anglia, East Midlands, North East, North West and Wales.

Habitat and ecology Sandy places, quarries, sand dunes, the edges of dune pools, coastal undercliffs, saltmarshes, sand flats, disturbed ground and ruderal habitats. Phytophagous. Associated with corn spurrey *Spergula arvensis*, sand spurrey *Spergularia rubra*, greater sea-spurrey *S. media*, lesser sea-spurrey *S. marina*, little mouse-ear *Cerastium semidecandarum* and possibly other Caryophyllaceae. Larvae probably occur at the roots of the foodplant, while adults are found at the base of the plants. Adults have been recorded in April, May, July, September and November.

Status Widely distributed and local in England and Wales, particularly on the coast. Often abundant where it occurs.

Threats Loss of habitat through reclamation, erosion, the construction of sea defences, afforestation and urban and holiday developments. This species may be further threatened by natural succession and scrub invasion.

Management and conservation Grazing or some other disturbance may be needed to maintain open conditions and encourage early successional stages.

GRYPUS EQUISETI — NOTABLE B
A weevil
Order COLEOPTERA   Family CURCULIONIDAE

Grypus equiseti (F., 1775). Formerly known as: *Grypidius equiseti* (F.).

Distribution England, Wales and Scotland.

Habitat and ecology Occurs in a variety of habitats including wetland, willow carr, disused gravel workings, grassland, field margins and roadside verges. Phytophagous. Associated with field horsetail *Equisetum arvense* and marsh horsetail *E. palustre*. Also recorded on great horsetail *E. telmateia*. The larvae feed in the stems. Adults occur at the base of the foodplants and have been recorded in March and from June to September.

Status Widespread but local. Possibly under-recorded because of its secretive nature.

Threats This species may be threatened by loss of habitat through conversion to other land use. Drainage, water abstraction schemes and natural succession may be further threats.

Management and conservation Water tables should be maintained at high levels. Grazing, cutting or some other disturbance, such as rotovation, on a rotational basis, is needed to maintain open conditions.

GYMNETRON BECCABUNGAE — NOTABLE A
A weevil
Order COLEOPTERA   Family CURCULIONIDAE

Gymnetron beccabungae (L., 1761).

Distribution Recorded from South Hampshire, North Hampshire, East Kent, West Kent, Surrey, Hertfordshire, Cheshire, East Inverness & Nairn and East Sutherland before 1970 and North Hampshire, West Norfolk, Northamptonshire, Glamorgan, Radnorshire, Carmarthenshire, Pembrokeshire, Cardiganshire, Denbighshire and East Sutherland from 1970 onwards.

Habitat and ecology Wetland. Phytophagous. Associated with brooklime *Veronica beccabunga* and marsh speedwell *V. scutellata*. On the Continent, this species has also been recorded from blue water-speedwell *V. anagallis-aquatica*. In Britain, adults have been found in June, July, September and October.

Status Local and with a disjunct distribution. Recorded in southern England with a few scattered records in other parts of England, throughout Wales, and very locally in Scotland. This species is difficult to identify and may be confused with other members of the genus, especially *G.veronicae*. Consequently, the exact status of this species is hard to assess.

Threats Drainage for reasons such as agricultural improvement and development is the primary cause of the loss of wetlands. Falling water tables because of water abstraction and river engineering schemes may also threaten this species.

Management and conservation Water tables should be maintained at high levels. Water bodies should be isolated from sources of eutrophication and pollution. Grazing or cutting, on a rotational basis, may be needed to maintain open conditions.

CURCULIONIDAE

Published sources Allen, A.A. (1989b), Bedwell, E.C. (1933), Champion, G.C. (1922), Collier, M.J. (1990), Walker, J.J. (1933).

GYMNETRON COLLINUM　　　**NOTABLE A**
A weevil
Order COLEOPTERA　　　Family CURCULIONIDAE

Gymnetron collinum (Gyllenhal, 1813). Formerly known as: *Gymnetron collinus* (Gyllenhal, 1813).

Distribution Recorded from Dorset, East Sussex, East Kent, West Kent, South Essex, Middlesex, Berkshire, West Suffolk, Cambridgeshire, Huntingdonshire, Cardiganshire and South Lancashire before 1970 and East Sussex, East Kent, South Essex, West Suffolk and Cardiganshire from 1970 onwards.

Habitat and ecology Calcareous grassland and disturbed ground, including pits and quarries. Phytophagous. Associated with common toadflax *Linaria vulgaris*. On the Continent, this species has also been recorded from pale toadflax *L. repens*. In Britain, larvae develop in galls on the roots of the foodplant in September, adults emerging about late September to early October. Adults have been recorded from April to June and from August to October.

Status Very local with a scattered distribution in England, recorded as far north as South Lancashire. Also noted in Wales. Recently recorded from just five vice-counties. This species is difficult to identify and may be confused with other members of the genus. Consequently, the exact status of this species is hard to assess.

Threats Loss of habitat through improvement and conversion to arable agriculture, development, the infilling of pits and quarries, and natural succession.

Management and conservation Disturbance, such as rotovation, is needed to maintain open conditions.

Published sources Allen, A.A. (1961), Allen, A.A. (1962a), Daltry, H.W. (1939), Donisthorpe, H.St J.K. (1937), Forster, H.W. (1951a), Fowler, W.W. (1891), Fowler, W.W. & Donisthorpe, H.St J.K. (1913).

GYMNETRON LINARIAE　　　**NOTABLE A**
A weevil
Order COLEOPTERA　　　Family CURCULIONIDAE

Gymnetron linariae (Panzer, 1795).

Distribution Recorded from South Hampshire, East Kent, West Kent, Surrey, Hertfordshire, Middlesex, Berkshire, West Suffolk, "Norfolk", Cambridgeshire, Cardiganshire, Merionethshire, South Lancashire and North-east Yorkshire before 1970 and South Hampshire, East Sussex, East Kent, South Essex, West Suffolk and Bedfordshire from 1970 onwards.

Habitat and ecology Calcareous grassland, disturbed ground and road-side verges. Phytophagous. Associated with common toadflax *Linaria vulgaris*. On the Continent, this species has also been recorded from pale toadflax *L. repens*. In Britain, larvae develop in galls on the roots of the foodplant in September, with adults emerging about late September to early October. Adults have been recorded in February, March, May, June, October and November.

Status Very local with a scattered distribution in England as far north as North-east Yorkshire. Also recorded in Wales. Recently recorded from only six vice-counties, all in southern and eastern England. This species is difficult to identify and may be confused with *G. antirrhini*. Consequently, the exact status of this species is hard to assess.

Threats Loss of habitat through improvement and conversion to arable agriculture and development. Natural succession may be a further threat.

Management and conservation Disturbance, such as rotovation, is needed to maintain open conditions.

Published sources Donisthorpe, H.St J.K. (1937), Fowler, W.W. (1891), Fowler, W.W. & Donisthorpe, H.St J.K. (1913).

GYMNETRON MELANARIUM　　　**NOTABLE B**
A weevil
Order COLEOPTERA　　　Family CURCULIONIDAE

Gymnetron melanarium (Germar, 1821). Formerly known as: *Gymnetron melanarius* (Germar, 1821).

Distribution South West, South, South East, East Anglia, East Midlands, West Midlands, North East and South Wales.

Habitat and ecology Grassland, damp meadows, field margins, road-side verges and possibly disturbed ground. Phytophagous. Associated with germander speedwell *Veronica chamaedrys*. On the Continent, this species has also been recorded from thyme-leaved speedwell *Veronica serpyllifolia* and heath speedwell *V. officinalis*. In Britain, adults have been found in March and from May to August.

Status Widespread and very local in southern England, with scattered records north to North-east Yorkshire. Also recorded in South Wales. This species is difficult to identify and may be confused with other members of the genus. Consequently, the exact status of this species is hard to assess.

Threats Loss of habitat through improvement and conversion to arable agriculture, and development. Natural succession, and the use of herbicides and pesticides are further threats.

Management and conservation Grazing, cutting or some other disturbance, such as rotovation, on a rotational basis, is needed to maintain open conditions.

GYMNETRON ROSTELLUM NOTABLE A
A weevil
Order COLEOPTERA Family CURCULIONIDAE

Gymnetron rostellum (Herbst, 1795). Formerly known as: *Gymnetron rostellium* (Herbst, 1795).

Distribution Recorded from South Devon, Isle of Wight, South Hampshire, East Sussex, East Kent, West Kent, Surrey, Berkshire, Oxfordshire, East Suffolk, East Norfolk and Cambridgeshire before 1970 and West Sussex, East Sussex, South Essex, Berkshire, West Suffolk, West Norfolk and Northamptonshire from 1970 onwards.

Habitat and ecology Disturbed ground, sand pits and probably road-side verges and field margins. Phytophagous. Possibly associated with *Matricaria* spp., cudweed *Filago*, and possibly also with germander speedwell *Veronica chamaedrys*. On the Continent, this species has also been recorded from *Achillea* spp., chamomile *Anthemis* and plantain *Plantago*. In Britain, larvae have been recorded in the flowerheads of *Matricaria*. Adults have been noted in April, May and July.

Status Very local. Only known in southern England and recorded as far north as Northamptonshire. Recently recorded from only six vice-counties. This species is difficult to identify and may be confused with other members of the genus. Consequently, the exact status of this species is hard to assess.

Threats Loss of habitat through improvement and conversion to arable agriculture, development and the infilling of pits. Natural succession, and the use of herbicides and pesticides are further threats.

Management and conservation Disturbance, such as rotovation, is needed to maintain open conditions.

Published sources Fowler, W.W. (1891), Fowler, W.W. & Donisthorpe, H.St J.K. (1913).

GYMNETRON VERONICAE NOTABLE B
A weevil
Order COLEOPTERA Family CURCULIONIDAE

Gymnetron veronicae (Germar, 1821). Formerly known as: *Gymnetron beccabungae var. nigrum* Hardy, 1852, *Gymnetron beccabungae var. veronicae* (Germar), *Gymnetron veronica* (Germar).

Distribution England, Dyfed-Powys, South East Scotland and South West Scotland.

Habitat and ecology Wetland. Phytophagous. Associated with brooklime *Veronica beccabunga* and occasionally figwort *Scrophularia*. On the Continent, this species has also been recorded from marsh speedwell *V. scutellata* and blue water-speedwell *V. anagallis-aquatica*. The larvae feed in galls on the flowers and flowerheads. In Britain, adults have been recorded from May to July.

Status Widespread but local in England, also recorded in southern Scotland. This species is difficult to identify and may be confused with other members of the genus, especially *G. beccabungae*. Consequently, the exact status of this species is hard to assess.

Threats Drainage for reasons such as agricultural improvement and development is the primary cause of the loss of wetlands. Falling water tables because of water abstraction and river engineering schemes, as well as pollution and eutrophication may threaten this species.

Management and conservation Water tables should be maintained at high levels. Ditch clearance should be undertaken on a rotational basis and should aim at maintaining the aquatic plant populations.

CURCULIONIDAE

GYMNETRON VILLOSULUM   NOTABLE B
A weevil
Order COLEOPTERA   Family CURCULIONIDAE

Gymnetron villosulum Gyllenhal, 1838. Formerly known as: *Gymnetron villosulus* Gyllenhal, 1838.

Distribution South East, South, South West, East Anglia, East Midlands and West Midlands.

Habitat and ecology Wetland. Phytophagous. Associated mainly with pink water-speedwell *Veronica catenata* and blue water-speedwell *Veronica anagallis-aquatica*. On the Continent, this species has also been recorded from marsh speedwell *V. scutellata* and brooklime *V. beccabunga*. In Britain, the larvae feed in galls on the inflorescences of the foodplants, with adults being recorded emerging in early July, early August and early October. Adults have been found from March to October.

Status Widespread but local in the southern half of England. This species is difficult to identify and may be confused with other members of the genus. Consequently, the exact status of this species is hard to assess.

Threats Drainage for reasons such as agricultural improvement and development is the primary cause of the loss of wetlands. Falling water tables because of water abstraction and river engineering schemes, as well as pollution and eutrophication may threaten this species.

Management and conservation Water tables should be maintained at high levels. Ditch clearance should be undertaken on a rotational basis and should aim to maintain the aquatic plant populations.

HYDRONOMUS ALISMATIS   NOTABLE B
A weevil
Order COLEOPTERA   Family CURCULIONIDAE

Hydronomus alismatis (Marsham, 1802). Formerly known as: *Bagous alismatis* (Marsham).

Distribution England, South Wales, North Wales, South East Scotland and South West Scotland.

Habitat and ecology Aquatic habitats such as ponds, lakes, ditches and dykes. Phytophagous. Associated with water-plantain *Alisma plantago-aquatica*, and possibly also other *Alisma* species and arrowhead *Sagittaria sagittifolia*. On the Continent, this species has also been recorded on floating water-plantain *Luronium natans*. In Britain, adults have been recorded from May to August.

Status Widespread but local in England, parts of Wales and southern Scotland.

Threats This species is threatened by drainage for reasons such as agricultural improvement and development. Falling water tables because of water abstraction schemes, the use of herbicides and pesticides, the infilling of ponds, and natural succession may also may also threaten this species.

Management and conservation Water tables should be maintained at high levels. Water bodies should be isolated from sources of eutrophication and pollution. Ditch and pond clearance should be undertaken on a rotational basis and should aim at maintaining aquatic plant populations.

HYLOBIUS TRANSVERSOVITTATUS
A weevil   **ENDANGERED**
Order COLEOPTERA   Family CURCULIONIDAE

Hylobius transversovittatus (Goeze, 1777).

Distribution Recorded from South Devon and North Somerset before 1970 and North Somerset from 1970 onwards.

Habitat and ecology Coastal undercliffs, grazing levels, and fenland. This species has been found on the ridges of loose peat between excavation trenches, and level areas where peat has previously been extracted. Phytophagous. Associated with purple-loosestrife *Lythrum salicaria*. Larvae feed in galls on the roots. Adults are found on the roots and at the base of the foodplant. Adults have been recorded in May, June and September.

Status Currently known from five 1km squares. Prior to the species' recent discovery on the Somerset Levels this beetle was only known from an example found in 1933 near Ashcott, North Somerset, and examples found between Seaton and Lyme Regis, South Devon.

Threats This species is threatened by drainage, peat cutting, and natural succession and scrub invasion.

Management and conservation Water tables should be maintained to keep the surface peat moist. Cutting or some other disturbance may be need to maintain open conditions.

Published sources Ashe, G.H. (1944b), Blair, K.G. (1945), Hodge, P.J. (1989), Shirt, D.B., ed. (1987).

HYPERA ARUNDINIS — EXTINCT
A weevil
Order COLEOPTERA   Family CURCULIONIDAE

Hypera arundinis (Paykull, 1792). Formerly known as: *Phytonomus arundinis* (Paykull).

Distribution Recorded from South Devon, South Hampshire, East Kent and "Norfolk".

Habitat and ecology Wetland. Phytophagous. Associated with semi-aquatic members of the Umbelliferae, in particular great water-parsnip *Sium latifolium*. On the Continent, this species has been noted on fine-leaved water-dropwort *Oenanthe aquatica* and lesser water-parsnip *Berula erecta*. The larvae probably feed externally on the foliage of the foodplant.

Status Presumed extinct. Last recorded in the 19th century.

Published sources Fowler, W.W. (1891), Shirt, D.B., ed. (1987).

HYPERA DAUCI — NOTABLE B
A weevil
Order COLEOPTERA   Family CURCULIONIDAE

Hypera dauci (Olivier, 1807). Formerly known as: *Hypera fasciculata* (Herbst, 1795), *Phytonomus dauci* (Olivier), *Phytonomus fasciculatus* (Herbst).

Distribution South West, South, South East, East Anglia, East Midlands, Wales, North East, North West and South East Scotland.

Habitat and ecology Sandy places in grassland, sand dunes and on disturbed ground. Also recorded on limestone grassland. Often found near the coast. Phytophagous. Associated with common stork's-bill *Erodium cicutarium*, the beetle requiring large plants in open situations. On the Continent, this species has noted on musk stork's-bill *E. moschatum*, round-leaved crane's-bill *Geranium rotundifolium* and dove's-foot crane's-bill *G. molle*. Larvae feed externally on the foliage of the foodplant. Adults are found at the base of plants and have been recorded from June to October.

Status Widespread and very local in England and Wales. Recently recorded from East Suffolk, north to Cumberland. There are no recent records for Scotland.

Threats Loss of sandy areas and unimproved grassland through improvement by reseeding, the application of fertilisers, conversion to arable agriculture, as well as through afforestation, urban and holiday developments, and natural succession and scrub invasion.

Management and conservation Disturbance, such as rotovation, is needed to maintain open conditions and encourage early successional stages.

Published sources Cooter, J. (1990c), Eyre, M.D. (1987), Fowler, W.W. (1891).

HYPERA DIVERSIPUNCTATA — RARE
A weevil
Order COLEOPTERA   Family CURCULIONIDAE

Hypera diversipunctata (Schrank, 1798). Formerly known as: *Hypera elongata* (Paykull, 1792), *Phytonomus diversipunctatus* (Schrank), *Phytonomus elongata* (Paykull).

Distribution Recorded from West Kent, East Suffolk, West Suffolk, East Norfolk, Nottinghamshire, Mid-west Yorkshire, Dumfriesshire, Selkirkshire and Midlothian before 1970 and West Norfolk, South-west Yorkshire and Dumfriesshire from 1970 onwards.

Habitat and ecology Recorded from an edge of a woodland, swept from a marshy area, and found amongst cut grass by the side of a river. Phytophagous. On the Continent, the species is associated with members of the Caryophyllaceae, particularly bog stitchwort *Stellaria alsine*, field mouse-ear *Cerastium arvense* and water-chickweed *Myosoton aquaticum*. The larvae feed externally on the foliage of the foodplant. In Britain, adults have been recorded in April and from July to September.

Status Old records suggest that this species had a scattered distribution in England, and was recorded as far north as Midlothian, Scotland. Recently recorded from only three, widely scattered, vice-counties.

Published sources Flint, J.H. (1986), Kevan, D.K. (1960), Owen, J.A. (1989a), Shirt, D.B., ed. (1987).

HYPERA FUSCOCINEREA — NOTABLE B
A weevil
Order COLEOPTERA   Family CURCULIONIDAE

Hypera fuscocinerea (Marsham, 1802). Formerly known as: *Hypera murina* (F., 1792), *Phytonomus fuscocinereus* (Marsham), *Phytonomus murinus* (F.).

Distribution England, South East Scotland and South West Scotland.

CURCULIONIDAE

Habitat and ecology Grassland, field margins, chalk pits and occasionally in wetland. Prefers open situations and primarily found on dry soils with some disturbance. Phytophagous. Associated with medick *Medicago* and also recorded from tufted vetch *Vicia cracca*. On the Continent, this species has been found on lucerne *Medicago sativa*, sickle medick *M. falcata*, ribbed melilot *Melilotus officinalis*, white melilot *M. alba*, white clover *Trifolium repens* and red clover *T. pratense*. The larvae feed externally on the foliage of the foodplant. In Britain, adults have been recorded from April to July.

Status Widespread but local in England and southern Scotland.

Threats Loss of unimproved grassland through improvement by reseeding or by the application of fertilisers, or by conversion to arable agriculture. The infilling of pits, the use of herbicides and pesticides, and natural succession and scrub invasion structure may also threaten this species.

Management and conservation Grazing or cutting, on a rotational basis, is needed to maintain open conditions.

HYPERA MELES　　　　　　　　　**NOTABLE A**
A weevil
Order COLEOPTERA　　　　Family CURCULIONIDAE

Hypera meles (F., 1792). Formerly known as: *Phytonomus meles* (F.).

Distribution Recorded from Isle of Wight, South Hampshire, Surrey, East Norfolk, Cambridgeshire, Glamorgan and Mid-west Yorkshire before 1970 and North Somerset, Dorset, South Hampshire, North Hampshire, East Sussex and East Kent from 1970 onwards.

Habitat and ecology Grassland, roadside verges and field margins. Phytophagous. Associated with white clover *Trifolium repens*, and possibly lucerne *Medicago sativa* and restharrow *Ononis*. On the Continent, this species has also been recorded from sickle medick *M. falcata*, ribbed melilot *Melilotus officinalis*, white melilot *M. alba*, red clover *T. pratense* and bird's-foot-trefoil *Lotus*. The larvae probably feed externally on the foliage of the foodplant. Adults have been recorded from April to September.

Status Status revised from RDB 3 (Rare) in Shirt (1987). Recorded in England from Dorset to Mid-west Yorkshire, also noted in South Wales. Recently recorded from only six vice-counties. This species is difficult to identify and may be confused with *H.*

plantaginis. Consequently, the exact status of this species is hard to assess.

Threats Loss of unimproved grassland through improvement by reseeding or by the application of fertilisers, or by conversion to arable agriculture. On verges, an intensive cutting regime is a threat. A lack of, or changes in grazing or cutting regimes, especially on grasslands, and the use of herbicides and pesticides may also threaten this species.

Management and conservation Grazing, cutting or some other disturbance, such as rotovation, on a rotational basis, is needed to maintain open conditions.

Published sources Allen, A.A. (1972e), Appleton, D. (1973a), Beare, T.H. (1922), Morris, M.G. (1982a), Shirt, D.B., ed. (1987).

HYPERA ONONIDIS INSUFFICIENTLY KNOWN
A weevil
Order COLEOPTERA　　　　Family CURCULIONIDAE

Hypera ononidis Chevrolat, 1863. Formerly known as: *Phytonomus ononidis* (Chevrolat).

Distribution Recorded from South Devon, North Devon and East Sussex before 1970 and Isle of Wight from 1970 onwards.

Habitat and ecology Coastal dunes, cliffs and shingle. Phytophagous. Associated with common restharrow *Ononis repens*. The larvae probably feed externally on the foliage of the foodplant. Adults have been recorded in May, June and September.

Status Not listed in the insect Red Data Book (Shirt, 1987). Only recorded between East Sussex and South Devon, and recently noted from just one vice-county. The taxonomic status of *H. ononidis* is in question, it is possible that this beetle may prove to be a form of *H. nigrirostris*. Further examination of previously determined examples is needed before this species can be confirmed as British.

Threats This species may be threatened by cliff stabilisation schemes, the construction of sea defences, gravel extraction, and urban and coastal developments. The degradation of suitable habitat through the invasion of scrub on stabilised areas is a further threat.

Management and conservation In areas of unstable cliff, occasional slippages are necessary to maintain habitat continuity. Large areas are required so that the population does not become isolated and subsequently threatened by individual landslips. Disturbance of coastal shingle should be avoided. On dunes some grazing and other disturbance may be desirable to maintain early successional stages and prevent the invasion of scrub.

Published sources Edmonds, T.H. (1930).

HYPERA PASTINACAE   **ENDANGERED**
A weevil
Order COLEOPTERA   Family CURCULIONIDAE

Hypera pastinacae (Rossi, 1790). Formerly known as: *Hypera tigrina* Boheman, 1834, *Phytonomus pastinacae* (Rossi).

Distribution Recorded from East Kent before 1970 and East Kent from 1970 onwards.

Habitat and ecology Coastal cliffs. Phytophagous. Associated with wild carrot *Daucus carota* subsp. *gummifer*. On the Continent, this species has also been recorded from wild parsnip *Pastinaca sativa*. The larvae feed externally in the incurved flower-heads of the foodplant. In Britain, pupae and newly emerged adults have been found within the fruiting heads of the foodplant in late August. Adults have been recorded in August and September.

Status Known from coastal cliffs in East Kent, where the species is well established.

Threats This species is threatened by coastal development, cliff stabilisation schemes and the construction of sea defences.

Management and conservation Large areas of cliff are required so that the population does not become isolated and subsequently threatened by individual landslips.

Published sources Fowler, W.W. (1891), Fowler, W.W. & Donisthorpe, H.St J.K. (1913), Shirt, D.B., ed. (1987).

LARINUS PLANUS   **NOTABLE B**
A weevil
Order COLEOPTERA   Family CURCULIONIDAE

Larinus planus (F., 1792). Formerly known as: *Larinus carlinae* (Olivier, 1807).

Distribution South East, South, South West, West Midlands, Dyfed-Powys and North Wales.

Habitat and ecology Grassland, usually near the coast. Also recorded inland. Phytophagous. Associated with creeping thistle *Cirsium arvense*, meadow thistle *C.*

dissectum, marsh thistle *C. palustre*, spear thistle *C. vulgare* and musk thistle *Carduus nutans*. On the Continent, this species has also been noted on cabbage thistle *C. oleraceum*, welted thistle *Carduus acanthoides*, greater knapweed *Centaurea scabiosa*, brown knapweed *C. jacea* and carline thistle *Carlina vulgaris*. In Britain, adults have been recorded from April to September.

Status Widespread but local in southern England, also recorded in West Midlands and parts of Wales.

Threats Loss of grassland through conversion to arable agriculture or other land use. Lack of, or changes in grazing or cutting regimes may threaten this species.

Management and conservation Grazing or some other disturbance is needed to maintain open conditions and encourage early successional stages.

LEIOSOMA OBLONGULUM   **NOTABLE B**
A weevil
Order COLEOPTERA   Family CURCULIONIDAE

Leiosoma oblongulum Boheman, 1842. Formerly known as: *Liosoma oblongulum* Boheman, 1842.

Distribution South West, South, South East, East Anglia, East Midlands, West Midlands, North West and Dyfed-Powys.

Habitat and ecology Broad-leaved woodland, coastal cliffs and possibly also grassland. Phytophagous. Probably associated with members of the Ranunculaceae. On the Continent, this species is associated with creeping buttercup *Ranunculus repens*, wood anemone *Anemone nemorosa* and columbine *Aquilegia vulgaris*. In Britain, it has been recorded in leaf-litter, damp moss and by sweeping roadside verges. Adults have been recorded from March to June.

Status Very local and recorded from southern England, north to Cumberland. Old records suggest that this species may have declined.

Threats This species may be threatened by the clear-felling of woodland and conversion to other land use, and the improvement of grassland and conversion to arable agriculture. Neglect and conversion to high forest, and scrub invasion may be further threats.

Management and conservation Grazing or cutting, on a rotational basis, is needed to maintain open conditions. Open glades and ride margins should be managed to retain a variety of vegetation structures.

CURCULIONIDAE

Published sources Atty, D.B. (1983), Fowler, W.W. (1891), Fowler, W.W. & Donisthorpe, H.St J.K. (1913), Read, R.W.J. (1980).

LEIOSOMA PYRENAEUM　　　　**VULNERABLE**
A weevil
Order COLEOPTERA　　　Family CURCULIONIDAE

Leiosoma pyrenaeum Brisout, 1866. Formerly known as: *Leiosoma pyrenaeum ssp. troglodytes* Rye, 1873, *Liosoma pyrenaeum* Rye, 1873, *Liosoma troglodytes* Rye, 1873.

Distribution Recorded from South Hampshire, East Sussex, East Kent and Surrey before 1970 and East Kent from 1970 onwards.

Habitat and ecology Calcareous grasslands, particularly downland. Phytophagous. Found at the roots of grass and in moss. Host-plants unknown in Britain, but probably consist of members of the Ranunculaceae. Adults have been recorded in April.

Status Status revised from RDB 3 (Rare) in Shirt (1987). Only known from southern and south-eastern England. Recently recorded from just one vice-county. Possibly under-recorded.

Threats Loss of calcareous grassland through improvement by reseeding or by the application of fertilisers, or by conversion to arable agriculture. Lack of, or changes in grazing or cutting regimes may threaten this species.

Management and conservation Grazing or cutting, on a rotational basis, is needed to maintain open conditions.

Published sources Fowler, W.W. (1891), Shirt, D.B., ed. (1987).

LEPYRUS CAPUCINUS　　　　　**EXTINCT**
A weevil
Order COLEOPTERA　　　Family CURCULIONIDAE

Lepyrus capucinus (Schaller, 1783). Formerly known as: *Lepyrus binotatus* (F., 1792).

Distribution Recorded from South Hampshire, North Hampshire, Surrey, Berkshire and Moray.

Habitat and ecology Wetland, willow carr and wet woodland. Phytophagous. Probably associated with *Salix* species. The species may be subterranean in its habits. Adults have been recorded in March.

Status Presumed extinct. There are only a few known occurences in the 19th century. Last recorded at Wellington College, Berkshire in 1897.

Published sources Fowler, W.W. (1891), Fowler, W.W. & Donisthorpe, H.St J.K. (1913), Shirt, D.B., ed. (1987).

LIMOBIUS BOREALIS　　　　　**NOTABLE A**
A weevil
Order COLEOPTERA　　　Family CURCULIONIDAE

Limobius borealis (Paykull, 1792). Formerly known as: *Limobius dissimilis* (Herbst, 1795).

Distribution Recorded from West Cornwall, South Devon, North Wiltshire, Dorset, North Hampshire, East Kent, West Kent, Surrey, Berkshire, Cambridgeshire, East Gloucestershire, Herefordshire, Caernarvonshire, Flintshire, Derbyshire, West Lancashire, North-east Yorkshire, Mid-west Yorkshire, North-west Yorkshire, Durham, South Northumberland, Cumberland, Renfrewshire, West Lothian and Kincardineshire before 1970 and North Wiltshire, West Gloucestershire, Durham and Roxburghshire from 1970 onwards.

Habitat and ecology Grassland, though possibly field margins and disturbed ground. Also recorded from a woodland. Phytophagous. Associated with *Geranium* species. Particularly found on meadow crane's-bill *G. pratense*, also herb robert *G. robertianum* and possibly bloody crane's-bill *G. sanguineum*. On the Continent, this species has also been noted on dove's-foot crane's-bill *G. molle*, hedgerow crane's-bill *G. pyrenaicum* and common stork's-bill *Erodium cicutarium*. In Britain, adults have been recorded from April to August.

Status Very local, though with a widely scattered distribution from West Gloucestershire to Roxburghshire in Scotland. Recently recorded from just four, widely scattered, vice-counties.

Threats This species has probably declined because of the loss of unimproved grassland through improvement by reseeding or by the application of fertilisers, or by conversion to arable agriculture. Lack of, or changes in grazing or cutting regimes, and the use of herbicides and pesticides may have contributed to this species' decline.

Management and conservation Grazing, cutting or some other disturbance, such as rotovation, on a rotational basis, may be needed to maintain open conditions.

Published sources Allen, A.A. (1973a), Atty, D.B. (1983), Eyre, M.D. (1987), Fowler, W.W. (1891), Fowler, W.W. & Donisthorpe, H.St J.K. (1913), Morris, M.G. (1974b).

LIMOBIUS MIXTUS　　　　　　**ENDANGERED**
A weevil
Order COLEOPTERA　　　Family CURCULIONIDAE

Limobius mixtus (Boheman, 1834).

Distribution Recorded from South Devon, Dorset, East Sussex, East Kent and West Suffolk before 1970 and East Sussex from 1970 onwards.

Habitat and ecology Sandy habitats, but not on very acid soils. Phytophagous. Associated with common stork's-bill *Erodium cicutarium*. Larvae feed externally on the foliage. Adults have been recorded in April, May and from July to September.

Status Status revised from RDB 2 (Vulnerable) in Shirt (1987). Only known from southern England. Recently recorded from just one site; Rye Harbour, East Sussex.

Threats This species is threatened by the infilling of gravel pits, tourist development and natural succession.

Management and conservation Management should aim at maintaining open conditions and encouraging early successional stages along pit edges. Rye Harbour is a Local Nature Reserve and has been designated as an SSSI.

Published sources Fowler, W.W. (1891), Fowler, W.W. & Donisthorpe, H.St J.K. (1913), Shirt, D.B., ed. (1987).

LIPARUS CORONATUS　　　　　**NOTABLE B**
A weevil
Order COLEOPTERA　　　Family CURCULIONIDAE

Liparus coronatus (Goeze, 1777).

Distribution South East, South, South West, East Anglia, East Midlands, West Midlands, North East and South Wales.

Habitat and ecology Wood-edge, field margins, gardens, undisturbed roadside verges and other grasslands. Phytophagous. Associated with members of the Umbelliferae, particularly cow parsley *Anthriscus sylvestris*. On the Continent, this species has also been found on wild carrot *Daucus carota*, bur chervil *A. caucalis*, chervil *Chaerophyllum* and wild parsnip *Pastinaca sativa*. Larvae feed in the roots of the foodplants, while adults occur on or at the base of the plants. In Britain, adults have been recorded from May to August.

Status Widespread but local in England, also recorded in South Wales.

Threats This species is threatened by the use of herbicides and pesticides. The loss of grassland through conversion to arable agriculture or other land use, and the lack of, or changes in grazing or cutting regimes may also threaten this species.

Management and conservation Open conditions conditions should be maintained.

LIPARUS GERMANUS　　　　　**VULNERABLE**
A weevil
Order COLEOPTERA　　　Family CURCULIONIDAE

Liparus germanus (L., 1758).

Distribution Recorded from East Sussex, East Kent, West Kent and Surrey before 1970 and East Kent and West Kent from 1970 onwards.

Habitat and ecology The margins of lanes, sheltered hedgebanks, ditches, field margins and possibly gardens. Phytophagous. Associated with hogweed *Heracleum sphondylium* and possibly wild angelica *Angelica sylvestris*. Larvae feed on the roots of the foodplant, while adults occur on or at the base of the plants. Adults have been recorded from June and August.

Status Extremely local and only known in south-eastern England. Recently recorded from just two vice-counties.

Threats This species may be threatened by development and change in land use, and the use of herbicides and pesticides. The lack of, or changes in grazing or cutting regimes could alter the vegetation structure which may in turn threaten this species.

Management and conservation Open conditions need to be maintained.

Published sources Duffield, C.A.W. (1921), Fowler, W.W. (1891), Fowler, W.W. & Donisthorpe, H.St J.K. (1913), Shirt, D.B., ed. (1987).

CURCULIONIDAE

LITODACTYLUS LEUCOGASTER   NOTABLE B
A weevil
Order COLEOPTERA   Family CURCULIONIDAE

Litodactylus leucogaster (Marsham, 1802). Formerly
known as: *Phytobius leucogaster* (Marsham).

Distribution England, Dyfed-Powys, South Wales and
Scotland.

Habitat and ecology Wetland and aquatic habitats,
including pools and brackish ditches. Phytophagous. On
the Continent, this species is associated with spiked
water-milfoil *Myriophyllum spicatum* and whorled
water-milfoil *M. verticillatum*. In Britain, adults have
been recorded from March to June, and in August and
September.

Status Widespread but local throughout Britain.

Threats This species is threatened by drainage for
reasons such as agricultural improvement and
development. Falling water tables because of water
abstraction schemes, the infilling of pools, the use of
herbicides and pesticides, and natural succession are
further threats.

Management and conservation Water tables should be
maintained at high levels. Water bodies should be
isolated from sources of eutrophication and pollution.
Ditch clearance should be undertaken on a rotational
basis and should aim at maintaining the aquatic plant
populations.

LIXUS ALGIRUS   ENDANGERED
A weevil
Order COLEOPTERA   Family CURCULIONIDAE

Lixus algirus (L., 1758).

Distribution Recorded from West Sussex, East Sussex
and West Kent before 1970.

Habitat and ecology Marshy and wet grassland.
Phytophagous. Associated with thistles *Cirsium* and
probably *Carduus* species. On the Continent, this
species has been recorded on creeping thistle *Cirsium
arvense*, marsh thistle *C. palustre*, common mallow
Malva sylvestris, hollyhock *Althaea rosea* and
Centaurea species. Larvae feed in the stems of the
foodplant. In Britain, adults have been recorded in
September.

Status Last recorded in 1923 at Fairlight, East Sussex.

Threats Loss of grassland through conversion to arable
agriculture or other land use. Drainage, and the lack of,

or changes in grazing or cutting regimes, and the use of
herbicides and pesticides may have contributed to this
species' decline.

Published sources Fowler, W.W. (1891), Fowler,
W.W. & Donisthorpe, H.St J.K. (1913), Mitford, R.S.
(1923), Shirt, D.B., ed. (1987).

LIXUS SCABRICOLLIS
A weevil   INSUFFICIENTLY KNOWN
Order COLEOPTERA   Family CURCULIONIDAE

Lixus scabricollis Boheman, 1843.

Distribution Recorded from West Kent from 1970
onwards.

Habitat and ecology Coastal shingle. Found on sea
beet *Beta vulgaris* subsp. *maritima*. On the Continent,
this species is associated with a variety of *Beta* species.
Adults have been recorded in September and October.

Status Not listed in the insect Red Data Book (Shirt,
1987). Discovered new to Great Britain in 1987 at a
site in West Kent. Probably a recent colonist or
possibly an accidental introduction.

Threats Uncertain, though this species may be
threatened by coastal developments and gravel
extraction.

Management and conservation Disturbance of coastal
shingle should be avoided.

LIXUS PARAPLECTICUS   ENDANGERED
A weevil
Order COLEOPTERA   Family CURCULIONIDAE

Lixus paraplecticus (L., 1758).

Distribution Recorded from North Somerset, South
Hampshire, West Sussex, East Kent, West Kent,
Surrey, Middlesex, Oxfordshire, East Suffolk, East
Norfolk, Cambridgeshire, Huntingdonshire, South-east
Yorkshire, South-west Yorkshire and Cumberland
before 1970.

Habitat and ecology Marshes, fens and the margins of
rivers and lakes. Phytophagous. Associated with
semi-aquatic members of the Umbelliferae, particularly
greater water-parsnip *Sium latifolium*, fine-leaved
water-dropwort *Oenanthe aquatica*, and possibly also
hemlock water-dropwort *O. crocata*. On the Continent,
this species has also been recorded from tubular
water-dropwort *O. fistulosa*. Larvae feed in the stems

of the foodplant. In Britain, adults have been recorded in April, July and August.

Status Widespread in southern England and East Anglia in the 19th and the early part of the 20th centuries. Last recorded in 1950 at Catcott Heath, North Somerset.

Threats This species has probably declined because of drainage of wetlands for reasons such as agricultural improvement and development. River engineering, including dredging, level regulation by damming and flood alleviation schemes, river pollution, as well as falling water tables because of water abstraction, and erosion because of boat traffic may also have contributed to this species' decline.

Management and conservation Water tables should be maintained at high levels. Water bodies should be isolated from sources of eutrophication and pollution.

Published sources Fowler, W.W. (1891), Fowler, W.W. & Donisthorpe, H.St J.K. (1913), Massee, A.M. (1940a), Shirt, D.B., ed. (1987), Wilson, W.A. (1958).

LIXUS VILIS   **ENDANGERED**
A weevil
Order COLEOPTERA   Family CURCULIONIDAE

Lixus vilis (Rossi, 1790). Formerly known as: *Lixus bicolor* Olivier, 1807.

Distribution Recorded from South Somerset, South Hampshire and East Kent before 1970.

Habitat and ecology Sand dunes and sandy places, particularly near the coast. Phytophagous. Associated with common storks-bill *Erodium cicutarium*. Larvae feed in the stems. Adults have been recorded in May and June.

Status Possibly extinct. Only known in southern England. Last recorded in 1905 at Deal, East Kent.

Threats This species has probably declined because of the loss of sandy habitat, particularly through urban and holiday development. The degradation of remaining suitable habitat by excessive disturbance of the vegetation through activities such as human trampling.

Published sources Fowler, W.W. (1891), Shirt, D.B., ed. (1987).

MAGDALIS BARBICORNIS   **NOTABLE A**
Pear weevil
Order COLEOPTERA   Family CURCULIONIDAE

Magdalis barbicornis (Latreille, 1804).

Distribution Recorded from North Somerset, South Hampshire, West Sussex, East Sussex, East Kent, West Kent, Surrey, Hertfordshire, Middlesex, Berkshire, Oxfordshire, East Suffolk, Cambridgeshire, Bedfordshire, Huntingdonshire, West Gloucestershire, Worcestershire, North Lincolnshire and Leicestershire & Rutland before 1970 and Dorset, West Sussex, East Sussex, West Kent, Surrey, Berkshire and Worcestershire from 1970 onwards.

Habitat and ecology Broad-leaved woodland, hedges, gardens and orchards. Phytophagous. Recorded from hawthorn, apple, medlar, pear and *Sorbus* species. On the Continent, this species has also been recorded from crab-apple. The larvae feed internally in twigs and branches. In Britain, adults have been recorded from late May to July and are probably short-lived.

Status Very local and widely distributed in southern England as far north as Worcestershire and North Lincolnshire.

Threats Loss of broad-leaved woodland through, for example, clear-felling and coniferisation. Neglect, and conversion to high forest, has led to increased shade and the loss of glades and broad sunny rides. The grubbing out of hedgerows and orchards, the mechanised trimming of hedgerows, and the use of pesticides may be further threats.

Management and conservation Open spaces in woodland need to be retained. Open glades and ride margins should be cut on rotation to retain a variety of vegetation structures.

Published sources Atty, D.B. (1983), Fowler, W.W. (1891), Fowler, W.W. & Donisthorpe, H.St J.K. (1913), Whitehead, P.F. (1990a).

MAGDALIS CARBONARIA   **NOTABLE B**
A weevil
Order COLEOPTERA   Family CURCULIONIDAE

Magdalis carbonaria (L., 1758).

Distribution England and Scotland.

Habitat and ecology Broad-leaved woodland and birch scrub. Phytophagous. Associated with birch. The larvae feed internally in dead twigs and branches. Adults have

CURCULIONIDAE

been recorded from April to July and are probably short-lived.

Status Widespread but local in England and Scotland, more frequent in the north. Not recorded in Wales. Often numerous where it occurs.

Indicator status Grade 2 in Garland (1983).

Threats Loss of broad-leaved woodland through, for example, clear-felling and coniferisation, and the grubbing out of scrub. Habitat loss, in particular, through the removal of dead wood, twigs and small branches from living trees. Neglect and conversion to high forest may be a further threat.

Management and conservation Gaps in the age structure of the tree population should be identified and regeneration encouraged to ensure the continuity of suitable habitat.

MAGDALIS CERASI   **NOTABLE B**
A weevil
Order COLEOPTERA   Family CURCULIONIDAE

Magdalis cerasi (L., 1758).

Distribution England.

Habitat and ecology Woodland, scrub and hedges. Phytophagous. Primarily associated with oak, though also found on species of Rosaceae, in particular blackthorn, hawthorn, pear, apple and rowan. Larvae feed internally in dead twigs and small branches. Adults have been recorded from May to August.

Status Widespread but local in central and southern England. More local in northern England.

Threats Loss of woodland through, for example, clear-felling and conversion to other land use. Habitat loss, in particular, through removal of dead wood and twigs from living trees and shrubs. Neglect and conversion to high forest, the grubbing out of scrub and hedgerows, and the mechanised trimming of hedgerows may be further threats.

Management and conservation Open glades and ride margins should be managed, on a rotational basis, to retain a variety of vegetation structures.

MAGDALIS DUPLICATA   **NOTABLE A**
A weevil
Order COLEOPTERA   Family CURCULIONIDAE

Magdalis duplicata Germar, 1818.

Distribution Recorded from Fife, Stirlingshire, Mid Perthshire, South Aberdeenshire, Moray and East Inverness & Nairn before 1970 and Mid-west Yorkshire, Cumberland, West Perthshire, Mid Perthshire, South Aberdeenshire, Moray and East Inverness & Nairn from 1970 onwards.

Habitat and ecology Native coniferous woodland and plantations. Phytophagous. Associated with Scots pine. The larvae feed internally in twigs and branches. Adults have been recorded from June to August and are probably short-lived.

Status Very local and predominantly Scottish in distribution. Probably a component of the native pine forest fauna which has spread, or been introduced, into other parts of Great Britain.

Threats Loss of native pine forest through, for example, clear-felling and conversion to other land use. Overgrazing, preventing regeneration, may be a further threat.

Management and conservation Gaps in the age structure of the tree population should be identified and regeneration encouraged to ensure the continuity of suitable habitat.

Published sources Alexander, K.N.A. & Grove, S.J. (1990), Allen, A.A. (1982a), Flint, J.H. (1988), Fowler, W.W. (1891), Fowler, W.W. & Donisthorpe, H.St J.K. (1913), Read, R.W.J. (1985).

MAGDALIS PHLEGMATICA   **NOTABLE A**
A weevil
Order COLEOPTERA   Family CURCULIONIDAE

Magdalis phlegmatica (Herbst, 1797).

Distribution Recorded from South Northumberland, Cumberland, Lanarkshire, West Lothian, Fife, Stirlingshire, Mid Perthshire, South Aberdeenshire, Moray and East Inverness & Nairn before 1970 and Mid-west Yorkshire, Peebleshire, Stirlingshire, West Perthshire, South Aberdeenshire, Moray and East Inverness & Nairn from 1970 onwards.

Habitat and ecology Native coniferous woodland and spruce plantation. Phytophagous. Associated with Scots pine. On the Continent, this species has been recorded on Norway spruce. The larvae feed internally in twigs and branches. In Britain, adults have been recorded from June to early August and are probably short-lived.

Status Very local and predominantly Scottish in distribution. Probably a component of the native pine forest fauna which has spread, or been introduced, into other parts of Britain.

Threats Loss of native pine forest through, for example, clear-felling and conversion to other land use. Overgrazing, preventing regeneration, may be a further threat.

Management and conservation Gaps in the age structure of the tree population should be identified and regeneration encouraged to ensure the continuity of suitable habitat.

Published sources Eyre, M.D. (1987), Fowler, W.W. & Donisthorpe, H.St J.K. (1913).

MECINUS CIRCULATUS NOTABLE B
A weevil
Order COLEOPTERA Family CURCULIONIDAE

Mecinus circulatus (Marsham, 1802).

Distribution South East, South, South West, East Anglia, East Midlands, West Midlands, South Wales and North Wales.

Habitat and ecology Grassland, chalk downland, undercliffs, cliff tops, sand dunes, disturbed ground and probably road-side verges and field margins. Predominantly coastal. Phytophagous. Associated with ribwort plantain *Plantago lanceolata* and buck's-horn plantain *P. coronopus*. Adults have been recorded in January, from April to June and in August and September.

Status Widespread but local in southern England, also recorded in parts of Wales.

Threats Loss of habitat through improvement and conversion to arable agriculture, urban and coastal developments, afforestation, the construction of sea defences and erosion. Natural succession, and the use of herbicides and pesticides may be a further threat.

Management and conservation Grazing, cutting or some other disturbance, such as rotovation, on a rotational basis, may be needed to maintain open conditions. In areas of undercliff, occasional slippages are necessary to maintain habitat continuity. Large areas of unstable cliff are required so that the population does not become isolated and subsequently threatened by individual landslips.

MECINUS COLLARIS NOTABLE B
A weevil
Order COLEOPTERA Family CURCULIONIDAE

Mecinus collaris Germar, 1821.

Distribution South East, South, South West, East Anglia, West Midlands, North East, North West, South Wales and South West Scotland.

Habitat and ecology Saltmarsh. Phytophagous. Associated with sea plantain *P. maritima*. On the Continent, this species has also been recorded from ribwort plantain *P. lanceolata*, hoary plantain *P. media* and greater plantain *P. major*. Larvae develop in galls at the top of the flowering stem below the inflorescence. In Britain, adults have been recorded as emerging from their galls in early July. Adults have been found in March, April and from June to September.

Status Widespread and very local around the coasts of England, also recorded in South Wales and South West Scotland. This species can suffer high levels of parasitism.

Threats Loss of saltmarsh through reclamation, erosion and the construction of sea defences. Overgrazing of saltmarshes could be a further threat to this species.

Management and conservation Grazing should not be introduced to a saltmarsh where there is no grazing at present.

MECINUS JANTHINUS NOTABLE A
A weevil
Order COLEOPTERA Family CURCULIONIDAE

Mecinus janthinus Germar, 1821.

Distribution Recorded from East Kent, West Kent, Surrey, South Essex and Middlesex before 1970 and East Kent, West Kent and South Essex from 1970 onwards.

Habitat and ecology Disturbed ground, grassland and road-side verges. Often on chalk. Phytophagous. Associated with common toadflax *Linaria vulgaris*. On the Continent, this species has also been recorded from small toadflax *Chaenorhinum minus*. In Britain, adults have been recorded from April to August.

Status Discovered in Britain in 1948. Only known from five vice-counties, and just three recently, all in south-eastern England.

CURCULIONIDAE

Threats Loss of grassland through improvement and conversion to arable agriculture, and development. Natural succession may be a further threat.

Management and conservation Some disturbance, such as rotovation, is needed to maintain open conditions and encourage early successional stages.

Published sources Allen, A.A. (1948d), Allen, A.A. (1951g), Allen, A.A. (1960c), Forster, H.W. (1951a), Forster, H.W. (1951b), George, R.S. (1953), Morris, M.G. (1960).

MESITES TARDII	NOTABLE B
A weevil	
Order COLEOPTERA	Family CURCULIONIDAE

Mesites tardii (Curtis, 1825). Formerly known as: *Mesites tardyi* (Curtis, 1825), *Rhopalomesites tardyi* (Curtis).

Distribution South East, South, South West, West Midlands, East Anglia, North East, North West, Wales, South West Scotland and North West Scotland.

Habitat and ecology Broad-leaved woodland and scrub, mainly near the coast. Phytophagous. In dead wood of a variety of tree species, including alder, birch, hawthorn, ash, holly, poplar, oak, willow, elm, beech, *Sorbus* and *Acer* species. On the Continent, the species has also been recorded from dead heath *Erica* and spurge *Euphorbia*. The beetle been found in dead wood on live trees and may possibly also attack living wood. This species may disperse utilising driftwood. In Britain, larvae have been found in mid August. Adults have been recorded from April to November.

Status Widespread but local in England, Wales and western Scotland.

Indicator status Grade 3 in Garland (1983). Grade 3 in Harding & Rose (1986).

Threats Loss of broad-leaved woodland through, for example, clear-felling and coniferisation. Habitat loss, in particular, through the felling of trees, removal of dead wood from living trees and the destruction or removal of standing and fallen dead wood for reasons such as forest hygiene, aesthetic tidiness, public safety or for use as fire wood. Beach parties and the clearance of driftwood may be a further threat.

Management and conservation Both fallen and standing dead timber, especially with the bark attached, should be retained. The removal of dead timber from trees should be avoided. Gaps in the age structure of the tree population should be identified and the

continuity of the appropriate dead wood habitat ensured by regeneration, suitable planting and possibly with pollarding.

MIARUS GRAMINIS	NOTABLE B
A weevil	
Order COLEOPTERA	Family CURCULIONIDAE

Miarus graminis (Gyllenhal, 1813).

Distribution South East, South, South West, East Anglia, East Midlands, West Midlands and North East.

Habitat and ecology Calcareous grassland and field margins. Phytophagous. Associated with clustered bellflower *Campanula glomerata*, and occasionally recorded from harebell *C. rotundifolia* and nettle-leaved bellflower *C. trachelium*. On the Continent, this species has also been recorded from large campanula *C. latifolia*. In Britain, adults have been recorded from May to August.

Status Widespread but local in England, although not recorded in North West England. Can be found in numbers where it occurs. This species is difficult to identify and may be confused with other members of the genus. Consequently, the exact status of this species is hard to assess.

Threats Loss of calcareous grassland through improvement and conversion to arable agriculture. Development, natural succession and the use of herbicides and pesticides are further threats.

Management and conservation Grazing or cutting, on a rotational basis, is needed to maintain open conditions.

MIARUS MICROS	RARE
A weevil	
Order COLEOPTERA	Family CURCULIONIDAE

Miarus micros (Germar, 1821).

Distribution Recorded from West Cornwall and East Cornwall before 1970 and West Cornwall from 1970 onwards.

Habitat and ecology Sandy grassland near the coast, sea cliffs, old walls, hedgebanks and chalk downland. Phytophagous. Associated with sheep's-bit *Jasione montana*. The larvae live gregariously and pupate in the capitula, adults emerging in early August. Adults have been recorded from June to September.

Status Only known from Cornwall. Recently recorded from just one vice-county. This species is difficult to identify and may be confused with other members of the genus. Consequently, the exact status of this species is hard to assess. Old records for Surrey are probably referable to *M. graminis*.

Threats Loss of habitat through natural succession, improvement and conversion to arable agriculture, and development may be threats to this species.

Management and conservation Grazing, cutting or some other disturbance, such as rotovation, on a rotational basis, may be needed to maintain open conditions.

Published sources Allen, A.A. (1989a), Butler, E.A. (1909), Fowler, W.W. (1891), Shirt, D.B., ed. (1987).

MIARUS PLANTARUM

A weevil   **INSUFFICIENTLY KNOWN**
Order COLEOPTERA   Family CURCULIONIDAE

Miarus plantarum (Germar, 1824).

Distribution Recorded from West Cornwall, North Wiltshire, Dorset, Isle of Wight, East Sussex, East Kent, West Kent, Surrey, North Essex, Hertfordshire, Berkshire, Oxfordshire, Buckinghamshire, East Norfolk, Cambridgeshire, Bedfordshire, Huntingdonshire, Northamptonshire, East Gloucestershire, West Gloucestershire and Leicestershire & Rutland before 1970 and South-west Yorkshire from 1970 onwards.

Habitat and ecology Grassland, road-side verges, field margins and hedgebanks. Phytophagous. Associated with bellflower *Campanula* and rampion *Phyteuma*. This species has been noted hibernating in apple trees. Adults have been recorded in January, from May to September and in November.

Status Not listed in the insect Red Data Book (Shirt, 1987). Old records suggest that this species was formerly more widespread and recorded throughout the southern half of England as far north as Leicestershire. There is only one recent record, that from South-west Yorkshire. This species is difficult to identify and may be confused with other members of the genus. Consequently, the exact status of this species is hard to assess.

Threats Uncertain, though this species may have declined through loss of grassland because of improvement and conversion to arable agriculture and development. Natural succession, and the use of herbicides and pesticides may also contributed to this species' decline.

Management and conservation Grazing or cutting, on a rotational basis, is needed to maintain open conditions.

Published sources Atty, D.B. (1983), Flint, J.H. (1988), Fowler, W.W. (1891), Fowler, W.W. & Donisthorpe, H.St J.K. (1913).

MONONYCHUS PUNCTUMALBUM NOTABLE A

A weevil
Order COLEOPTERA   Family CURCULIONIDAE

Mononychus punctumalbum (Herbst, 1784). Formerly known as: *Mononychus pseudacori* (F., 1792).

Distribution Recorded from South Devon, North Somerset, Dorset, Isle of Wight and South Hampshire before 1970 and North Somerset, South Wiltshire, Dorset and Isle of Wight from 1970 onwards.

Habitat and ecology Coastal cliffs, though occasionally also found in wetland. Phytophagous. Associated with stinking iris *Iris foetidissima*, though also found on yellow flag *I. pseudacorus*. Larvae develop in the seed pods. Adults are frequently found on the flowers of wild carrot and other plants. Larvae have been recorded in August and September, with pupae occurring in September. Adults have been found from May to August and in November.

Status Very local and known only from southern England. Recently recorded from just four vice-counties.

Threats This species is threatened by cliff stabilisation schemes and the construction of sea defences. Coastal developments may reduce the amount of available habitat. Activities that accelerate or reduce the rate of erosion should be avoided. The degradation of suitable habitat through natural succession and the invasion of scrub on stabilised areas, and in wetlands, drainage and water abstraction schemes are further threats.

Management and conservation Occasional slippages are necessary to maintain habitat continuity. Large areas of unstable cliff are required so that the population does not become isolated and subsequently threatened by individual landslips. Water tables should be maintained at high levels. The presence of nectar sources such as umbellifers and composite herbs may also be important for this species.

Published sources Eagles, T.R. (1952), Eagles, T.R. (1953c), Fowler, W.W. (1891), Fowler, W.W. & Donisthorpe, H.St J.K. (1913), Jones, R.A. (1977b).

CURCULIONIDAE

NOTARIS AETHIOPS
NOTABLE A
A weevil
Order COLEOPTERA   Family CURCULIONIDAE

Notaris aethiops (F., 1792). Formerly known as:
Erirrhinus aethiops (F.).

Distribution Recorded from North-east Yorkshire,
Mid-west Yorkshire, Cumberland, Dumfriesshire,
Kirkcudbrightshire, Renfrewshire, Roxburghshire,
Berwickshire, Stirlingshire, Mid Perthshire, South
Aberdeenshire, Moray and East Inverness & Nairn
before 1970 and Mid Perthshire and Moray from 1970
onwards.

Habitat and ecology Wetland. Phytophagous.
Associated with branched bur-reed *Sparganium
erectum*. Adults have been recorded from April to
August.

Status Very local. Primarily Scottish, though also
recorded in northern England.

Threats Drainage for reasons such as agricultural
improvement and development. Falling water tables
because of water abstraction and river engineering
schemes may also threaten this species.

Management and conservation Water tables should be
maintained at high levels. Water bodies should be
isolated from sources of eutrophication and pollution.
Clearance should be undertaken on a rotational basis
and should aim at encouraging a variety of early
successional stages and ensuring the presence of the
foodplant.

Published sources Fowler, W.W. (1891), Fowler,
W.W. & Donisthorpe, H.St J.K. (1913), Read, R.W.J.
(1981).

NOTARIS BIMACULATUS
NOTABLE B
A weevil
Order COLEOPTERA   Family CURCULIONIDAE

Notaris bimaculatus (F., 1787). Formerly known as:
Erirrhinus bimaculatus (F.).

Distribution England, Wales, South East Scotland and
South West Scotland.

Habitat and ecology Wetland and river banks.
Phytophagous. Associated with reed canary-grass
Phalaris arundinacea, common reed *Phragmites
australis*, bulrush *Typha latifolia*, and possibly sedge
Carex. Larvae develop in the stems of the foodplant.

Adults have been recorded from April to July and in
September.

Status Widespread but local. Recorded throughout
Great Britain as far north as southern Scotland.

Threats Drainage for reasons such as agricultural
improvement and development. Falling water tables
because of water abstraction and river engineering
schemes may also threaten this species.

Management and conservation Water tables should be
maintained at high levels. Water bodies should be
isolated from sources of eutrophication and pollution.
Clearance should be undertaken on a rotational basis
and should aim at maintaining a variety of early
successional stages and ensuring the presence of the
foodplant.

NOTARIS SCIRPI
NOTABLE B
A weevil
Order COLEOPTERA   Family CURCULIONIDAE

Notaris scirpi (F., 1792). Formerly known as:
Erirrhinus scirpi (F.).

Distribution South East, South, East Anglia, East
Midlands, West Midlands, North East, North West and
Wales.

Habitat and ecology Wetland. Phytophagous.
Associated with lesser pond-sedge *Carex acutiformis*
and bulrush *Typha latifolia*. On the Continent, this
species has been recorded from sedges and club-rushes.
Larvae develop at the roots of the foodplant. In Britain,
adults have been recorded in February, April, from
June to August and in October.

Status Widespread but local. Distributed throughout
England and Wales, although not recorded in South
West England.

Threats Drainage for reasons such as agricultural
improvement and development. Falling water tables
because of water abstraction and river engineering
schemes may also threaten this species.

Management and conservation Water tables should be
maintained at high levels. Water bodies should be
isolated from sources of eutrophication and pollution.
Clearance should be undertaken on a rotational basis
and should aim at maintaining a variety of early
successional stages and ensuring the presence of the
foodplant.

OMIAMIMA MOLLINA — NOTABLE A
A weevil
Order COLEOPTERA — Family CURCULIONIDAE

Omiamima mollina (Boheman, 1834). Formerly known as: *Omias mollinus* Boheman, 1834.

Distribution Recorded from South Hampshire, South Essex, North Essex, Middlesex, East Gloucestershire, Worcestershire, Staffordshire, North Lincolnshire, Derbyshire, West Lancashire, South-east Yorkshire, North-east Yorkshire, South-west Yorkshire, Mid-west Yorkshire, Durham, Cumberland, Lanarkshire, Roxburghshire and Berwickshire before 1970 and Northamptonshire, South Lincolnshire, Derbyshire, South-east Yorkshire, North-east Yorkshire, South-west Yorkshire, Mid-west Yorkshire, Durham, South Northumberland and Lanarkshire from 1970 onwards.

Habitat and ecology Grassland, chalk quarries, woodland and a fen. Phytophagous and probably polyphagous. On the Continent, this species has been associated with bulbous buttercup *Ranunculus bulbosus*. In Britain, most examples have been recorded at the roots of plants and under stones. Adults have been recorded from April to August.

Status Status revised from RDB 3 (Rare) in Shirt (1987). Formerly recorded from South Hampshire north to Berwickshire in Scotland. Recently recorded with a scattered distribution from Northamptonshire north to Lanarkshire.

Threats Loss of unimproved grassland through improvement by reseeding or by the application of fertilisers, or by conversion to arable agriculture. Lack of, or changes in grazing or cutting regimes, the infilling of quarries, and clear-felling and conversion to other land use may be further threats.

Management and conservation Grazing or cutting, on a rotational basis, may be needed to maintain open conditions. Open spaces in woodland need to be retained. Open glades and ride margins should be managed to retain a variety of vegetation structures.

Published sources Atty, D.B. (1983), Drane, A.B. (1985c), Eyre, M.D. (1987), Flint, J.H. (1988), Fowler, W.W. (1891), Fowler, W.W. & Donisthorpe, H.St J.K. (1913).

ORTHOCHAETES INSIGNIS — NOTABLE B
A weevil
Order COLEOPTERA — Family CURCULIONIDAE

Orthochaetes insignis (Aubé, 1863). Formerly known as: *Orthocaetes insignis* (Aubé, 1863).

Distribution South East, South, South West, South Wales and North Wales.

Habitat and ecology Chalk pits, cliff tops, undercliffs, coastal shingle, gardens and disturbed ground. Particularly in sandy or chalky places. Phytophagous and very polyphagous. The larvae are leaf-miners. Adults generally occur at the base of plants, such as mayweed and stonecrop, and have been recorded in June, September, November and December.

Status Local and widely distributed in southern England and parts of Wales. Possibly under-recorded.

Threats This species is threatened by changes in land use such as through development, and natural succession.

Management and conservation Except on coastal shingle, which can be left unmanaged, grazing, cutting or some other disturbance is needed to maintain open conditions.

ORTHOCHAETES SETIGER — NOTABLE B
A weevil
Order COLEOPTERA — Family CURCULIONIDAE

Orthochaetes setiger (Beck, 1817). Formerly known as: *Orthocaetes setiger* (Beck, 1817).

Distribution England, Dyfed-Powys, North Wales, South East Scotland and South West Scotland.

Habitat and ecology Sand dunes, sand quarry, calcareous grassland, coastal shingle, disturbed ground and a garden. Phytophagous and polyphagous. Adults generally occur at the base of plants and, in Britain, have been recorded from January to May, in July and from October to December.

Status Widespread but local, particularly in southern England, though noted as far north as southern Scotland. Probably under-recorded.

Threats Loss of calcareous grassland through improvement by reseeding or by the application of fertilisers, or by conversion to arable agriculture. This species may be further threatened by development, afforestation and a lack of, or changes in grazing or cutting regimes leading to scrub invasion.

CURCULIONIDAE

Management and conservation Except on coastal shingle, which can be left unmanaged, grazing, cutting or some other disturbance is needed to maintain open conditions and encourage early successional stages.

OTIORHYNCHUS AUROPUNCTATUS
A weevil   **ENDANGERED**
Order COLEOPTERA   Family CURCULIONIDAE

Otiorhynchus auropunctatus Gyllenhal, 1834. Formerly known as: *Otiorrhynchus auropunctatus* Gyllenhal, 1834.

Distribution Recorded from West Ross before 1970.

Habitat and ecology Montane habitats. Phytophagous and polyphagous. Adults probably occur on and at the base of shrubs and herbs, while the larvae feed on the roots of plants. An adult has been found amongst moss. Adults have been recorded in June and August.

Status Extremely local, with very few records. First recorded in Great Britain in 1964 from Stac Polly (Pollaidh), West Ross.

Management and conservation Stac Polly is part of Inverpolly NNR.

Published sources Morris, M.G. (1972), Shirt, D.B., ed. (1987).

OTIORHYNCHUS DESERTUS   NOTABLE B
A weevil
Order COLEOPTERA   Family CURCULIONIDAE

Otiorhynchus desertus Rosenhauer, 1847. Formerly known as: *Otiorrhynchus desertus* Rosenhauer, 1847, *Otiorrhynchus muscorum* Brisout, 1863.

Distribution South East, South West, East Anglia, East Midlands, West Midlands, North East, North West, Wales and Scotland.

Habitat and ecology Heather moorland and sandy areas. Recorded between a saltmarsh and a sand dune, and probably occurs on disturbed ground and grassland. Phytophagous and probably polyphagous. Larvae probably feed on the roots of plants. Adults have been found under stones, at plant roots and probably also occur in litter under plants. Adults are nocturnal and have been recorded from April to September.

Status Widespread but local. Possibly more frequent in the northern parts of Great Britain.

Threats This species is threatened by urban and holiday developments, and improvement and conversion to arable agriculture. Afforestation, natural succession and scrub invasion are further threats.

Management and conservation Grazing, cutting or some other disturbance, such as rotovation, on a rotational basis, may be needed to maintain open conditions.

OTIORHYNCHUS LIGUSTICI   VULNERABLE
A weevil
Order COLEOPTERA   Family CURCULIONIDAE

Otiorhynchus ligustici (L., 1758). Formerly known as: *Otiorrhynchus ligustici* (L., 1758).

Distribution Recorded from Isle of Wight, North Hampshire, Surrey, South Essex, East Suffolk, East Norfolk, East Gloucestershire, Shropshire, Leicestershire & Rutland, Derbyshire, South Lancashire, Durham and Midlothian before 1970 and South Devon and Isle of Wight from 1970 onwards.

Habitat and ecology Maritime cliff, grassland and possibly also disturbed ground. Phytophagous. The species is polyphagous on a variety of plants, with a possible preference for kidney vetch *Anthyllis vulneraria*. The larvae feed on the roots of plants. Adults have been recorded from March to July. This species is parthenogenetic.

Status Old records suggest that this species was formerly widespread and distributed from southern England north to Midlothian, Scotland. Recently recorded from only two vice-counties, both in southern England.

Threats This species is threatened by cliff stabilisation schemes and the construction of sea defences. Coastal developments may reduce the amount of available habitat. Activities that accelerate or reduce the rate of erosion should be avoided. Loss of unimproved grassland through improvement by reseeding or by the application of fertilisers, or by conversion to arable agriculture. Lack of, or changes in grazing or cutting regimes, and natural succession and scrub invasion may threaten this species.

Management and conservation Occasional slippages are necessary to maintain habitat continuity. Large areas of unstable cliff are required so that the population does not become isolated and subsequently threatened by individual landslips. On grassland, grazing or cutting is needed to maintain open conditions.

Published sources Atty, D.B. (1983), Eyre, M.D. (1987), Fowler, W.W. (1891), Fowler, W.W. & Donisthorpe, H.St J.K. (1913), Morris, M.G. (1965a), Shirt, D.B., ed. (1987), Turner, M. (1976).

OTIORHYNCHUS MORIO   **INDETERMINATE**
A weevil
Order COLEOPTERA   Family CURCULIONIDAE

Otiorhynchus morio (F., 1781). Formerly known as: *Otiorrhynchus morio* (F., 1781).

Distribution Recorded from West Sutherland before 1970.

Habitat and ecology Montane habitats. Phytophagous and polyphagous. The larvae probably feed on the roots of plants. Adults have been recorded in June.

Status Status revised from RDB3 (Rare) in Shirt (1987). Last recorded in 1901 on the shores of Loch Assynt, West Sutherland.

Published sources Fowler, W.W. & Donisthorpe, H.St J.K. (1913), Kidson-Taylor, J. (1906), Shirt, D.B., ed. (1987).

OTIORHYNCHUS RAUCUS   **NOTABLE B**
A weevil
Order COLEOPTERA   Family CURCULIONIDAE

Otiorhynchus raucus (F., 1777). Formerly known as: *Otiorrhynchus raucus* (F., 1777).

Distribution South East, South, South West, East Anglia, East Midlands, North East and South Wales.

Habitat and ecology Sand pits, disturbed ground, a cliff edge, woodland and possibly also gardens and allotments. The species appears to prefer loose sandy or chalky soils. Phytophagous and polyphagous. The larvae feed on the roots of plants while adults tend to be found at the base of plants and in litter. Adults have been recorded from March to May, July and in September.

Status Widespread but local in southern England. Also recorded in North East England and South Wales.

Threats This species may be threatened by urban development, the infilling of pits, the use of pesticides and herbicides, and natural succession and scrub invasion.

Management and conservation Management should aim at maintaining open conditions.

OTIORHYNCHUS SCABER   **NOTABLE B**
A weevil
Order COLEOPTERA   Family CURCULIONIDAE

Otiorhynchus scaber (L., 1758). Formerly known as: *Otiorrhynchus scaber* (L., 1758), *Otiorrhynchus septentrionis* (Herbst, 1795).

Distribution Scotland.

Habitat and ecology Woodland, sand dunes and wetland. Phytophagous and polyphagous. Recorded from birch, pine, at the base of marram and in litter and moss, including *Sphagnum*. Parthenogenetic. Adults have been recorded from June to August.

Status Only recorded in Scotland. A record for Bedfordshire in 1947 is known to be incorrect.

Threats Uncertain, though this species may be threatened by the clear-felling of woodland and conversion to other land use. Further threats may be through coastal developments and drainage.

PACHYTYCHIUS HAEMATOCEPHALUS
A weevil   **ENDANGERED**
Order COLEOPTERA   Family CURCULIONIDAE

Pachytychius haematocephalus (Gyllenhal, 1836).

Distribution Recorded from Dorset and South Hampshire before 1970 and South Hampshire from 1970 onwards.

Habitat and ecology Exposed herb-rich grassland on steep south-facing slopes. Phytophagous. Associated with common bird's-foot-trefoil *Lotus corniculatus*. The larvae develop in the pods, feeding on the unripe seeds. Adults occur at the roots of grass tussocks growing immediately adjacent to patches of the foodplant. Adults have been recorded in February, June and from August to October.

Status Extremely local. Recorded for over a hundred years from a single site in South Hampshire. A record for North Wiltshire requires confirmation.

Threats This species may be threatened by natural succession and any change in land use.

Management and conservation Grazing or cutting, on a rotational basis, is needed to maintain open

CURCULIONIDAE

conditions. There is no statutory protection for this species' only site.

Published sources Fowler, W.W. (1891), Morris, M.G. (1974a), Shirt, D.B., ed. (1987).

PHLOEOPHAGUS GRACILIS   **EXTINCT**
A weevil
Order COLEOPTERA   Family CURCULIONIDAE

Phloeophagus gracilis (Rosenhauer, 1856). Formerly known as: *Rhyncolus gracilis* osenhauer.

Distribution Recorded from Surrey, Middlesex, West Gloucestershire, Warwickshire, Nottinghamshire and South Lancashire.

Habitat and ecology Broad-leaved woodland. Phytophagous. Larvae and adults occur in dead wood, with records from rotten beech, birch twigs, holly trunks and hawthorn. Adults have been recorded in March.

Status Presumed extinct. Last recorded in 1897 from Ealing, Middlesex.

Published sources Atty, D.B. (1983), Fowler, W.W. (1891), Fowler, W.W. & Donisthorpe, H.St J.K. (1913), Shirt, D.B., ed. (1987).

PHLOEOPHAGUS TRUNCORUM   **NOTABLE A**
A weevil
Order COLEOPTERA   Family CURCULIONIDAE

Phloeophagus truncorum (Germar, 1824). Formerly known as: *Rhyncolus truncorum* (Germar), *Stereocorynes truncorum* (Germar).

Distribution Recorded from North Somerset, South Hampshire, West Kent, Surrey, South Essex, Hertfordshire, Berkshire, East Suffolk, Cambridgeshire, East Gloucestershire and Leicestershire & Rutland before 1970 and South Hampshire, West Sussex, East Kent, South Essex, Berkshire, Herefordshire and Worcestershire from 1970 onwards.

Habitat and ecology Ancient broad-leaved woodland and pasture-woodland. Phytophagous. Associated with dead wood of oak, beech, ash, maple, willow, poplar and apple. On the Continent, this species has been found on Norway spruce. In Britain, adults have been recorded in February and from April to December.

Status Local. Only recorded in the southern half of England, as far north as Herefordshire.

Indicator status Grade 1 in Harding & Rose (1986).

Threats Loss of broad-leaved woodland and parkland through, for example, clear-felling and changes to other land use. Habitat loss, in particular, through the felling of ancient trees, removal of dead wood from living trees and the destruction or removal of standing and fallen dead wood for reasons such as forest hygiene, aesthetic tidiness, public safety or for use as fire wood.

Management and conservation Ancient trees and both fallen and standing dead timber, especially with the bark attached, should be retained. The removal of dead timber from ancient trees should be avoided. Gaps in the age structure of the tree population should be identified and the continuity of the appropriate dead wood habitat ensured by regeneration, suitable planting and possibly with pollarding.

Published sources Fowler, W.W. & Donisthorpe, H.St J.K. (1913), Harding, P.T. (1978), Harding, P.T. & Rose, F. (1986), Owen, J.A. (1990a), Whitehead, P.F. (1990b).

PHYLLOBIUS VESPERTINUS   **NOTABLE B**
A weevil
Order COLEOPTERA   Family CURCULIONIDAE

Phyllobius vespertinus (F., 1792). Formerly known as: *Phyllobius artemisiae* Desbrochers, 1873.

Distribution South East, East Anglia, East Midlands, North East and North West.

Habitat and ecology Predominantly coastal and recorded from saltmarsh, coastal shingle and estuaries. Also noted inland from a chalk pit. Phytophagous and probably polyphagous. The larvae probably feed on the roots of the foodplant. Adults have been recorded from April to June.

Status Recorded from East Kent to South-east Yorkshire, as well as from South Lancashire and Cumberland.

Threats Loss of saltmarsh through reclamation, erosion and the construction of sea defences. Overgrazing of saltmarshes could be a further threat to this species. This species may be further threatened by gravel extraction and coastal developments.

Management and conservation Grazing should not be introduced to a saltmarsh where there is no grazing at present. Disturbance of coastal shingle should be avoided.

PHYTOBIUS CANALICULATUS   NOTABLE B
A weevil
Order COLEOPTERA   Family CURCULIONIDAE

Phytobius canaliculatus Fahraeus, 1843.

Distribution England, Wales, South East Scotland, South West Scotland and North East Scotland.

Habitat and ecology Wetland. Phytophagous. Associated with water-milfoil *Myriophyllum*. Larvae probably feed externally on the foodplant. Adults have been recorded from April to June, and in August and September.

Status Widespread but local in Great Britain, not recorded in North West Scotland.

Threats This species is threatened by drainage for reasons such as agricultural improvement and development. Falling water tables because of water abstraction schemes, the use of herbicides and pesticides, and natural succession are further threats.

Management and conservation Water tables should be maintained at high levels. Water bodies should be isolated from sources of eutrophication and pollution. Clearance should be undertaken on a rotational basis and should aim at maintaining aquatic and semi-aquatic plant populations.

PHYTOBIUS COMARI   NOTABLE B
A weevil
Order COLEOPTERA   Family CURCULIONIDAE

Phytobius comari (Herbst, 1795).

Distribution England, Dyfed-Powys, North Wales and South West Scotland.

Habitat and ecology Wetland. Phytophagous. Associated with marsh cinquefoil *Potentilla palustris*, though also recorded at the roots and feeding on the leaves of purple-loosestrife *Lythrum salicaria*. On the Continent, this species has also been recorded from lady's-mantle *Alchemilla vulgaris*, great burnet *Sanguisorba officinalis* and redshank *Polygonum persicaria*. In Britain, eggs are laid on the surface of leaves of the foodplant from late April until early August. Larvae feed externally on the leaves (predominantly the underside) of the foodplant and have been recorded from late April until mid September. Pupation occurs, following three larval instars, from mid June until late September, with adult

weevils being recorded in February and from April to September.

Status Widespread but local in England, parts of Wales, and also recorded in South West Scotland.

Threats This species is threatened by drainage for reasons such as agricultural improvement and development. Falling water tables because of water abstraction, river engineering schemes, and natural succession may also threaten this species.

Management and conservation Water tables should be maintained at high levels. Water bodies should be isolated from sources of eutrophication and pollution. Grazing or cutting, on a rotational basis, is needed to maintain open conditions.

PHYTOBIUS MURICATUS   NOTABLE A
A weevil
Order COLEOPTERA   Family CURCULIONIDAE

Phytobius muricatus Brisout, 1867. Formerly known as: *Phytobius quadrinodosus* sensu Fowler, 1891 not (Gyllenhal, 1813).

Distribution Recorded from Dorset, South Hampshire, East Sussex, West Kent, Hertfordshire, East Norfolk, Cambridgeshire, West Gloucestershire, Glamorgan, Cheshire, North-east Yorkshire, North-west Yorkshire and Cumberland before 1970 and North Devon, North Somerset, South Hampshire, North Essex and Carmarthenshire from 1970 onwards.

Habitat and ecology Wetland, including wet meadows. Phytophagous. Host-plant uncertain. Adults have been recorded in January and from April to June.

Status Very local and recently recorded from only five, widely scattered, vice-counties. Old records suggest that this species was widely distributed from southern England to Cumberland. This species is difficult to identify and may be confused with other members of the genus. Consequently, the exact status of this species is hard to assess. There is much confusion with the old records for this species.

Threats Drainage for reasons such as agricultural improvement and development. Falling water tables because of water abstraction, river engineering schemes, and natural succession may also threaten this species.

Management and conservation Water tables should be maintained at high levels. Water bodies should be isolated from sources of eutrophication and pollution.

CURCULIONIDAE

Grazing or cutting, on a rotational basis, is needed to maintain open conditions.

Published sources Allen, A.A. (1946a), Fowler, W.W. (1891), Fowler, W.W. & Donisthorpe, H.St J.K. (1913).

PHYTOBIUS OLSSONI   **RARE**
A weevil
Order COLEOPTERA   Family CURCULIONIDAE

Phytobius olssoni Israelson, 1972. Formerly confused under: *Phytobius quadrituberculatus* (F., 1787).

Distribution Recorded from Berkshire before 1970 and West Sussex and Cardiganshire from 1970 onwards.

Habitat and ecology Wetland and damp ground, particularly in cart-ruts etc.. Phytophagous. Associated with water-purslane *Peplis portula*. On the Continent, larvae have been recorded feeding externally on the leaves of the foodplant. In Britain, adults have been recorded from May to September.

Status Recorded new to Britain in 1982 from three examples collected in 1927 from Easthampstead, Berkshire. Discovered at a second site, in West Sussex, in 1984. Extremely difficult to separate from the common *P. quadrituberculatus* and requiring the dissection of the male for conclusive identification. Consequently, the exact status of this species is hard to assess.

Threats Drainage for reasons such as agricultural improvement and development. Falling water tables because of water abstraction, river engineering schemes, and natural succession may also threaten this species.

Management and conservation Water tables should be maintained at high levels. Grazing, cutting or some other disturbance, on a rotational basis, is needed to maintain open conditions.

Published sources Boyce, D.C. (1990), Johnson, C. (1982), Shirt, D.B., ed. (1987).

PHYTOBIUS QUADRICORNIGER   NOTABLE A
A weevil
Order COLEOPTERA   Family CURCULIONIDAE

Phytobius quadricorniger Colonnelli,1986. Formerly known as: Phytobius quadricornis (Gyllenhal, 1813).

Distribution Recorded from North Devon, East Kent, Surrey, Hertfordshire, Berkshire, East Suffolk, East Norfolk, West Norfolk and Warwickshire before 1970 and North Hampshire and West Norfolk from 1970 onwards.

Habitat and ecology Wetland. Phytophagous. Associated with pale persicaria *Polygonum lapathifolium*. On the Continent, this species has also been recorded from amphibious bistort *P. amphibium* v. *terrestre*. Larvae probably feed externally on the foodplant. In Britain, adults have been found from May to September.

Status Very local and recently recorded from only two vice-counties. Old records show that this species has been more widely recorded in southern England and noted as far north as Warwickshire.

Threats Drainage for reasons such as agricultural improvement and development. Falling water tables because of water abstraction, and natural succession may also threaten this species.

Management and conservation Water tables should be maintained at high levels.

Published sources Allen, A.A. (1946a), Collier, M.J. (1988a), Fowler, W.W. (1891), Pope, R.D. (1969).

PHYTOBIUS QUADRINODOSUS   NOTABLE A
A weevil
Order COLEOPTERA   Family CURCULIONIDAE

Phytobius quadrinodosus (Gyllenhal, 1813). Formerly known as: *Rhinoncus denticollis* (Gyllenhal, 1837).

Distribution Recorded from South Devon, North Wiltshire, Isle of Wight, South Hampshire, West Sussex, East Sussex, East Kent, West Kent, Surrey, Hertfordshire, Berkshire, Oxfordshire, Cambridgeshire, East Gloucestershire, Glamorgan and Pembrokeshire before 1970 and South Wiltshire, West Sussex, East Sussex, East Kent, South Essex, North Essex, Northamptonshire and Leicestershire & Rutland from 1970 onwards.

Habitat and ecology Calcareous grassland, particularly downland. Phytophagous. Possibly associated with *Polygonum* species. Larvae probably feed externally on the foodplant(s). Adults have been found at the roots of grass, and have been recorded from March to September.

Status Status revised from RDB 3 (Rare) in Shirt (1987). Very local and distributed in southern England as far north as Leicestershire. There are old records from southern Wales. The historical distribution may be in doubt because of confusion with the very similar *P. muricatus*.

Threats Loss of calcareous grassland through improvement by reseeding or by the application of fertilisers, or by conversion to arable agriculture. Lack of, or changes in grazing or cutting regimes may threaten this species.

Management and conservation Grazing or cutting, on a rotational basis, is needed to maintain open conditions.

Published sources Atty, D.B. (1983), Fowler, W.W. (1891), Fowler, W.W. & Donisthorpe, H.St J.K. (1913), Morris, M.G. & Rispin, W.E. (1988).

PHYTOBIUS WALTONI	NOTABLE B

A weevil
Order COLEOPTERA   Family CURCULIONIDAE

Phytobius waltoni Boheman, 1843.

Distribution South East, South, South West, East Anglia, North West and Wales.

Habitat and ecology Fens and lake margins, damp areas in woodland, particularly cart-ruts and gullies, and also cliff grassland and probably other wetland habitats. Phytophagous. Associated with water-pepper *Polygonum hydropiper*. On the Continent, this species has also been found on tasteless water-pepper *P. mite*. Larvae feed externally on the foodplant. In Britain, the complete life-cycle from egg to adult is very short, and can last just 18 days (J. Parry, pers. comm.). Adults have been recorded from June to August.

Status Widespread but local in southern England and Wales. Also recorded in North West England.

Threats Drainage, falling water tables because of water abstraction schemes, natural succession and overgrazing may threaten this species.

Management and conservation Water tables should be maintained at high levels. In woodlands, open glades and ride margins should be cut on rotation to retain a variety of vegetation structures. Some grazing or cutting is needed to maintain open conditions.

PISSODES VALIDIROSTRIS	RARE

A weevil
Order COLEOPTERA   Family CURCULIONIDAE

Pissodes validirostris (Sahlberg, 1834).

Distribution Recorded from Fife, Mid Perthshire, Kincardineshire, South Aberdeenshire, Banffshire, Moray, East Inverness & Nairn and West Ross before 1970 and Moray from 1970 onwards.

Habitat and ecology Coniferous woodland. Phytophagous. Associated with Scots pine. Larvae feed in the cones and occasionally in the shoots. Adults have been recorded in June and July.

Status Extremely local and only recorded in Scotland. recently recorded from just one vice-county. A component of the native pine forest fauna. This species is difficult to identify and may be confused with *P. castaneus*. Consequently, the exact status of this species is hard to assess.

Threats Loss of native pine forest through, for example, clear-felling and conversion to other land use. Overgrazing, preventing regeneration, may be a further threat.

Management and conservation Gaps in the age structure of the tree population should be identified and regeneration encouraged to ensure the continuity of suitable habitat.

Published sources Allen, A.A. (1970b), Beare, T.H. (1930b), Bevan, D. (1971), Shirt, D.B., ed. (1987).

PLINTHUS CALIGINOSUS	NOTABLE A

Hop root weevil
Order COLEOPTERA   Family CURCULIONIDAE

Plinthus caliginosus (F., 1775). Formerly known as: *Epipolaeus caliginosus* (F.).

Distribution Recorded from South Wiltshire, South Hampshire, East Sussex, East Kent, West Kent, Surrey, Middlesex, Herefordshire and Worcestershire before 1970 and East Sussex, East Kent, West Kent, Worcestershire and Derbyshire from 1970 onwards.

Habitat and ecology Hop fields, and probably hedgebanks and other sites with the foodplant. Particularly on chalky soils, though also on clay and sand. Phytophagous. Associated with hops *Humulus lupulus*, though also recorded from a wide range of rootstocks of herbaceous and woody plants and from dead wood of both deciduous and coniferous trees. The larvae and adults are found at the roots of plants, in

CURCULIONIDAE

moss and in leaf litter. Eggs have been recorded between mid August and April, the larval period probably lasting from nine to 18 months. Pupae have been found from June to September. Adults have been recorded all year round.

Status Recorded in England from South Wiltshire, north to Derbyshire. This species has been noted as a pest of hops.

Threats This species may be threatened by the clear-felling of woodland and conversion to other land use, and the grubbing out and mechanised cutting of hedgerows. Neglect, and conversion to high forest, and the use of pesticides may be a further threat.

Management and conservation Hedgerows, and open glades and ride margins in woodland, should be cut on rotation to retain a variety of vegetation structures.

Published sources Collingwood, C.A. (1954), Fowler, W.W. (1891).

POLYDRUSUS CONFLUENS   **NOTABLE B**
A weevil
Order COLEOPTERA   Family CURCULIONIDAE

Polydrusus confluens Stephens, 1831. Formerly known as: *Polydrosus confluens* Stephens, 1831.

Distribution South East, South, South West, East Anglia, North East, North West, South Wales and North West Scotland.

Habitat and ecology Grassland, heathland and scrub. Phytophagous. Associated with broom *Cytisus scoparius*, gorse *Ulex europaeus*, and dwarf gorse *U. minor*, and possibly *Genista* species. Flightless. Adults have been recorded from May to August.

Status Widespread but local in southern and northern England. Also recorded in South Wales and North West Scotland.

Threats Much heathland and grassland has been lost, or fragmented, through changes in land use, mainly by conversion to arable, forestry and urban development. This species may be further threatened by the grubbing out of scrub.

Management and conservation Management should aim for a diversity of successional stages, ranging from open ground to mature scrub, preferably by grazing or by scraping or possibly through cutting.

POLYDRUSUS FLAVIPES   **NOTABLE B**
A weevil
Order COLEOPTERA   Family CURCULIONIDAE

Polydrusus flavipes (Degeer, 1775). Formerly known as: *Polydrosus flavipes* (Degeer, 1775).

Distribution England.

Habitat and ecology Open and coppiced broad-leaved woodland, also pasture-woodland. Phytophagous. Associated with young oak and aspen, possibly also with hazel, beech, birch and hawthorn. This beetle has been noted from mature oaks in parkland. Adults have been recorded from May to September.

Status Widespread but local in England. This species is difficult to identify and may be confused with other members of the genus. Consequently, the exact status of this species is hard to assess.

Threats Loss of broad-leaved woodland and parkland through, for example, clear-felling and coniferisation. Neglect, cessation of the coppice cycle, and conversion to high forest, has led to increased shade and the loss of glades and broad sunny rides.

Management and conservation Open spaces in woodland need to be retained. Open spaces and ride margins should be cut on rotation to retain a variety of vegetation structures.

POLYDRUSUS MARGINATUS   **VULNERABLE**
A weevil
Order COLEOPTERA   Family CURCULIONIDAE

Polydrusus marginatus Stephens, 1831. Formerly known as: *Metallites marginatus* (Stephens), *Polydrosus marginatus* (Stephens).

Distribution Recorded from South Hampshire, West Sussex, East Sussex, East Kent, West Kent, Hertfordshire, Berkshire, Oxfordshire, Buckinghamshire, Shropshire and Leicestershire & Rutland before 1970 and West Kent from 1970 onwards.

Habitat and ecology Broad-leaved woodland and possibly scrub. Also recorded from heathland. Phytophagous. Recorded from birch, hazel and broom. On the Continent, this species has been noted from oak and members of the Rosaceae. Flightless. In Britain, adults have been recorded from April to June.

Status Not listed in the insect Red Data Book (Shirt, 1987). Old records suggest that this species was recorded in southern England, north to Salop and

290

Leicestershire. Recently recorded from only one vice-county.

Threats Loss of broad-leaved woodland through, for example, clear-felling and coniferisation. Neglect, and conversion to high forest, has led to increased shade and the loss of glades and broad sunny rides. The grubbing out of scrub may be a further threat.

Management and conservation Open spaces in woodland need to be retained. Open glades and ride margins should be cut on rotation to retain a variety of vegetation structures. On heathland management should aim for a diversity of successional stages preferably by grazing or by rotational cutting, scraping or burning.

Published sources Allen, A.A. (1940a), Allen, A.A. (1962c), Buck, F.D. (1960b), Fowler, W.W. (1891), Fowler, W.W. & Donisthorpe, H.St J.K. (1913).

POLYDRUSUS MOLLIS   NOTABLE B
A weevil
Order COLEOPTERA   Family CURCULIONIDAE

Polydrusus mollis (Ström, 1768). Formerly known as: *Polydrosus mollis* (Ström, 1768), *Polydrusus micans* (F., 1792).

Distribution England, South Wales, Dyfed-Powys, South East Scotland and South West Scotland.

Habitat and ecology Broad-leaved woodland, and hazel scrub in limestone dales. Phytophagous. Associated with hazel and also recorded from oak, birch and sallow. The beetle has been found on small oaks in grassland, in oak scrub, in open coppiced woodland and in shaded situations in fairly dense woodland. Parthenogenetic. Adults have been recorded from April to June.

Status Widespread but local in England, parts of Wales and southern Scotland.

Threats Loss of broad-leaved woodland through, for example, clear-felling and coniferisation. This species may be further threatened by neglect and shading.

Management and conservation Open glades and ride margins should be cut on rotation to retain a variety of vegetation structures.

POLYDRUSUS PULCHELLUS   NOTABLE B
A weevil
Order COLEOPTERA   Family CURCULIONIDAE

Polydrusus pulchellus Stephens, 1831. Formerly confused under: *Polydrusus chrysomela* (Olivier, 1807).

Distribution England, South Wales, Dyfed-Powys and South West Scotland.

Habitat and ecology Saltmarsh, coastal shingle and estuaries. Phytophagous, feeding on a variety of saltmarsh plants, including sea wormwood *Artemisia maritima*, sea-purslane *Halimione portulacoides* and sea beet *Beta maritima*, possibly also buck's-horn plantation *Plantago coronopus*. Larvae probably feed on the roots of the foodplants. Adults have been recorded from May to August.

Status Widespread but local in England, also recorded in parts of Wales and South West Scotland.

Threats Loss of habitat through reclamation, erosion, the construction of sea defences and gravel extraction. Overgrazing of saltmarshes could be a further threat to this species.

Management and conservation Grazing should not be introduced to a saltmarsh where there is no grazing at present. Disturbance of coastal shingle should be avoided.

POLYDRUSUS SERICEUS   NOTABLE A
A weevil
Order COLEOPTERA   Family CURCULIONIDAE

Polydrusus sericeus (Schaller, 1783). Formerly known as: *Polydrosus sericeus* (Schaller, 1783).

Distribution Recorded from South Wiltshire, South Hampshire, North Hampshire and West Kent before 1970 and North Somerset, South Wiltshire, North Hampshire, East Sussex and West Kent from 1970 onwards.

Habitat and ecology Broad-leaved woodland rides, clearings and wood-edge. Phytophagous. Recorded from hazel, oak, alder, birch, cherry and wild rose. Adults have been recorded from May to September.

Status Very local in southern England from West Kent to North Somerset. Can be abundant where found.

Threats Loss of broad-leaved woodland through, for example, clear-felling and coniferisation. Neglect, and conversion to high forest, has led to increased shade and the loss of glades and broad sunny rides.

CURCULIONIDAE

Management and conservation Open spaces in woodland need to be retained. Open glades and ride margins should be cut on rotation to retain a variety of vegetation structures.

Published sources Allen, A.A. (1977a), Fowler, W.W. (1891), Fowler, W.W. & Donisthorpe, H.St J.K. (1913), Hodge, P.J. (1989), Hodge, P.J. (1990), Morris, M.G. (1978a), Morris, M.G. (1978b).

PROCAS ARMILLATUS INDETERMINATE
A weevil
Order COLEOPTERA Family CURCULIONIDAE

Procas armillatus (F., 1801).

Distribution Recorded from South Devon, East Sussex, East Kent, West Kent, Surrey, North Essex, Hertfordshire, "Norfolk", West Gloucestershire, Nottinghamshire and North Northumberland before 1970.

Habitat and ecology Recorded from the margins of a field of oats and clover and from moorland. Phytophagous. The beetle has been recorded at the roots of herbs and grasses and from under stones. May be predominantly subterranean. Most records are for May, although the beetle has also been recorded in March, April and June.

Status Status revised from RDB 3 (Rare) in Shirt (1987). Very local with very few records. Apart from an abundance of the beetle at the edge of a field near Brighton, East Sussex in May 1930, and smaller numbers in the following years, perhaps as late as 1950, most other records for the species appear to be of single individuals. Records of three examples of this species for Cumberland have recently been published, though these are probably referable to *P. granulicollis*.

Threats Uncertain, though this species has probably declined through development and change in land use.

Published sources Allen, A.A. (1971b), Cox, L.G. (1930), Fowler, W.W. (1891), Jennings, F.B. (1906), Kenward, H.K. (1990), Read, R.W.J. (1989a), Read, R.W.J. (1989b), Shirt, D.B., ed. (1987), Walker, J.J. (1930a).

PROCAS GRANULICOLLIS INDETERMINATE
A weevil **ENDEMIC**
Order COLEOPTERA Family CURCULIONIDAE

Procas granulicollis Walton, 1848. Formerly known as: *Procas armillatus var. granulicollis* Walton.

Distribution Recorded from Cumberland and Kirkcudbrightshire before 1970 and Cardiganshire from 1970 onwards.

Habitat and ecology Recorded in a birch woodland, a south-east facing sessile oak woodland and from woodlands on north-facing valley slopes. This species has been found by sweeping in an area of young birches growing closely together to give a completely closed canopy. This was on moist and peaty soil, with a thin layer of leaf litter. Also found in woodland glades containing climbing corydalis *Corydalis claviculata* by sieving bracken litter and from moss, deep leaf litter and humus in a wood. Adults have been recorded in April, May and August.

Status Endemic. Status revised from RDB Appendix in Shirt (1987). Formerly regarded as probably only a variety of *P. armillatus*. However, the discovery of a single example, and first record for over 120 years, in 1968 in Kirkcudbrightshire, has enabled a re-evaluation of this species. Prior to the discovery of this beetle in 1968, the species was only known from four very old specimens, only one of these bearing data. It is very likely that all four examples are from the same area, which is just across the Solway Firth from the Kirkcudbrightshire site. A recent published record of three examples of *P. armillatus* found in an oakwood in Cumberland probably refer to *P. granulicollis*, but confirmation is required. This is also the case for a single specimen recently collected from oakwood leaf litter in Cardiganshire. The record from Brighton, Sussex is unlikely and could be the result of confusion over labelling.

Threats Uncertain, though this species may be threatened by clear-felling and change in land use.

Published sources Fowler, W.W. (1891), Kenward, H.K. (1990), Read, R.W.J. (1989a), Read, R.W.J. (1989b), Shirt, D.B., ed. (1987).

PSELACTUS SPADIX   NOTABLE B
A weevil
Order COLEOPTERA   Family CURCULIONIDAE

Pselactus spadix (Herbst, 1795). Formerly known as: *Codiosoma spadix* (Herbst), *Phloeophagia spadix* (Herbst).

Distribution South East, South, South West, East Anglia, East Midlands, North East and South Wales.

Habitat and ecology In groynes, driftwood and old timber on the coast. Primarily coastal. Phytophagous. Found in rotten wood. This species has been found in old railway sleepers on the shoreline. Adults have been recorded in January, from April to June and from August to October.

Status Widespread but local in southern and eastern England. Also recorded in South Wales.

Threats This species may be threatened by coastal development, the clearing up of driftwood and beach parties. Pollution may be a further threat.

PSEUDOSTYPHLUS PILLUMUS   NOTABLE A
A weevil
Order COLEOPTERA   Family CURCULIONIDAE

Pseudostyphlus pillumus (Gyllenhal, 1836). Formerly known as: *Pseudostyphlus pilumnus* (Gyllenhal, 1836).

Distribution Recorded from South Hampshire, East Kent, West Kent, Surrey, South Essex, Middlesex, Oxfordshire, Buckinghamshire and Cambridgeshire before 1970 and East Sussex, West Kent, South Essex and Berkshire from 1970 onwards.

Habitat and ecology Sand dunes, coastal undercliffs, disturbed ground, field margins, and along path edges. Chiefly in sandy places. Phytophagous. Associated, possibly exclusively, with scented mayweed *Matricaria recutita*. On the Continent, this species has also benn found on scentless mayweed *Tripleurospermum maritimum* subsp *inodorum*. In Britain, adults occur at the roots and under the basal leaves and have been recorded from May to July.

Status Very local and known from southern England as far north as Cambridgeshire.

Threats This species may be threatened by loss of habitat through afforestation, urban and holiday development, and natural succession. The degradation of remaining habitat by excessive disturbance of the vegetation through activities such as motorbike access, horse-riding and human trampling.

Management and conservation Some grazing, rotational cutting or other disturbance, such as rotovation, is necessary to maintain open conditions and encourage early successional stages.

Published sources Fowler, W.W. (1891), Fowler, W.W. & Donisthorpe, H.St J.K. (1913).

RHINOCYLLUS CONICUS   NOTABLE A
A weevil
Order COLEOPTERA   Family CURCULIONIDAE

Rhinocyllus conicus (Froelich, 1792). Formerly known as: *Rhinocyllus latirostris* (Latreille, 1804).

Distribution Recorded from West Cornwall, South Devon, South Wiltshire, Dorset, Isle of Wight, South Hampshire, West Sussex, East Sussex, East Kent and Cambridgeshire before 1970 and South Devon, Dorset and Isle of Wight from 1970 onwards.

Habitat and ecology Grassland, particularly on calcareous soils. Usually found near the coast, though also recorded inland. Phytophagous. Associated with creeping thistle *Cirsium arvense*, marsh thistle *C. palustre*, spear thistle *C. vulgare* and musk thistle *Carduus nutans*. On the Continent, this species has also been recorded on common knapweed *Centaurea nigra*. In Britain, adults have been recorded from April to September.

Status Very local and only known from southern England. Recently recorded from just three vice-counties.

Threats Loss of grassland through conversion to arable agriculture or other land use. Lack of, or changes in grazing or cutting regimes may threaten this species.

Management and conservation Grazing or some other disturbance may be needed to maintain open conditions and encourage early successional stages.

Published sources Drane, A.B. (1990), Fowler, W.W. (1891), Morris, M.G. (1983).

RHINONCUS ALBICINCTUS   ENDANGERED
A weevil
Order COLEOPTERA   Family CURCULIONIDAE

Rhinoncus albicinctus Gyllenhal, 1837.

Distribution Recorded from Berkshire before 1970 and Berkshire from 1970 onwards.

CURCULIONIDAE

Habitat and ecology Known from the shore of a single lake in Britain. Phytophagous. Associated with amphibious bistort *Polygonum amphibium* v. *natans*. The larvae have been recorded in root-nodules on the foodplant. Adults have been recorded in June, July, and August, sitting on floating leaves of the foodplant. The adults overwinter on dry land.

Status Recorded new to Britain in 1972 from a site in Berkshire.

Threats This species is threatened by the clearance of vegetation from the banks and shallow water for lake maintenance and the interest of angling. Natural succession may be a further threat.

Management and conservation Water tables should be maintained at high levels. Water bodies should be isolated from sources of eutrophication and pollution. Lake clearance should be undertaken on a rotational basis and should aim at maintaining the foodplant population.

Published sources Allen, A.A. (1973d), Morris, M.G. (1975), Shirt, D.B., ed. (1987).

RHYNCHAENUS CALCEATUS
A weevil   **INSUFFICIENTLY KNOWN**
Order COLEOPTERA Family CURCULIONIDAE

Rhynchaenus calceatus Germar, 1821.

Distribution Recorded from Surrey, Oxfordshire and Bedfordshire before 1970.

Habitat and ecology Birch woodland. Biology uncertain, but appears to be associated with birch.

Status Not listed in the insect Red Data Book (Shirt, 1987). Possibly only a form of *R. testaceus*.

Threats Uncertain, though loss of birch woodland through clear-felling and conversion to other land use may be a threat.

Published sources Allen, A.A. (1988a).

RHYNCHAENUS DECORATUS
A weevil   **INDETERMINATE**
Order COLEOPTERA Family CURCULIONIDAE

Rhynchaenus decoratus (Germar, 1821). Formerly known as: *Orchestes decoratus* Germar.

Distribution Recorded from West Suffolk and Dumfriesshire before 1970 and North Hampshire from 1970 onwards.

Habitat and ecology Fenland, and willow carr along river margins. Phytophagous. Associated with purple willow and creeping willow. On the Continent, this species has also been recorded from almond willow, crack willow and black poplar. The larvae are leaf-miners. In Britain, adults have been recorded from April to June and in August.

Status Status revised from RDB 3 (Rare) in Shirt (1987). Formerly only known from just two records in circa 1879 and 1905. Found in 1970, and subsequently in 1972, on purple willow along a riverside near Stockbridge, North Hampshire. These are the only recent records.

Threats Uncertain, though drainage, falling water tables because of water abstraction, river engineering schemes, and the removal of the species host-shrubs may threaten this species.

Management and conservation Water tables should be maintained at high levels. Gaps in the age structure of the host-shrub population should be identified and the continuity of the appropriate habitat ensured through regeneration.

Published sources Morris, M.G. (1973a), Shirt, D.B., ed. (1987).

RHYNCHAENUS FOLIORUM **NOTABLE A**
A weevil
Order COLEOPTERA Family CURCULIONIDAE

Rhynchaenus foliorum (Müller, 1764). Formerly known as: *Orchestes foliorum* (Müller), *Orchestes saliceti* (Paykull, 1792).

Distribution Recorded from West Cornwall, South Devon, North Somerset, South Hampshire, East Sussex, West Kent, Surrey, Middlesex, Berkshire, Oxfordshire, East Suffolk, West Suffolk, East Norfolk, Herefordshire, Glamorgan, North Lincolnshire, Leicestershire & Rutland, Derbyshire, South Lancashire, North-east Yorkshire, South-west Yorkshire, North-west Yorkshire, Durham, Westmorland & North Lancashire, Cumberland,

Dumfriesshire, Lanarkshire, Roxburghshire, "Perth", Argyll Main, East Ross and Orkney before 1970 and East Norfolk, Caernarvonshire, Denbighshire, North Lincolnshire, Mid-west Yorkshire and Ayrshire from 1970 onwards.

Habitat and ecology Wetland, including fen and carr. Phytophagous. Associated with grey willow, white willow, goat willow and osier. The larvae are leaf-miners. Adults have been recorded in May and June.

Status Very local and recently recorded from only six, widely scattered, vice-counties. Old records show that this species has been widely recorded from southern England through to Orkney in Scotland.

Threats Drainage for reasons such as agricultural improvement and development is the primary cause of the loss of wetlands. Falling water tables because of water abstraction and river engineering schemes may also threaten this species.

Management and conservation Water tables should be maintained at high levels. Gaps in the age structure of the tree population should be identified and the continuity of the appropriate habitat ensured by regeneration, suitable planting and possibly pollarding.

Published sources Eyre, M.D. (1987), Fowler, W.W. (1891), Fowler, W.W. & Donisthorpe, H.St J.K. (1913).

RHYNCHAENUS IOTA   **NOTABLE B**
A weevil
Order COLEOPTERA   Family CURCULIONIDAE

Rhynchaenus iota (F., 1787). Formerly known as: *Orchestes iota* (F.).

Distribution South East, South, South West and East Anglia.

Habitat and ecology Wetland, particularly wet heath and humid heath. Phytophagous. Associated with bog myrtle *Myrica gale*. On the Continent, this species has also been recorded from alder, silver birch, downy birch and *Salix* species. The larvae are leaf-miners. In Britain, adults have been recorded from May to August.

Status Widespread and very local in southern and south-eastern England.

Threats Loss of habitat through improvement and conversion to arable agriculture, development and afforestation. Drainage, water abstraction schemes, and heath fires may be further threats.

Management and conservation Water tables should be maintained at a high level. On heathland, management should aim for a diversity of successional stages and include mature heath.

RHYNCHAENUS POPULI
A weevil   **INSUFFICIENTLY KNOWN**
Order COLEOPTERA   Family CURCULIONIDAE

Rhynchaenus populi (F., 1792). Formerly known as: *Orchestes populi* (F.).

Distribution Recorded from East Kent and "London district" before 1970 and East Kent from 1970 onwards.

Habitat and ecology Broad-leaved woodland and probably wetland. Phytophagous. Associated with grey poplar and a single example has been recorded on sallow. On the Continent, this species has been recorded from white willow, crack willow, bay willow, osier, black poplar and Lombardy poplar. In Britain, adults have been recorded in July and October.

Status Not listed in the insect Red Data Book (Shirt, 1987). Formerly of doubtful status in Britain, this weevil was added to the British list from an example recorded in 1952 at Canterbury, East Kent. A further example was found at Sandwich, East Kent in 1970. The species has since been noted on at least two other occasions, both in East Kent.

Threats Uncertain, though this species is probably threatened by the loss of woodland through clear-felling and conversion to other land use. Neglect and conversion to high forest may be a further threat.

Management and conservation Open glades and ride margins should be cut on rotation to retain a variety of vegetation structures.

Published sources Parry, J.A. (1981), Stephens, J.F. (1839).

RHYNCHAENUS PRATENSIS   **NOTABLE B**
A weevil
Order COLEOPTERA   Family CURCULIONIDAE

Rhynchaenus pratensis (Germar, 1821). Formerly known as: *Orchestes pratensis* Germar.

Distribution South East, South, South West, East Anglia, East Midlands and West Midlands.

Habitat and ecology Grassland, woodland clearings and rides on calcareous soils, also recorded in wetland.

CURCULIONIDAE

Phytophagous. Associated with common knapweed *Centaurea nigra*. The larvae are leaf-miners. Adults have been recorded from May to September.

Status Widespread but local in the southern half of England, more local in south-western England.

Threats Loss of habitat through improvement and conversion to arable agriculture, clear-felling and conversion to other land use, and development. Natural succession is a further threat.

Management and conservation Grazing or cutting, on a rotational basis, is needed to maintain open conditions. Open glades and ride margins should be managed to retain a variety of vegetation structures.

RHYNCHAENUS PSEUDOSTIGMA
A weevil   **INSUFFICIENTLY KNOWN**
Order COLEOPTERA   Family CURCULIONIDAE

Rhynchaenus pseudostigma Temperé, 1982.

Distribution Recorded from Westmorland & North Lancashire before 1970 and South Wiltshire from 1970 onwards.

Habitat and ecology Probably associated with deciduous woodland and scrub. Virtually unknown. Phytophagous, almost certainly a leafminer, as are other species in the genus. Probable hosts are birch and *Salix* species.

Status Not listed in the insect Red Data Book (Shirt, 1987). Recognised as a distinct species in 1982 when this species was separated from *R. stigma*.

Published sources Morris, M.G. (1987).

RHYNCHAENUS TESTACEUS   **VULNERABLE**
A weevil
Order COLEOPTERA   Family CURCULIONIDAE

Rhynchaenus testaceus (Müller, 1776). Formerly known as: *Orchestes scutellaris* (F., 1801), *Orchestes testaceus* (Müller).

Distribution Recorded from Dorset, Isle of Wight, East Kent, West Kent, Surrey, Oxfordshire, East Suffolk, West Suffolk, East Norfolk, Bedfordshire, Huntingdonshire, Glamorgan, Nottinghamshire, Cheshire, South Lancashire, South-west Yorkshire, Durham, North Northumberland, Cumberland, Dumfriesshire, Berwickshire, South Aberdeenshire and

East Inverness & Nairn before 1970 and West Norfolk and Huntingdonshire from 1970 onwards.

Habitat and ecology Alder carr, and probably alder woodland along river margins. Phytophagous. Associated with alder, and possibly also with grey alder. Adults have been recorded from wild cherry, though it is doubtful that this is a typical foodplant. Adults have been recorded from late April to June.

Status Not listed in the insect Red Data Book (Shirt, 1987). Old records show that this species was widely noted in southern England, with scattered records as far north as Easterness, Scotland. There are only two recent records.

Threats Uncertain, though this species may have declined through loss of habitat because of drainage, water abstraction schemes, and clear-felling and conversion to other land use.

Management and conservation Water tables should be maintained at high levels. Gaps in the age structure of the tree population should be identified and the continuity of the appropriate habitat ensured by regeneration and possibly suitable planting.

Published sources Collier, M.J. (1988a), Collier, M.J. (1989), Eyre, M.D. (1987), Fowler, W.W. (1891), Fowler, W.W. & Donisthorpe, H.St J.K. (1913).

RUTIDOSOMA GLOBULUS   **NOTABLE A**
A weevil
Order COLEOPTERA   Family CURCULIONIDAE

Rutidosoma globulus (Herbst, 1795). Formerly known as: *Rhytidosoma globulus* (Herbst, 1795), *Rhytidosomus globulus* (Herbst), *Rutidosoma globula* (Herbst).

Distribution Recorded from South Devon, South Hampshire, East Sussex, East Kent, West Kent, Surrey, Hertfordshire, Middlesex, Berkshire, East Suffolk, West Suffolk, Warwickshire, South Lincolnshire, North Lincolnshire and South Northumberland before 1970 and South Hampshire, North Hampshire, East Kent and Dunbartonshire from 1970 onwards.

Habitat and ecology Broad-leaved woodland. Phytophagous. Associated with aspen and grey poplar, and has been found in leaf litter. Possibly associated with several *Populus* species. On the Continent, this species has also been recorded on white poplar. In Britain, adults have been recorded from May to September.

Status Very local and recently recorded from four vice-counties, three of these in southern England, the remaining one being Dunbarton, Scotland. Old records show that this species has been widely recorded in England.

Threats Loss of broad-leaved woodland through, for example, clear-felling and coniferisation. Neglect and conversion to high forest, has led to increased shade and the loss of glades and broad sunny rides. The selective removal of aspen may also be a further threat to this species.

Management and conservation Open glades and ride margins should be cut on rotation to retain a variety of vegetation structures.

Published sources Fowler, W.W. (1891), Fowler, W.W. & Donisthorpe, H.St J.K. (1913).

SIBINIA ARENARIAE	NOTABLE B

A weevil
Order COLEOPTERA   Family CURCULIONIDAE

Sibinia arenariae Stephens, 1831.

Distribution South East, South, South West, East Anglia and Dyfed-Powys.

Habitat and ecology Grassland, coastal cliffs and the drier parts of saltmarshes. This species has only been recorded from coastal habitats. Phytophagous. Associated with rock spurrey *Spergularia rupestris*, lesser sea-spurrey *S. marina* and sand spurrey *S. rubra*. On the Continent, this species has also been recorded from greater sea-spurrey *S. media*. Adults are found at the base of the foodplant. In Britain, adults have been recorded from April to June, and in August and September.

Status Widespread but local along the coasts of southern England. Also recorded from parts of Wales. There may be confusion over identification of this species as it is transposed with *S. primitus* in the key in Joy (1932).

Threats Loss of habitat through reclamation, erosion, improvement and conversion to arable agriculture, coastal developments and natural succession.

Management and conservation Some grazing, rotational cutting or other disturbance, such as rotovation, may be needed to maintain open conditions.

SIBINIA POTENTILLAE	NOTABLE B

A weevil
Order COLEOPTERA   Family CURCULIONIDAE

Sibinia potentillae Germar, 1824. Formerly known as: *Gymnetron lloydi* Donisthorpe, 1929.

Distribution South East, South, South West, East Anglia, East Midlands and West Midlands.

Habitat and ecology Disturbed ground, particularly on sandy soils. Phytophagous. Associated with corn spurrey *Spergula arvensis*. On the Continent, this species has also been found on *S. morisonii*. Adults occur on, under and near the foodplant. In Britain, adults have been recorded in April, June, August and September.

Status Widespread but local in the southern half of England.

Threats This species is threatened by natural succession.

Management and conservation Disturbance, such as rotavation, is necessary to maintain open conditions.

SIBINIA PRIMITUS	NOTABLE B

A weevil
Order COLEOPTERA   Family CURCULIONIDAE

Sibinia primitus (Herbst, 1795). Formerly known as: *Sibinia primita* (Herbst, 1795), *Sibinia signata* (Gyllenhal, 1813).

Distribution South East, South, South West, East Anglia, East Midlands, West Midlands, North East, South Wales and Dyfed-Powys.

Habitat and ecology Disturbed ground, sand pits, coastal shingle, heathland, downland and gardens. Particularly on sandy soils. Phytophagous. Associated with pearlwort *Sagina*, also recorded from rock sea-spurrey *Spergularia rupicola* and possibly associated with chickweed *Stellaria* and mouse-ear *Cerastium*. Adults of this species have been found on the flowers of golden rod *Solidago canadensis* growing in a garden. Adults have been recorded in May and August.

Status Widespread but local in the southern half of England, and also recorded in parts of Wales. Recorded as far north as North Lincolnshire. There may be confusion over identification of this species as it is transposed in the key in Joy (1932).

CURCULIONIDAE

Threats Loss of habitat through improvement and conversion to arable agriculture, development, afforestation, gravel extraction, the infilling of pits, and natural succession.

Management and conservation Except on coastal shingle, which should be left undisturbed, disturbance, such as rotovation, is needed to maintain open conditions.

SIBINIA SODALIS   **NOTABLE A**
A weevil
Order COLEOPTERA   Family CURCULIONIDAE

Sibinia sodalis Germar, 1824.

Distribution Recorded from West Cornwall, South Devon, North Devon, Dorset, South Hampshire, Glamorgan and Pembrokeshire before 1970 and North Devon, South Hampshire, Glamorgan and Pembrokeshire from 1970 onwards.

Habitat and ecology Coastal habitats, predominantly cliff tops and shingle. Phytophagous. Associated with thrift *Armeria maritima*. Adults generally occur at the base of the foodplant and have been recorded from April to June and in September.

Status Primarily southern and western in distribution. Very local and recently recorded from only four vice-counties.

Threats This species is threatened by coastal developments, erosion and gravel extraction

Published sources De La Garde, P. (1906a), Fowler, W.W. (1891), Fowler, W.W. & Donisthorpe, H.St J.K. (1913), Henderson, M.K. (1989).

SITONA CINERASCENS
A weevil   **INSUFFICIENTLY KNOWN**
Order COLEOPTERA   Family CURCULIONIDAE

Sitona cinerascens (Fahraeus, 1840). Formerly known as: *Sitona cambricus var. cinerascens* (Fahraeus), *Sitona cambricus?* sensu auct. Brit. partim not Stephens, 1831, *Sitones cambricus var. cinerascens* (Fahraeus).

Distribution Recorded from a single example of unknown origin.

Status Not listed in the insect Red Data Book (Shirt, 1987). Very closely allied to *S. puberulus* and only known from a single example in the Stephens Collection at the British Museum (Nat. Hist.). This

example is without data, though it is considered to be of British origin.

SITONA GEMELLATUS   **ENDANGERED**
A weevil
Order COLEOPTERA   Family CURCULIONIDAE

Sitona gemellatus Gyllenhal, 1834. Formerly known as: *Sitones gemellatus* (Gyllenhal).

Distribution Recorded from South Devon before 1970 and Dorset from 1970 onwards.

Habitat and ecology Undercliffs on sandy substrates, and coastal shingle. Phytophagous. Most frequently found under restharrow *Ononis repens*, also associated with black medick *Medicago lupulina* and probably other Leguminosae. On the Continent, this species has also been found on greater bird's-foot-trefoil *Lotus uliginosus* and meadow vetchling *Lathyrus pratensis*. The larvae are root-feeders. In Britain, adults have been recorded from April to June and in August.

Status Only known from a very few sites on the south coast of England, and recently only from one site. First recorded in 1908 from Sidmouth, South Devon.

Threats Activities that accelerate or reduce the rate of erosion should be avoided. The degradation of suitable habitat through natural succession and the invasion of scrub on stabilised areas may be a threat.

Management and conservation Occasional slippages are necessary to maintain habitat continuity. Large areas of unstable cliff are required so that the population does not become isolated and subsequently threatened by individual landslips.

Published sources Allen, J.W. & Nicholson, G.W. (1924b), Fryer, J.C.F. & Fryer, H.F. (1923a), Shirt, D.B., ed. (1987).

SITONA MACULARIUS   **NOTABLE B**
A weevil
Order COLEOPTERA   Family CURCULIONIDAE

Sitona macularius (Marsham, 1802). Formerly known as: *Sitona crinitus* (Herbst, 1795), *Sitones crinitus* (Herbst).

Distribution England, South Wales, North Wales, South East Scotland, South West Scotland and North West Scotland.

Habitat and ecology Grassland, field margins in cultivated land, quarries and disturbed ground, particularly on chalky soils. Phytophagous. The larvae feed on roots and root nodules. Adults have been recorded from sainfoin *Onobrychis sativa*, wild liquorice *Astragalus glycyphyllos*, bird's-foot-trefoil *Lotus*, hairy tare *Vicia hirsuta* and smooth tare *V. tetrasperma*. On the Continent, the species has been associated with common vetch *Vicia sativa*, lucerne *Medicago sativa*, black medick *M. lupulina*, red clover *Trifolium pratense* and cultivated pea *Pisum sativum*. In Britain, adults have been recorded from January to September.

Status Widespread but local in Britain. Possibly declining. This species is difficult to identify and may be confused with other members of the genus. Consequently, the exact status of this species is hard to assess.

Threats This species may be threatened by the loss of grassland through conversion to arable agriculture or other land use. The infilling of quarries, the use of herbicides and pesticides, and natural succession and scrub invasion may be further threats.

Management and conservation Grazing, cutting or some other disturbance, such as rotovation, on a rotational basis, is needed to maintain open conditions and encourage early successional stages.

SITONA ONONIDIS　　　　　**NOTABLE B**
A weevil
Order COLEOPTERA　　　　Family CURCULIONIDAE

Sitona ononidis Sharp, 1866. Formerly known as: *Sitona suturalis var. ononidis* Sharp, *Sitones ononidis* (Sharp).

Distribution South, South West, South East, East Anglia, West Midlands, North West and North East.

Habitat and ecology Grassland, coastal cliffs, disturbed ground and quarries. Phytophagous. Associated with common restharrow *Ononis repens* and spiny restharrow *O. spinosa*. On the Continent, this species has also been recorded from meadow vetchling *Lathyrus pratensis*, tuberous pea *L. tuberosus* and tufted vetch *Vicia cracca*. Larvae feed on the roots of plants. In Britain, adults have been recorded in July, September and October.

Status Widespread but local in England.

Threats This species is threatened by the loss of unimproved grassland through improvement and conversion to arable agriculture, coastal developments,

cliff stabilisation schemes, the infilling of quarries, and natural succession and scrub invasion.

Management and conservation Grazing, cutting or some other disturbance, such as rotovation, on a rotational basis, is needed to maintain open conditions and encourage early successional stages.

SITONA PUBERULUS
A weevil　　　　**INSUFFICIENTLY KNOWN**
Order COLEOPTERA　　　　Family CURCULIONIDAE

Sitona puberulus Reitter, 1903. Formerly confused under: *Sitona cambricus* Stephens, 1831, *Sitones cambricus* sensu Fowler, 1891 partim not (Stephens).

Distribution Recorded from South Hampshire, Kirkcudbrightshire and Clyde Islands before 1970.

Habitat and ecology Uncertain, but probably associated with grassland and possibly sand dunes. Phytophagous. Possibly associated with bird's-foot-trefoil *Lotus* species. Adults have been recorded in July and August.

Status Not listed in the insect Red Data Book (Shirt, 1987). Only known from seven examples. Possibly under-recorded due to confusion with *S. cambricus*. Last recorded in 1946 from near Creetown, Kirkcudbrightshire.

Published sources Allen, A.A. (1965a), Kevan, D.K. (1963a).

SITONA WATERHOUSEI　　　　**NOTABLE B**
A weevil
Order COLEOPTERA　　　　Family CURCULIONIDAE

Sitona waterhousei Walton, 1846. Formerly known as: *Sitones waterhousei* (Walton).

Distribution South East, South, South West, East Midlands, West Midlands, North East, North West, South Wales and North Wales.

Habitat and ecology Coastal undercliffs, calcareous grassland and possibly coastal shingle and quarries near the coast. Phytophagous. The larvae probably feed on the roots of plants. Adults are associated with common bird's-foot-trefoil *Lotus corniculatus* and with narrow-leaved bird's-foot-trefoil *L. tenuis*. On the Continent, the species has also been found on greater bird's-foot-trefoil *L. uliginosus* and black medick *Medicago lupulina*. In Britain, adults have been recorded from February to September.

CURCULIONIDAE

Status Widespread but local in England, although not recorded in East Anglia. Also noted in parts of Wales. Often abundant where found.

Threats This species is threatened by cliff stabilisation schemes and the construction of sea defences. Coastal developments may reduce the amount of available habitat. Activities that accelerate or reduce the rate of erosion should be avoided. Loss of calcareous grassland through improvement and conversion to arable agriculture, gravel extraction, the infilling of quarries, and the degradation of suitable habitat through natural succession and the invasion of scrub on stabilised areas are further threats.

Management and conservation In unstable cliff habitats, occasional slippages are necessary to maintain habitat continuity. Large areas of cliff are required so that the population does not become isolated and subsequently threatened by individual landslips. The disturbance of coastal shingle should be avoided. In areas of grassland, grazing or cutting may be needed to maintain open conditions.

SMICRONYX COECUS **RARE**
A weevil
Order COLEOPTERA   Family CURCULIONIDAE

Smicronyx coecus (Reich, 1797).

Distribution Recorded from North Devon, Dorset, Isle of Wight, East Kent and West Kent before 1970 and North Devon, East Kent and Surrey from 1970 onwards.

Habitat and ecology Grassland, heathland, coastal shingle, cliff tops and other habitats in which the foodplants occur. Phytophagous. Associated with dodder *Cuscuta epithymum* and with greater dodder *C. europaea*. Adults have been recorded from May to July and in September.

Status Very local and only known from southern England. Recently recorded from only three vice-counties, all in southern England. This species is difficult to identify and may be confused with *S. jungermanniae*. Consequently, the exact status of this species is hard to assess.

Threats Loss of habitat through gravel extraction and development. This species may be further threatened by afforestation, improvement by reseeding or by the application of fertilisers, or by conversion to arable agriculture, and natural succession.

Management and conservation Disturbance of coastal shingle should be avoided. Grazing or cutting, on a rotational basis, may be needed to maintain open conditions.

Published sources Blair, K.G. (1935), Edwards, J. (1910), Fowler, W.W. & Donisthorpe, H.St J.K. (1913), Shirt, D.B., ed. (1987), Walker, J.J. (1894).

SMICRONYX JUNGERMANNIAE **NOTABLE B**
A weevil
Order COLEOPTERA   Family CURCULIONIDAE

Smicronyx jungermanniae (Reich, 1797).

Distribution South East, South, South West, East Anglia and East Midlands.

Habitat and ecology Grassland, heathland, coastal shingle and probably other habitats in which the foodplants occur. Phytophagous. Associated with dodder *Cuscuta epithymum* and with greater dodder *C. europaea*, usually on common gorse *Ulex europaeus* and broom *Cytisus scoparius*. Adults have been recorded from April to September.

Status Widespread but local, known only in the southern half of England. This species is difficult to identify and may be confused with *S. coecus*.

Threats Loss of habitat through improvement and conversion to arable agriculture, afforestation, development and gravel extraction. Natural succession may be a further threat.

Management and conservation Disturbance of coastal shingle should be avoided. Grazing or cutting, on a rotational basis, may be needed to maintain open conditions.

SMICRONYX REICHI **RARE**
A weevil
Order COLEOPTERA   Family CURCULIONIDAE

Smicronyx reichi (Gyllenhal, 1836). Formerly known as: *Smicronyx reichei* (Gyllenhal, 1836), *Smicronyx seripilosus* Tournier, 1874.

Distribution Recorded from South Devon, North Devon, North Somerset, Dorset, Isle of Wight, West Sussex, East Kent, West Kent, Surrey, Berkshire, Bedfordshire and West Gloucestershire before 1970 and East Sussex and Surrey from 1970 onwards.

Habitat and ecology Calcareous grassland. Phytophagous. Associated with common centaury *Centaurium erythraea*, yellow-wort *Blackstonia perfoliata* and possibly autumn gentian *Gentianella* *germanica*, and not with dodder *Cuscuta* like other members of the genus. On the Continent, this species has also been associated with Chiltern gentian *G. germanica*. In Britain, adults have emerged in early September from seedheads of common century. Adults have also been recorded in January and from May to August.

Status Not listed in the insect Red Data Book (Shirt, 1987). Very local and recently recorded from just two vice-counties. Old records suggest that this species was formerly more widespread in southern England. Possibly under-recorded.

Threats This species may have declined through the loss of calcareous grassland through improvement by reseeding or by the application of fertilisers, or by conversion to arable agriculture. Lack of, or changes in grazing or cutting regimes may also have contributed to this species' decline.

Management and conservation Grazing or cutting, on a rotational basis, is needed to maintain open conditions.

Published sources Atty, D.B. (1983), Blair, K.G. (1935).

STENOCARUS UMBRINUS　　　**NOTABLE B**
A weevil
Order COLEOPTERA　　　Family CURCULIONIDAE

Stenocarus umbrinus (Gyllenhal, 1837). Formerley known as: *Coeliodes cardui* sensu Fowler, 1891 not (Herbst, 1784), *Stenocarus fuliginosus* (Marsham, 1802).

Distribution England, South Wales, South East Scotland, South West Scotland and North East Scotland.

Habitat and ecology Roadside verges, field margins and disturbed ground, chiefly on base-rich or sandy soils. Has been found in unsprayed, arable fields and on new construction sites. Phytophagous. Associated with poppies *Papaver* species. On the Continent, this species has been recorded from common poppy *P. rhoeas*. Larvae feed in the root and rootstock. In Britain, adults have been recorded from March to November.

Status Widespread but local. Recorded throughout England, parts of Wales and throughout Scotland, except North West Scotland.

Threats This species is threatened by change in land use, natural succession and the use of herbicides and pesticides.

Management and conservation Disturbance is essential to maintain open conditions and encourage early successional stages.

STROPHOSOMA FABER　　　**NOTABLE B**
A weevil
Order COLEOPTERA　　　Family CURCULIONIDAE

Strophosoma faber (Herbst, 1784). Formerly known as: *Strophosomus faber* (Herbst, 1784).

Distribution England and Wales.

Habitat and ecology Heathland and grassland in sandy or gravelly situations. Particularly on open ground and on cliffs with a short turf. Phytophagous. Usually found at the roots of plants, particularly sheep's sorrel *Rumex acetosella*. On the Continent, this species has been associated with sheep's-fescue *Festuca ovina*, meadow fescue *F. pratensis*, wall barley *Hordeum murinum* and mouse-ear hawkweed *Hieracium pilosella*. In Britain, adults have been recorded in January, February and from April to October.

Status Widespread but local in England and Wales.

Threats Loss of habitat through improvement and conversion to arable agriculture, development and afforestation. Natural succession may be a further threat.

Management and conservation Some grazing, rotational cutting or other disturbance, such as rotovation, may be needed to maintain open conditions.

STROPHOSOMA FULVICORNE　**VULNERABLE**
A weevil
Order COLEOPTERA　　　Family CURCULIONIDAE

Strophosoma fulvicorne Walton,J., 1846. Formerly confused under: *Strophosoma curvipes* (Thomson, 1868), *Strophosoma fulvicornis* sensu auct. not Walton, 1846.

Distribution Recorded from Dorset and South Hampshire before 1970 and Dorset from 1970 onwards.

Habitat and ecology Heathland, particularly dune-heath, and probably also sand dunes. Phytophagous and probably associated with heath and heather. Adults have been recorded in May, June,

CURCULIONIDAE

September and October. Probably hibernates as an adult.

Status Status revised from RDB 3 (Rare) in Shirt (1987). Only known from a small area of heathland and dune heath in Purbeck, Dorset and from an old record at Ringwood, South Hampshire.

Threats Much heathland has been lost, or fragmented, through changes in land use, mainly by conversion to arable, forestry and urban development. Further habitat degradation has been through the cessation of traditional heathland management practices.

Management and conservation Management should aim for a diversity of successional stages from bare ground to mature heath, preferably by grazing, rotational cutting, or scraping.

Published sources Drane, A.B. (1990), Fowler, W.W. (1891), Sharp, D. (1912a), Shirt, D.B., ed. (1987).

TANYMECUS PALLIATUS   **NOTABLE B**
A weevil
Order COLEOPTERA   Family CURCULIONIDAE

Tanymecus palliatus (F., 1787).

Distribution England, South Wales, South East Scotland and South West Scotland.

Habitat and ecology Hedgebanks, roadside verges, grassland, undercliffs and possibly also woodland. Phytophagous. The larvae feed on the roots of plants. Adults are associated with thistles, nettles, greater burdock *Arctium lappa*, greater knapweed *Centaurea scabiosa* and black knapweed *C. nigra*. On the Continent, this species has been found on orache *Atriplex*, beet *Beta vulgaris*, bindweed *Convolvulus*, *Polygonum*, and members of the Leguminosae. In Britain, adults have been recorded from May to July.

Status Widespread but local. Recorded from southern England north to West Perthshire in Scotland.

Threats Loss of grassland through conversion to arable agriculture or other land use. Lack of, or changes in grazing or cutting regimes, and the use of pesticides and herbicides may threaten this species. The construction of sea defences and cliff stabilisation schemes may be further threats.

Management and conservation Grazing, cutting or some other disturbance, such as rotovation, on a rotational basis, is needed to maintain open conditions and encourage early successional stages.

TAPINOTUS SELLATUS   **NOTABLE A**
A weevil
Order COLEOPTERA   Family CURCULIONIDAE

Tapinotus sellatus (F., 1794).

Distribution Recorded from North Hampshire, Surrey, East Norfolk, West Norfolk and Huntingdonshire before 1970 and Dorset, North Hampshire, Surrey, Berkshire and East Norfolk from 1970 onwards.

Habitat and ecology Wetland and lake margins. Phytophagous. Associated with yellow-loosestrife *Lysimachia vulgaris*. Adults have been recorded from April to September.

Status Very local and recently recorded from only five vice-counties, all in southern England.

Threats Drainage for reasons such as agricultural improvement and development. Falling water tables because of water abstraction, river engineering schemes, and natural succession may also threaten this species.

Management and conservation Water tables should be maintained at high levels. Water bodies should be isolated from sources of eutrophication and pollution. Grazing or cutting, on a rotational basis, is needed to maintain open conditions.

Published sources Allen, A.A. (1972d), Allen, J.W. (1935), Bedwell, E.C. (1932b), Halstead, A. (1989), Luff, M.L. (1974), Morris, M.G. (1982b), Thouless, H.J. (1921).

THRYOGENES SCIRRHOSUS   **NOTABLE B**
A weevil
Order COLEOPTERA   Family CURCULIONIDAE

Thryogenes scirrhosus (Gyllenhal, 1836). Formerly known as: *Erirrhinus scirrhosus* (Gyllenhal).

Distribution South East, South, South West, East Anglia, East Midlands, West Midlands and North East.

Habitat and ecology Wetland. Phytophagous. Associated with club-rushes, bur-reed and possibly sedges. Adults have been recorded in March, from May to July, September and in December.

Status Widespread but local in England, though not known from North West England.

Threats Drainage for reasons such as agricultural improvement and development. Falling water tables because of water abstraction and river engineering schemes may also threaten this species.

Management and conservation Water tables should be maintained at high levels. Water bodies should be isolated from sources of eutrophication and pollution. Clearance should be undertaken on a rotational basis and should aim at maintaining early successional stages and ensuring the presence of the foodplant.

TRACHODES HISPIDUS　　　　**NOTABLE B**
A weevil
Order COLEOPTERA　　　Family CURCULIONIDAE

Trachodes hispidus (L., 1758).

Distribution South West, South, South East, West Midlands, East Midlands, North West, South Wales, Dyfed-Powys and South West Scotland.

Habitat and ecology Broad-leaved woodland, pasture-woodland and possibly hedgerows. Phytophagous. Recorded from leaf-litter and under bark, particularly on oak. Also recorded in small dead oak branches, from dead hazel branches, and on living and dead hornbeam. Adults have been noted from April to September.

Status Widespread but local in England, recorded as far north as Wigtownshire in Scotland.

Indicator status Grade 3 in Harding & Rose (1986).

Threats Loss of broad-leaved woodland and parkland through, for example, clear-felling and coniferisation. Habitat loss, in particular, through the felling of trees, removal of dead wood from living trees and the destruction or removal of standing and fallen dead wood for reasons such as forest hygiene, aesthetic tidiness, public safety or for use as fire wood. The grubbing-out of hedgerows may be a further threat.

Management and conservation Both fallen and standing dead timber, especially with the bark attached, should be retained. The removal of dead timber from trees should be avoided. Gaps in the age structure of the tree population should be identified and the continuity of the appropriate dead wood habitat ensured by regeneration, suitable planting and possibly with pollarding.

TRACHYPHLOEUS ALTERNANS　　**NOTABLE B**
A weevil
Order COLEOPTERA　　　Family CURCULIONIDAE

Trachyphloeus alternans Gyllenhal, 1834.

Distribution England, North Wales, South East Scotland, South West Scotland and North East Scotland.

Habitat and ecology Grassland, coastal cliffs and quarries, particularly on calcareous soils. Phytophagous. On the Continent, this species is associated with common rock-rose *Helianthemum nummularium*. Possibly also associated with this plant in Britain, as well as plantain *Plantago*. Larvae probably feed on the roots of plants, while adults are found at the base of plants. Parthenogenetic. Adults have been recorded from May to October.

Status Widespread but local in England and recorded as far north as Kincardineshire in Scotland. Also recorded from north Wales. This species is difficult to identify and may be confused with other members of the genus. Consequently, the exact status of this species is hard to assess.

Threats Improvement of calcareous grassland by reseeding or the application of fertilisers, or by conversion to arable agriculture. Lack of, or changes in grazing or cutting regimes, urban and holiday developments, and the infilling of quarries may be further threats.

Management and conservation Grazing or cutting, on a rotational basis, is needed to maintain open conditions.

TRACHYPHLOEUS ARISTATUS　　**NOTABLE B**
A weevil
Order COLEOPTERA　　　Family CURCULIONIDAE

Trachyphloeus aristatus (Gyllenhal, 1827).

Distribution England, Dyfed-Powys, South East Scotland and South West Scotland.

Habitat and ecology Grassland, coastal cliffs and quarries, possibly with a preference for calcareous soils. Phytophagous and probably polyphagous. The larvae possibly feed on the roots of plants. Adults have been recorded at the roots of plants and in litter. This species has also been found under buck's-horn plantain *Plantago coronopus* and may be associated with other plantain *Plantago*. Parthenogenetic. Adults have been recorded throughout the year.

CURCULIONIDAE

Status Widespread but local in England and recorded in southern Scotland and mid Wales. This species is difficult to identify and may be confused with other members of the genus. Consequently, the exact status of this species is hard to assess.

Threats Improvement of grassland by reseeding or the application of fertilisers, or by conversion to arable agriculture. Lack of, or changes in grazing or cutting regimes, and the infilling of quarries may be further threats.

Management and conservation Some grazing, rotational cutting or other disturbance, such as rotovation, is needed to maintain open conditions.

TRACHYPHLOEUS ASPERATUS   NOTABLE B
A weevil
Order COLEOPTERA   Family CURCULIONIDAE

Trachyphloeus asperatus Boheman, 1843. Formerly known as: *Trachyphloeus olivieri* Bedel, 1883, *Trachyphloeus squamulatus* sensu (Olivier, 1808) not (Herbst, 1795).

Distribution England and South East Scotland.

Habitat and ecology On both sandy and chalky soils. Found on disturbed ground, sand dunes, chalk pits, heathland with short turf and coastal cliffs. Phytophagous and probably polyphagous. The larvae probably feed on the roots of plants. Adults are found at the base of plants and have been recorded from sheep's sorrel *Rumex acetosella* and common bird's-foot-trefoil *Lotus corniculatus*. Parthenogenetic. Adults have been recorded from March to November.

Status Widespread but local in England, also recorded in South East Scotland. This species is difficult to identify and may be confused with other members of the genus. Consequently, the exact status of this species is hard to assess.

Threats This species may be threatened by the loss of habitat through urban and holiday development, afforestation, the infilling of chalk pits and natural succession and scrub invasion.

Management and conservation Some grazing, rotational cutting or some other disturbance, such as rotovation, is needed to maintain open conditions.

TRACHYPHLOEUS DIGITALIS   NOTABLE A
A weevil
Order COLEOPTERA   Family CURCULIONIDAE

Trachyphloeus digitalis (Gyllenhal, 1827).

Distribution Recorded from Isle of Wight, East Kent, West Kent, Berkshire, West Suffolk and "Lincs" before 1970 and East Kent and Northamptonshire from 1970 onwards.

Habitat and ecology Dry banks, pits and coastal cliffs. On calcareous soils. Phytophagous and probably polyphagous. The larvae probably feed on the roots of plants. Adults have been recorded at the base of rock-rose *Helianthemum* and bird's-foot-trefoil *Lotus*, though probably also occur on other plants. Parthenogenetic. Adults have been recorded in May, July and August.

Status Formerly with a scattered distribution from the Isle of Wight to Lincolnshire. Recently recorded from only two vice-counties.

Threats Loss of calcareous grassland through improvement by reseeding or by the application of fertilisers, or by conversion to arable agriculture. Lack of, or changes in grazing or cutting regimes, and the infilling of pits and quarries may be further threats to this species.

Management and conservation Grazing, cutting or some other disturbance, such as rotovation, on a rotational basis, is needed to maintain open conditions.

Published sources Allen, A.A. (1982b), Donisthorpe, H.St J.K. (1948), Newberry, E.A. (1913).

TRACHYPHLOEUS LATICOLLIS   NOTABLE A
A weevil
Order COLEOPTERA   Family CURCULIONIDAE

Trachyphloeus laticollis Boheman, 1843.

Distribution Recorded from North Devon, South Somerset and North Somerset before 1970 and West Cornwall, Cardiganshire, Westmorland & North Lancashire, Cumberland, Dumfriesshire, Wigtownshire and Berwickshire from 1970 onwards.

Habitat and ecology Rough heathy ground and coastal habitats, including stony grassland. Particularly on sandy soils. Phytophagous and probably polyphagous. The larvae probably feed on the roots of plants. Adults probably occur at the base of various plants, possibly with a preference for plantain *Plantago*, sorrel *Rumex* or bird's-foot-trefoil *Lotus*. Parthenogenetic. Adults have been recorded in March, April, June, August, September and October.

Status Recorded in south western and north western England, and also mid Wales and southern Scotland. Recently recorded from just seven vice-counties. This species is difficult to identify and may be confused with other members of the genus. Consequently, the exact status of this species is hard to assess.

Threats Much heathland has been lost, or fragmented, through changes in land use, mainly by conversion to arable, forestry and urban development. Further habitat degradation has been through the cessation of traditional heathland management practices. This species may be further threatened by coastal developments and through erosion in areas frequented by large numbers of walkers.

Management and conservation Management should aim for a diversity of successional stages from bare ground to mature heath, preferably by grazing or by rotational cutting, scraping or burning. Where there has been a cessation of management and scrub is invading, public pressure through walking may be the only factor maintaining open conditions.

Published sources Fowler, W.W. (1891), Fowler, W.W. & Donisthorpe, H.St J.K. (1913).

TRACHYPHLOEUS SPINIMANUS   NOTABLE B
A weevil
Order COLEOPTERA   Family CURCULIONIDAE

Trachyphloeus spinimanus Germar, 1824.

Distribution South East, South, South West, East Anglia, East Midlands and Dyfed-Powys.

Habitat and ecology Disturbed ground, grassland, coastal cliffs and chalk pits. Nearly always on chalky soils, though also recorded from a sand pit. Phytophagous and probably polphagous. The larvae probably feed on the roots of plants. Parthenogenetic. Adults have been found on ribwort plantain *Plantago lanceolata* and have been recorded from March to October.

Status Widespread but local in southern England. Also recorded in mid Wales. This species is difficult to identify and may be confused with other members of the genus. Consequently, the exact status of this species is hard to assess.

Threats Loss of calcareous grassland through improvement by reseeding or by the application of fertilisers, or by conversion to arable agriculture. Lack

of, or changes in grazing or cutting regimes, coastal developments, and the infilling of pits may also threaten this species.

Management and conservation Grazing, cutting or some other disturbance, such as rotovation, on a rotational basis, is needed to maintain open conditions and encourage early successional stages.

TRICHOSIROCALUS BARNEVILLEI
A weevil   **NOTABLE B**
Order COLEOPTERA   Family CURCULIONIDAE

Trichosirocalus barnevillei (Brisout, 1866). Formerly known as: *Ceuthorhynchidius barnevillei* Brisout, *Ceuthorhynchidius chevrolati* Crotch, 1865, *Ceuthorrhynchidius chevrolati* Crotch, 1865.

Distribution South East, South, East Anglia and East Midlands.

Habitat and ecology Disturbed ground, coastal shingle, Breck grassland and established grassland on disturbed sites. Phytophagous. Associated with yarrow *Achillea millefolium*. On the Continent, this beetle is also recorded from *Tanacetum* spp. and chamomile *Anthemis*. In Britain, adults have been recorded in March and from May to September.

Status Local in southern and eastern England.

Threats This species is threatened by gravel extraction, afforestation, urban and tourist development, and improvement and conversion to arable agriculture. Natural succession is a further threat.

Management and conservation Disturbance of coastal shingle should be avoided. Grazing, cutting or some other disturbance, such as rotovation, on a rotational basis, is needed to maintain open conditions and encourage early successional stages.

TRICHOSIROCALUS DAWSONI   NOTABLE B
A weevil
Order COLEOPTERA   Family CURCULIONIDAE

Trichosirocalus dawsoni (Brisout, 1869). Formerly known as: *Ceuthorhynchidius dawsoni* (Brisout), *Ceuthorrhynchidius dawsoni* (Brisout).

Distribution South East, South, South West, East Anglia, North East, North West, Dyfed-Powys, North Wales and South West Scotland.

Habitat and ecology Coastal cliffs, undercliffs and saltmarshes. Also found inland in chalk pits.

CURCULIONIDAE

Phytophagous. Associated with buck's-horn plantain *Plantago coronopus* and also sea plantain *P. maritima*. Adults have been recorded from March to September.

Status Widespread but local in southern and northern England. Also recorded in parts of Wales and South West Scotland. Often numerous where found.

Threats This species is threatened by coastal developments, natural succession and the loss of saltmarsh through reclamation, erosion and the construction of coastal defences. The infilling of chalk pits may be a further threat.

Management and conservation Grazing should not be introduced to a saltmarsh where there is no grazing at present. In areas of unstable cliff, occasional slippages are necessary to maintain habitat continuity. Large areas of unstable cliff are required so that the population does not become isolated and subsequently threatened by individual landslips.

TRICHOSIROCALUS HORRIDUS   NOTABLE A
A weevil
Order COLEOPTERA   Family CURCULIONIDAE

Trichosirocalus horridus (Panzer, 1801). Formerly known as: *Ceuthorhynchidius horridus* (Panzer), *Ceuthorrhynchidius horridus* (Panzer).

Distribution Recorded from West Cornwall, South Devon, North Devon, Dorset, Isle of Wight, West Sussex, East Sussex, East Kent, West Kent, Surrey, Hertfordshire, Berkshire, Oxfordshire, Buckinghamshire, East Suffolk, West Suffolk, "Norfolk", Cambridgeshire, East Gloucestershire, West Gloucestershire, North Lincolnshire, Leicestershire & Rutland, North-east Yorkshire, Durham, North Northumberland, East Lothian, Midlothian and Fife before 1970 and Dorset, Isle of Wight, West Sussex and East Sussex from 1970 onwards.

Habitat and ecology Disturbed ground and woodland rides. Phytophagous. Associated with spear thistle *Cirsium vulgare* and musk thistle *Carduus nutans*, probably occasionally on other species of thistle. On the Continent, this species has been noted from cotton thistle *Onopordum acanthium*. In Britain, adults have been recorded in March and from May to September.

Status Very local and recently recorded from only four vice-counties, all in southern England. Old records show that this species has been more widely recorded in England and noted as far north as Fife, Scotland.

Threats This species is probably threatened by loss of habitat through development and improvement, and conversion to arable agriculture. Natural succession may also have contributed to this species decline.

Management and conservation Grazing, cutting or some other disturbance, such as rotovation, on a rotational basis, is needed to maintain open conditions and encourage early successional stages.

Published sources Atty, D.B. (1983), Crowson, R.A. (1970), Eyre, M.D. (1987), Fowler, W.W. (1891), Fowler, W.W. & Donisthorpe, H.St J.K. (1913).

TRICHOSIROCALUS RUFULUS   NOTABLE A
A weevil
Order COLEOPTERA   Family CURCULIONIDAE

Trichosirocalus rufulus (Dufour, 1851). Formerly known as: *Ceuthorhynchidius rufulus* (Dufour), *Ceuthorrhynchidius rufulus* (Dufour).

Distribution Recorded from Dorset, West Sussex, East Sussex, East Kent, West Kent, Surrey, Berkshire, Oxfordshire, West Suffolk and Cambridgeshire before 1970 and Dorset, East Sussex, East Kent and West Kent from 1970 onwards.

Habitat and ecology Disturbed ground, particularly on sandy and chalky soils. Also long established grassland such as on coastal cliffs. Phytophagous. Associated with species of plantain. On the Continent, this species has been recorded from ribwort plantain *P. lanceolata* and sea plantain *P. maritima*. In Britain, adults have been recorded in March, and from May to October.

Status Very local and recently recorded from only four vice-counties, all in south-eastern England. Old records show that this species has been more widely recorded and found as far north as Cambridgeshire. There is an unconfirmed record from South-west Yorkshire.

Threats This species is threatened by improvement and conversion to arable agriculture, urban and tourist development, and natural succession.

Management and conservation Grazing, cutting or some other disturbance, such as rotovation, on a rotational basis, is needed to maintain open conditions and encourage early successional stages.

Published sources Allen, A.A. (1981b), Allen, A.A. (1989d), Fowler, W.W. (1891).

TROPIPHORUS ELEVATUS   NOTABLE B
A weevil
Order COLEOPTERA   Family CURCULIONIDAE

Tropiphorus elevatus (Herbst, 1795). Formerly known as: *Tropiphorus carinatus* sensu (Müller, 1776) not (L., 1767).

Distribution South East, South, South West, East Anglia, East Midlands, North East, North West, South East Scotland, South West Scotland and North East Scotland.

Habitat and ecology Broad-leaved woodland and pasture-woodland. Phytophagous. Associated with field layer plants and possibly with dog's mercury *Mercurialis perennis*. Parthenogenetic. Adults have been recorded from April to July.

Status Widespread but local in England, with isolated records from Scotland as far north as the Shetlands.

Threats Loss of broad-leaved woodland through, for example, clear-felling and coniferisation.

Management and conservation Open spaces in woodland should to be retained. Open glades and ride margins should be cut on rotation to retain a variety of vegetation structures.

TROPIPHORUS OBTUSUS   NOTABLE A
A weevil
Order COLEOPTERA   Family CURCULIONIDAE

Tropiphorus obtusus (Bonsdorff, 1785).

Distribution Recorded from Derbyshire, Cheshire, South Lancashire, Westmorland & North Lancashire, Cumberland, Dumfriesshire, Lanarkshire, Fife, Mid Perthshire, Angus, South Aberdeenshire, Moray, Clyde Islands, South Ebudes, North Ebudes and Shetland before 1970 and Cheshire, Mid-west Yorkshire, North-west Yorkshire, South Northumberland, Lanarkshire, Moray and Shetland from 1970 onwards.

Habitat and ecology Open and well vegetated ground, including a sand dune. Occasionally found in woodland. Phytophagous and possibly polyphagous. This species has been found in moss and flood refuse. Parthenogenetic. Adults have been recorded from March to October.

Status Very local with a northern distribution, recently recorded from Cheshire north to Shetland in Scotland.

Threats Uncertain, though the loss of habitat through improvement by reseeding or by the application of

fertilisers, or by conversion to arable agriculture may threaten this species. The clear-felling of woodland and conversion to other land use, afforestation, and the lack of, or changes in grazing or cutting regimes, may be further threats.

Management and conservation Uncertain, though grazing or cutting may be needed to maintain open conditions.

Published sources Fitton, M.G. (1965), Fowler, W.W. (1891), Fowler, W.W. & Donisthorpe, H.St J.K. (1913).

TROPIPHORUS TERRICOLA   NOTABLE B
A weevil
Order COLEOPTERA   Family CURCULIONIDAE

Tropiphorus terricola (Newman, 1838). Formerly known as: *Tropiphorus tomentosus* (Marsham, 1802).

Distribution England, South Wales, South East Scotland, South West Scotland and North East Scotland.

Habitat and ecology Herbaceous vegetation in open situations, limestone dales and sand dunes. Phytophagous and possibly polyphagous. Recorded from dog's mercury *Mercurialis perennis*. Parthenogenetic. Adults have been recorded from April to July and in December.

Status Widespread but local in England and parts of Scotland, also recorded in South Wales.

Threats Loss of habitat through improvement and conversion to arable agriculture, development and afforestation. Natural succession may be a further threat.

Management and conservation Some grazing, rotational cutting or other disturbance is needed to maintain open conditions.

TYCHIUS CRASSIROSTRIS
A weevil   **INSUFFICIENTLY KNOWN**
Order COLEOPTERA   Family CURCULIONIDAE

Tychius crassirostris Kirsch, 1871.

Distribution Recorded from Dorset before 1970.

Habitat and ecology Only recorded from coastal undercliff in Britain. Phytophagous. Probably associated with melilot *Melilotus* and medick *Medicago*. On the Continent, this species has been associated with white melilot *Melilotus alba*, tall

CURCULIONIDAE

melilot *M. altissima*, ribbed melilot *M. officinalis* and lucerne *Medicago sativa*. In Britain, adults have been recorded in May and June.

Status Status revised from RDB 3 (Rare) in Shirt (1987). Only known from a small number of examples recorded in 1926 and 1927 from Charmouth, Dorset. This species is difficult to identify and may be confused with other members of the genus. Consequently, the exact status of this species is hard to assess.

Threats Uncertain, though this species may be threatened by by cliff stabilisation schemes and the construction of sea defences. Coastal developments may reduce the amount of available habitat. Activities that accelerate or reduce the rate of erosion should be avoided. The degradation of suitable habitat through natural succession and the invasion of scrub on stabilised areas may be a further threat.

Published sources Shirt, D.B., ed. (1987).

TYCHIUS LINEATULUS	NOTABLE A

A weevil
Order COLEOPTERA   Family CURCULIONIDAE

Tychius lineatulus Stephens, 1831.

Distribution Recorded from North Devon, South Wiltshire, Dorset, Isle of Wight, South Hampshire, West Sussex, East Sussex, East Kent, West Kent, South Essex, Berkshire, Cambridgeshire, Bedfordshire, East Gloucestershire, South-east Yorkshire and North-east Yorkshire before 1970 and Dorset, East Sussex and East Kent from 1970 onwards.

Habitat and ecology Grassland, road-side verges and disturbed ground. Phytophagous. Associated with zigzag clover *Trifolium medium*. Also recorded from red clover *T. pratense* and kidney vetch *Anthyllis vulneraria*. Adults have been found in January, from April to September, and in November.

Status Very local and recently recorded from only three vice-counties, all in southern England. Old records show that this species has been widely recorded in southern England, and had a scattered distribution as far north as North-east Yorkshire. This species is difficult to identify and may be confused with other members of the genus. Consequently, the exact status of this species is hard to assess.

Threats This species is threatened by improvement and conversion to arable agriculture, development and natural succession.

Management and conservation Grazing or cutting, on a rotational basis, is needed to maintain open conditions.

Published sources Atty, D.B. (1983), Fowler, W.W. (1891), Fowler, W.W. & Donisthorpe, H.St J.K. (1913), Morris, M.G. (1974a).

TYCHIUS PARALLELUS	NOTABLE A

A weevil
Order COLEOPTERA   Family CURCULIONIDAE

Tychius parallelus (Panzer, 1794). Formerly known as: *Tychius venustus* sensu auct. not (F., 1781).

Distribution Recorded from South Devon, South Hampshire, East Sussex, East Kent, West Kent, Surrey, North Essex, Berkshire, West Suffolk, "Norfolk", Glamorgan, Moray and East Inverness & Nairn before 1970 and East Kent, West Suffolk and West Norfolk from 1970 onwards.

Habitat and ecology Grassland, woodland, roadside verges, coastal shingle and probably disturbed ground. Phytophagous. Associated with broom *Cytisus scoparius*. On the Continent, this species may also be associated with dyer's greenweed *Genista tinctoria*. The larvae feed on unripe seeds in the pods. In Britain, adults have been recorded in April, May and July.

Status Very local and recently recorded from only three vice-counties, all in south-eastern England. Old records suggest that this species was formerly more widespread, though had a disjunct distribution being recorded in southern England, and Elgin and Easterness in Scotland.

Threats This species is threatened by loss of woodland and conversion to other land use, and the loss of grassland through improvement and conversion to arable agriculture. Gravel extraction, coastal developments, and natural succession may be further threats.

Management and conservation Except on coastal shingle, which should be left undisturbed, grazing, cutting or some other disturbance may be needed to maintain open conditions.

Published sources Collier, M.J. (1990), Fowler, W.W. (1891).

TYCHIUS POLYLINEATUS
A weevil   **INSUFFICIENTLY KNOWN**
Order COLEOPTERA   Family CURCULIONIDAE

Tychius polylineatus (Germar, 1824).

Distribution Recorded from East Sussex, Berkshire and Cambridgeshire before 1970.

Habitat and ecology Chalk downland. Phytophagous. Associated with red clover *Trifolium pratense* and possibly common bird's-foot-trefoil *Lotus corniculatus*. On the Continent, this species has also been noted from zigzag clover *T. medium*, knotted clover *T. striatum* and hare's-foot clover *T. arvense*. The larvae are gall formers.

Status Status revised from RDB 3 (Rare) in Shirt (1987). Only known from very few vice-counties in southern England and with very few records. Last recorded in 1909 from East Sussex. This species is difficult to identify and may be confused with other members of the genus. Consequently, the exact status of this species is hard to assess.

Threats Loss of calcareous grassland through improvement by reseeding or by the application of fertilisers, or by conversion to arable agriculture. Lack of, or changes in grazing or cutting regimes may also have threatened this species.

Published sources Fowler, W.W. (1891), Fowler, W.W. & Donisthorpe, H.St J.K. (1913), Shirt, D.B., ed. (1987).

TYCHIUS PUSILLUS   NOTABLE B
A weevil
Order COLEOPTERA   Family CURCULIONIDAE

Tychius pusillus Germar, 1842. Formerly known as: *Tychius pygmaeus* Brisout, 1860.

Distribution South East, South, South West, East Anglia and West Midlands.

Habitat and ecology Grassland, field margins, road-side verges and disturbed ground. Phytophagous. Associated with lesser trefoil *Trifolium dubium* and possibly with other species of clover. On the Continent, this species has also been associated with hop trefoil *T. campestre* and strawberry clover *T. fragiferum*. In Britain, adults have been recorded from May to July and in September.

Status Widespread but local in southern England.

Threats Loss of grassland through improvement and conversion to arable agriculture, development and natural succession. This species may also be threatened by the use of herbicides and pesticides.

Management and conservation Grazing, cutting or some other disturbance, such as rotovation, on a rotational basis, is needed to maintain open conditions.

TYCHIUS QUINQUEPUNCTATUS VULNERABLE
A weevil
Order COLEOPTERA   Family CURCULIONIDAE

Tychius quinquepunctatus (L., 1758). Formerly known as: *Aoromius quinquepunctatus* (L.), *Aoromius 5-punctatus* (L.).

Distribution Recorded from South Devon, South Hampshire, East Sussex, Surrey, Middlesex, West Norfolk and Cambridgeshire before 1970 and West Norfolk and Glamorgan from 1970 onwards.

Habitat and ecology Sand dunes, grassland, heathland, and woodland rides. Phytophagous. Associated with bitter vetch *Lathyrus montanus* and narrow-leaved vetch *Vicia sativa* subsp. *nigra*. On the Continent, this species has also been recorded from bush vetch *V. sepium*, common vetch *V. sativa*, broad bean *V. faba* and field pea *Pisum sativum*. In Britain, adults have been found in May and June.

Status Only known from southern England and South Wales. Recently recorded from just two vice-counties. A species that has declined in many areas.

Threats This species has probably declined through loss of habitat and conversion to other land use. Natural succession, and in some areas overgrazing, may have also contributed to this species' decline. The degradation of remaining habitat by excessive disturbance of the vegetation through activities such as motorbike access, horse-riding and human trampling may be a threat.

Management and conservation Grazing, cutting or some other disturbance, such as rotovation, on a rotational basis, may be necessary to maintain open conditions.

Published sources Collier, M.J. (1990), De La Garde, P. (1906b), Drane, A.B. (1990), Fowler, W.W. (1891), Fowler, W.W. & Donisthorpe, H.St J.K. (1913), Hodge, P. (1987), Shirt, D.B., ed. (1987).

CURCULIONIDAE

TYCHIUS SCHNEIDERI   NOTABLE B
A weevil
Order COLEOPTERA   Family CURCULIONIDAE

Tychius schneideri (Herbst, 1795).

Distribution South East, South, South West, East Anglia, East Midlands and Dyfed-Powys.

Habitat and ecology Calcareous grassland, cliff tops and slopes, and shingle banks. Phytophagous. Associated with kidney vetch *Anthyllis vulneraria*. Adults have been recorded from April to August and in October.

Status Widespread but local in southern England. Also recorded from Wales. This species is difficult to identify and may be confused with other members of the genus. Consequently, the exact status of this species is hard to assess.

Threats This species is threatened by the construction of sea defences and cliff stabilisation schemes, and the loss of calcareous grassland through improvement and conversion to arable agriculture. Development, gravel extraction, and natural succession may be further threats.

Management and conservation Grazing, cutting or some other disturbance, such as rotovation, on a rotational basis, may be needed to maintain open conditions.

TYCHIUS SQUAMULATUS   NOTABLE B
A weevil
Order COLEOPTERA   Family CURCULIONIDAE

Tychius squamulatus Gyllenhal, 1836. Formerly known as: *Tychius flavicollis* sensu auct. Brit. partim not Stephens, 1831.

Distribution South East, South, South West, East Anglia, East Midlands, North West and Wales.

Habitat and ecology Grassland, chalk downland, field margins, coastal habitats including sand dunes, and possibly also on disturbed ground. Particularly found on calcareous soils, though also on sandy soils. Phytophagous. Associated with common bird's-foot-trefoil *Lotus corniculatus* and also found at the roots of kidney vetch *Anthyllis vulneraria*. The larvae feed in the pods of the foodplant. Adults have been recorded in January, March, May, June, August and October.

Status Widespread but local in southern England, also recorded in North West England and throughout Wales.

This species is difficult to identify and may be confused with other members of the genus. Consequently, the exact status of this species is hard to assess.

Threats Loss of habitat through improvement by reseeding or by the application of fertilisers, or by conversion to arable agriculture. Afforestation, development, natural succession, and the use of herbicides and pesticides may also threaten this species.

Management and conservation Grazing, cutting or some other disturbance, such as rotovation, on a rotational basis, is needed to maintain open conditions.

TYCHIUS TIBIALIS   NOTABLE A
A weevil
Order COLEOPTERA   Family CURCULIONIDAE

Tychius tibialis Boheman, 1843.

Distribution Recorded from South Devon, Isle of Wight, South Hampshire, East Sussex, East Kent, Surrey, South Essex, Oxfordshire, West Norfolk, Cambridgeshire, Huntingdonshire, Worcestershire and Leicestershire & Rutland before 1970 and Isle of Wight, South Hampshire, East Sussex, East Kent, South Essex, East Norfolk and West Norfolk from 1970 onwards.

Habitat and ecology Primarily coastal. Grassland, coastal habitats and probably disturbed ground. Particularly on sandy soils. Phytophagous. On the Continent, this species has been associated with hop trefoil *Trifolium campestre* and knotted clover *T. striatum*. In Britain, adults have been recorded in May and from July to September.

Status Very local. Only known from southern England and recorded as far north as Leicestershire. Recently recorded from only seven vice-counties. This species is difficult to identify and may be confused with other members of the genus. Consequently, the exact status of this species is hard to assess.

Threats Loss of habitat through improvement and conversion to arable agriculture, development, afforestation, and natural succession.

Management and conservation Grazing, cutting or some other disturbance, such as rotovation, on a rotational basis, is needed to maintain open conditions.

Published sources Fowler, W.W. (1891), Fowler, W.W. & Donisthorpe, H.St J.K. (1913).

ZACLADUS EXIGUUS   NOTABLE B
A weevil
Order COLEOPTERA   Family CURCULIONIDAE

Zacladus exiguus (Olivier, 1807). Formerly known as: *Allodactylus exiguus* (Olivier), *Coeliodes exiguus* (Olivier).

Distribution England and South Wales.

Habitat and ecology Roadside verges and coastal cliffs, though also recorded in sand-dunes and other coastal sites. Associated with several small-flowered crane's-bill *Geranium* species. The larvae are thought to feed on the roots of the foodplant. Adults have been recorded from May to July.

Status Widespread but local in England and parts of Wales. Possibly more frequent in southern England than in the rest of its range in Great Britain.

Threats The use of herbicides and pesticides are a threat to this species. Cliff stabilisation schemes, the construction of sea defences, and natural succession and the invasion of scrub may be further threats.

Management and conservation Large areas of cliff are required so that the population does not become isolated and subsequently threatened by individual landslips. Grazing, cutting or some other disturbance, such as rotovation, on a rotational basis, is necessary to maintain open conditions.

ANTHRENUS PIMPINELLAE   EXTINCT

Order COLEOPTERA   Family DERMESTIDAE

Anthrenus pimpinellae (F., 1775).

Distribution Recorded from "Devon", West Kent, "London district" and "Suffolk" before 1970.

Habitat and ecology This beetle has been found on a bough of a maple. On the Continent, this species has been recorded from bird's nests and also feeding on pollen and nectar of various flowers, in particular umbellifers and also on *Spiraea*. It has also been recorded from a variety of indoor situations where it may occasionally be a pest, feeding on stored products and carpets.

Status Not listed in the insect Red Data Book (Shirt, 1987). Presumed extinct. Possibly over-looked or mis-identified, because of confusion with other, more common members of the genus. The record for the "London district" given in the distribution section may refer to the West Kent record.

Published sources Fowler, W.W. (1889).

ANTHRENUS SCROPHULARIAE   EXTINCT

Order COLEOPTERA   Family DERMESTIDAE

Anthrenus scrophulariae (L., 1758). Formerly known as: *Anthrenus scropulariae* (L., 1758).

Distribution Recorded from Dorset, "London district" and Midlothian before 1970.

Habitat and ecology Adults apparently feed only on pollen and nectar. On the Continent, this species has been recorded from flowers of a variety of plants. Larvae have been found feeding on carrion, in bird's nests and in bee's nests. It has also been recorded from a variety of indoor situations where it may occasionally be a pest, feeding on stored products and carpets.

Status Not listed in the insect Red Data Book (Shirt, 1987). Presumed extinct. There is an unconfirmed record from Dauddyfryn, Merionethshire. Possibly over-looked or mis-identified, because of confusion with other, more common members of the genus.

Published sources Fowler, W.W. (1889).

CTESIAS SERRA   NOTABLE B

Order COLEOPTERA   Family DERMESTIDAE

Ctesias serra (F., 1792). Formerly known as: *Tiresias serra* (F.).

Distribution South East, South, South West, East Midlands, East Anglia, West Midlands, North East, Dyfed-Powys, South West Scotland and South East Scotland.

Habitat and ecology Ancient broad-leaved woodland, pasture-woodland and isolated trees. Recorded under the loose, dry bark of broad-leaved trees, with records from oak, elm, ash and willow. Adults and larvae are frequently found amongst spider's webs where they probably feed on trapped insect remains and detritus. Larvae have been recorded in March, April and November and probably occur all year round. Adults have been noted from April to June.

Status Widespread but local. Recorded throughout Great Britain apart from northern Scotland.

Indicator status Grade 1 in Garland (1983). Grade 3 in Harding & Rose (1986).

DERMESTIDAE

Threats Loss of broad-leaved woodland, parkland and isolated trees through, for example, clear-felling and conversion to other land use. Habitat loss, in particular, through the felling of ancient trees, removal of dead wood from living trees and the destruction or removal of standing and fallen dead wood for reasons such as forest hygiene, aesthetic tidiness, public safety or for use as fire wood.

Management and conservation Ancient trees and both fallen and standing dead timber, especially with the bark attached, should be retained. The removal of dead timber from ancient trees should be avoided. Gaps in the age structure of the tree population should be identified and the continuity of the appropriate dead wood habitat ensured by suitable planting and possibly with pollarding.

GLOBICORNIS NIGRIPES — ENDANGERED

Order COLEOPTERA   Family DERMESTIDAE

Globicornis nigripes (F., 1792).

Distribution Recorded from Berkshire and East Gloucestershire before 1970 and Berkshire and Buckinghamshire from 1970 onwards.

Habitat and ecology Ancient broad-leaved woodland and pasture-woodland. This species probably develops under bark or in dead wood. On the Continent, larvae of *G. nigripes* have been found in dead wood in the spring, adults emerging by mid April. In Britain, adults may be pollen feeders as many records are from flowers, particularly hogweed, hedge-parsley and also on *Spiraea*. Also recorded by sweeping grass under old oaks and, on at least one occasion, have been found on the foliage of an old oak. Adults have been noted from May to July.

Status Apart from an old and undated record from Tewkesbury, East Gloucestershire and a recent record from Slough, Buckinghamshire, *G. nigripes* has only been recorded from Windsor Forest, Windsor Great Park and the surrounding area, Berkshire.

Indicator status Grade 2 in Harding & Rose (1986).

Threats Loss of broad-leaved woodland and parkland through, for example, clear-felling and coniferisation. Habitat loss, in particular, through the felling of ancient trees, removal of dead wood from living trees and the destruction or removal of standing and fallen dead wood for reasons such as forest hygiene, aesthetic tidiness, public safety or for use as fire wood.

Management and conservation Ancient trees and both fallen and standing dead timber, especially with the bark attached, should be retained. The removal of dead timber from ancient trees should be avoided. Gaps in the age structure of the tree population should be identified and the continuity of the appropriate dead wood habitat ensured by suitable planting and possibly with pollarding. The presence of nectar sources such as umbellifers, may also be particularly important for this species. Windsor Forest and parts of Windsor Great Park are notified as SSSIs.

Published sources Harding, P.T. (1978), Harding, P.T. & Rose, F. (1986), Shirt, D.B., ed. (1987).

MEGATOMA UNDATA — NOTABLE B

Order COLEOPTERA   Family DERMESTIDAE

Megatoma undata (L., 1758).

Distribution South East, South, East Anglia, East Midlands, West Midlands, North East, North West and South Wales.

Habitat and ecology Woodland, pasture-woodland and isolated trees and bushes. Adults and larvae have been recorded from under the bark of dead wood on trees, with larval records from oak and sweet chestnut and adult records from oak, sweet chestnut, goat willow, field maple, hawthorn and sycamore. Possibly associated with older trees. This species has also been recorded on posts and garden fences, and has been found in the nests and larval burrows of a variety of insects and amongst spider's webs, where it probably feeds on dead insect remains and detritus. Adults have been recorded at flowers, such as hawthorn and crab-apple, probably feeding on pollen. Larvae have been recorded in March and July. Adults have been noted in February and from April to July.

Status Widespread but local in England, also reported in South Wales.

Indicator status Grade 3 in Garland (1983).

Threats Loss of broad-leaved woodland and parkland through, for example, clear-felling and coniferisation. Habitat loss, in particular, through the felling of ancient trees, removal of dead wood from living trees and the destruction or removal of standing and fallen dead wood for reasons such as forest hygiene, aesthetic tidiness, public safety or for use as fire wood.

Management and conservation Ancient trees and both fallen and standing dead timber, especially with the bark attached, should be retained. The removal of dead timber from ancient trees should be avoided. Gaps in the age structure of the tree population should be identified and the continuity of the appropriate dead wood habitat ensured by suitable planting and possibly with pollarding. The presence of nectar sources such as hawthorn, may also be particularly important for this species.

TRINODES HIRTUS **RARE**

Order COLEOPTERA Family DERMESTIDAE

Trinodes hirtus (F., 1781).

Distribution Recorded from South Devon, Surrey, Berkshire, Oxfordshire, East Suffolk and Cheshire before 1970 and Surrey, Berkshire and East Suffolk from 1970 onwards.

Habitat and ecology Ancient broad-leaved woodland and pasture-woodland. Adults and larvae have been found amongst spider's webs beneath loose, dry bark, mainly of oak, where they feed on the trapped remains of insects. Adults have been beaten from the boughs of old oak trees and recorded by sweeping under oaks. This species has also been recorded from an old, live sweet chestnut and Scots pine. Larvae have been found in April (adults emerging in July). Adults have been recorded from April to July.

Status Very local, this species has recently been recorded from just three vice-counties in southern and south-eastern England.

Indicator status Grade 1 in Harding & Rose (1986).

Threats Loss of broad-leaved woodland and parkland through, for example, clear-felling and coniferisation. Habitat loss, in particular, through the felling of ancient trees, removal of dead wood from living trees and the destruction or removal of standing and fallen dead wood for reasons such as forest hygiene, aesthetic tidiness, public safety or for use as fire wood.

Management and conservation Ancient trees and both fallen and standing dead timber, especially with the bark attached, should be retained. The removal of dead timber from ancient trees should be avoided. Gaps in the age structure of the tree population should be identified and the continuity of the appropriate dead wood habitat ensured by suitable planting and possibly with pollarding.

Published sources Fowler, W.W. (1889), Fowler, W.W. & Donisthorpe, H.St J.K. (1913), Garland, S.P. (1983), Harding, P.T. (1978), Harding, P.T. & Rose, F. (1986), Nash, D.R. (1990), Shirt, D.B., ed. (1987).

DRILUS FLAVESCENS **NOTABLE A**

Order COLEOPTERA Family DRILIDAE

Drilus flavescens (Fourcroy, 1785).

Distribution Recorded from Isle of Wight, South Hampshire, North Hampshire, East Sussex, East Kent, West Kent, Surrey and Berkshire before 1970 and Isle of Wight, North Hampshire, West Sussex, East Sussex, East Kent and West Kent from 1970 onwards.

Habitat and ecology Grassland, woodland rides and coastal shingle. Almost exclusively on chalk. Females are larviform and apterous. Larvae, and probably adults, are predatory on snails (Mollusca), particularly *Helicella itala, Cernuella virgata, Candidula intersecta, Monacha cantiana, Trichia striolata, Oxychilus cellarius, Cepaea nemoralis* and *Helix aspersa*. The preferred host snail species may vary with the distribution of the beetle. The larva feeds during the summer by invading and eating successively larger host snails, and may take two seasons to become fully grown. Three or four snails are usually devoured in a season, the larva moulting within the empty shell. By about mid September the larva stops feeding, moults and remains within its host-shell to pass the winter. Pupation and adult emergence occurs around the following May. Adults have been recorded from May to July.

Status Local in southern England from the Isle of Wight to East Kent.

Threats Grassland improvement and conversion to arable agriculture cause the main losses of calcareous grassland. Changes in grazing or cutting regimes may also adversely affect this species. Coastal shingle is threatened by shingle extraction and disturbance through practices such as motorbike riding and trampling.

Management and conservation Grazing or cutting, on a rotational basis, is needed to maintain open conditions on downland sites. Disturbance of coastal shingle sites should be avoided.

Published sources Alexander, K.N.A. & Clements, D.K. (1988), Crawshay, L.R. (1903), Fowler, W.W. (1890).

ELATERIDAE

ADRASTUS RACHIFER   RARE
A click beetle
Order COLEOPTERA   Family ELATERIDAE

Adrastus rachifer (Fourcroy, 1785). Formerly known
as: *Adrastus pusillus* (F.).

Distribution Recorded from East Kent before 1970 and
East Kent from 1970 onwards. A map is given in
Mendel (1988).

Habitat and ecology Grassland, primarily calcareous.
Biology uncertain. Adults have been found in August
by beating hedges and by sweeping grass on a cliff-top.

Status Extremely local, being known only from the
east coast and neighbouring areas of East Kent. In
certain localities this species may be quite common.
This species is widespread on the Continent.

Threats Loss of calcareous downland either by
improvement through reseeding or fertilisation or
through conversion to arable agriculture or other land
use. Lack of grazing or changes in grazing regimes
could alter the vegetation structure which in turn may
affect this species.

Management and conservation Grazing or cutting, on
a rotational basis, is needed to maintain open
conditions.

Published sources Fowler, W.W. (1890), Fowler,
W.W. & Donisthorpe, H.St J.K. (1913), Mendel, H.
(1988a).

AGRIOTES SORDIDUS   RARE
A click beetle
Order COLEOPTERA   Family ELATERIDAE

Agriotes sordidus (Illiger, 1807).

Distribution Recorded from South Devon, Dorset, Isle
of Wight, South Hampshire, East Kent, West Kent,
Surrey, South Essex, North Essex, Middlesex,
Pembrokeshire and West Lancashire before 1970 and
Isle of Wight and Berkshire from 1970 onwards. A
map is given in Mendel (1988).

Habitat and ecology Banks of tidal rivers, coastal and
estuarine habitats, including under stones on a beach.
Biology unknown, though the larvae probably develop
in the soil at the roots of plants. Adults have been
recorded from April to July.

Status Very local and possibly declining. Records are
centred around the Isle of Wight, the Thames estuary
and Essex coast, with scattered records to south Wales.

Threats Loss of natural and semi-natural coastal and
estuarine habitat due the building of sea defences and
industrial and urban development. Much remaining
habitat has been degraded by recreational pressures.

Management and conservation Management should
aim to maintain the tidal influence and open conditions.

Published sources Atty, D.B. (1983), Fowler, W.W.
(1890), Mendel, H. (1988a).

AMPEDUS CARDINALIS   VULNERABLE
A click beetle
Order COLEOPTERA   Family ELATERIDAE

Ampedus cardinalis (Schiödte, 1865). Formerly known
as: *Elater coccinatus* Rye, 1867, *Elater praeustus* sensu
auct. Brit. partim not F., 1792.

Distribution Recorded from West Sussex, South Essex,
Middlesex, Berkshire, Herefordshire and
Nottinghamshire before 1970 and West Sussex, West
Kent, Surrey, South Essex, Berkshire, Herefordshire
and Nottinghamshire from 1970 onwards. A map is
given in Mendel (1988).

Habitat and ecology Ancient broad-leaved woodland
and pasture-woodland. Appears to be exclusively
associated with oak, breeding predominantly in
red-rotten wood. Adults have been recorded from May
to July, and have been found in their pupal cells from
September to April.

Status Associated with a declining habitat and known
from very few sites.

Indicator status Grade 1 in Harding & Rose (1986).
Listed in Speight (1989) Appendix 1.

Threats Loss of broad-leaved woodland and parkland
through, for example, clear-felling and coniferisation.
Habitat loss, in particular, through the felling of ancient
trees, removal of dead wood from living trees and the
destruction or removal of standing and fallen dead
wood for reasons such as forest hygiene, aesthetic
tidiness, public safety or for use as fire wood.

Management and conservation Ancient trees and both
fallen and standing dead timber, especially with the
bark attached, should be retained. The removal of dead
timber from ancient trees should be avoided. Gaps in
the age structure of the tree population should be
identified and the continuity of the appropriate dead
wood habitat ensured by suitable planting and possibly
with pollarding.

Published sources Allen, A.A. (1966), Allen, A.A. (1990b), Copestake, D.R. (1990a), Fowler, W.W. & Donisthorpe, H.St J.K. (1913), Garland, S.P. (1983), Hammond, P.M. (1979), Harding, P.T. (1978), Harding, P.T. & Rose, F. (1986), Mendel, H. (1988a), Shirt, D.B., ed. (1987), Speight, M.C.D. (1989).

AMPEDUS CINNABARINUS   **RARE**
A click beetle
Order COLEOPTERA   Family ELATERIDAE

Ampedus cinnabarinus (Eschsholtz, 1829). Formerly known as: *Elater cinnabarinus* Eschscholtz, *Elater lythropterus* (Germar, 1844).

Distribution Recorded from North Devon, South Wiltshire, South Hampshire, West Sussex, Surrey, South Essex, East Gloucestershire, West Gloucestershire, Glamorgan, Merionethshire and Nottinghamshire before 1970 and South Hampshire, West Sussex, West Gloucestershire and Monmouthshire from 1970 onwards. A map is given in Mendel (1988).

Habitat and ecology Ancient broad-leaved woodland, though also recorded from alder carr. Associated with beech, oak, birch, alder and occasionally pine. Probably breeds in other trees. Larvae develop in rotten wood. Adults have been recorded from March to July and also in September, though the main period of adult emergence is probably from mid April to mid May. Eclosion occurs about late August.

Status Possibly declining with few recent records. Centres of distribution are the woodlands on the West Sussex and Hampshire county boundaries; the New Forest, South Hampshire; and the Forest of Dean and Wye Valley woodlands.

Indicator status Grade 1 in Harding & Rose (1986).

Threats Loss of broad-leaved woodland through, for example, clear-felling and coniferisation. Habitat loss, in particular, through the felling of ancient trees, removal of dead wood from living trees and the destruction or removal of standing and fallen dead wood for reasons such as forest hygiene, aesthetic tidiness, public safety or for use as fire wood.

Management and conservation Ancient trees and both fallen and standing dead timber, especially with the bark attached, should be retained. The removal of dead timber from ancient trees should be avoided. Gaps in the age structure of the tree population should be identified and the continuity of the appropriate dead wood habitat ensured by suitable planting and possibly with pollarding.

Published sources Allen, A.A. (1966), Allen, A.A. (1990b), Atty, D.B. (1983), Black, J.E. (1924), Cooter, J. (1970), Fowler, W.W. (1890), Garland, S.P. (1983), Harding, P.T. (1978), Harding, P.T. & Rose, F. (1986), Horton, G.A.N. (1980), Horton, G.A.N. (1989), Mendel, H. (1988a), Mendel, H. (1990b), Moore, D. (1989), Shirt, D.B., ed. (1987), Walker, J.J. (1927), Walker, J.J. (1932c).

AMPEDUS ELONGANTULUS   **NOTABLE A**
A click beetle
Order COLEOPTERA   Family ELATERIDAE

Ampedus elongantulus (F., 1787). Formerly known as: *Elater elongatulus* (F., 1787).

Distribution Recorded from East Cornwall, South Hampshire, West Sussex, West Kent, Surrey, Berkshire, Buckinghamshire and Northamptonshire before 1970 and North Somerset, North Wiltshire, South Wiltshire, Isle of Wight, South Hampshire, North Hampshire, West Sussex, East Sussex, East Kent, Surrey, Berkshire and Oxfordshire from 1970 onwards. A map is given in Mendel (1988).

Habitat and ecology Ancient broad-leaved woodland. Associated with oak, beech and pine. Larvae are found in dead wood, mainly red-rotten stumps. Adults have been recorded from March to July, with the main period of adult emergence probably occurring from mid April to mid May.

Status Very local, with little evidence of a decline. Widely distributed in southern and south-eastern England.

Indicator status Grade 3 in Harding & Rose (1986).

Threats Loss of broad-leaved woodland through, for example, clear-felling and coniferisation. Habitat loss, in particular, through the felling of ancient trees, removal of dead wood from living trees and the destruction or removal of stumps, standing and fallen dead wood for reasons such as forest hygiene, aesthetic tidiness, public safety or for use as fire wood.

Management and conservation Ancient trees and both fallen and standing dead timber, especially with the bark attached, should be retained. The removal of dead timber from ancient trees should be avoided. Gaps in the age structure of the tree population should be identified and the continuity of the appropriate dead wood habitat ensured by suitable planting and possibly with pollarding.

Published sources Allen, A.A. (1966), Fowler, W.W. (1890), Fowler, W.W. & Donisthorpe, H.St J.K. (1913),

ELATERIDAE

Harding, P.T. (1978), Harding, P.T. & Rose, F. (1986), Mendel, H. (1988a), Osborne, P.J. (1951).

AMPEDUS NIGERRIMUS　　　　**ENDANGERED**
A click beetle
Order COLEOPTERA　　　　Family ELATERIDAE

Ampedus nigerrimus (Lacordaire, 1835). Formerly known as: *Elater aethiops* sensu Fowler, 1890 not Boisduval and Lacordaire, 1835, *Elater nigerrimus* Lacordaire.

Distribution Recorded from Berkshire before 1970 and Berkshire from 1970 onwards.

Habitat and ecology Ancient broad-leaved woodland and pasture-woodland. Breeds exclusively in decayed oak, particularly in red-rotten wood of trunks, logs, large boughs and stumps. Adults have been recorded in March, April, June and October. The main period of adult emergence is probably during April and May.

Status Only known from Windsor Forest, Berkshire. Records listed for other British localities have yet to be confirmed.

Indicator status Grade 1 in Harding & Rose (1986). Listed in Speight (1989) Appendix 1.

Threats Loss of broad-leaved woodland and parkland through, for example, clear-felling and coniferisation. Habitat loss, in particular, through the felling of ancient trees, removal of dead wood from living trees and the destruction or removal of stumps, standing and fallen dead wood for reasons such as forest hygiene, aesthetic tidiness, public safety or for use as fire wood.

Management and conservation Ancient trees and both fallen and standing dead timber, especially with the bark attached, should be retained. The removal of dead timber from ancient trees should be avoided. Gaps in the age structure of the tree population should be identified and the continuity of the appropriate dead wood habitat ensured by suitable planting and possibly with pollarding. Windsor Forest is notified as an SSSI.

Published sources Allen, A.A. (1966), Harding, P.T. (1978), Harding, P.T. & Rose, F. (1986), Mendel, H. (1988a), Shirt, D.B., ed. (1987), Speight, M.C.D. (1989).

AMPEDUS NIGRINUS　　　　**NOTABLE B**
A click beetle
Order COLEOPTERA　　　　Family ELATERIDAE

Ampedus nigrinus (Herbst, 1784). Formerly known as: *Elater nigrinus* Herbst.

Distribution West Midlands, North East, North West, South East Scotland, North East Scotland, North West Scotland, South West Scotland and Dyfed-Powys. A map is given in Mendel (1988).

Habitat and ecology Woodland. Primarily associated with Scots pine. On the Continent, this species is known to be associated with hardwoods, this may also be the case in this country. The larvae occur in dead wood. Adults have been recorded in May, June and August, though the main period of adult emergence is probably in May and June.

Status Very local. Predominantly Scottish, its distribution being centred on areas of native pine forest. There are scattered records in England, from as far south as the Forest of Dean, Worcestershire.

Threats Loss of native pine forest through, for example, clear-felling or conversion to plantation forest. Habitat loss in the remaining areas through the felling of ancient trees and the destruction or removal of standing and fallen dead wood for reasons such as forest hygiene, aesthetic tidiness, public safety or for use as fire wood.

Management and conservation Ancient trees and both standing and fallen dead timber, especially with the bark attached, should be retained. Gaps in the age structure of the tree population should be identified and regeneration encouraged to ensure the continuity of dead wood habitats.

AMPEDUS POMORUM　　　　**NOTABLE B**
A click beetle
Order COLEOPTERA　　　　Family ELATERIDAE

Ampedus pomorum (Herbst, 1784). Formerly known as: *Elater ferrugatus* Boisduval and Lacordaire, 1835, *Elater pomorum* Herbst.

Distribution East Midlands, West Midlands, North East, North West and Scotland. A map is given in Mendel (1988).

Habitat and ecology Ancient broad-leaved woodland and birch woodland on bogs. Associated mainly with birch, though also with oak, ash and possibly on other tree species. Larvae develop in decayed, red-rotten wood. Adults have been recorded in February and from

April to June, with the main period of adult emergence probably occurring from mid April to mid May.

Status Very local. A widely scattered distribution from the Forest of Dean, Worcestershire through northern England to northern and western Scotland.

Indicator status Grade 1 in Garland (1983). Grade 3 in Harding & Rose (1986).

Threats Loss of broad-leaved woodland through, for example, clear-felling and coniferisation. Habitat loss, in particular, through the felling of ancient trees, removal of dead wood from living trees and the destruction or removal of standing and fallen dead wood for reasons such as forest hygiene, aesthetic tidiness, public safety or for use as fire wood.

Management and conservation Ancient trees and both fallen and standing dead timber, especially with the bark attached, should be retained. The removal of dead timber from ancient trees should be avoided. Gaps in the age structure of the tree population should be identified and the continuity of the appropriate dead wood habitat ensured by suitable planting and possibly with pollarding.

AMPEDUS QUERCICOLA　　　　　**NOTABLE B**
A click beetle
Order COLEOPTERA　　　　　Family ELATERIDAE

Ampedus quercicola (du Buysson, 1887). Formerly known as: *Ampedus pomonae* (Stephens, 1830), *Elater miniatus* Gorham, 1892.

Distribution South East, South, East Anglia and East Midlands. A map is given in Mendel (1988).

Habitat and ecology Ancient broad-leaved woodland and birch woodland on fenland sites. Recorded from ash, beech, birch and oak, but probably associated with a number of other broad-leaved trees. A single example has been reared from spruce. The larvae occur in dead wood. Adults have been recorded in February, April, May and September with the main period of adult emergence probably occurring from mid April to mid May.

Status Very local. The main centres of distribution are the New Forest, Hampshire and parts of the East Midlands. This species has been reported from west Wales, West Midlands and South Scotland, there are, as yet, no confirmed records from any of these areas.

Indicator status Grade 1 in Harding & Rose (1986).

Threats Loss of broad-leaved woodland and parkland through practices, for example, clear-felling and coniferisation. Habitat loss, in particular, through the felling of ancient trees, removal of dead wood from living trees and the destruction or removal of standing and fallen dead wood for reasons such as forest hygiene, aesthetic tidiness, public safety or for use as fire wood.

Management and conservation Ancient trees and both fallen and standing dead timber, especially with the bark attached, should be retained. The removal of dead timber from ancient trees should be avoided. Gaps in the age structure of the tree population should be identified and the continuity of the appropriate dead wood habitat ensured by suitable planting and possibly with pollarding.

AMPEDUS RUFICEPS　　　　　**ENDANGERED**
A click beetle
Order COLEOPTERA　　　　　Family ELATERIDAE

Ampedus ruficeps (Mulsant and Guillebeau, 1855). Formerly known as: *Elater ruficeps* Mulsant and Guillebeau.

Distribution Recorded from Berkshire before 1970 and Berkshire from 1970 onwards.

Habitat and ecology Ancient broad-leaved woodland and pasture-woodland. In Britain, larvae and adults have been recorded from damp wood mould and red-rotten wood in cavities in decaying ancient oaks. On the Continent, the beetle is also associated with beech. Adults have been recorded in April and May.

Status Only known in Britain from Windsor Great Park, Berkshire. The beetle was first discovered in 1938 from a single adult in wood mould in a cavity high in an old oak. The species was not rediscovered until 1986 when adults and larvae were found, again in wood mould of decaying oak. Another colony of the beetle has since been discovered in 1987, also in wood mould in an ancient oak. The original tree in which this species was first discovered no longer exists.

Indicator status Grade 1 in Harding & Rose (1986). Listed in Speight (1989) Appendix 1.

Threats Loss of broad-leaved woodland and parkland through, for example, clear-felling and coniferisation. Habitat loss, in particular, through the felling of ancient trees, removal of dead wood from living trees and the destruction or removal of standing and fallen dead wood for reasons such as forest hygiene, aesthetic tidiness, public safety or for use as fire wood.

ELATERIDAE

Management and conservation Ancient trees and standing dead timber, especially with the bark attached, should be retained. The removal of dead timber from ancient trees should be avoided. Gaps in the age structure of the tree population should be identified and the continuity of the appropriate dead wood habitat ensured by suitable planting and possibly with pollarding. Parts of Windsor Great Park are notified as an SSSI.

Published sources Allen, A.A. (1966), Harding, P.T. (1978), Harding, P.T. & Rose, F. (1986), Mendel, H. (1988a), Porter, D.A. (1989), Shirt, D.B., ed. (1987), Speight, M.C.D. (1989).

AMPEDUS RUFIPENNIS　　　　　**VULNERABLE**
A click beetle
Order COLEOPTERA　　　　　Family ELATERIDAE

Ampedus rufipennis (Stephens, 1830). Formerly known as: *Elater rufipennis* Stephens.

Distribution Recorded from East Kent, Surrey, Berkshire, East Norfolk and Herefordshire before 1970 and Berkshire, Herefordshire and Worcestershire from 1970 onwards. A map is given in Mendel (1988).

Habitat and ecology Ancient broad-leaved woodland and pasture-woodland. Associated predominantly with beech, but occasionally with elm, birch, ash, oak and sycamore. Larvae develop in rotten wood in trunks, logs and boughs, and more rarely stumps. Adults have been recorded in April, May and September, though the main period of adult emergence is probably from mid April to mid May.

Status Extremely local, known from very few scattered localities where its occurrence is erratic. Recent records include Windsor Forest and Windsor Great Park, Berkshire, Moccas Park, Herefordshire and Bredon Hill, Worcestershire.

Indicator status Grade 1 in Harding & Rose (1986).

Threats Loss of broad-leaved woodland and parkland through, for example, clear-felling and coniferisation. Habitat loss, in particular, through the felling of ancient trees, removal of dead wood from living trees and the destruction or removal of stumps, standing and fallen dead wood for reasons such as forest hygiene, aesthetic tidiness, public safety or for use as fire wood.

Management and conservation Ancient trees and both fallen and standing dead timber, especially with the bark attached, should be retained. The removal of dead timber from ancient trees should be avoided. Gaps in the age structure of the tree population should be

identified and the continuity of the appropriate dead wood habitat ensured by suitable planting and possibly with pollarding. Windsor Forest and parts of Windsor Great Park are notified as SSSIs, Moccas Park and Bredon Hill are NNRs.

Published sources Allen, A.A. (1966), Allen, A.A. (1990b), Donisthorpe, H.St J.K. (1925d), Harding, P.T. (1978), Harding, P.T. & Rose, F. (1986), Mendel, H. (1988a), Moore, D. (1989), Shirt, D.B., ed. (1987).

AMPEDUS SANGUINEUS　　　　　**EXTINCT**
A click beetle
Order COLEOPTERA　　　　　Family ELATERIDAE

Ampedus sanguineus (L., 1758). Formerly known as: *Elater sanguineus* L.

Distribution Recorded from South Wiltshire and South Hampshire.

Habitat and ecology Probably associated with dead wood. On the Continent mainly, but not exclusively, found in coniferous wood, including heaps of sawdust.

Status Presumed extinct. Last recorded in 1830 from the New Forest, South Hampshire. Stephens' record of *A. sanguineus* from Bagley Wood, Berkshire should be deleted as it refers to *A. rufipennis* (A. A. Allen, pers.comm.).

Published sources Allen, A.A. (1966), Allen, A.A. (1990b), Mendel, H. (1988a).

AMPEDUS SANGUINOLENTUS　　　**NOTABLE A**
A click beetle
Order COLEOPTERA　　　　　Family ELATERIDAE

Ampedus sanguinolentus (Schrank, 1776). Formerly known as: *Elater sanguinolentus* Schrank.

Distribution Recorded from South Somerset, South Wiltshire, South Hampshire, North Hampshire, West Kent, Surrey, South Essex, Berkshire, Oxfordshire, East Suffolk, Worcestershire and Caernarvonshire before 1970 and North Somerset, Dorset, South Hampshire, West Sussex, East Sussex and Surrey from 1970 onwards. A map is given in Mendel (1988).

Habitat and ecology Heathland and woodland on acid soils. Associated with birch, though possibly also on other tree species such as pine and sallow. Larvae develop in dead wood, particularly stumps. Adults have been recorded under bark, in stumps, from pine branches, from gorse bushes and at the roots of heather. A single adult has been found under a stone.

They may also visit flowers such as hawthorn blossom. Adults have been found in March and from May to August.

Status Local. Primarily southern, its distribution being centred on the New Forest, Hampshire.

Threats Loss of broad-leaved woodland through, for example, clear-felling and coniferisation. Loss of heathland or fragmentation through changes in land use, such as conversion to arable agriculture, forestry and urban development. Habitat loss, in particular, through the removal of dead wood from living trees and the destruction or removal of stumps, standing and fallen dead wood for reasons such as forest hygiene, aesthetic tidiness, public safety or for use as fire wood.

Management and conservation Birch stumps, fallen and standing dead timber, especially with the bark attached, should be retained. Gaps in the age structure of the tree population should be identified and the continuity of the appropriate dead wood habitat ensured. The presence of nectar sources such as hawthorn, umbellifers and composite herbs may also be particularly important for this species. Much of the New Forest is notified as an SSSI.

Published sources Allen, A.A. (1966), Fowler, W.W. (1890), Fowler, W.W. & Donisthorpe, H.St J.K. (1913), Mendel, H. (1988a), Mendel, H. (1990b).

AMPEDUS TRISTIS	**VULNERABLE**
A click beetle	
Order COLEOPTERA	Family ELATERIDAE

Ampedus tristis (L., 1758). Formerly known as: *Elater tristis* L.

Distribution Recorded from Mid Perthshire, Moray and East Inverness & Nairn before 1970 and South Aberdeenshire from 1970 onwards. A map is given in Mendel (1988).

Habitat and ecology Coniferous woodland. Associated with Scots pine, spruce and doubtfully or very rarely with birch. Larvae occur in dead wood. Adults have mainly been found in decaying spruce logs (though apparently not in stumps) and in Scots pine and birch. Adults have been recorded from May to July and in October. The main period of adult emergence is probably in May and June.

Status Status revised from RDB 3 (Rare) in Shirt (1987). Extremely local with recent records only from South Aberdeenshire. Exclusively Scottish and probably a component of the native pine forest fauna.

Threats Loss of native pine forest through, for example, clear-felling and conversion to plantation forest. Habitat loss in the remaining areas through the felling of ancient trees and the destruction or removal of standing and fallen dead wood for reasons such as forest hygiene, aesthetic tidiness, public safety or for use as fire wood.

Management and conservation Ancient trees and both standing and fallen dead timber, especially with the bark attached, should be retained. Gaps in the age structure of the tree population should be identified and regeneration encouraged to ensure the continuity of the dead wood habitats.

Published sources Fergusson, A. (1935), Fowler, W.W. (1890), Harwood, P. (1944), Mendel, H. (1988a), Shirt, D.B., ed. (1987).

ANOSTIRUS CASTANEUS	**ENDANGERED**
A click beetle	
Order COLEOPTERA	Family ELATERIDAE

Anostirus castaneus (L., 1758). Formerly known as: *Corymbites castaneus* (L.).

Distribution Recorded from Isle of Wight, East Norfolk, West Gloucestershire, Monmouthshire, Mid-west Yorkshire and Durham before 1970 and Isle of Wight and Mid-west Yorkshire from 1970 onwards. A map is given in Mendel (1988).

Habitat and ecology Grassland and heathland. This species has also been recorded from sparsely vegetated sand at the base of a coastal cliff. There are several coastal records, though the species has also been found inland. The beetle breeds in the soil at the roots of plants. Larvae are predatory at the roots of isolated tufts of grass and in sand virtually devoid of plant material. Adults have been found from March to May.

Status Extremely local, with only two modern sites. The species appears to be extremely localized even within a larger area of apparently suitable habitat. Old records suggest that the species was established on the Durham coast and in the Forest of Dean area, Monmouthshire and West Gloucestershire. It is possible that as yet undiscovered colonies still occur in these areas.

Threats Agricultural improvement through reseeding and the application of fertiliser, and urban and holiday developments. Changes in grazing regimes may also adversely affect this species.

Management and conservation Some degree of disturbance is desirable to maintain the early

ELATERIDAE

successional stages and open conditions and prevent the invasion of scrub.

Published sources Appleton, D. (1973b), Atty, D.B. (1983), Mendel, H. (1988a), Shirt, D.B., ed. (1987).

ATHOUS CAMPYLOIDES **NOTABLE B**
A click beetle
Order COLEOPTERA Family ELATERIDAE

Athous campyloides Newman, 1833. Formerly known as: *Athous difformis* Boisduval and Lacordaire, 1835, *Orthathous difformis* (Boisduval and Lacordaire).

Distribution South East, South, South West and East Midlands. A map is given in Mendel (1988).

Habitat and ecology Grassland, road-sides verges, gardens, coastal cliffs, a disused chalk pit and a disused clay pit. The beetle probably breeds in the soil at the roots of plants. The species is crepuscular and gregarious. Adults have been recorded from June to August.

Status Local in south-east England, but with a scatter of records to West Cornwall. A record for Mid-west Yorkshire requires confirmation. The beetle may be under-recorded due to its crepuscular habits.

Threats Uncertain, though improvement or conversion to arable agriculture cause the main loss of much unimproved grassland. Changes in grazing and cutting regimes can lead to a change in the vegetation structure which in turn may adversely affect this species.

Management and conservation Grazing or cutting, on a rotational basis, is needed to maintain open contions.

ATHOUS SUBFUSCUS **RARE**
A click beetle
Order COLEOPTERA Family ELATERIDAE

Athous subfuscus (Müller, 1764).

Distribution Recorded from Shetland before 1970 and Surrey, Orkney and Shetland from 1970 onwards. A map is given in Mendel (1988).

Habitat and ecology Grassland and heathland. Adults have been found under heather and have been swept from grass, bracken and low growing vegetation. The beetle breeds in the soil at the roots of plants. Adults have been recorded in June and July.

Status Formerly known only from the Orkney and Shetland Isles. The species has also recently been found to be locally common in a very few localities in Surrey where it is thought to have been either introduced or a recent colonist (H. Mendel pers.comm.). This beetle may possibly be spreading in southern England.

Threats Agricultural improvement through reseeding and the application of fertiliser, and conversion to arable cause the main loss of much unimproved grassland. Changes in grazing regimes can lead to a change in the vegetation structure in turn may adversely affect this species.

Management and conservation Grazing or cutting, on a rotational basis, is needed to maintain open conditions.

Published sources Fowler, W.W. (1890), Mendel, H. (1988a), Prance, D.A. (1985), Shirt, D.B., ed. (1987).

CARDIOPHORUS ASELLUS **NOTABLE B**
A click beetle
Order COLEOPTERA Family ELATERIDAE

Cardiophorus asellus Erichson, 1840.

Distribution South East, South, South West, East Anglia, East Midlands, North East, South Wales and North Wales. A map is given in Mendel (1988).

Habitat and ecology Sand dunes, particularly dunes supporting a scattering of marram and with significant amounts of moss and lichens. Also found on heathland and sandy areas inland. Adults, pupae and larvae are found in the soil. Adults have also been found by sweeping vegetation. The main period of adult activity is in May when egg laying probably takes place. Apart from a lack of records for June and July, larvae appear to be present in the soil all year round. Pupation occurs in late summer with the new generation of adults remaining in the ground until the following spring. Adults and larvae possibly feed on the roots of plants including heather.

Status Local, with a scattered distribution as far north as Anglesey in the west and Humberside in the east.

Threats Loss of dune and heathland habitat, particularly through changes in land use to arable agriculture and forestry, as well as through urban and holiday development. Degradation of habitat by a change in traditional management practices, excessive disturbance of the vegetation through activities such as motorbike access, horse-riding and human trampling.

Management and conservation On dunes some grazing and other disturbance may be desirable to

320

maintain early successional stages and prevent the invasion of scrub and rank grassland. Heathland management should aim to maintain a diversity of successional stages from bare ground to mature heathland, preferably by animal grazing or by cutting and scraping.

CARDIOPHORUS ERICHSONI   VULNERABLE
A click beetle
Order COLEOPTERA   Family ELATERIDAE

Cardiophorus erichsoni du Buysson, 1901. Formerly known as: *Cardiophorus rufipes* (Fourcroy).

Distribution Recorded from West Cornwall and North Devon before 1970 and West Cornwall, South Devon, North Devon and Pembrokeshire from 1970 onwards. A map is given in Mendel (1988).

Habitat and ecology Coastal habitats, including gorse heath on rocky cliffs. There are records for over 80 years from the same abandoned gardens on the Isle of Lundy, North Devon. The beetle probably breeds in the soil at the roots of plants. Adults have been recorded in June.

Status Not listed in the insect Red Data Book (Shirt, 1987). Extremely local. Apparently restricted to a very few coastal sites in south-western England and southern Wales.

Threats Agricultural improvement is one of the main causes of the loss of grassland. This species is further threatened by coastal developments. Changes in grazing regimes and alterations in the levels of gorse burning may also adversely affect this species.

Management and conservation Grazing or cutting, on a rotational basis, is needed to maintain open conditions. Where appropriate, traditional gorse burning management should be continued.

Published sources Alexander, K.N.A. (1987), Fowler, W.W. (1890), Mendel, H. (1988a), Payne, K.G. (1977).

CARDIOPHORUS GRAMINEUS   EXTINCT
A click beetle
Order COLEOPTERA   Family ELATERIDAE

Cardiophorus gramineus (Scopoli, 1763). Formerly known as: *Cardiophorus thoracicus* (F., 1775).

Distribution Recorded from "Somerset", South Wiltshire, South Hampshire, "London district", Berkshire, "Norfolk" and Cumberland.

Habitat and ecology Broad-leaved woodland. On the Continent, this species is associated with oak.

Status Presumed extinct. Last recorded between 1835 and 1845 from Windsor Forest, Berkshire.

Published sources Allen, A.A. (1966), Fowler, W.W. (1890), Mendel, H. (1988a), Shirt, D.B., ed. (1987).

CARDIOPHORUS RUFICOLLIS   EXTINCT
A click beetle
Order COLEOPTERA   Family ELATERIDAE

Cardiophorus ruficollis (L., 1758).

Distribution Recorded from "London district" and "Norfolk".

Habitat and ecology Broad-leaved woodland. Associated with oak.

Status Presumed extinct. Last recorded in 19th century. Records listed by Stephens (1827-1833) are from the 'neighbourhood' of London and from Norfolk.

Published sources Fowler, W.W. (1890), Mendel, H. (1988a), Shirt, D.B., ed. (1987).

CTENICERA PECTINICORNIS   NOTABLE A
A click beetle
Order COLEOPTERA   Family ELATERIDAE

Ctenicera pectinicornis (L., 1758). Formerly known as: *Corymbites pectinicornis* (L.).

Distribution Recorded from North Essex, Berkshire, Bedfordshire, West Gloucestershire, Monmouthshire, Worcestershire, Warwickshire, Staffordshire, Caernarvonshire, Denbighshire, North Lincolnshire, Leicestershire & Rutland, Nottinghamshire, Cheshire, South Lancashire, South-east Yorkshire, South-west Yorkshire, Mid-west Yorkshire, Durham, Westmorland & North Lancashire and Selkirkshire before 1970 and North Somerset, Warwickshire, Breconshire, Radnorshire, North-east Yorkshire, North Northumberland and Cumberland from 1970 onwards. A map is given in Medel (1988).

Habitat and ecology Lush grassland in old hay meadows. The beetle breeds in the soil at the roots of plants. Adults have been recorded in May and June.

Status Possibly declining. Very local, with a widely scattered distribution through central and northern England, Wales, to southern Scotland.

ELATERIDAE

Threats Loss of unimproved grassland through improvement by the application of fertilisers, reseeding, drainage or conversion to arable land.

Management and conservation Maintain traditional hay meadow management.

Published sources Anon, (1962), Atty, D.B. (1983), Carr, J.W. (1916), Denison-Roebuck, W. (1900), Eyre, M.D. & Sheppard, D.A. (1982), Fitter, A. & Smith, C.J. (1979), Fowler, W.W. (1890), Fowler, W.W. & Donisthorpe, H.St J.K. (1913), Mendel, H. (1988a).

DICRONYCHUS EQUISETI	VULNERABLE
A click beetle	
Order COLEOPTERA	Family ELATERIDAE

Dicronychus equiseti (Herbst, 1784). Formerly known as: *Cardiophorus equiseti* (Herbst).

Distribution Recorded from North Devon, Dorset, Glamorgan, Carmarthenshire and Pembrokeshire before 1970 and Glamorgan and Carmarthenshire from 1970 onwards. A map is given in Mendel (1988).

Habitat and ecology Sand dunes. Adults occur in the sand at the roots of plants, particularly marram. Adults have been recorded in April and May.

Status Not listed in the insect Red Data Book (Shirt, 1987). Restricted to a very few coastal sites in south-west England and south Wales.

Threats Loss of dune habitat, particularly through urban and holiday development. Degradation of habitat by excessive disturbance of the vegetation through activities such as motorbike access, horse-riding and human trampling.

Management and conservation On dunes some grazing and other disturbance may be desirable to maintain the early successional stages and prevent the invasion of scrub and rank grassland.

Published sources Cooter, J. (1990c), Fowler, W.W. (1890), Fowler, W.W. & Donisthorpe, H.St J.K. (1913), Mendel, H. (1988a).

ELATER FERRUGINEUS	ENDANGERED
A click beetle	
Order COLEOPTERA	Family ELATERIDAE

Elater ferrugineus L., 1758. Formerly known as: *Ludius ferrugineus* (L.).

Distribution Recorded from East Kent, West Kent, Surrey, Middlesex, Berkshire, East Suffolk, Cambridgeshire and Glamorgan before 1970 and Berkshire and East Suffolk from 1970 onwards. A map is given in Mendel (1988).

Habitat and ecology Ancient broad-leaved woodland and pasture-woodland. Breeds in decayed, rotten wood, red-rotten wood and mould in the trunks and boughs of old trees, chiefly elm, beech, ash and grey poplar. The larvae are often found in rot-holes where there has been a birds nest. Pupation occurs in late June or early July with the main period of adult emergence probably occurring in July and August. It is unlikely that the adults hibernate.

Status Extremely local. Formerly more widespread in southern England, with a record for south Wales. Regularly recorded this century only from Windsor Forest and Windsor Great Park, Berkshire, though recently (1987) discovered at a new locality in Suffolk (H. Mendel pers.comm.).

Indicator status Grade 1 in Harding & Rose (1986). Listed in Speight (1989) Appendix 1.

Threats Loss of broad-leaved woodland and parkland through, for example, clear-felling and coniferisation. Habitat loss, in particular, through the felling of ancient trees, removal of dead wood from living trees and the destruction or removal of standing and fallen dead wood for reasons such as forest hygiene, aesthetic tidiness, public safety or for use as fire wood. The widespread removal of dead elms in recent years following the epidemic of Dutch elm disease may have destroyed unknown colonies and also resulted in the loss of suitable breeding sites.

Management and conservation Ancient trees and both standing and fallen dead timber, especially with the bark attached, should be retained. Gaps in the age structure of the tree population should be identified and the continuity of the appropriate dead wood habitat ensured by suitable planting and possibly with pollarding. Windsor Forest and parts of Windsor Great Park are notified as SSSIs.

Published sources Allen, A.A. (1966), Donisthorpe, H.St J.K. (1945b), Fowler, W.W. (1890), Fowler, W.W. & Donisthorpe, H.St J.K. (1913), Harding, P.T. (1978), Harding, P.T. & Rose, F. (1986), Mendel, H. (1988a),

Mendel, H. (1989), Shirt, D.B., ed. (1987), Speight, M.C.D. (1989), Tyler, P.S. (1959), Van Emden, F.I. (1945), Verdcourt, B. (1983).

FLEUTIAUXELLUS MARITIMUS — NOTABLE A
A click beetle
Order COLEOPTERA — Family ELATERIDAE

Fleutiauxellus maritimus (Curtis, 1840). Formerly known as: *Cryptohypnus maritimus* (Curtis), *Hypnoidus maritimus* (Curtis).

Distribution Recorded from Monmouthshire, Breconshire, Caernarvonshire, Denbighshire, West Lancashire, Mid-west Yorkshire, North-west Yorkshire, Cumberland, Dumfriesshire, Ayrshire, Lanarkshire, South Aberdeenshire, East Inverness & Nairn and Argyll Main before 1970 and Radnorshire, Carmarthenshire, Cardiganshire, South-east Yorkshire, South Northumberland, Dumfriesshire, Selkirkshire, Moray, Dunbartonshire and West Ross from 1970 onwards. A map is given in Mendel (1988).

Habitat and ecology River shingle and lakeside shingle. Adults and larvae live in open areas of river shingle where there are very few or no plants. Adults are found under larger pebbles, among finer particles and on the surface of the shingle. Larvae occur among the shingle and may be predatory. Adults have been recorded from May and June.

Status A northern and western species. Very local, though possibly under-recorded.

Threats River engineering, including dredging, level regulation by damming and flood alleviation. In some areas colonisation by Himalayan balsam *Impatiens glandulifera* can reduce the available habitat. Livestock access can damage the shingle structure. River pollution is a further threat.

Management and conservation An under-valued habitat. Shingle tends to be mobile and relies on the free-flow of river and stream systems. Activities that hinder this flow should be prevented.

Published sources Boyce, D.C. & Fowles, A.P. (1988), Cooper, B.A. (1947), Day, F.H. (1909), Fowler, W.W. (1890), Fowler, W.W. & Donisthorpe, H.St J.K. (1913), Fowles, A.P. (1989), Mendel, H. (1988a), Morgan, I.K. (1988), Morgan, I.K. (1990).

FLEUTIAUXELLUS QUADRIPUSTULATUS
A click beetle — NOTABLE A
Order COLEOPTERA — Family ELATERIDAE

Fleutiauxellus quadripustulatus (F., 1792). Formerly known as: *Cryptohypnus quadripustulatus* (F.), *Hypnoidus quadripustulatus* (F.).

Distribution Recorded from North Devon, West Kent, Surrey, South Essex, North Essex, Hertfordshire, Middlesex, Berkshire, Oxfordshire, East Suffolk, Cambridgeshire, Bedfordshire, East Gloucestershire, West Gloucestershire, Worcestershire, Warwickshire, Staffordshire, Shropshire, Caernarvonshire, North Lincolnshire, Leicestershire & Rutland, Nottinghamshire, Derbyshire, West Lancashire, South-east Yorkshire, North-east Yorkshire, South-west Yorkshire, North-west Yorkshire, South Northumberland and North Northumberland before 1970 and West Sussex, Oxfordshire, East Norfolk, East Gloucestershire, Herefordshire, Worcestershire, Warwickshire and Leicestershire & Rutland from 1970 onwards. A map is given in Mendel (1988).

Habitat and ecology Fields and field margins by water courses, fens and probably other wetland habitats. Larvae have been recorded in abundance from grass fields. The larvae probably develop and pupate in the soil at the roots of plants. Adults have been recorded from April to August and in late winter. Pupation occurs from mid March to mid May with a new generation of adults emerging in April and May.

Status Declining. Very local, with a widely scattered distribution from southern England through to North Northumberland.

Threats Agricultural improvement through reseeding and the application of fertilisers, and conversion to arable cause the main losses of unimproved grazing land. Falling water tables due to river improvements make the long term future of many remaining water meadows and fenland pastures uncertain. Scrub invasion or changes in grazing regimes could adversely affect this species.

Management and conservation Water tables should be maintained at high levels. Water bodies should be isolated from sources of eutrophication and pollution. Grazing or cutting, on a rotational basis, is needed to maintain open conditions.

Published sources Atty, D.B. (1983), Fowler, W.W. (1890), Fowler, W.W. & Donisthorpe, H.St J.K. (1913), McClenaghan, I. (1989), Mendel, H. (1988a), Roebuck, A. & Broadbent, L. (1945).

ELATERIDAE

HARMINIUS UNDULATUS NOTABLE B
A click beetle
Order COLEOPTERA   Family ELATERIDAE

Harminius undulatus (Degeer, 1774). Formerly known as: *Athous undulatus* (Degeer).

Distribution North East and Scotland. A map is given in Mendel (1988).

Habitat and ecology Birch woodland. Larvae are usually found in fallen birch trunks living under the bark or in the subcortical wood. They take 4-5 years to develop. Pupation is in spring or early summer. Adults are mature about three weeks later and have been found from May to July. They are very seldomly found outside the tree.

Status Status revised from RDB 3 (Rare) in Shirt (1987). Local. Not as rare as formerly thought. Predominantly Scottish, this species has also been reported in northern England.

Threats Loss of birch woodland through practices such as clear-felling, conversion to plantation forest and through lack of regeneration due to overgrazing. Habitat loss in the remaining areas through the felling of trees and the destruction or removal of standing and fallen dead wood for reasons such as forest hygiene, aesthetic tidiness, public safety or for use as fire wood.

Management and conservation Both standing and fallen dead timber, especially with the bark attached, should be retained. Gaps in the age structure of the tree population should be identified and regeneration encouraged to ensure the continuity of the dead wood habitats.

ISCHNODES SANGUINICOLLIS NOTABLE A
A click beetle
Order COLEOPTERA   Family ELATERIDAE

Ischnodes sanguinicollis (Panzer, 1793).

Distribution Recorded from South Devon, South Somerset, North Somerset, West Sussex, East Kent, West Kent, Surrey, South Essex, North Essex, Hertfordshire, Middlesex, Berkshire, Oxfordshire, East Suffolk and Derbyshire before 1970 and North Somerset, South Hampshire, East Sussex, North Essex, Berkshire, East Suffolk, East Norfolk and Cambridgeshire from 1970 onwards. A map is given in Mendel (1988).

Habitat and ecology Ancient broad-leaved woodland and pasture-woodland. Appears to be mainly associated with ash and elm, though also found on other tree species, notably beech and occasionally oak. Remains have also been found under the bark of field maple. Larvae develop in soft, decaying wood and wood mould. Adults have been recorded in January, April to July, September and October. The main period of adult emergence is probably during April and May.

Status Widespread but local in southern and south-eastern England.

Indicator status Grade 2 in Harding & Rose (1986). Listed in Speight (1989) Appendix 1.

Threats Loss of broad-leaved woodland and parkland through, for example, clear-felling and coniferisation. Habitat loss, in particular, through the felling of ancient trees, removal of dead wood from living trees and the destruction or removal of standing and fallen dead wood for reasons such as forest hygiene, aesthetic tidiness, public safety or for use as fire wood.

Management and conservation Ancient trees and both fallen and standing dead timber, especially with the bark attached, should be retained. The removal of dead timber from ancient trees should be avoided. Gaps in the age structure of the tree population should be identified and the continuity of the appropriate dead wood habitat ensured by suitable planting and possibly with pollarding.

Published sources Allen, A.A. (1966), Forster, H.W. (1952a), Fowler, W.W. (1890), Fowler, W.W. & Donisthorpe, H.St J.K. (1913), Garland, S.P. (1983), Harding, P.T. (1978), Harding, P.T. & Rose, F. (1986), Imms, A.D. (1927), Mendel, H. (1988a), Mendel, H. & Owen, J.A. (1990), Speight, M.C.D. (1989), Wilson, W.A. (1958).

LACON QUERCEUS ENDANGERED
A click beetle
Order COLEOPTERA   Family ELATERIDAE

Lacon querceus (Herbst, 1784).

Distribution Recorded from Berkshire before 1970 and Berkshire from 1970 onwards.

Habitat and ecology Ancient broad-leaved woodland and pasture-woodland. Breeds exclusively in dry, flaky red-rotten oak in dead trunks (both standing and fallen) and large boughs, but apparently not in stumps. Often found in association with the larvae of *Mycetophagus piceus* (Mycetophagidae). Eclosion occurs about late August. Adults have been recorded from March to May, July and September with the main period of emergence probably occurring during March and April.

Status Only known from Windsor Forest and Windsor Great Park, Berkshire.

Indicator status Grade 1 in Harding & Rose (1986). Listed in Speight (1989) Appendix 1.

Threats This species is threatened by the felling of ancient trees, removal of dead wood from living trees and the destruction or removal of standing and fallen dead wood for reasons such as aesthetic tidiness, public safety or for use as fire wood.

Management and conservation Ancient trees and both fallen and standing dead timber, especially with the bark attached, should be retained. The removal of dead timber from ancient trees should be avoided. Gaps in the age structure of the tree population should be identified and the continuity of the appropriate dead wood habitat ensured by suitable planting and possibly with pollarding. Windsor Forest and parts of Windsor Geat Park are notified as SSSIs.

Published sources Allen, A.A. (1966), Harding, P.T. (1978), Harding, P.T. & Rose, F. (1986), Mendel, H. (1988a), Owen, J.A. (1990a), Shirt, D.B., ed. (1987), Speight, M.C.D. (1989).

LIMONISCUS VIOLACEUS　　　　**ENDANGERED**
Violet click beetle
Order COLEOPTERA　　　　Family ELATERIDAE

Limoniscus violaceus (Müller, 1821).

Distribution Recorded from Berkshire and East Gloucestershire before 1970 and Berkshire and Worcestershire from 1970 onwards. A map is given in Mendel (1988).

Habitat and ecology Ancient broad-leaved woodland and pasture-woodland. Appears to be primarily associated with beech, though the species has also been found in an ash stump. Larvae develop in a mixture of wood and leaf mould in hollow beech trees. They are predatory and possibly also feed on the remains of dead insects. Adults have been found in similar habitat. They have been recorded in April and May, and in their pupal chambers in February. Adults are thought to be primarily nocturnal. They have been noted at hawthorn blossom, though the time of day is unknown. Information from rearing in near natural conditions coupled to field observations suggest the following life cycle: The eggs are laid in fissures in the lining of the breeding cavity. After hatching the larvae descend to the substrate to begin a life amongst the leaf and wood mould. The larval stage lasts a little over two years, pupating in July or August of the second year. The pupal stage lasts only a matter of weeks, adults

remaining in the pupal chamber until the following spring.

Status Within Windsor Forest, Berkshire this beetle is apparently restricted to just three or four trees. Recently it was discovered in Worcestershire. An adult has been recorded in Windsor Great Park, Berkshire. An earlier record for Tewkesbury, assumed to be a site in East Gloucestershire may refer to one in Worcestershire. In 1988 *L.violaceus* was given protection by its addition to Schedule 5 of the Wildlife and Countryside Act 1981.

Indicator status Grade 1 in Harding & Rose (1986). Listed in Speight (1989) Appendix 1.

Threats Loss of broad-leaved woodland and parkland through, for example, clear-felling and coniferisation. Habitat loss, in particular, through the felling of ancient trees, removal of dead wood from living trees and the destruction or removal of standing and fallen dead wood for reasons such as forest hygiene, aesthetic tidiness, public safety or for use as fire wood. The species was first noted at its Windsor site in 1937. By 1966 the three or four trees in which it had been found had gone, due to felling. The species was refound in a single tree in 1972, and has now been noted from three or four trees. The felling of just a few trees could have serious consequences for the species continued existence in Britain.

Management and conservation Ancient trees and both fallen and standing dead timber, especially with the bark attached, should be retained. The removal of dead timber from ancient trees should be avoided. Gaps in the age structure of the tree population should be identified and the continuity of the appropriate dead wood habitat ensured by suitable planting and pollarding. The presence of nectar sources such as hawthorn may also be important for this species. Windsor Forest and parts of Windsor Great Park are notified as SSSIs. Whitten (1990), who was not aware of the recent discovery in Worcestershire, suggested translocation to suitable trees on other sites. Such an operation would risk damaging depletion of the existing colonies with no guarantee of success.

Published sources Allen, A.A. (1966), Harding, P.T. (1978), Harding, P.T. & Rose, F. (1986), Mendel, H. (1988a), Mendel, H. & Owen, J.A. (1990), Shirt, D.B., ed. (1987), Speight, M.C.D. (1989), Whitten, A.J. (1990).

ELATERIDAE

MEGAPENTHES LUGENS   ENDANGERED
A click beetle
Order COLEOPTERA   Family ELATERIDAE

Megapenthes lugens (Redtenbacher, 1842).

Distribution Recorded from South Hampshire, Surrey, South Essex, Middlesex and Berkshire before 1970 and South Hampshire and Berkshire from 1970 onwards. A map is given in Mendel (1988).

Habitat and ecology Ancient broad-leaved woodland and pasture-woodland. Breeds in decaying elm and also beech. The larvae feed in fairly hard, dry rotten wood. A larva was found in rotten wood lining a cavity in a large section of a fallen beech tree. The adults are more often found on flowers, particularly hawthorn blossom. A single adult has been found on holly and a further one on nettle flowers. Adults have been recorded in February (in pupal cells), March and May. The main period of adult emergence is probably from mid April to mid May.

Status Extremely local and possibly declining. Recent records emanate from Windsor Forest, Berkshire and the New Forest, Hampshire.

Indicator status Grade 1 in Harding & Rose (1986). Listed in Speight (1989) Appendix 1.

Threats Loss of broad-leaved woodland and parkland through, for example, clear-felling and coniferisation. Habitat loss, in particular, through the felling of ancient trees, removal of dead wood from living trees and the destruction or removal of standing and fallen dead wood for reasons such as forest hygiene, aesthetic tidiness, public safety or for use as fire wood. The widespread removal of dead elms in recent years following the epidemic of Dutch elm disease may well have destroyed unknown colonies, and certainly resulted in a loss of potential breeding sites.

Management and conservation Ancient trees and both fallen and standing dead timber, especially with the bark attached, should be retained. The removal of dead timber from ancient trees should be avoided. Gaps in the age structure of the tree population should be identified and the continuity of the appropriate dead wood habitat ensured by suitable planting and possibly with pollarding. The presence of nectar sources such as hawthorn may also be particularly important for this species. Windsor Forest and much of the New Forest are notified as SSSIs.

Published sources Allen, A.A. (1964b), Allen, A.A. (1966), Atty, D.B. (1983), Harding, P.T. (1978), Harding, P.T. & Rose, F. (1986), Mendel, H. (1988a),

Owen, J.A. (1990a), Shirt, D.B., ed. (1987), Speight, M.C.D. (1989).

MELANOTUS PUNCTOLINEATUS
A click beetle   **ENDANGERED**
Order COLEOPTERA   Family ELATERIDAE

Melanotus punctolineatus (Pelerin, 1829). Formerly known as: *Ectinus punctolineatus* (Pelerin).

Distribution Recorded from East Kent, Surrey, Middlesex and Glamorgan before 1970 and East Kent from 1970 onwards. A map is given in Mendel (1988).

Habitat and ecology Sand dunes and grassland, particularly coastal grassland on sandy soils. Larvae live in sand at the roots of marram. Adults have been found from April to June and in August, probably also occurring from autumn to spring.

Status Status revised from RDB 3 (Rare) in Shirt (1987). Reliably known this century only from coastal sites at Deal, Dover, Pegwell Bay and Littlestone, East Kent. Most recently recorded in 1986 from a sandpit at Deal, East Kent, this being the first record of this species since its occurrence at Littlestone, East Kent in 1950.

Threats Loss of dune and sandy grassland, particularly to urban and holiday development, as well as through improvement or conversion to arable agriculture. Degradation of habitat by excessive disturbance of the vegetation through activities such as motorbike access, horse-riding and human trampling. The Deal site for this species has recently been destroyed by in-filling.

Management and conservation On dunes and sandy grassland some grazing and other disturbance may be desirable to maintain the early successional stages and prevent the invasion of scrub.

Published sources Fowler, W.W. (1890), Mendel, H. (1988a), Owen, J.A. (1990a), Shirt, D.B., ed. (1987).

NEGASTRIUS PULCHELLUS   VULNERABLE
A click beetle
Order COLEOPTERA   Family ELATERIDAE

Negastrius pulchellus (L., 1761). Formerly known as: *Cryptohypnus pulchellus* (L.), *Hypnoidus pulchellus* (L.).

Distribution Recorded from East Inverness & Nairn before 1970 and Moray and Argyll Main from 1970 onwards. A map is given in Mendel (1988).

Habitat and ecology Adults are associated with river shingle and have been recorded from June to August. Larvae may be predatory amongst river shingle.

Status Status revised from RDB 3 (Rare) in Shirt (1987). Possibly declining. Exclusively Scottish and extremely local, though may be common where found.

Threats River engineering, including dredging, level regulation by damming and flood alleviation. Livestock access can damage shingle structure. River pollution is a further threat.

Management and conservation An under-valued habitat. Shingle tends to be mobile and relies on the free-flow of river and stream systems. Activities that hinder this flow should be prevented. Adjacent scrub and rank grassland may be important for the species to escape late summer floods.

Published sources Fowler, W.W. (1890), Fowler, W.W. & Donisthorpe, H.St J.K. (1913), Mendel, H. (1988a), Shirt, D.B., ed. (1987).

NEGASTRIUS SABULICOLA   VULNERABLE
A click beetle
Order COLEOPTERA   Family ELATERIDAE

Negastrius sabulicola (Boheman, 1853). Formerly known as: *Cryptohypnus sabulicola* (Boheman), *Hypnoidus sabulicola* (Boheman).

Distribution Recorded from West Gloucestershire, Herefordshire, Shropshire, North-west Yorkshire, North Northumberland, Cumberland and Dumfriesshire before 1970 and Herefordshire, Radnorshire and Carmarthenshire from 1970 onwards. A map is given in Mendel (1988).

Habitat and ecology River shingle. Adults, and probably larvae, live in open areas of river shingle where there are very few or no plants. Adults are found beneath larger pebbles, among finer particles and on the surface of the shingle. Larvae probably occur in the shingle and may be predatory. Adults have been recorded in April and June.

Status Status revised from RDB 3 (Rare) in Shirt (1987). A species with a primarily western distribution, possibly under-recorded.

Threats River engineering, including dredging, level regulation by damming and flood alleviation. In some areas colonisation by Himalayan balsam *Impatiens glandulifera* can reduce the available habitat. Livestock access can damage shingle structure. River pollution is a further threat.

Management and conservation An under-valued habitat. Shingle tends to be mobile and relies on the free-flow of river and stream systems. Activities that hinder this flow should be prevented.

Published sources Atty, D.B. (1983), Cooper, B.A. (1945), Day, F.H. (1909), Fowler, W.W. (1890), Mendel, H. (1988a), Shirt, D.B., ed. (1987), Tomlin, J.R.le B. (1949).

PROCRAERUS TIBIALIS   RARE
A click beetle
Order COLEOPTERA   Family ELATERIDAE

Procraerus tibialis (Boisduval and Lacordaire, 1835). Formerly known as: *Megapenthes tibialis* (Boisduval and Lacordaire).

Distribution Recorded from North Wiltshire, South Wiltshire, South Hampshire, West Sussex, West Kent, Surrey, South Essex, Hertfordshire, Middlesex, Berkshire, Buckinghamshire, East Suffolk, Herefordshire, Worcestershire, Leicestershire & Rutland and Nottinghamshire before 1970 and Surrey, South Essex, North Essex, Hertfordshire, Berkshire, West Suffolk, Northamptonshire, West Gloucestershire and Herefordshire from 1970 onwards. A map is given in Mendel (1988).

Habitat and ecology Ancient broad-leaved woodland and pasture-woodland. Larvae develop in dead wood in hollow and decayed oak, beech and hornbeam. An adult was recorded in dead wood from a hedgerow ash tree. Adults have been noted in March, May, June and July with the main period of adult emergence probably occurring in May.

Status Status revised from RDB 2 (Vulnerable) in Shirt (1987). Widely distributed, but local, in southern and central England.

Indicator status Grade 1 in Harding & Rose (1986).

Threats Loss of broad-leaved woodland and parkland through, for example, clear-felling and coniferisation. Habitat loss, in particular, through the felling of ancient trees, removal of dead wood from living trees and the destruction or removal of standing and fallen dead wood for reasons such as forest hygiene, aesthetic tidiness, public safety or for use as fire wood.

Management and conservation Ancient trees and both fallen and standing dead timber, especially with the bark attached, should be retained. The removal of dead timber from ancient trees should be avoided. Gaps in the age structure of the tree population should be identified and the continuity of the appropriate dead

ELATERIDAE

wood habitat ensured by suitable planting and possibly with pollarding.

Published sources Allen, A.A. (1955b), Allen, A.A. (1966), Allen, A.A. (1971a), Allen, A.A. (1981a), Cooter, J. (1978), Drane, A.B. (1985b), Foster, A.P. (1988), Garland, S.P. (1983), Harding, P.T. (1978), Harding, P.T. & Rose, F. (1986), Mendel, H. (1988a), Owen, J.A. (1990a), Shirt, D.B., ed. (1987).

SELATOSOMUS ANGUSTULUS   RARE
A click beetle
Order COLEOPTERA   Family ELATERIDAE

Selatosomus angustulus (Kiesenwetter, 1858). Formerly known as: *Corymbites angustulus* (Kiesenwetter).

Distribution Recorded from Shropshire and Montgomeryshire before 1970 and North Somerset, Shropshire, Radnorshire and Montgomeryshire from 1970 onwards. A map is given in Mendel (1988).

Habitat and ecology Lush grassland, particularly along river banks. The beetle probably breeds at the roots of grasses and other plants. Numerous adults have been found by sweeping riverside vegetation. Adults have been recorded in June.

Status Possibly under-recorded. The distribution of this species appears to be centred on central Wales and the Welsh border.

Threats Agricultural improvement through reseeding and the application of fertiliser. Falling water tables due to river improvements. Changes in grazing regimes may also adversely affect this species.

Management and conservation Water tables should be maintained. Grazing or cutting, on a rotational basis, is needed to maintain open conditions.

Published sources Hignett, J. (1940), Mendel, H. (1988a), Shirt, D.B., ed. (1987).

SELATOSOMUS BIPUSTULATUS   NOTABLE B
A click beetle
Order COLEOPTERA   Family ELATERIDAE

Selatosomus bipustulatus (L., 1767). Formerly known as: *Calambus bipustulatus* (L.), *Corymbites bipustulatus* (L.).

Distribution South East, South, South West, East Anglia, East Midlands, West Midlands and Dyfed-Powys. A map is given in Mendel (1988).

Habitat and ecology Broad-leaved woodland and pasture-woodland. Associated with oak, willow, alder, hawthorn and probably other tree species. Larvae develop in dead wood. Adults have been recorded from April to July.

Status Possibly declining. Widespread, but local, in southern and central England, also recorded in mid Wales.

Indicator status Grade 1 in Garland (1983). Grade 3 in Harding & Rose (1986).

Threats Loss of broad-leaved woodland and parkland through, for example, clear-felling and coniferisation. Habitat loss, in particular, through the felling of ancient trees, removal of dead wood from living trees and the destruction or removal of standing and fallen dead wood for reasons such as forest hygiene, aesthetic tidiness, public safety or for use as fire wood.

Management and conservation Ancient trees and both standing and fallen dead timber, especially with the bark attached, should be retained. Gaps in the age structure of the tree population should be identified and the continuity of the appropriate dead wood habitat ensured by suitable planting and possibly with pollarding.

SELATOSOMUS CRUCIATUS   EXTINCT
A click beetle
Order COLEOPTERA   Family ELATERIDAE

Selatosomus cruciatus (L., 1758).

Distribution Recorded from Berkshire.

Habitat and ecology Broad-leaved woodland. Biology unknown, though the larvae probably develop in the soil.

Status Presumed extinct. Last recorded in c.1840 from Windsor Forest, Berkshire.

Published sources Allen, A.A. (1966), Mendel, H. (1988a), Shirt, D.B., ed. (1987).

SELATOSOMUS IMPRESSUS — NOTABLE B
A click beetle
Order COLEOPTERA — Family ELATERIDAE

Selatosomus impressus (F., 1792). Formerly known as: *Corymbites impressus* (F.).

Distribution North East, North West, Scotland, West Midlands and North Wales. A map is given in Mendel (1988).

Habitat and ecology Woodland. Biology unknown, though the larvae probably develop in the soil. Adults have been found on pine, birch and oak. Adults have been recorded from June to August.

Status Widespread, but local. Scotland and northern England, with old records from north Wales.

Threats Uncertain, though it is possible that overgrazing of the woodland flora and changes in vegetation structure due to 20th century management practices and the neglect of woodlands may influence the distribution of this beetle.

SELATOSOMUS NIGRICORNIS — RARE
A click beetle
Order COLEOPTERA — Family ELATERIDAE

Selatosomus nigricornis (Panzer, 1799). Formerly known as: *Corymbites metallicus* (Paykull, 1800), *Corymbites nigricornis* (Panzer).

Distribution Recorded from "Somerset", South Hampshire, East Sussex, East Kent, West Kent, Surrey, South Essex, North Essex, Hertfordshire, Middlesex, Oxfordshire, East Suffolk, East Norfolk, Cambridgeshire, Bedfordshire, East Gloucestershire, West Gloucestershire, Glamorgan, South Lincolnshire, North Lincolnshire, Leicestershire & Rutland, Derbyshire, South-east Yorkshire, North-east Yorkshire, South-west Yorkshire and Mid-west Yorkshire before 1970 and South Wiltshire, South Hampshire, West Sussex, Huntingdonshire, East Gloucestershire, West Gloucestershire and Worcestershire from 1970 onwards. A map is given in Mendel (1988).

Habitat and ecology Wet broad-leaved woodland and wetland. Larvae apparently develop in waterlogged soil. Adults have been recorded from sallows in swampy places in woods, and also from oak, birch and from a beech log. Adults have been recorded in May and June.

Status Possibly declining and very local. This species has a scattered distibution through southern England to northern England with an old record in south Wales.

Threats Loss of wet broad-leaved woodland through, for example, clear-felling and conversion to other land use. The vegetation structure of many remaining woodlands has been considerably altered by 20th century management practices. The vegetation communities in the wetter areas of broad-leaved woodlands are likely to be adversely affected by nutrient enrichment, pollution and falling water tables, due to water abstraction and drainage.

Management and conservation Water tables should be maintained at a high level to ensure that the wetter areas do not dry out. Any management should take into account the value of wetland habitats within woodlands.

Published sources Atty, D.B. (1983), Fowler, W.W. (1890), Fowler, W.W. & Donisthorpe, H.St J.K. (1913), Mendel, H. (1988a), Welch, R.C. (1981).

SYNAPTUS FILIFORMIS — ENDANGERED
A click beetle
Order COLEOPTERA — Family ELATERIDAE

Synaptus filiformis (F., 1781).

Distribution Recorded from Surrey, East Gloucestershire, West Gloucestershire and Monmouthshire before 1970 and South Somerset from 1970 onwards. A map is given in Mendel (1988).

Habitat and ecology River banks and canal margins. Biology uncertain, though this beetle has been found by sweeping long grass under hedges and willows along river banks and canal margins. Larvae may develop in dead wood. Adults have been recorded in May and June.

Status Status revised from RDB 3 (Rare) in Shirt (1987). Known from very few records, the only modern record being from South Somerset.

Threats Dead wood may be an important part of this species' life cycle and changes in management affecting the supply of this habitat could affect the beetle's survival.

Management and conservation Gaps in the age structure of trees should be identified and the continuity of dead wood ensured by suitable pollarding or planting.

Published sources Atty, D.B. (1983), Fowler, W.W. (1890), Fowler, W.W. & Donisthorpe, H.St J.K. (1913), Mendel, H. (1988a), Payne, K.G. (1977), Shirt, D.B., ed. (1987).

ENDOMYCHIDAE

LYCOPERDINA BOVISTAE   RARE

Order COLEOPTERA   Family ENDOMYCHIDAE

Lycoperdina bovistae (F., 1792).

Distribution Recorded from South Devon, Isle of Wight, West Sussex, East Sussex, West Kent, Surrey, Berkshire, East Suffolk, "Norfolk", East Gloucestershire and Leicestershire & Rutland before 1970 and South Hampshire, East Sussex and West Norfolk from 1970 onwards.

Habitat and ecology Recorded from deep shade in a mixed woodland and from a number of other woodland sites, though probably also occurs in more open situations such as heathland, grassland and quarries. Associated with puff-ball fungi, with records from *Lycoperdon bovista*, and *L. perlatum*. Also recorded from *L. pyriforme* growing on timber. Adults have been recorded in January, September and October.

Status Not listed in the insect Red Data Book (Shirt, 1987). Very local. Old records suggest that this species was widely distributed in southern England and recorded as far north as Leicestershire. Recently recorded from just three vice-counties.

Threats Uncertain, though loss of habitat through conversion to other land use may threaten this species. Natural succession may be a further threat.

Management and conservation Grazing or cutting may be needed to maintain open conditions. In woodland, management should aim to retain a variety of vegetation structures.

Published sources Atty, D.B. (1983), Fowler, W.W. (1889), Fowler, W.W. & Donisthorpe, H.St J.K. (1913), Hooper, M. (1978).

LYCOPERDINA SUCCINCTA   VULNERABLE

Order COLEOPTERA   Family ENDOMYCHIDAE

Lycoperdina succincta (L., 1767).

Distribution Recorded from West Suffolk and West Norfolk before 1970 and West Suffolk from 1970 onwards.

Habitat and ecology Open, sandy ground in Breckland. The adults and larvae are found in various puff-ball fungi. This species has been recorded from *Lycoperdon gemmatum* and *L. caelatum*. On the Continent, this beetle has also been found in *Bovista nigrescens*. In Britain, larvae and pupae have been

noted at the beginning of May (adults emerging in early June) and larvae have also been found in October. Adults have been recorded from September to November and are known to overwinter in decaying puff-balls.

Status Extremely local and only known from a few sites in the Brecklands. Recently recorded from just one vice-county.

Threats Loss of Breck grassland through improvement and conversion to arable agriculture, development and afforestation. Natural succession is a further threat.

Management and conservation Grazing, cutting or some other disturbance is needed to maintain open conditions.

Published sources Mendel, H. (1989), Nicholson, G.W. (1916), Shirt, D.B., ed. (1987), Taylor, S. (1943).

SYMBIOTES LATUS   NOTABLE B

Order COLEOPTERA   Family ENDOMYCHIDAE

Symbiotes latus Redtenbacher, 1849.

Distribution South East, South, East Anglia, East Midlands and West Midlands.

Habitat and ecology Broad-leaved woodland. Associated with elm, especially *Scolytus* (Scolytidae) infested bark. Also recorded from a dead ash, from beech, under the bark of a dead poplar and from turkey oak. Possiby also associated with fungi on wood. Adults have been recorded from February to August.

Status Widespread but local in southern and central England, not known from South West England.

Indicator status Grade 3 in Harding & Rose (1986).

Threats Loss of broad-leaved woodland through, for example, clear-felling and coniferisation. Habitat loss, in particular, through the felling of ancient trees, removal of dead wood from living trees and the destruction or removal of standing and fallen dead wood for reasons such as forest hygiene, aesthetic tidiness, public safety or for use as fire wood.

Management and conservation Ancient trees and both fallen and standing dead timber, especially with the bark attached, should be retained. The removal of dead timber from ancient trees should be avoided. Gaps in the age structure of the tree population should be identified and the continuity of the appropriate dead wood habitat ensured by regeneration, suitable planting and possibly with pollarding.

TRIPLAX LACORDAIRII   RARE

Order COLEOPTERA   Family EROTYLIDAE

Triplax lacordairii Crotch, 1870. Formerly known as: *Triplax lacordairei* Crotch, 1870.

Distribution Recorded from Dorset, South Hampshire, East Kent, West Kent, Surrey and Worcestershire before 1970 and Isle of Wight, South Hampshire, East Kent and South Essex from 1970 onwards.

Habitat and ecology Ancient broad-leaved woodland and pasture-woodland. Adults and probably larvae occur in fungi on trees, with records from the fungus *Pleurotus*, from a bracket fungus on a fallen beech, and also from ash and elm trees. Adults have been recorded June to September.

Status Only known from southern England and recently recorded from just four vice-counties. An old record for Berkshire is in error.

Indicator status Grade 3 in Harding & Rose (1986).

Threats Loss of broad-leaved woodland and parkland through, for example, clear-felling and coniferisation. Habitat loss, in particular, through the felling of fungus-infected trees, removal of dead wood from living trees and the destruction or removal of standing and fallen dead wood for reasons such as forest hygiene, aesthetic tidiness, public safety or for use as fire wood.

Management and conservation Fungus-infected trees and both fallen and standing dead timber, especially with the bark attached, should be retained. The removal of dead timber from fungus-infected trees should be avoided. Gaps in the age structure of the tree population should be identified and the continuity of the appropriate dead wood habitat ensured by regeneration, suitable planting and possibly with pollarding.

Published sources Allen, A.A. (1965c), Fowler, W.W. (1889), Harding, P.T. (1978), Harding, P.T. & Rose, F. (1986), Prance, D.A. (1988), Shirt, D.B., ed. (1987), Skidmore, P. (1962).

TRIPLAX SCUTELLARIS   RARE

Order COLEOPTERA   Family EROTYLIDAE

Triplax scutellaris Charpentier, 1825.

Distribution Recorded from Cheshire and Durham before 1970 and North-east Yorkshire and South Northumberland from 1970 onwards.

Habitat and ecology Ancient broad-leaved woodland. Recorded in the fungus *Pleurotus*, and in fungi growing on elm and holly. Adults have been recorded from March to May, and in July and September.

Status Only known from four vice-counties, all in northern England. Recently recorded from just two vice-counties.

Indicator status Grade 1 in Garland (1983). Grade 3 in Harding & Rose (1986).

Threats Loss of broad-leaved woodland through, for example, clear-felling and coniferisation. Habitat loss, in particular, through the felling of fungus-infected trees, removal of dead wood from living trees and the destruction or removal of standing and fallen dead wood for reasons such as forest hygiene, aesthetic tidiness, public safety or for use as fire wood.

Management and conservation Fungus-infected trees and both fallen and standing dead timber, especially with the bark attached, should be retained. The removal of dead timber from fungus-infected trees should be avoided. Gaps in the age structure of the tree population should be identified and the continuity of the appropriate dead wood habitat ensured by regeneration, suitable planting and possibly with pollarding.

Published sources Aubrook, E.W. (1973), Garland, S.P. (1983), Harding, P.T. (1978), Harding, P.T. & Rose, F. (1986), Luff, M.L. & Walker, M. (1981), Shirt, D.B., ed. (1987).

TRITOMA BIPUSTULATA   NOTABLE A

Order COLEOPTERA   Family EROTYLIDAE

Tritoma bipustulata F., 1775. Formerly known as: *Cyrtotriplax bipustulata* (F.).

Distribution Recorded from Dorset, South Hampshire, North Hampshire, West Sussex, West Kent, Surrey, South Essex, East Suffolk, Huntingdonshire, East Gloucestershire, Glamorgan, Nottinghamshire, North-east Yorkshire, South-west Yorkshire and South

EUCNEMIDAE

Northumberland before 1970 and South Wiltshire, South Hampshire, West Sussex, East Sussex, Berkshire, Nottinghamshire and Westmorland & North Lancashire from 1970 onwards.

Habitat and ecology Ancient broad-leaved woodland and pasture-woodland. Recorded from an open, damp deciduous wood, containing birch and also from thin woodland growing on limestone. Occurs in fungi on trees. Recorded from a fungus on beech, in the fungus *Trametes* on oak, and under bark and in an oak stump. Adults have been recorded from April to August.

Status Widespread and very local. Recently recorded from seven vice-counties from South Wiltshire to Westmorland. There is an old record for South Wales.

Indicator status Grade 2 in Garland (1983). Grade 3 in Harding & Rose (1986).

Threats Loss of broad-leaved woodland and parkland through, for example, clear-felling and coniferisation. Habitat loss, in particular, through the felling of fungus-infected trees, removal of dead wood from living trees and the destruction or removal of standing and fallen dead wood for reasons such as forest hygiene, aesthetic tidiness, public safety or for use as fire wood.

Management and conservation Fungus-infected trees and both fallen and standing dead timber, especially with the bark attached, should be retained. The removal of dead timber from fungus-infected trees should be avoided. Gaps in the age structure of the tree population should be identified and the continuity of the appropriate dead wood habitat ensured by regeneration, suitable planting and possibly with pollarding.

Published sources Atty, D.B. (1983), Fowler, W.W. (1889), Fowler, W.W. & Donisthorpe, H.St J.K. (1913), Garland, S.P. (1983), Harding, P.T. (1978), Harding, P.T. & Rose, F. (1986), Johnson, C. (1990), Williams, S.A. (1975).

DIRHAGUS PYGMAEUS   RARE
A false click beetle
Order COLEOPTERA   Family EUCNEMIDAE

Dirhagus pygmaeus (F., 1792). Formerly known as: *Dirrhagus pygmaeus* (F., 1792), *Microrrhagus pygmaeus* (F.).

Distribution Recorded from South Devon, Isle of Wight, South Hampshire, West Sussex, East Kent, West Kent, West Gloucestershire, North-east Yorkshire, East Inverness & Nairn, West Ross and East Ross

before 1970 and East Cornwall, North Somerset, South Hampshire, West Sussex, East Kent, East Gloucestershire, Herefordshire, Westmorland & North Lancashire, Kirkcudbrightshire and East Inverness & Nairn from 1970 onwards. A map is given in Mendel (1988).

Habitat and ecology Broad-leaved woodland and pasture-woodland. Larvae probably develop in dead wood. Adults have been found in rotten beech and in a dry, rotten birch log. This species has been recorded by sweeping, particularly from bracken and often under oak and beech trees. Also found in oak leaf litter. Adults have been recorded June to August.

Status Widely scattered and very local in England and Wales. Recently recorded from West Cornwall to East Kent and north to Easterness.

Indicator status Grade 1 in Garland (1983). Grade 3 in Harding & Rose (1986).

Threats Loss of broad-leaved woodland and parkland through, for example, clear-felling and coniferisation. Habitat loss, in particular, may be through the felling of ancient trees, removal of dead wood from living trees and the destruction or removal of standing and fallen dead wood for reasons such as forest hygiene, aesthetic tidiness, public safety or for use as fire wood.

Management and conservation Ancient trees and both fallen and standing dead timber, especially with the bark attached, should be retained. Dead timber should not be removed from ancient trees. Gaps in the age structure of the tree population should be identified and the continuity of the appropriate dead wood habitat ensured by regeneration, suitable planting and possibly with pollarding.

Published sources Allen, A.A. (1966), Allen, A.A. (1973c), Atty, D.B. (1983), Cooper, B.A. (1946), Davidson, A. (1956), Fowler, W.W. (1890), Fowler, W.W. & Donisthorpe, H.St J.K. (1913), Garland, S.P. (1983), Harding, P.T. (1978), Harding, P.T. & Rose, F. (1986), Hodge, P.J. (1990), Mendel, H. (1988a), Orton, P.D. (1988), Shirt, D.B., ed. (1987), Welch, R.C. (1983b).

EUCNEMIS CAPUCINA   ENDANGERED
A false click beetle
Order COLEOPTERA   Family EUCNEMIDAE

Eucnemis capucina Ahrens, 1812.

Distribution Recorded from South Hampshire and Berkshire before 1970 and South Hampshire and

Berkshire from 1970 onwards. A map is given in Mendel (1988).

Habitat and ecology Ancient broad-leaved woodland and pasture-woodland. Larvae develop in rotten wood and under bark. Mainly associated with beech, but also recorded from ash and probably associated with other deciduous trees. Pupae have been found in mould beneath a fallen beech branch in March. Adults have been recorded in April, June and July.

Status Only known from Windsor Forest and Windsor Great Park, Berkshire and the New Forest, South Hampshire, and only from a small number of examples.

Indicator status Grade 1 in Harding & Rose (1986).

Threats Loss of broad-leaved woodland and parkland through, for example, clear-felling and coniferisation. Habitat loss, in particular through the felling of ancient trees, removal of dead wood from living trees and the destruction or removal of standing and fallen dead wood for reasons such as forest hygiene, aesthetic tidiness, public safety or for use as fire wood.

Management and conservation Ancient trees and both fallen and standing dead timber, especially with the bark attached, should be retained. Dead timber should not be removed from ancient trees. Gaps in the age structure of the tree population should be identified and the continuity of the appropriate dead wood habitat ensured by regeneration, suitable planting and possibly with pollarding. Windsor Forest, parts of Windsor Great Park and much of the New Forest are notified as SSSIs.

Published sources Allen, A.A. (1966), Appleton, D. (1972a), Harding, P.T. (1978), Harding, P.T. & Rose, F. (1986), Mendel, H. (1988a), Shirt, D.B., ed. (1987).

HYLIS CARINICEPS   **ENDANGERED**
A false click beetle
Order COLEOPTERA   Family EUCNEMIDAE

Hylis cariniceps (Reitter, 1902).

Distribution Recorded from South Hampshire before 1970 and Dorset from 1970 onwards. A map is given in Mendel (1988).

Habitat and ecology Broad-leaved woodland. Probably associated with ancient beech. The larvae probably develop in dead wood. Adults have been found in July.

Status Only known in Great Britain from two examples; a female found in 1966 near some old beech trees in Mallard Wood, New Forest, South Hampshire

and a second example found, post 1976, on Brownsea Island, Dorset.

Threats Uncertain, though this species is probably threatened by the loss of broad-leaved woodland through, for example, clear-felling and coniferisation. Habitat loss, in particular, may be through the felling of ancient trees, removal of dead wood from living trees and the destruction or removal of standing and fallen dead wood for reasons such as forest hygiene, aesthetic tidiness, public safety or for use as fire wood.

Management and conservation Ancient trees and both fallen and standing dead timber, especially with the bark attached, should be retained. Dead timber should not be removed from ancient trees. Gaps in the age structure of the tree population should be identified and the continuity of the appropriate dead wood habitat ensured by regeneration, suitable planting and possibly with pollarding.

Published sources Mendel, H. (1988a), Shirt, D.B., ed. (1987).

HYLIS OLEXAI   **RARE**
A false click beetle
Order COLEOPTERA   Family EUCNEMIDAE

Hylis olexai (Palm, 1955)

Distribution Recorded from West Kent and Surrey before 1970 and South Hampshire and West Sussex from 1970 onwards. A map is given in Mendel (1988).

Habitat and ecology Broad-leaved woodland. Associated with beech and probably hornbeam. Possibly also associated with other broad-leaved tree species. Larvae develop in dead wood. Adults have been recorded from July to September.

Status Only known from South and South East England. Recently recorded from only two vice-counties.

Threats Loss of broad-leaved woodland through, for example, clear-felling and coniferisation. Habitat loss, in particular, through the felling of ancient trees, removal of dead wood from living trees and the destruction or removal of standing and fallen dead wood for reasons such as forest hygiene, aesthetic tidiness, public safety or for use as fire wood.

Management and conservation Ancient trees and both fallen and standing dead timber, especially with the bark attached, should be retained. Dead timber should not be removed from ancient trees. Gaps in the age

EUCNEMIDAE

structure of the tree population should be identified and the continuity of the appropriate dead wood habitat ensured by regeneration, suitable planting and possibly with pollarding.

Published sources Allen, A.A. (1954f), Buck, F.D. (1960b), Mendel, H. (1988a), Shirt, D.B., ed. (1987).

MELASIS BUPRESTOIDES　　　　**NOTABLE B**
A false click beetle
Order COLEOPTERA　　　　Family EUCNEMIDAE

Melasis buprestoides (L., 1761).

Distribution South East, South, East Anglia, East Midlands, West Midlands, North East, North West and South Wales. A map is given by Mendel (1988).

Habitat and ecology Broad-leaved woodland and pasture woodland. Associated with beech, oak, birch, ash and probably other deciduous trees. A single adult has also been found in the fungus *Daldinia concentrica*. Larvae develop in hard, dead wood and found in standing trunks, stumps and fallen branches, and make flat (1mm thick) yellow coloured galleries across the grain (J. Parry pers.comm.). Adults have been recorded in May, June and August.

Status Widespread but local in England, though not recorded in South West England. This species may be under-recorded because of its secretive nature.
ces Fowler, W.W. (1890), Hodge, P.J. (1990), Jessop, L. (1986).

GEOTRUPES VERNALIS　　　　**NOTABLE B**
A dor beetle
Order COLEOPTERA　　　　Family GEOTRUPIDAE

Geotrupes vernalis (L., 1758).

Distribution England, Wales and Scotland.

Habitat and ecology Grassland and heather moorland, particularly on dry, sandy soils. This species has been found in sheep dung. Also in fox dung and under dead birds. The larva develops in a burrow. Adults have been found throughout summer and autumn.

Status Widespread and very local. Declining, now possibly absent in south-eastern England.

Threats This species is probably threatened by the loss of its habitat through grassland improvement, conversion to arable agriculture, forestry and urban development. Lack of, or a change in grazing regimes is a further threat.

Management and conservation An opportunistic and very mobile species requiring a continuity of dung availability. Because of this, this species is likely to require large areas of suitable habitat for its continued survival. Management should aim for a diversity of successional stages, preferably by animal grazing.

ODONTEUS ARMIGER　　　　**NOTABLE A**
A dor beetle
Order COLEOPTERA　　　　Family GEOTRUPIDAE

Odonteus armiger (Scopoli, 1772). Formerly known as: *Odontaeus armiger* (Scopoli, 1772), *Odontaeus mobilicornis* (F., 1775).

Distribution Recorded from North Somerset, South Wiltshire, Dorset, Isle of Wight, South Hampshire, North Hampshire, West Sussex, East Sussex, East Kent, West Kent, Surrey, Hertfordshire, Berkshire, Oxfordshire, Buckinghamshire, West Suffolk, East Norfolk, West Norfolk, Cambridgeshire, Northamptonshire, East Gloucestershire and West Gloucestershire before 1970 and South Hampshire, East Kent, West Kent, Berkshire, Buckinghamshire, West Suffolk, West Norfolk, Bedfordshire and Radnorshire from 1970 onwards.

Habitat and ecology Grassland and heathland on chalky or sandy soils. Subterranean, occasionally found in and around rabbit burrows. Possibly feeds on subterranean fungi. This species has been found under dry cow dung and sheep droppings. Adults fly in the evening in hot weather, in cooler weather they have been noted flying in the afternoon. Adults have been found from May to November, though most records are from June and July. A high proportion of records of this beetle are from light traps.

Status Status revised from RDB 3 (Rare) in Shirt (1987). Very local and widely distributed in southern Great Britain. Virtually all records are south of a line from the Wash to the Bristol Channel, though with a recent record from Radnorshire, Wales. Probably under-recorded because of its secretive habits.

Threats Loss of calcareous grassland and heathland through agricultural improvement by reseeding or by the application of fertilisers, by conversion to arable or through forestry. Further habitat degradation through the cessation of traditional heathland and downland management pratices. Lack of, or changes in the grazing or cutting regimes could alter the vegetation structure which may in turn threaten this species.

Management and conservation Grazing or cutting, on a rotational basis, is needed to maitain open conditions.

Published sources Albertini, M. & Hall, P. (1988),
Ansorge, E. (1958), Ansorge, E. (1962), Ansorge, E.
(1963), Appleton, D. (1972b), Ashford, R.W. (1961),
Atty, D.B. (1983), Bedwell, E.C. (1932a), Chinery, M.
(1962), Duffield, C.A.W. (1925), Eagles, T.R. (1948a),
Eeles, W.J. (1961), Foster, A.P. (1988), Fowler, W.W.
(1890), Fowler, W.W. & Donisthorpe, H.St J.K. (1913),
Grant, M.H. (1888), Harman, T.W. (1977), Hobby,
B.M. (1939), Jessop, L. (1986), Jones-Walters, L.M.
(1984), Parsons, M.S. (1988), Sankey, J.H.P. (1950),
Saunders, C.J. (1936a), Shirt, D.B., ed. (1987), Stott,
C.E. (1925), Wilkinson, W. (1953).

HETEROCERUS FUSCULUS   NOTABLE B

Order COLEOPTERA   Family HETEROCERIDAE

Heterocerus fusculus Kiesenwetter, 1843.

Distribution South, South West, East Anglia and North
West.

Habitat and ecology The base of sea-cliffs. Possibly
detritivorous. The beetle lives in shallow galleries in
muddy ground close to the high tide mark. Adults have
been recorded from May to July.

Status Widespread but local in southern England, also
recorded from North West England.

Threats This species is threatened by cliff stabilisation
schemes and the construction of sea defences. Coastal
developments may reduce the amount of available
habitat. Activities that accelerate or reduce the rate of
erosion should be avoided. The degradation of suitable
habitat through natural succession on stabilised areas is
a further threat.

Management and conservation Occasional slippages
are necessary to maintain habitat continuity. Large
areas of unstable cliff are required so that the
population does not become isolated and subsequently
threatened by individual landslips.

HETEROCERUS HISPIDULUS   RARE

Order COLEOPTERA   Family HETEROCERIDAE

Heterocerus hispidulus Kiesenwetter, 1843.

Distribution Recorded from East Sussex and East Kent
before 1970 and East Sussex, East Kent and East
Suffolk from 1970 onwards.

Habitat and ecology Water-filled sand pits and gravel
pits. Probably detritivorous occurring in shallow

galleries just below the surface of the sand, close to the
waters edge. Adults have been recorded from April to
June and from August to October.

Status Recorded new to Great Britain in 1969 and now
known from three vice-counties, all in South East
England and East Anglia.

Threats This species is threatened by the infilling of
gravel pits, tourist development and natural succession.

Management and conservation Management should
aim at maintaining open conditions and encouraging
early successional stages along pit edges.

Published sources Clarke, R.O.S. (1973), Collier, M.J.
(1988a), Collier, M.J. (1988b), Hodge, P.J. (1990),
Shirt, D.B., ed. (1987).

ABRAEUS GRANULUM   NOTABLE A

Order COLEOPTERA   Family HISTERIDAE

Abraeus granulum Erichson, 1839.

Distribution Recorded from South Devon, South
Hampshire, West Sussex, West Kent, Surrey, South
Essex, Berkshire, Oxfordshire, Cambridgeshire,
Herefordshire, Warwickshire and Nottinghamshire
before 1970 and Surrey, Berkshire, East Suffolk, East
Gloucestershire, Derbyshire, Cheshire and South
Lancashire from 1970 onwards.

Habitat and ecology Ancient broad-leaved woodland
and pasture-woodland. Probably predatory. Found in
rotten wood and recorded from beech, oak, ash, wych
elm and English elm. This beetle has been found in the
company of the ant *Lasius brunneus* (Hymenoptera).
Adults have been recorded in January, February, April,
June, July and November.

Status Widely distributed in England but very local.
Recently recorded from seven vice-counties and found
as far north as South Lancashire.

Indicator status Grade 1 in Harding & Rose (1986).

Threats Loss of broad-leaved woodland and parkland
through, for example, clear-felling and coniferisation.
Habitat loss, in particular, through the felling of ancient
trees, removal of dead wood from living trees and the
destruction or removal of standing and fallen dead
wood for reasons such as forest hygiene, aesthetic
tidiness, public safety or for use as fire wood.

Management and conservation Ancient trees, and
both fallen and standing dead timber, especially with

335

HISTERIDAE

the bark attached, should be retained. The removal of dead timber from ancient trees should be avoided. Gaps in the age structure of the tree population should be identified and the continuity of the appropriate dead wood habitat ensured by regeneration, suitable planting and possibly with pollarding.

Published sources Collier, M.J. (1988a), Garland, S.P. (1983), Halstead, D.G.H. (1963), Hammond, P.M. (1979), Harding, P.T. (1978), Harding, P.T. & Rose, F. (1986).

ACRITUS HOMOEOPATHICUS   RARE

Order COLEOPTERA   Family HISTERIDAE

Acritus homoeopathicus Wollaston, 1857.

Distribution Recorded from Dorset, South Hampshire, East Kent, Surrey, South Essex and Huntingdonshire before 1970 and South Hampshire, West Sussex and East Sussex from 1970 onwards.

Habitat and ecology Found on burnt ground. Associated with the fungus *Pyronema confluens*. Also noted under burnt bark. Adults have been recorded in April, May, July and September.

Status Only recorded in southern England and found as far north as Huntingdonshire. Recently recorded from just three vice-counties.

Threats The cessation of burning as a management technique in certain habitats could be a threat to this species. The tidying of burnt material may be a further threat.

Management and conservation Small areas of burnt timber should be left undisturbed following a burn.

Published sources Cooter, J. (1973), Forster, H.W. (1953), Halstead, D.G.H. (1963), Massee, A.M. (1950), Shirt, D.B., ed. (1987), Welch, R.C. (1968).

AELETES ATOMARIUS   RARE

Order COLEOPTERA   Family HISTERIDAE

Aeletes atomarius (Aubé, 1842). Formerly known as: *Acritus atomarius* (Aubé).

Distribution Recorded from South Hampshire, North Hampshire, Berkshire, Oxfordshire, Herefordshire and Worcestershire before 1970 and South Hampshire, North Hampshire, Berkshire, West Gloucestershire, Herefordshire, Derbyshire and North-east Yorkshire from 1970 onwards.

Habitat and ecology Ancient broad-leaved woodland and pasture-woodland. Found in dead wood, under bark and in stumps. Mainly recorded from beech, though also from elm, ash, the red rotten heartwood of a fallen oak and probably other trees. It has been found in the burrows of the lesser stag-beetle *Dorcus parallelepipedus* (Lucanidae) and in dead wood damaged by the beetle *Sinodendron cylindricum* (Lucanidae), though this association may be coincidental. On the Continent, this species has been Found in association with the ant *Lasius brunneus* (Hymenoptera). In Britain, adults have been recorded in September and December.

Status Widely distributed but very local in England. This species has been recently recorded from seven vice-counties and found as far north as North-east Yorkshire. Possibly under-recorded because of its secretive habits and extremely small size.

Indicator status Grade 1 in Harding & Rose (1986).

Threats Loss of broad-leaved woodland and parkland through, for example, clear-felling and coniferisation. Habitat loss, in particular, through the felling of ancient trees, removal of dead wood from living trees and the destruction or removal of standing and fallen dead wood for reasons such as forest hygiene, aesthetic tidiness, public safety or for use as fire wood.

Management and conservation Ancient trees, and both fallen and standing dead timber, especially with the bark attached, should be retained. The removal of dead timber from ancient trees should be avoided. Gaps in the age structure of the tree population should be identified and the continuity of the appropriate dead wood habitat ensured by regeneration, suitable planting and possibly with pollarding.

Published sources Alexander, K.N.A. & Clements, D.K. (1988), Allen, A.A. (1951d), Flint, J.H. (1988), Halstead, D.G.H. (1963), Harding, P.T. (1978), Harding, P.T. & Rose, F. (1986), Marsh, R.J. (1988), Shirt, D.B., ed. (1987).

BAECKMANNIOLUS DIMIDIATUS   NOTABLE B

Order COLEOPTERA   Family HISTERIDAE

Baeckmanniolus dimidiatus (Illiger, 1807). Formerly known as: *Baeckmanniolus dimidiatus ssp. maritimus* (Stephens, 1830), *Baeckmanniolus maritimus* (Stephens), *Pachylopus maritimus* (Stephens), *Saprinus maritimus* (Stephens).

Distribution South East, South, South West, East Anglia, East Midlands, North East, North West, Wales, South East Scotland and South West Scotland.

Habitat and ecology Sandy coastal habitats, particularly along the strandline and on dunes. There is an old inland record from Torksey, North Lincolnshire. Found under driftwood, in decaying seaweed, and also in dung and once in dead sedge. Adults have been recorded from April to August.

Status Widespread but local in Great Britain and recorded as far north as southern Scotland.

Threats Coastal developments, the construction of sea-defences and the tidying up of beaches. Coastal pollution may be a further threat.

Management and conservation Grazing or some other disturbance may be needed to maintain open conditions in dune habitats.

EPIERUS COMPTUS INSUFFICIENTLY KNOWN

Order COLEOPTERA   Family HISTERIDAE

Epierus comptus Erichson, 1834.

Distribution Recorded from South Wiltshire from 1970 onwards.

Habitat and ecology Ancient broad-leaved woodland. Recorded from under the bark of a mature, fallen beech in August.

Status Status revised from RDB 3 (Rare) in Shirt (1987). Still only known from a single example recorded in 1980 from South Wiltshire.

Threats Uncertain, though the loss of broad-leaved woodland through, for example, clear-felling and coniferisation may be a threat. Habitat loss, in particular, may be through the felling of ancient trees, removal of dead wood from living trees and the destruction or removal of standing and fallen dead wood for reasons such as forest hygiene, aesthetic tidiness, public safety or for use as fire wood.

Management and conservation Uncertain, though ancient trees, and both fallen and standing dead timber, especially with the bark attached, should be retained. The removal of dead timber from ancient trees should be avoided. Gaps in the age structure of the tree population should be identified and the continuity of the appropriate dead wood habitat ensured by regeneration, suitable planting and possibly with pollarding.

Published sources Nash, D.R. (1982c), Shirt, D.B., ed. (1987).

GNATHONCUS BUYSSONI   NOTABLE A

Order COLEOPTERA   Family HISTERIDAE

Gnathoncus buyssoni Auzat, 1917.

Distribution Recorded from South Hampshire, "Sussex", East Kent, West Kent, Hertfordshire, Berkshire, "Norfolk", Warwickshire and South Lancashire before 1970 and Surrey, Cardiganshire, South Lincolnshire, Leicestershire & Rutland, Cheshire and South Lancashire from 1970 onwards.

Habitat and ecology Woodland and pasture-woodland. Recorded from bird's nests, fungi, a squirrel's drey and rat droppings. On the Continent, this species has been noted from the fungi *Polyporus* and *Poria*. In Britain, adults have been recorded from June to August and in December.

Status Widespread and very local. Recorded from southern England, with scattered records north to South Lancashire. Also noted in mid Wales. Recently noted in six vice-counties. This species is difficult to identify and may be confused with other members of the genus. Consequently, the exact status of this species is hard to assess.

Threats Loss of woodland through, for example, clear-felling and conversion to other land use. Habitat loss, in particular, through the felling of fungus-infected trees, removal of dead wood from living trees and the destruction or removal of standing and fallen dead wood for reasons such as forest hygiene, aesthetic tidiness, public safety or for use as fire wood.

Management and conservation Fungus-infected trees, and both fallen and standing dead timber, especially with the bark attached, should be retained. The removal of dead timber from fungus-infected trees should be avoided. Gaps in the age structure of the tree population should be identified and the continuity of

HISTERIDAE

the appropriate habitat ensured by regeneration, suitable planting and possibly with pollarding.

Published sources Bedwell, E.C. (1943b), Boyce, D.C. (1989), Boyce, D.C. (1990), Halstead, D.G.H. (1963), Wood, O.E. (1964).

GRAMMOSTETHUS MARGINATUS NOTABLE B

Order COLEOPTERA   Family HISTERIDAE

Grammostethus marginatus (Erichson, 1834). Formerly known as: *Hister marginatus* Erichson, *Margarinotus marginatus* (Erichson).

Distribution South East, South, East Anglia, East Midlands, North East, North West, South Wales and South West Scotland.

Habitat and ecology Woodland. Mainly found in mole's nests, though also found in rabbit's burrows, horse dung and once in grass cuttings. Adults have been recorded from February to July.

Status Widespread but local in England. Also recorded in South Wales and South West Scotland. Possibly under-recorded because of this species' secretive habits.

Threats Uncertain, though the clear-felling of woodland and conversion to other land use may be a threat to this species.

HALACRITUS PUNCTUM
INSUFFICIENTLY KNOWN
Order COLEOPTERA   Family HISTERIDAE

Halacritus punctum (Aubé, 1842). Formerly known as: *Acritus punctum* (Aubé).

Distribution Recorded from East Cornwall, North Somerset, Dorset, Isle of Wight and South Hampshire before 1970 and North Devon from 1970 onwards.

Habitat and ecology On the sea-shore, in sand under seaweed just above the high-water level. Adults have been recorded in June.

Status Not listed in the insect Red Data Book (Shirt, 1987). Only known from the coast of southern England. Recently recorded from just one vice-county. There is a published record for Leicestershire. This species is probably under-recorded because of its secretive habits and very small size.

Threats Coastal developments, the construction of sea defences, recreational pressure and the tidying of beaches. Pollution may be a further threat.

Published sources Fowler, W.W. (1889), Fowler, W.W. & Donisthorpe, H.St J.K. (1913), Halstead, D.G.H. (1963).

HETAERIUS FERRUGINEUS   INDETERMINATE

Order COLEOPTERA   Family HISTERIDAE

Hetaerius ferrugineus (Olivier, 1789).

Distribution Recorded from South Hampshire, Surrey and Middlesex before 1970.

Habitat and ecology Probably associated with woodland and grassland. Found in the nests of the ants, *Formica fusca* and *F. sanguinea* (Hymenoptera). Adults have been recorded in April and May.

Status Status revised from RDB 3 (Rare) in Shirt (1987). Only known from three vice-counties, all in southern England. Last recorded in 1954 from Hum, near Christchurch, South Hampshire.

Threats Uncertain, though the loss of woodland through clear-felling and conversion to other land use, and the loss of grassland through improvement and conversion to arable agriculture, devorestation may be threats to this species.

Published sources Brown, S.C.S. (1971), Fowler, W.W. (1889), Fowler, W.W. & Donisthorpe, H.St J.K. (1913), Halstead, D.G.H. (1963), Shirt, D.B., ed. (1987).

HISTER BISSEXSTRIATUS   NOTABLE B

Order COLEOPTERA   Family HISTERIDAE

Hister bissexstriatus F., 1801.

Distribution South East, East Anglia, East Midlands, North East and North West.

Habitat and ecology Grassland. Found in cow dung.

Status Widespread but local in England.

Threats Because of its dependence on dung this species is likely to require large areas of suitable habitat for its continued survival. Agricultural improvement and development are the causes of the main losses of unimproved pasture.

Management and conservation An opportunistic and mobile species requiring a continuity of dung availability. Grazing is needed to keep the habitat open and provide suitable conditions for this species.

HISTER ILLIGERI   EXTINCT

Order COLEOPTERA   Family HISTERIDAE

Hister illigeri Duftschmid, 1805. Formerly known as: *Hister sinuatus* sensu Illiger, 1798 not F., 1792).

Distribution Recorded from South Devon, West Kent, Worcestershire and Glamorgan.

Habitat and ecology Probably associated with dung and/or carrion.

Status There are 19th century records from four vice-counties, though it is possible that this species may not be British.

Published sources Fowler, W.W. (1889), Shirt, D.B., ed. (1987).

HISTER QUADRIMACULATUS
                    INSUFFICIENTLY KNOWN
Order COLEOPTERA   Family HISTERIDAE

Hister quadrimaculatus L., 1758.

Distribution Recorded from Dorset, Isle of Wight, South Hampshire, East Kent, West Kent, Surrey and North Essex before 1970 and East Kent from 1970 onwards.

Habitat and ecology On or near the coast and rarely found inland. Found in dung and carrion. Adults have been recorded from April to June and in August and December.

Status Status revised from RDB 2 (Vulnerable) in Shirt (1987). Recorded from Dorset to North Essex. Most records are from Kent, and recently recorded only from East Kent.

Threats Because of its dependence on dung/carrion this species is likely to require large areas of suitable habitat for its continued survival. Loss of habitat through coastal developments, the construction of coastal defences and gravel extraction. Loss of grazing in coastal areas may be a further threat.

Management and conservation An opportunistic and very mobile species requiring a continuity of

dung/carrion availability. The disturbance of coastal shingle should be avoided.

Published sources Duffy, E.A.J. (1945), Fowler, W.W. (1889), Fowler, W.W. & Donisthorpe, H.St J.K. (1913), Halstead, D.G.H. (1963), Parsons, M.S. (1989), Shirt, D.B., ed. (1987), Whicher, L.S. (1952b).

HISTER QUADRINOTATUS   EXTINCT

Order COLEOPTERA   Family HISTERIDAE

Hister quadrinotatus Scriba, 1790.

Distribution Recorded from "London district" and West Gloucestershire.

Habitat and ecology Probably associated with dung and/or carrion.

Status There are 19th century records from Bristol and near London, though it is possible that this species may not be British.

Published sources Fowler, W.W. (1889), Shirt, D.B., ed. (1987).

HYPOCACCUS METALLICUS   RARE

Order COLEOPTERA   Family HISTERIDAE

Hypocaccus metallicus (Herbst, 1792). Formerly known as: *Saprinus metallicus* (Herbst).

Distribution Recorded from Dorset, East Sussex, East Kent, East Norfolk, West Norfolk and North Lincolnshire before 1970 and East Sussex, East Kent and West Norfolk from 1970 onwards.

Habitat and ecology Coastal sandhills and shingle banks. Found in dung and carrion on bare sand, and also found under driftwood. Adults have been noted crawling over bare sand on warm spring days. Adults have been recorded from April to August.

Status Status revised from RDB 2 (Vulnerable) in Shirt (1987). Reported along the coasts from Dorset to South Lincolnshire. Recently recorded from just three vice-counties.

Threats Because of its dependence on dung/carrion this species is likely to require large areas of suitable habitat for its continued survival. Loss of dune habitat, particularly through afforestation, urban and holiday development. The degradation of remaining sandhill habitat by excessive disturbance of the vegetation through activities such as motorbike access and human trampling. The tidying of beaches may be a further threat.

Management and conservation An opportunistic and mobile species requiring a continuity of dung/carrion availability. On dunes some grazing and other disturbance may be desirable to maintain the early successional stages, prevent the invasion of scrub and provide dung. The disturbance of coastal shingle should be avoided.

Published sources Collier, M.J. (1988a), Fowler, W.W. (1889), Halstead, D.G.H. (1963), Shirt, D.B., ed. (1987).

HYPOCACCUS RUGICEPS NOTABLE A

Order COLEOPTERA Family HISTERIDAE

Hypocaccus rugiceps (Duftschmid, 1805). Formerly known as: *Saprinus quadristriatus* sensu (Hoffmann, 1803) not (Thunberg, 1794), *Saprinus 4-striatus* sensu (Hoffmann, 1803) not (Thunberg, 1794).

Distribution Recorded from Dorset, Merionethshire, Anglesey, Cheshire, South Lancashire, West Lancashire, Westmorland & North Lancashire, Cumberland, Ayrshire, Renfrewshire, Clyde Islands, South Ebudes and Mid Ebudes before 1970 and Carmarthenshire, Merionethshire, Anglesey, South Lancashire and Cumberland from 1970 onwards.

Habitat and ecology Coastal sandhills and dunes. Found in dung and carrion. Adults have been recorded from April to August.

Status Status revised from RDB 2 (Vulnerable) in Shirt (1987). Widely distributed but very local in Great Britain, recorded from Dorset to Mid Ebudes in Scotland. Recently recorded from along the Welsh and North West England coasts and found in a total of five vice-counties.

Threats Because of its dependence on carrion this species is likely to require large areas of suitable habitat for its continued survival. Loss of dune habitat, particularly through afforestation, urban and holiday development. The degradation of remaining habitat by excessive disturbance of the vegetation through activities such as motorbike access and human trampling.

Management and conservation An opportunistic and mobile species requiring a continuity of dung/carrion availability. On dunes some grazing and other disturbance may be desirable to maintain the early

successional stages, prevent the invasion of scrub and provide dung.

Published sources Angus, R.B. (1964), Fowler, W.W. (1889), Halstead, D.G.H. (1963), Shirt, D.B., ed. (1987).

HYPOCACCUS RUGIFRONS NOTABLE B

Order COLEOPTERA Family HISTERIDAE

Hypocaccus rugifrons (Paykull, 1798). Formerly known as: *Saprinus rugifrons* (Paykull).

Distribution South East, South, South West, East Anglia, East Midlands, West Midlands, North East, South Wales and North Wales.

Habitat and ecology Primarily found on coastal sandhills, though also recorded from sandy places inland. Adults have been recorded in June. Found in dung and carrion.

Status Widespread and local in England and parts of Wales. Not recorded in North West England.

Threats Because of its dependence on dung/carrion this species is likely to require large areas of suitable habitat for its continued survival. Loss of dune and sandy habitats, particularly through afforestation, urban and holiday development. The degradation of remaining habitat by excessive disturbance of the vegetation through activities such as motorbike access, horse-riding and human trampling.

Management and conservation An opportunistic and mobile species requiring a continuity of dung/carrion availability. Some grazing or other disturbance, such as rotovation, on a rotational basis, may be desirable to maintain the early successional stages and prevent the invasion of scrub.

MYRMETES PICEUS NOTABLE B

Order COLEOPTERA Family HISTERIDAE

Myrmetes piceus (Paykull, 1809).

Distribution South East, South, East Anglia, East Midlands, West Midlands, North East and North East Scotland.

Habitat and ecology Woodland. Found in the nests of the wood ants, *Formica rufa* and *F. lugubris* (Hymenoptera). Adults have been recorded from June to August.

Status Widespread but local in England, though absent from South West and North West England. Also recorded in North East Scotland.

Threats Loss of broad-leaved woodland through, for example, clear-felling and coniferisation. Neglect and conversion to high forest has led to increased shade and the loss of glades and broad sunny rides.

Management and conservation Open glades and ride margins should be managed to retain a variety of vegetation structures.

ONTHOPHILUS PUNCTATUS
                              INSUFFICIENTLY KNOWN
Order COLEOPTERA   Family HISTERIDAE

Onthophilus punctatus (Müller, 1776). Formerly known as: *Onthophilus globulosus* sensu Fowler, 1889 not (Olivier, 1789), *Onthophilus sulcatus* (Fourcroy, 1785).

Distribution Recorded from South Devon, Isle of Wight, Surrey, Berkshire, West Suffolk, East Norfolk, West Norfolk, Bedfordshire and Nottinghamshire before 1970 and West Kent, South Essex and West Suffolk from 1970 onwards.

Habitat and ecology Uncertain, though possibly associated with wooded situations and sandy soils. Mainly found in mole's nests, though also recorded from a rabbit burrow. Adults have been recorded in February and March.

Status Not listed in the insect Red Data Book (Shirt, 1987). Widely distributed but very local in southern England and recorded as far north as Nottinghamshire. Recently recorded from just three vice-counties. Possibly under-recorded because of its secretive habits.

Threats Uncertain, though the clear-felling of woodland and conversion to other land use may be a threat to this species. The loss of Breckland through improvement and conversion to arable agriculture, afforestation and development may be a further threat.

Management and conservation In Breckland, grazing or some other disturbance, such as rotovation, may be needed to maintain open conditions.

Published sources Fowler, W.W. (1889), Fowler, W.W. & Donisthorpe, H.St J.K. (1913), Halstead, D.G.H. (1963).

PARALISTER OBSCURUS   **ENDANGERED**
Order COLEOPTERA   Family HISTERIDAE

Paralister obscurus (Kugelann, 1792). Formerly known as: *Hister stercorarius* Hoffmann, *Margarinotus stercorarius* (Hoffmann), *Paralister stercorarius* (Hoffmann).

Distribution Recorded from West Cornwall, South Devon, North Devon, South Hampshire, "Norfolk", Glamorgan and South Lancashire before 1970.

Habitat and ecology Sandhills and also recorded inland. Found in dung. The adult has been recorded in June.

Status Status revised from RDB 2 (Vulnerable) in Shirt (1987). No recent records. This species had a widely scattered distribution in England and was recorded from Glamorgan in South Wales. Last found in 1947 from Colyton, South Devon. A record for London requires confirmation.

Threats Loss of dune habitat, particularly through afforestation, urban and holiday development. The degradation of remaining habitat by excessive disturbance of the vegetation through activities such as human trampling may have contributed to this species' decline.

Published sources Fowler, W.W. (1889), Fowler, W.W. & Donisthorpe, H.St J.K. (1913), Halstead, D.G.H. (1963), Shirt, D.B., ed. (1987).

PAROMALUS PARALLELEPIPEDUS
                              ENDANGERED
Order COLEOPTERA   Family HISTERIDAE

Paromalus parallelepipedus (Herbst, 1792). Formerly known as: *Microlomalus parallelepipedus* (Herbst), *Micromalus parallelopipedus* (Herbst), *Paromalus parallelopipedus* (Herbst).

Distribution Recorded from South Hampshire and East Kent before 1970.

Habitat and ecology Woodland. Found under bark and possibly also in dead wood. Adults have been recorded in July.

Status Only known from a few examples from the New Forest, South Hampshire, where it was last noted 1910, and a single example which was found in 1952 in Pennipot Wood, Canterbury, East Kent.

HISTERIDAE

Threats Uncertain, though this species may be threatened by the loss of broad-leaved woodland through, for example, clear-felling and coniferisation. Habitat loss, in particular, may be through the felling of ancient trees, removal of dead wood from living trees and the destruction or removal of standing and fallen dead wood for reasons such as forest hygiene, aesthetic tidiness, public safety or for use as fire wood.

Management and conservation Much of the New Forest has been notified as an SSSI.

Published sources Allen, A.A. (1971c), Fowler, W.W. & Donisthorpe, H.St J.K. (1913), Halstead, D.G.H. (1963), Shirt, D.B., ed. (1987).

PLEGADERUS DISSECTUS   NOTABLE B

Order COLEOPTERA   Family HISTERIDAE

Plegaderus dissectus Erichson, 1839.

Distribution South East, South, East Anglia, East Midlands, West Midlands, North West and Dyfed-Powys.

Habitat and ecology Ancient broad-leaved woodland, pasture-woodland and wooded fens. Probably predatory. In old decaying trees, stumps and logs, and under bark. Recorded mainly from beech and oak, though also from elm, poplar, a red-rotten birch log and horse-chestnut. Adults have been recorded from March to October.

Status Widespread and local in England and parts of Wales.

Indicator status Grade 2 in Harding & Rose (1986).

Threats Loss of broad-leaved woodland and parkland through, for example, clear-felling and coniferisation. Habitat loss, in particular, through the felling of ancient trees, removal of dead wood from living trees and the destruction or removal of standing and fallen dead wood for reasons such as forest hygiene, aesthetic tidiness, public safety or for use as fire wood.

Management and conservation Ancient trees, and both fallen and standing dead timber, especially with the bark attached, should be retained. The removal of dead timber from ancient trees should be avoided. Gaps in the age structure of the tree population should be identified and the continuity of the appropriate dead wood habitat ensured by regeneration, suitable planting and possibly with pollarding.

SAPRINUS CUSPIDATUS   NOTABLE B

Order COLEOPTERA   Family HISTERIDAE

Saprinus cuspidatus Ihssen, 1949. Formerly confused under: *Saprinus nitidulus* (F., 1801), *Saprinus semistratus* (Scriba, 1790).

Distribution South East, South, South West, East Anglia, East Midlands, West Midlands, North West, Wales and South West Scotland.

Habitat and ecology Coastal sandhills, though possibly also in other situations. Found in carrion and rarely in dung. Adults have been recorded in May to September.

Status Widespread but local in England and Wales, also recorded in South West Scotland. This species is difficult to identify and may be confused with other members of the genus. Consequently, the exact status of this species is hard to assess.

Threats Because of its dependence on dung/carrion this species is likely to require large areas of suitable habitat for its continued survival. Loss of dune habitat, particularly through afforestation, urban and holiday development. The degradation of remaining habitat by excessive disturbance of the vegetation through activities such as motorbike access and human trampling.

Management and conservation An opportunistic and mobile species requiring a continuity of dung/carrion availability. On dunes some grazing and other disturbance may be desirable to maintain the early successional stages and prevent the invasion of scrub.

SAPRINUS IMMUNDUS   NOTABLE B

Order COLEOPTERA   Family HISTERIDAE

Saprinus immundus (Gyllenhal, 1827). Formerly known as: *Saprinus aeneus var. immundus* (Gyllenhal).

Distribution South East, East Anglia and East Midlands.

Habitat and ecology Coastal sandhills and sandy coasts. Found in dung and carrion. This species has been recorded from dog dung. Adults have been recorded in August and September.

Status Only known from southern and eastern England. Can be common where found.

Threats Because of its dependence on dung/carrion this species is likely to require large areas of suitable habitat for its continued survival. Loss of dune habitat, particularly through afforestation, urban and holiday development. The degradation of remaining habitat by excessive disturbance of the vegetation through activities such as motorbike access, horse-riding and human trampling.

Management and conservation An opportunistic and very mobile species requiring a continuity of dung/carrion availability. On dunes some grazing and other disturbance may be desirable to maintain the early successional stages and prevent the invasion of scrub.

SAPRINUS SUBNITESCENS   **EXTINCT**

Order COLEOPTERA   Family HISTERIDAE

Saprinus subnitescens Bickhardt, 1909. Formerly confused under: *Saprinus nitidulus* (F., 1801), *Saprinus semistriatus*.

Distribution Recorded from West Sussex.

Habitat and ecology Probably associated with carrion and/or dung. The adult has been recorded in August.

Status Presumed extinct. Only known in Great Britain from two examples; one without data, the other recorded in 1892 from Colgate, West Sussex.

Published sources Halstead, D.G.H. (1963), Shirt, D.B., ed. (1987).

SAPRINUS VIRESCENS
                          INSUFFICIENTLY KNOWN
Order COLEOPTERA   Family HISTERIDAE

Saprinus virescens (Paykull, 1798).

Distribution Recorded from West Cornwall, South Devon, Dorset, Isle of Wight, North Hampshire, East Sussex, East Kent, West Kent, Surrey, Berkshire, East Suffolk, East Norfolk, Cambridgeshire, Bedfordshire, Huntingdonshire, East Gloucestershire, Glamorgan, North Lincolnshire, Nottinghamshire, South-east Yorkshire, North-east Yorkshire and Mid-west Yorkshire before 1970 and Cardiganshire from 1970 onwards.

Habitat and ecology Marshes, well vegetated river banks and water-meadows. Predatory on the larvae of species of *Phaedon* (Chrysomelidae) and probably found on water-cress *Nasturtium officinale*. This

species has also been noted in carrion and hibernating under oak bark. Adults have been recorded from April to June and in August and September.

Status Not listed in the insect Red Data Book (Shirt, 1987). Formerly widespread and recorded in southern England, with scattered records north to North-east Yorkshire. Recently recorded from just one vice-county, this in mid Wales.

Threats Drainage for reasons such as agricultural improvement and development is the primary cause of the loss of wetlands. Falling water tables because of water abstraction and river engineering schemes may also threaten this species.

Management and conservation Water tables should be maintained at high levels. Grazing or cutting, on a rotational basis, may be needed to maintain open conditions.

Published sources Allen, A.A. (1973b), Atty, D.B. (1983), Fowler, W.W. (1889), Fowler, W.W. & Donisthorpe, H.St J.K. (1913), Halstead, D.G.H. (1963), Russell, H.M. (1953).

TERETRIUS FABRICII   **ENDANGERED**

Order COLEOPTERA   Family HISTERIDAE

Teretrius fabricii Mazur, 1972. Formerly known as: *Teretrius picipes* sensu (F., 1792) not (Olivier, 1789).

Distribution Recorded from East Kent, West Kent, Surrey, Berkshire, East Suffolk, East Norfolk, West Gloucestershire and Glamorgan before 1970.

Habitat and ecology The majority of records are from fresh oak palings, though probably associated with broad-leaved woodland. Larvae and probably adults prey on the immature stages of the beetles *Lyctus brunneus* and *L. linearis* (Lyctidae) and possibly other bostrichoid beetles. Adults have been recorded in June and July.

Status Formerly recorded over a wide area of south-eastern England, with isolated records from West Gloucestershire and South Wales. Last recorded in 1936 from Walderslade, East Kent.

Threats This species may have declined through the loss of broad-leaved woodland through, for example, clear-felling and coniferisation. Habitat loss, in particular, may have been through the felling of ancient trees, removal of dead wood from living trees and the destruction or removal of standing and fallen dead

HYPOCOPRIDAE

wood for reasons such as forest hygiene, aesthetic tidiness, public safety or for use as fire wood.

Published sources Allen, A.A. (1963), Bedwell, E.C. (1907), Fowler, W.W. (1889), Halstead, D.G.H. (1963), Shirt, D.B., ed. (1987).

HYPOCOPRUS LATRIDIOIDES
INDETERMINATE
Order COLEOPTERA   Family HYPOCOPRIDAE

Hypocoprus latridioides Motschulsky, 1839. Formerly known as: *Hypocoprus lathridioides* Motschulsky, 1839, *Hypocoprus quadricollis* Reitter, 1877.

Distribution Recorded from East Sussex and West Suffolk before 1970.

Habitat and ecology Open, sandy areas. Recorded under cow and sheep's dung, and possibly noted from under dead fowl. On the Continent, this species has been found under dry dung and also in the nest of the ant *Formica rufa* (Hymenoptera). The adult has been recorded in August.

Status Not listed in the insect Red Data Book (Shirt, 1987). Only recorded on three occasions, the last being in 1902 from Camber, East Sussex.

Published sources Butler, E.A. (1903), Fowler, W.W. (1889), Fowler, W.W. & Donisthorpe, H.St J.K. (1913).

LAMPROHIZA SPLENDIDULA   EXTINCT
A glow-worm
Order COLEOPTERA   Family LAMPYRIDAE

Lamprohiza splendidula (L., 1767).

Distribution Recorded from East Kent before 1970.

Habitat and ecology This species is almost certainly predatory. At least one and possibly both British examples were beaten from a hedge. On the Continent, this species is found in damp meadows, along streeams, inwoodland glades, along paths and in gardens.

Status Not listed in the insect Red Data Book (Shirt, 1987). Presumed extinct. Known from just two examples recorded in 1884 from East Kent. This species may be an extremely rare native, it is equally possible that the records are the result of a chance importation. This species can be confused with the common glow-worm *Lampyris noctiluca*.

Published sources Allen, A.A. (1989c).

PHOSPHAENUS HEMIPTERUS   ENDANGERED
A glow-worm
Order COLEOPTERA   Family LAMPYRIDAE

Phosphaenus hemipterus (Goeze, 1777).

Distribution Recorded from South Hampshire, West Sussex and East Sussex before 1970.

Habitat and ecology In Britain, usually recorded in gardens and churchyards where it frequents walls, tombstones and rockeries etc.. Almost certainly predatory, possibly on snails. The males can be active during the day. The females are larviform and are faintly luminescent at dusk. Adults have been recorded in June and July.

Status Only known from a very small number of records, the most recent being in 1961 from Chelwood Gate, Ashdown Forest, East Sussex.

Threats This species has suffered from the attention of collectors.

Published sources Fowler, W.W. (1890), Fowler, W.W. & Donisthorpe, H.St J.K. (1913), Shirt, D.B., ed. (1987).

LIMNICHUS PYGMAEUS   NOTABLE A

Order COLEOPTERA   Family LIMNICHIDAE

Limnichus pygmaeus (Sturm, 1807).

Distribution Recorded from South Devon, Dorset, Isle of Wight, East Sussex, East Kent, West Kent, South Essex, East Norfolk, Cambridgeshire and North Lincolnshire before 1970 and Dorset, South Hampshire, North Lincolnshire and Mid-west Yorkshire from 1970 onwards.

Habitat and ecology Wetland, fens, bogs and coastal habitats. Also in sandy or chalky places. Usually found crawling on mud or damp ground, though has also been found in sedge litter and in moss. Adults have been recorded in August and September.

Status Old records indicate that this species was formerly widespread in England and recorded from South Devon, north to North Lincolnshire. Recently recorded from just four vice-counties, including Mid-west Yorkshire.

Threats Drainage for reasons such as agricultural improvement and development. Falling water tables because of water abstraction and river engineering schemes, coastal developments and natural succession may be further threats to this species.

Management and conservation Water tables should be maintained at high levels. Grazing, cutting or some other disturbance, on a rotational basis, may be needed to maintain open conditions.

Published sources Fowler, W.W. (1889), Fowler, W.W. & Donisthorpe, H.St J.K. (1913).

LUCANUS CERVUS   **NOTABLE B**
Stag beetle
Order COLEOPTERA   Family LUCANIDAE

Lucanus cervus (L., 1758).

Distribution England and North Wales. An unpublished distribution map has been prepared by BRC.

Habitat and ecology Broad-leaved woodland, pasture-woodland and gardens. The larvae take at least three and a half years to develop and feed in the decaying roots of old stumps and logs, especially elm, lime and beech. Larvae have also been recorded from a compost heap. Adults are active in the evening, probably only the males fly regularly, and occasionally large swarms are reported. Adults feed on fruit and exudate on trees, and have been recorded from May to August, though also noted in September and November. Adults have been noted at light.

Status Local in south-eastern England, also recorded in South Wales, South West England, and central and northern England. Declining, and recently recorded only in southern England, from Dorset to West Suffolk, most frequent in south London and the New Forest. *L. cervus* is listed on Appendix III of the Bern Convention (the Convention of European Wildlife and Natural Habitats).

Threats Loss of broad-leaved woodland and parkland through, for example, clear-felling and coniferisation. Habitat loss, in particular, through the felling of ancient trees, removal of dead wood from living trees and the destruction or removal of standing and fallen dead wood for reasons such as forest hygiene, aesthetic tidiness, public safety or for use as fire wood. Collecting has been considered a major threat in Europe.

Management and conservation Ancient trees and both fallen and standing dead timber, especially with the bark attached, should be retained. The removal of dead timber from ancient trees should be avoided. Gaps in the age structure of the tree population should be identified and the continuity of the appropriate dead wood habitat ensured by regeneration, suitable planting and possibly with pollarding.

PLATYCERUS CARABOIDES   **EXTINCT**

Order COLEOPTERA   Family LUCANIDAE

Platycerus caraboides (L., 1758). Formerly known as: *Systenocerus caraboides* (L.).

Distribution Recorded from Berkshire, Oxfordshire, Buckinghamshire and South Aberdeenshire.

Habitat and ecology Associated with woodland. On the Continent, this beetle is considered to be an "Urwald" species (associated with primeval or virgin forest). Probably develops in dead wood. Adults have been recorded in June and July.

Status Presumed extinct. Last recorded pre-1830 from Oxfordshire.

Published sources Allen, A.A. (1967e), Fowler, W.W. (1890), Shirt, D.B., ed. (1987).

DICTYOPTERA AURORA   **NOTABLE B**
A net-winged beetle
Order COLEOPTERA   Family LYCIDAE

Dictyoptera aurora (Herbst, 1784). Formerly known as: *Dictyopterus aurora* (Herbst, 1784), *Eros aurora* (Herbst).

Distribution Scotland.

Habitat and ecology Coniferous woodland. Associated with Scots pine. Larvae probably occur in dead wood and under bark. Adults have been found in an old pine stump, under pine bark and under pine chips, and have been recorded in March and from May to July. Adults are short-lived.

Status Local in Highland Scotland. The beetle is a component of the native pine forest fauna. Records outside this distribution are probably the result of chance introduction or casual importation.

Threats Loss of native pine forest through practices such as clear-felling or conversion to plantation forest. Habitat loss in the remaining areas through the felling of ancient trees and the destruction or removal of standing and fallen dead wood for reasons such as

LYCIDAE

forest hygiene, aesthetic tidiness, public safety or for use as fire wood.

Management and conservation Ancient trees and both standing and fallen dead timber, especially with the bark attached, should be retained. Dead wood should not be removed from trees. Gaps in the age structure of the tree population should be identified and regeneration encouraged to ensure the continuity of the dead wood habitats.

PLATYCIS COSNARDI   **INDETERMINATE**
A net-winged beetle
Order COLEOPTERA   Family LYCIDAE

Platycis cosnardi (Chevrolat, 1829). Formerly known as: *Dictyopterus cosnardi* (Chevrolat).

Distribution Recorded from West Sussex and West Gloucestershire before 1970 and West Sussex from 1970 onwards.

Habitat and ecology Ancient, broad-leaved woodland. Found in rotten wood and under bark, possibly associated with beech. The larvae may be predatory. An adult was swept from the herb layer under beech trees. Adults have been recorded in May and June.

Status Status revised from RDB 1 (Endangered) in Shirt (1987). Only known from four examples; the first two were recorded in 1944 from a garden on the Staunton road, near Monmouth, West Gloucestershire, another in 1969 from near Goodwood, West Sussex and most recently in 1984 at Duncton Chalk Pit, West Sussex.

Indicator status Grade 1 in Harding & Rose (1986).

Threats Loss of broad-leaved woodland through, for example, clear-felling and coniferisation. Habitat loss, in particular, through the felling of ancient trees, removal of dead wood from living trees and the destruction or removal of standing and fallen dead wood for reasons such as forest hygiene, aesthetic tidiness, public safety or for use as fire wood.

Management and conservation Ancient trees and both fallen and standing dead timber, especially with the bark attached, should be retained. Dead timber should not be removed from ancient trees. Gaps in the age structure of the tree population should be identified and the continuity of the appropriate dead wood habitat ensured by regeneration and suitable planting.

Published sources Airy Shaw, H.K. (1944), Cooter, J. (1969b), Harding, P.T. & Rose, F. (1986), Porter, D.A. (1987), Shirt, D.B., ed. (1987).

PLATYCIS MINUTA   **NOTABLE B**
A net-winged beetle
Order COLEOPTERA   Family LYCIDAE

Platycis minuta (F., 1787). Formerly known as: *Platycis minutus* (F., 1787).

Distribution England and South Wales.

Habitat and ecology Ancient, broad-leaved woodland and pasture-woodland. Associated predominantly with beech, though also on ash oak and probably other tree species. Larvae develop in dead wood and may be predatory. Adults are usually found inside rotten logs of birch and beech, and by sweeping herbage. Adults have been recorded from August to October, and are short-lived.

Status Widespread but local in England, becoming more local in the north. Also recorded in South Wales.

Indicator status Grade 1 in Garland (1983). Grade 3 in Harding & Rose (1986).

Threats Loss of broad-leaved woodland and parkland through, for example, clear-felling and coniferisation. Habitat loss, in particular, through the felling of ancient trees, removal of dead wood from living trees and the destruction or removal of standing and fallen dead wood for reasons such as forest hygiene, aesthetic tidiness, public safety or for use as fire wood.

Management and conservation Ancient trees and both fallen and standing dead timber, especially with the bark attached, should be retained. Dead timber should not be removed from ancient trees. Gaps in the age structure of the tree population should be identified and the continuity of the appropriate dead wood habitat ensured by regeneration, suitable planting and possibly with pollarding.

PYROPTERUS NIGRORUBER   **NOTABLE A**
A net-winged beetle
Order COLEOPTERA   Family LYCIDAE

Pyropterus nigroruber (Degeer, 1774). Formerly known as: *Dictyopterus affinis* (Paykull, 1799), *Pyropterus affinis* (Paykull).

Distribution Recorded from Nottinghamshire, North-east Yorkshire, South-west Yorkshire, Mid-west Yorkshire, Moray, East Inverness & Nairn and West

Ross before 1970 and North Lincolnshire, Derbyshire, South-west Yorkshire, Mid-west Yorkshire, South Aberdeenshire, Moray and East Inverness & Nairn from 1970 onwards.

Habitat and ecology Ancient, broad-leaved woodland and pasture-woodland. Found in rotten wood and under bark. Adults and larvae have been noted in rotten beech, and larvae have been recorded in May from birch. The species is almost certainly associated with other trees, including pine. Larvae may be predatory. Adults have been found sitting on foliage and have been recorded in July and August. Adults are very short-lived.

Status Status revised from RDB 3 (Rare) in Shirt (1987). This species is very local and has two population centres. One from Nottinghamshire to North-east Yorkshire, the other in northern Scotland. Surrey is listed on the strength of a single example from Box Hill, this is almost certainly a casual importation (A.A. Allen pers. comm.).

Indicator status Grade 1 in Garland (1983). Grade 2 in Harding & Rose (1986).

Threats Loss of broad-leaved woodland and parkland through, for example, clear-felling and coniferisation. Habitat loss, in particular, through the felling of ancient trees, removal of dead wood from living trees and the destruction or removal of standing and fallen dead wood for reasons such as forest hygiene, aesthetic tidiness, public safety or for use as fire wood.

Management and conservation Ancient trees and both fallen and standing dead timber, especially with the bark attached, should be retained. Dead timber should not be removed from ancient trees. Gaps in the age structure of the tree population should be identified and the continuity of the appropriate dead wood habitat ensured by regeneration, suitable planting and possibly with pollarding.

Published sources Buck, F.D. (1938), Donisthorpe, H.St J.K. (1938b), Fowler, W.W. (1890), Fowler, W.W. & Donisthorpe, H.St J.K. (1913), Garland, S.P. (1983), Harding, P.T. (1978), Harding, P.T. & Rose, F. (1986), Shirt, D.B., ed. (1987).

LYCTUS LINEARIS   **NOTABLE B**
A powder-post beetle
Order COLEOPTERA   Family LYCTIDAE

Lyctus linearis (Goeze, 1777). Formerly known as: *Lyctus canaliculatus* F., 1792, *Lyctus fuscus* (L., 1767).

Distribution South East, South, East Anglia, East Midlands, West Midlands, North East and North West.

Habitat and ecology Ancient, broad-leaved woodland. This species has also been found in timber yards and in buildings. Develops in dead wood, particularly that of oak, beech and ash. Often found in fresh oak palings. Adults have been recorded from June to August.

Status Widespread but local in England, not recorded in South West England.

Threats This species is threatened by the loss of broad-leaved woodland through, for example, clear-felling and coniferisation. Habitat loss, in particular, through the felling of ancient trees, removal of dead wood from living trees and the destruction or removal of standing and fallen dead wood for reasons such as forest hygiene, aesthetic tidiness, public safety or for use as fire wood.

Management and conservation Ancient trees and both fallen and standing dead timber, especially with the bark attached, should be retained. Dead timber should not be removed from ancient trees. Gaps in the age structure of the tree population should be identified and the continuity of the appropriate dead wood habitat ensured by regeneration, suitable planting and possibly with pollarding.

HYLECOETUS DERMESTOIDES   **NOTABLE B**

Order COLEOPTERA   Family LYMEXYLIDAE

Hylecoetus dermestoides (L., 1761).

Distribution East Midlands, West Midlands, North East, North West, Dyfed-Powys and Scotland.

Habitat and ecology Ancient broad-leaved woodland and coniferous woodland. The larvae develop in very hard dead wood, particularly trunks and stumps. This species has been recorded from birch, oak, beech, ash, wych elm, holly and pine. Males occasionally swarm. Pupae have been found in May. Adults have been recorded from May to July and are very short-lived.

Status Widespread but local. Recorded from the Midlands through northern England and parts of Wales to northern Scotland.

LYMEXYLIDAE

Indicator status Grade 2 in Garland (1983). Grade 3 in Harding & Rose (1986).

Threats Loss of broad-leaved woodland and native pine forest through, for example, clear-felling and coniferisation. Habitat loss, in particular, through the felling of ancient trees, removal of dead wood from living trees and the destruction or removal of standing and fallen dead wood for reasons such as forest hygiene, aesthetic tidiness, public safety or for use as fire wood.

Management and conservation Ancient trees and both fallen and standing dead timber, especially with the bark attached, should be retained. The removal of dead timber from ancient trees should be avoided. Gaps in the age structure of the tree population should be identified and the continuity of the appropriate dead wood habitat ensured by encouraging regeneration, suitable planting and possibly with pollarding.

LYMEXYLON NAVALE   VULNERABLE

Order COLEOPTERA   Family LYMEXYLIDAE

Lymexylon navale (L., 1758). Formerly known as: *Limexylon navale* (L.).

Distribution Recorded from South Hampshire, Hertfordshire, Berkshire, Cheshire and South Lancashire before 1970 and South Hampshire, Surrey, Berkshire and Herefordshire from 1970 onwards.

Habitat and ecology Ancient broad-leaved woodland and pasture-woodland. Breeds in dead oak. The larvae bore into the seasoned timber of dead standing oaks, usually at some distance above ground. Local populations are often restricted to individual trees. Adults fly in the evening and occasionally swarm. Adults have been recorded from late June to September.

Status Possibly declining. Very local and known from very few sites. Recently recorded from just four vice-counties. Possibly under-recorded owing to its habit of frequenting dead timber well above ground level. This species is occassionally imported with timber.

Indicator status Grade 1 in Harding & Rose (1986).

Threats Loss of broad-leaved woodland and parkland through, for example, clear-felling and coniferisation. Habitat loss, in particular, through the felling of ancient trees, removal of dead wood from living trees and the destruction or removal of standing and fallen dead

wood for reasons such as forest hygiene, aesthetic tidiness, public safety or for use as fire wood.

Management and conservation Ancient trees and both fallen and standing dead timber, especially with the bark attached, should be retained. The removal of dead timber from ancient trees should be avoided. Gaps in the age structure of the tree population should be identified and the continuity of the appropriate dead wood habitat ensured by suitable planting and possibly with pollarding.

Published sources Bedwell, E.C. (1926a), Donisthorpe, H.St J.K. (1926b), Harding, P.T. (1978), Harding, P.T. & Rose, F. (1986), Menzies, I.S. (1990), Shirt, D.B., ed. (1987).

ABDERA AFFINIS   ENDANGERED
A false darkling beetle
Order COLEOPTERA   Family MELANDRYIDAE

Abdera affinis (Paykull, 1799). Formerly known as: *Carida affinis* (Paykull).

Distribution Recorded from Moray before 1970 and Mid Perthshire from 1970 onwards.

Habitat and ecology Woodland. In fungi on trees. Recorded from a fungus on birch and from a fungus-infected birch stump. Adults have been recorded in June and July.

Status Only one recent record, in 1982 from mid Perthshire. Prior to this, it was last recorded in 1909 from Nethy Bridge, Elgin.

Threats Loss of woodland through, for example, clear-felling or conversion to plantation forest. Habitat loss in the remaining areas through the felling of fungus-infected trees and the destruction or removal of standing and fallen dead wood for reasons such as forest hygiene, aesthetic tidiness, public safety or for use as fire wood.

Management and conservation Fungus-infected trees and both standing and fallen dead timber, especially with the bark attached, should be retained. Gaps in the age structure of the tree population should be identified and regeneration encouraged to ensure the continuity of the appropriate habitat.

Published sources Fowler, W.W. & Donisthorpe, H.St J.K. (1913), Shirt, D.B., ed. (1987).

ABDERA BIFLEXUOSA   NOTABLE B
A false darkling beetle
Order COLEOPTERA   Family MELANDRYIDAE

Abdera biflexuosa (Curtis, 1829). Formerly known as: *Abdera bifasciata* sensu Fowler, 1891 not (Marsham, 1802).

Distribution South East, South, South West, East Anglia, East Midlands, West Midlands, North East and Dyfed-Powys.

Habitat and ecology Ancient broad-leaved woodland and parkland. In dead wood, probably breeding in twigs, with records from oak, ash and found under lime. Also recorded in a fungus on alder. Adults have been recorded in April and from June to August.

Status Widespread but local in England, also recorded in Wales.

Indicator status Grade 2 in Garland (1983). Grade 3 in Harding & Rose (1986).

Threats Loss of broad-leaved woodland and parkland through, for example, clear-felling and coniferisation. Habitat loss, in particular, through the felling of ancient trees, removal of dead wood from living trees and the destruction or removal of standing and fallen dead wood for reasons such as forest hygiene, aesthetic tidiness, public safety or for use as fire wood.

Management and conservation Ancient trees and both fallen and standing dead timber, especially with the bark attached, should be retained. The removal of dead timber from ancient trees should be avoided. Gaps in the age structure of the tree population should be identified and the continuity of the appropriate dead wood habitat ensured by regeneration, suitable planting and possibly with pollarding.

ABDERA FLEXUOSA   NOTABLE B
A false darkling beetle
Order COLEOPTERA   Family MELANDRYIDAE

Abdera flexuosa (Paykull, 1799).

Distribution South East, South, East Anglia, East Midlands, West Midlands, North East, North West, Dyfed-Powys, North Wales and Scotland.

Habitat and ecology Woodland, pasture-woodland, and wooded heaths and bogs. In fungi on trees, particularly the fungus *Polyporus radiatus* on alder. Also on willow, and recorded under oak bark and on a fungus-infected oak fence post. Adults have been recorded in May and June.

Status Widespread but local in Great Britain.

Indicator status Grade 2 in Garland (1983).

Threats Loss of woodland and parkland through, for example, clear-felling and conversion to other land use. Habitat loss, in particular, through the felling of fungus-infected trees, removal of dead wood from living trees and the destruction or removal of standing and fallen dead wood for reasons such as forest hygiene, aesthetic tidiness, public safety or for use as fire wood.

Management and conservation Fungus-infected trees and both fallen and standing dead timber, especially with the bark attached, should be retained. The removal of dead timber from trees should be avoided. Gaps in the age structure of the tree population should be identified and the continuity of the appropriate habitat ensured by regeneration, suitable planting and possibly with pollarding.

ABDERA QUADRIFASCIATA   NOTABLE A
A false darkling beetle
Order COLEOPTERA   Family MELANDRYIDAE

Abdera quadrifasciata (Curtis, 1829).

Distribution Recorded from South Hampshire, West Kent, Surrey, South Essex, Middlesex, Berkshire, Huntingdonshire, West Gloucestershire, Herefordshire, Shropshire, Leicestershire & Rutland and Cheshire before 1970 and Surrey, Huntingdonshire, Herefordshire, Worcestershire and Cheshire from 1970 onwards.

Habitat and ecology Ancient broad-leaved woodland and parkland. In rotten wood, particularly hornbeam and oak. Also recorded from beech, horse-chestnut and by beating birch. Adults have been noted from June to September.

Status Widely distributed and very local in England, recorded from South Hampshire to Cheshire. Recently noted from just five vice-counties.

Indicator status Grade 1 in Harding & Rose (1986).

Threats Loss of broad-leaved woodland and parkland through, for example, clear-felling and coniferisation. Habitat loss, in particular, through the felling of ancient trees, removal of dead wood from living trees and the destruction or removal of standing and fallen dead wood for reasons such as forest hygiene, aesthetic tidiness, public safety or for use as fire wood.

MELANDRYIDAE

Management and conservation Ancient trees and both fallen and standing dead timber, especially with the bark attached, should be retained. The removal of dead timber from ancient trees should be avoided. Gaps in the age structure of the tree population should be identified and the continuity of the appropriate dead wood habitat ensured by regeneration, suitable planting and possibly with pollarding.

Published sources Hammond, P.M. (1979), Harding, P.T. (1978), Harding, P.T. & Rose, F. (1986), Menzies, I.S. (1990).

ABDERA TRIGUTTATA   **NOTABLE A**
A false darkling beetle
Order COLEOPTERA   Family MELANDRYIDAE

Abdera triguttata (Gyllenhal, 1810).

Distribution Recorded from West Suffolk, Herefordshire, Lanarkshire, South Aberdeenshire, Moray, East Inverness & Nairn and North Ebudes before 1970 and West Perthshire and East Inverness & Nairn from 1970 onwards.

Habitat and ecology Probably associated with coniferous woodland. Found under the flakey bark of pine and recorded from the canopy of an oak. Adults have been recorded in June and July.

Status Primarily restricted to Scotland, though there are also old records from West Suffolk and Herefordshire in England.

Threats Loss of woodland through, for example, clear-felling and conversion to other land use. Habitat loss in the remaining areas may be through the felling of trees and the destruction or removal of standing and fallen dead wood for reasons such as forest hygiene, aesthetic tidiness, public safety or for use as fire wood.

Management and conservation Trees and both standing and fallen dead timber, especially with the bark attached, should be retained. Gaps in the age structure of the tree population should be identified and regeneration encouraged to ensure the continuity of the dead wood habitats.

Published sources Buck, F.D. (1957c), Buck, F.D. (1958), Fowler, W.W. (1891), Fowler, W.W. & Donisthorpe, H.St J.K. (1913).

ANISOXYA FUSCULA   **NOTABLE A**
A false darkling beetle
Order COLEOPTERA   Family MELANDRYIDAE

Anisoxya fuscula (Illiger, 1798).

Distribution Recorded from Isle of Wight, South Hampshire, East Kent, West Kent, Surrey, "Essex", Hertfordshire, Middlesex, Berkshire, Oxfordshire, East Suffolk, Cambridgeshire, Huntingdonshire, Warwickshire and Merionethshire before 1970 and North Somerset, Surrey, Berkshire, West Suffolk, West Norfolk, Worcestershire, Glamorgan, Leicestershire & Rutland and South-west Yorkshire from 1970 onwards.

Habitat and ecology Ancient broad-leaved woodland, more open areas in the Breckland and also recorded in suburban gardens. Larvae develop in the dead twigs of ash, willow, beech, field maple and lilac. This beetle has also been recorded from grey poplar. Adults generally occur in the canopy and have been recorded from May and to October.

Status Status revised from RDB 3 (Rare) in Shirt (1987). Widespread and very local, recorded from the Isle of Wight, north to South-west Yorkshire, also recorded in parts of Wales. Recently recorded from nine vice-counties.

Indicator status Grade 3 in Harding & Rose (1986).

Threats Loss of broad-leaved woodland through, for example, clear-felling and coniferisation. Habitat loss, in particular, through the felling of ancient trees, removal of dead wood from living trees and the destruction or removal of standing and fallen dead wood for reasons such as forest hygiene, aesthetic tidiness, public safety or for use as fire wood.

Management and conservation Ancient trees and both fallen and standing dead timber, especially with the bark attached, should be retained. The removal of dead timber from ancient trees should be avoided. Gaps in the age structure of the tree population should be identified and the continuity of the appropriate dead wood habitat ensured by regeneration, suitable planting and possibly with pollarding.

Published sources Alexander, K.N.A. & Clements, D.K. (1988), Atty, D.B. (1983), Copestake, D.R. (1990a), Fowler, W.W. (1891), Fowler, W.W. & Donisthorpe, H.St J.K. (1913), Harding, P.T. (1978), Harding, P.T. & Rose, F. (1986), Jones, R.A. (1988), Massee, A.M. (1944), Mendel, H. (1989), Shirt, D.B., ed. (1987), Whitehead, P.F. (1990a), Williams, B.S. (1924).

CONOPALPUS TESTACEUS   **NOTABLE B**
A false darkling beetle
Order COLEOPTERA   Family MELANDRYIDAE

Conopalpus testaceus (Olivier, 1790).

Distribution England.

Habitat and ecology Ancient broad-leaved woodland, wooded bogs, hedgerows, and on farmland and orchards. In dead boughs and small branches. Recorded from rotten oak and also from hazel, apple and once from a beech bough. A larva has been found in November. Adults have been recorded from May to August and may visit wild flowers, particularly umbellifers. This species has been recorded at a mercury vapour light trap.

Status Widespread but local in England.

Indicator status Grade 1 in Garland (1983). Grade 3 in Harding & Rose (1986).

Threats Loss of broad-leaved woodland and parkland through, for example, clear-felling and coniferisation. Habitat loss, in particular, through the felling of trees, removal of dead wood from living trees and the destruction or removal of standing and fallen dead wood for reasons such as forest hygiene, aesthetic tidiness, public safety or for use as fire wood. The use of herbicides and pesticides may be a further threat to this species.

Management and conservation Trees and both fallen and standing dead timber, especially with the bark attached, should be retained. The removal of dead timber from trees should be avoided. Gaps in the age structure of the tree population should be identified and the continuity of the appropriate dead wood habitat ensured by regeneration, suitable planting and possibly with pollarding. The presence of nectar sources such as umbellifers may also be particularly important for this species.

HALLOMENUS BINOTATUS   **NOTABLE B**
A false darkling beetle
Order COLEOPTERA   Family MELANDRYIDAE

Hallomenus binotatus (Quensel, 1790). Formerly known as: *Hallomenus humeralis* Panzer, 1793.

Distribution South East, South, East Anglia, East Midlands, West Midlands, North East, North West and Scotland.

Habitat and ecology Ancient woodland and pasture-woodland. In fungus-infected wood and in fungi on wood. Recorded from the fungi *Polyporus* , *Trametes* and *Laetiporus sulphureus*. Noted from oak, beech, birch, aspen, under pine bark and on the fungus infected wood of a felled Scots pine. This species has been recorded at a light trap. Larvae have been found in April. Adults have been noted from May to September.

Status Widespread but local in England and Scotland, not known in Wales.

Indicator status Grade 3 in Garland (1983). Grade 3 in Harding & Rose (1986).

Threats Loss of woodland and parkland through, for example, clear-felling and conversion to other land use. Habitat loss, in particular, through the felling of fungus-infected trees, removal of dead wood from living trees and the destruction or removal of standing and fallen dead wood for reasons such as forest hygiene, aesthetic tidiness, public safety or for use as fire wood.

Management and conservation Fungus-infected trees and both fallen and standing dead timber, especially with the bark attached, should be retained. The removal of dead timber from fungus-infected trees should be avoided. Gaps in the age structure of the tree population should be identified and the continuity of the appropriate habitat ensured by regeneration, suitable planting and possibly with pollarding.

HYPULUS QUERCINUS   **VULNERABLE**
A false darkling beetle
Order COLEOPTERA   Family MELANDRYIDAE

Hypulus quercinus (Quensel, 1790).

Distribution Recorded from South Devon, North Somerset, Dorset, West Sussex, West Kent, Surrey, Middlesex, Cambridgeshire and Huntingdonshire before 1970 and Bedfordshire, Huntingdonshire, Northamptonshire and Pembrokeshire from 1970 onwards.

Habitat and ecology Ancient broad-leaved woodland and ancient hazel coppice. In the decaying wood of oak, hazel and birch. On the Continent, the larva has also been recorded from ash. In Britain, adults have been recorded from April to June.

Status Old records indicate that this species was widely distributed from South Devon, north to Huntingdonshire. Recently recorded from four vice-counties, including Northamptonshire, and Pembrokeshire in Wales.

MELANDRYIDAE

Indicator status Grade 2 in Harding & Rose (1986).

Threats Loss of broad-leaved woodland through, for example, clear-felling and coniferisation. Habitat loss, in particular, through the felling of ancient trees, removal of dead wood from living trees and the destruction or removal of standing and fallen dead wood for reasons such as forest hygiene, aesthetic tidiness, public safety or for use as fire wood. The loss of ancient hazel coppice through neglect may also be a threat to this species.

Management and conservation Ancient trees and both fallen and standing dead timber, especially with the bark attached, should be retained. The removal of dead timber from ancient trees should be avoided. Gaps in the age structure of the tree population should be identified and the continuity of the appropriate dead wood habitat ensured by regeneration, suitable planting and possibly with pollarding.

Published sources Allen, A.A. (1947b), Harding, P.T. (1978), Harding, P.T. & Rose, F. (1986), Shirt, D.B., ed. (1987).

MELANDRYA BARBATA   ENDANGERED
A false darkling beetle
Order COLEOPTERA   Family MELANDRYIDAE

Melandrya barbata (F., 1787). Formerly known as: *Melandrya dubia* sensu auct. Brit. not (Schaller, 1783).

Distribution Recorded from South Hampshire before 1970 and South Hampshire and Surrey from 1970 onwards.

Habitat and ecology Ancient broad-leaved woodland. In decaying wood, primarily that of oak and beech. Adults have been recorded in May and June.

Status Only three recent records; one in 1971 from Surrey, the other two from the New Forest, South Hampshire. Otherwise only known from old records from the New Forest. Old records for Oxfordshire and Berkshire are considered doubtful.

Indicator status Grade 1 in Harding & Rose (1986).

Threats Loss of broad-leaved woodland through, for example, clear-felling and coniferisation. Habitat loss, in particular, through the felling of ancient trees, removal of dead wood from living trees and the destruction or removal of standing and fallen dead wood for reasons such as forest hygiene, aesthetic tidiness, public safety or for use as fire wood.

Management and conservation Ancient trees and both fallen and standing dead timber, especially with the bark attached, should be retained. The removal of dead timber from ancient trees should be avoided. Gaps in the age structure of the tree population should be identified and the continuity of the appropriate dead wood habitat ensured by regeneration, suitable planting and possibly with pollarding. Much of the New Forest has been notified as an SSSI, it is also designated as a National Park.

Published sources Harding, P.T. (1978), Harding, P.T. & Rose, F. (1986), Shirt, D.B., ed. (1987).

MELANDRYA CARABOIDES   NOTABLE B
A false darkling beetle
Order COLEOPTERA   Family MELANDRYIDAE

Melandrya caraboides (L., 1761).

Distribution England, Wales, South East Scotland and South West Scotland.

Habitat and ecology Ancient broad-leaved woodland, pasture-woodland, wooded heaths and bogs, and river margins. In rotten wood, particularly stumps, possibly with a preference for willow. Also recorded from oak, beech, birch, hornbeam, ash, elm, cultivated plum and cherry, and probably on a number of other tree species. Adults have been recorded from April to July.

Status Widespread but local in Great Britain, not known from northern Scotland.

Indicator status Grade 2 in Garland (1983). Grade 3 in Harding & Rose (1986).

Threats Loss of broad-leaved woodland and parkland through, for example, clear-felling and coniferisation. Habitat loss, in particular, through the felling of trees, removal of dead wood from living trees and the destruction or removal of standing and fallen dead wood for reasons such as forest hygiene, aesthetic tidiness, public safety or for use as fire wood. River engineering and water abstraction schemes may also be a threat to this species.

Management and conservation Trees and both fallen and standing dead timber, especially with the bark attached, should be retained. The removal of dead timber from trees should be avoided. Gaps in the age structure of the tree population should be identified and the continuity of the appropriate dead wood habitat ensured by regeneration, suitable planting and possibly with pollarding.

ORCHESIA MICANS   NOTABLE B
A false darkling beetle
Order COLEOPTERA   Family MELANDRYIDAE

Orchesia micans (Panzer, 1793).

Distribution South East, South, East Anglia, East Midlands, West Midlands, North East, North West, Wales, South East Scotland, South West Scotland and North East Scotland.

Habitat and ecology Woodland and pasture-woodland. In bracket fungi on trees. Recorded from *Polyporus hispidus* on ash, *P. radiatus* on alder and bracket-fungi on beech. Larvae have been recorded in April. Adults have been noted in March, July, August and October.

Status Widespread and very local in Great Britain.

Threats Loss of woodland and parkland through, for example, clear-felling and conversion to other land use. Habitat loss, in particular, through the felling of fungus-infected trees, removal of dead wood from living trees and the destruction or removal of standing and fallen dead wood for reasons such as forest hygiene, aesthetic tidiness, public safety or for use as fire wood.

Management and conservation Fungus-infected trees and both fallen and standing dead timber, especially with the bark attached, should be retained. The removal of dead timber from fungus-infected trees should be avoided. Gaps in the age structure of the tree population should be identified and the continuity of the appropriate dead wood habitat ensured by regeneration, suitable planting and possibly with pollarding.

ORCHESIA MINOR   NOTABLE B
A false darkling beetle
Order COLEOPTERA   Family MELANDRYIDAE

Orchesia minor Walker, 1836. Formerly known as: *Clinocara tetratoma* (Thomson, 1864).

Distribution England, Dyfed-Powys and Scotland.

Habitat and ecology Woodland and pasture-woodland. In fungi on trees, particularly *Polyporus*, and in rotten wood. Recorded from hazel, lime, beech and under pine log bark. This species has been found by beating oak and also recorded from a bird's nest. Adults have been recorded in January and from April to September.

Status Widespread and very local in Great Britain.

Threats Loss of woodland and parkland through, for example, clear-felling and coniferisation. Habitat loss, in particular, through the felling of fungus-infected trees, removal of dead wood from living trees and the destruction or removal of standing and fallen dead wood for reasons such as forest hygiene, aesthetic tidiness, public safety or for use as fire wood.

Management and conservation Fungus-infected trees and both fallen and standing dead timber, especially with the bark attached, should be retained. The removal of dead timber from fungus-infected trees should be avoided. Gaps in the age structure of the tree population should be identified and the continuity of the appropriate dead wood habitat ensured by regeneration, suitable planting and possibly with pollarding.

OSPHYA BIPUNCTATA   RARE
A false darkling beetle
Order COLEOPTERA   Family MELANDRYIDAE

Osphya bipunctata (F., 1775).

Distribution Recorded from West Kent, South Essex, Berkshire, Oxfordshire, East Suffolk, Cambridgeshire, Bedfordshire, Huntingdonshire, Northamptonshire, East Gloucestershire and North-east Yorkshire before 1970 and North Somerset, West Kent, South Essex, Buckinghamshire, Cambridgeshire, Huntingdonshire, Northamptonshire and East Gloucestershire from 1970 onwards.

Habitat and ecology Ancient broad-leaved woodland. Adults are usually found on the flowers of hawthorn, and have also been recorded from dog rose, wayfaring tree, guelder-rose and field maple. Larvae probably develop in dead wood. Adults have been recorded in May and June.

Status Widespread and very local in England. Recorded from West Kent, north to North-east Yorkshire.

Threats Loss of broad-leaved woodland through, for example, clear-felling and coniferisation. Habitat loss, in particular, through the felling of trees, removal of dead wood from living trees and the destruction or removal of standing and fallen dead wood for reasons such as forest hygiene, aesthetic tidiness, public safety or for use as fire wood. Neglect and conversion to high forest may also be a threat to this species.

Management and conservation Trees and both fallen and standing dead timber, especially with the bark attached, should be retained. The removal of dead timber from trees should be avoided. Gaps in the age

MELANDRYIDAE

structure of the tree and shrub population should be identified and the continuity of the appropriate habitat ensured by regeneration or possibly suitable planting. Open glades and ride margins should be cut on rotation to retain a variety of vegetation structures.

Published sources Atty, D.B. (1983), Carter, I.S. (1980), Forster, H.W. (1952b), Fowler, W.W. & Donisthorpe, H.St J.K. (1913), Hunter, F.A. (1955), McClenaghan, I. (1988), Shirt, D.B., ed. (1987), Stott, N. (1981).

PHLOIOTRYA VAUDOUERI   NOTABLE B
A false darkling beetle
Order COLEOPTERA   Family MELANDRYIDAE

Phloiotrya vaudoueri Mulsant, 1856. Formerly known as: *Phloeotrya rufipes* Mulsant, 1856, *Phloiotrya rufipes* sensu auct. Brit. not (Gyllenhal, 1810).

Distribution England.

Habitat and ecology Ancient broad-leaved woodland and parkland. Larvae develop in dead sapwood, mainly that of oak though also found in sweet chestnut. Adults have been recorded from beech, ash and hornbeam. Adults have been recorded from May to July.

Status Widespread but local in England.

Indicator status Grade 1 in Garland (1983). Grade 2 in Harding & Rose (1986).

Threats Loss of broad-leaved woodland and parkland through, for example, clear-felling and coniferisation. Habitat loss, in particular, through the felling of ancient trees, removal of dead wood from living trees and the destruction or removal of standing and fallen dead wood for reasons such as forest hygiene, aesthetic tidiness, public safety or for use as fire wood.

Management and conservation Ancient trees and both fallen and standing dead timber, especially with the bark attached, should be retained. The removal of dead timber from ancient trees should be avoided. Gaps in the age structure of the tree population should be identified and the continuity of the appropriate dead wood habitat ensured by regeneration, suitable planting and possibly with pollarding.

XYLITA LAEVIGATA   NOTABLE A
A false darkling beetle
Order COLEOPTERA   Family MELANDRYIDAE

Xylita laevigata (Hellenius, 1786). Formerly known as: *Xylita buprestoides* (F., 1792).

Distribution Recorded from Midlothian, Mid Perthshire, South Aberdeenshire, Moray and East Inverness & Nairn before 1970 and South Aberdeenshire, East Inverness & Nairn, West Inverness, Argyll Main, West Ross and East Ross from 1970 onwards.

Habitat and ecology Native pine forest and conifer plantations. In rotten wood and under bark. Larvae have been recorded in June and July, and have been found in very dry wood in a dead, standing pine.

Status Only known in Scotland and recently recorded from just six vice-counties.

Threats Loss of native pine forest through, for example, clear-felling and conversion to other land use. Habitat loss in the remaining areas through the felling of ancient trees and the destruction or removal of standing and fallen dead wood for reasons such as forest hygiene, aesthetic tidiness, public safety or for use as fire wood.

Management and conservation Ancient trees and both standing and fallen dead timber, especially with the bark attached, should be retained. Gaps in the age structure of the tree population should be identified and regeneration encouraged to ensure the continuity of the dead wood habitats.

Published sources Fowler, W.W. (1891), Hunter, F.A. (1977), Owen, J.A. (1990a).

ZILORA FERRUGINEA   NOTABLE B
A false darkling beetle
Order COLEOPTERA   Family MELANDRYIDAE

Zilora ferruginea (Paykull, 1798).

Distribution Scotland.

Habitat and ecology Coniferous woodland. In fungi on trees. Recorded in *Trichaptum abietinum* on Scots pine, also under pine bark and from a cut pine bough. The pupa has been found in July (adult emerged in August). Adults have been recorded from June to August.

Status Very local and only known from Scotland.

Threats Loss of native pine forest through, for example, clear-felling and conversion to other land use. Habitat loss in the remaining areas through the felling of ancient and fungus-infected trees and the destruction or removal of standing and fallen dead wood for reasons such as forest hygiene, aesthetic tidiness, public safety or for use as fire wood.

Management and conservation Ancient and fungus-infected trees and both standing and fallen dead timber, especially with the bark attached, should be retained. Gaps in the age structure of the tree population should be identified and regeneration encouraged to ensure the continuity of the dead wood habitats.

APALUS MURALIS   **ENDANGERED**
An oil beetle
Order COLEOPTERA   Family MELOIDAE

Apalus muralis (Forster, 1771). Formerly known as: *Sitaris muralis* (Forster).

Distribution Recorded from "Devon", South Hampshire, "Kent", Surrey, Oxfordshire, West Gloucestershire and Warwickshire before 1970.

Habitat and ecology In and about the nests of mason bees and other bees (Hymenoptera), mainly in old walls and occasionally in the ground or in banks. The larvae feed on the bee's brood. In Britain, it is probably chiefly associated with *Anthophora plumipes* and *A. retusa*, though it has also been recorded from a *Bombus terrestris* nest. Adults have been noted in August and September.

Status No recent records. Formerly with a widely scattered distribution in southern England. Not uncommon in the Oxford area from the early part of this century up until the mid-1940's. Last recorded in 1969 from Wheatley, Oxford.

Threats The habitat of this species in the Oxford area has been destroyed.

Published sources Atty, D.B. (1983), Fowler, W.W. (1891), Fowler, W.W. & Donisthorpe, H.St J.K. (1913), Parmenter, L. (1956), Shirt, D.B., ed. (1987).

MELOE AUTUMNALIS   **ENDANGERED**
An oil beetle
Order COLEOPTERA   Family MELOIDAE

Meloe autumnalis Olivier, 1792.

Distribution Recorded from South Devon, Dorset, East Kent, West Kent, North Essex and Cambridgeshire before 1970.

Habitat and ecology Primarily coastal. Sandy cliffs and grassland. The larvae are parasitic, probably on bees (Hymenoptera) of the genera *Anthophora* and *Osmia*. Adults occur mainly in the autumn though have also been recorded in March.

Status Status revised from RDB 3 (Rare) in Shirt (1987). Virtually all of the few known records are for the 19th century, the only 'recent' record being in 1952 from Holland-on-Sea, North Essex.

Threats Uncertain, though this species may have declined because of grassland improvement or conversion to arable agriculture. Cliff stabilisation and coastal defence schemes may also have played a part in this beetles decline.

Published sources Allen, A.A. (1962b), Shirt, D.B., ed. (1987).

MELOE BREVICOLLIS   **ENDANGERED**
An oil beetle
Order COLEOPTERA   Family MELOIDAE

Meloe brevicollis Panzer, 1793.

Distribution Recorded from East Cornwall, South Devon, Dorset, Isle of Wight, South Hampshire, East Sussex, East Kent, West Kent, Surrey, Berkshire, Merionethshire and Derbyshire before 1970.

Habitat and ecology Sandy heathland and coastal cliffs. The larvae are parasitic, probably on bees (Hymenoptera) of the genera *Anthophora* and *Osmia*. Adults have been recorded from April and May.

Status Status revised from RDB 3 (Rare) in Shirt (1987). No recent records. Last recorded in 1948 from Chailey Common, East Sussex.

Threats This species may have declined because of the loss and fragmentation of heathland through changes in land use, primarily to arable agriculture, forestry, urban development and through lack of traditional management. Cliff stabilisation and coastal defence may also have played a part in this beetles decline.

MELOIDAE

Published sources Allen, A.A. (1962b), Fowler, W.W. (1891), Fowler, W.W. & Donisthorpe, H.St J.K. (1913), Shirt, D.B., ed. (1987).

MELOE CICATRICOSUS   **ENDANGERED**
An oil beetle
Order COLEOPTERA   Family MELOIDAE

Meloe cicatricosus Leach, 1811. Formerly known as: *Meloe cicatricosa* Leach, 1811.

Distribution Recorded from South Devon, East Kent and South Essex before 1970.

Habitat and ecology Coastal grassland. The larvae are parasitic, probably on bees (Hymenoptera) of the genera *Anthophora* and *Osmia*. Adults generally occur in the early spring.

Status Status revised from RDB 3 (Rare) in Shirt (1987). Virtually all the few known records are from the 19th century. Last recorded in 1906 from Margate, East Kent.

Threats Uncertain, though this species may have declined because of grassland improvement or conversion to arable agriculture.

Published sources Fowler, W.W. (1891), Shirt, D.B., ed. (1987).

MELOE RUGOSUS   **RARE**
An oil beetle
Order COLEOPTERA   Family MELOIDAE

Meloe rugosus Marsham, 1802. Formerly known as: *Meloe rugosa* Marsham, 1802.

Distribution Recorded from South Devon, East Kent, South Essex, Berkshire, Oxfordshire, East Gloucestershire and Derbyshire before 1970 and North Somerset, East Gloucestershire, West Gloucestershire and Worcestershire from 1970 onwards.

Habitat and ecology Grassland, including a grassy area in a garden and an area on Oolitic limestone, and a sandpit. The larvae are parasitic on bees (Hymenoptera) of the genera *Anthophora* and *Osmia*. Adults have been noted feeding on the foliage of highly toxic plant species. Recent observations have shown adults to be active in the autumn with females ovipositing in December. Adults seek shelter in severe waether. Spring records are probably the result of adults overwintering.

Status Possibly declining. Records indicate a scattered distribution throughout southern England, recorded as far north as Derbyshire. All modern records are from the West Midlands and North Somerset.

Threats Uncertain, though this species is probably threatened by grassland improvement or conversion to arable agriculture.

Management and conservation Cutting or grazing is needed to maintain open conditions.

Published sources Alexander, K.N.A. (1989), Alexander, K.N.A. & Grove, S.J. (1990), Atty, D.B. (1983), Brown, D.G. (1977), Fowler, W.W. (1891), Shirt, D.B., ed. (1987), Walker, J.J. (1928a), Whitehead, P.F. (1989a), Whitehead, P.F. (1990a).

MELOE VARIEGATUS   **EXTINCT**
An oil beetle
Order COLEOPTERA   Family MELOIDAE

Meloe variegatus Donovan, 1793. Formerly known as: *Meloe variegata* Donovan, 1793.

Distribution Recorded from "Hants" and East Kent before 1970.

Habitat and ecology Grassland near the coast. The larvae are parasitic, probably on bees (Hymenoptera) of the genera *Anthophora* and *Osmia*. Adults occur in the early spring.

Status Status revised from RDB 3 (Rare) in Shirt (1987). Presumed extinct. No known records this century. Last recorded in 1882 from Margate, East Kent.

Published sources Fowler, W.W. (1891), Shirt, D.B., ed. (1987).

MELOE VIOLACEUS   **NOTABLE B**
Oil beetle
Order COLEOPTERA   Family MELOIDAE

Meloe violaceus Marsham, 1802. Formerly known as: *Meloe violacea* Marsham, 1802.

Distribution England, South Wales, Dyfed-Powys and Scotland.

Habitat and ecology Grassland, heathland and grassy areas in moorland. The larvae are parasitic, probably on bees (Hymenoptera) of the genera *Anthophora* and *Osmia*. Adults have been noted feeding on watercress, round-leaved crowfoot, creeping buttercup and lesser spearwort. Adults have been recorded from March to May.

Status Widespread but local, though predominantly a northern and western species.

Threats Uncertain, though this species is possibly threatened by loss of its habitat through afforestation, agricultural improvement and conversion to arable.

Management and conservation Management should concentrate on maintaining a diversity of succesional stages, preferably by animal grazing or by rotational cutting, scraping or in certain cases burning.

APLOCNEMUS NIGRICORNIS   NOTABLE A
A false soldier beetle
Order COLEOPTERA   Family MELYRIDAE

Aplocnemus nigricornis (F., 1792). Formerly known as: *Haplocnemus nigricornis* (F., 1792).

Distribution Recorded from South Hampshire, East Sussex, East Kent, Hertfordshire, Middlesex, Huntingdonshire, East Gloucestershire, Leicestershire & Rutland, Nottinghamshire, Mid-west Yorkshire and Midlothian before 1970 and East Kent, South Essex, East Suffolk, Huntingdonshire, Radnorshire, North-east Yorkshire and Roxburghshire from 1970 onwards.

Habitat and ecology Ancient broad-leaved woodland and pasture-woodland. Also found in mixed and coniferous woodland. This species has been found by beating and sweeping beneath trees, a single example was noted by beating bird cherry. Adults have been recorded from April to July.

Status Very local and widely scattered throughout Great Britain, being recorded as far north as Midlothian. Few recent records. Possibly not as uncommon as records suggest due to the difficulty in distinguishing this species from *A. pini*.

Indicator status Grade 1 in Garland (1983). Grade 3 in Harding & Rose (1986).

Threats Uncertain, though this species is probably threatened by the loss of broad-leaved woodland and parkland through, for example, clear-felling and conversion to other land use.

Published sources Allen, A.A. (1951a), Atty, D.B. (1983), Crowson, R.A. (1962), Crowson, R.A. (1983b), Flint, J.H. (1988), Fowler, W.W. (1890), Garland, S.P. (1983), Harding, P.T. (1978), Harding, P.T. & Rose, F. (1986), Nash, D.R. (1975).

APLOCNEMUS PINI   NOTABLE B
A false soldier beetle
Order COLEOPTERA   Family MELYRIDAE

Aplocnemus pini (Redtenbacher, 1849). Formerly known as: *Haplocnemus impressus* sensu Fowler, 1890 not (Marsham, 1802), *Haplocnemus pini* sensu Fowler, 1890 not (Marsham, 1802).

Distribution England, South East Scotland and North East Scotland.

Habitat and ecology Ancient broad-leaved woodland and pasture-woodland. Also recorded from coniferous woodland and possibly orchards. Found under bark, in decayed wood and by sweeping. Recorded from oak, pine, willow and hawthorn, possibly also associated with apple, pear and elm. Adults have been recorded in April and June.

Status Widely distributed and local. Apparently not known from Wales. This species is difficult to distinguish from *A. nigricornis*.

Indicator status Grade 3 in Harding & Rose (1986).

Threats Uncertain, though this species is probably threatened by the loss of broad-leaved woodland and parkland through, for example, clear-felling and conversion to other land use.

Management and conservation Uncertain, though ancient trees and both fallen and standing dead timber, especially with the bark attached, should be retained. The removal of dead timber from ancient trees should be avoided. Gaps in the age structure of the tree population should be identified and the continuity of the appropriate dead wood habitat ensured by regeneration, suitable planting and possibly with pollarding.

AXINOTARSUS PULICARIUS   ENDANGERED
A false soldier beetle
Order COLEOPTERA   Family MELYRIDAE

Axinotarsus pulicarius (F., 1777).

Distribution Recorded from East Sussex, East Kent, West Kent, Surrey and North Essex before 1970.

Habitat and ecology Grassland and probably coastal shingle. Most records are on or near the coast. Larvae probably develop in stems or at the roots of plants. Adults occur on grasses, herbage and flowers. Adults have been recorded in June.

357

MELYRIDAE

Status Status revised from RDB 2 (Vulnerable) in Shirt (1987). No recent records. Apparently last recorded in 1923 from Wivenhoe near Colchester, North Essex. Any modern records of *A. pulicarius* should be checked carefully as this species is easily be confused with *A. marginalis*, a recent, and spreading, colonist.

Threats Uncertain, though loss of coastal grassland through agricultural improvement, conversion to arable and urban development may have been responsible for this species decline.

Published sources Fowler, W.W. (1890), Fowler, W.W. & Donisthorpe, H.St J.K. (1913), Shirt, D.B., ed. (1987).

CERAPHELES TERMINATUS   NOTABLE A
A false soldier beetle
Order COLEOPTERA   Family MELYRIDAE

Cerapheles terminatus (Ménétriés, 1832). Formerly known as: *Anthocomus terminatus* (Ménétriés).

Distribution Recorded from Surrey, East Norfolk, Cambridgeshire and Huntingdonshire before 1970 and Dorset, East Norfolk, Cambridgeshire, Huntingdonshire and Glamorgan from 1970 onwards.

Habitat and ecology Fens, saltmarshes and other wetland habitats. Adults have been swept from vegetation. Adults have been recorded in June.

Status Very local, with a scattered distribution in southern England ranging to South Wales.

Threats Drainage for reasons such as agricultural improvement is the primary cause of the loss of fenland and other wetland habitats. Saltmarshes are threatened by reclamation and the construction of sea defences.

Management and conservation Water tables should be maintained at high levels. On fenland sites grazing or cutting is needed to maintain open conditions. On saltmarshes, grazing should not be introduced on sites where there is no grazing at present.

Published sources Fowler, W.W. (1890), Fowler, W.W. & Donisthorpe, H.St J.K. (1913).

DASYTES COERULEUS
A false soldier beetle   INSUFFICIENTLY KNOWN
Order COLEOPTERA   Family MELYRIDAE

Dasytes coeruleus (Degeer, 1774).

Distribution Recorded from Anglesey from 1970 onwards.

Habitat and ecology Unknown.

Status Status uncertain. Not listed in the insect Red Data Book (Shirt, 1987). Only known from one specimen taken in Anglesey before 1975 (but probably post 1970) by a student.

Published sources Johnson, C. (1975b).

DASYTES NIGER   NOTABLE A
A false soldier beetle
Order COLEOPTERA   Family MELYRIDAE

Dasytes niger (L., 1761).

Distribution Recorded from North Wiltshire, South Hampshire, West Sussex, Surrey and Berkshire before 1970 and South Wiltshire, South Hampshire and West Sussex from 1970 onwards.

Habitat and ecology Woodland and downland. Larvae probably develop in dead wood. Adults are usually found visiting flowers in more open situations such as downland, hedge-banks and railway margins, and have been recorded on common rockrose. Larvae have been found in November. Adults have been recorded from May to -July.

Status Very local and apparently confined to a small number of counties in southern England. There is a 19th century record of this species from Midlothian which, until authenticated, must be regarded as doubtful.

Threats The loss of herb rich areas adjacent to woodland. Loss of woodland through, for example, clear-felling and conversion to other land use. This species may be further threatened by the removal of dead wood from living trees and shrubs and the destruction or removal of standing and fallen dead wood for reasons such as forest hygiene, aesthetic tidiness, public safety or for use as firewood.

Management and conservation Ancient trees and both fallen and standing dead timber, especially with the bark attached, should be retained. The removal of dead timber from ancient trees should be avoided. Gaps in the age structure of the tree population should be identified and the continuity of the appropriate dead wood habitat ensured by suitable planting and possibly with pollarding. The presence of nectar sources such as umbellifers and composite herbs may be particularly important for this species.

Published sources Allen, A.A. (1959d), Easton, A.M. (1965), Fowler, W.W. (1890), Fowler, W.W. & Donisthorpe, H.St J.K. (1913), Holford, N.A. (1968).

DASYTES PLUMBEUS   **NOTABLE B**
A false soldier beetle
Order COLEOPTERA   Family MELYRIDAE

Dasytes plumbeus (Müller, 1776). Formerly known as: *Dasytes oculatus* sensu Fowler, 1890 not Kiesenwetter, 1867.

Distribution South East, South West, East Anglia, East Midlands, North East, North West and South Wales.

Habitat and ecology This species has been noted from grassland in derelict chalk pits, cliff tops, grazing levels, fen edge, neutral grassland on a railway cutting and possibly woodland. Found by sweeping grassland and also by beating oak. Adults have been recorded in June and early July.

Status Uncertain because of former confusion over nomenclature combined with the difficulty in reliably distinguishing this species from its close relative *D. puncticollis*. This is likely to lead to under-recording of this species. From specimens so far identified it would appear that *D. plumbeus* is the commoner of the two species (R.S. Key pers. comm.).

Indicator status Grade 2 in Garland (1983).

DASYTES PUNCTICOLLIS   **NOTABLE B**
A false soldier beetle
Order COLEOPTERA   Family MELYRIDAE

Dasytes puncticollis Reitter, 1888. Formerly known as: *Dasytes flavipes* (Olivier, 1790).

Distribution South East, South West and East Anglia.

Habitat and ecology Grassland, especially coastal grassland. Adults have been recorded in June and July.

Status The exact status of this species is uncertain because of former confusion over nomenclature combined with difficulty in reliably distinguishing the beetle from its close relative *D. plumbeus*. This is likely to lead to under-recording of this species. From specimens so far identified it would appear that *D.*

puncticollis is the rarer of the two species (R.S. Key pers. comm.). Possibly noted in the East Midlands region.

DOLICHOSOMA LINEARE   **NOTABLE B**
A false soldier beetle
Order COLEOPTERA   Family MELYRIDAE

Dolichosoma lineare (Rossi, 1794).

Distribution South East, South West, East Anglia and East Midlands.

Habitat and ecology Saltmarshes, tidal creeks and coastal grassland. Adults are usually swept from vegetation. Adults have been recorded from June to September.

Status Local in southern and south-eastern England, as far north as Lincolnshire.

Threats Loss of saltmarsh through reclamation and the construction of sea defences. Overgrazing of saltmarshes could be a further threat to this species. The agricultural improvement of coastal grassland.

Management and conservation Grazing should not be introduced to a site where there is no grazing at present.

EBAEUS PEDICULARIUS   **EXTINCT**
A false soldier beetle
Order COLEOPTERA   Family MELYRIDAE

Ebaeus pedicularius (L., 1758).

Distribution Recorded from "Devon", "Bristol district" and Berkshire.

Status Presumed extinct. Last recorded in the 19th century.

Published sources Donisthorpe, H.St J.K. & Tomlin J.R. le B. (1934), Shirt, D.B., ed. (1987).

HYPEBAEUS FLAVIPES   **ENDANGERED**
Moccas beetle
Order COLEOPTERA   Family MELYRIDAE

Hypebaeus flavipes (F., 1787). Formerly known as: *Ebaeus abietinus* sensu auct. Brit. not Abeille, 1869.

Distribution Recorded from Herefordshire before 1970 and Herefordshire from 1970 onwards.

MELYRIDAE

Habitat and ecology Ancient broad-leaved woodland and pasture-woodland. Associated with oak. In Germany, this species has been noted from hornbeam. In Britain, larvae probably develop in the red-rotten wood. Adults occur in the canopy and have been found on the grass under host oaks. Adults have been recorded in June and July.

Status Only known from very few trees in Moccas Park, Herefordshire, where it was first discovered in 1934.

Indicator status Grade 1 in Harding & Rose (1986).

Threats Habitat loss through natural death, the felling of ancient trees, removal of dead wood from living trees and the destruction or removal of standing and fallen dead wood for reasons such as forest hygiene, aesthetic tidiness, public safety or for use as fire wood.

Management and conservation Ancient trees and both fallen and standing dead timber, especially with the bark attached, should be retained. The removal of dead timber from ancient trees should be avoided. Gaps in the age structure of the tree population should be identified and the continuity of the appropriate dead wood habitat ensured by suitable planting and possibly with pollarding. Moccas Park is an NNR.

Published sources Blair, K.G. (1943), Cooter, J. (1990d), Donisthorpe, H.St J.K. & Tomlin J.R. le B. (1934), Harding, P.T. (1978), Harding, P.T. & Rose, F. (1986), Shirt, D.B., ed. (1987).

MALACHIUS AENEUS   RARE
A false soldier or malachite beetle
Order COLEOPTERA   Family MELYRIDAE

Malachius aeneus (L., 1758).

Distribution Recorded from South Devon, North Devon, North Somerset, Dorset, Isle of Wight, South Hampshire, East Sussex, East Kent, West Kent, Surrey, South Essex, Hertfordshire, Berkshire, East Gloucestershire, Warwickshire, Glamorgan, Derbyshire, Cheshire, South-east Yorkshire, Durham and South Northumberland before 1970 and North Somerset, South Hampshire, West Kent and Hertfordshire from 1970 onwards.

Habitat and ecology Grassland and grassy areas in woodland. Larvae are probably predators. Adults frequent flowers, especially *Ranunculus* spp., and have been swept from vegetation. Adults have been recorded in May and June.

Status Formerly local and widespread in England. This beetle has declined and has recently been recorded from very few sites in southern England.

Threats Grasslands are threatened by agricultural improvement, conversion to arable and reseeding. Changes in grazing regimes may also adversely affect this species.

Management and conservation Grazing or cutting, on a rotational basis, is needed to maintain open conditions.

Published sources Atty, D.B. (1983), Fowler, W.W. (1890), Fowler, W.W. & Donisthorpe, H.St J.K. (1913), Hodge, P.J. (1989), Hodge, P.J. (1990), Shirt, D.B., ed. (1987).

MALACHIUS BARNEVILLEI   RARE
A false soldier or malachite beetle
Order COLEOPTERA   Family MELYRIDAE

Malachius barnevillei Puton, 1865.

Distribution Recorded from West Norfolk before 1970 and West Norfolk from 1970 onwards.

Habitat and ecology Coastal sandhills. Adults have been found on marram and on the flowers of field bindweed and ragwort. Adults have been recorded from June to August.

Status Only known from coastal sandhills on the north Norfolk coast.

Threats Loss of habitat, particularly through urban and holiday development, and afforestation. The degradation of other areas by excessive disturbance of the vegetation through activities such as motorbike access, horse-riding and human trampling.

Management and conservation Some grazing and other disturbance may be desirable to maintain the early successional stages and prevent the invasion of scrub.

Published sources Anon, (1935), Collier, M.J. (1988a), Fowler, W.W. & Donisthorpe, H.St J.K. (1913), Shirt, D.B., ed. (1987).

MALACHIUS MARGINELLUS   NOTABLE B
A false soldier or malachite beetle
Order COLEOPTERA   Family MELYRIDAE

Malachius marginellus Olivier, 1790. Formerly known as: *Malachius angustimarginalis* Donisthorpe, 1933, *Malachius elegans* sensu auct. Brit. not (Fourcroy, 1785), *Malachius pseudosardous* Reclairie and va der Wiel, 1932.

Distribution South East, South, South West, East Anglia, East Midlands, West Midlands and South East Scotland.

Habitat and ecology Primarily coastal. This species has been recorded from coastal shingle, riverside vegetation and grassland. Adults have been found on flowers, with records from hogweed, tansy and *Hieracium* spp.. Adults have been recorded from May to July.

Status Widely distributed and local in the southern half of England. This beetle has also been recorded from Peeblesshire in Scotland.

Threats The loss of coastal grassland habitats through agricultural improvements and conversion to arable. Coastal shingle is threatened by gravel extraction, disturbance and urban and tourist development. The construction of sea defences and river engineering schemes may also affect this species.

Management and conservation Except on coastal shingle, which can be left unmanaged, grazing or cutting may be required to maintain open conditions.

MALACHIUS VULNERATUS   RARE
A false soldier or malachite beetle
Order COLEOPTERA   Family MELYRIDAE

Malachius vulneratus Abeille, 1891.

Distribution Recorded from East Kent and West Kent before 1970 and East Kent, West Kent, South Essex and North Essex from 1970 onwards.

Habitat and ecology Saltmarshes, tidal creeks and coastal grazing marshes. Adults have been found by sweeping vegetation. Adults have been recorded from May to July.

Status Very local and apparently confined to coastal sites on the Thames Estuary.

Threats Loss of saltmarsh through reclamation and the construction of sea defences. Overgrazing of saltmarshes could be a further threat to this species.

Management and conservation Grazing should not be introduced to a site where there is no grazing at present.

Published sources Allen, A.A. (1949a), Fowler, W.W. & Donisthorpe, H.St J.K. (1913), Shirt, D.B., ed. (1987).

SPHINGINUS LOBATUS
A false soldier beetle   INSUFFICIENTLY KNOWN
Order COLEOPTERA   Family MELYRIDAE

Sphinginus lobatus (Olivier, 1790).

Distribution Recorded from South Hampshire from 1970 onwards.

Habitat and ecology Grassy banks under and near oak trees, also recorded from an oak on a roadside verge. Probably breeds in dead twigs of oak and possibly other trees. Adults have been recorded in June and July.

Status Not listed in the insect Red Data Book (Shirt, 1987). Recorded new to Great Britain from examples found in 1982 at Titchfield Common, South Hampshire. Since found at other sites close to the original site. The beetle is very small and is likely to be overlooked.

Threats Uncertain, though the clear-felling of the host-trees, the use of herbicides and pesticides, and natural succession are possible threats.

Management and conservation Gaps in the age structure of the tree population should be determined and continuity of suitable habitat ensured by regenration or planting.

Published sources Allen, A.A. (1984b).

MORDELLA HOLOMELAENA
     INSUFFICIENTLY KNOWN
Order COLEOPTERA   Family MORDELLIDAE

Mordella holomelaena Apfelbeck, 1914. Formerly confused under: *Mordella aculeata* L., 1758.

Distribution Recorded from North Somerset, South Wiltshire, Dorset, West Sussex, East Kent, West Kent, Surrey, Hertfordshire, East Suffolk, Huntingdonshire, East Gloucestershire, West Gloucestershire, Herefordshire and Glamorgan before 1970 and South Hampshire, East Gloucestershire and West Gloucestershire from 1970 onwards.

MORDELLIDAE

Habitat and ecology Ancient broad-leaved woodland. Larvae develop in either dead wood or plant stems, probably the latter. Adults are usually found on flowers, particularly hogweed, though with also noted on buttercup, bramble, wild carrot, guelder-rose and hawthorn blossom. Adults have been recorded from May to August.

Status Not listed in the insect Red Data Book (Shirt, 1987). Recently added to the British list and previously confused with *M. aculeata* (a species which on present evidence does not occur in Great Britain). Only known in southern England as far north as Herefordshire and also noted in South Wales. Recently recorded from just three vice-counties. Because of the confusion over the taxonomy and indentity of this species, its exact status is uncertain.

Indicator status Status for *M. aculeata* was Grade 3 in Harding & Rose (1986).

Threats Loss of broad-leaved woodland and parkland through, for example, clear-felling and coniferisation. Habitat loss, in particular, may be through the felling of ancient trees, removal of dead wood from living trees and the destruction or removal of standing and fallen dead wood for reasons such as forest hygiene, aesthetic tidiness, public safety or for use as fire wood. Neglect and conversion to high forest may be a further threat.

Management and conservation Uncertain, though ancient trees, and both fallen and standing dead timber, especially with the bark attached, should be retained. The removal of dead timber from ancient trees should be avoided. Gaps in the age structure of the tree population should be identified and the continuity of the appropriate habitat ensured by regeneration, suitable planting and possibly with pollarding. The presence of nectar sources such as hawthorn, umbellifers and composite herbs may also be particularly important for this species.

Published sources Atty, D.B. (1983), Batten, R. (1986), Buck, F.D. (1954), Fowler, W.W. (1891), Fowler, W.W. & Donisthorpe, H.St J.K. (1913), Harding, P.T. (1978), Harding, P.T. & Rose, F. (1986).

MORDELLA LEUCASPIS
INSUFFICIENTLY KNOWN
Order COLEOPTERA   Family MORDELLIDAE

Mordella leucaspis Kuester, 1849. Formerly confused under: *Mordella aculeata* L., 1757.

Distribution Recorded from South Wiltshire and Herefordshire before 1970.

Habitat and ecology Uncertain, though probably associated with ancient broad-leaved woodland. Larvae probably develop in plant stems, while adults are probably frequent flowers.

Status Not listed in the insect Red Data Book (Shirt, 1987). Recently added to the British list from examples previously confused with *M. aculeata* (which on presnt evidence does not occur in Great Britain). So far, only known from two old examples. Because of the confusion over the taxonomy and identity of this species, its exact status is uncertain.

Indicator status Status for *M. aculeata* was Grade 3 in Harding & Rose (1986).

Threats Uncertain, though this species may be threatened by the loss of woodland through, for example, clear-felling and conversion to other land use. Habitat loss, in particular, may be through the felling of trees, removal of dead wood from living trees and the destruction or removal of standing and fallen dead wood for reasons such as forest hygiene, aesthetic tidiness, public safety or for use as fire wood. Neglect and conversion to high forest may be a further threat.

Management and conservation Uncertain, though trees, and both fallen and standing dead timber, especially with the bark attached, should be retained. The removal of dead timber from trees should be avoided. Gaps in the age structure of the tree population should be identified and the continuity of the appropriate dead wood habitat ensured by regeneration, suitable planting and possibly with pollarding. The presence of nectar sources such as umbellifers and composite herbs may also be particularly important for this species.

Published sources Batten, R. (1986).

MORDELLISTENA ACUTICOLLIS
INSUFFICIENTLY KNOWN
Order COLEOPTERA   Family MORDELLIDAE

Mordellistena acuticollis Schilsky, 1895. Formerly confused under: *Mordellistena parvula* (Gyllenhal, 1827).

Distribution Recorded from West Kent from 1970 onwards.

Habitat and ecology Wood margins. Larvae are possibly associated with creeping thistle *Cirsium arvense*. Adults have been swept from long grass mixed with nettles and young trees, and from short grass mixed with a few docks under an oak tree. The adult has been recorded in July.

Status Not listed in the insect Red Data Book (Shirt, 1987). Recently added to the British list through two examples recorded from Shooters Hill in West Kent in 1984 and 1985. This species is difficult to identify and may be confused with other members of the genus. Consequently, the exact status of this species is hard to assess.

Indicator status Grade 3 in Harding & Rose (1986).

Threats Shooters Hill is part of Oxleas Woods SSSI which is threatened by the construction of a road. Natural succession may be a further threat to this species.

Management and conservation Some cutting or other disturbance, on a rotational basis, may be needed to maintain open conditions. Shooters Hill is part of Oxleas Woods SSSI.

Published sources Allen, A.A. (1986), Batten, R. (1986).

MORDELLISTENA BREVICAUDA
INSUFFICIENTLY KNOWN
Order COLEOPTERA   Family MORDELLIDAE

Mordellistena brevicauda (Boheman, 1849).

Distribution Recorded from East Kent, Surrey, South Essex, Berkshire and Oxfordshire before 1970 and East Sussex, South Essex and Berkshire from 1970 onwards.

Habitat and ecology Recorded from calcareous grassland, with most records from chalk downland, though also recorded from limestone soils. Possibly found on other soil types. The larvae develop in either dead wood or plant stems, probably the latter. Adults have been recorded in August.

Status Not listed in the insect Red Data Book (Shirt, 1987). Very local and found in southern and south-eastern England. Because of taxonomic confusion the exact status of this species is hard to assess.

Threats Uncertain, though this species may be threatened by the loss of calcareous grassland through improvement and conversion to arable agriculture. Natural succession may be a further threat.

Management and conservation Grazing or cutting, on a rotational basis, may be needed to maintain open conditions.

Published sources Batten, R. (1986).

MORDELLISTENA HUMERALIS
INSUFFICIENTLY KNOWN
Order COLEOPTERA   Family MORDELLIDAE

Mordellistena humeralis (L., 1758). Formerly confused under: *Mordellistena variegata* (F., 1798).

Distribution Recorded from Huntingdonshire before 1970 and West Kent from 1970 onwards.

Habitat and ecology Broad-leaved woodland. Larvae develop in either dead wood or plant stems, probably the latter. Adults probably frequent flowers such as umbellifers. Adults have been recorded in August.

Status Not listed in the insect Red Data Book (Shirt, 1987). Confirmed records exist for just two vice-counties. There are published records for Cambridgeshire, East Suffolk, North Essex and Surrey which require confirmation. This species is difficult to identify and has been confused with *M. variegata* and *M. neuwaldeggiana*. Consequently, its exact status is uncertain. Also see the data sheet for *M. neuwaldeggiana*.

Threats Loss of broad-leaved woodland through, for example, clear-felling and coniferisation. Habitat loss may be through the felling of trees, removal of dead wood from living trees and the destruction or removal of standing and fallen dead wood for reasons such as forest hygiene, aesthetic tidiness, public safety or for use as fire wood. Neglect and conversion to high forest may be a further threat.

Management and conservation Uncertain, though trees, and both fallen and standing dead timber, especially with the bark attached, should be retained. The removal of dead timber from ancient trees should be avoided. Gaps in the age structure of the tree population should be identified and the continuity of the appropriate dead wood habitat ensured by regeneration, suitable planting and possibly with pollarding. The presence of nectar sources such as umbellifers and composite herbs may also be particularly important for this species.

Published sources Allen, A.A. (1989g), Batten, R. (1986), Cox, D. (1950), Donisthorpe, H.St J.K. (1938a), Easton, A.M. (1948), Pope, R.D. (1969).

MORDELLIDAE

MORDELLISTENA NANULOIDES
INSUFFICIENTLY KNOWN
Order COLEOPTERA   Family MORDELLIDAE

Mordellistena nanuloides Ermisch, 1967. Formerly confused under: *Mordellistena parvula* (Gyllenhal, 1827).

Distribution Recorded from East Kent before 1970 and West Kent from 1970 onwards.

Habitat and ecology Saltmarshes and other maritime habitats where the foodplant occurs. Associated, possibly exclusively, with sea wormwood *Artemisia maritima*. Adults occur on the plants, the larvae probably developing in the stems. The adult has been recorded in August.

Status Not listed in the insect Red Data Book (Shirt, 1987). Recently identified as a separate species in this country and split from *M. parvula*. Only known from the Isle of Sheppey, East Kent where it used to occur commonly on the foodplant, and recently found on the Isle of Grain, West Kent. Because of taxonomic confusion the exact status of this species is hard to assess.

Threats Loss of saltmarsh through reclamation, erosion and the construction of sea defences. Overgrazing of saltmarshes could be a further threat to this species.

Management and conservation Grazing should not be introduced to a saltmarsh where there is no grazing at present.

Published sources Allen, A.A. (1986), Batten, R. (1986), Owen, J.A. (1989a), Owen, J.A. (1990b).

MORDELLISTENA NEUWALDEGGIANA
INSUFFICIENTLY KNOWN
Order COLEOPTERA   Family MORDELLIDAE

Mordellistena neuwaldeggiana (Panzer, 1796). Formerly known as: *Mordellistena brunnea* (F., 1801), *Mordellistena newaldeggiana* (F., 1801).

Distribution Recorded from South Hampshire, Berkshire, Oxfordshire and Buckinghamshire before 1970 and West Kent and South Essex from 1970 onwards.

Habitat and ecology Woodland and pasture-woodland. On the Continent, this species is mainly recorded from wood-edges. Larvae develop in either dead wood or plant stems, probably the latter. The adult has been noted on the flowers of hogweed. In Britain, adults have been recorded in August.

Status Not listed in the insect Red Data Book (Shirt, 1987). There are confirmed records for only six vice-counties, and only two recently. There are published records for Cambridgeshire, East Suffolk, Huntingdonshire, North Essex and Surrey which require confirmation. There are also unconfirmed recent records from Dorset and East Sussex, though these may be referable to *M. humeralis*. This species is difficult to identify and may be confused with other members of the genus. Consequently, the exact status of this species is hard to assess.

Threats Uncertain, though the loss of woodland and parkland through, for example, clear-felling and conversion to other land use. Habitat loss may be through the felling of trees, removal of dead wood from living trees and the destruction or removal of standing and fallen dead wood for reasons such as forest hygiene, aesthetic tidiness, public safety or for use as fire wood. Neglect and conversion to high forest may be a further threat.

Management and conservation Uncertain, though trees, and both fallen and standing dead timber, especially with the bark attached, should be retained. The removal of dead timber from trees should be avoided. Gaps in the age structure of the tree population should be identified and the continuity of the appropriate dead wood habitat ensured by regeneration, suitable planting and possibly with pollarding. The presence of nectar sources such as umbellifers and composite herbs may also be particularly important for this species.

Published sources Allen, A.A. (1989g), Batten, R. (1986), Cox, D. (1950), Donisthorpe, H.St J.K. (1938a), Easton, A.M. (1948), Pope, R.D. (1969), Steele, R.C. & Welch, R.C. (eds.) (1973).

MORDELLISTENA PARVULA
INSUFFICIENTLY KNOWN
Order COLEOPTERA   Family MORDELLIDAE

Mordellistena parvula (Gyllenhal, 1827).

Distribution Recorded from West Cornwall, West Sussex, East Kent, Surrey, Hertfordshire and "Norfolk" before 1970 and East Sussex and West Norfolk from 1970 onwards.

Habitat and ecology Chalk downland, Breck grassland and coastal cliffs. Larvae probably develop in plant stems. Adults appear to be mainly associated with mugwort, though have also been recorded from yarrow. Adults have been noted in July.

Status Not listed in the insect Red Data Book (Shirt, 1987). Only recorded in southern England as far north as West Norfolk. Recently recorded from just two vice-counties. There is also a published record for West Suffolk which requires confirmation. Because of taxanomic confusion the exact status of this species is hard to assess.

Threats Uncertain, though the loss of chalk and Breck grassland through improvement and conversion to arable agriculture, afforestation and development may threaten this species. Cliff stabilisation schemes and natural succession may be further threats.

Management and conservation Grazing or cutting, on a rotational basis, may be needed to maintain open conditions.

Published sources Allen, A.A. (1986), Batten, R. (1986), Jones, R.A. (1984).

MORDELLISTENA PARVULOIDES
INSUFFICIENTLY KNOWN
Order COLEOPTERA   Family MORDELLIDAE

Mordellistena parvuloides Ermisch, 1956. Formerly confused under: *Mordellistena parvula* (Gyllenhal, 1827).

Distribution Recorded from West Kent from 1970 onwards.

Habitat and ecology Broad-leaved woodland. Recorded from oak woodland. The adult has been noted in July.

Status Not listed in the insect Red Data Book (Shirt, 1987). Recently added to the British list on the strength of one example found in 1985 from Shooters Hill in West Kent. This species is difficult to identify and may be confused with other members of the genus. Consequently, the exact status of this species is hard to assess.

Threats Shooters Hill is part of Oxleas Woods SSSI which is threatened by the construction of a road.

Management and conservation Shooters Hill is part of Oxleas Woods SSSI.

Published sources Allen, A.A. (1986), Batten, R. (1986).

MORDELLISTENA PSEUDOPUMILA
INSUFFICIENTLY KNOWN
Order COLEOPTERA   Family MORDELLIDAE

Mordellistena pseudopumila Ermisch, 1962. Formerly confused under: *Mordellistena pumila* (Gyllenhal, 1810).

Distribution Recorded from Berkshire before 1970.

Habitat and ecology Chalk downland. Larvae probably develop in the stems of plants. The adult has been recorded in May.

Status Not listed in the insect Red Data Book (Shirt, 1987). Recently added to the British list on the strength of three examples found in 1935 from Streatley, Berkshire. This species is difficult to identify and may be confused with other members of the genus. Consequently, the exact status of this species is hard to assess.

Threats Uncertain, though the loss of chalk downland through improvement and conversion to arable agriculture, and development may threaten this species.

Management and conservation Grazing or cutting, on a rotational basis, may be needed to maintain open conditions.

Published sources Batten, R. (1986).

TOMOXIA BUCEPHALA   NOTABLE A
A tumbling flower beetle
Order COLEOPTERA   Family MORDELLIDAE

Tomoxia bucephala Costa, 1854. Formerly known as: *Tomoxia biguttata* (Gyllenhal, 1827).

Distribution Recorded from South Wiltshire, Dorset, South Hampshire, North Hampshire, West Sussex, Surrey, Berkshire and Durham before 1970 and North Wiltshire, South Hampshire, West Sussex, East Sussex, East Kent, Surrey, Hertfordshire and Berkshire from 1970 onwards.

Habitat and ecology Ancient broad-leaved woodland and pasture-woodland. Larvae develop in rotten wood, particularly trunks and stumps of beech. Adults have been recorded sitting, running and flying on and off stumps and logs. Adults have been recorded from June to August.

Status Status revised from RDB 3 (Rare) in Shirt (1987). Possibly spreading. Primarily recorded from southern England as far north as Berkshire, though this species has also been noted in Durham. Recently recorded from eight vice-counties. This species can be confused with *Varimorda villosa*.

Indicator status Grade 1 in Harding & Rose (1986).

Threats Loss of broad-leaved woodland and parkland through, for example, clear-felling and coniferisation. Habitat loss, in particular, through the felling of ancient trees, removal of dead wood from living trees and the destruction or removal of standing and fallen dead wood for reasons such as forest hygiene, aesthetic tidiness, public safety or for use as fire wood.

Management and conservation Ancient trees, and both fallen and standing dead timber, especially with the bark attached, should be retained. The removal of dead timber from ancient trees should be avoided. Gaps in the age structure of the tree population should be identified and the continuity of the appropriate dead wood habitat ensured by regeneration, suitable planting and possibly with pollarding.

Published sources Fowler, W.W. (1891), Godfrey, A. (1989), Halstead, A. (1989), Harding, P.T. (1978), Harding, P.T. & Rose, F. (1986), Menzies, I.S. (1990), Moore, D. (1989), Nash, D.R. (1971), Nicholson, G.W. (1931), Onslow, N. (1989), Shirt, D.B., ed. (1987).

VARIIMORDA VILLOSA   NOTABLE B
A tumbling flower beetle
Order COLEOPTERA   Family MORDELLIDAE

Variimorda villosa (Schrank, 1781). Formerly known as: *Mordella fasciata* F., 1775, *Mordella villosa* (Schrank).

Distribution South East, South, South West, East Anglia, East Midlands, West Midlands and South Wales.

Habitat and ecology Ancient broad-leaved woodland and pasture-woodland. Larvae develop in either dead wood or plant stems, probably the latter. Adults are usually found on flowers, particularly umbellifers. Adults have been recorded from May to September.

Status Widespread and local in the southern half of England, also recorded in South Wales.

Indicator status Grade 3 in Harding & Rose (1986).

Threats Loss of broad-leaved woodland and parkland through, for example, clear-felling and coniferisation. Habitat loss, in particular, may be through the felling of trees, removal of dead wood from living trees and the destruction or removal of standing and fallen dead wood for reasons such as forest hygiene, aesthetic

tidiness, public safety or for use as fire wood. Neglect and conversion to high forest may be a further threat.

Management and conservation Uncertain, though trees, and both fallen and standing dead timber, especially with the bark attached, should be retained. The removal of dead timber from ancient trees should be avoided. Gaps in the age structure of the tree population should be identified and the continuity of the appropriate dead wood habitat ensured by regeneration, suitable planting and possibly with pollarding. The presence of nectar sources such as umbellifers and composite herbs may also be particularly important for this species.

MYCETOPHAGUS FULVICOLLIS   EXTINCT
A fungus beetle
Order COLEOPTERA   Family MYCETOPHAGIDAE

Mycetophagus fulvicollis F., 1792.

Distribution Recorded from Mid Perthshire before 1970.

Habitat and ecology Coniferous woodland. Probably in dead wood and under bark. Possibly also in fungi. The adult has been recorded in June.

Status Not listed in the insect Red Data Book (Shirt, 1987). Presumed extinct. Only known from two records from the Dall Sawpit, Black Wood of Rannoch, Mid Perthshire, in 1865 and 1870.

Published sources Crowson, R.A. (1960), Fowler, W.W. (1889), Fowler, W.W. & Donisthorpe, H.St J.K. (1913).

MYCETOPHAGUS PICEUS   NOTABLE B
A fungus beetle
Order COLEOPTERA   Family MYCETOPHAGIDAE

Mycetophagus piceus (F., 1777).

Distribution South East, South, East Anglia, East Midlands, West Midlands, North East, North West, South Wales, Dyfed-Powys and South West Scotland.

Habitat and ecology Ancient broad-leaved woodland, pasture-woodland and also in more open country, such as Breckland. Lives and breeds in red-rotten heart wood of oak attacked by the fungus *Laetiporus sulphureus*. Also recorded under oak bark, under beech, in a bracket fungus on elm, in oyster fungi, and probably in other fungi and other trees attacked by *L. sulphureus*. This species has also been recorded from a compost heap. Adults have been recorded in June and from August to October.

Status Widespread but local in England and Wales, also recorded in South West Scotland.

Indicator status Grade 2 in Garland (1983). Grade 3 in Harding & Rose (1986).

Threats Loss of broad-leaved woodland and parkland through, for example, clear-felling and coniferisation. Habitat loss, in particular, through the felling of ancient and fungus-infected trees, removal of dead wood from living trees and the destruction or removal of standing and fallen dead wood for reasons such as forest hygiene, aesthetic tidiness, public safety or for use as fire wood.

Management and conservation Ancient and fungus-infected trees and both fallen and standing dead timber, especially with the bark attached, should be retained. The removal of dead timber from ancient and fungus-infected trees should be avoided. Gaps in the age structure of the tree population should be identified and the continuity of the appropriate habitat ensured by regeneration, suitable planting and possibly with pollarding.

MYCETOPHAGUS POPULI   NOTABLE A
A fungus beetle
Order COLEOPTERA   Family MYCETOPHAGIDAE

Mycetophagus populi F., 1798.

Distribution Recorded from South Hampshire, South Essex, Berkshire, East Suffolk, East Norfolk, Herefordshire, South Lincolnshire, Nottinghamshire, Mid-west Yorkshire and Lanarkshire before 1970 and West Suffolk, Northamptonshire, East Gloucestershire, Worcestershire, Leicestershire & Rutland and Derbyshire from 1970 onwards.

Habitat and ecology Broad-leaved woodland, pasture-woodland and probably isolated trees. Under bark and in soft, wet, decaying fungus-infected wood. Recorded in numbers from an elm stump and in numbers from an ash boundary pollard. Adults have been recorded in February, April, May, September and October.

Status Very local. Widely distributed and recorded from South Hampshire to Lanarkshire in Scotland. Recently recorded from seven vice-counties.

Threats Loss of broad-leaved woodland and parkland through, for example, clear-felling and coniferisation. Habitat loss, in particular, through the felling of ancient

and fungus-infected trees, removal of dead wood from living trees and the destruction or removal of standing and fallen dead wood for reasons such as forest hygiene, aesthetic tidiness, public safety or for use as fire wood.

Management and conservation Ancient and fungus-infected trees, and both fallen and standing dead timber, especially with the bark attached, should be retained. The removal of dead timber from ancient and fungus-infected trees should be avoided. Gaps in the age structure of the tree population should be identified and the continuity of the appropriate dead wood habitat ensured by regeneration, suitable planting and possibly with pollarding.

Published sources Crowson, R.A. (1960), Drane, A.B. (1990), Fowler, W.W. (1889), Fowler, W.W. & Donisthorpe, H.St J.K. (1913), Mendel, H. (1989).

MYCETOPHAGUS QUADRIGUTTATUS
A fungus beetle   NOTABLE A
Order COLEOPTERA   Family MYCETOPHAGIDAE

Mycetophagus quadriguttatus Müller, 1821.

Distribution Recorded from South Devon, South Hampshire, West Sussex, East Kent, West Kent, Surrey, Middlesex, Berkshire, West Suffolk and Lanarkshire before 1970 and South Hampshire, Berkshire and West Suffolk from 1970 onwards.

Habitat and ecology Synanthropic and found in granaries, food stores etc., though this species has also been recorded from fungi on trees and in tree hollow debris. Usually found associated with stored products such as grain, corn and peas, though it has been recorded from mouldy hay in a stable. Also found in beef-steak fungus *Fistulina hepatica* growing on beech. Adults have been recorded in June and July.

Status Widely distributed in England and recorded as far north as Lanarkshire in Scotland. Recently recorded from just three vice-counties, all in southern England.

Threats In the wild this species may be threatened by loss of broad-leaved woodland through, for example, clear-felling and coniferisation. Habitat loss, in particular, through the felling of fungus-infected trees, removal of dead wood from living trees and the destruction or removal of standing and fallen dead wood for reasons such as forest hygiene, aesthetic tidiness, public safety or for use as fire wood.

Management and conservation In the wild, fungus-infected trees and both fallen and standing dead timber, especially with the bark attached, should be retained. The removal of dead timber from fungus-infected trees should be avoided. Gaps in the age structure of the tree population should be identified and the continuity of the appropriate habitat ensured by regeneration, suitable planting and possibly with pollarding.

Published sources Crowson, R.A. (1960), Drane, A.B. (1990), Fowler, W.W. (1889), Fowler, W.W. & Donisthorpe, H.St J.K. (1913), Mendel, H. (1989).

MYCTERUS CURCULIOIDES   **EXTINCT**

Order COLEOPTERA   Family MYCTERIDAE

Mycterus curculioides (F., 1781). Formerly known as: *Mycterus curculionoides* (F., 1781).

Distribution Recorded from "Devon" and Oxfordshire.

Habitat and ecology Probably open areas. Possibly associated with thistles and umbellifers.

Status Presumed extinct. Last recorded in 1882 from Oxfordshire.

Published sources Fowler, W.W. (1891), Shirt, D.B., ed. (1987).

CHRYSANTHIA NIGRICORNIS   **ENDANGERED**
A flower beetle
Order COLEOPTERA   Family OEDEMERIDAE

Chrysanthia nigricornis (Westhoff, 1881).

Distribution Recorded from South Aberdeenshire from 1970 onwards.

Habitat and ecology Coniferous woodland. Adults have been swept from heather in open canopy Scots pine forest in August. Larvae have been found in August from the heart-wood of a sodden old pine branch (5 cm thick) lying beneath tufts of moss and heather.

Status Only known from two sites in South Aberdeenshire. This beetle was discovered new to Great Britain in 1971. The species may be a component of the native pine forest fauna.

Threats Loss of native pine forest through, for example, clear-felling or conversion to plantation forest. Habitat loss in the remaining areas through the felling of ancient trees and the destruction or removal of standing and fallen dead wood for reasons such as

forest hygiene, aesthetic tidiness, public safety or for use as fire wood.

Management and conservation Ancient trees and both standing and fallen dead timber, especially with the bark attached, should be retained. Gaps in the age structure of the tree population should be identified and regeneration encouraged to ensure the continuity of the dead wood habitats.

Published sources Shirt, D.B., ed. (1987), Skidmore, P. (1973).

ISCHNOMERA CAERULEA   **RARE**
A flower beetle
Order COLEOPTERA   Family OEDEMERIDAE

Ischnomera caerulea (L., 1758). Formerly known as: *Ischnomera coerulea* (L.).

Distribution Recorded from South Hampshire, Surrey and Berkshire before 1970 and Surrey, Berkshire and Herefordshire from 1970 onwards.

Habitat and ecology Ancient broad-leaved woodland and pasture-woodland. Probably associated with dead wood, examples have been recorded from dead elm and rotten wood removed from a rot hole. On the Continent, the species may be exclusively associated with oak. In Britain, adults have been noted at hawthorn blossom. Adults have been recorded from April to June.

Status This species has recently been recognised as a separate species and was formerly confused with *I. cyanea*. Because of this, this data sheet must remain provisional until more collected material has been checked. This species, however, would appear the less frequently encountered of the two, and on current information has only been recorded from four vice-counties. Old records for *I. caerulea* needed to be treated with caution as these are more likely to refer to *I. cyanea*.

Threats Loss of broad-leaved woodland and parkland through, for example, clear-felling and coniferisation. Habitat loss, in particular, may be through the felling of ancient trees, removal of dead wood from living trees and the destruction or removal of standing and fallen dead wood for reasons such as forest hygiene, aesthetic tidiness, public safety or for use as fire wood.

Management and conservation Ancient trees and both fallen and standing dead timber, especially with the bark attached, should be retained. The removal of dead timber from ancient trees should be avoided. Gaps in the age structure of the tree population should be identified and the continuity of the appropriate dead wood habitat ensured by regeneration, suitable planting and possibly with pollarding. The presence of nectar sources such as hawthorn may also be particularly important for this species.

Published sources Allen, A.A. (1988d), Mendel, H. (1990a).

ISCHNOMERA CINERASCENS   VULNERABLE
A flower beetle
Order COLEOPTERA   Family OEDEMERIDAE

Ischnomera cinerascens (Pandelle, 1867).

Distribution Recorded from Herefordshire before 1970 and Buckinghamshire, Herefordshire and North-east Yorkshire from 1970 onwards.

Habitat and ecology Ancient broad-leaved woodland, pasture-woodland and secondary woodland. Adults have been beaten from very large wych elms and it is possible that the larvae develop in dead wood of this tree species. Adults have been recorded in May and June.

Status Described new to Great Britain from a few examples from Duncombe Park in North-east Yorkshire in 1979. Subsequent examination of material of the closely related *I. caerulea* resulted in a record from Moccas Park, Herefordshire in 1965, where it has since been refound. Recently, it has also been found in Buckinghamshire.

Indicator status Grade 1 in Garland (1983). Grade 1 in Harding & Rose (1986).

Threats Loss of broad-leaved woodland and parkland through, for example, clear-felling and coniferisation. Habitat loss, in particular, through the felling of ancient trees, removal of dead wood from living trees and the destruction or removal of standing and fallen dead wood, for reasons such as forest hygiene, aesthetic tidiness, public safety or for use as fire wood, will probably threaten this species. Dutch elm disease and the removal of dead and dying elms may be a further threat to this beetle.

Management and conservation Ancient trees and both fallen and standing dead timber, especially with the bark attached, should be retained. The removal of dead timber from ancient trees should be avoided. Gaps in the age structure of the tree population should be identified and the continuity of the appropriate dead wood habitat ensured by suitable planting and possibly with pollarding. Moccas Park and Duncombe Park are NNRs.

Published sources Alexander, K.N.A. & Clements, D.K. (1988), Garland, S.P. (1983), Harding, P.T. (1978), Harding, P.T. & Rose, F. (1986), Shirt, D.B., ed. (1987), Skidmore, P. & Hunter, F.A. (1980).

ISCHNOMERA CYANEA   NOTABLE B
A flower beetle
Order COLEOPTERA   Family OEDEMERIDAE

Ischnomera cyanea (F., 1787). Formerly confused under: *Ischnomera caerulea* (L., 1758), *Ischnomera coerulea* (L.).

Distribution South East, South, South West, East Anglia, East Midlands, West Midlands, North East and Wales.

Habitat and ecology Ancient broad-leaved woodland, pasture-woodland, downland and hedgerows. Larvae develop in rotten wood of elm, willow, beech, ivy and probably oak. Adults have been beaten from the foliage of oak, field maple, elm and hawthorn. Also recorded from an old, dead ash and swept from grassy, herb-rich woodland rides. The adults are often found on the flowers of hogweed and hawthorn. Larvae have been recorded in May. Adults overwinter in the pupal chambers and have been noted from April to August.

Status This species has recently been recognised as a separate species from *I. caerulea*. In Great Britain, *I. cyanea* appears to be the more familiar insect of the species split. However, until more collected material has been checked this data sheet, and the species' status for conservation purposes, must remain provisional.

Indicator status Prior to the species split *I. caerulea* was listed as Grade 1 in Garland (1983) and also Grade 3 in Harding & Rose (1986), it was also listed in Speight (1989) Appendix 1.

Threats Loss of broad-leaved woodland and parkland through, for example, clear-felling and coniferisation. Habitat loss, in particular, through the felling of ancient trees, removal of dead wood from living trees and the destruction or removal of standing and fallen dead wood for reasons such as forest hygiene, aesthetic tidiness, public safety or for use as fire wood.

Management and conservation Ancient trees and both fallen and standing dead timber, especially with the bark attached, should be retained. The removal of dead timber from ancient trees should be avoided. Gaps in the age structure of the tree population should be identified and the continuity of the appropriate dead wood habitat ensured by suitable planting and possibly with pollarding. The presence of nectar sources such as

OEDEMERIDAE

hawthorn and umbellifers, may also be particularly important for this species.

ISCHNOMERA SANGUINICOLLIS NOTABLE B
A flower beetle
Order COLEOPTERA   Family OEDEMERIDAE

Ischnomera sanguinicollis (F., 1787). Formerly known as: *Asclera sanguinicollis* (F.).

Distribution South West, South, South East, East Anglia, West Midlands, East Midlands, North East, Dyfed-Powys and .

Habitat and ecology Ancient broad-leaved woodland, pasture-woodland and hedgerows. Larvae develop in dead wood and have been bred from elm. Adults have been beaten from the flowers and foliage of hawthorn, field maple, sycamore, oak and lime. The adults have also been found on the flowers of guelder-rose, crab apple, chestnut and ox-eye daisy, and the foliage of wych elm. Adults have been recorded from April to July.

Status Widespread but local. Recorded throughout the southern half of England, being found as far north as North-east Yorkshire. This beetle has also been noted in Radnorshire, Wales.

Indicator status Grade 1 in Garland (1983). Grade 1 in Harding & Rose (1986).

Threats Loss of broad-leaved woodland and parkland through, for example, clear-felling and coniferisation. Habitat loss, in particular, through the felling of ancient trees, removal of dead wood from living trees and the destruction or removal of standing and fallen dead wood for reasons such as forest hygiene, aesthetic tidiness, public safety or for use as fire wood. Dutch elm disease and the removal of dead and dying elms may be a further threat to this species.

Management and conservation Ancient trees and both fallen and standing dead timber, especially with the bark attached, should be retained. The removal of dead timber from ancient trees should be avoided. Gaps in the age structure of the tree population should be identified and the continuity of the appropriate dead wood habitat ensured by suitable planting and possibly with pollarding. The presence of nectar sources such as hawthorn and composite herbs, may also be particularly important for this species.

OEDEMERA VIRESCENS   VULNERABLE
A flower beetle
Order COLEOPTERA   Family OEDEMERIDAE

Oedemera virescens (L., 1767).

Distribution Recorded from East Norfolk, East Gloucestershire, West Gloucestershire and North-east Yorkshire before 1970 and North-east Yorkshire from 1970 onwards.

Habitat and ecology Broad-leaved woodland, pasture-woodland and wood edges. Larval biology unknown but probably associated with dead wood. Adults have been recorded in June.

Status Status revised from RDB 3 (Rare) in Shirt (1987). Possibly declining. This beetle appears to be restricted to a small area in three 10km squares in North-east Yorkshire.

Threats Uncertain, though this species is probably threatened by the loss of broad-leaved woodland and parkland through, for example, clear-felling and coniferisation. Habitat loss, in particular, may be through the felling of ancient trees, removal of dead wood from living trees and the destruction or removal of standing and fallen dead wood for reasons such as forest hygiene, aesthetic tidiness, public safety or for use as fire wood.

Management and conservation Ancient trees and both fallen and standing dead timber, especially with the bark attached, should be retained. The removal of dead timber from ancient trees should be avoided. Gaps in the age structure of the tree population should be identified and the continuity of the appropriate dead wood habitat ensured by suitable planting and possibly with pollarding.

Published sources Atty, D.B. (1983), Buck, F.D. (1954), Shirt, D.B., ed. (1987).

ONCOMERA FEMORATA   NOTABLE B
A flower beetle
Order COLEOPTERA   Family OEDEMERIDAE

Oncomera femorata (F., 1792).

Distribution South East, South, South West, East Anglia, North West, South Wales and North Wales.

Habitat and ecology Possibly associated with wooded areas and hedgerows. Larval biology unknown. Adults have been recorded from the blossom of sallow and ivy. This species has been noted at mercury vapour light and has been found in a shed during the winter months. Adults are nocturnal and have been recorded in March, April, June and from September to November.

Status Widespread and very local. Recorded throughout southern England and with a scattered distribution through the the rest of England. This species was noted in the past from southern Scotland and parts of Wales. Possibly under-recorded.

Threats Uncertain, though this species may be threatened by the loss of woodland and hedgerows through practices such as uprooting and clear-felling. The mechanised trimming of hedgerows may be a further threat.

OSTOMA FERRUGINEUM   **ENDANGERED**

Order COLEOPTERA   Family PELTIDAE

Ostoma ferrugineum (L., 1758). Formerly known as: *Ostoma ferruginea* (L., 1758).

Distribution Recorded from South Aberdeenshire and East Inverness & Nairn before 1970 and South Aberdeenshire from 1970 onwards.

Habitat and ecology Ancient native pine forest. The larvae feed in the heartwood and sapwood of Scots pine that has been extensively rotted by the fungus *Phaeolus schweinitzii*, while adults are usually found under pine bark. Larvae have been found in early April, these pupated in late May. Adults have been recorded from April to June.

Status First recorded in Great Britain in 1952. Extremely local and only known from two vice-counties, and only one recently.

Threats Loss of native pine forest through practices such as clear-felling or conversion to plantation forest. Habitat loss in the remaining areas through the felling of ancient and fungus-infected trees and the destruction or removal of standing and fallen dead wood for reasons such as forest hygiene, aesthetic tidiness, public safety or for use as fire wood.

Management and conservation Ancient and fungus-infected trees, and both standing and fallen dead timber, especially with the bark attached, should be retained. Gaps in the age structure of the tree population should be identified and regeneration encouraged to ensure the continuity of the appropriate habitats.

Published sources Hammond, P.M., Smith, K.G.V., Else, G.R. & Allen, G.W. (1989), Shirt, D.B., ed. (1987).

THYMALUS LIMBATUS   **NOTABLE B**

Order COLEOPTERA   Family PELTIDAE

Thymalus limbatus (F., 1787).

Distribution South East, South, South West, East Midlands, West Midlands, North East, North West, Dyfed-Powys, North Wales, South West Scotland, North East Scotland and North West Scotland.

Habitat and ecology Ancient broad-leaved woodland and pasture-woodland. Occurs under bark, particularly that of oak and beech. Also under the bark of alder and pine, and in bracket fungi, particularly on birch. Larvae have been recorded in January, February and June. Adults have been noted from February to September.

Status Widespread but local in Great Britain.

Indicator status Grade 1 in Garland (1983). Grade 3 in Harding & Rose (1986).

Threats Loss of broad-leaved woodland and parkland through, for example, clear-felling and coniferisation. Habitat loss, in particular, through the felling of trees, removal of dead wood from living trees and the destruction or removal of standing and fallen dead wood for reasons such as forest hygiene, aesthetic tidiness, public safety or for use as fire wood.

Management and conservation Trees, and both fallen and standing dead timber, especially with the bark attached, should be retained. The removal of dead timber from trees should be avoided. Gaps in the age structure of the tree population should be identified and the continuity of the appropriate dead wood habitat ensured by regeneration, suitable planting and possibly with pollarding.

OLIBRUS FLAVICORNIS
          INSUFFICIENTLY KNOWN
Order COLEOPTERA   Family PHALACRIDAE

Olibrus flavicornis (Sturm, 1807). Formerly known as: *Olibrus helveticus* Rye, 1876.

Distribution Recorded from Isle of Wight, East Sussex, East Kent, Surrey, Buckinghamshire and "Suffolk" before 1970.

Habitat and ecology Probably associated with grassland and coastal habitats. On the Continent, this species is apparently associated with autumn hawkbit *Leontodon autumnalis*. Larvae probably develop in the flower head of the foodplant, while the adults feed on

PHALACRIDAE

pollen. In Britain, the adults have been recorded in June and July.

Status Not listed in the insect Red Data Book (Shirt, 1987). Only known from southern England and recorded as far north as Suffolk. Last recorded in 1950 from Camber, East Sussex. This species is difficult to identify and may be confused with other members of the genus.

Threats Uncertain, though this species may have declined because of the loss of habitat through improvement and conversion to arable agriculture, afforestation, development and natural succession.

Published sources Allen, A.A. (1971d), Fowler, W.W. (1889), Fowler, W.W. & Donisthorpe, H.St J.K. (1913), Thompson, R.T. (1958).

OLIBRUS MILLEFOLII   NOTABLE B

Order COLEOPTERA   Family PHALACRIDAE

Olibrus millefolii (Paykull, 1800).

Distribution South East, South, South West, East Anglia and East Midlands.

Habitat and ecology Heathland, grassland, and probably hedge-banks. Occurs on yarrow. Larvae probably develop in the flower heads. The adults feed on pollen and have been recorded in May, June and August.

Status Widespread and very local in southern England. Most frequent in the Brecklands. This species is difficult to identify and may be confused with other members of the genus. Consequently, the exact status of this species is hard to assess.

Threats Loss or fragmentation of habitat through improvement and conversion to arable agriculture, afforestation and development. Natural succession is a further threat.

Management and conservation Grazing or cutting, on a rotational basis, is needed to maintain open conditions.

OLIBRUS PYGMAEUS   NOTABLE B

Order COLEOPTERA   Family PHALACRIDAE

Olibrus pygmaeus (Sturm, 1807).

Distribution South East, South, South West, East Anglia, East Midlands and North East.

Habitat and ecology Marshland, probably also grassland and disturbed ground. Recorded from common cudweed *Filago vulgaris*. On the Continent, this species is also associated with narrow-leaved cudweed *F. gallica*, hawkbit *Leontodon* and hawk's-beard *Crepis*. Larvae probably develop in the flower heads. The adults feed on pollen and, in Britain, they have been recorded in June, August and October.

Status Widespread and very local in England, not recorded in North West England or the West Midlands. This species is difficult to identify and may be confused with other members of the genus. Consequently, the exact status of this species is hard to assess.

Threats Loss of habitat through drainage and water abstraction schemes. Improvement and conversion to arable agriculture, development, and natural succession are further threats.

Management and conservation Grazing or cutting, on a rotational basis, is needed to maintain open conditions. In areas of marshland, water tables should be maintained at high levels.

PHALACRUS BRUNNIPES   NOTABLE A
A smut beetle
Order COLEOPTERA   Family PHALACRIDAE

Phalacrus brunnipes Brisout, 1863. Formerly known as: *Phalacrus championi* Guillebeau, 1892.

Distribution Recorded from South Hampshire, "Sussex", East Kent, West Kent, Surrey and Cambridgeshire before 1970 and East Kent, South Essex, North Essex, East Suffolk, South Aberdeenshire, North Aberdeenshire and East Inverness & Nairn from 1970 onwards.

Habitat and ecology Saltmarshes and also in other coastal habitats, along river margins, and occasionally inland on chalk hills and in woodland. Also recorded from a disused railway embankment. Adults are usually found on flowers, and in grass tussocks during the winter. The larvae feed on the spores of smut fungi on grasses and sedges. Eggs are probably laid towards the end of June in the inflorescences of grasses etc., and hatch after about three to five days. The larvae feed for about three to four weeks and then fall to the ground where they innovate earthen cells which they line with silk in which they pupate. Adults have been recorded in January, February, from May to July and in November.

Status Widely distributed and very local in England, and recorded as far north as Easterness in Scotland. Recently recorded from just six vice-counties. Can be common where found.

Threats Loss of saltmarsh through reclamation, erosion and the construction of sea defences. This species may also be threatened by river engineering schemes, improvement and conversion to arable agriculture, development, clear-felling and conversion to other land use and natural succession.

Management and conservation Grazing or cutting may be needed to maintian open conditions. Grazing should not be introduced to a saltmarsh where there is no grazing at present.

Published sources Allen, A.A. (1952a), Owen, J.A. (1988a), Owen, J.A. (1988b), Thompson, R.T. (1958).

STILBUS ATOMARIUS
INSUFFICIENTLY KNOWN
Order COLEOPTERA   Family PHALACRIDAE

Stilbus atomarius (L., 1767).

Distribution Recorded from "Sussex", West Kent, East Suffolk, East Norfolk, Cambridgeshire and Bedfordshire before 1970.

Habitat and ecology Marshland, fens and ditch sides. Possibly associated with reedmace *Typha*.

Status Not listed in the insect Red Data Book (Shirt, 1987). Last recorded in 1946 from Hipsey Spinney, Bedfordshire. This species is difficult to identify and may be confused with other members of the genus.

Threats Uncertain, though this species may have declined through loss of habitat because of drainage and water abstraction schemes.

Published sources Fowler, W.W. (1889), Fowler, W.W. & Donisthorpe, H.St J.K. (1913), Thompson, R.T. (1958).

PHLOIOPHILUS EDWARDSI   NOTABLE B
Order COLEOPTERA   Family PHLOIOPHILIDAE

Phloiophilus edwardsi Stephens, 1830. Formerly known as: *Phloeophilus edwardsi* Stephens, 1830, *Phloiophilus edwardsii* Stephens, 1830.

Distribution England, Dyfed-Powys, North Wales, South East Scotland, South West Scotland and North East Scotland.

Habitat and ecology Ancient broad-leaved woodland and pasture-woodland. Associated with fungi. The species has been recorded from *Corticium quercinum* growing on dead branches of oak, beech, hazel, birch and probably also on other trees, as well as from *Phlebia merismoides* growing on oak and hazel. The beetle has also been recorded from holly and under the rotten bark of beech. The eggs are laid in autumn, with the larvae feeding from early winter to between late April and July. This species primarily pupates in the ground, though pupae have also been found under bark (in August). Adults emerge from the pupal cells during late September and early October. Adults are found during the winter months, and have also been recorded in May, June and August.

Status Widespread but local in Great Britain.

Indicator status Grade 1 in Garland (1983). Grade 3 in Harding & Rose (1986).

Threats Loss of broad-leaved woodland and parkland through, for example, clear-felling and coniferisation. Habitat loss, in particular, through the felling of fungus-infected trees, removal of dead wood from living trees and the destruction or removal of standing and fallen dead wood for reasons such as forest hygiene, aesthetic tidiness, public safety or for use as fire wood.

Management and conservation Fungus-infected trees and both fallen and standing dead timber, especially with the bark attached, should be retained. The removal of fungus-infected timber from ancient trees should be avoided. Gaps in the age structure of the tree population should be identified and the continuity of the appropriate dead wood habitat ensured by regeneration, suitable planting and possibly with pollarding.

PLATYPODIDAE

PLATYPUS CYLINDRUS   NOTABLE B
Oak pin-hole borer
Order COLEOPTERA   Family PLATYPODIDAE

Platypus cylindrus (F., 1792).

Distribution South East, South, South West, East
Anglia, East Midlands, West Midlands, North East,
Dyfed-Powys and North Wales.

Habitat and ecology Ancient broad-leaved woodland
and pasture-woodland. Associated with oak, beech and
ash. Also recorded from sweet chestnut. Adults and
larvae live in galleries in dead wood which extend deep
into the heartwood, feeding on fungi which they
'culture' in their burrows. Larvae have been recorded
in January and February. Adults have been noted from
April to October.

Status Status revised from RDB 3 (Rare) in Shirt
(1987). Widepsread and local in England and Wales.

Indicator status Grade 3 in Harding & Rose (1986).

Threats Loss of broad-leaved woodland and parkland
through, for example, clear-felling and coniferisation.
Habitat loss, in particular, through the felling of ancient
trees, removal of dead wood from living trees and the
destruction or removal of standing and fallen dead
wood for reasons such as forest hygiene, aesthetic
tidiness, public safety or for use as fire wood.

Management and conservation Ancient trees and both
fallen and standing dead timber, especially with the
bark attached, should be retained. The removal of dead
timber from ancient trees should be avoided. Gaps in
the age structure of the tree population should be
identified and the continuity of the appropriate dead
wood habitat ensured by regeneration, suitable planting
and possibly with pollarding.

PLATYPUS PARALLELUS   INDETERMINATE

Order COLEOPTERA   Family PLATYPODIDAE

Platypus parallelus (F., 1801). Formerly known as:
suturalis (Gyllenhal).

Distribution Recorded from West Kent before 1970
and West Kent from 1970 onwards.

Habitat and ecology A wood-borer. Adults have been
noted in July and August. This species has been
recorded at mercury vapour light traps.

Status Not listed in the insect Red Data Book (Shirt,
1987). Known from just three examples. The first was
noted near Sydenham in 1832, the second at
Blackheath in 1973 and the third at Charlton in 1983.

Published sources Allen, A.A. (1976b), Allen, A.A.
(1985a).

EUBRIA PALUSTRIS   RARE

Order COLEOPTERA   Family PSEPHENIDAE

Eubria palustris Germar, 1818.

Distribution Recorded from South Devon, Dorset, East
Sussex, Berkshire, Oxfordshire, East Norfolk,
North-east Yorkshire and Durham before 1970 and
Dorset, Caernarvonshire and Anglesey from 1970
onwards.

Habitat and ecology Wetland, fens, marshes and from
wet flushes on soft-rock cliffs. Recorded from
brookweed *Samolus valerandi* and from sallow
blossom. Adults have been recorded in early spring and
in June.

Status A widely scattered distribution in England and
Wales and recorded from South Devon as far north as
Durham. Recently recorded from just three
vice-counties.

Threats Drainage for reasons such as agricultural
improvement, cliff stabilisation schemes, the
construction of sea defences and coastal developments.
In areas of soft-rock cliffs activities that accelerate or
reduce the rate of erosion should be avoided. The
degradation of suitable habitat through natural
succession is a further threat.

Management and conservation Water tables should be
maintained at high levels. In areas of soft-rock cliff,
occasional slippages are necessary to maintain habitat
continuity. Large areas of unstable cliff are required so
that the population does not become isolated and
subsequently threatened by individual landslips.

Published sources Drane, A.B. (1990), Fowler, W.W.
(1890), Fowler, W.W. & Donisthorpe, H.St J.K. (1913),
Shirt, D.B., ed. (1987).

PTINUS LICHENUM — RARE
A spider beetle
Order COLEOPTERA — Family PTINIDAE

Ptinus lichenum Marsham, 1802.

Distribution Recorded from Dorset, East Kent, West Kent, Surrey, South Essex, Hertfordshire, Middlesex, Berkshire, "Norfolk", West Gloucestershire and Dumfriesshire before 1970 and East Sussex, East Gloucestershire and South-west Yorkshire from 1970 onwards.

Habitat and ecology Woodland. Possibly associated with dead wood. The beetle has been found on lichen covered fence posts, and also recorded from sycamore. Adults have been noted in March, May, June, and November.

Status Not listed in the insect Red Data Book (Shirt, 1987). Recorded from southern England as far north as Norfolk, also noted in South-west Yorkshire and Dumfriesshire, Scotland. Recently recorded from just three vice-counties.

Threats Uncertain, though this species may be threatened through, for example, clear-felling and conversion to other land use. Habitat loss, in particular, maybe through the removal of dead wood from living trees and the destruction or removal of standing and fallen dead wood for reasons such as forest hygiene, aesthetic tidiness, public safety or for use as fire wood.

Management and conservation Trees and both fallen and standing dead timber, especially with the bark attached, should be retained. The removal of dead timber from trees should be avoided. Gaps in the age structure of the tree population should be identified and the continuity of the appropriate habitat ensured by regeneration, suitable planting and possibly with pollarding.

Published sources Atty, D.B. (1983), Flint, J.H. (1988), Fowler, W.W. (1890), Fowler, W.W. & Donisthorpe, H.St J.K. (1913).

PTINUS PALLIATUS — NOTABLE A
A spider beetle
Order COLEOPTERA — Family PTINIDAE

Ptinus palliatus Perris, 1847. Formerly known as: *Ptinus germanus* sensu F., 1781 not Goeze, 1777 ?L., 1767.

Distribution Recorded from "Devon", East Sussex, West Kent, Surrey, South Essex, North Essex, Berkshire, East Suffolk, "Norfolk" and South

Northumberland before 1970 and West Sussex, Berkshire and East Suffolk from 1970 onwards.

Habitat and ecology Ancient broad-leaved woodland and exposed situations near the coast. Associated with dry, rotten dead wood, with records from oak and from under the bark of a dead beech tree. It has also been reared from old fence posts. Adults have been recorded in March, May and June.

Status Widely scattered in England, with records from southern England, north to Norfolk. Also reported in South Northumberland. Recently recorded from only three vice-counties.

Indicator status Grade 3 in Harding & Rose (1986).

Threats Loss of broad-leaved woodland through, for example, clear-felling and coniferisation. Habitat loss, in particular, through the felling of ancient trees, removal of dead wood from living trees and the destruction or removal of standing and fallen dead wood for reasons such as forest hygiene, aesthetic tidiness, public safety or for use as fire wood. Coastal developments may be a further threat.

Management and conservation Ancient trees and both fallen and standing dead timber, especially with the bark attached, should be retained. The removal of dead timber from ancient trees should be avoided. Gaps in the age structure of the tree population should be identified and the continuity of the appropriate dead wood habitat ensured by regeneration, suitable planting and possibly with pollarding.

Published sources Fowler, W.W. (1890), Fowler, W.W. & Donisthorpe, H.St J.K. (1913), Harding, P.T. (1978), Harding, P.T. & Rose, F. (1986), Owen, J.A. (1990a), Pope, R.D. (1988).

PTINUS SEXPUNCTATUS — NOTABLE B
A spider beetle
Order COLEOPTERA — Family PTINIDAE

Ptinus sexpunctatus Panzer, 1792.

Distribution South East, South, South West, East Anglia, East Midlands, West Midlands, North West and South East Scotland.

Habitat and ecology Possibly associated with dead wood. Recorded from the nests of bees, particularly *Osmia rufa* (Hymenoptera), an old house-martin's nest, under the peeling bark of a fence post, from ivy, from blossom and indoors. Occasionally recorded at mercury vapour light traps. Adults have been recorded in February and from May to August.

PTINIDAE

Status Widespread but local in England. Also recorded in South East Scotland.

Threats Uncertain, though this species maybe be threatened by the loss of dead wood through the felling of trees, removal of dead wood from living trees and the destruction or removal of standing and fallen dead wood for reasons such as forest hygiene, aesthetic tidiness, public safety or for use as fire wood.

Management and conservation Ancient trees and both fallen and standing dead timber, especially with the bark attached, should be retained. The removal of dead timber from ancient trees should be avoided. Gaps in the age structure of the tree population should be identified and the continuity of the appropriate dead wood habitat ensured by regeneration, suitable planting and possibly with pollarding.

PTINUS SUBPILOSUS　　　　　　**NOTABLE B**
A spider beetle
Order COLEOPTERA　　　　　　Family PTINIDAE

Ptinus subpilosus Sturm, 1837.

Distribution South East, South, South West, East Midlands, West Midlands, North Wales, South East Scotland, South West Scotland and North East Scotland.

Habitat and ecology Ancient broad-leaved woodland and pasture-woodland. In old, hollow trees and under bark. Mainly recorded from oak, though also found under the bark of sycamore and on one occasion under the bark of plane. In Scotland, this species has been recorded from old Scots pine. It has also been found in the nests of the ants *Lasius brunneus* and *L. fuliginosus* (Hymenoptera). Adults have been recorded from February to April and in June and July.

Status Widespread but local in Great Britain, not recorded from northern England.

Indicator status Grade 1 in Garland (1983). Grade 2 in Harding & Rose (1986).

Threats Loss of broad-leaved woodland and parkland through, for example, clear-felling and coniferisation. Habitat loss, in particular, through the felling of ancient trees, removal of dead wood from living trees and the destruction or removal of standing and fallen dead wood for reasons such as forest hygiene, aesthetic tidiness, public safety or for use as fire wood.

Management and conservation Ancient trees and both fallen and standing dead timber, especially with the bark attached, should be retained. The removal of dead

timber from ancient trees should be avoided. Gaps in the age structure of the tree population should be identified and the continuity of the appropriate dead wood habitat ensured by regeneration, suitable planting and possibly with pollarding.

PYROCHROA COCCINEA　　　　　**NOTABLE B**
Black-headed cardinal beetle
Order COLEOPTERA　　　　Family PYROCHROIDAE

Pyrochroa coccinea (L., 1761).

Distribution South East, South, East Anglia, East Midlands, West Midlands, North East, North West and Wales.

Habitat and ecology Ancient broad-leaved woodland and pasture-woodland. In dead wood and under the bark of oak, beech and elm. Also found breeding in walnut. Adults are often found on flowers. Adults have been recorded from April to June.

Status Widely distributed and local in England and Wales.

Indicator status Grade 1 in Garland (1983). Grade 3 in Harding & Rose (1986).

Threats Loss of broad-leaved woodland and parkland through, for example, clear-felling and coniferisation. Habitat loss, in particular, through the felling of ancient trees, removal of dead wood from living trees and the destruction or removal of standing and fallen dead wood for reasons such as forest hygiene, aesthetic tidiness, public safety or for use as fire wood.

Management and conservation Ancient trees and both fallen and standing dead timber, especially with the bark attached, should be retained. The removal of dead timber from ancient trees should be avoided. Gaps in the age structure of the tree population should be identified and the continuity of the appropriate dead wood habitat ensured by suitable planting and possibly with pollarding.

SCHIZOTUS PECTINICORNIS　　　**NOTABLE A**
A cardinal beetle
Order COLEOPTERA　　　　Family PYROCHROIDAE

Schizotus pectinicornis (L., 1758). Formerly known as: *Pyrochroa pectinicornis* (L.).

Distribution Recorded from Herefordshire, Breconshire, South Aberdeenshire, Moray, East Inverness & Nairn, Mid Ebudes, West Ross and East Ross before 1970 and Herefordshire, Breconshire,

Radnorshire, East Inverness & Nairn, West Inverness, North Ebudes, West Ross and East Ross from 1970 onwards.

Habitat and ecology Broad leaved woodland. In dead wood and under the bark of oak and birch. Adults have been recorded in May and June.

Status Status revised from RDB 3 (Rare) in Shirt (1987). Restricted to central Wales and Herefordshire, the Western Isles and the Highlands of Scotland.

Threats Loss of broad-leaved woodland through, for example, clear-felling and coniferisation. Habitat loss, in particular, through the felling of ancient trees, removal of dead wood from living trees and the destruction or removal of standing and fallen dead wood for reasons such as forest hygiene, aesthetic tidiness, public safety or for use as fire wood.

Management and conservation Ancient trees and both standing and fallen dead timber, especially with the bark attached, should be retained. Gaps in the age structure of the tree population should be identified and where appropriate regeneration or planting and possibly pollarding encouraged to ensure the continuity of the dead wood habitats.

Published sources Allen, A.A. (1956a), Fowler, W.W. (1891), Shirt, D.B., ed. (1987).

PYTHO DEPRESSUS	NOTABLE A
Order COLEOPTERA	Family PYTHIDAE

Pytho depressus (L., 1767).

Distribution Recorded from Mid Perthshire and Moray before 1970 and South Aberdeenshire, East Inverness & Nairn and West Inverness from 1970 onwards.

Habitat and ecology Coniferous woodland. Adults and larvae live under the bark of Scots pine. Larvae have been recorded in June and July. Adults have been noted in July.

Status Exclusively Scottish, probably a component of the native pine forest fauna.

Indicator status Listed in Speight (1989) Appendix 1.

Threats Loss of native pine forest through practices such as clear-felling or conversion to plantation forest. Habitat loss in the remaining areas through the felling of ancient trees and the destruction or removal of standing and fallen dead wood for reasons such as

forest hygiene, aesthetic tidiness, public safety or for use as fire wood.

Management and conservation Ancient trees and both standing and fallen dead timber, especially with the bark attached, should be retained. Gaps in the age structure of the tree population should be identified and regeneration encouraged to ensure the continuity of the dead wood habitats.

Published sources Owen, J.A. (1990a), Speight, M.C.D. (1989).

CYANOSTOLUS AENEUS	NOTABLE A
Order COLEOPTERA	Family RHIZOPHAGIDAE

Cyanostolus aeneus (Richter, 1820). Formerly known as: *Rhizophagus aeneus* Richter, *Rhizophagus caeruleipennis* Sahlberg, 1837, *Rhizophagus coeruleipennis* Sahlberg, 1837.

Distribution Recorded from South Devon, North Devon, South Hampshire, West Sussex, East Sussex, West Kent, Surrey, Worcestershire, Breconshire, Derbyshire, South Lancashire, North-east Yorkshire, Durham, Dumfriesshire and Lanarkshire before 1970 and West Sussex, Worcestershire, Breconshire, Radnorshire, Carmarthenshire and Derbyshire from 1970 onwards.

Habitat and ecology Wet woodland, river margins and probably other wetland habitats. Has occurred away from water. Found at sap, under and in crevices in bark, usually in damp places. Also found on floating or partially submerged timber. Recorded under the bark of elm, beech, oak and occasionally associated with alder, birch, sweet chestnut, pine and apple. The beetle is probably predatory on bark beetles (Scolytidae) of the genera *Xyleborus*, *Scolytus* and *Hylesinus*. Adults have been recorded from April to July and in September.

Status Status revised from RDB 3 (Rare) in Shirt (1987). Old records indicate that this species was widely distributed and recorded from South Devon to Lanarkshire in Scotland. Recently recorded in six vice-counties.

Threats Loss of broad-leaved woodland through, for example, clear-felling and coniferisation. Habitat loss, in particular, through the felling of trees, removal of dead wood from living trees and the destruction or removal of standing and fallen dead wood for reasons such as forest hygiene, aesthetic tidiness, public safety or for use as fire wood. Drainage and water abstraction schemes may be further threats to this species.

RHIZOPHAGIDAE

Management and conservation Trees and both fallen and standing dead timber, especially with the bark attached, should be retained. The removal of dead timber from trees should be avoided. Gaps in the age structure of the tree population should be identified and the continuity of the appropriate habitat ensured by regeneration, suitable planting and possibly with pollarding. Water tables should be maintained at high levels.

Published sources Airy-Shaw, H.K. (1946), Allen, A.A. (1955a), Ashe, G.H. (1934), Blair, K.G. (1936), Fowler, W.W. (1889), Fowler, W.W. & Donisthorpe, H.St J.K. (1913), Gimingham, C.T. (1930), Peacock, E.R. (1977), Shirt, D.B., ed. (1987), Walker, J.J. (1930b).

MONOTOMA ANGUSTICOLLIS   **RARE**

Order COLEOPTERA   Family RHIZOPHAGIDAE

Monotoma angusticollis (Gyllenhal, 1827). Formerly known as: *Monotoma formicetorum* Thomson, 1863.

Distribution Recorded from South Somerset, Dorset, Isle of Wight, South Hampshire, North Hampshire, West Sussex, East Sussex, East Kent, West Kent, Surrey, North Essex, Hertfordshire, Middlesex, Berkshire, Oxfordshire, East Norfolk, East Gloucestershire, Herefordshire, Worcestershire, Warwickshire, Staffordshire, Shropshire, Breconshire, North Lincolnshire, Leicestershire & Rutland, North-east Yorkshire, Durham, South Northumberland, Westmorland & North Lancashire and Cumberland before 1970 and East Sussex, Surrey and Cardiganshire from 1970 onwards.

Habitat and ecology Woodland. Myrmecophilous in the nests of *Formica rufa*, *F. lugubris* and *F. aquilonia* (Hymenoptera). Occasionally found with *M. conicicollis*. Adults have been recorded in January, February, April and December.

Status Old records indicate that this species was widespread in England and recorded as far north as Cumberland. Recently recorded from just three vice-counties, two in southern England and one in mid Wales. This species is difficult to identify and may be confused with other members of the genus. Consequently, the exact status of this species is hard to assess.

Threats Loss of broad-leaved woodland through, for example, clear-felling and conversion to other land use. Neglect and conversion to high forest may be a further threat.

Management and conservation Open glades and ride margins should be cut on rotation to retain a variety of vegetation structures.

Published sources Atty, D.B. (1983), Fowler, W.W. (1889), Fowler, W.W. & Donisthorpe, H.St J.K. (1913), Peacock, E.R. (1977), Shirt, D.B., ed. (1987).

RHIZOPHAGUS NITIDULUS   **NOTABLE B**

Order COLEOPTERA   Family RHIZOPHAGIDAE

Rhizophagus nitidulus (F., 1798).

Distribution England, Dyfed-Powys, North Wales and Scotland.

Habitat and ecology Ancient broad-leaved woodland and pasture-woodland. Adults occur under bark, with records from oak, beech, hornbeam, ash, rowan, birch, and sycamore. Also recorded at sap. Larvae develop in dead wood, with a record of a larva from a southern beech log. Larvae have been found in June. Adults have been noted from April to October.

Status Widespread but local in Great Britain, possibly increasing.

Indicator status Grade 2 in Garland (1983). Grade 3 in Harding & Rose (1986).

Threats Loss of broad-leaved woodland and parkland through, for example, clear-felling and coniferisation. Habitat loss, in particular, through the felling of ancient trees, removal of dead wood from living trees and the destruction or removal of standing and fallen dead wood for reasons such as forest hygiene, aesthetic tidiness, public safety or for use as fire wood.

Management and conservation Ancient trees and both fallen and standing dead timber, especially with the bark attached, should be retained. The removal of dead timber from ancient trees should be avoided. Gaps in the age structure of the tree population should be identified and the continuity of the appropriate dead wood habitat ensured by regeneration, suitable planting and possibly with pollarding.

RHIZOPHAGUS OBLONGICOLLIS
ENDANGERED
Order COLEOPTERA   Family RHIZOPHAGIDAE

Rhizophagus oblongicollis Blatch and Horner, 1892.
Formerly known as: *Rhizophagus simplex* sensu auct.
Brit. not Reitter, 1884.

Distribution Recorded from Surrey, South Essex,
Hertfordshire, Berkshire, Oxfordshire, Staffordshire,
Nottinghamshire and North-east Yorkshire before 1970
and Surrey, Berkshire and East Gloucestershire from
1970 onwards.

Habitat and ecology Ancient broad-leaved woodland
and pasture-woodland. Occurs under the bark of oak
stumps or logs, or in fungi on logs. There is one record
from beech. Larvae develop in dead wood. Adults have
been recorded in March, April, June, July, September
and October.

Status Old records indicate that this species had a
widely scattered distribution in England and was
recorded from Surrey, north to North-east Yorkshire.
Recently recorded from three vice-counties.

Indicator status Grade 1 in Harding & Rose (1986).

Threats Loss of broad-leaved woodland and parkland
through, for example, clear-felling and coniferisation.
Habitat loss, in particular, through the felling of ancient
and fungus-infected trees, removal of dead wood from
living trees and the destruction or removal of standing
and fallen dead wood for reasons such as forest
hygiene, aesthetic tidiness, public safety or for use as
fire wood.

Management and conservation Ancient and
fungus-infected trees and both fallen and standing dead
timber, especially with the bark attached, should be
retained. The removal of dead timber from ancient and
fungus-infected trees should be avoided. Gaps in the
age structure of the tree population should be identified
and the continuity of the appropriate habitat ensured by
regeneration, suitable planting and possibly with
pollarding.

Published sources Garland, S.P. (1983), Harding, P.T.
(1978), Harding, P.T. & Rose, F. (1986), Peacock, E.R.
(1977), Shirt, D.B., ed. (1987).

RHIZOPHAGUS PARVULUS   RARE
Order COLEOPTERA   Family RHIZOPHAGIDAE

Rhizophagus parvulus (Paykull, 1800).

Distribution Recorded from East Inverness & Nairn
before 1970 and East Inverness & Nairn from 1970
onwards.

Habitat and ecology Broad-leaved woodland. Occurs
under the bark of deciduous trees. Found on sappy
stumps of fallen silver birch, also recorded from oak
and in bracket fungi. Adults have been noted in June.

Status Only known from a small area of the Highlands
of Scotland.

Threats Loss of woodland through, for example,
clear-felling or conversion to plantation forest. Habitat
loss in the remaining areas through the felling of
ancient and fungus-infected trees and the destruction or
removal of standing and fallen dead wood for reasons
such as forest hygiene, aesthetic tidiness, public safety
or for use as fire wood.

Management and conservation Ancient and
fungus-infected trees and both standing and fallen dead
timber, especially with the bark attached, should be
retained. Gaps in the age structure of the tree
population should be identified and regeneration
encouraged to ensure the continuity of the appropriate
habitat.

Published sources Johnson, C. (1962b), Shirt, D.B.,
ed. (1987).

RHIZOPHAGUS PICIPES   NOTABLE A
Order COLEOPTERA   Family RHIZOPHAGIDAE

Rhizophagus picipes (Olivier, 1790). Formerly known
as: *Rhizophagus politus* (Hellwig, 1792).

Distribution Recorded from South Devon, South
Hampshire, East Kent, West Kent, Surrey, South Essex,
Berkshire, Oxfordshire, West Gloucestershire,
Monmouthshire, Worcestershire, Warwickshire,
Staffordshire, Nottinghamshire, Cheshire, South
Lancashire, West Lancashire, Westmorland & North
Lancashire, South Aberdeenshire and East Inverness &
Nairn before 1970 and West Sussex, Worcestershire,
Cheshire, South Lancashire, South-east Yorkshire,
South-west Yorkshire and Mid-west Yorkshire from
1970 onwards.

SALPINGIDAE

Habitat and ecology Woodland and pasture-woodland. Occurs at sap and under the bark of pine, oak, beech, ash, poplar, alder, sycamore and once in an apple stump. Probably occurs in association with other trees. Also found in fungi and rotting vegetation. Adults have been recorded from May to July and in September.

Status Status revised from RDB 3 (Rare) in Shirt (1987). Old records indicate that this species has been widely distributed in Great Britain and recorded from South Devon to Easterness in Scotland. Recently recorded from seven vice-counties.

Threats Loss of woodland and parkland through, for example, clear-felling and coniferisation. Habitat loss, in particular, through the felling of trees, removal of dead wood from living trees and the destruction or removal of standing and fallen dead wood for reasons such as forest hygiene, aesthetic tidiness, public safety or for use as fire wood.

Management and conservation Trees and both fallen and standing dead timber, especially with the bark attached, should be retained. The removal of dead timber from trees should be avoided. Gaps in the age structure of the tree population should be identified and the continuity of the appropriate habitat ensured by regeneration, suitable planting and possibly with pollarding.

Published sources Atty, D.B. (1983), Fowler, W.W. (1889), Fowler, W.W. & Donisthorpe, H.St J.K. (1913), Peacock, E.R. (1977), Shirt, D.B., ed. (1987).

LISSODEMA CURSOR — NOTABLE A

Order COLEOPTERA — Family SALPINGIDAE

Lissodema cursor (Gyllenhal, 1813). Formerly known as: *Lissodema kirkae* Donisthorpe, 1925.

Distribution Recorded from East Sussex, East Kent, West Kent, Surrey, North Essex, Hertfordshire, Berkshire, Oxfordshire, West Norfolk, Cambridgeshire, Herefordshire, Derbyshire and Cheshire before 1970 and East Sussex, West Kent, South Essex, Northamptonshire and Leicestershire & Rutland from 1970 onwards.

Habitat and ecology Ash woodland, pasture-woodland, wooded fens and isolated ash trees. Associated with the dead top-most branches of ash, the larvae probably developing in the dead wood. Adults are usually found under ash trees, especially following severe winds. Adults have been recorded from June to September.

Status Recorded from East Kent to Cheshire in England. Recently recorded from just five vice-counties. Probably under-recorded.

Threats Loss of broad-leaved woodland and parkland through, for example, clear-felling and conversion to other land use. Habitat loss, in particular, through the felling of trees and the removal of dead wood from living trees. for reasons such as forest hygiene, aesthetic tidiness, public safety or for use as fire wood.

Management and conservation Trees, especially with the bark attached, should be retained. Gaps in the age structure of the tree population should be identified and regeneration encouraged to ensure the continuity of the appropriate habitat.

Published sources Allen, A.A. (1978b), Blenkarn, S.A. (1922), Cox, D. (1951), Donisthorpe, H.St J.K. (1925c), Fowler, W.W. (1891), Fowler, W.W. & Donisthorpe, H.St J.K. (1913), Grensted, L.W. (1949), Lloyd, R.W. (1946), Massee, A.M. (1945b), Saunders, C.J. (1936b), Walker, J.J. (1936), Williams, S.A. (1972).

LISSODEMA QUADRIPUSTULATA — NOTABLE B

Order COLEOPTERA — Family SALPINGIDAE

Lissodema quadripustulata (Marsham, 1802). Formerly known as: *Lissodema quadripustulatum* (Marsham, 1802).

Distribution South East, South, South West, East Anglia, East Midlands, West Midlands and North East.

Habitat and ecology Woodland, pasture-woodland and hedgerow trees. In dead wood, under bark, and occasionally found under or near the host-trees. Possibly preferring elm, this species has been recorded from beech, field maple, hazel, oak, hawthorn, holly, sycamore and pine, and probably occurs on a variety of other tree species. The larvae probably develop in dead wood. Adults have been recorded from June to September.

Status Widespread but local in England, though not recorded in North West England.

Threats Loss of broad-leaved woodland and parkland through, for example, clear-felling and conversion to other land use. Habitat loss, in particular, through the felling of trees, removal of dead wood from living trees and the destruction or removal of standing and fallen dead wood for reasons such as forest hygiene, aesthetic tidiness, public safety or for use as fire wood.

Management and conservation Trees and both fallen and standing dead timber, especially with the bark attached, should be retained. The removal of dead timber from trees should be avoided. Gaps in the age structure of the tree population should be identified and the continuity of the appropriate dead wood habitat ensured by regeneration, suitable planting and possibly with pollarding.

RABOCERUS FOVEOLATUS   NOTABLE A

Order COLEOPTERA   Family SALPINGIDAE

Rabocerus foveolatus (Ljungh, 1824). Formerly known as: *Rabocerus bishopi* Sharp, 1909, *Salpingus mutilatus* sensu Fowler, 1891 not Beck, 1817, *Sphaeriestes foveolatus* (Ljungh).

Distribution Recorded from Surrey, East Gloucestershire, West Gloucestershire, North-east Yorkshire, South-west Yorkshire, Moray and Clyde Islands before 1970 and South Hampshire, Surrey, Radnorshire, South-west Yorkshire, Durham and Argyll Main from 1970 onwards.

Habitat and ecology Woodland. In dead wood and under bark, with records from beech and pine. The larvae probably develop in dead wood. Adults have been recorded in February, March, from May to July, September and October.

Status Widely distributed and very local in Great Britain, and recorded north to the Clyde Isles. This species is difficult to identify and may be confused with *R. gabrieli*. Consequently, the exact status of this species is hard to assess.

Threats Loss of woodland through, for example, clear-felling and conversion to other land use. Habitat loss, in particular, through the felling of trees, removal of dead wood from living trees and the destruction or removal of standing and fallen dead wood for reasons such as forest hygiene, aesthetic tidiness, public safety or for use as fire wood.

Management and conservation Trees and both standing and fallen dead timber, especially with the bark attached, should be retained. Gaps in the age structure of the tree population should be identified and regeneration, or possibly planting, encouraged to ensure the continuity of the appropriate habitat.

Published sources Atty, D.B. (1983), Fowler, W.W. (1891), Fowler, W.W. & Donisthorpe, H.St J.K. (1913), Walsh, G.B. & Rimington, F.C. (1953).

RABOCERUS GABRIELI   NOTABLE B

Order COLEOPTERA   Family SALPINGIDAE

Rabocerus gabrieli Gerhardt, 1901. Formerly known as: *Salpingus foveolatus* sensu Fowler, 1891 not Ljungh, 1824, *Sphaeriestes gabrieli* (Gerhardt).

Distribution South East, West Midlands, North East, North West, Dyfed-Powys and Scotland.

Habitat and ecology Woodland. In dead wood and under bark. Recorded from beech, elm, alder and pine. Larvae probably develop in dead wood. Adults have been recorded in February, March, May, June and from August to November.

Status Widespread but local in northern Great Britain, very local in southern England. This species is difficult to identify and may be confused with *R. foveolatus*. Consequently, the exact status of this species is hard to assess.

Threats Loss of woodland through, for example, clear-felling and conversion to other land use. Habitat loss, in particular, through the felling of trees, removal of dead wood from living trees and the destruction or removal of standing and fallen dead wood for reasons such as forest hygiene, aesthetic tidiness, public safety or for use as fire wood.

Management and conservation Trees and both standing and fallen dead timber, especially with the bark attached, should be retained. Gaps in the age structure of the tree population should be identified and regeneration, or possibly planting, encouraged to ensure the continuity of the appropriate habitat.

SCAPHISOMA ASSIMILE   INDETERMINATE

Order COLEOPTERA   Family SCAPHIDIIDAE

Scaphisoma assimile Erichson, 1845. Formerly known as: *Scaphosoma assimile* Erichson.

Distribution Recorded from Surrey before 1970 and East Kent from 1970 onwards.

Habitat and ecology Woodland. This species has been found in fungus-infected dead wood and fungus infected plant litter. Adults have been recorded in March.

Status Not listed in the insect Red Data Book (Shirt, 1987). Extremely local and known from very few examples.

SCAPHIDIIDAE

Threats Uncertain, though this species may be threatened by the clear-felling of woodland and conversion to other land use. Habitat loss, in particular, may be through the felling of fungus infected trees, removal of dead wood from living trees and the destruction or removal of standing and fallen dead wood for reasons such as forest hygiene, aesthetic tidiness, public safety or for use as fire wood.

Management and conservation Fungus-infected trees and both fallen and standing dead timber, especially with the bark attached, should be retained. The removal of dead timber from trees should be avoided. Gaps in the age structure of the tree population should be identified and the continuity of the appropriate habitat ensured by suitable regeneration, planting and possibly with pollarding.

Published sources Fowler, W.W. (1889), Philp, E.G. (1990).

SCAPHISOMA BOLETI	NOTABLE B
Order COLEOPTERA	Family SCAPHIDIIDAE

Scaphisoma boleti (Panzer, 1793).

Distribution England, Dyfed-Powys, South East Scotland and South West Scotland.

Habitat and ecology Broad-leaved woodland and pasture-woodland. Also found in a disused quarry, a raised bog, a fen, a riverbank and a garden. This species may have a preference for old dark hornbeam woodland in the south-east. Recorded from an oyster fungus *Pleurotus* spp., a polypore *Polyporus* spp., and a gill fungus, as well as in fungus infected rotten willow and a fungus on a dead elm stump. The beetle has also been noted from Atlantic cedar in a garden. Adults have been recorded from January to March and from May to November.

Status Widespread but local in England and southern Scotland, also recorded in mid-Wales.

Indicator status Grade 2 in Garland (1983).

Threats Loss of broad-leaved woodland and parkland through, for example, clear-felling and conversion to other land use. Habitat loss, in particular through the felling of fungus-infected trees, removal of dead wood from living trees and the destruction or removal of standing and fallen dead wood for reasons such as forest hygiene, aesthetic tidiness, public safety or for use as fire wood. This species may be further threatened by the infilling of quarries and drainage of wetland sites.

Management and conservation Fungus-infected trees and both fallen and standing dead timber, especially with the bark attached, should be retained. Dead timber should not be removed from trees. Gaps in the age structure of the tree population should be identified and the continuity of the appropriate habitat ensured by suitable planting and possibly with pollarding.

SCAPHIUM IMMACULATUM	
	INSUFFICIENTLY KNOWN
Order COLEOPTERA	Family SCAPHIDIIDAE

Scaphium immaculatum (Olivier, 1790).

Distribution Recorded from East Kent before 1970.

Habitat and ecology Uncertain, though probably in open ground or woodland in Britain. Recorded from sand dunes in Holland. In Germany this species has been found in moss, plant litter and rotting fungi. In Britain, the beetle is probably associated with rotting vegetation, plant litter and probably found in situations such as at the roots of plants and under bark or in fungi. Adults have been recorded in April, May, July and September.

Status Status revised from RDB 1 (Endangered) in Shirt (1987). Known from only eighteen examples, four recorded by P. Harwood in 1918, and fourteen found by E.C. Bedwell between 1921 and 1936, all near St Margaret's Bay, East Kent. This species may only be a chance immigrant to these shores which occasionally establishes a temporary population.

Published sources Shirt, D.B., ed. (1987).

AEGIALIA RUFA	**ENDANGERED**
A dung beetle	
Order COLEOPTERA	Family SCARABAEIDAE

Aegialia rufa (F., 1792). Formerly known as: *Rhysothorax rufus* (F.).

Distribution Recorded from Merionethshire, Cheshire and South Lancashire before 1970 and Glamorgan and South Lancashire from 1970 onwards.

Habitat and ecology Coastal sand dunes. Feeds on debris in sand. Adults probably occur at the roots of marram and other dune plants. Adults have been recorded from May to July.

Status Extremely local. Restricted to the coastal sand dunes between the rivers Ribble and Dee in the vicinity of Liverpool, Cheshire and South Lancashire, with a single record from a site in Glamorganshire. There is one old record for Barmouth, Merionethshire.

Threats Loss of dune habitat, particularly through afforestation, urban and holiday development. The degradation of remaining habitat by excessive disturbance of the vegetation through activities such as motorbike access, horse-riding and human trampling.

Management and conservation On dunes some grazing and other disturbance may be desirable to maintain the early successional stages and prevent the invasion of scrub.

Published sources Jessop, L. (1986), Shirt, D.B., ed. (1987), Williams, S.A. (1968).

AEGIALIA SABULETI   NOTABLE B
A dung beetle
Order COLEOPTERA   Family SCARABAEIDAE

Aegialia sabuleti (Panzer, 1796). Formerly known as: *Psammoporus sabuleti* (Panzer).

Distribution South East, West Midlands, North East, North West, South Wales, South East Scotland, South West Scotland and North East Scotland.

Habitat and ecology In sandy areas, usually on the sandy banks of rivers and streams. Also noted at sites well away from water. Possibly associated with decaying vegetable matter, algae and fungi. It has been found under stones. Adults have been recorded in April, June and July.

Status Very local. Primarily a northern and western species.

Threats River engineering, including dredging, level regulation by damming and flood alleviation. In some areas colonisation by Himalayan balsam *Impatiens glandulifera* can reduce the available habitat. River pollution is a further threat.

Management and conservation Shingle and sand banks are an under-valued habitat. They tend to be mobile and rely on the free-flow of river and stream systems. Activities that hinder this flow should be prevented.

AMPHIMALLON OCHRACEUS   NOTABLE A
A chafer
Order COLEOPTERA   Family SCARABAEIDAE

Amphimallon ochraceus (Knoch, 1801). Formerly known as: *Amphimallon ochraceum* (Knoch, 1801), *Amphimallus ochraceus* (Knoch, 1801), *Rhizotrogus ochraceus* (Knoch).

Distribution Recorded from West Cornwall, Surrey, Berkshire, West Gloucestershire, Glamorgan, Pembrokeshire, Caernarvonshire and Anglesey before 1970 and North Devon, East Sussex, Glamorgan, Pembrokeshire, Cardiganshire, Caernarvonshire and Anglesey from 1970 onwards.

Habitat and ecology Predominantly coastal. Frequents cliffs and downland, also found inland on downland. Requires areas of unimproved, undisturbed grassland. Adults have been found on the foliage of herbs and grasses. They only appear to be active for a few hours at around mid-day, when they fly with rapid, weaving movements above and among herbage. Adults have been recorded from June to August.

Status Very local. This species has a scattered distribution along the Welsh coast and the south-west peninsula of England to East Sussex.

Threats Loss of grassland through agricultural improvement by reseeding or by the application of fertilisers, or by conversion to arable. Lack of, or changes in grazing or cutting regimes could alter the vegetation structure which may in turn threaten this species.

Management and conservation Grazing or cutting, on a rotational basis, is needed to maintain open conditions.

Published sources Allen, A.A. (1978c), Allen, A.A. (1978d), Boyce, D.C. (1989), Halstead, A.J. (1988), Jessop, L. (1986).

APHODIUS BREVIS   ENDANGERED
A dung beetle
Order COLEOPTERA   Family SCARABAEIDAE

Aphodius brevis Erichson, 1848. Formerly known as: *Ammoecius brevis* (Erichson, 1848).

Distribution Recorded from South Lancashire before 1970.

Habitat and ecology Coastal dunes. *A. brevis* has found in various sorts of dung, especially that of rabbit, usually partly dried out. This beetle excavates burrows about 4 cm long into which it retreats in dry weather. Also recorded from partly dry cow dung. Adults have been recorded in May and June.

Status No modern records. Only reliably known from the coastal dunes in South Lancashire, and last recorded in 1962 from Ainsdale. The 1956 record from the River Wharfe, at Castley Ford, near Pool, Mid-west Yorkshire is a misidentification. There have been identification problems with this species which has been confused with *Aegialia sabuleti*.

Threats This species may have declined because of the loss of dune habitat, particularly through afforestation, urban and holiday development. The degradation of remaining habitat by excessive disturbance of the vegetation through activities such as motorbike access, horse-riding and human trampling.

Management and conservation An opportunistic and mobile species requiring a continuity of dung availability. Because of this, this species is likely to require large areas of suitable habitat for its continued survival. On dunes some grazing and other disturbance may be desirable to maintain the early successional stages and prevent the invasion of scrub. Part of the area is covered by Ainsdale Sand Dunes and Cabin Hill NNRs.

Published sources Flint, J.H. (1957), Jessop, L. (1986), Shirt, D.B., ed. (1987).

APHODIUS COENOSUS   **NOTABLE B**
A dung beetle
Order COLEOPTERA   Family SCARABAEIDAE

Aphodius coenosus (Panzer, 1798). Formerly known as: *Aphodius tristis* Zenker in Panzer, 1801.

Distribution South East, South, South West, East Anglia, East Midlands, West Midlands, North East and Wales.

Habitat and ecology Dry, exposed, sunny areas on sandy soils. In various types of dung with records from that of horse, sheep, cow and rabbit. Adults have been found from March to May.

Status Widespread but local in England and Wales.

Threats This species may be threatened by practices such as grassland improvement or conversion to arable agriculture. Lack of, or a change in grazing regimes may be a further threat.

Management and conservation An opportunistic and mobile species requiring a continuity of dung

availability. Because of this, this species is likely to require large areas of suitable habitat for its continued survival. Grazing is needed to maintain open conditions and provide a suitable food source.

APHODIUS CONSPURCATUS   **NOTABLE B**
A dung beetle
Order COLEOPTERA   Family SCARABAEIDAE

Aphodius conspurcatus (L., 1758).

Distribution South East, South, East Anglia, East Midlands, West Midlands, North East, North West, Wales and South East Scotland.

Habitat and ecology Woodland, rarely found in exposed habitats. Recorded from horse dung, also found in cow dung. Adults have been noted in February, March, from August to October and in December.

Status Local in England and Wales, very local in Scotland and southern England.

Threats This species may be threatened by the loss of woodland through, for example, clear-felling and conversion to other land use.

Management and conservation An opportunistic and mobile species requiring a continuity of dung availability. Because of this, this species is likely to require large areas of suitable habitat for its continued survival. Grazing is needed to maintain a suitable food source, this should be managed to ensure woodland regeneration.

APHODIUS CONSPUTUS   **RARE**
A dung beetle
Order COLEOPTERA   Family SCARABAEIDAE

Aphodius consputus Creutzer, 1799.

Distribution Recorded from "Devon", East Sussex, East Kent, West Kent, South Essex, Middlesex and Cambridgeshire before 1970 and North Somerset, East Sussex and East Kent from 1970 onwards.

Habitat and ecology Downland, heathland and wet or marshy meadows. In various types of dung with records from that of horse, sheep and cow. Also recorded from a dead rabbit. Adults have been found from March to May and from September to December.

Status Not listed in the insect Red Data Book (Shirt, 1987). Very local and possibly declining. Only known in southern England and recently recorded from just three vice-counties. Can be common where found.

Threats Loss of unimproved grazing land through improvement or conversion to arable agriculture. Lack of, or a change in grazing regimes may be a further threat.

Management and conservation An opportunistic and mobile species requiring a continuity of dung availability. Because of this, this species is likely to require large areas of suitable habitat for its continued survival. Grazing is needed to maintain open conditions and provide a suitable food source.

Published sources Fowler, W.W. (1890), Fowler, W.W. & Donisthorpe, H.St J.K. (1913), Hammond, P.M. (1979), Jessop, L. (1986).

APHODIUS DISTINCTUS **NOTABLE B**
A dung beetle
Order COLEOPTERA Family SCARABAEIDAE

Aphodius distinctus (Müller, 1776). Formerly known as: *Aphodius inquinatus* (Herbst, 1783), *Aphodius melanostictus* sensu auct. Brit. not Schmidt.

Distribution South East, South, East Anglia, West Midlands, East Midlands, North East, North West and South East Scotland.

Habitat and ecology Grazing land, particularly on sandy soils. This species has been recorded from fens and downland. In various types of dung and decomposing vegetable material, though possibly with a preference for horse dung. This beetle has also been found in a rabbit burrow and a rabbit carcass. Adults have been recorded from April to June and from August to October.

Status Very local. Formerly not uncommon. This species has periods of comparative abundance.

Threats This species may be threatened by the loss of unimproved grazing land through improvement or conversion to arable agriculture. Lack of, or a change in grazing regimes is a further threat.

Management and conservation An opportunistic and mobile species requiring a continuity of dung availability. Because of this, this species is likely to require large areas of suitable habitat for its continued survival. Grazing is needed to maintain open conditions and provide a suitable food source.

APHODIUS FASCIATUS **NOTABLE B**
A dung beetle
Order COLEOPTERA Family SCARABAEIDAE

Aphodius fasciatus (Olivier, 1789). Formerly known as: *Aphodius foetidus* (F., 1792) nec (Herbst, 1783), *Aphodius putridus* (Herbst, 1789) nec (Fourcroy, 1785), *Aphodius tenellus* Say, 1823.

Distribution South East, West Midlands, North East, North West, Dyfed-Powys and North East Scotland.

Habitat and ecology Possibly associated with woodland. Has been found on moorland. In various types of dung with records from that of deer, sheep and horse. Adults have been found from April to October and December, though mainly in the autumn.

Status Very local, primarily a northern species.

Threats This species may be threatened by the loss of woodland through, for example, clear-felling and conversion to other land use. Lack of, or a change in grazing regimes is a further threat.

Management and conservation An opportunistic and mobile species requiring a continuity of dung availability. Because of this, this species is likely to require large areas of suitable habitat for its continued survival. Grazing is needed to maintain a suitable food source.

APHODIUS LIVIDUS **ENDANGERED**
A dung beetle
Order COLEOPTERA Family SCARABAEIDAE

Aphodius lividus (Olivier, 1789).

Distribution Recorded from East Kent, West Kent, Hertfordshire, Oxfordshire, East Suffolk, East Norfolk, Cambridgeshire, Huntingdonshire, Anglesey, "Lincs" and South Northumberland before 1970 and West Kent from 1970 onwards.

Habitat and ecology Probably associated with pasture. In various types of dung, in manure and compost heaps. Adults have been recorded in June, August, September and November.

Status Status revised from RDB 3 (Rare) in Shirt (1987). One recent record. Formerly widespread, with scattered records throughout England, also noted from Anglesey, Wales.

Threats Loss of unimproved grazing land through improvement or conversion to arable agriculture. Lack of, or a change in grazing regimes may have also contributed to a decline in this species.

Management and conservation An opportunistic and mobile species requiring a continuity of dung, and possibly rotting vegetation, availability. Because of this, this species is likely to require large areas of suitable habitat for its continued survival. Grazing is needed to maintain open conditions and provide a suitable food source.

Published sources Fowler, W.W. (1890), Fowler, W.W. & Donisthorpe, H.St J.K. (1913), Jessop, L. (1986), Shirt, D.B., ed. (1987), Walker, J.J. (1928b), Williams, B.S. (1928a).

APHODIUS NEMORALIS   NOTABLE A
A dung beetle
Order COLEOPTERA   Family SCARABAEIDAE

Aphodius nemoralis Erichson, 1848.

Distribution Recorded from Merionethshire, Westmorland & North Lancashire, Cumberland, "Perth", South Aberdeenshire, Moray and Dunbartonshire before 1970 and Cumberland, West Perthshire, Angus, South Aberdeenshire and East Inverness & Nairn from 1970 onwards.

Habitat and ecology Shaded conditions, primarily in coniferous woodland. Mainly in deer dung, though also in sheep dung. Adults have been recorded in the spring and rarely in the autumn.

Status Very local and primarily Scottish in distribution, with few records for northern England and Wales.

Threats The loss of woodland through, for example, clear-felling and conversion to other land use may be a threat to this species. Lack of, or a change in grazing regimes, including the establishment of deer exclosures, may be a further threat.

Management and conservation An opportunistic and mobile species requiring a continuity of dung availability. Because of this, this species is likely to require large areas of suitable habitat for its continued survival. Grazing is needed to provide a suitable food source, this should be managed to ensure woodland regeneration.

Published sources Jessop, L. (1986), Owen, J.A. (1988a).

APHODIUS NIGER   ENDANGERED
A dung beetle
Order COLEOPTERA   Family SCARABAEIDAE

Aphodius niger (Panzer, 1796).

Distribution Recorded from South Hampshire before 1970 and South Hampshire from 1970 onwards.

Habitat and ecology Pond and ditch margins. Feeds on decaying matter in damp soil and mud at the sides of a pond frequented by cattle and horses. This beetle has been found at the roots of grass under and adjacent to dung on wet mud, as well as under sods on the bed of the same pond after it had dried out. Adults have been recorded in April, July and September.

Status Currently only known from one pond in the New Forest, South Hampshire

Threats This species would be threatened by drainage, infilling, excavating and the polluting of its sole site. Lack of access to the pond for cattle and horses may be a further threat.

Management and conservation Every effort should be made to ensure the continued existence of this pond and its immediate surroundings. Current conditions should be maintained i.e. variable water level. Grazing of the surrounding area may be needed to maintain open conditions.

Published sources Jessop, L. (1986), Shirt, D.B., ed. (1987).

APHODIUS PAYKULLI   NOTABLE B
A dung beetle
Order COLEOPTERA   Family SCARABAEIDAE

Aphodius paykulli Bedel, 1908. Formerly known as: *Aphodius tessulatus* (Paykull, 1798).

Distribution South, South East, South West, East Anglia, West Midlands, North East, North West, South Wales, South East Scotland and North East Scotland.

Habitat and ecology Downland and heathland. This beetle has also been recorded from arable land on calcareous soil. In various types of dung, including that of cow. Adults have been recorded in January, April, October and December.

Status Widespread and very local. Possibly under-recorded as this is a winter species.

Threats Loss of grassland and heathland through agricultural improvement by reseeding or by the application of fertilisers or through afforestation and urban development. Lack of, or a change in grazing regimes may threaten this species.

Management and conservation An opportunistic and mobile species requiring a continuity of dung availability. Because of this, this species is likely to require large areas of suitable habitat for its continued survival. Management should aim for a diversity of successional stages, preferably by animal grazing.

APHODIUS PLAGIATUS   **NOTABLE B**
A dung beetle
Order COLEOPTERA   Family SCARABAEIDAE

Aphodius plagiatus (L., 1767).

Distribution South East, South, South West, East Anglia, East Midlands, North East, North West and Wales.

Habitat and ecology Sandhills, sand dunes, saltmarshes and damp places near the coast. Associated with small fungi growing in damp hollows on sandhills. It has also been found in rotting seaweed on a saltmarsh. Adults have been recorded from March to June and in November.

Status Widespread and very local. Apparently not known in Scotland.

Threats Loss of dune habitat, particularly through afforestation, urban and holiday development. The degradation of remaining habitat by excessive disturbance of the vegetation through activities such as motorbike access, horse-riding and human trampling. This species may be further threatened by the loss of saltmarsh through reclamation, erosion and the construction of sea defences.

Management and conservation On dunes some grazing or other disturbance may be desirable to maintain the early successional stages and prevent the invasion of scrub.

APHODIUS PORCUS   **NOTABLE B**
A dung beetle
Order COLEOPTERA   Family SCARABAEIDAE

Aphodius porcus (F., 1792).

Distribution South East, South, South West, East Anglia, West Midlands, South Wales, North East, North West, South East Scotland and South West Scotland.

Habitat and ecology Grassland, including south-facing slopes on chalk downland and sandy pastures. This beetle has been shown to be a cuckoo parasite in the burrows of *Geotrupes stercorarius* (Geotrupidae). Adults are often found in dung, with records from that of horse and cow. Adults have been recorded in September and October. This is an autumn species which may overwinter in the burrows of *Geotrupes*.

Status Widespread and very local. Recorded throughout Great Britain except the far north of Scotland

Threats This species is probably threatened by the loss of unimproved grassland through agricultural improvement by reseeding or by the application of fertilisers, or by conversion to arable. Lack of, or a change in grazing regimes may be a further threat.

Management and conservation An opportunistic and mobile species requiring a continuity of dung availability. Because of this, this species is likely to require large areas of suitable habitat for its continued survival. Grazing is needed to maintain open conditions and to provide a suitable food source.

APHODIUS PUTRIDUS   **NOTABLE B**
A dung beetle
Order COLEOPTERA   Family SCARABAEIDAE

Aphodius putridus (Fourcroy, 1785). Formerly known as: *Aphodius arenarius* sensu (Olivier, 1789) not (F., 1787), *Aphodius rhododactylus* (Marsham, 1802), *Plagiogonus arenarius* sensu (Olivier) not (F.).

Distribution South East, South, East Anglia, East Midlands, North East, North West and South Wales.

Habitat and ecology Sand dunes. Also known from downland and other open situations on dry, chalky or sandy soils. In sheep, horse and cow dung. Recorded in rabbit dung at the entrance to their burrows and in vegetable debris. Adults have been found in March and June.

Status Possibly declining. Now very local, particularly outside south-eastern England.

Threats This species may be threatened by the loss of dune habitat, particularly through afforestation, urban and holiday development. The degradation of remaining habitat by excessive disturbance of the vegetation through activities such as motorbike access, and human trampling. Loss of calcareous grassland through agricultural improvement by reseeding or by the application of fertilisers, or by conversion to arable.

SCARABAEIDAE

Lack of, or a change in grazing regimes may be a further threat.

Management and conservation An opportunistic and mobile species requiring a continuity of dung availability. Because of this, this species is likely to require large areas of suitable habitat for its continued survival. Some grazing is needed to maintain open conditions, as well as to providing a suitable food source.

APHODIUS QUADRIMACULATUS
A dung beetle   **ENDANGERED**
Order COLEOPTERA   Family SCARABAEIDAE

Aphodius quadrimaculatus (L., 1761).

Distribution Recorded from North Somerset, East Kent, Surrey, Berkshire, East Suffolk, "Norfolk", West Gloucestershire, Glamorgan, Cheshire and Midlothian before 1970 and East Sussex from 1970 onwards.

Habitat and ecology Dry pastures, including chalk downland. In sheep and cow dung. Adults have been recorded from April to May and in July.

Status Status revised from RDB 3 (Rare) Shirt (1987). Formerly mainly recorded from East Kent though with scattered records throughout Great Britain, noted as far north as Midlothian, Scotland. Recently recorded from just one vice-county.

Threats This species may be threatened by the loss of unimproved grassland through improvement or conversion to arable agriculture. Lack of, or a change in grazing regimes may be a further threat.

Management and conservation An opportunistic and mobile species requiring a continuity of dung availability. Because of this, this species is likely to require large areas of suitable habitat for its continued survival. Grazing is needed to maintain open conditions and to provide a suitable food source.

Published sources Eagles, T.R. (1949), Fowler, W.W. (1890), Fowler, W.W. & Donisthorpe, H.St J.K. (1913), Jessop, L. (1986), Shirt, D.B., ed. (1987).

APHODIUS SCROFA   **EXTINCT**
A dung beetle
Order COLEOPTERA   Family SCARABAEIDAE

Aphodius scrofa (F., 1787).

Distribution Recorded from East Cornwall, South Lancashire and Renfrewshire.

Habitat and ecology Dry, exposed situations on sandy ground. In sheep, horse and cow dung.

Status Presumed extinct. Last recorded in the 19th century.

Published sources Fowler, W.W. (1890), Jessop, L. (1986), Shirt, D.B., ed. (1987).

APHODIUS SORDIDUS   **NOTABLE A**
A dung beetle
Order COLEOPTERA   Family SCARABAEIDAE

Aphodius sordidus (F., 1775).

Distribution Recorded from "Devon", South Hampshire, West Kent, Surrey, South Essex, Middlesex, East Suffolk, West Suffolk, East Norfolk, West Gloucestershire, Glamorgan, North Lincolnshire, Cheshire, West Lancashire, South-west Yorkshire, Cumberland, Midlothian and Fife before 1970 and West Sussex, West Norfolk, East Gloucestershire, West Gloucestershire, Caernarvonshire and Kirkcudbrightshire from 1970 onwards.

Habitat and ecology Dry, sandy or chalky areas, often near the coast. Prefers exposed habitats. In various types of dung, including that of horse. Adults have been recorded in June, August and September.

Status Formerly more widespread, now very local in Great Britain with very few recent and widely scattered records.

Threats This species may be threatened by the loss of unimproved grassland through improvement or conversion to arable agriculture. Lack of, or a change in grazing regimes may be a further threat.

Management and conservation An opportunistic and mobile species requiring a continuity of dung availability. Because of this, this species is likely to require large areas of suitable habitat for its continued survival. Grazing is needed to maintain open conditions and provide a suitable food source.

Published sources Atty, D.B. (1983), Day, F.H.
(1952), Fowler, W.W. (1890), Fowler, W.W. &
Donisthorpe, H.St J.K. (1913), Jessop, L. (1986).

APHODIUS SUBTERRANEUS ENDANGERED
A dung beetle
Order COLEOPTERA   Family SCARABAEIDAE

Aphodius subterraneus (L., 1758). Formerly known as:
Colobopterus subterraneus (L.).

Distribution Recorded from South Hampshire, North
Hampshire, East Kent, West Kent, Warwickshire,
South-east Yorkshire and Cumberland before 1970 and
West Kent from 1970 onwards.

Habitat and ecology Possibly associated with grazing
land, downland and cliffs. In various types of dung,
carrion, compost and other debris. Adults have been
recorded from April to September.

Status Status revised from RDB 3 (Rare) in Shirt
(1987). Formerly more widespread with a scattered
distribution in England, recorded as far north as
Cumberland. Prior to a single recent record in West
Kent this beetle was last recorded in 1948 from
Eversley, North Hampshire.

Threats This species has possibly declined because of
practices such as grassland improvement or conversion
to arable agriculture.

Management and conservation An opportunistic and
mobile species requiring a continuity of resource
availability i.e. dung and carrion. Because of this, this
species is likely to require large areas of suitable
habitat for its continued survival. The presence of
grazing is needed to maintain open conditions and
provide a suitable food source.

Published sources Day, F.H. (1952), Jessop, L. (1986),
Shirt, D.B., ed. (1987), Whicher, L.S. (1948).

APHODIUS ZENKERI NOTABLE B
A dung beetle
Order COLEOPTERA   Family SCARABAEIDAE

Aphodius zenkeri Germar, 1813.

Distribution South East, South, South West, East
Anglia, East Midlands, West Midlands, North West,
North East and North Wales.

Habitat and ecology Woodland and deer parks. This
species has also been noted on downland.
Predominantly in deer dung. Occasionally in other

types of dung, though usually where deer are present.
Adults have been recorded from July to October,
though the main period of adult emergence is probably
in August.

Status Widespread and very local. Recorded
throughout southern England and north to Yorkshire
and Cheshire. This species has also been noted in
North Wales.

Threats This species is probably threatened by the loss
of pasture-woodland and woodland through, for
example, clear-felling and conversion to other land use.
Lack of, or a change in grazing regimes is a further
threat.

Management and conservation An opportunistic and
mobile species requiring a continuity of dung
availability. Because of this, this species is likely to
require large areas of suitable habitat for its continued
survival. Grazing by deer is needed to maintain a
suitable food source. Deer populations should be
managed to ensure woodland regeneration.

BRINDALUS PORCICOLLIS EXTINCT
A dung beetle
Order COLEOPTERA   Family SCARABAEIDAE

Brindalus porcicollis (Illiger, 1803). Formerly known
as: *Psammobius porcicollis* (Illiger), *Psammodius
porcicollis* (Illiger).

Distribution Recorded from East Cornwall and
Glamorgan before 1970.

Habitat and ecology Sandy places on the coast. In
vegetable debris, under stones and at the roots of low
herbage, particularly restharrow. Adults have been
recorded in September.

Status Status revised from RDB 1+ (Endangered,
beleived extinct) in Shirt (1987). Presumed extinct.
Only known from two localities: Pyle, Glamorganshire,
where it was last recorded, in numbers, in 1899, and
Whitsand Bay, near Plymouth, East Cornwall, where a
few examples were found between 1875 and 1897. It
has not been seen since.

Published sources Jessop, L. (1986), Shirt, D.B., ed.
(1987).

SCARABAEIDAE

CETONIA CUPREA   **NOTABLE B**
A chafer
Order COLEOPTERA   Family SCARABAEIDAE

Cetonia cuprea F., 1775. Formerly known as: *Cetonia floricola* Herbst, 1790, *Potosia cuprea* F., 1775.

Distribution South, North East, North West and Scotland.

Habitat and ecology Woodland and open birch scrub. Larvae develop in nests of wood ants *Formica* spp. (Hymenoptera). Adults have been found on flowers such as umbellifers and the blossom of cherry and lilac. Also noted at exuding sap of damaged trees. Larvae collected in June 1977 emerged as adults in August 1978. Adults have been found in late spring and summer.

Status Local. Apart from three records from the New Forest, South Hampshire, this beetle has only been recorded from the north of England and Scotland.

Threats Loss of woodland through, for example, clear-felling and conversion to other land use. The cessation of management practices that encourage a varied ride edge structure and open conditions.

Management and conservation Open glades and ride margins should be cut on rotation to retain a variety of vegetation structures.

COPRIS LUNARIS   **ENDANGERED**
Horned dung beetle
Order COLEOPTERA   Family SCARABAEIDAE

Copris lunaris (L., 1758).

Distribution Recorded from North Somerset, South Hampshire, West Sussex, East Kent, West Kent, Surrey, Berkshire, East Suffolk and Staffordshire before 1970.

Habitat and ecology Well-drained, unploughed pastures on chalk or sandy soils. The adults co-operate in excavating an oblique or vertical tunnel up to 10-20 cm deep, under cow (or horse) dung, leaving a large cast on the surface. A large terminal brood chamber is prepared and furnished with four to seven brood balls of dung, and only one egg is laid in each ball. The female remains in the brood chamber until the new adults emerge three to four months later. Adults are usually seen from mid May to July and fly at dusk on warm evenings.

Status No recent records. Formerly recorded from a number of localities in central and southern England,

though the majority of examples appear to have been found in the neighbourhood of Box Hill, Dorking, Godalming and Guildford in Surrey. Last recorded in 1955 from Juniper Hall Field Centre, near Box Hill, Surrey.

Threats This species may be have declined because of the loss of suitable grassland through improvement or conversion to arable agriculture. Lack of, or a change in grazing regimes may be a further threat. This species may have suffered in the Box Hill area because of the attentions of collectors.

Management and conservation An opportunistic and mobile species requiring a continuity of dung availability. Because of this, this species is likely to require large areas of suitable habitat for its continued survival. Grazing is needed to maintain open conditions and to provide a suitable food source.

Published sources Eagles, T.R. (1948b), Fowler, W.W. (1890), Fowler, W.W. & Donisthorpe, H.St J.K. (1913), Jessop, L. (1986), Miles, P.M. (1942), Shirt, D.B., ed. (1987).

DIASTICTUS VULNERATUS   **VULNERABLE**
A dung beetle
Order COLEOPTERA   Family SCARABAEIDAE

Diastictus vulneratus (Sturm, 1805).

Distribution Recorded from West Suffolk before 1970 and West Suffolk and West Norfolk from 1970 onwards.

Habitat and ecology Dry, open heathy areas on sandy soils. In entrances to rabbit burrows, under stones, in moss and in ground litter. Probably feeds on vegetable debris. Adults have been recorded from April to early July and November.

Status Extremely local. Only known from the Breckland area of Suffolk and Norfolk.

Threats Much Breckland has been lost, or fragmented, through changes in land use, mainly by conversion to arable agriculture and forestry. Habitat degradation through the cessation of traditional management leading to loss through invasion by bracken and scrub is a further threat.

Management and conservation Management should aim for a diversity of successional stages from bare ground to mature breck grassland, preferably by animal grazing or by rotational cutting and scraping.

Published sources Jessop, L. (1986), Shirt, D.B., ed. (1987).

EUHEPTAULACUS SUS   **ENDANGERED**
A dung beetle
Order COLEOPTERA   Family SCARABAEIDAE

Euheptaulacus sus (Herbst, 1783). Formerly known as: *Aphodius sus* (Herbst), *Heptaulacus sus* (Herbst).

Distribution Recorded from South Devon, North Devon, North Somerset, Dorset, East Sussex, East Kent, East Norfolk, Glamorgan and Cheshire before 1970 and North Devon and West Lothian from 1970 onwards.

Habitat and ecology Dry, sandy pastures. In dung. Adults have been recorded in August.

Status Status revised from RDB 3 (Rare) in Shirt (1987). There are old scattered records throughout southern England ranging as far north as Cheshire, also noted in South Wales. Recently recorded from two vice-counties, including West Lothian in Scotland.

Threats This species may be threatened by the loss of unimproved grassland through improvement or conversion to arable agriculture. Lack of, or a change in grazing regimes may be a further threat.

Management and conservation An opportunistic and mobile species requiring a continuity of dung availability. Because of this, this species is likely to require large areas of suitable habitat for its continued survival. Grazing is needed to maintain open conditions and to provide a suitable food source.

Published sources Britton, E.B. (1956), Crowson, R.A. (1986), Crowson, R.A. (1987), Fowler, W.W. (1890), Jessop, L. (1986), Shirt, D.B., ed. (1987).

EUHEPTAULACUS VILLOSUS   **NOTABLE A**
A dung beetle
Order COLEOPTERA   Family SCARABAEIDAE

Euheptaulacus villosus Gyllenhal, 1806. Formerly known as: *Aphodius villosus* (Gyllenhal), *Heptaulacus villosus* (Gyllenhal).

Distribution Recorded from North Devon, North Somerset, South Wiltshire, Isle of Wight, East Sussex, East Kent, West Kent, Surrey, Hertfordshire, Berkshire, Oxfordshire, "Suffolk", Merionethshire, Caernarvonshire, North Lincolnshire, Cheshire, South Lancashire, West Lancashire, North-east Yorkshire, Mid-west Yorkshire, Berwickshire, East Lothian and

Fife before 1970 and North Somerset, South Hampshire, East Sussex, West Norfolk, Northamptonshire, Glamorgan, South-west Yorkshire and Angus from 1970 onwards.

Habitat and ecology Sand dunes and meadows near the sea on sandy or chalky soils. At the roots of plants, in vegetable matter, in dung and occasionally beaten or swept from vegetation. *E. villosus* may be associated with rabbit burrows. Adults may be short-lived as nearly all the records are for June or July, though it has also been recorded in May and August.

Status Widespread and very local. Most northern records are from coastal, sandy localities, the southern records are primarily from inland sites. Possibly under-recorded because of the short lifespan of the adult.

Threats This beetle may be threatened by the loss of unimproved grassland through improvement or conversion to arable agriculture and the loss of dune habitat, particularly through afforestation, urban and holiday development. Lack of, or a change in grazing regimes may be a further threat.

Management and conservation Because of its dependence on dung etc. this species is likely to require large areas of suitable habitat for its continued survival. Grazing is needed to maintain open conditions.

Published sources Allen, A.A. (1954d), Appleton, D. (1972b), Collier, M.J. (1990), Foster, A.P. (1988), Fowler, W.W. (1890), Fowler, W.W. & Donisthorpe, H.St J.K. (1913), Hodge, P.J. (1979), Jessop, L. (1986), Kevan, D.K. (1964), Lee, J. (1980), Welch, R.C. (1978).

GNORIMUS NOBILIS   **VULNERABLE**
A chafer
Order COLEOPTERA   Family SCARABAEIDAE

Gnorimus nobilis (L., 1758).

Distribution Recorded from North Devon, South Hampshire, West Sussex, East Kent, West Kent, Surrey, South Essex, Middlesex, Oxfordshire, Buckinghamshire, East Norfolk, West Gloucestershire, Herefordshire, Worcestershire and Cumberland before 1970 and South Hampshire, West Gloucestershire and Worcestershire from 1970 onwards.

Habitat and ecology Woodland and orchards. Larvae develop in wood mould, particularly of that of plum, but also apple, pear, damson, cherry and willow. Adults have been found on the flowers of hogweed and dog rose. Once recorded from oak. Larvae have been found

SCARABAEIDAE

in April. Development probably takes at least two years. Reared larvae produced pupae in February. In captivity adults have emerged in March. In the wild adults have been noted from April to September. The main period of adult emergence is probably in June.

Status Status revised from RDB 3 (Rare) in Shirt (1987). Very local and declining. Predominantly noted in south-east England, though there are recent records for only three vice-counties, one in southern England and two in the West Midlands.

Threats Loss of broad-leaved woodland and old orchards through, for example, clear-felling and conversion to other land use. Habitat loss, in particular, through the felling of ancient trees, removal of dead wood from living trees and the destruction or removal of standing and fallen dead wood for reasons such as aesthetic tidiness, public safety or for use as fire wood.

Management and conservation Ancient trees and both fallen and standing dead timber, especially with the bark attached, should be retained. The removal of dead timber from ancient trees should be avoided. Gaps in the age structure of the tree population should be identified and the continuity of the appropriate dead wood habitat ensured by suitable planting and possibly with pollarding. The presence of nectar sources such as umbellifers, may also be particularly important for this species.

Published sources Allen, A.A. (1947c), Allen, A.A. (1949b), Atty, D.B. (1983), Fowler, W.W. (1890), Fowler, W.W. & Donisthorpe, H.St J.K. (1913), Grensted, L.W. (1945), Griffith, C.F. (1947), Jessop, L. (1986), Massee, A.M. (1940b), Massee, A.M. (1946), Owen, J.A. (1990a), Shirt, D.B., ed. (1987), Smith, K.G.V. (1948).

GNORIMUS VARIABILIS   ENDANGERED
A chafer
Order COLEOPTERA   Family SCARABAEIDAE

Gnorimus variabilis (L., 1758).

Distribution Recorded from West Kent, Surrey and Berkshire before 1970 and Berkshire from 1970 onwards.

Habitat and ecology Ancient broad-leaved woodland. Larvae feed in the red heart wood and damp wood mould of old oak trees and also in oak logs which can be up to 20 years old. Adults are secretive and are occasionally found resting on trunks or flying. Larvae occur all year round and probably take at least 3 to 4 years to develop. Adults have been noted in June, July

and rarely in August. Populations are very localised and can be restricted to just a single tree.

Status Formerly known last century and very early this century from a small number of localities in the vicinity of London. Since the early part of this century the beetle has only been recorded from Windsor Forest, Berkshire.

Indicator status Grade 1 in Harding & Rose (1986).

Threats Loss of parkland and isolated trees through, for example, clear-felling and development. Habitat loss, in particular, through the felling of ancient trees, removal of dead wood from living trees and the destruction or removal of standing and fallen dead wood for reasons such as forest hygiene, aesthetic tidiness, public safety or for use as fire wood.

Management and conservation Ancient trees and both fallen and standing dead timber, especially with the bark attached, should be retained. The removal of dead timber from ancient trees should be avoided. Gaps in the age structure of the tree population should be identified and the continuity of the appropriate dead wood habitat ensured by suitable planting and possibly with pollarding. Windsor Forest is notified as an SSSI.

Published sources Harding, P.T. (1978), Harding, P.T. & Rose, F. (1986), Jessop, L. (1986), Owen, J.A. (1990a), Shirt, D.B., ed. (1987).

HEPTAULACUS TESTUDINARIUS
A dung beetle   ENDANGERED
Order COLEOPTERA   Family SCARABAEIDAE

Heptaulacus testudinarius (F., 1775). Formerly known as: *Aphodius testudinarius* (F.).

Distribution Recorded from South Hampshire, Surrey, South Essex, Middlesex, Glamorgan and "Yorks" before 1970 and South Hampshire from 1970 onwards.

Habitat and ecology Sandy areas. In dry dung, with a record from horse dung, and also in rotten vegetable material. The species has been reported as overwintering in the burrows of *Geotrupes mutator* (Geotrupidae). Adults have been recorded in April and May.

Status Status revised from RDB 3 (Rare) in Shirt (1987). One recent record from South Hampshire. Previously known from widely scattered records from Surrey north to Yorkshire.

Threats This species may be threatened by the loss of unimproved grassland through agricultural improvement and conversion to arable. Lack of, or a change in grazing regimes may be a further threat.

Management and conservation An opportunistic and mobile species requiring a continuity of dung availability. Because of this, this species is likely to require large areas of suitable habitat for its continued survival. Grazing is needed to maintain open conditions and to provide a suitable food source.

Published sources Fowler, W.W. (1890), Fowler, W.W. & Donisthorpe, H.St J.K. (1913), Jessop, L. (1986), Shirt, D.B., ed. (1987).

MELOLONTHA HIPPOCASTANI
A cockchafer   **INSUFFICIENTLY KNOWN**
Order COLEOPTERA   Family SCARABAEIDAE

Melolontha hippocastani F., 1801.

Distribution Recorded from Herefordshire, Merionethshire, Derbyshire, Durham, Westmorland & North Lancashire, Mid Perthshire, Moray, East Inverness & Nairn and Dunbartonshire before 1970.

Habitat and ecology Woodland. Larvae are probably root-feeders. This species has been beaten from hawthorn and various tree species. The adult has been noted flying at sunset. Adults have been recorded from June to August.

Status Not listed in the insect Red Data Book (Shirt, 1987). This species may be over-looked as it is difficult to identify and may be confused with *M. melolontha*. Identified examples indicate that this species may have a northern distribution.

Threats Uncertain, though this species may be threatened by the loss of woodland through, for example, clear-felling and conversion to other land use.

Management and conservation Open glades and ride margins should be cut on rotation to retain a variety of vegetation structures.

Published sources Jessop, L. (1986), Wood, J.J. (1899).

OMALOPLIA RURICOLA   NOTABLE B
A chafer
Order COLEOPTERA   Family SCARABAEIDAE

Omaloplia ruricola (F., 1775). Formerly known as: *Homaloplia ruricola* (F., 1775)

Distribution South East, South, East Anglia, East Midlands and West Midlands.

Habitat and ecology In dry, calcareous areas. Possibly myrmecophilous. Adults occur on the foliage of herbs and grasses. They are active throughout the day and will fly in hot sunshine. Adults have been recorded in June and July.

Status Very local in the southern half of England. Like most chafers, populations fluctuate from year to year.

Threats Loss of calcareous grassland through agricultural improvement by reseeding or by the application of fertilisers, or by conversion to arable. Lack of, or changes in grazing or cutting regimes could alter the vegetation structure which may in turn threaten this species.

Management and conservation Grazing or cutting, on a rotational basis, is needed to maintain open conditions.

ONTHOPHAGUS FRACTICORNIS
A dung beetle   **INSUFFICIENTLY KNOWN**
Order COLEOPTERA   Family SCARABAEIDAE

Onthophagus fracticornis (Preyssler, 1790).

Distribution Recorded from South Hampshire before 1970.

Habitat and ecology Possibly associated with pasture and wet heathland. In dung.

Status Not listed in the insect Red Data Book (Shirt, 1987). Known from just a few old specimens without data and a single female in 1957 from Matley Bog, New Forest, South Hampshire. There is not a single undisputed male specimen with full data from Great Britain. This species is difficult to identify and has been confused with *O. similis*.

Threats It is probable that a lack of, or changes in grazing regimes may be a threat to this species

Management and conservation This is probably an opportunistic and mobile species which requires a continuity of dung availability. Because of this, this species is likely to need large areas of suitable habitat

SCARABAEIDAE

for its continued survival. Grazing is needed to maintain open conditions and to provide a suitable food source.

Published sources Jessop, L. (1986).

ONTHOPHAGUS NUCHICORNIS   NOTABLE A
A dung beetle
Order COLEOPTERA   Family SCARABAEIDAE

Onthophagus nuchicornis (L., 1758).

Distribution Recorded from North Devon, West Sussex, East Sussex, East Kent, West Kent, Surrey, South Essex, West Suffolk, East Norfolk, West Norfolk, Huntingdonshire, Cheshire and North-east Yorkshire before 1970 and North Devon, East Sussex, West Suffolk, West Norfolk and Glamorgan from 1970 onwards.

Habitat and ecology Sand dunes and sandy areas inland. In various types of dung, with records from that of sheep, horse, rabbit and dog. Adults have been recorded in May, June and August.

Status Very local and possibly declining. This species has a widely scattered distribution through southern England and recorded as far north as North-east Yorkshire. Also noted in Glamorganshire, Wales. This species is difficult to identify and has been confused with *O. similis*. Consequently, the exact status of this species is hard to assess.

Threats This species may be threatened by the loss of dune habitat, particularly through afforestation, urban and holiday development, and the loss of sandy pasture through agricultural improvement and conversion to arable. Lack of, or a change in grazing regimes may be a further threat.

Management and conservation An opportunistic and mobile species requiring a continuity of dung availability. Because of this, this species is likely to require large areas of suitable habitat for its continued survival. Grazing is needed to maintain open conditions, as well as to provide a suitable food source.

Published sources Jessop, L. (1986).

ONTHOPHAGUS NUTANS   ENDANGERED
A dung beetle
Order COLEOPTERA   Family SCARABAEIDAE

Onthophagus nutans (F., 1787). Formerly known as: *Onthophagus verticicornis* (Laicharting, 1781).

Distribution Recorded from North Somerset, Dorset, West Kent, South Essex and Glamorgan before 1970.

Habitat and ecology Possibly associated with pastureland. In dung. Adults have been recorded in May.

Status Status revised from RDB Appendix (Extinct) in Shirt (1987). Only known this century from a single example recorded in 1926 from Milborne St. Andrew, Dorset.

Threats The reasons for this species decline are uncertain, though the loss of unimproved grassland through improvement or conversion to arable agriculture could be contributory factors. Lack of, or changes in grazing regimes may also have been a further threat.

Management and conservation An opportunistic and mobile species requiring a continuity of dung availability. Because of this, this species is likely to require large areas of suitable habitat for its continued survival.

Published sources Allen, A.A. (1965b), Fowler, W.W. (1890), Jessop, L. (1986), Shirt, D.B., ed. (1987).

ONTHOPHAGUS TAURUS   EXTINCT
A dung beetle
Order COLEOPTERA   Family SCARABAEIDAE

Onthophagus taurus (Schreber, 1759).

Distribution Recorded from South Devon, South Hampshire and Oxfordshire before 1970.

Habitat and ecology Pastureland. In dung. Adults have been recorded in August and October.

Status Not listed in the insect Red Data Book (Shirt, 1987). Presumed extinct. Apparently recorded on only four occasions in Great Britain, the last being circa 1867 from Exmouth, South Devon. On the Continent, *O. taurus* has been found to be show a degree of migratory behaviour. It is possible that this beetle was never a true native breeding species in Great Britain but occurred here through occasional stray immigrants from the Continent or the Channel Islands.

Published sources Allen, A.A. (1967e), Jessop, L. (1986).

ONTHOPHAGUS VACCA　　　　NOTABLE B
A dung beetle
Order COLEOPTERA　　　Family SCARABAEIDAE

Onthophagus vacca (L., 1767).

Distribution South East, South and South West.

Habitat and ecology Pastureland, especially in low-lying areas in river valleys. In cow, horse and sheep dung. Adults have been recorded in May and June.

Status Local and possibly declining. A southern species.

Threats Loss of unimproved grazing land through agricultural improvement by reseeding or by the application of fertilisers, or by conversion to arable. Lack of, or changes in grazing regimes may be a further threat.

Management and conservation An opportunistic and mobile species requiring a continuity of dung availability. Because of this, this species is likely to require large areas of suitable habitat for its continued survival. Grazing is needed to maintain open conditions and to provide a suitable food source.

PLEUROPHORUS CAESUS　　　　EXTINCT
A dung beetle
Order COLEOPTERA　　　Family SCARABAEIDAE

Pleurophorus caesus (Creutzer in Panzer, 1796). Formerly known as: *Psammobius caesus* (Creutzer).

Distribution Recorded from West Cornwall, "Bristol district", Glamorgan and South Lancashire.

Habitat and ecology Dry, sandy areas on the coast. In plant debris and probably at the roots of plants.

Status Presumed extinct and possibly never a true native. Last recorded in 1890 from the Isles of Scilly, West Cornwall.

Published sources Fowler, W.W. (1890), Jessop, L. (1986), Johnson, C. (1962a), Shirt, D.B., ed. (1987).

POLYPHYLLA FULLO　　　　EXTINCT
Pine chafer
Order COLEOPTERA　　　Family SCARABAEIDAE

Polyphylla fullo (L., 1758).

Distribution Recorded from East Sussex and East Kent.

Habitat and ecology Coastal sandhills. Larvae feed at the roots of plants. Adults have been recorded from June to August and in November.

Status Presumed extinct. Recorded on a number of occasions from coastal sites in East Kent up to about the mid-19th century. A single example recorded from St Leonards, East Sussex in 1902 probably represents an adventive. The species occurs on the sandhills of the French, Dutch and Belgium coasts. It is possible that all records of this species in Great Britain are the result of chance immigration.

Published sources Allen, A.A. (1967e), Allen, A.A. (1970c), Jessop, L. (1986), Shirt, D.B., ed. (1987).

PSAMMODIUS ASPER　　　　NOTABLE A
A dung beetle
Order COLEOPTERA　　　Family SCARABAEIDAE

Psammodius asper (F., 1775). Formerly known as: *Psammobius sulcicollis* (Illiger, 1801).

Distribution Recorded from North Devon, North Somerset, East Kent, West Kent, West Norfolk, West Gloucestershire, Glamorgan, South Lancashire and East Perthshire before 1970 and East Sussex, East Kent, Glamorgan, Pembrokeshire and South Lancashire from 1970 onwards.

Habitat and ecology Sandy coastal areas and inland gravel pits. Adults occur in and on sand, at the roots of plants and have been noted under seaweed. Adults have been recorded from April to June and in August and October.

Status Widespread and very local. Recorded from southern England and scattered northwards to East Perthshire, Scotland, though recently known only as far north as South Lancashire. Also noted in South Wales.

Threats Loss of dune habitat, particularly through afforestation, urban and holiday development. The degradation of remaining habitat by excessive disturbance of the vegetation through activities such as motorbike access, horse-riding and human trampling. Infilling, and the invasion of scrub at the edges of gravel pits may be a further threat.

SCARABAEIDAE

Management and conservation On dunes and at the edges of gravel pits some grazing or other disturbance may be desirable to maintain the early successional stages and prevent the invasion of scrub.

Published sources Fowler, W.W. (1890), Fowler, W.W. & Donisthorpe, H.St J.K. (1913), Jessop, L. (1986).

RHYSSEMUS GERMANUS   **EXTINCT**
A dung beetle
Order COLEOPTERA   Family SCARABAEIDAE

Rhyssemus germanus (L., 1767).

Distribution Recorded from "Bristol district", Glamorgan and South Lancashire.

Habitat and ecology Dry, sandy areas. Typically beside rivers. In decomposing plant debris and at the roots of plants.

Status Presumed extinct and possibly never a true native. Last recorded in the 19th century.

Published sources Fowler, W.W. (1890), Jessop, L. (1986), Johnson, C. (1962a), Shirt, D.B., ed. (1987).

CYPHON KONGSBERGENSIS   **NOTABLE A**

Order COLEOPTERA   Family SCIRTIDAE

Cyphon kongsbergensis Munster, 1924.

Distribution Recorded from Carmarthenshire, Cardiganshire, Montgomeryshire, Merionethshire, Caernarvonshire, West Inverness and West Ross from 1970 onwards.

Habitat and ecology Found in *Sphagnum* bogs. This species has been recorded from poorly drained, low level moorland. Occurs mainly in open bogs, though also from wooded margins of bogs. Larvae of *Cyphon* species are semi-aquatic in wet moss. Adults have been recorded in July and August.

Status Recorded new to Great Britain in 1981 from Westerness. Now recorded from seven vice-counties, though only known from Wales and western Scotland. This species is difficult to identify and may be confused with other members of the genus. Consequently, the exact status of this species is hard to assess.

Threats Drainage for reasons such as agricultural improvement and development. Falling water tables because of water abstraction and river engineering schemes may also threaten this species.

Management and conservation Water tables should be maintained at high levels.

Published sources Owen, J.A. (1988a), Owen, J.A. (1989b), Skidmore, P. (1985b).

CYPHON PUBESCENS   **NOTABLE B**

Order COLEOPTERA   Family SCIRTIDAE

Cyphon pubescens (F., 1792). Formerly confused under: *Cyphon variabilis* (Thunberg, 1787).

Distribution South East, South, East Anglia, West Midlands, North West, North Wales and North East Scotland.

Habitat and ecology Ponds, bogs and probably other wetland habitats. Recorded from grass tussocks (probably over-wintering sites). Adults probably also occur amongst other herbage in damp or wet places, and have been recorded in April and June.

Status Status revised from RDB 3 (Rare) in Shirt (1987). Widely distributed and local. Recorded from West Kent, north to Easterness in Scotland. This species is difficult to identify and may be confused with other members of the genus. Consequently, the exact status of this species is hard to assess.

Threats Drainage for reasons such as agricultural improvement and development. Falling water tables because of water abstraction and river engineering schemes, and the infilling of ponds may also threaten this species.

Management and conservation Water tables should be maintained at high levels.

CYPHON PUNCTIPENNIS   **NOTABLE A**

Order COLEOPTERA   Family SCIRTIDAE

Cyphon punctipennis Sharp, 1872. Formerly known as: *Cyphon variabilis var. nigriceps* sensu Kloet and Hincks, 1945 ?not Kiesenwetter, 1860.

Distribution Recorded from Cumberland, "Solway district", "Tay district", "Dee district" and Kintyre before 1970 and East Norfolk, Breconshire, Cardiganshire, Caernarvonshire, Moray and North Ebudes from 1970 onwards.

Habitat and ecology Fens, blanket bog and other damp places. Found on bushes and young trees. Adults have been recorded in May, June and September.

Status Very local. Recorded from East Norfolk, parts of Wales, to North Ebudes in Scotland. This species is difficult to identify and may be confused with other members of the genus. Consequently, the exact status of this species is hard to assess.

Threats Drainage for reasons such as agricultural improvement and development. Falling water tables because of water abstraction and river engineering schemes may also threaten this species.

Management and conservation Water tables should be maintained at high levels.

Published sources Moseley, K.A. (1979).

ELODES ELONGATA　　　　**INDETERMINATE**

Order COLEOPTERA　　　　Family SCIRTIDAE

Elodes elongata Tournier, 1868.

Distribution Recorded from Berkshire before 1970 and Herefordshire from 1970 onwards.

Habitat and ecology Probably associated with damp places. Recorded from tree foliage and noted from hawthorn by a pond. Adults have been recorded in June.

Status Status revised from RDB 3 (Rare) in Shirt (1987). Only known from two examples; the first recorded in 1934 in Windsor Great Park, Berkshire, and the second found in 1981 at Brampton Bryon Park, Herefordshire. This species is difficult to identify and requires dissection for identification.

Threats Uncertain, though this species may be threatened by drainage for reasons such as agricultural improvement and development. Falling water tables because of water abstraction and river engineering schemes may be further threats.

Management and conservation Water tables should be maintained at high levels.

HYDROCYPHON DEFLEXICOLLIS　**NOTABLE B**

Order COLEOPTERA　　　　Family SCIRTIDAE

Hydrocyphon deflexicollis (Muller, 1821).

Distribution South East, South West, East Anglia, East Midlands, West Midlands, North East, North West, Dyfed-Powys, North Wales and Scotland.

Habitat and ecology Gravelly stream margins and shingle banks. Possibly occurs in other damp or wet situations. Found under a stone on gravelly ground, and possibly also occurs amongst herbage. Adults have been recorded in July and August.

Status Widespread but local in Great Britain.

Threats River engineering, including dredging, level regulation by damming and flood alleviation schemes. In some areas colonisation by Himalayan balsam *Impatiens glandulifera* can reduce the available habitat. Livestock access can damage shingle structure and disturb vegetation communities. River pollution may be a further threat.

Management and conservation An under-valued habitat. Shingle banks tend to be mobile and rely on the free flow of river and stream systems. Activities that hinder this flow should be avoided. Adjacent scrub and rank grassland may be important for the species in winter and to escape late summer floods.

PRIONOCYPHON SERRICORNIS　**NOTABLE B**

Order COLEOPTERA　　　　Family SCIRTIDAE

Prionocyphon serricornis (Müller, 1821).

Distribution England, Dyfed-Powys and South East Scotland.

Habitat and ecology Ancient broad-leaved woodland and pasture-woodland. Associated with wet or water filled rot-holes in trees. The larva is probably predatory on fly (Diptera) and other larvae that live in rot-holes. Adults have been beaten from oak, lime, found under beech bark and from under beech. Larvae have been found in August. Adults have been recorded from June to August and have been noted at mercury vapour light traps.

Status Status revised from RDB 3 (Rare) in Shirt (1987). Widespread but local in England, also recorded in South East Scotland.

SCIRTIDAE

Indicator status Grade 1 in Garland (1983). Grade 2 in Harding & Rose (1986).

Threats Loss of broad-leaved woodland and parkland through, for example, clear-felling and coniferisation. Habitat loss, in particular, through the felling of ancient trees, removal of dead wood from living trees and the destruction or removal of standing and fallen dead wood for reasons such as forest hygiene, aesthetic tidiness, public safety or for use as fire wood.

Management and conservation Ancient trees and both fallen and standing dead timber, especially with the bark attached, should be retained. The removal of dead timber from ancient trees should be avoided. Gaps in the age structure of the tree population should be identified and the continuity of the appropriate dead wood habitat ensured by regeneration, suitable planting and possibly with pollarding.

SCIRTES ORBICULARIS   NOTABLE A

Order COLEOPTERA   Family SCIRTIDAE

Scirtes orbicularis (Panzer, 1793).

Distribution Recorded from "Devon", Isle of Wight, East Sussex, East Kent, West Kent, Surrey, South Essex, East Suffolk and East Norfolk before 1970 and South Somerset, North Somerset, West Sussex, East Sussex, East Kent, East Norfolk and Monmouthshire from 1970 onwards.

Habitat and ecology Wetland, marshes, grazing levels and marshy dykes. Occurs amongst wetland herbage, including sedges, reedmace and bur-reeds. Adults have been recorded from June to August.

Status Status revised from RDB 3 (Rare) in Shirt (1987). Recorded in southern England as far north as East Norfolk, also known from Monmouthshire in Wales. Recently recorded from seven vice-counties.

Threats Drainage for reasons such as agricultural improvement and development. Falling water tables because of water abstraction and river engineering schemes may also threaten this species.

Management and conservation Water tables should be maintained at high levels. Pollution and eutrophication may be further threats.

Published sources Fowler, W.W. (1890), Fowler, W.W. & Donisthorpe, H.St J.K. (1913), Shirt, D.B., ed. (1987).

DRYOCOETINUS ALNI   NOTABLE A
A bark beetle
Order COLEOPTERA   Family SCOLYTIDAE

Dryocoetinus alni (Georg, 1856). Formerly known as: *Dryocaetes alni* (Georg, 1856), *Dryocoetes alni* (Georg).

Distribution Recorded from Surrey, Worcestershire, Warwickshire, Staffordshire, Cheshire, South Lancashire, North-east Yorkshire, Mid-west Yorkshire, Durham, North Northumberland, Ayrshire, Lanarkshire, Berwickshire, Midlothian, Stirlingshire, East Inverness & Nairn and Dunbartonshire before 1970 and East Sussex, Surrey, Herefordshire, Cheshire, South Lancashire, South-west Yorkshire and West Perthshire from 1970 onwards.

Habitat and ecology Broad-leaved woodland. Occurs in the bark of dead alder, beech and hazel. Adults have been recorded in February, May, June and September.

Status Status revised from RDB 3 (Rare) in Shirt (1987). Widely distributed and very local. Recorded from East Sussex, north to West Perthshire. There are published records for Kircudbrightshire, Scotland, though these are referable to *D. villosus*.

Threats Loss of broad-leaved woodland through, for example, clear-felling and coniferisation. Habitat loss, in particular, through the felling of trees, removal of dead wood from living trees and the destruction or removal of standing and fallen dead wood for reasons such as forest hygiene, aesthetic tidiness, public safety or for use as fire wood.

Management and conservation Trees and both fallen and standing dead timber, especially with the bark attached, should be retained. The removal of dead timber from trees should be avoided. Gaps in the age structure of the tree population should be identified and the continuity of the appropriate dead wood habitat ensured by regeneration, suitable planting and possibly with pollarding.

Published sources Allen, A.A. (1973c), Clemons, L. (1983), Flint, J.H. (1988), Fowler, W.W. (1891), Fowler, W.W. & Donisthorpe, H.St J.K. (1913), Shirt, D.B., ed. (1987), White, I. (1983).

ERNOPORUS CAUCASICUS — ENDANGERED
A bark beetle
Order COLEOPTERA — Family SCOLYTIDAE

Ernoporus caucasicus Lindemann, 1876.

Distribution Recorded from Northamptonshire, Herefordshire and Leicestershire & Rutland before 1970 and Northamptonshire, Herefordshire and Derbyshire from 1970 onwards.

Habitat and ecology Ancient broad-leaved woodland and pasture-woodland. Occurs in the bark of dead branches of lime, mainly small-leaved lime though also found in common lime. Adults have been recorded from the canopy foliage. Populations of this beetle appear to be restricted to just one or two trees where the species occurs. Larvae have been found in June. Adults have been recorded in February and from May to July.

Status Known from a very few sites. Recently recorded in just three vice-counties.

Indicator status Grade 1 in Garland (1983). Grade 1 in Harding & Rose (1986).

Threats Loss of broad-leaved woodland and parkland through, for example, clear-felling and coniferisation. Habitat loss, in particular, through the felling of ancient trees, removal of dead wood from living trees and the destruction or removal of standing and fallen dead wood for reasons such as forest hygiene, aesthetic tidiness, public safety or for use as fire wood.

Management and conservation Ancient trees and standing dead timber, especially with the bark attached, should be retained. The removal of dead timber from ancient trees should be avoided. Gaps in the age structure of the tree population should be identified and the continuity of the appropriate dead wood habitat ensured by regeneration or suitable planting.

Published sources Cooter, J. (1980b), Drane, A.B. (1985a), Garland, S.P. (1983), Harding, P.T. (1978), Harding, P.T. & Rose, F. (1986), Shirt, D.B., ed. (1987).

ERNOPORUS FAGI — NOTABLE A
A bark beetle
Order COLEOPTERA — Family SCOLYTIDAE

Ernoporus fagi (F., 1798). Formerly known as: *Cryphalus fagi* (F.), *Ernopocerus fagi* (F.).

Distribution Recorded from South Hampshire, West Sussex, East Sussex, West Kent, Surrey, South Essex, Middlesex, Berkshire, "Norfolk" and West Gloucestershire before 1970 and East Sussex, South Essex, Berkshire, East Suffolk, Herefordshire and Leicestershire & Rutland from 1970 onwards.

Habitat and ecology Ancient broad-leaved woodland and pasture-woodland. Occurs under the bark of recently dead or dying twigs and small, thin boughs of beech. On the Continent, this species is also associated with oak and birch. In Britain, adults have been recorded from February to August and in December.

Status Widespread and very local in southern England, recorded as far north as Leicestershire. Recently recorded from seven vice-counties.

Indicator status Grade 3 in Harding & Rose (1986).

Threats Loss of broad-leaved woodland and parkland through, for example, clear-felling and coniferisation. Habitat loss, in particular, through the felling of ancient trees, removal of dead wood from living trees and the destruction or removal of standing and fallen dead wood for reasons such as forest hygiene, aesthetic tidiness, public safety or for use as fire wood.

Management and conservation Ancient trees and both fallen and standing dead timber, especially with the bark attached, should be retained. The removal of dead timber from ancient trees should be avoided. Gaps in the age structure of the tree population should be identified and the continuity of the appropriate dead wood habitat ensured by regeneration, suitable planting and possibly with pollarding.

Published sources Atty, D.B. (1983), Fowler, W.W. (1891), Fowler, W.W. & Donisthorpe, H.St J.K. (1913), Halstead, A. (1989), Harding, P.T. (1978), Harding, P.T. & Rose, F. (1986).

SCOLYTIDAE

ERNOPORUS TILIAE   **ENDANGERED**
A bark beetle
Order COLEOPTERA   Family SCOLYTIDAE

Ernoporus tiliae (Panzer, 1793). Formerly known as:
Cryphalus tiliae (Panzer).

Distribution Recorded from West Gloucestershire,
Worcestershire, Shropshire, Merionethshire, "Lincs",
Leicestershire & Rutland and Durham before 1970 and
Mid-west Yorkshire from 1970 onwards.

Habitat and ecology Broad-leaved woodland. Occurs
in the bark of recently dead or dying twigs and small,
thin boughs of small-leaved lime and possibly other
Tilia species. Adults have been recorded in February,
June, August and December.

Status Status revised from RDB 3 (Rare) in Shirt
(1987). Only one recent record. Old records indicate
that this species had a widely scattered distribution
from West Gloucestershire to Durham.

Threats Loss of broad-leaved woodland through, for
example, clear-felling and coniferisation. Habitat loss,
in particular, through the felling of trees, removal of
dead wood from living trees and the destruction or
removal of standing and fallen dead wood for reasons
such as forest hygiene, aesthetic tidiness, public safety
or for use as fire wood.

Management and conservation Trees and both fallen
and standing dead timber, especially with the bark
attached, should be retained. The removal of dead
timber from trees should be avoided. Gaps in the age
structure of the tree population should be identified and
the continuity of the appropriate dead wood habitat
ensured by regeneration, suitable planting and possibly
with pollarding.

Published sources Allen, A.A. (1969b), Cooter, J.
(1980b), Fowler, W.W. (1891), Fowler, W.W. &
Donisthorpe, H.St J.K. (1913), Shirt, D.B., ed. (1987).

KISSOPHAGUS HEDERAE   **NOTABLE B**
A bark beetle
Order COLEOPTERA   Family SCOLYTIDAE

Kissophagus hederae (Schmitt, 1843). Formerly known
as: *Cissophagus hederae* (Schmitt, 1843).

Distribution South West, South, South East, East
Midlands, West Midlands, East Anglia, North East,
South Wales and North Wales.

Habitat and ecology Woodland, pasture-woodland and
isolated trees. Found in the dead stems of ivy. Adults
have been recorded in June and November.

Status Widely distributed and very local in England
and Wales. Recorded from South Devon, north to
North-east Yorkshire.

Threats Loss of broad-leaved woodland and parkland
through, for example, clear-felling and conversion to
other land use. The selective removal of ivy may be a
particular threat to this species.

Management and conservation In woodland, open
glades and ride margins should be cut on rotation to
retain a variety of vegetation structures.

LEPERISINUS ORNI   **NOTABLE B**
A bark beetle
Order COLEOPTERA   Family SCOLYTIDAE

Leperisinus orni (Fuchs, 1906).

Distribution South East, South, South West, East
Anglia, West Midlands, North East and North West.

Habitat and ecology Broad-leaved woodland and
pasture-woodland. Associated with dead or cut slender
branches of ash, constructing galleries under the bark.
Adults have been recorded from April to June and from
August to October.

Status Widespread but local in England.

Threats Loss of broad-leaved woodland and parkland
through, for example, clear-felling and coniferisation.
Habitat loss, in particular, through the felling of trees,
removal of dead wood from living trees and the
destruction or removal of standing and fallen dead
wood for reasons such as forest hygiene, aesthetic
tidiness, public safety or for use as fire wood.

Management and conservation Trees and both fallen
and standing dead timber, especially with the bark
attached, should be retained. The removal of dead
timber from trees should be avoided. Gaps in the age
structure of the tree population should be identified and
the continuity of the appropriate habitat ensured by
regeneration or possibly through suitable planting.

SCOLYTIDAE

PITYOGENES QUADRIDENS   NOTABLE A
A bark beetle
Order COLEOPTERA   Family SCOLYTIDAE

Pityogenes quadridens (Hartig, 1834).

Distribution Recorded from North Lincolnshire, Mid Perthshire, South Aberdeenshire, Moray, East Inverness & Nairn and Orkney before 1970 and Mid-west Yorkshire, West Perthshire, Mid Perthshire, Moray and East Inverness & Nairn from 1970 onwards.

Habitat and ecology Coniferous woodland. Breeds under the bark of small branches of Scots pine, also recorded from Norway spruce and fir. Adults have been recorded from May to August.

Status Status revised from RDB 3 (Rare) in Shirt (1987). Very local and apart from records for North Lincolnshire and Mid-west Yorkshire, this species is only known from central and northern Scotland. Probably only native in Scotland.

Threats Loss of woodland through, for example, clear-felling or conversion to other land use. Habitat loss in the remaining areas through the felling of trees and the destruction or removal of standing and fallen dead wood for reasons such as forest hygiene, aesthetic tidiness, public safety or for use as fire wood.

Management and conservation Trees and both standing and fallen dead timber, especially with the bark attached, should be retained. Gaps in the age structure of the tree population should be identified and regeneration encouraged to ensure the continuity of the dead wood habitats.

Published sources Fowler, W.W. (1891), Shirt, D.B., ed. (1987).

PITYOGENES TREPANATUS   NOTABLE A
A bark beetle
Order COLEOPTERA   Family SCOLYTIDAE

Pityogenes trepanatus (Nördlinger, 1848).

Distribution Recorded from Surrey, Berkshire, East Perthshire and Moray before 1970 and East Suffolk, West Norfolk and East Inverness & Nairn from 1970 onwards.

Habitat and ecology Coniferous woodland. Associated mainly with Scots pine, though also with Norway spruce and fir. Adults and larvae live in the bark and smaller branches. Adults have been recorded in May, July, September and October.

Status Status revised from RDB 3 (Rare) in Shirt (1987). Very local and with a widely scattered distribution in Great Britain known from parts of southern England, East Anglia and central and northern Scotland. Probably only native in Scotland.

Threats Loss of woodland through, for example, clear-felling or conversion to other land use. Habitat loss in the remaining areas through the felling of trees and the destruction or removal of standing and fallen dead wood for reasons such as forest hygiene, aesthetic tidiness, public safety or for use as fire wood.

Management and conservation Trees and both standing and fallen dead timber, especially with the bark attached, should be retained. Gaps in the age structure of the tree population should be identified and regeneration encouraged to ensure the continuity of the dead wood habitats.

Published sources Allen, A.A. (1951e), Allen, A.A. (1975a), Beare, T.H. (1937), Fowler, W.W. & Donisthorpe, H.St J.K. (1913), Nash, D.R. (1982a), Shirt, D.B., ed. (1987).

PITYOPHTHORUS LICHTENSTEINI   RARE
A bark beetle
Order COLEOPTERA   Family SCOLYTIDAE

Pityophthorus lichtensteini (Ratzeburg, 1837). Formerly known as: *Pityophthorus lichtensteinii* (Ratzeburg, 1837).

Distribution Recorded from Mid Perthshire, South Aberdeenshire, Moray and East Inverness & Nairn before 1970 and South-east Yorkshire and South Aberdeenshire from 1970 onwards.

Habitat and ecology Coniferous woodland. Breeds in the bark of twigs of pine. Adults have been recorded from cut pine tops and have been recorded from June to August.

Status Very local and recently recorded in just two vice-counties. Apart from South-east Yorkshire, this species is only known in Scotland. There is an unconfirmed record for Suffolk. Probably only native in Scotland. This species is difficult to identify and may be confused with other members of the family. Consequently, the exact status of this species is hard to assess.

Threats Loss of woodland through, for example, clear-felling and conversion to other land use. Habitat

SCOLYTIDAE

loss in the remaining areas through the felling of trees and the destruction or removal of standing and fallen dead wood for reasons such as forest hygiene, aesthetic tidiness, public safety or for use as fire wood.

Management and conservation Trees and both standing and fallen dead timber, especially with the bark attached, should be retained. Gaps in the age structure of the tree population should be identified and regeneration encouraged to ensure the continuity of the appropriate habitat.

Published sources Allen, A.A. (1990a), Flint, J.H. (1988), Fowler, W.W. (1891), Shirt, D.B., ed. (1987).

SCOLYTUS MALI	NOTABLE B
Large fruit bark beetle	
Order COLEOPTERA	Family SCOLYTIDAE

Scolytus mali (Bechstein, 1805). Formerly known as: *Eccoptogaster mali* Bechstein, *Scolytus pruni* (Ratzeburg, 1837).

Distribution South East, South, East Anglia, East Midlands, West Midlands, North East, North West and South Wales.

Habitat and ecology Orchards and woodland. Associated with apple, pear, hawthorn, cultivated and wild cherry, plum, blackthorn, rowan and elm. The larvae develop in galleries in the sapwood just under the bark, where they feed on the living wood. Larvae have been found in February (emerged November). Adults have been recorded from May to July, and in October and November.

Status Widespread but local in England, also recorded in South Wales.

Threats Loss of broad-leaved woodland through, for example, clear-felling and coniferisation. Neglect and conversion to high forest, and the grubbing out of orchards are further threats to this species.

Management and conservation Open glades and ride margins should be cut on rotation to retain a variety of vegetation structures.

SCOLYTUS RATZEBURGI	NOTABLE B
A bark beetle	
Order COLEOPTERA	Family SCOLYTIDAE

Scolytus ratzeburgi Janson, 1856. Formerly known as: *Eccoptogaster ratzeburgi* (Janson).

Distribution South East Scotland, North East Scotland and North West Scotland.

Habitat and ecology Birch woodland. Associated with birch. The larvae develop in galleries just under the bark. Usually found in dead or dying branches, though also in stumps, where they feed on dead wood. Adults have been recorded from June to August.

Status Known from central Scotland, north to West Sutherland. There is a record for South Hampshire in England that requires confirmation.

Threats Loss of birch woodland through practices such as clear-felling or conversion to plantation forest. Habitat loss in the remaining areas through the felling of trees and the destruction or removal of standing and fallen dead wood for reasons such as forest hygiene, aesthetic tidiness, public safety or for use as fire wood.

Management and conservation Trees and both standing and fallen dead timber, especially with the bark attached, should be retained. Gaps in the age structure of the tree population should be identified and regeneration encouraged to ensure the continuity of the appropriate habitats.

TAPHRORYCHUS BICOLOR	NOTABLE A
A bark beetle	
Order COLEOPTERA	Family SCOLYTIDAE

Taphrorychus bicolor (Herbst, 1793).

Distribution Recorded from Dorset, South Hampshire, West Sussex, East Sussex, West Kent, Surrey, South Essex, Middlesex and Berkshire before 1970 and South Hampshire, West Sussex, East Sussex, West Kent, Surrey, South Essex and Berkshire from 1970 onwards.

Habitat and ecology Broad-leaved woodland and pasture-woodland. Found in the bark of dead beech and only rarely in other trees. On the Continent, this species is associated with a variety of trees including beech, hornbeam, oak, birch, elm, hazel, aspen, sycamore and walnut. In Britain, larvae have been found in May and June. Adults have been recorded from April to October.

Status Restricted to southern and south-eastern England. Recently recorded from seven vice-counties.

Threats Loss of broad-leaved woodland and parkland through, for example, clear-felling and coniferisation Habitat loss, in particular, through the felling of ancient trees, removal of dead wood from living trees and the destruction or removal of standing and fallen dead wood for reasons such as forest hygiene, aesthetic tidiness, public safety or for use as fire wood.

Management and conservation Ancient trees and both fallen and standing dead timber, especially with the bark attached, should be retained. The removal of dead timber from ancient trees should be avoided. Gaps in the age structure of the tree population should be identified and the continuity of the appropriate dead wood habitat ensured by regeneration, suitable planting and possibly with pollarding.

Published sources Allen, A.A. (1951b), Cooter, J. (1971), Fowler, W.W. (1891).

TOMICUS MINOR   **RARE**
A bark beetle
Order COLEOPTERA   Family SCOLYTIDAE

Tomicus minor (Hartig, 1834). Formerly known as: *Blastophagus minor* (Hartig), *Myelophilus minor* (Hartig).

Distribution Recorded from Dorset, "Perth", Angus, Kincardineshire, South Aberdeenshire, Moray, West Inverness and "Ross" before 1970 and Dorset and East Inverness & Nairn from 1970 onwards.

Habitat and ecology Coniferous woodland. Associated with small branches of Scots pine, though also in Norway spruce and possibly other introduced conifer species, tunnelling under the bark. Adults have been recorded in July.

Status Very local. Primarily recorded in Scotland, though also noted in Dorset. Recently noted from two vice-counties. Probably native only in Scotland, the Dorset population may be the result of an introduction or recent colonisation.

Threats Loss of native pine forest through, for example, clear-felling and conversion to other land use. Habitat loss in the remaining areas through the felling of trees and the destruction or removal of standing and fallen dead wood for reasons such as forest hygiene, aesthetic tidiness, public safety or for use as fire wood.

Management and conservation Trees and both standing and fallen dead timber, especially with the bark attached, should be retained. Gaps in the age structure of the tree population should be identified and

regeneration encouraged to ensure the continuity of the appropriate habitat.

Published sources Fowler, W.W. (1891), Owen, J.A. (1990a), Shirt, D.B., ed. (1987).

TRIOTEMNUS CORYLI   **ENDANGERED**
A bark beetle
Order COLEOPTERA   Family SCOLYTIDAE

Triotemnus coryli (Perris, 1855). Formerly known as: *Dryocaetes coryli* (Perris, 1855), *Dryocoetes coryli* (Perris), *Lymantor coryli* (Perris).

Distribution Recorded from South Hampshire, West Kent, Surrey, North Essex, East Norfolk and Worcestershire before 1970.

Habitat and ecology Broad-leaved woodland. Occurs in or under the bark of dead, dry branches, mainly of hazel and hornbeam. Also recorded in oak, pear and buckthorn. Adults have been noted in September.

Status Not listed in the insect Red Data Book (Shirt, 1987). No recent records. Only known from southern England and recorded as far north as East Norfolk.

Threats Loss of broad-leaved woodland through, for example, clear-felling and coniferisation. Habitat loss, in particular, through the felling of ancient trees, removal of dead wood from living trees and the destruction or removal of standing and fallen dead wood for reasons such as forest hygiene, aesthetic tidiness, public safety or for use as fire wood.

Management and conservation Ancient trees and both fallen and standing dead timber, especially with the bark attached, should be retained. The removal of dead timber from ancient trees should be avoided. Gaps in the age structure of the tree population should be identified and the continuity of the appropriate dead wood habitat ensured by regeneration, suitable planting and possibly with pollarding.

Published sources Fowler, W.W. (1891), Fowler, W.W. & Donisthorpe, H.St J.K. (1913).

SCOLYTIDAE

TRYPOPHLOEUS ASPERATUS   NOTABLE A
A bark beetle
Order COLEOPTERA   Family SCOLYTIDAE

Trypophloeus asperatus (Gyllenhal, 1813). Formerly known as: *Cryphalus asperatus* (Gyllenhal), *Cryphalus binodulus* (Ratzeburg, 1837).

Distribution Recorded from Dorset, West Kent, Surrey, South Essex, Hertfordshire, Middlesex, West Suffolk, Monmouthshire, Worcestershire, Cheshire and North-east Yorkshire before 1970 and West Kent, East Suffolk, West Suffolk and South Lancashire from 1970 onwards.

Habitat and ecology Broad-leaved woodland. Occurs in recently dead, small or slender branches of *Populus*, particularly aspen, though also recorded from balsam poplar. Pupae have been noted in May. Adults have been recorded from May to August.

Status Status revised from RDB 3 (Rare) in Shirt (1987). Very local and widely distributed in England. Recorded from Dorset, north to North-east Yorkshire, also noted in Wales. Recently found in just four vice-counties.

Threats Loss of broad-leaved woodland through, for example, clear-felling and coniferisation. Habitat loss, in particular, through the felling of trees, removal of dead wood from living trees and the destruction or removal of standing and fallen dead wood for reasons such as forest hygiene, aesthetic tidiness, public safety or for use as fire wood.

Management and conservation Trees and both fallen and standing dead timber, especially with the bark attached, should be retained. The removal of dead timber from trees should be avoided. Gaps in the age structure of the tree population should be identified and the continuity of the appropriate dead wood habitat ensured by regeneration, suitable planting and possibly with pollarding.

Published sources Allen, A.A. (1977b), Fowler, W.W. (1891), Hammond, P.M. (1979), Shirt, D.B., ed. (1987), Williams, B.S. (1928b).

TRYPOPHLOEUS GRANULATUS   EXTINCT
A bark beetle
Order COLEOPTERA   Family SCOLYTIDAE

Trypophloeus granulatus (Ratzeburg, 1837). Formerly known as: *Cryphalus granulatus* (Ratzeburg).

Distribution Recorded from Surrey.

Habitat and ecology Broad-leaved woodland. Associated with *Populus*. The adult has been recorded in June.

Status Presumed extinct and possibly never native. Known from just a single example found in 1867 near Surbiton, Surrey.

Published sources Fowler, W.W. (1891), Shirt, D.B., ed. (1987).

XYLEBORUS DISPAR   NOTABLE B
A bark beetle
Order COLEOPTERA   Family SCOLYTIDAE

Xyleborus dispar (F., 1792). Formerly known as: *Anisandrus dispar* (F.).

Distribution South East, South, East Midlands, West Midlands and North East.

Habitat and ecology Ancient broad-leaved woodland, though also on non-native trees and conifers. Associated with dead wood of oak, beech, elm, hazel, holly, pear, plum, cherry, apple, aspen, sweet chestnut, Norway maple, Scots pine and also *Magnolia soulangiana*. The larvae feed on a symbiotic fungus that lines the beetles' brood tunnels in the solid wood. Adults have been recorded from April to June and from August to October.

Status Status revised from RDB 3 (Rare) in Shirt (1987). Widespread but local in England.

Indicator status Grade 2 in Garland (1983). Grade 3 in Harding & Rose (1986).

Threats Loss of woodland through, for example, clear-felling and conversion to other land use. Habitat loss, in particular, through the felling of trees, removal of dead wood from living trees and the destruction or removal of standing and fallen dead wood for reasons such as forest hygiene, aesthetic tidiness, public safety or for use as fire wood

Management and conservation Trees and both fallen and standing dead timber, especially with the bark attached, should be retained. The removal of dead timber from trees should be avoided. Gaps in the age structure of the tree population should be identified and the continuity of the appropriate dead wood habitat ensured by regeneration, suitable planting and possibly with pollarding.

XYLEBORUS DRYOGRAPHUS   NOTABLE B
A bark beetle
Order COLEOPTERA   Family SCOLYTIDAE

Xyleborus dryographus (Ratzeburg, 1837). Formerly known as: *Anisandrus dryographus* (Ratzeburg), *?Taphrorychus villifrons* sensu Donisthorpe, 1924 not (Dufour, 1843), *Xyleborus dryophagus* sensu Donisthorpe, 1924 not (Dufour, 1843).

Distribution South West, South, South East, East Anglia, East Midlands, South Wales and Dyfed-Powys.

Habitat and ecology Ancient broad-leaved woodland and pasture-woodland. Associated with oak and sweet chestnut, though also recorded from beech and elm. The larvae feed on a symbiotic fungus that lines the beetles' brood tunnels. Adults have been found under bark and have been recorded from March to July.

Status Very local and widely distributed in southern England, recorded as far north as South Lincolnshire. Also recorded in parts of Wales.

Indicator status Grade 3 in Harding & Rose (1986).

Threats Loss of broad-leaved woodland and parkland through, for example, clear-felling and coniferisation. Habitat loss, in particular, through the felling of ancient trees, removal of dead wood from living trees and the destruction or removal of standing and fallen dead wood for reasons such as forest hygiene, aesthetic tidiness, public safety or for use as fire wood.

Management and conservation Ancient trees and both fallen and standing dead timber, especially with the bark attached, should be retained. The removal of dead timber from ancient trees should be avoided. Gaps in the age structure of the tree population should be identified and the continuity of the appropriate dead wood habitat ensured by regeneration, suitable planting and possibly with pollarding.

XYLOTERUS SIGNATUS   NOTABLE B
A bark beetle
Order COLEOPTERA   Family SCOLYTIDAE

Xyloterus signatus (F., 1792). Formerly known as: *Trypodendron quercus* sensu Fowler, 1891 ?not Eichhoff, *Trypodendron signatum* (F.), *Xyloterus signatum* (F.).

Distribution South, South West, East Midlands, West Midlands, North East, North West, Dyfed-Powys, North Wales and North East Scotland.

Habitat and ecology Ancient broad-leaved woodland and pasture-woodland. Breeds in hard, dead wood of oak, beech, birch and sycamore. This species has once been found in *Acacia dealbata*. The larvae feed on a symbiotic fungus that lines the beetles' brood tunnels in the solid wood. Adults have been found from March to September.

Status Status revised from RDB 3 (Rare) in Shirt (1987). Widespread but local in England and Wales, also recorded in North East Scotland.

Indicator status Grade 1 in Garland (1983). Grade 3 in Harding & Rose (1986).

Threats Loss of broad-leaved woodland and parkland through, for example, clear-felling and coniferisation. Habitat loss, in particular, through the felling of ancient trees, removal of dead wood from living trees and the destruction or removal of standing and fallen dead wood for reasons such as forest hygiene, aesthetic tidiness, public safety or for use as fire wood.

Management and conservation Ancient trees and both fallen and standing dead timber, especially with the bark attached, should be retained. The removal of dead timber from ancient trees should be avoided. Gaps in the age structure of the tree population should be identified and the continuity of the appropriate dead wood habitat ensured by regeneration, suitable planting and possibly with pollarding.

ANASPIS BOHEMICA
                        INSUFFICIENTLY KNOWN
Order COLEOPTERA   Family SCRAPTIIDAE

Anaspis bohemica Schilsky, 1898.

Distribution Recorded from Moray before 1970 and East Inverness & Nairn from 1970 onwards.

Habitat and ecology Probably associated with woodland. The larvae probably develop in dead wood.

405

SCRAPTIIDAE

Adults have been noted from broom and dead pine branches. The adult has been recorded in June.

Status Not listed in the insect Red Data Book (Shirt, 1987). Only known from a few examples found at Nethy Bridge, Elgin in 1951 and a single female from Coylumbridge, Easterness in 1986. In the past, this species may have been confused with other members of the genus. Consequently, the exact status of this species is hard to assess.

Threats Uncertain, though this species may be threatened by the loss of woodland through, for example, clear-felling and conversion to plantation forest. Habitat loss in the remaining areas may be through the felling of ancient trees and the destruction or removal of standing and fallen dead wood for reasons such as forest hygiene, aesthetic tidiness, public safety or for use as fire wood.

Management and conservation Ancient trees, and both standing and fallen dead timber, especially with the bark attached, should be retained. Gaps in the age structure of the tree population should be identified and regeneration encouraged to ensure the continuity of suitable habitat.

Published sources Allen, A.A. (1975c), Buck, F.D. (1954), Owen, J.A. (1988d).

ANASPIS MELANOSTOMA
INSUFFICIENTLY KNOWN
Order COLEOPTERA   Family SCRAPTIIDAE

Anaspis melanostoma Costa, 1854.

Distribution Recorded from South Hampshire, West Kent, South Essex and Cumberland before 1970.

Habitat and ecology Broad-leaved woodland. Larvae probably develop in dead wood. Adults have been found on hawthorn flowers and once on bramble blossom. Adults have been recorded from May to July.

Status Status revised from RDB 3 (Rare) in Shirt (1987). Only known from four, widely scattered, vice-counties in England. Last recorded in 1966 from Denny Wood, New Forest, South Hampshire. A record for North Somerset requires confirmation. This species is difficult to identify and may be confused with other members of the genus. Consequently, the exact status of this species is hard to assess.

Threats Uncertain, though this species may be threatened by the loss of broad-leaved woodland through, for example, clear-felling and coniferisation. Habitat loss, in particular, may be through the felling

of trees, removal of dead wood from living trees and the destruction or removal of standing and fallen dead wood for reasons such as forest hygiene, aesthetic tidiness, public safety or for use as fire wood.

Management and conservation Trees, and both fallen and standing dead timber, especially with the bark attached, should be retained. The removal of dead timber from trees should be avoided. Gaps in the age structure of the tree population should be identified and the continuity of the appropriate dead wood habitat ensured by regeneration, suitable planting and possibly with pollarding. The presence of nectar sources such as hawthorn may also be particularly important for this species.

Published sources Buck, F.D. (1954), Donisthorpe, H.St J.K. (1930b), Fowler, W.W. (1889), Fowler, W.W. (1891), Fowler, W.W. & Donisthorpe, H.St J.K. (1913), Shirt, D.B., ed. (1987).

ANASPIS SCHILSKYANA   INDETERMINATE

Order COLEOPTERA   Family SCRAPTIIDAE

Anaspis schilskyana Csiki, 1915.

Distribution Recorded from Oxfordshire before 1970 and Herefordshire from 1970 onwards.

Habitat and ecology Ancient broad-leaved woodland and pasture-woodland. On the Continent, larvae have been recorded in half-dry, red-rotten oak wood in January, the adults ecloding in June. On emergence they have been found visiting the flowers of hawthorn. In Britain, the beetle has been beaten from oak foliage. Adults are possibly very short-lived, being recorded only in early June.

Status Status revised from RDB 1 (Endangered) in Shirt (1987). Recognised as a British species in 1975. Only known from five examples from just two localities. Recently recorded only from Moccas Park, Herefordshire. This species can be difficult to identify and may be confused with other members of the genus. Consequently, the exact status of this species is hard to assess.

Indicator status Grade 1 in Harding & Rose (1986).

Threats Loss of broad-leaved woodland and parkland through, for example, clear-felling and coniferisation. Habitat loss, in particular, through the felling of ancient trees, removal of dead wood from living trees and the destruction or removal of standing and fallen dead wood for reasons such as forest hygiene, aesthetic tidiness, public safety or for use as fire wood.

Management and conservation Ancient trees, and both fallen and standing dead timber, especially with the bark attached, should be retained. The removal of dead timber from ancient trees should be avoided. Gaps in the age structure of the tree population should be identified and the continuity of the appropriate dead wood habitat ensured by regeneration, suitable planting and possibly with pollarding. The presence of nectar sources such as hawthorn may also be particularly important for this species. Moccas Park is an NNR.

Published sources Allen, A.A. (1975c), Cooter, J. (1990d), Harding, P.T. (1978), Harding, P.T. & Rose, F. (1986), Owen, J.A. (1982), Shirt, D.B., ed. (1987).

ANASPIS SEPTENTRIONALIS	EXTINCT
	ENDEMIC
Order COLEOPTERA	Family SCRAPTIIDAE

Anaspis septentrionalis Champion, 1891.

Distribution Recorded from East Inverness & Nairn before 1970.

Habitat and ecology Woodland. Larvae probably develop in dead wood. The adult has been recorded in July.

Status Endemic. Not listed in the insect Red Data Book (Shirt, 1987). Presumed extinct. Only known from a pair found in 1876 from Aviemore, Easterness.

Published sources Champion, G.C. (1891), Donisthorpe, H.St J.K. (1930b).

ANASPIS THORACICA	NOTABLE A
Order COLEOPTERA	Family SCRAPTIIDAE

Anaspis thoracica (L., 1758). Formerly known as: *Anaspis latipalpis* Schilsky, 1895.

Distribution Recorded from East Sussex, East Kent, Surrey, Berkshire, Buckinghamshire, West Norfolk, Cheshire, South Lancashire and Dumfriesshire before 1970 and West Kent, Northamptonshire, South-west Yorkshire and North Northumberland from 1970 onwards.

Habitat and ecology Woodland, wood margins etc. Larvae probably develop in dead wood. Adults have been found on blossom and have been recorded in June and July.

Status Widespread but very local. Recorded from southern England to Dumfriesshire in Scotland. The female of this species is difficult to identify and may be confused with other members of the genus. Consequently, the exact status of this species is hard to assess.

Threats Loss of woodland through, for example, clear-felling and conversion to other land use. Habitat loss, in particular, may be through the felling of trees, removal of dead wood from living trees and the destruction or removal of standing and fallen dead wood for reasons such as forest hygiene, aesthetic tidiness, public safety or for use as fire wood.

Management and conservation Trees, and both fallen and standing dead timber, especially with the bark attached, should be retained. The removal of dead timber from trees should be avoided. Gaps in the age structure of the tree population should be identified and the continuity of the appropriate dead wood habitat ensured by regeneration, suitable planting and possibly with pollarding. The presence of nectar sources such as hawthorn may also be particularly important for this species.

Published sources Allen, A.A. (1975c), Easton, A.M. (1948), Johnson, C., Robinson, N.A. & Stubbs, A. (1977).

SCRAPTIA DUBIA	EXTINCT
Order COLEOPTERA	Family SCRAPTIIDAE

Scraptia dubia (Olivier, 1790).

Distribution Recorded from Dorset before 1970.

Habitat and ecology Ancient broad-leaved woodland. Larvae develop in decaying wood. Possibly associated with flowers such as hawthorn, canopy foliage and hollow trees. The adult is probably short-lived. The adult has been recorded in June.

Status Not listed in the insect Red Data Book (Shirt, 1987). Presumed extinct. Reliably known from just a single example, found in 1842 from Glanvilles Wootton, Dorset. The species was apparently recorded from Windsor Forest, Berkshire last century, and from other sites this century, though it is probable that these records refer to another member of the genus.

Indicator status Grade 1 in Harding & Rose (1986).

Published sources Fowler, W.W. (1891), Harding, P.T. (1978), Harding, P.T. & Rose, F. (1986).

SCRAPTIIDAE

SCRAPTIA FUSCULA　　　　　ENDANGERED

Order COLEOPTERA　　　　Family SCRAPTIIDAE

Scraptia fuscula Müller, 1821.

Distribution Recorded from Surrey and Berkshire before 1970 and Berkshire from 1970 onwards.

Habitat and ecology Ancient broad-leaved woodland and pasture-woodland. Larvae develop in red-rotten wood, almost certainly that of oak, while adults are found on canopy foliage and inside hollow ancient oaks. The adults has also been found on flowers in gardens. Adults have been recorded from June to August; the males probably being very short-lived.

Status Not listed in the insect Red Data Book (Shirt, 1987). Apart from the first record of *S. fuscula* in Great Britain, at Ripley, Surrey, this species has only reliably been recorded from Windsor Forest and Windsor Great Park. There are published records for Sherwood Forest, Nottinghamshire and Moccas Park, Herefordshire. These, and other records outside the Windsor Forest area, probably refer to *S. testacea* with which it was formerly confused.

Indicator status Grade 1 in Harding & Rose (1986).

Threats Loss of broad-leaved woodland and parkland through, for example, clear-felling and coniferisation. Habitat loss, in particular, through the felling of ancient trees, removal of dead wood from living trees and the destruction or removal of standing and fallen dead wood for reasons such as forest hygiene, aesthetic tidiness, public safety or for use as fire wood.

Management and conservation Ancient trees, and both fallen and standing dead timber, especially with the bark attached, should be retained. The removal of dead timber from ancient trees should be avoided. Gaps in the age structure of the tree population should be identified and the continuity of the appropriate dead wood habitat ensured by regeneration, suitable planting and possibly with pollarding. Windsor Forest and parts of Windsor Great Park are notified as SSSIs.

Published sources Allen, A.A. (1940b), Carr, J.W. (1916), Garland, S.P. (1983), Hallett, H.M. (1951), Harding, P.T. (1978), Harding, P.T. & Rose, F. (1986).

SCRAPTIA TESTACEA　　　　　　RARE

Order COLEOPTERA　　　　Family SCRAPTIIDAE

Scraptia testacea Allen, 1940. Formerly confused under: *Scraptia fuscula* Müller, 1821.

Distribution Recorded from South Hampshire, Surrey, South Essex, Hertfordshire, Middlesex, Berkshire, Huntingdonshire, Herefordshire, Worcestershire, Nottinghamshire and Cumberland before 1970 and East Sussex, Berkshire, West Suffolk, Herefordshire and Nottinghamshire from 1970 onwards.

Habitat and ecology Ancient broad-leaved woodland and pasture-woodland. Larvae develop in red-rotten wood, principally oak, though also beech and hawthorn. Adults are found on canopy foliage and inside hollow trees, and have also been recorded from elm and by sweeping under mature limes. Adults have been recorded from June to August; the males being very short-lived.

Status Not listed in the insect Red Data Book (Shirt, 1987). Widely distributed but very local in England and recorded as far north as Cumberland. Recently recorded from five vice-counties.

Indicator status Grade 1 in Harding & Rose (1986).

Threats Loss of broad-leaved woodland and parkland through, for example, clear-felling and coniferisation. Habitat loss, in particular, through the felling of ancient trees, removal of dead wood from living trees and the destruction or removal of standing and fallen dead wood for reasons such as forest hygiene, aesthetic tidiness, public safety or for use as fire wood.

Management and conservation Ancient trees, and both fallen and standing dead timber, especially with the bark attached, should be retained. The removal of dead timber from ancient trees should be avoided. Gaps in the age structure of the tree population should be identified and the continuity of the appropriate dead wood habitat ensured by regeneration, suitable planting and possibly with pollarding.

Published sources Allen, A.A. (1940b), Buck, F.D. (1954), Cooter, J. (1990d), Garland, S.P. (1983), Harding, P.T. (1978), Harding, P.T. & Rose, F. (1986).

ACLYPEA OPACA NOTABLE A
Beet carrion beetle
Order COLEOPTERA Family SILPHIDAE

Aclypea opaca (L., 1758). Formerly known as:
Blitophaga opaca (L.), *Silpha opaca* L.

Distribution Recorded from South Devon, North
Devon, East Kent, West Kent, Surrey, "Suffolk",
"Norfolk", Cambridgeshire, East Gloucestershire, West
Gloucestershire, Warwickshire, Staffordshire,
Glamorgan, Radnorshire, Cardiganshire, North
Lincolnshire, Nottinghamshire, South Lancashire,
North-east Yorkshire, South-west Yorkshire, Mid-west
Yorkshire, South Northumberland, Lanarkshire, "Tweed
district", "Forth district", "Tay district", "Dee district",
"Moray district" and Clyde Islands before 1970 and
South Devon, North Somerset, East Sussex, North-east
Yorkshire, Mid-west Yorkshire and Durham from 1970
onwards.

Habitat and ecology Probably not habitat specific.
This species has been recorded from cultivated fields
and a wetland site. Phytophagous at the roots of plants,
and has been found damaging beet and turnip crops in
cultivated fields. It has also been swept from
water-dropwort, recorded from the flowers of rowan
and found in a rotten stump. Adults have been recorded
from May to August.

Status Declining, from the records this species was
formerly widespread throughout Great Britain. There
are recent records from very few vice-counties in
southern England and northern England.

Threats Uncertain, though modern intensive
agricultural practices are likely to be a threat to this
species.

Published sources Allen, A.A. (1981c), Atty, D.B.
(1983), Fowler, W.W. (1889), Fowler, W.W. &
Donisthorpe, H.St J.K. (1913), Hodge, P.J. (1990).

ACLYPEA UNDATA ENDANGERED

Order COLEOPTERA Family SILPHIDAE

Aclypea undata (Müller, 1776). Formerly known as:
Silpha reticulata F., 1787.

Distribution Recorded from North Somerset, Surrey,
"Norfolk", Cambridgeshire, Glamorgan,
Nottinghamshire and North-east Yorkshire before 1970.

Habitat and ecology Probably not habitat specific.
Phytophagous, almost certainly at the roots of plants.
Adults have been recorded in May and June.

Status Status revised from RDB 3 (Rare) in Shirt
(1987). No recent records. Last recorded in 1936 from
Buttercrambe Woods, North-east Yorkshire.

Published sources Fowler, W.W. (1889), Fowler,
W.W. & Donisthorpe, H.St J.K. (1913), Shirt, D.B., ed.
(1987).

DENDROXENA QUADRIMACULATA
NOTABLE B
Order COLEOPTERA Family SILPHIDAE

Dendroxena quadrimaculata (Scopoli, 1772). Formerly
known as: *Dendroxena quadripunctata* sensu auct. not
(L., 1758), *Silpha quadripunctata* sensu Fowler, 1889
not (L.), *Xylodrepa quadripunctata* sensu auct. not (L.).

Distribution England and Dyfed-Powys.

Habitat and ecology Mainly oak woodland. Adults and
probably larvae occur in the canopy of oak and
occasionally other tree species, where they feed on
caterpillars. It has also been found under bark. This
beetle appears to have a preference for trees heavily
infested with caterpillars. Adults have been recorded
primarily in June, though also in March and May.

Status Widespread but local.

Threats Loss of broad-leaved woodland through, for
example, clear-felling and coniferisation.

NICROPHORUS INTERRUPTUS NOTABLE B
A burying or sexton beetle
Order COLEOPTERA Family SILPHIDAE

Nicrophorus interruptus Stephens, 1830. Formerly
known as: *Necrophorus interruptus* Stephens, 1830.

Distribution England, North Wales and Dyfed-Powys.

Habitat and ecology Probably not habitat specific. In
carrion. Adults have been recorded from June to
October.

Status Possibly declining. Widespread but local in
England and Wales.

Threats This is and opportunistic and highly mobile
species requiring a continuity of carrion availability.

SILPHIDAE

Management and conservation Because of its
dependence on carrion this species is likely to require
large areas of suitable habitat for its continued survival.

NICROPHORUS VESTIGATOR	NOTABLE A
A burying or sexton beetle	
Order COLEOPTERA	Family SILPHIDAE

Nicrophorus vestigator Herschel, 1807. Formerly
known as: *Necrophorus vestigator* Herschel, 1807.

Distribution Recorded from South Devon, North
Devon, North Somerset, Dorset, South Hampshire,
West Sussex, East Kent, Surrey, Oxfordshire, West
Suffolk, East Norfolk, West Norfolk, Cambridgeshire,
Huntingdonshire, North-east Yorkshire and Durham
before 1970 and West Suffolk, East Norfolk, West
Norfolk and South Lincolnshire from 1970 onwards.

Habitat and ecology Probably not habitat specific.
Most recent records are from areas with sandy soils. In
carrion. Adults have been recorded in May and August.

Status Declining, with very few recent records.
Formerly widely distributed in England as far north as
Durham. Now virtually restricted to the Brecklands of
East Anglia.

Threats This is a mobile and opportunistic species
requiring a continuity of carrion availability.

Management and conservation Because of its
dependence on carrion this species is likely to require
large areas of suitable habitat for its continued survival.

Published sources Collier, M.J. (1988a), Fowler, W.W.
(1889), Fowler, W.W. & Donisthorpe, H.St J.K. (1913).

SILPHA CARINATA	ENDANGERED
Order COLEOPTERA	Family SILPHIDAE

Silpha carinata Herbst, 1783.

Distribution Recorded from South Hampshire before
1970 and South Wiltshire from 1970 onwards.

Habitat and ecology The margins of broad-leaved
woodland. Recorded from carrion and pitfall traps
baited with dead fish, and also from damp straw, moss
and under a stone. In captivity *S. carinata* was found to
eat a variety of both plant and dead animal material.
Adults probably occur all year round.

Status Extremely localised, modern records exist for
just three 1 km grid squares in South Wiltshire, where
it was last recorded in 1977. Formerly known in Great
Britain from just six examples recorded pre-1839 from
Winchester, South Hampshire.

Threats Uncertain, though the loss of broad-leaved
woodland through, for example, clear-felling and
coniferisation may well be a threat to this species. This
beetle is opportunistic and is likely to require a
continuity of resource availability such as carrion.

Management and conservation As this species
appears to be associated with carrion this species is
likely to require large areas of suitable habitat for its
continued survival.

Published sources Fowler, W.W. (1889), Fowler,
W.W. & Donisthorpe, H.St J.K. (1913), Shirt, D.B., ed.
(1987).

SILPHA OBSCURA	VULNERABLE
Order COLEOPTERA	Family SILPHIDAE

Silpha obscura L., 1758.

Distribution Recorded from "Devon", Dorset, Isle of
Wight, South Hampshire, East Sussex, East Kent,
Surrey, South Essex, Cambridgeshire, West
Gloucestershire, North Northumberland and
Cumberland before 1970 and "Cornwall" and South
Somerset from 1970 onwards.

Habitat and ecology Possibly associated with sandy or
chalky areas. Doubtfully or only rarely feeds on
carrion, possibly predatory on molluscs. Usually found
under stones and at the roots of plants. Adults have
been recorded from April to June and more rarely in
August and September.

Status Not listed in the insect Red Data Book (Shirt,
1987). Declining, records suggest that this species was
formerly widespread in England and was recorded as
far north as Cumberland. Modern records exist for just
three vice-counties, all in southern England.

Threats Uncertain, though this species may be
threatened by the loss of calcareous and sandy habitats
through agricultural improvement and conversion to
arable.

Published sources Atty, D.B. (1983), Fowler, W.W.
(1889), Fowler, W.W. & Donisthorpe, H.St J.K. (1913)

SILPHA TYROLENSIS   NOTABLE B

Order COLEOPTERA   Family SILPHIDAE

Silpha tyrolensis Laicharting, 1781. Formerly known as. *Silpha nigrita* Creutzer, 1799.

Distribution South West, East Midlands, West Midlands, North East, North West, Dyfed-Powys, North Wales and Scotland.

Habitat and ecology Probaly not habitat specific. Doubtfully or rarely feeding on carrion, possibly predatory on molluscs. This species usually occurs under stones and at the roots of plants. Adults have been recorded in April and May.

Status Widespread but local. Predominantly a northern and western species.

THANATOPHILUS DISPAR   ENDANGERED

Order COLEOPTERA   Family SILPHIDAE

Thanatophilus dispar (Herbst, 1793). Formerly known as: *Silpha dispar* Herbst.

Distribution Recorded from North Essex, East Suffolk, "Norfolk", Caernarvonshire, Leicestershire & Rutland, Derbyshire, Mid-west Yorkshire, Durham, Renfrewshire and Fife before 1970.

Habitat and ecology Probably not habitat specific. In carrion. This beetle has been found in fish carcasses washed up on the shores of Loch Leven (R.A. Crowson pers. comm.). Adults have been recorded in June.

Status Status revised from RDB 3 (Rare) in Shirt (1987). Declining, records indicate that this species formerly had a widely scattered distribution in the country. There are old records and one recent record from the shores of Loch Leven, it is uncertain whether this species has occurred in just one or both the following vice-counties; Main Argyll or Westerness.

Threats This is a mobile and opportunistic species which requires a continuity of carrion availability.

Management and conservation Because of its dependence on carrion this species is likely to require large areas of suitable habitat for its continued survival.

Published sources Fowler, W.W. (1889), Fowler, W.W. & Donisthorpe, H.St J.K. (1913), Nelson, J.M. (1983), Shirt, D.B., ed. (1987).

SILVANOPRUS FAGI   ENDANGERED

Order COLEOPTERA   Family SILVANIDAE

Silvanoprus fagi (Guerin-Meneville, 1844). Formerly known as: *Silvanus fagi* Guerin Meneville, *Silvanus similis* Erichson, 1846.

Distribution Recorded from South Hampshire, West Kent and Surrey before 1970.

Habitat and ecology Woodland. Recorded from under beech and pine bark. Adults have been recorded in July.

Status Status revised from RDB 3 (Rare) in Shirt (1987). Only recorded on just four occasions; from Esher, Surrey (last century and in 1966), from Cobham Park, West Kent (a single example, last century), and a single example, from Mark Ash, New Forest, South Hampshire.

Threats Loss of woodland through, for example, clear-felling and conversion to other land use. Habitat loss, in particular, through the felling of ancient trees, removal of dead wood from living trees and the destruction or removal of standing and fallen dead wood for reasons such as forest hygiene, aesthetic tidiness, public safety or for use as fire wood.

Management and conservation Ancient trees and both fallen and standing dead timber, especially with the bark attached, should be retained. The removal of dead timber from ancient trees should be avoided. Gaps in the age structure of the tree population should be identified and the continuity of the appropriate dead wood habitat ensured by regeneration, suitable planting and possibly with pollarding.

Published sources Fowler, W.W. (1889), Shirt, D.B., ed. (1987).

SILVANUS BIDENTATUS   NOTABLE B

Order COLEOPTERA   Family SILVANIDAE

Silvanus bidentatus (F., 1792).

Distribution South, South East, East Anglia, East Midlands, West Midlands, North West, North East and South West Scotland.

Habitat and ecology Ancient broad-leaved woodland, pasture-woodland and also recorded from conifers. Adults and probably also larvae occur under the bark of a variety of trees, with records from oak, beech,

411

sweet chestnut and Scots pine. Adults have been recorded from March to November.

Status Status revised from RDB 3 (Rare) in Shirt (1987). Known from southern England, north to Renfrewshire in Scotland. Formerly very local, this species appears to have increased in distribution during the past 30 years and is still possibly spreading.

Indicator status Grade 2 in Harding & Rose (1986).

Threats Loss of woodland through, for example, clear-felling and conversion to other land use. Habitat loss, in particular, through the felling of ancient trees, removal of dead wood from living trees and the destruction or removal of standing and fallen dead wood for reasons such as forest hygiene, aesthetic tidiness, public safety or for use as fire wood.

Management and conservation Ancient trees and both fallen and standing dead timber, especially with the bark attached, should be retained. The removal of dead timber from ancient trees should be avoided. Gaps in the age structure of the tree population should be identified and the continuity of the appropriate dead wood habitat ensured by regeneration, suitable planting and possibly with pollarding.

SPHAERITES GLABRATUS   **RARE**

Order COLEOPTERA   Family SPHAERITIDAE

Sphaerites glabratus (F., 1792).

Distribution Recorded from North-east Yorkshire, Mid-west Yorkshire, North Northumberland, Cumberland, Berwickshire, West Perthshire, Mid Perthshire, South Aberdeenshire, Moray and East Inverness & Nairn before 1970 and Cardiganshire, South Aberdeenshire and East Inverness & Nairn from 1970 onwards.

Habitat and ecology Probably associated with woodland. Recorded from under the bark of dead trees, at sap, in dung, in a hen house and in carrion, including a record about 15cm deep in the soil beneath deer corpses. Also recorded from fungi, including a polypore fungus on a birch stump, the gills of an agaric fungus, in a rotting *Boletus luteus* fungus and other decaying fungi. Adults have been recorded from May to July and in September.

Status A northern species. Primarily recorded in Scotland, though with a few records in northern England, found as far south as Mid-west Yorkshire, and also noted in mid Wales. Recently recorded from three vice-counties.

Threats Uncertain, though this species may be threatened by the loss of woodland through, for example, clear-felling and conversion to other land use. Habitat loss, in particular, may be through the felling of trees, removal of dead and fungus-infected wood from living trees and the destruction or removal of standing and fallen dead wood for reasons such as forest hygiene, aesthetic tidiness, public safety or for use as fire wood.

Management and conservation Trees, and both fallen and standing dead timber, especially with the bark attached, should be retained. The removal of dead or fungus-infected timber from trees should be avoided. Gaps in the age structure of the tree population should be identified and the continuity of the appropriate habitat ensured by regeneration, suitable planting and possibly with pollarding.

Published sources Boyce, D.C. (1989), Fowler, W.W. (1891), Fowler, W.W. & Donisthorpe, H.St J.K. (1913), Thompson, M.L. (1923).

SPHINDUS DUBIUS   **NOTABLE B**
A dry fungus beetle
Order COLEOPTERA   Family SPHINDIDAE

Sphindus dubius (Gyllenhal, 1808).

Distribution England and North East Scotland.

Habitat and ecology Woodland and pasture-woodland. On powdery myxomycete fungi (slime-moulds) on trees and occasionally found under bark. Recorded from oak, beech, pine and horse-chestnut. Adults have been noted from May to September.

Status Widespread but local in England, also recorded in North East Scotland.

Threats Loss of woodland and parkland through, for example, clear-felling and conversion to other land use. Habitat loss, in particular, through the felling of trees, removal of dead wood from living trees and the destruction or removal of standing and fallen dead wood for reasons such as forest hygiene, aesthetic tidiness, public safety or for use as fire wood.

Management and conservation Trees and both fallen and standing dead timber, especially with the bark attached, should be retained. The removal of dead timber from trees should be avoided. Gaps in the age structure of the tree population should be identified and the continuity of the appropriate dead wood habitat ensured by regeneration, suitable planting and possibly with pollarding.

BOLITOPHAGUS RETICULATUS   RARE
A darkling beetle
Order COLEOPTERA   Family TENEBRIONIDAE

Bolitophagus reticulatus (L., 1767).

Distribution Recorded from Mid Perthshire, South Aberdeenshire, East Inverness & Nairn, West Inverness, West Ross and East Ross before 1970 and Mid Perthshire, East Perthshire, East Inverness & Nairn, West Inverness and Argyll Main from 1970 onwards.

Habitat and ecology Birch woodland and isolated trees. In the bracket fungus *Fomes fomentarius* on birch. Adults have been recorded from May to August.

Status Widespread but local in the Highlands of Scotland.

Threats Loss of birch woodland through, for example, clear-felling or conversion to plantation forest. Habitat loss in the remaining areas through the felling of fungus infected trees and the destruction or removal of standing and fallen dead wood for reasons such as forest hygiene, aesthetic tidiness, public safety or for use as fire wood.

Management and conservation Fungus infected birches and both standing and fallen dead timber should be retained. Gaps in the age structure of the tree population should be identified and regeneration encouraged to ensure the continuity of suitable habitat.

Published sources Brendel, M.J.D. (1975), Crowson, R.A. (1983a), Owen, J.A. (1990a), Shirt, D.B., ed. (1987).

CORTICEUS UNICOLOR   RARE
A darkling beetle
Order COLEOPTERA   Family TENEBRIONIDAE

Corticeus unicolor Piller and Mitterpacher, 1783. Formerly known as: *Hypophlaeus castaneus* (F., 1790), *Hypophlaeus unicolor* (Piller and Mitterpacher), *Hypophloeus castaneus* (Piller and Mitterpacher), *Hypophloeus unicolor* (Piller and Mitterpacher).

Distribution Recorded from South Devon, South Hampshire, Nottinghamshire, Cheshire, South Lancashire, South-east Yorkshire and South-west Yorkshire before 1970 and Herefordshire, Nottinghamshire, Derbyshire and South-west Yorkshire from 1970 onwards.

Habitat and ecology Ancient broad-leaved woodland and pasture-woodland. Also noted amongst birch on raised bogs and heathland. Under the decaying bark of beech, oak, elm and birch, also noted in the decaying wood of birch. Adults have been recorded in March and April and from June to August.

Status The centre of this species distribution is in an area ranging from Sherwood Forest, Nottinghamshire to Thorne, South-west Yorkshire. Elsewhere this species is extremely local and recorded from very few vice-counties in England.

Indicator status Grade 1 in Garland (1983). Grade 2 in Harding & Rose (1986).

Threats Loss of broad-leaved woodland and parkland through, for example, clear-felling and coniferisation. Habitat loss, in particular, through the felling of ancient trees, removal of dead wood from living trees and the destruction or removal of standing and fallen dead wood for reasons such as forest hygiene, aesthetic tidiness, public safety or for use as fire wood.

Management and conservation Ancient trees and both fallen and standing dead timber, especially with the bark attached, should be retained. The removal of dead timber from ancient trees should be avoided. Gaps in the age structure of the tree population should be identified and the continuity of the appropriate dead wood habitat ensured by suitable planting and possibly with pollarding.

Published sources Brendel, M.J.D. (1975), Garland, S.P. (1983), Harding, P.T. (1978), Harding, P.T. & Rose, F. (1986), Shirt, D.B., ed. (1987).

CRYPTICUS QUISQUILIUS   NOTABLE B
A darkling beetle
Order COLEOPTERA   Family TENEBRIONIDAE

Crypticus quisquilius (L., 1761).

Distribution South East, South, South West, East Anglia, East Midlands, West Midlands, North East, Dyfed-Powys, North Wales and South West Scotland.

Habitat and ecology Almost exclusively coastal, though with a few records inland. Principally on sand dunes, heathlands and sandpits. Also recorded from coastal shingle and alluvial soils on bare ground at field edges. Probably at the roots of plants and under stones, as well as in bare ground. Adults have been recorded in June.

Status Widespread but local along the coasts of England and Wales, also recorded from South West Scotland. Possibly declining.

TENEBRIONIDAE

Threats Loss of dune habitat, particularly through afforestation, urban and holiday development and the loss of coastal shingle through activities such as shingle extraction. The degradation of remaining habitat by excessive disturbance of the vegetation through activities such as motorbike access and human trampling.

Management and conservation On dunes some grazing and other disturbance may be desirable to maintain the early successional stages and prevent the invasion of scrub. Disturbance of coastal shingle sites should be avoided.

CYLINDRINOTUS PALLIDUS　　　　**NOTABLE B**
A darkling beetle
Order COLEOPTERA　　　Family TENEBRIONIDAE

Cylindrinotus pallidus (Curtis, 1830). Formerly known as: *Cylindronotus pallidus* (Curtis, 1830), *Helops pallidus* (Curtis).

Distribution South East, South, South West, East Anglia, West Midlands, North East, North West and Wales.

Habitat and ecology Sand dunes and sandy areas on the coast. Usually found at the roots of marram. Also found under orache. and possibly under saltwort. The beetle has been swept from marram at night. Adults have been recorded in August and October.

Status Widespread but local along the coasts of England and Wales.

Threats Loss of dune and sandy habitat, particularly through afforestation, urban and holiday development. The degradation of remaining habitat by excessive disturbance of the vegetation through activities such as motorbike access, horse-riding and human trampling.

Management and conservation On dunes and sandy areas some grazing and other disturbance may be desirable to maintain the early successional stages and prevent the invasion of scrub.

DIAPERIS BOLETI　　　　**VULNERABLE**
A darkling beetle
Order COLEOPTERA　　　Family TENEBRIONIDAE

Diaperis boleti (L., 1758).

Distribution Recorded from Dorset, South Hampshire, East Sussex, East Suffolk, Nottinghamshire and Cumberland before 1970 and East Sussex, East Kent and Huntingdonshire from 1970 onwards.

Habitat and ecology Birch and damp woodland. Adults, pupae and larvae are found in the bracket fungus *Piptoporus betulinus* on birch. This species has also been found in the bracket fungus *Polyporus squamosus* on poplar. Possibly restricted to fungi that are large, dry and old. On the Continent, this beetle has been reported from a number of other bracket fungi species including *Fomes fomentarius*, *Laetiporus sulphureus* and *Coriolus versicolor*. The adults and larvae apparently feed on the soft, fleshy part of the fungus just above the gills, pupation occurring in a roomy excavation within the fungus. In Britain, larvae and pupae have been recorded in October. Adults have been noted in June and from August to October. The life cycle takes one year.

Status Very local with modern records from only three vice-counties. Primarily recorded from southern and south-eastern England. This species has been found as far north as Nottinghamshire and Cumberland.

Threats Loss of woodland through, for example, clear-felling or conversion to plantation forest. Habitat loss in the remaining areas through the felling of fungus infected trees and the destruction or removal of standing and fallen dead wood for reasons such as forest hygiene, aesthetic tidiness, public safety or for use as fire wood.

Management and conservation Fungus infected trees and both standing and fallen dead timber should be retained. Gaps in the age structure of the tree population should be identified and regeneration encouraged to ensure the continuity of the appropriate habitat.

Published sources Brendel, M.J.D. (1975), Harwood, P. (1956), Hodge, P.J. (1990), Sankey, J.H.P. (1956), Shirt, D.B., ed. (1987).

ELEDONA AGRICOLA　　　　**NOTABLE B**
A darkling beetle
Order COLEOPTERA　　　Family TENEBRIONIDAE

Eledona agricola (Herbst, 1783). Formerly known as: *Eledona agaricola* (Herbst, 1783), *Heledona agaricola* (Herbst, 1783).

Distribution England, South Wales and Dyfed-Powys.

Habitat and ecology Ancient broad-leaved woodland, pasture-woodland and riverside trees. Predominantly in the bracket fungus *Laetiporus sulphureus* on oak, beech, willow and yew. Also recorded from the bracket fungus *Polyporus squamosus* and the giant polypore *Meripilus giganteus*. A colony of the species was kept in captivity on the bracket fungus *Piptoporus betulinus*

(I. McClenaghan pers. comm.). Adults have been recorded in January, May and from August to November.

Status Widespread but local throughout England and recorded in South Wales. This species can be abundant where found. There is an unconfirmed record of this species taken in 1892 from Carnsalloch Wood near Dumfries, Dumfriesshire.

Indicator status Grade 2 in Garland (1983). Grade 3 in Harding & Rose (1986).

Threats Loss of broad-leaved woodland and parkland through, for example, clear-felling and coniferisation. Habitat loss, in particular, through the felling of fungus infected trees, removal of dead wood from living trees and the destruction or removal of standing and fallen dead wood for reasons such as forest hygiene, aesthetic tidiness, public safety or for use as fire wood.

Management and conservation Fungus infected trees and both fallen and standing dead timber, especially with the bark attached, should be retained. The removal of dead timber from ancient trees should be avoided. Gaps in the age structure of the tree population should be identified and the continuity of the appropriate habitat ensured by suitable planting and possibly with pollarding.

HELOPS CAERULEUS   **NOTABLE B**
A darkling beetle
Order COLEOPTERA   Family TENEBRIONIDAE

Helops caeruleus (L., 1758). Formerly known as: *Helops coeruleus* (L., 1758).

Distribution South East, South, South West, East Anglia, East Midlands, West Midlands, North West and South Wales.

Habitat and ecology Woodland and pasture-woodland. Primarily found near the coast. This species has also been noted in urban areas. Found in dead and dying trees, usually in the latter stages of decay. This beetle has been noted principally from oak, though also from elm, beech, willow and rarely pine. Also recorded in old railway sleepers, in old groyne timber and in other prepared timber such as window frames. The larvae feed on dead wood. Adults have been found in April, July and August. Adults appear to be primarily nocturnal.

Status Local, though occassionally not uncommon where found. As this species is also found in prepared timber some records may represent introductions.

Threats Loss of woodland through, for example, clear-felling and conversion to other land use. Habitat loss, in particular, through the felling of ancient trees, removal of dead wood from living trees and the destruction or removal of standing and fallen dead wood for reasons such as forest hygiene, aesthetic tidiness, public safety or for use as fire wood.

Management and conservation Ancient trees and both fallen and standing dead timber, especially with the bark attached, should be retained. The removal of dead timber from ancient trees should be avoided. Gaps in the age structure of the tree population should be identified and the continuity of the appropriate dead wood habitat ensured by suitable planting and possibly with pollarding.

LAGRIA ATRIPES   **ENDANGERED**
A darkling beetle
Order COLEOPTERA   Family TENEBRIONIDAE

Lagria atripes Mulsant and Guillebeau, 1855.

Distribution Recorded from South Hampshire and East Kent before 1970.

Habitat and ecology Woodland. Has been found by beating chestnut and aspen, and by sweeping in grassy areas. Adults have been recorded in June and July.

Status Not listed in the insect Red Data Book (Shirt, 1987). Only known from three localities: the New Forest, South Hampshire (last recorded circa 1901); Blean Woods, East Kent (last recorded 1948) and Orlestone Forest, East Kent (last recorded 1963).

Threats Uncertain, though this species may be threatened by the loss of broad-leaved woodland through, for example, clear-felling and coniferisation.

Management and conservation Uncertain, though effort should initially ensure the prevention of further habitat loss.

Published sources Allen, A.A. (1948d), Buck, F.D. (1954), Henderson, J.L. (1957).

TENEBRIONIDAE

MYCETOCHARA HUMERALIS   NOTABLE A
A darkling beetle
Order COLEOPTERA   Family TENEBRIONIDAE

Mycetochara humeralis (F., 1787). Formerly known as: *Mycetochares bipustulata* (Illiger, 1794).

Distribution Recorded from South Hampshire, East Kent, West Kent, Surrey, South Essex, Cambridgeshire, Glamorgan, Nottinghamshire and Cheshire before 1970 and North Wiltshire, South Hampshire, Berkshire, Northamptonshire, Worcestershire, South Lincolnshire, Nottinghamshire and Derbyshire from 1970 onwards.

Habitat and ecology Ancient broad-leaved woodland. Found in dead wood and under bark, with records from oak, beech, maple, sycamore, poplar, small-leaved lime and cherry. Adults have been recorded in May and June.

Status Very local and with a widely scattered distribution in England as far north as Cheshire. This species has been recorded from south Wales in the past.

Indicator status Grade 3 in Harding & Rose (1986).

Threats Loss of broad-leaved woodland and parkland through, for example, clear-felling and coniferisation. Habitat loss, in particular, through the felling of ancient trees, removal of dead wood from living trees and the destruction or removal of standing and fallen dead wood for reasons such as forest hygiene, aesthetic tidiness, public safety or for use as fire wood.

Management and conservation Ancient trees and both fallen and standing dead timber, especially with the bark attached, should be retained. The removal of dead timber from ancient trees should be avoided. Gaps in the age structure of the tree population should be identified and the continuity of the appropriate dead wood habitat ensured by suitable planting and possibly with pollarding.

Published sources Alexander, K.N.A. & Clements, D.K. (1988), Fowler, W.W. (1891), Fowler, W.W. & Donisthorpe, H.St J.K. (1913), Garland, S.P. (1983), Harding, P.T. (1978), Harding, P.T. & Rose, F. (1986), MacKechnie-Jarvis, C. (1976), Whitehead, P.F. (1990a).

OMOPHLUS RUFITARSIS   ENDANGERED
A darkling beetle
Order COLEOPTERA   Family TENEBRIONIDAE

Omophlus rufitarsis (Leske, 1785). Formerly known as: *Omophlus armeriae* (Curtis, 1836).

Distribution Recorded from Dorset before 1970 and Dorset from 1970 onwards.

Habitat and ecology This beetle inhabits the interface between a coastal shingle bank and saltmarsh. Adults have been found on the flowers and at the roots of thrift. Also under wet seaweed at the edge of tidal pools in an area of saltmarsh. Adults have been recorded in June and July.

Status Only known from a small area of Chesil Beach, Dorset where it was recently rediscovered. There is an unconfirmed record from the New Forest, Hampshire.

Threats Shingle beaches are sensitive to the effects of trampling, motorbike access etc., which damages any vegetated sections causing accumulated humus to erode. Part of the only British site for this species is used as a car park.

Management and conservation Disturbance of the shingle on Chesil Beach should be avoided. The area where this beetle is found is in the Chesil and Fleet SSSI.

Published sources Buck, F.D. (1954), Cooter, J. (1990b), Fowler, W.W. (1891), Shirt, D.B., ed. (1987).

OPATRUM SABULOSUM   NOTABLE B
A darkling beetle
Order COLEOPTERA   Family TENEBRIONIDAE

Opatrum sabulosum (L., 1758).

Distribution South East, South, South West, East Anglia, East Midlands, North West and Wales.

Habitat and ecology Almost exclusively coastal. Principally on sand dunes, coastal shingle and at the base of cliffs. There are a small number of records inland, including at least one from chalk downland. Usually found at the roots of plants, under stones and also in bare sandy areas. Adults have been record from May, June and September.

Status Widespread but local along the coasts of England and Wales.

Threats Loss of coastal habitats due to the building of sea defences. The loss of dune habitat, particularly through afforestation, urban and holiday development and the loss of coastal shingle through activities such as shingle extraction. The degradation of remaining habitat by excessive disturbance through activities such as motorbike access and human trampling.

Management and conservation On dunes some grazing and other disturbance may be desirable to maintain the early successional stages and prevent the invasion of scrub. Disturbance of coastal shingle sites should be avoided.

PLATYDEMA VIOLACEUM — ENDANGERED
A darkling beetle
Order COLEOPTERA   Family TENEBRIONIDAE

Platydema violaceum (F., 1790). Formerly known as: *Platydema dytiscoides* sensu Fowler, 1891 not (Rossi, 1790).

Distribution Recorded from South Hampshire and Surrey before 1970.

Habitat and ecology Broad-leaved woodland. On the Continent, this beetle has been recorded from the jews-ear fungus *Auricularia auricula-judae* and the fungus *A. mesenterica* on elder and elm. Also noted from under the fungoid bark of rotten beech and oak. The larvae and adults are found in the outer, more rotten parts of *Auricularia*, pupation occurring in the fungus. In Britain, this species has only been recorded from under the bark of a felled oak and at light. Adults have been noted in August.

Status Only known this century from seven examples recorded in 1901 from the New Forest, South Hampshire and from one example (at light) in 1957 from the Juniper Hall Field Centre, near Dorking, Surrey.

Threats Uncertain, though loss of broad-leaved woodland through, for example, clear-felling and coniferisation may be a threat to this species. Habitat loss, in particular, through the felling of fungus infected trees and shrubs and the destruction or removal of standing and fallen dead wood for reasons such as forest hygiene, aesthetic tidiness, public safety or for use as fire wood.

Management and conservation Fungus infected trees and shrubs and both fallen and standing dead and fungoid timber should be retained.

Published sources Sankey, J. (1957), Shirt, D.B., ed. (1987).

PRIONYCHUS ATER — NOTABLE B
A darkling beetle
Order COLEOPTERA   Family TENEBRIONIDAE

Prionychus ater (F., 1775). Formerly known as: *Eryx ater* (F.).

Distribution South East, South, South West, East Anglia, East Midlands, West Midlands, Dyfed-Powys and South Wales.

Habitat and ecology Ancient broad-leaved woodland, pasture-woodland, secondary woodland, orchards and stream and river margins. In dead and decaying trees with records from oak, ash, birch, elm, beech, willow and apple. The larvae develop in wood mould. This species has been noted at mercury vapour light trap. Adults have been recorded from June to September.

Status Widely distributed and local over the southern half of Great Britain. Not as rare as formerly thought. Often found in the larval state.

Indicator status Grade 3 in Harding & Rose (1986).

Threats Loss of broad-leaved woodland and parkland through, for example, clear-felling and coniferisation. Habitat loss, in particular, through the felling of ancient trees, removal of dead wood from living trees and the destruction or removal of standing and fallen dead wood for reasons such as forest hygiene, aesthetic tidiness, public safety or for use as fire wood.

Management and conservation Ancient trees and both fallen and standing dead timber, especially with the bark attached, should be retained. The removal of dead timber from ancient trees should be avoided. Gaps in the age structure of the tree population should be identified and the continuity of the appropriate dead wood habitat ensured by suitable planting and possibly with pollarding.

PRIONYCHUS MELANARIUS — VULNERABLE
A darkling beetle
Order COLEOPTERA   Family TENEBRIONIDAE

Prionychus melanarius (Germar, 1813). Formerly known as: *Prionychus farimairei* sensu auct. Brit. not Reiche, 1861.

Distribution Recorded from West Sussex, East Suffolk and Nottinghamshire before 1970 and West Sussex, East Suffolk, East Gloucestershire and Nottinghamshire from 1970 onwards.

Habitat and ecology Ancient broad-leaved woodland and pasture-woodland. Usually found as larvae. The

417

TENEBRIONIDAE

adults and larvae are found in dry frass under loose bark and in rotten wood. In dead and dying trees with records from beech, birch, ash, oak, elm and walnut. The larvae develop in wood mould. Larvae have been recorded in March and April (emerged May), May (emerged June), August and September. Adults have been reported from May to August. Adults are nocturnal.

Status Very local with very few, widely scattered sites.

Indicator status Grade 1 in Harding & Rose (1986).

Threats Loss of broad-leaved woodland and parkland through, for example, clear-felling and coniferisation. Habitat loss, in particular, through the felling of ancient trees, removal of dead wood from living trees and the destruction or removal of standing and fallen dead wood for reasons such as forest hygiene, aesthetic tidiness, public safety or for use as fire wood.

Management and conservation Ancient trees and both fallen and standing dead timber, especially with the bark attached, should be retained. The removal of dead timber from ancient trees should be avoided. Gaps in the age structure of the tree population should be identified and the continuity of the appropriate dead wood habitat ensured by suitable planting and possibly with pollarding.

Published sources Bedwell, E.C. (1923), Garland, S.P. (1983), Harding, P.T. (1978), Harding, P.T. & Rose, F. (1986), Morris, M.G. (1978b), Owen, J.A. (1990a), Shirt, D.B., ed. (1987).

PSEUDOCISTELA CERAMBOIDES NOTABLE B
A darkling beetle
Order COLEOPTERA Family TENEBRIONIDAE

Pseudocistela ceramboides (L., 1758). Formerly known as: *Cistela ceramboides* (L.), *Gonodera ceramboides* (L.).

Distribution South West, South, South East, East Anglia, West Midlands, East Midlands and South Wales.

Habitat and ecology Ancient broad-leaved woodland and pasture-woodland. Found in dead and dying trees, with records from oak and beech. The larvae develop in dead wood, particularly red-rotten wood and wood mould in hollow trees. Adults have been found in hollow trees and also beaten from oak, hawthorn and probably also from beech. Adults have been noted at mercury vapour light. Larvae have been recorded in May. Adults have been found from April to August.

Status Local and widespread throughout the southern half of England. Recently recorded in Monmouthshire, South Wales.

Indicator status Grade 2 in Harding & Rose (1986).

Threats Loss of broad-leaved woodland and parkland through, for example, clear-felling and coniferisation. Habitat loss, in particular, through the felling of ancient trees, removal of dead wood from living trees and the destruction or removal of standing and fallen dead wood for reasons such as forest hygiene, aesthetic tidiness, public safety or for use as fire wood.

Management and conservation Ancient trees and both fallen and standing dead timber, especially with the bark attached, should be retained. The removal of dead timber from ancient trees should be avoided. Gaps in the age structure of the tree population should be identified and the continuity of the appropriate dead wood habitat ensured by suitable planting and possibly with pollarding.

SCAPHIDEMA METALLICUM NOTABLE B
A darkling beetle
Order COLEOPTERA Family TENEBRIONIDAE

Scaphidema metallicum (F., 1792).

Distribution England, South Wales and North Wales.

Habitat and ecology This species has been found in a wide variety of habitats including broad-leaved woodland, allotments, estuarine beaches and coastal shingle. Usually found in moist conditions. Recorded under the bark of elm, hornbeam, chestnut, broom, gorse and probably on other tree and shrub species. This species has been found in moss, leaf-litter, dry rotten hawthorn and under driftwood. Adults have been recorded from February to June.

Status Widespread but local, recorded throughout England and noted in both South and North Wales.

Threats Uncertain, though this species may be threatened by the loss of woodland and hedgerows through, for example, clear-felling and uprooting. This species could be further threatened by the removal of driftwood from beaches.

Management and conservation In woodlands, old trees and both fallen and standing dead timber, especially with the bark attached, should be retained. The removal dead and decaying wood in other habitats, including driftwood on the strandline, should be avoided.

TETRATOMA ANCORA   NOTABLE B

Order COLEOPTERA   Family TETRATOMIDAE

Tetratoma ancora F., 1791.

Distribution South East, South West, East Anglia, East Midlands, North East, North West, Dyfed-Powys, North Wales and Scotland.

Habitat and ecology Ancient woodland. Adults have been found on trunks, and also in tree foliage, under bark, and in fungi on other trees. Recorded from oak, beech, hazel, pine and from under hornbeam. Larvae have been recorded from under the encrusting fruit-bodies of the fungus *Corticium quercinum* on dead oak branches, and are probably associated with this fungus on hazel and also possibly on other trees. Adults have been noted in association with the fungus *Collybia radicata*. Larvae have been found in September. Adults have been recorded in March, May and June.

Status Widespread but local in Great Britain, more local in southern England.

Indicator status Grade 2 in Garland (1983). Grade 3 in Harding & Rose (1986).

Threats Loss of woodland through, for example, clear-felling and conversion to other land use. Habitat loss, in particular, through the felling of ancient and fungus-infected trees, removal of dead wood from living trees and the destruction or removal of standing and fallen dead wood for reasons such as forest hygiene, aesthetic tidiness, public safety or for use as fire wood.

Management and conservation Ancient and fungus-infected trees, and both standing and fallen dead timber, especially with the bark attached, should be retained. Gaps in the age structure of the tree population should be identified and regeneration, or possibly planting and pollarding, encouraged to ensure the continuity of suitable habitat.

TETRATOMA DESMARESTI   NOTABLE A

Order COLEOPTERA   Family TETRATOMIDAE

Tetratoma desmaresti Latreille, 1807. Formerly known as: *Tetratoma desmarestii* Latreille, 1807.

Distribution Recorded from South Hampshire, West Sussex, East Kent, West Kent, Surrey, Hertfordshire, Berkshire, East Suffolk, "Norfolk", Bedfordshire, Herefordshire, Nottinghamshire, Cheshire, North-east Yorkshire, Cumberland, Kirkcudbrightshire and Lanarkshire before 1970 and North Somerset, Buckinghamshire, East Suffolk, South Lincolnshire and South Lancashire from 1970 onwards.

Habitat and ecology Ancient broad-leaved woodland and pasture-woodland. Adults are usually found in the canopy and under the bark of oak, in dead wood and fungi on oak. Possibly also associated with willow. Larvae have been recorded in April from a yellow encrusting fungus on oak, probably in the genus *Stereum*, while a larva and pupa have been recorded in moss under oaks, in June. A probable life-cycle has been suggested as follows; eggs laid in the autumn, larvae developing over the winter and going to ground to pupate in late spring or early summer, adults probably remaining some time in the pupal cells and emerging in September to October. Adults have been recorded in January and from September to December.

Status Old records indicate that this species was formerly widely distributed in England and recorded north to Lanarkshire in Scotland. Recently recorded from just five vice-counties.

Indicator status Grade 3 in Garland (1983). Grade 3 in Harding & Rose (1986).

Threats Loss of broad-leaved woodland and parkland through, for example, clear-felling and coniferisation. Habitat loss, in particular, through the felling of ancient and fungus-infected trees, removal of dead wood from living trees and the destruction or removal of standing and fallen dead wood for reasons such as forest hygiene, aesthetic tidiness, public safety or for use as fire wood.

Management and conservation Ancient and fungus-infected trees, and both fallen and standing dead timber, especially with the bark attached, should be retained. The removal of dead timber from ancient and fungus-infected trees should be avoided. Gaps in the age structure of the tree population should be identified and the continuity of the appropriate habitat ensured by regeneration, suitable planting or possibly through pollarding.

Published sources Crowson, R.A. (1963), Fowler, W.W. (1891), Fowler, W.W. & Donisthorpe, H.St J.K. (1913), Garland, S.P. (1983), Harding, P.T. (1978), Harding, P.T. & Rose, F. (1986).

THROSCIDAE

AULONOTHROSCUS BREVICOLLIS   RARE

Order COLEOPTERA   Family THROSCIDAE

Aulonothroscus brevicollis (Bonvouloir, 1859).
Formerly known as: *Trixagus brevicollis* (Bonvouloir).

Distribution Recorded from South Devon, Surrey, Berkshire and Herefordshire before 1970 and Berkshire, East Suffolk, West Suffolk and Herefordshire from 1970 onwards. A map is given in Mendel (1988).

Habitat and ecology Pasture-woodland and, rarely, broad-leaved woodland. Associated with oak. Adults are usually found in the canopy. Larvae probably develop in dead wood. Adults havce been recorded in April and from June to August.

Status Very local, with a scattered distribution in southern England. This species is difficult to identify and may be confused with other members of the family. Consequently, the exact status of this species is hard to assess.

Indicator status Grade 1 in Harding & Rose (1986).

Threats Loss of parkland and broad-leaved woodland through, for example, clear-felling and coniferisation. Habitat loss, in particular, may be through the felling of ancient trees, removal of dead wood from living trees and the destruction or removal of standing and fallen dead wood for reasons such as forest hygiene, aesthetic tidiness, public safety or for use as fire wood.

Management and conservation Ancient trees and both fallen and standing dead timber, especially with the bark attached, should be retained. The removal of dead timber from ancient trees should be avoided. Gaps in the age structure of the tree population should be identified and the continuity of the appropriate dead wood habitat ensured by regeneration, suitable planting and possibly through pollarding.

Published sources Ashe, G.H. (1942), Cooter, J. (1990d), Harding, P.T. (1978), Harding, P.T. & Rose, F. (1986), Mendel, H. (1985), Mendel, H. (1988a), Shirt, D.B., ed. (1987).

TRIXAGUS ELATEROIDES   RARE

Order COLEOPTERA   Family THROSCIDAE

Trixagus elateroides (Heer, 1841). Formerly known as: *Throscus elateroides* (Heer).

Distribution Recorded from East Kent, West Kent, South Essex, Hertfordshire and Herefordshire before 1970 and East Sussex, East Kent, Surrey, South Essex and North Essex from 1970 onwards. A map is given in Mendel (1988).

Habitat and ecology Predominantly recorded from coastal and estuarine habitats, including coastal shingle and saltmarsh. Also noted inland in pasture-woodland. Most records are from the roots of plants and grass tussocks. Adults have been recorded in April, May and August.

Status Not listed in the insect Red Data Book (Shirt, 1987). Very local and primarily a species of south-eastern England. Can be numerous in some sites. This species is difficult to identify and may be confused with other members of the family. Consequently, the exact status of this species is hard to assess.

Threats This species may be threatened by urban and holiday developments, the construction of coastal defences, gravel extraction, reclamation, and erosion.

Management and conservation Disturbance of coastal shingle should be avoided. Management should aim at maintaining open conditions.

Published sources Fowler, W.W. (1890), Hallett, H.M. (1951), Lloyd, R.W. (1949), Mendel, H. (1988a).

TROX PERLATUS   ENDANGERED
A hide beetle
Order COLEOPTERA   Family TROGIDAE

Trox perlatus Goeze, 1777. Formerly known as: *Trox hispidus* sensu auct. Brit. not (Pontoppidan, 1763).

Distribution Recorded from "Devon" and Dorset before 1970.

Habitat and ecology Coastal cliffs. In carrion. Adults have been recorded in March, April, August and October.

Status First listed as a British species in 1860. Two further examples had been found, although the identity of one was not realised until a later date. In 1929 and 1930 small numbers of this beetle were found on the skins of two very young dead lambs on the cliffs above Worbarrow Bay, Dorset. *T. perlatus* does not appear to have been recorded since.

Threats This is an opportunistic and mobile species requiring a continuity of carrion availability.

Management and conservation Because of its dependence on carrion this species is likely to require large areas of suitable habitat for its continued survival.

Published sources Allen, A.A. (1967e), Harwood, P. (1929), Jessop, L. (1986), Shirt, D.B., ed. (1987).

TROX SABULOSUS   **NOTABLE A**
A hide beetle
Order COLEOPTERA   Family TROGIDAE

Trox sabulosus (L., 1758).

Distribution Recorded from "Devon", Dorset, South Hampshire, West Sussex, East Sussex, East Kent, West Kent, Surrey, Berkshire, Oxfordshire, East Suffolk, Worcestershire, Staffordshire, Leicestershire & Rutland and Cumberland before 1970 and West Sussex, South Essex, East Suffolk, Glamorgan, Lanarkshire and East Inverness & Nairn from 1970 onwards.

Habitat and ecology Sandy habitats, in particular sandy heathland. In dry carcasses. Recorded from a dead sheep. Adults have been recorded in May, June and August.

Status Widespread and very local. Possibly declining, this species was formerly known throughout southern England with scattered records as far north as Cumberland. Recently recorded from very few vice-counties, including two in Scotland and one from South Wales.

Threats The loss and fragmentation of heathland and sandy areas through changes in land use, mainly by conversion to arable, forestry and urban development, may be a threat to this species. Further habitat degradation has been through the cessation of traditional heathland management practices.

Management and conservation An opportunistic and mobile species requiring a continuity of carrion availability. Because of this, this species is likely to require large areas of suitable habitat for its continued survival. Management should aim for a diversity of successional stages, preferably by animal grazing or by rotational cutting, scraping or burning.

Published sources Bates, F. (1896), Day, F.H. (1909), Foster, G.N. & Young, R.J. (1978), Fowler, W.W. (1890), Fowler, W.W. & Donisthorpe, H.St J.K. (1913), Jessop, L. (1986), Jones, R.A. (1988)

NEMOZOMA ELONGATUM   **RARE**

Order COLEOPTERA   Family TROGOSSITIDAE

Nemozoma elongatum (L., 1761). Formerly known as: *Nemosoma elongatum* (L., 1761).

Distribution Recorded from North Somerset, West Kent, Oxfordshire, Bedfordshire, West Gloucestershire, Herefordshire, Warwickshire, Leicestershire & Rutland, Nottinghamshire and Cheshire before 1970 and Cambridgeshire, Northamptonshire, Worcestershire and Derbyshire from 1970 onwards.

Habitat and ecology Woodland, pasture-woodland and isolated groups of trees. Predatory. Occurs in dead wood, mainly in the burrows of the bark beetle *Acrantus vittatus* (Scolytidae) in elm, though also recorded from those of the beetle *Leperisinus varius* (Scolytidae). Also recorded from old palings. Adults have been recorded in March, July and August.

Status Very local. Widely scattered in England from North Somerset to Cheshire. Recently recorded from only four vice-counties.

Threats Loss of broad-leaved woodland and parkland through, for example, clear-felling and coniferisation. Habitat loss, in particular, through the felling of trees, removal of dead wood from living trees and the destruction or removal of standing and fallen dead wood for reasons such as forest hygiene, aesthetic tidiness, public safety or for use as fire wood.

Management and conservation Trees and both fallen and standing dead timber, especially with the bark attached, should be retained. The removal of dead timber from trees should be avoided. Gaps in the age structure of the tree population should be identified and the continuity of the appropriate dead wood habitat ensured by regeneration, suitable planting and possibly with pollarding.

Published sources Atty, D.B. (1983), Fowler, W.W. (1889), Roche, P.J.L. (1944), Shirt, D.B., ed. (1987), Tozer, D. (1944).

REFERENCES QUOTED IN DATA SHEETS

AIRY-SHAW, H.K. 1944. *Dictyopterus (Platycis) cosmandi* Chevr. (Col., Cantharidae, Lycidae) new to Britain. *Entomologist's Monthly Magazine, 80:* 204-205.

AIRY-SHAW, H.K. 1946. *Rhizophagus aeneus* Richter (Col., Rhizophagidae) in Surrey. *Entomologist's Monthly Magazine, 82:* 137.

AITKEN, J.F. 1988. Coleoptera new to Skye. *Entomologist's Monthly Magazine, 124:* 146.

ALBERTINI, M. & HALL, P. 1988. 1987 Annual Exhibition. Imperial College, London SW7 - 24th October 1987. Coleoptera. *British Journal of Entomology and Natural History, 1:* 38.

ALDRIDGE, R.J.W. & POPE, R.D. 1986. The British species of *Bruchidius* Schilsky (Coleoptera: Bruchidae). *Entomologist's Gazette, 37:* 181-193.

ALEXANDER, K.N.A. 1979. *Rhagonycha elongata* Fall. (Col., Cantharidae) in Kirkcudbright. *Entomologist's Monthly Magazine, 114:* 20.

ALEXANDER, K.N.A. 1986. Two new localities for *Anchonidium unguiculare* (Aubé) (Col., Curculionidae) in West Cornwall. *Entomologist's Monthly Magazine, 122:* 36.

ALEXANDER, K.N.A. 1987. Some notable Coleoptera from Pembrokeshire, including *Ceutorhynchus sulcicollis* (Payk.) new to Wales. *Entomologist's Monthly Magazine, 123:* 44.

ALEXANDER, K.N.A. 1989. *Meloe rugosus* Marsham (Col., Meloidae) in Gloucestershire. *Entomologist's Monthly Magazine, 125:* 127.

ALEXANDER, K.N.A. & CLEMENTS, D.K. 1988. 1987 Annual Exhibition. Imperial College, London SW7 - 24th October 1987. Coleoptera. *British Journal of Entomology and Natural History, 1:* 38-39.

ALEXANDER, K.N.A. & GROVE, S.J. 1990. 1989 Annual Exhibition. Imperial College, London SW7 - 28th October 1989. Coleoptera. *British Journal of Entomology and Natural History, 3:* 83.

ALLEN, A.A. 1937. *Malthodes crassicornis* Mäk. in Berkshire. *Entomologist's Monthly Magazine, 73:* 191.

ALLEN, A.A. 1940a. Abnormal food-plant of *Polydrosus confluens* St. (Col., Curculionidae). *Entomologist's Monthly Magazine, 76:* 40.

ALLEN, A.A. 1940b. *Scraptia testacea* nom. nov. and *S. fuscula* Müll. (Col., Scraptiidae). *Entomologist's Monthly Magazine, 76:* 56-58.

ALLEN, A.A. 1942. A recent capture of *Malthodes brevicollis* Payk. (*nigellus* Kies.) (Col., Cantharidae). *Entomologist's Monthly Magazine, 78:* 117.

ALLEN, A.A. 1945a. *Ceuthorrhynchus chalybaeus* Germ. (Col., Curculionidae) in Surrey, and its specific characters. *Entomologist's Monthly Magazine, 81:* 58.

ALLEN, A.A. 1945b. *Apion millum* Bach (Col., Curculionidae), etc., in Sussex. *Entomologist's Monthly Magazine, 81:* 78.

ALLEN, A.A. 1946a. *Phytobius quadricornis* Gyll. and *P. muricatus* Bris. (Col., Curculionidae) in Herts. *Entomologist's Monthly Magazine, 82:* 12.

ALLEN, A.A. 1946b. *Rhynchites opthalmicus* Steph. (Col., Curculionidae) in Monmouthshire. *Entomologist's Monthly Magazine, 82:* 93.

ALLEN, A.A. 1947a. *Agrilus sinuatus* Ol. (Col., Buprestidae) new to Hertfordshire. *Entomologist's Monthly Magazine, 83:* 8.

ALLEN, A.A. 1947b. *Hypulus quercinus* Quens. (Col., Melandryidae) not extinct in Britain. *Entomologist's Monthly Magazine, 83:* 9.

ALLEN, A.A. 1947c. *Gnorimus nobilis* L. (Col., Scarabeidae) in Sussex. *Entomologist's Monthly Magazine, 83:* 80.

ALLEN, A.A. 1947d. *Apion lemoroi* Bris. (Col., Curculionidae) in Cambridgeshire. *Entomologist's Monthly Magazine, 83:* 147.

ALLEN, A.A. 1947e. The three fumitory-weevils (Col., Curculionidae) at Windsor. (*Ceuthorrhynchus nigrinus* Marsh.; *Ceuthorrhynchus quercicola* Payk.; *Ceuthorrhynchus mixtus* Muls.). *Entomologist's Monthly Magazine, 83:* 152.

ALLEN, A.A. 1948a. Recent abundance of *Chrysomela* (=*Melasoma*) *populi* L. and *C. tremula* F. (Col.). *Entomologist's Monthly Magazine, 84:* 34.

ALLEN, A.A. 1948b. *Caenocara bovistae* Hoff. (Col., Anobiidae) in Essex. *Entomologist's Monthly Magazine, 84:* 45.

REFERENCES

ALLEN, A.A. 1948c. *Bagous cylindrus* Payk. (Col., Curculionidae), etc., in Middlesex. *Entomologist's Monthly Magazine, 84*: 46.

ALLEN, A.A. 1948d. Two species of Coleoptera new to Britain, in Kent. (*Lagria atripes; Mecinus janthinus*). *Entomologist's Monthly Magazine, 84*: 287-288.

ALLEN, A.A. 1949a. *Baris scolopacea* Germ. (Col., Curculionidae) at Port Victoria, north Kent. *Entomologist's Monthly Magazine, 85*: 61.

ALLEN, A.A. 1949b. *Gnorimus nobilis* L. (Col., Scarabaeidae) in Darenth Wood, Kent. *Entomologist's Monthly Magazine, 85*: 63.

ALLEN, A.A. 1950a. A second English capture of *Trechus subnotatus* Dej. (Col., Carabidae). *Entomologist's Monthly Magazine, 86*: 38.

ALLEN, A.A. 1950b. *Cassida nebulosa* L. (Col., Chrysomelidae) in Surrey and Cambs. *Entomologist's Monthly Magazine, 86*: 43.

ALLEN, A.A. 1950c. Recent abundance of certain Halticini (Col., Chrysomelidae). *Entomologist's Monthly Magazine, 86*: 49.

ALLEN, A.A. 1950d. *Coeliodes nigritarsis* Hartmann (Col., Curculionidae) in Scotland; an addition to the British fauna. *Entomologist's Monthly Magazine, 86*: 88-89.

ALLEN, A.A. 1950e. Coleoptera on Canvey Island, including two species new to Essex. *Entomologist's Monthly Magazine, 86*: 324.

ALLEN, A.A. 1950f. Recent captures of *Amara* spp. (Col., Carabidae). *Entomologist's Monthly Magazine, 86*: 344.

ALLEN, A.A. 1951a. *Haplocnemus nigricornis* F. (Col., Dasytidae) in Kent. *Entomologist's Monthly Magazine, 87*: 27.

ALLEN, A.A. 1951b. *Taphrorychus bicolor* Herbst (Coleoptera, Scolytidae) in Kent and Surrey. *Entomologist's Monthly Magazine, 87*: 31.

ALLEN, A.A. 1951c. *Cis coluber* Ab. (Col., Cisidae) new to Kent. *Entomologist's Monthly Magazine, 87*: 34.

ALLEN, A.A. 1951d. *Acritus atomarius* Aubé (Col., Histeridae) in Herefordshire, and a diagnostic note. *Entomologist's Monthly Magazine, 87*: 59.

ALLEN, A.A. 1951e. *Pityogenus trepanatus* Nördl. (Col., Scolytidae) spreading in England. *Entomologist's Monthly Magazine, 87*: 115-116.

ALLEN, A.A. 1951f. *Malthodes crassicornis* Mäkl. (Col., Cantharidae) etc., in Epping Forest, Essex. *Entomologist's Monthly Magazine, 87*: 214.

ALLEN, A.A. 1951g. A second West Kent record of *Mecinus janthinus* Germ. (Col., Curculionidae). *Entomologist's Monthly Magazine, 87*: 270.

ALLEN, A.A. 1952a. *Phalacrus substriatus* Gyll. (Col., Phalacridae) in Kent; and a few discrepancies etc., relating to other species. *Entomologist's Monthly Magazine, 88*: 18.

ALLEN, A.A. 1952b. A not generally recognised sexual colour-difference in *Leptura scutellata* F. (Col., Cerambycidae). *Entomologist's Monthly Magazine, 88*: 192.

ALLEN, A.A. 1952c. *Harpalus anxius* Dufts. (Carabidae) in South-east London. *Entomologist's Record and Journal of Variation, 64*: 263.

ALLEN, A.A. 1952d. *Apion difficile* Hbst. (Curculionidae) in Surrey and Hants. *Entomologist's Record and Journal of Variation, 64*: 294-295.

ALLEN, A.A. 1953. *Dromius insignis* Lucas (Carabidae) under bark: an unusual habitat. *Entomologist's Record and Journal of Variation, 65*: 121-122.

ALLEN, A.A. 1954a. *Langelandia anophthalma* Aubé (Col., Colydiidae) in South London. *Entomologist's Monthly Magazine, 90*: 42.

ALLEN, A.A. 1954b. A possibly unrecorded habit of *Donacia impressa* Payk. (Col., Chrysomelidae). *Entomologist's Monthly Magazine, 90*: 56.

ALLEN, A.A. 1954c. *Colon latum* Kr. (Col., Cholevidae), etc. at Wychwood, Oxon. *Entomologist's Monthly Magazine, 90*: 144.

ALLEN, A.A. 1954d. *Heptaulacus villosus* Gyll. (Col., Scarabaeidae) in a new West Kent locality. *Entomologist's Monthly Magazine, 90*: 193.

ALLEN, A.A. 1954e. *Perileptus areolatus* Gzr. (Col., Carabidae) in Herefordshire. *Entomologist's Monthly Magazine, 90*: 227.

ALLEN, A.A. 1954f. *Hypocoelus procerulus* Mannh. (Col., Eucnemidae, Anelastini) in Kent and Surrey: a tribe, genus and species new to Britain. *Entomologist's Monthly Magazine, 90*: 228-230.

ALLEN, A.A. 1954g. A melanic form of *Podagrica fuscicornis* L. (Col., Chrysomelidae). *Entomologist's Monthly Magazine*, *90*: 247.

ALLEN, A.A. 1955a. *Rhizophagus aeneus* Richt. (Col., Rhizophagidae) new to Kent. *Entomologist's Monthly Magazine*, *91*: 29.

ALLEN, A.A. 1955b. *Procraerus tibialis* Lac. (Col., Elateridae), *Hylotrupes bajulus* L. and *Obrium brunneum* F. (Cerambycidae), etc., in Hants. *Entomologist's Monthly Magazine*, *91*: 140.

ALLEN, A.A. 1955c. A note concerning the biology of two Carabid beetles (Agonini). (*Agonum sexpunctatum* L.; *Synuchus nivalis* Panz.). *Entomologist's Monthly Magazine*, *91*: 142.

ALLEN, A.A. 1955d. The Attelabidae of Darenth Wood, Kent. *Entomologist's Record and Journal of Variation*, *67*: 67-69.

ALLEN, A.A. 1955e. Notes on some Longicornia from Herefordshire. *Entomologist's Record and Journal of Variation*, *67*: 88-89.

ALLEN, A.A. 1956a. *Schizotus pectinicornis* L. (Col., Pyrochroidae) and other beetles in Breconshire. *Entomologist's Monthly Magazine*, *92*: 196.

ALLEN, A.A. 1956b. *Pediacus depressus* Hbst. (Col., Cucujidae) in Berks. and Sussex; with a summary of its British history. *Entomologist's Monthly Magazine*, *92*: 212.

ALLEN, A.A. 1956c. *Longitarsus quadriguttatus* Pont. (Col., Chrysomelidae) and its foodplants. *Entomologist's Monthly Magazine*, *92*: 218.

ALLEN, A.A. 1957a. *Antomaria alpina* Heer (Col., Cryptophagidae) in East and West Kent. *Entomologist's Monthly Magazine*, *93*: 18.

ALLEN, A.A. 1957b. A few records of *Badister* (*Baudia*) *anomalus* Perris and *dilatus* Chaud. (Col., Carabidae). *Entomologist's Monthly Magazine*, *93*: 20.

ALLEN, A.A. 1957c. *Ceuthorhynchus suturellus* Gyll. (Col., Curculionidae) in Sussex. *Entomologist's Monthly Magazine*, *93*: 88.

ALLEN, A.A. 1957d. An old Middlesex capture of *Strangalia aurulenta* F. (Col., Cerambycidae). *Entomologist's Monthly Magazine*, *93*: 91.

ALLEN, A.A. 1957e. *Pediacus depressus* Hbst. (Col., Cucujidae): an addendum. *Entomologist's Monthly Magazine*, *93*: 113

ALLEN, A.A. 1958. *Agrilus pannonicus* Dill. & Mitt. (Col., Buprestidae) in Herts. *Entomologist's Monthly Magazine*, *94*: 52.

ALLEN, A.A. 1959a. *Aderus brevicornis* Perris (Col., Aderidae) recaptured in Windsor Forest; with a survey of its British history and further notes. *Entomologist's Monthly Magazine*, *95*: 120.

ALLEN, A.A. 1959b. *Bradycellus distinctus* Dej. (Col., Carabidae) new to Kent, and *B. sharpi* Joy new to Essex. *Entomologist's Monthly Magazine*, *95*: 273.

ALLEN, A.A. 1959c. Abundance and habits of *Luperus circumfusus* Marsh. (Chrysomelidae) in a Surrey locality; and a submelanic form. *Entomologist's Record and Journal of Variation*, *71*: 141-142.

ALLEN, A.A. 1959d. *Dasytes niger* L. (Melyridae) on the Chilterns. *Entomologist's Record and Journal of Variation*, *71*: 142.

ALLEN, A.A. 1960a. *Longitarsus waterhousei* Kuts. (Col., Chrysomelidae) in South-east London. *Entomologist's Monthly Magazine*, *96*: 21.

ALLEN, A.A. 1960b. A note on the foodplant of *Apion flavimanum* Gyll. (Col., Apionidae). *Entomologist's Monthly Magazine*, *96*: 166.

ALLEN, A.A. 1960c. *Mecinus janthinus* Germ. (Col., Curculionidae) in Surrey. *Entomologist's Monthly Magazine*, *96*: 212.

ALLEN, A.A. 1960d. A new capture of *Cryptocephalus 10- maculatus* L. (Col., Chrysomelidae) in Scotland. *Entomologist's Monthly Magazine*, *96*: 271.

ALLEN, A.A. 1961. *Apion semivittatum* Gyll. and *Gymnetron collinum* Gyll. (Col., Curculionidae) in Kent and West Sussex. *Entomologist's Monthly Magazine*, *97*: 21.

ALLEN, A.A. 1962a. A reputed twentieth-century (?) occurrence of *Rhynchites bacchus* L. (Col., Curculionidae) in Kent. *Entomologist's Monthly Magazine*, *98*: 50.

ALLEN, A.A. 1962b. An additional record of *Meloe autumnalis* Ol. for Dorset, and of *M.brevicollois* Panz. (Col., Meloidae) for Cornwall. *Entomologist's Monthly Magazine*, *98*: 106.

REFERENCES

ALLEN, A.A. 1962c. *Polydrusus confluens* Steph. (Col., Curculionidae) recorded in error from oak. *Entomologist's Monthly Magazine, 98*: 108.

ALLEN, A.A. 1963. The occurrence of *Teretrius pipcipes* F. (Col., Histeridae) at Oxshott, Surrey; with short notes on the other British records. *Entomologist's Monthly Magazine, 99*: xix.

ALLEN, A.A. 1964a. The genus *Synchita* Hellw. (Col., Colydiidae) in Britain; with an addition to the fauna and a new synonomy. *Entomologist's Monthly Magazine, 100*: 36-42.

ALLEN, A.A. 1964b. *Megapenthes lugens* Rebt. (Col., Elateridae) in Hants., Gloucs., etc., with additional notes. *Entomologist's Monthly Magazine, 100*: 95-96.

ALLEN, A.A. 1964c. *Harpalus honestus* Duft. (Col., Carabidae) confirmed as British. *Entomologist's Monthly Magazine, 100*: 155-157.

ALLEN, A.A. 1965a. *Sitona puberulus* Reitt. (Col., Curculionidae) in Ireland and the Isle of Arran; and a comment on the British status of *S. cinerascens* Fahr. *Entomologist's Monthly Magazine, 101*: 19.

ALLEN, A.A. 1965b. Is *Onthophagus nutans* F. (Col., Scarabaeidae) still taken in Britain?. *Entomologist's Monthly Magazine, 101*: 30.

ALLEN, A.A. 1965c. *Triplax lacordairei* Crotch (Col., Erotylidae) in Dorset and Worcs., with brief remarks on the other British species. *Entomologist's Monthly Magazine, 101*: 46-48.

ALLEN, A.A. 1965d. Notes on records of two species of *Bembidion* (Col., Carabidae). (*Bembidion lunatum* Duft.; *Bembidion octomaculatum* Gze.). *Entomologist's Monthly Magazine, 101*: 177.

ALLEN, A.A. 1966. The rarer Sternoxia (Col.) of Windsor Forest. *Entomologist's Record and Journal of Variation, 78*: 14-23.

ALLEN, A.A. 1967a. Two new species of *Longitarsus* Latr. (Col., Chrysomelidae) in Britain. *Entomologist's Monthly Magazine, 103*: 75-82.

ALLEN, A.A. 1967b. A note on *Strangalia aurulenta* F. (Col., Cerambycidae) in Sussex. *Entomologist's Monthly Magazine, 103*: 236.

ALLEN, A.A. 1967c. *Dorytomus affinis* Payk. (Col., Curculionidae) in Kent and notes on its British allies. *Entomologist's Monthly Magazine, 103*: 264-267.

ALLEN, A.A. 1967d. An inquiry into the British status of the genus *Trichodes* Hbst. (Col., Cleridae). *Entomologist's Record and Journal of Variation, 79*: 54-58.

ALLEN, A.A. 1967e. A review of the status of certain Scarabaeoidea (Col.) in the British fauna; with the addition to our list of *Onthophagus similis* Scriba. *Entomologist's Record and Journal of Variation, 79*: 201-206.

ALLEN, A.A. 1969a. On the sexual characters of *Aderus brevicornis* Perris (Col., Aderidae); with an additional record. *Entomologist's Monthly Magazine, 105*: 164.

ALLEN, A.A. 1969b. *Ernoporus caucasicus* Lind. and *Leperisinus orni* Fuchs. (Col., Scolytidae) in Britain. *Entomologist's Monthly Magazine, 105*: 245-9.

ALLEN, A.A. 1970a. An overlooked Sussex record of *Crytocephalus 10-maculatus* L. (Col., Chrysomelidae), *C. biguttatus* Scop. in Surrey. *Entomologist's Monthly Magazine, 106*: 120.

ALLEN, A.A. 1970b. A note on *Pissodes validirostris* Gyll. (Col., Curculionidae) and on the British range of *P. pini* L. *Entomologist's Monthly Magazine, 106*: 204.

ALLEN, A.A. 1970c. The British status of certain Scarabaeoidea (Col.): supplementary notes and corrections. (*Platycerus caraboides* L.; *Onthophagus taurus* Schreb.;*Polyphylla fullo* L.). *Entomologist's Record and Journal of Variation, 82*: 91.

ALLEN, A.A. 1970d. Revisional notes on the British species of *Orthoperus* Steph. (Col., Corylophidae). *Entomologist's Record and Journal of Variation, 82*: 112-120.

ALLEN, A.A. 1971a. *Procraerus tibialis* Lac. (Col., Elateridae) in Wilts. and Herts. *Entomologist's Monthly Magazine, 107*: 12.

ALLEN, A.A. 1971b. *Procas armillatus* F. (Col., Curculionidae) new to Devonshire. *Entomologist's Monthly Magazine, 107*: 52.

ALLEN, A.A. 1971c. *Microlomalus parallelepipedus* Hbst. (Col., Histeridae) in Kent. *Entomologist's Monthly Magazine, 107*: 80.

ALLEN, A.A. 1971d. A few records of *Olibrus* spp. (Col., Phalacridae). *Entomologist's Monthly Magazine, 107*: 126.

426

ALLEN, A.A. 1972a. *Strangalia revestita* L. (Col., Cerambycidae) in Surrey; with a synopsis of its British history. *Entomologist's Monthly Magazine, 108*: 22.

ALLEN, A.A. 1972b. A second British capture of *Leptura rufa* Brulle (Col., Cerambycidae); and remarks on *L. sanguinolenta* L. in Britain. *Entomologist's Monthly Magazine, 108*: 92.

ALLEN, A.A. 1972c. *Hydrothassa hannoverana* F. (Col., Chrysomelidae) in the New Forest, Hants. *Entomologist's Monthly Magazine, 108*: 101.

ALLEN, A.A. 1972d. *Tapinotus sellatus* F. (Col., Curculionidae) in Berkshire. *Entomologist's Monthly Magazine, 108*: 208.

ALLEN, A.A. 1972e. A contribution to the knowledge of *Phytonomus meles* F. (Col., Curculionidae) in Britain. *Entomologist's Record and Journal of Variation, 84*: 110-113.

ALLEN, A.A. 1973a. *Limobius borealis* Payk. (Col., Curculionidae) in Wilts. and Hants. *Entomologist's Monthly Magazine, 109*: 61.

ALLEN, A.A. 1973b. An ecological note on *Saprinus virescens* Payk. (Col., Histeridae). *Entomologist's Monthly Magazine, 109*: 131.

ALLEN, A.A. 1973c. Previous Scottish records, etc., of *Dryocoetes alni* Georg (Coleoptera: Scolytidae) and *Dirhagus pygmaeus* F. (Eucnemidae):. *Entomologist's Monthly Magazine, 109*: 140.

ALLEN, A.A. 1973d. *Rhinoncus albicinctus* Gyll. (Col., Curculionidae) new to Britain. *Entomologist's Monthly Magazine, 109*: 188-190.

ALLEN, A.A. 1974. Notes on British Cionini (Col.) mainly arising out of Mr. Cunningham's findings in the Portsmouth area. *Entomologist's Record and Journal of Variation, 86*: 265-269.

ALLEN, A.A. 1975a. *Pityogenes trepanatus* Nördl. (Col., Scolytidae) in Norfolk. *Entomologist's Monthly Magazine, 111*: 22.

ALLEN, A.A. 1975b. The second Kent record of *Anthicus bimaculatus* Ill. (Col., Anthicidae) and the first for *Bledius pallipes* Grav. (Staphylinidae), etc. *Entomologist's Monthly Magazine, 111*: 228.

ALLEN, A.A. 1975c. Two species of *Anaspis* (Col.: Mordellidae) new to Britain; with a consideration of the status of *A. hudsoni* Donis., etc. *Entomologist's Record and Journal of Variation, 87*: 269- 274.

ALLEN, A.A. 1976a. *Thalassophilus longicornis* (Sturm) (Col., Carabidae) in Caernarvonshire. *Entomologist's Monthly Magazine, 112*: 242.

ALLEN, A.A. 1976b. *Platypus parallelus* F. (= *micus* & Steph.) (Col., Scolytidae) recaptured in Britain after 150 years. *Entomologist's Record and Journal of Variation, 88*: 57-58

ALLEN, A.A. 1976c. Notes on some British Chrysomelidae (Col.) including amendments and additions to the list. *Entomologist's Record and Journal of Variation, 88*: 220-225.

ALLEN, A.A. 1976d. Notes on some Briitsh Chrysomelidae (Col.) including ammendments and additions to the list. *Entomologist's Record and Journal of Variation, 88*: 294-299.

ALLEN, A.A. 1977a. *Polydrusus sericeus* Schall. (Col.: Curculionidae) in Sussex. *Entomologist's Record and Journal of Variation, 89*: 182.

ALLEN, A.A. 1977b. *Trypophloeus asperatus* Gyll. (Col.: Scolytidae) in S.E. London. *Entomologist's Record and Journal of Variation, 89*: 185.

ALLEN, A.A. 1977c. A recent occurrence of *Phyllotreta vittata* F. (= *sinuata* auct. Brit.) (Col.: Chrysomelidae). *Entomologist's Record and Journal of Variation, 89*: 332.

ALLEN, A.A. 1977d. *Agonum gracilipes* Duft. (Col.: Carabidae) in Sussex, and its deletion from the Irish list. *Entomologist's Record and Journal of Variation, 89*: 343.

ALLEN, A.A. 1978a. *Longitarsus rutilus* (Ill.) (Col., Chrysomelidae) in east Cornwall. *Entomologist's Monthly Magazine, 114*: 62.

ALLEN, A.A. 1978b. *Lissodema cursor* (Gyll.) (Col., Salpingidae) in suburban Kent. *Entomologist's Monthly Magazine, 114*: 101.

ALLEN, A.A. 1978c. *Amphimallon ochraceum* Knoch (Col.: Scarabaeidae): recent captures in Wales and Sussex, with a brief survey of earlier records. *Entomologist's Record and Journal of Variation, 90*: 17.

ALLEN, A.A. 1978d. *Amphimallon ochraceum* Knoch (Col. Scarabeidae) an addendum. *Entomologist's Record and Journal of Variation, 90*: 278.

ALLEN, A.A. 1979. *Apion semivittatum* Gyll. (Col., Apionidae) in N.E. London. *Entomologist's Record and Journal of Variation, 91*: 328-329.

REFERENCES

ALLEN, A.A. 1981a. A recent Essex find of *Procraerus tibilis* Lac. (Col.: Elateridae). *Entomologist's Record and Journal of Variation, 93*: 43.

ALLEN, A.A. 1981b. *Ceuthorynchidius rufulus* Duf., etc., (Col.: Curculionidae) in S.E. London. *Entomologist's Record and Journal of Variation, 93*: 141.

ALLEN, A.A. 1981c. *Aclypea opaca* L. (Col.: Silphidae) in West Kent. *Entomologist's Record and Journal of Variation, 93*: 147.

ALLEN, A.A. 1981d. Notes, mainly diagnostic, on *Ceuthorhynchus pulvinatus* Gyll. (Col.: Curculionidae). *Entomologist's Record and Journal of Variation, 93*: 234-235.

ALLEN, A.A. 1982a. *Magdalis violacea* L. (Col.: Curculionidae): correction of a record. *Entomologist's Record and Journal of Variation, 94*: 120-121.

ALLEN, A.A. 1982b. A note on two British *Trachyphloeus* spp. (Col., Curculionidae). (*Trachyploeus scabriculus* L.; *Trachyphloeus digitalis* Gyll.). *Entomologist's Record and Journal of Variation, 94*: 129.

ALLEN, A.A. 1984a. *Amara nitida* Stm. (Col., Carabidae) in Surrey. *Entomologist's Monthly Magazine, 120*: 222.

ALLEN, A.A. 1984b. A genus and species of Malachiinae (Col., Melyridae) new to Britain. (*Sphinginus lobatus* Ol.). *Entomologist's Record and Journal of Variation, 96*: 243-244.

ALLEN, A.A. 1985a. *Platypus parallelus* (F.) (Col., Scolytidae) again captured at light in S.E. London. *Entomologist's Monthly Magazine, 121*: 141.

ALLEN, A.A. 1985b. A fourth capture of *Aderus brevicornis* Perris (Col.) at Windsor. *Entomologist's Record and Journal of Variation, 97*: 34-35.

ALLEN, A.A. 1985c. *Brachinus sclopeta* F. (Col., Carabidae): two captures in the present century. *Entomologist's Record and Journal of Variation, 97*: 137-139.

ALLEN, A.A. 1986. On the British species of *Mordellistena* Costa (Col.: Mordellidae) resembling *parvula* Gyll. *Entomologist's Record and Journal of Variation, 98*: 47-50.

ALLEN, A.A. 1988a. *Rhynchaenus calceatus* Germ. (Col., Curculionidae), an addition to the British list. *Entomologist's Monthly Magazine, 124*: 147-148.

ALLEN, A.A. 1988b. Notes on *Agrilus pannonicus* Pill. & Mitt. (Col.: Buprestidae) in 1985. *Entomologist's Record and Journal of Variation, 100*: 25-28.

ALLEN, A.A. 1988c. *Dorcatoma serra* Panz. (Col.: Anobiidae) in W.Norfolk. *Entomologist's Record and Journal of Variation, 100*: 46.

ALLEN, A.A. 1988d. A fourth species of *Ischnomera* Steph. (Col.: Oedemeridae) in Britain. *Entomologist's Record and Journal of Variation, 100*: 199-202.

ALLEN, A.A. 1988e. Two notable garden beetles. *Entomologist's Record and Journal of Variation, 100*: 277.

ALLEN, A.A. 1989a. *Miarus micros* (Germ.) (Col., Curculionidae): a problem of foodplant. *Entomologist's Monthly Magazine, 125*: 3.

ALLEN, A.A. 1989b. *Gymnetron beccabungae* (L.) (Col., Curculionidae) in Scotland. *Entomologist's Monthly Magazine, 125*: 51.

ALLEN, A.A. 1989c. *Lamprohiza splendidula* (L.) (Col., Lampyridae) taken in Kent in 1884. *Entomologist's Monthly Magazine, 125*: 182.

ALLEN, A.A. 1989d. Beetles and bugs on a Thame-side wall in autumn. *Entomologist's Record and Journal of Variation, 101*: 47-49.

ALLEN, A.A. 1989e. *Bembidion* (*Lymnaeum*) *nigropiceum* Marsh. (Col: Carabidae) in West Kent. *Entomologist's Record and Journal of Variation, 101*: 55.

ALLEN, A.A. 1989f. A brief history of *Carabus intricatus* L. (Col; Carabidae) in Britain, with special reference to its present-day status. *Entomologist's Record and Journal of Variation, 101*: 113-115.

ALLEN, A.A. 1989g. Uncommon Heteromera (Col.) from a S.E. London wood. *Entomologist's Record and Journal of Variation, 101*: 233-234.

ALLEN, A.A. 1990a. *Pityophthorus lichtensteini* (Ratz.) (Col., Scolytidae) in Speyside. *Entomologist's Monthly Magazine, 126*: 19.

ALLEN, A.A. 1990b. Notes on, and a key to, the often-confused British species of *Ampedus* Germ. (Col.: Elateridae), with corrections of some erroneous records. *Entomologist's Record and Journal of Variation, 102*: 121-127.

ALLEN, A.A. 1990c. The earliest British capture of *Cis dentatus* Mell (Col.: Cisidae); with diagnostic notes. *Entomologist's Record and Journal of Variation, 102*: 179-180.

ALLEN, A.A. 1990d. *Cossonus linearis* F. (Col.: Curculionidae) in Surrey and West Kent. *Entomologist's Record and Journal of Variation, 102*: 300.

ALLEN, A.A. 1990e. Two interesting Carabid captures (Col.) in S.E.London (W.Kent). *Entomologist's Record and Journal of Variation, 102*: 305-306.

ALLEN, A.A. & LLOYD, R.W. 1951. *Pyrrhidium sanguineum* L. (Col., Cerambycidae) as a British species. *Entomologist's Monthly Magazine, 87*: 157-8.

ALLEN, J.W. 1935. *Tapinotus sellatus* in Surrey. *Entomologist's Monthly Magazine, 71*: 67.

ALLEN, J.W. & NICHOLSON, G.W. 1924a. *Tachys micros* Fisch. (*Gregarius* Chaud.), an addition to the list of British Coleoptera. *Entomologist's Monthly Magazine, 60*: 225.

ALLEN, J.W. & NICHOLSON, G.W. 1924b. *Sitona gemellatus* Gyll. in South Devon. *Entomologist's Monthly Magazine, 60*: 231.

ALLEN, J.W. & NICHOLSON, G.W. 1924c. *Chlaenius schranki* Dufts. (*nitidulus* Schrank) on the Dorset coast. *Entomologist's Monthly Magazine, 60*: 231.

ALLEN, S.E. 1953. Coleoptera of Woolmer Bog. *Bulletin of the Amateur Entomological Society, 12*: 6-7.

ANGUS, R.B. 1964. Some Coleoptera from Cumberland, Westmorland and the northern part of Lancashire. *Entomologist's Monthly Magazine, 100*: 61-69.

ANON, 1924. Society: The South London Entomological and Natural History Society: August 14th. *Entomologist's Record and Journal of Variation, 36*: 147.

ANON, 1925. Exhibition of objects of natural history other than Lepidoptera. *Proceedings of the South London Entomological and Natural History Society, 1925-1926*: 107-111.

ANON, 1932. Reviews and notices of books: An annotated list of the Coleoptera of the Isle of

Sheppey by J.J.Walker. *Entomologist's Record and Journal of Variation, 44*: 147-148.

ANON, 1933. Miscellaneous observations. *Transactions of the Norfolk and Norwich Naturalists' Society, 13*: 505-508.

ANON, 1935. Miscellaneous observations. *Transactions of the Norfolk and Norwich Naturalists' Society, 14*: 106-108.

ANON, 1939. Handlist of the Coleoptera of the Marlborough district (10 miles radius). *Report of the Marlborough College Natural History Society, 87* (1938): 54-86.

ANON, 1960. Insect notes: Coleoptera. *Field Naturalist, 5*: 27-28.

ANON, 1962. 588th meeting of the Yorkshire Naturalists' Union. Meltham V.C. 63. *Yorkshire Naturalists' Union Circular, 614.*.

ANON, 1972. Annual exhibition: 30th October 1971. *Proceedings and Transactions of the British Entomological and Natural History Society, 5*: 21-26.

ANSORGE, E. 1958. *Odontaeus armiger* (Scopoli) in Buckinghamshire (Coleoptera, Scarabaeidae). *Entomologist's Gazette, 9*: 101.

ANSORGE, E. 1962. *Odontaeus armiger* (Scopoli) in Buckinghamshire (Coleoptera, Scarabaeidae). *Entomologist's Gazette, 13*: 177.

ANSORGE, E. 1963. *Odontaeus armiger* (Scopoli) (Coleoptera, Scarabaeidae) in Buckinghamshire. *Entomologist's Gazette, 14*: 162.

APPLETON, D. 1969a. *Tropideres sepicola* (F.) and *Choragus sheppardi* Kirby (Col., Platystomidae) in the New Forest. *Entomologist's Monthly Magazine, 105*: 47.

APPLETON, D. 1969b. *Pterostichus aterrimus* (Herbst) (Col., Carabidae) in the New Forest. *Entomologist's Monthly Magazine, 105*: 179.

APPLETON, D. 1971. *Zyras haworthi* (Steph.) (Col., Staphylinidae) in the New Forest. *Entomologist's Monthly Magazine, 107*: 256.

APPLETON, D. 1972a. *Eucnemis capucina* Ahr. (Col., Eucnemidae) in the New Forest. *Entomologist's Monthly Magazine, 108*: 2.

REFERENCES

APPLETON, D. 1972b. *Heptaulacus villosus* (Gyll.) (Col., Scarabaeidae) in south Hants. *Entomologist's Monthly Magazine, 108*: 87.

APPLETON, D. 1973a. *Phytonomus meles* (F.) (Col., Curculionidae) in south Hants. *Entomologist's Monthly Magazine, 109*: 78.

APPLETON, D. 1973b. *Corymbites castaneus* (L.) (Col., Elateridae) in the Isle of Wight. *Entomologist's Monthly Magazine, 109*: 202.

APPLETON, D. 1986. *Baris analis* (Olivier) (Col., Curculionidae) rediscovered in Isle of Wight. *Entomologist's Monthly Magazine, 122*: 232.

ASHE, G.H. 1922. *Caenocara bovistae* in Caernarvonshire. *Entomologist's Monthly Magazine, 58*: 230.

ASHE, G.H. 1934. A note on *Rhizophagus aeneus* Richt. *Entomologist's Monthly Magazine, 70*: 94.

ASHE, G.H. 1942. *Trixagus (=Throscus) brevicollis* Bonv. (Col., Trixagidae), a species new to Britain. *Entomologist's Monthly Magazine, 78*: 287.

ASHE, G.H. 1944a. Devonshire and Sussex Coleoptera for 1942-43. *Entomologist's Monthly Magazine, 80*: 70.

ASHE, G.H. 1944b. *Hylobius transversovittatus* Goeze (= *fatuus* Rossi) (Col., Curculionidae) new to Britain. *Entomologist's Monthly Magazine, 80*: 287.

ASHE, G.H. 1952. Coleoptera at Nethy Bridge, Inverness-shire. *Entomologist's Monthly Magazine, 88*: 166.

ASHFORD, R.W. 1961. *Odontaeus armiger* (Scop.)(Col., Scarabaeidae) in Berks. *Entomologist's Monthly Magazine, 97*: 236.

ASMOLE, N.P. *and others*. 1983. Insects and spiders on snowfields in the Cairngorms, Scotland. *Journal of Natural History, 17*: 599-613.

ATTY, D.B. 1983. *Coleoptera of Gloucestershire.* Cheltenham, D.B. Atty.

AUBROOK, E.W. 1963. Entomology: Coleoptera. *Naturalist, 88*: 17-18.

AUBROOK, E.W. 1970. *Cis dentatus* Mell. (Col. Cisidae): an addition to the British list. *Entomologist, 103*: 250-251.

AUBROOK, E.W. 1972. Entomology: Coleoptera. *Naturalist, 97*: 25-27.

AUBROOK, E.W. 1973. *Triplax scutellaris* Charp. v. *gyllenhali* Crotch (Col., Erotylidae). *Entomologist's Monthly Magazine, 109*: 88.

BAGNALL, R.S. 1905. Notes and further additions to the Coleoptera of Northumberland & Durham. *Entomologist's Record and Journal of Variation, 17*: 331-333.

BAINES, J.M. 1952. *Hippodamia 13-punctata* L. (Coccinellidae) at Hastings. *Entomologist, 85*: 262.

BATES, F. 1896. The Coleoptera of Bradgate Park. *Transactions of the Leicester Literary and Philosophical Society, 4*: 170-176.

BATTEN, R. 1986. A review of the British Mordellidae (Coleoptera). *Entomologist's Gazette, 37*: 225-235.

BEARE, T.H. 1912. *Thanasimus rufipes*, Brahm: a beetle new to Britian. *Entomologist's Monthly Magazine, 48*: 255-257.

BEARE, T.H. 1922. *Hypera meles* and other Coleoptera in a lucerne field at Wicken. *Entomologist's Monthly Magazine, 58*: 249.

BEARE, T.H. 1927. *Dorytomus hirtipennis* Bed. in Suffolk. *Entomologist's Monthly Magazine, 63*: 12.

BEARE, T.H. 1930a. *Apion minimum* Hbst., an addition to the Rochester list of Coleoptera. *Entomologist's Monthly Magazine, 66*: 215-216.

BEARE, T.H. 1930b. *Pissodes validirostris* Gyll., a British insect. *Entomologist's Monthly Magazine, 66*: 274.

BEARE, T.H. 1934a. Capture of *Cathormiocerus britannicus* Blair. *Entomologist's Monthly Magazine, 70*: 54.

BEARE, T.H. 1934b. *Bagous lutosus* Gyll. at Fleet. *Entomologist's Monthly Magazine, 70*: 66.

BEARE, T.H. 1937. *Pityogenes trepanatus* Nördl. at Dulnain Bridge, N.B. *Entomologist's Monthly Magazine, 73*: 259.

BEDWELL, E.C. 1907. Is *Teretrius picipes*, F., parasitic on *Lyctus canaliculatus*, F., as well as on *L. brunneus*. *Entomologist's Monthly Magazine, 43*: 275.

BEDWELL, E.C. 1923. *Prionychus (Eryx) fairmairei* Reiche; a southern record. *Entomologist's Monthly Magazine, 59*: 236-237.

BEDWELL, E.C. 1926a. *Lymeylon navale* L. at Windsor. *Entomologist's Monthly Magazine*, : 62: 240.

BEDWELL, E.C. 1926b. *Teredus cylindricus* Ol. and *Cryptocephalus quercetI* Suffr. at Windsor in July 1925. *Entomologist's Monthly Magazine*, 62: 240-241.

BEDWELL, E.C. 1932a. *Odontaeus mobilicornis* Fab. at Coulsdon, Surrey. *Entomologist's Monthly Magazine, 68*: 188.

BEDWELL, E.C. 1932b. A new variety of *Tapinotus sellatus* F. in Hampshire. *Entomologist's Monthly Magazine, 68*: 277.

BEDWELL, E.C. 1933. Localities for *Gymnetron beccabungae* L. *Entomologist's Monthly Magazine, 69*: 189-190.

BEDWELL, E.C. 1936a. *Anchonidium unguiculare* Aubé in Cornwall. *Entomologist's Monthly Magazine, 72*: 220.

BEDWELL, E.C. 1936b. A belated British record of *Agelastica alni* L. *Entomologist's Monthly Magazine, 72*: 257-258.

BEDWELL, E.C. 1943a. *Tachys micros* Frsch. (Col., Carabidae) in Sussex. *Entomologist's Monthly Magazine, 79*: 208.

BEDWELL, E.C. 1943b. *Gnathoncus buyssoni* Auzat (Col., Histeridae) in Surrey. *Entomologist's Monthly Magazine, 79*: 223.

BENSON, R.B. 1967. (Footnote). *Entomologist's Monthly Magazine, 103*: 72.

BEVAN, D. 1971. Notes on *Pissodes validirostris* Gyll. and *P. pini* L. (Col., Curculionidae). *Entomologist's Monthly Magazine, 107*: 90.

BÍLÝ, S. 1982. *The Buprestidae (Coleoptera) of Fennoscandia and Denmark*. Klampenborg, Scandinavian Science Press. (Fauna Entomologica Scandinavica, Volume 10).

BÍLÝ, S. & MEHL, O. 1989. *Longhorn beetles (Coleoptera, Cerambycidae) of Fennoscandia and Denmark*. Leiden, E.J. Brill. (Fauna Entomologica Scandinavica, Volume 22).

BLACK, J.E. 1924. Coleoptera in the New Forest. *Entomologist's Monthly Magazine, 60*: 36-37.

BLAIR, K.G. 1935. *Smicronyx reichi* Gyll. with notes on other species of the genus (Col.). *Entomologist's Monthly Magazine, 71*: 127-130.

BLAIR, K.G. 1936. *Rhizophagus aeneus* Richt. in Tilgate Forest. *Entomologist's Monthly Magazine, 72*: 209.

BLAIR, K.G. 1943. *Hypobaous flavipos* F. (not *Ebaous abietinus* Abielle) (Col., Malachiidae) in Britain: a correction. *Entomologist's Monthly Magazine, 79*: 16.

BLAIR, K.G. 1945. *Hylobius transversovittatus* Goeze (Col., Curculionidae) in Somerset. *Entomologist's Monthly Magazine, 81*: 10.

BLAIR, K.G. 1950. Some northern British Coleoptera, including *Nebria nivalis* Payk. a Carabid species not hitherto known from Britain. *Entomologist's Monthly Magazine, 86*: 219-220.

BLENKARN, S.A. 1922. *Lissodema cursor* Gyll., etc., at Box Hill. *Entomologist's Monthly Magazine, 58*: 65.

BOOTH, R.G. 1977. *Langelandia anophthalma* Aubé (Col., Colydiidae) in the West Country. *Entomologist's Monthly Magazine, 113*: 112.

BOOTH, R.G. 1978. *Apion sicardi* Desbr. (Col., Apionidae) and *Crepidocera* (ie *Crepidodera*) *impressa* (F.) (Col., Chrysomelidae) in Hampshire. *Entomologist's Monthly Magazine, 114*: 102.

BOYCE, D.C. 1987. Beetles of the Tanybwlch area, Aberystwyth (22/580800), VC46. *Dyfed Invertebrate Group Newsletter*, No. 7: 3.

BOYCE, D.C. 1988. Deadwood beetles in Ceredigion (VC46). *Dyfed Invertebrate Group Newsletter*, No. 11: 13-15.

BOYCE, D.C. 1989. Coleoptera recording in Ceredigion during 1988. *Dyfed Invertebrate Group Newsletter*, No. 12: 15-18.

BOYCE, D.C. 1990. Coleoptera recording in Ceredigion (VC46) during 1989. *Dyfed Invertebrate Group Newsletter*, No. 16: 16-21.

BOYCE, D.C. & FOWLES, A.P. 1988. Coleoptera. Ceredigion Coleoptera records. *Dyfed Invertebrate Group Newsletter*, No. 9: 17-19.

BRENDEL, M.J.D. 1975. Coleoptera: Tenebrionidae. *Handbooks for the Identification of British Insects, 5*(10): 1-22.

431

REFERENCES

BRITTEN, H. 1937. A further note on *Agelastica alni* L. *Entomologist's Monthly Magazine, 73*: 108-109.

BRITTEN, H. 1943. The Coleoptera of the Isle of Man. *North Western Naturalist, 18*:. 73-87.

BRITTON, E.B. 1956. Coleoptera: Scarabaeoidea. *Handbooks for the Identification of British Insects, 5*(11): 1-53.

BROMLEY, P.J. 1947. Biological observations on *Chrysomela tremula* F. (Col., Chrysomelidae) at Oxford. *Entomologist's Monthly Magazine, 83*: 57-58.

BROWN, D.G. 1977. *Meloe rugosus* Marsham (Col., Meloidae) at Bath, Avon. *Entomologist's Monthly Magazine, 113*: 94.

BROWN, E.S. 1948a. *Calosoma inquisitor* L. (Col.) in Hertfordshire. *Entomologist's Monthly Magazine, 84*: 124.

BROWN, E.S. 1948b. Recent abundance of *Chrysomela* (*Melasoma*) *populi* L. and *C. tremula* F. (Col.). *Entomologist's Monthly Magazine, 84*: 138.

BROWN, S.C.S. 1954. *Feronia kugelanni* Panz. (Col., Carabidae) in the New Forest, Hants. *Entomologist's Monthly Magazine, 90*: 22.

BROWN, S.C.S. 1971. *Hetaerius ferrugineus* Bh. (Col., Histeridae) in Hants. *Entomologist's Monthly Magazine, 107*: 72.

BUCK, F.D. 1938. *Cassida vittata*, Vl. and *Dictyopterus affinis*, Pk. *Entomologist's Record and Journal of Variation, 50*: 23.

BUCK, F.D. 1949. *Agonum quadripunctatum* Degeer and *Feronia angustata* Dufts. (Col., Carabidae) in Surrey. *Entomologist's Monthly Magazine, 85*: 132.

BUCK, F.D. 1953. *Hippodmia 13-punctata* (L.) (Col., Coccinellidae) its environment and pabulum. *Entomologist, 86*: 105-106.

BUCK, F.D. 1954. Coleoptera: Lagriidae to Meloidae. *Handbooks for the Identification of British Insects, 5*(9): 1-30.

BUCK, F.D. 1955. A provisional list of the Coleoptera of Epping Forest. *Entomologist's Monthly Magazine, 91*: 174-192.

BUCK, F.D. 1956. Society: The South London Entomological and Natural History Society: June 14th, 1956. *Entomologist's Monthly Magazine, 92*: 335.

BUCK, F.D. 1957a. *Dorcatoma serra* Panz. (Col., Anobiidae) and its Braconid (Hym.) parasite in Norfolk. *Entomologist's Monthly Magazine, 93*: 245.

BUCK, F.D. 1957b. Society: The South London Entomological and Natural History Society: July 11th, 1957. *Entomologist's Monthly Magazine, 93*: 255.

BUCK, F.D. 1957c. *Abdera triguttata* (Gyll.) (Col., Melandryidae) in Suffolk. *Entomologist's Monthly Magazine, 93*: 280.

BUCK, F.D. 1958. Society: The South London Entomological and Natural History Society: March 27th, 1958. *Entomologist's Monthly Magazine, 94*: xxvii.

BUCK, F.D. 1959a. Society: The South London Entomological and Natural History Society: February 12th, 1959. *Entomologist's Monthly Magazine, 95*: 15.

BUCK, F.D. 1959b. Society: The South London Entomological and Natural History Society: September 24th, 1959. *Entomologist's Monthly Magazine, 95*: 39-40.

BUCK, F.D. 1960a. Society: The South London Entomological and Natural History Society: March 24th, 1960. *Entomologist's Monthly Magazine, 96*: 19.

BUCK, F.D. 1960b. Society: The South London Entomological and Natural History Society: April 28th, 1960. *Entomologist's Monthly Magazine, 96*: 23-24.

BUCK, F.D. 1962. A provisional list of the Coleoptera in Wood Walton Fen, Hunts. *Transactions and Proceedings of the South London Entomological and Natural History Society, (1961)*: 93-117.

BUCKLAND, P.C. & JOHNSON, C. 1983. *Curimopsis nigrita* (Palm) (Coleoptera: Byrrhidae) from Thorne Moors, South Yorkshire. *Naturalist, 108*: 153-154.

BUTLER, E.A. 1903. *Hypocoprus* in East Sussex. *Entomologist's Monthly Magazine, 39*: 301.

BUTLER, E.A. 1909. On the *Miarus micros* of British catalogues, together with a table of the British species of the genus. *Entomologist's Monthly Magazine, 45*: 99-102.

CAMPBELL, J.M. 1985. *Oxford Museums: Techical paper no. 8: An atlas of Oxfordshire ladybirds.* Woodstock, Oxfordshire County Council.

CARR, J.W. 1916. *The invertebrate fauna of Nottinghamshire.* Nottingham. J. & H. Bell.

CARTER, A.E.J. 1924. *Pachyta sexmaculata* L. at Pitlochry. *Entomologist's Monthly Magazine, 60*: 138.

CARTER, I.S. 1980. *Osphya bipuncta* (F.) (Col. Melandryidae) in Avon. *Entomologist's Monthly Magazine, 116*: 256.

CHAMPION, G.C. 1891. Description of a new species of *Anaspis* from Scotland, with some remarks on the black species occurring in Britain. *Entomologist's Monthly Magazine, 27*: 104-105.

CHAMPION, G.C. 1897. The *Lema erichsoni*, Suffr., of British collections: synonymical note. *Entomologist's Monthly Magazine, 33*: 135-136.

CHAMPION, G.C. 1922. *Gymnetron squamicolle* Reitt. in Hants and Surrey. *Entomologist's Monthly Magazine, 58*: 277.

CHAMPION, G.C. 1924. Notes on the autumn *Anthonomi*. *Entomologist's Monthly Magazine, 60*: 74-76.

CHAMPION, G.C. 1926. *Prionus coriarius* L. captured by a small bird. *Entomologist's Monthly Magazine, 62*: 219.

CHINERY, M. 1962. *Odontaeus armiger* (Scop.) (Col., Scarabaeidae) in Surrey. *Entomologist's Monthly Magazine, 98*: 79.

CHITTY, A.J. 1904. Collecting (chiefly Coleoptera) in old hedges near Faversham, Kent. *Entomologist's Monthly Magazine, 40*: 100-103.

CLARKE, R.O.S. 1973. Coleoptera, Heteroceridae. *Handbooks for the Identification of British Insects, 5*(2c): 1-15.

CLEMONS, L. 1983. *Gronops inaequalis* Boheman (Col.: Curculionidae): a weevil new to Britain. *Entomologist's Record and Journal of Variation, 95*: 213-215.

COLLIER, M.J. 1988a. 1987 Annual Exhibition. Imperial College, London SW7 - 24th October 1987. Coleoptera. *British Journal of Entomology and Natural History, 1*: 39-40.

COLLIER, M.J. 1988b. *Dyschirius obscurus* Gyll. and *Nebria livida* (L.) (Col. Carabidae) in a Suffolk sand pit. *Entomologist's Monthly Magazine, 124*: 254.

COLLIER, M.J. 1989. *Rhynchaenus testaceus* (Müll.) (Col., Curculionidae) in Norfolk. *Entomologist's Monthly Magazine, 125*: 167.

COLLIER, M.J. 1990. 1989 Annual Exhibition. Imperial College, London SW7 - 28th October 1989. Coleoptera. *British Journal of Entomology and Natural History, 3*: 83-84.

COLLINGWOOD, C.A. 1953. *Trichodes alvearius* F. in England. *Entomologist's Record and Journal of Variation, 65*: 301.

COLLINGWOOD, C.A. 1954. The biology of *Epipolaeus caliginosus* F. (Col., Curculionidae). *Entomologist's Monthly Magazine, 90*: 169-172.

COLLINGWOOD, C.A. 1957. *Myrmechixenus subterraneus* Chev. (Col., Colydiidae) in Derbyshire. *Entomologist's Monthly Magazine, 93*: 142.

COLLINS, J. 1923. Captures of *Rhynchites* in the Oxford district in 1923. *Entomologist's Monthly Magazine, 59*: 202-203.

COOPER, B.A. 1945. *Cryptohypnus sabulicola* Boh. (Col., Elateridae) in North Riding, Yorkshire. *Entomologist's Monthly Magazine, 81*: 133.

COOPER, B.A. 1946. *Dirhagus pygmaeus* F. (Col., Eucnemidae) in North Riding, Yorkshire. *Entomologist's Monthly Magazine, 82*: 16.

COOPER, B.A. 1947. *Cryptohypnus maritimus* Curt. (Col., Elateridae) in North Riding, Yorkshire. *Entomologist's Monthly Magazine, 83*: 199.

COOTER, J. 1966. *Strangalia aurulenta* (F.), (Col., Cerambycidae) in West Sussex. *Entomologist's Monthly Magazine, 102*: 228.

COOTER, J. 1969a. *Trachys pumilus* Ill. (Col., Bupestridae) and *Dasytes niger* (L.) (Col., Melyridae) in West Sussex. *Entomologist's Monthly Magazine, 105*: 84.

COOTER, J. 1969b. *Platycis cosnardi* Chevr. (Col., Lycidae) the third British record. *Entomologist's Monthly Magazine, 105*: 171.

COOTER, J. 1970. *Elater cinnabarinus* Eschscholtz (=*lythropterus* Germar) (Col., Elateridae) in West Sussex. *Entomologist's Monthly Magazine, 106*: 120.

REFERENCES

COOTER, J. 1971. *Taphrorychus bicolor* (Herbst) (Col., Scolytidae) in West Sussex. *Entomologist's Monthly Magazine, 107*: 78.

COOTER, J. 1972. *Cicones variegatus* Hellwig (Col., Colydidae) in West Sussex (at last!). *Entomologist's Monthly Magazine, 108*: 18.

COOTER, J. 1973. *Acritus homeopathicus* Wollaston (Col., Histeridae) in West Sussex. *Entomologist's Monthly Magazine, 109*: 122.

COOTER, J. 1977a. *Orthoperus nitidulus* Allen (Col., Corylophidae) in Richmond Park, Surrey. *Entomologist's Monthly Magazine, 112*: 90.

COOTER, J. 1977b. *Bagous czwalinai* Seidlitz (Col., Curculionidae), a second locality. *Entomologist's Monthly Magazine, 113*: 167.

COOTER, J. 1977c. A good year for Cerambycidae (Col.). *Entomologist's Monthly Magazine, 113*: 240.

COOTER, J. 1977d. *Pediacus depressus* (Herbst., 1797) (Col.: Cucujidae) new to Scotland. *Entomologist's Record and Journal of Variation, 89*: 339-340.

COOTER, J. 1978. *Procraerus tibialis* Bois. & Lac., and other beetles in Moccas Park, Herefordshire. *Entomologist's Record and Journal of Variation, 90*: 24.

COOTER, J. 1980a. A further note on *Pyrrhidium sanguineum* (L.) (Col., Cerambycidae). *Entomologist's Monthly Magazine, 116*: 104.

COOTER, J. 1980b. A note on *Ernopus caucasicus* Lind. (Col., Scolytidae) in Britain. *Entomologist's Monthly Magazine, 116*: 112.

COOTER, J. 1982. Yet another note on *Pyrrhidium sanguineum* (L.) (Col., Cerambycidae). *Entomologist's Monthly Magazine, 118*: 54.

COOTER, J. 1989. An annotated list of beetles collected during the course of a field meeting held in Dorset, June 2nd to 5th, 1989: A Supplement to the Coleopterists Society Newsletter (unpublished).

COOTER, J. 1990a. Three species of Coleoptera new to Dorset. *Entomologist's Gazette, 41*: 31-32.

COOTER, J. 1990b. *Omophlus rufitarsis* (Leske, 1785) (Coleoptera: Alleculidae) in Dorset. *Entomologist's Gazette, 41*: 33-34.

COOTER, J. 1990c. Some uncommon beetles captured at Merthyr Mawr, Glamorganshire. *Entomologist's Monthly Magazine, 126*: 32.

COOTER, J. 1990d. Some beetles from Moccas Park, Herefordshire. *Entomologist's Monthly Magazine, 126*: 70.

COOTER, J. & OWEN, J.A. 1978. Note on the occurrence of *Amara quenseli* (Schoenherr) (Col., Carabidae) in East Inverness-shire. *Entomologist's Monthly Magazine, 114*: 30.

COOTER, J. & WELCH, R.C. 1981. The Coleoptera of Moccas Park, Herefordshire. Unpublished report to the Nature Conservancy Council.

COPESTAKE, D.R. 1990a. 1989 Annual Exhibition. Imperial College, London SW7 - 28 October 1989. Coleoptera. *British Journal of Entomology and Natural History, 3*: 84.

COPESTAKE, D.R. 1990b. 1989 Annual Exhibition. Imperial College, London SW7 - 28 October 1989. Coleoptera. *British Journal of Entomology and Natural History, 3*: 84.

COX, D. 1947. *Lytta vesicatoria* L. (Col., Meloidae) and *Zeugophora flavicollis* Marsh. (Col., Chrysomelidae) in Essex. *Entomologist's Monthly Magazine, 83*: 104.

COX, D. 1948. Food-plants of *Cryptocephalus 6-punctatus* L. (Col., Chrysomelidae). *Entomologist's Monthly Magazine, 84*: 185.

COX, D. 1950. Coleoptera from the Colchester district, N.E.Essex. *Entomologist's Monthly Magazine, 86*: 142-143.

COX, D. 1951. *Lissodema cursor* L. and *L.quadripustulatum* Marsh. (Col., Pithidae) in Essex. *Entomologist's Monthly Magazine, 87*: 271.

COX, L.G. 1921. *Nebria livida* F. at Mundesley, Norfolk. *Entomologist's Monthly Magazine, 57*: 233.

COX, L.G. 1930. *Procas armillatus* F. in abundance near Brighton. *Entomologist's Monthly Magazine, 66*: 231.

CRAWSHAY, L.R. 1903. On the life history of *Drilus flavescens*, Rossi. *Transactions of the Royal Entomological Society of London, 51*: 39-45.

CRIBB, J. 1954. The species of *Plateumaris* and *Donacia* (Col., Chrysomelidae) in Sussex. *Entomologist's Monthly Magazine, 90*: 80.

434

CROSSLEY, R. & NORRIS, A. 1975. *Bembidion humerale* Sturm (Col. Carabidae) new to Britain. *Entomologist's Monthly Magazine, 111*: 59-60.

CROWSON, R.A. 1960. Observations on Scottish Mycetophagidae (Col.). *Entomologist's Monthly Magazine, 96*: 244.

CROWSON, R.A. 1962. Observations on Coleoptera in Scottish oak woods. *Glasgow Naturalist, 18*: 177-195.

CROWSON, R.A. 1963. Observations on British Tetratomidae (Col.), with a key to the larvae. *Entomologist's Monthly Magazine, 99*: 82-86.

CROWSON, R.A. 1970. Coleoptera from East Lothian, including a species of *Philonthus* (Staphylinidae) apparently new to Britain. *Entomologist's Monthly Magazine, 106*: 95-96.

CROWSON, R.A. 1976. Some records of Chrysomelidae (Col.) from Ayrshire. *Entomologist's Monthly Magazine, 112*: 76.

CROWSON, R.A. 1983a. Old woodland beetles at Kindrogan. *Glasgow Naturalist, 20*: 368.

CROWSON, R.A. 1983b. Beetles from Jed Forest. *Glasgow Naturalist, 20*: 368-369.

CROWSON, R.A. 1986. Some records of Coleoptera from Dalmeny Park, West Lothian. *Glasgow Naturalist, 21*: 225-227.

CROWSON, R.A. 1987. Some records of Coleoptera from Dalmeny Park, West Lothian, Scotland. *Entomologist's Monthly Magazine, 123*: 14.

DALTRY, H.W. 1939. *Gymnetron collinum* Gyll. (Col., Curculionidae) in East Kent. *Entomologist's Monthly Magazine, 75*: 59.

DAVIDSON, A. 1956. *Dirhagus pygmaeus* F. (Col., Eucnemidae) new to Scotland. *Entomologist's Monthly Magazine, 92*: 149.

DAVIDSON, W.F. 1954. *Leistus montanus* Steph. *Entomologist's Record and Journal of Variation, 66*: 230-231.

DAVIDSON, W.F. 1960a. Mountain and moorland beetles of the North West. *Changing Scene, 2*: 56-60.

DAVIDSON, W.F. 1960b. *Amara nitida* Sturm. (Coleoptera) in Cumberland. *Entomologist's Gazette, 11*: 36.

DAVIDSON, W.F. 1961. Notes on Cumberland and Westmorland Coleoptera. *Entomologist's Monthly Magazine, 97*: 15-21.

DAVIES, R.D. & BURROWS, J.W. 1952. *Saperda scalaris* L. (Col., Cerambycidae) breeding in Caernarvon. *Entomologist's Monthly Magazine, 88*: 205.

DAY, F.H. 1909. The Coleoptera of Cumberland. *Transactions of the Carlisle Natural History Society*, Four parts plus supplements in volumes 1 to 5.

DAY, F.H. 1930. *Lebia crux-minor* L. in Cumberland. *Entomologist's Monthly Magazine, 66*: 139.

DAY, F.H. 1932. Coleoptera, Hemiptera, etc., in Ashdown Forest. *Entomologist's Monthly Magazine, 68*: 38-39.

DAY, F.H. 1943. *Leistus montanus* Steph. (Col., Carabidae) in Cumberland. *Entomologist's Monthly Magazine, 79*: 251.

DAY, F.H. 1952. The genus *Aphodius* Ill. in Cumberland. *Entomologist's Record and Journal of Variation, 64*: 125-126.

DE LA GARDE, P. 1906a. *Sibinia sodalis*, Germ., and *Apion filirostre*, Kirby, in Devonshire. *Entomologist's Monthly Magazine, 42*: 180.

DE LA GARDE, P. 1906b. Coleoptera in Devon. *Entomologist's Monthly Magazine, 42*: 230-231.

DENISON-ROEBUCK, W. 1900. Insects. Unpublished manuscript for Victoria County History. Westmorland.

DE WORMS, C.G.M. & PAYNE, R.M. 1952. Sectional reports: Entomology. *London Naturalist, 32*: 127-128.

DINNAGE, H. 1945. *Cassida nebulosa* L. (Col., Chrysomelidae) and other Coleoptera in Surrey. *Entomologist's Monthly Magazine, 81*: 239.

DINNAGE, H. 1952. Coleoptera at Horsham, Sussex. *Entomologist's Monthly Magazine, 88*: 179.

DOLLING, W.R. 1974. The first record of *Apion dispar* Germar (Col., Curculionidae) in Britain. *Entomologist's Monthly Magazine, 110*: 181.

DONISTHORPE, H.ST J.K. 1918. *Caenocara subglobosa* Muls., a species of Coleoptera new to Britain. *Entomologist's Monthly Magazine, 54*: 55-56.

REFERENCES

DONISTHORPE, H.ST J.K. 1921. *Baris scolopacea* Germ. in Sussex. *Entomologist's Monthly Magazine, 57*: 153-154.

DONISTHORPE, H.ST J.K. 1924. *Anthonomus cinctus* Kollar in Surrey. *Entomologist's Monthly Magazine, 60*: 138.

DONISTHORPE, H.ST J.K. 1925a. *Dryophthorus corticalis* Pk., a genus and species of Coleoptera new to Britain. *Entomologist's Monthly Magazine, 61*: 182.

DONISTHORPE, H.ST J.K. 1925b. *Agrilus sinuatus* Ol. in Surrey. *Entomologist's Monthly Magazine, 61*: 206.

DONISTHORPE, H.ST J.K. 1925c. *Lissodema kirkae* n.sp. a species of Coleoptera new to science. *Entomologist's Record and Journal of Variation, 37*: 106.

DONISTHORPE, H.ST J.K. 1925d. *Elutella rufipennis*, Stephens, a distinct species. *Entomologist's Record and Journal of Variation, 37*: 124-128.

DONISTHORPE, H.ST J.K. 1926a. A day's collecting in Hainault Forest. *Entomologist's Monthly Magazine, 62*: 39.

DONISTHORPE, H.ST J.K. 1926b. A few notes on some Windsor beetles. *Entomologist's Monthly Magazine, 62*: 263-265.

DONISTHORPE, H.ST J.K. 1928. *Dorcatoma dresdensis* Hbst. and *D.serra* Pz., two new British insects. With notes on the other British species of the genus. *Entomologist's Monthly Magazine, 64*: 196-199.

DONISTHORPE, H.ST J.K. 1930a. A few notes on *Myrmechixenus* (=*Myrmecoxenus*) *subterraneus* Chevr. *Entomologist's Monthly Magazine, 66*: 153-155.

DONISTHORPE, H.ST J.K. 1930b. The British species of *Anaspis* Geoffroy. *Entomologist's Monthly Magazine, 66*: 249-252.

DONISTHORPE, H.ST J.K. 1935. *Anthonomus rufus* Schon. (Curculionidae, Col.), a beetle new to Ireland. *Entomologist's Monthly Magazine, 71*: 101.

DONISTHORPE, H.ST J.K. 1936a. A note on *Langelandia anophthalma* Aubé. *Entomologist's Monthly Magazine, 72*: 229.

DONISTHORPE, H.ST J.K. 1936b. *Agelastica alni* L. : another record. *Entomologist's Monthly Magazine, 72*: 279.

DONISTHORPE, H.ST J.K. 1937. *Gymnetron collinum* Gyll. and *G. linariae* Panz. near Hendon. *Entomologist's Monthly Magazine, 73*: 258.

DONISTHORPE, H.ST J.K. 1938a. Coleoptera. *In: The Victoria history of the county of Cambridgeshire and the Isle of Ely*, ed. by L.F. Salzman, 104-137. London, Oxford University Press.

DONISTHORPE, H.ST J.K. 1938b. *Cassida vittata*, Vill. inland, etc. *Entomologist's Record and Journal of Variation, 50*: 44.

DONISTHORPE, H.ST J.K. 1939a. *A preliminary list of the Coleoptera of Windsor Forest*. London. Nathaniel Lloyd.

DONISTHORPE, H.ST J.K. 1939b. *Cossonus linearis* (Col., Curculionidae); a species of Coleoptera new to Britain. *Entomologist's Monthly Magazine, 75*: 203.

DONISTHORPE, H.ST J.K. 1942a. *Agrilus viridus* L. and *Anthaxia nitidula* L. (Col., Buprestidae) in the New Forest. *Entomologist's Monthly Magazine, 78*: 247.

DONISTHORPE, H.ST J.K. 1942b. *Dorcatoma dresdensis* Hbst. (Col., Anobiidae) in Cambridgeshire. *Entomologist's Record and Journal of Variation, 54*: 105.

DONISTHORPE, H.ST J.K. 1944. *Longitarsus aeruginosus*, Foudr. (Col., Chrysomelidae), a genuine British species. *Entomologist's Record and Journal of Variation, 56*: 93.

DONISTHORPE, H.ST J.K. 1945a. New localities for four rare British beetles. *Entomologist's Monthly Magazine, 81*: 120.

DONISTHORPE, H.ST J.K. 1945b. *Ludius ferrugineus*, L., ab. *occitanicus*, Villers (Col., Elateridae), an aberration new to the British list. *Entomologist's Record and Journal of Variation, 57*: 97.

DONISTHORPE, H.ST J.K. 1948. A new species of *Trachyphloeus* Germ (Col., Curculionidae) from Britain. *Entomologist's Monthly Magazine, 84*: 50.

DONISTHORPE, H.ST J.K. & TOMLIN J.R.LE B. 1934. *Ebaeus abietinus* Abeille (Malachiidae, Col.), a beetle new to Britain. *Entomologist's Monthly Magazine, 70*: 198-199.

DRANE, A.B. 1978. *Cossonus linearis* (F.) and *C. parallelpipedus* (Herbst) (Col., Curculionidae) occurring together in a willow at Wicken Fen Nature Reserve, Cambs. *Entomologist's Monthly Magazine, 114,* £00.

DRANE, A.B. 1985a. A second Northants. locality for *Ernoporus caucasicus* Lindenmann (Col., Scolytidae) and notes on some other beetles. *Entomologist's Monthly Magazine, 121*: 107.

DRANE, A.B. 1985b. *Procraerus tibialis* (Boisd. & Lacord.) (Col., Elateridae) in Northants. *Entomologist's Monthly Magazine, 121*: 166.

DRANE, A.B. 1985c. *Omias mollinus* Boheman (Col., Curculionidae) in Grimsthorpe Park, S. Lincs. *Entomologist's Monthly Magazine, 121*: 184.

DRANE, A.B. 1990. 1989 Annual Exhibition. Imperial College, London SW7 - 28th October 1989. Coleoptera. *British Journal of Entomology and Natural History, 3*: 84-85.

DRUMMOND, D.C. 1952. *Macquartia tenebricosa* ab. *nitida* Meig. (Dipt., Tachinidae) bred from *Chrysolina graminis* L. (Col., Chrysomelidae). *Entomologist's Monthly Magazine, 88*: 46.

DRUMMOND, D.C. 1956. *Hydrothassa hannoveriana* (F.) (Col., Chrysomelidae) in Orkney. *Entomologist's Monthly Magazine, 92*: 368.

DUFFIELD, C.A.W. 1921. A note on *Liparus germanus* L. *Entomologist's Monthly Magazine, 57*: 142-143.

DUFFIELD, C.A.W. 1922. *Anthonomus cinctus* Kollar in Kent. *Entomologist's Monthly Magazine, 58*: 37.

DUFFIELD, C.A.W. 1925. *Odontaeus armiger* Scop. in Kent. *Entomologist's Monthly Magazine, 61*: 225.

DUFFY, E.A.J. 1945. The Coleopterous fauna of the Hants. - Surrey border. *Entomologist's Monthly Magazine, 81*: 169-179.

EAGLES, T.R. 1946. Society: South London Entomological and Natural History Society: August 28th: 1946. *Entomologist's Monthly Magazine, 82*: 255.

EAGLES, T.R. 1948a. Society: South London Entomological and Natural History Society: September 8th, 1948. *Entomologist's Monthly Magazine, 84*: xl.

EAGLES, T.R. 1948b. Societies: South London Entomological and Natural History Society: May 26th, 1948. *Entomologist's Monthly Magazine, 84*: 168.

EAGLES, T.R. 1948c. Society: South London Entomological and Natural History Society; June 9th, 1948. *Entomologist's Monthly Magazine, 84*: 185.

EAGLES, T.R. 1949. Society: South London Entomological and Natural History Society: 24th August, 1949. *Entomologist's Monthly Magazine, 85*: xxxv.

EAGLES, T.R. 1951. Societies; The South London Entomological and Natural History Society, April, 1951. *Entomologist, 84*: 167-168.

EAGLES, T.R. 1952. Society: South London Entomological and Natural History Society: August 27th, 1952. *Entomologist's Monthly Magazine, 88*: 251.

EAGLES, T.R. 1953a. Society: South London Entomological and Natural History Society: May 27th, 1953. *Entomologist's Monthly Magazine, 89*: 235.

EAGLES, T.R. 1953b. Society: South London Entomological and Natural History Society: August 12th, 1953. *Entomologist's Monthly Magazine, 89*: 264.

EAGLES, T.R. 1953c. Society: South London Entomological and Natural History Society: September 23rd, 1953. *Entomologist's Monthly Magazine, 89*: 288.

EAGLES, T.R. 1955a. Society: South London Entomological and Natural History Society: April 27th, 1955. *Entomologist's Monthly Magazine, 91*: 168.

EAGLES, T.R. 1955b. South London Entomological and Natural History Society: August 25th, 1955. *Entomologist's Monthly Magazine, 91*: 288.

EAGLES, T.R. 1955c. South London Entomological and Natural History Society: September 15th, 1955. *Entomologist's Monthly Magazine, 91*: 288.

EASTON, A.M. 1946a. *Apion millum* Bach (Col., Curculionidae) in three counties. *Entomologist's Monthly Magazine, 82*: 24.

EASTON, A.M. 1946b. *Apion lemoroi* Brisout (Col., Curculionidae); a species new to Britain. *Entomologist's Monthly Magazine, 82*: 130-131.

REFERENCES

EASTON, A.M. 1947. The Coleoptera of flood-refuse; a comparison of samples from Surrey and Oxfordshire. *Entomologist's Monthly Magazine, 83*: 113-115.

EASTON, A.M. 1948. The Coleoptera of Bookham Common: Sternoxia, Teredilia, Heteromera, Longicornia. *London Naturalist, 27*: 61-65.

EASTON, A.M. 1965. A further record of *Dasytes niger* (L.) (Col., Dasytidae) in Hampshire, south. *Entomologist's Monthly Magazine, 101*: 30.

ECCLES, T.M. & BOWESTEAD, S. 1987. *Anommatus diecki* Reitter (Coleoptera: Cerylonidae) new to Britain. *Entomologist's Gazette, 38*: 225-227.

EDMONDS, T.H. 1930. *Hypera nigrirostris* var. *ononinis* Stevens of the British list. *Entomologist's Monthly Magazine, 66*: 41.

EDMONDS, T.H. 1934. A new species of *Tachys* (Coleoptera, Carabidae), from the New Forest, new to science. (*Tachys piceus* n. sp.). *Entomologist's Monthly Magazine, 70*: 7-9.

EDWARDS, J. 1893. Fauna and flora of Norfolk. Part XII, Coleoptera. *Transactions of the Norfolk and Norwich Naturalists' Society, 5*: 427-508.

EDWARDS, J. 1910. On the British species of *Smicronyx*, Schonherr. *Entomologist's Monthly Magazine, 46*: 132-135.

EDWARDS, J.E. 1917. On *Rhynchites ophthalmicus* Stephens, with a table of the British species of that genus. *Entomologist's Monthly Magazine, 53*: 22-26.

EELES, W.J. 1961. *Odontaeus armiger* (Scop.) (Col., Scarabaeidae) in South Oxon. *Entomologist's Monthly Magazine, 97*: 236.

ELLIS, E.A. 1943. Miscellaneous observations: Beetle grubs damaging telephone poles. *Transactions of the Norfolk and Norwich Naturalists' Society, 15*: 432-433.

ELSE, G.R. 1970. *Leptura sanguinolenta* L. (Col., Cerambycidae) in Inverness-shire. *Entomologist's Monthly Magazine, 106*: 173.

EYRE, M.D. 1987. A checklist of the weevils of Northumberland and County Durham. *Recording News, No. 9*: 6-12.

EYRE, M.D. & COX, M.L. 1987. The Chrysomelidae (leaf beetles) of Northumberland and County Durham. *Recording News, No. 10*: 14-20.

EYRE, M.D. & SHEPPARD, D.A. 1982. Literature records of Invertebrates from Gibside, County Durham. Unpublished report.

FERGUSSON, A. 1922. *Leistus montanus* Steph. in Arran. *Entomologist's Monthly Magazine, 58*: 250.

FERGUSSON, A. 1935. *Elater tristis* L. in Scotland. *Entomologist's Monthly Magazine, 71*: 18.

FERRY, R.S. 1952. Longicorns from North Hertfordshire. *Entomologist's Record and Journal of Variation, 64*: 29-30.

FERRY, R.S. 1953. In search of *Strangalia aurulenta* Fabricius. *Entomologist's Record and Journal of Variation, 65*: 26-28.

FINCHER, F. 1947. *Lasiorhynchites* (=*Rhynchites*) *ophthalmicus* Steph. (Col., Curculionidae) in Worcestershire. *Entomologist's Monthly Magazine, 83*: 153.

FINCHER, F. 1955. *Bytiscus populi* (L.) in Worcestershire. *Entomologist's Record and Journal of Variation, 67*: 69.

FITTER, A. & SMITH, C.J. 1979. *A wood in Ascam - a study in woodland conservation.* York, Sessions.

FITTON, M.G. 1965. The distribution of *Tropiphorus obtusus* Bons. (Col., Curculionidae). *Bulletin of the Amateur Entomological Society, 24*: 55-56.

FLINT, J.H. 1943a. Longicornia (Col.) in Nottinghamshire. *Entomologist's Monthly Magazine, 79*: 199.

FLINT, J.H. 1943b. *Calosoma inquisitor* L. (Col., Carabidae) in Cumberland. *Entomologist's Monthly Magazine, 79*: 223.

FLINT, J.H. 1946. Species of genus *Anthicus* (Col., Carabidae) in Essex. *Entomologist's Monthly Magazine, 82*: 219.

FLINT, J.H. 1947. Carabidae (Col.) in South Essex. *Entomologist's Monthly Magazine, 83*: 244-245.

FLINT, J.H. 1957. *Aphodius brevis* Er. (Col., Scarabaeidae) in Yorkshire. *Entomologist's Monthly Magazine, 93*: 12.

FLINT, J.H. 1973. *Hydrothassa hannoveriana* (F.) (Col., Chrysomelidae) in Yorkshire and Durham. *Entomologist's Monthly Magazine, 109*: 147.

FLINT, J.H. 1984a. A second Yorkshire record of *Agonum gracilipes* (Duft.) (Col., Carabidae). *Entomologist's Record and Journal of Variation, 96*: 30.

FLINT, J.H. 1984b. Entomological Reports for 1973-83: Coleoptera. Part 1, Carabidae. *Naturalist, 109*: 116-120.

FLINT, J.H. 1986. Entomological reports for 1973-1983 Coleoptera: Part 2, Haliplidae - Scolytidae. *Naturalist, 111*: 25-30.

FLINT, J.H. 1988. Entomological reports for 1984-1986. Coleoptera. *Naturalist, 113*: 69-72.

FORDHAM, W.J. 1942. More Yorkshire beetles. *Naturalist, 67*: 92.

FORSTER, H.W. 1951a. A further record of *Mecinus janthinus* Germ. (Col., Curculionidae). *Entomologist's Monthly Magazine, 87*: 206.

FORSTER, H.W. 1951b. Some interesting beetles from South Essex marshes. *Entomologist's Monthly Magazine, 87*: 223.

FORSTER, H.W. 1952a. *Ischnodes sanguinicollis* Panz. (Col., Elateridae), etc. in Epping Forest, Essex. *Entomologist's Monthly Magazine, 88*: 151.

FORSTER, H.W. 1952b. *Osphya bipunctata* F. (Col., Melandryidae) discovered in a new locality. *Entomologist's Monthly Magazine, 88*: 164.

FORSTER, H.W. 1953. *Acritus homoeopathicus* Woll. (Col., Histeridae) and *Micropeplus tessurula* Ct. (Col., Micropeplidae) in Epping Forest. *Entomologist's Monthly Magazine, 88*: 50.

FOSTER, A.P. 1988. 1987 Annual Exhibition. Imperial College, London SW7 - 24 October 1987. Coleoptera. *British Journal of Entomology and Natural History, 1*: 40.

FOSTER, A.P. 1989. 1988 Annual Exhibition. Imperial College, London SW7 - 19th November 1988. Coleoptera. *British Journal of Entomology and Natural History, 2*: 47-48.

FOSTER, G.N. & YOUNG, R.J. 1978. *Ptinomorphus imperialis* (L.) (Col., Anobiidae) and *Trox sabulosus* (L.) (Col., Trogidae) in Lanarkshire. *Entomologist's Monthly Magazine, 114*: 202.

FOWLER, W.W. 1889. On the species of the genus *Anaspis*, Geoffrey, with description of a new species. *Entomologist's Monthly Magazine, 25*: 331-338.

FOWLER, W.W. 1887. The Coleoptera of the British Islands. Volume 1. London, Reeve.

FOWLER, W.W. 1888. The Coleoptera of the British Islands. Volume 2. London, Reeve.

FOWLER, W.W. 1889. The Coleoptera of the British Islands. Volume 3. London, Reeve.

FOWLER, W.W. 1890. The Coleoptera of the British Islands. Volume 4. London, Reeve.

FOWLER, W.W. 1891. *The Coleoptera of the British Islands*. Volume 5. London, Reeve.

FOWLER, W.W. & DONISTHORPE, H.ST J.K. 1913. *The Coleoptera of the British Islands*. Volume 6. London, Reeve.

FOWLES, A.P. 1989. The Coleoptera of shingle banks on the River Ystwyth, Dyfed. *Entomologist's Record and Journal of Variation, 101*: 209-221.

FOWLES, A.P. & MORGAN, I.K. 1987. The lost world: Discoveries in Dyfed. *Dyfed Invertebrate Group Newsletter*, No. 6: 8-9.

FRASER, F.C. 1949. *Cicindela germanica* L. (Col., Carabidae) on the mainland, Hampshire. *Entomologist's Monthly Magazine, 85*: 126.

FRASER, M.G. 1948. *Saperda scalaris* L. (Col., Cerambycidae) in Cheshire. *Entomologist's Monthly Magazine, 84*: 120.

FRYER, J.C.F. 1929a. *Ceuthorrhynchidius pulvinatus* Gyll. *Entomologist's Monthly Magazine, 65*: 64.

FRYER, J.C.F. 1929b. The food-plant of *Ceuthorrhynchus angulosus* Bohem. *Entomologist's Monthly Magazine, 65*: 65.

FRYER, J.C.F. & FRYER, H.F. 1923a. *Sitona gemellatus* Gyll. in Britain. *Entomologist's Monthly Magazine, 59*: 80-81.

FRYER, J.C.F. & FRYER, H.F. 1923b. *Dibolia cynoglossi* Koch in Cambridgeshire. *Entomologist's Monthly Magazine, 59*: 89.

FRYER, J.C.F. & FRYER, H.F. 1923c. *Chrysolina marginata* Linn. and its food-plant. *Entomologist's Monthly Magazine, 59*: 89.

GARLAND, S.P. 1983. Beetles as primary woodland indicators. *Sorby Record, 21*: 3-38.

REFERENCES

GARLAND, S.P. & LEE, J. 1981. *Nebria nivalis* (Payk.) from Snowdon, Gwynedd. *Entomologist's Monthly Magazine, 117*: 141.

GEORGE, R.S. 1953. *Mecinus janthinus* Germ. (Col., Curculionidae) in Essex. *Entomologist's Monthly Magazine, 89*: 125.

GILBERT, O. 1958. Beetles of south-west Anglesey. *Entomologist's Monthly Magazine, 94*: 133.

GILMOUR, E.F. 1946. Two rare British Carabidae (Col.). (*Dromius quadrisignatus* Dej.; *Bembidion obliquum* Sturm.). *Entomologist's Monthly Magazine, 82*: 44.

GIMINGHAM, C.T. 1930. *Rhizophagus aeneus* Richt. (*coeruleipennis* Sahl.) in Sussex. *Entomologist's Monthly Magazine, 66*: 230.

GODFREY, A. 1989. *Tomoxia bucephala* Costa (Mordellidae), *Bitoma crenata* (F.) (Colydiidae) and other local Coleoptera from Whippendell Wood, Hertfordshire. *Entomologist's Monthly Magazine, 125*: 252.

GRADWELL, G.R. 1953. Prey of *Calosoma clathratus jansoni* Kraatz (Col., Carabidae) in Sutherland. *Entomologist's Monthly Magazine, 89*: 78.

GRANT, M.H. 1888. *Odontaeus mobilicornis* in the Isle of Wight. *Entomologist, 21*: 92.

GREEN, M.E. 1972. *Pyrrhidium sanguineum* (L.) (Col., Cerambycidae) established as an indigenous species. *Entomologist's Monthly Magazine, 108*: 65.

GREENSLADE, P.J.M. 1963a. A concentration of Carabidae (Coleoptera) at Fleet in Dorset. *Entomologist's Monthly Magazine, 99*: 46-48.

GREENSLADE, P.J.M. 1963b. Aggregation in British Carabidae (Coleoptera). *Entomologist's Monthly Magazine, 99*: 202.

GRENSTED, L.W. 1931. *Agonum scitulum* Dej. in Hampshire. *Entomologist's Monthly Magazine, 67*: 142.

GRENSTED, L.W. 1945. *Gnorimus nobilis* L. (Col., Scarabaeidae) in Herefordshire. *Entomologist's Monthly Magazine, 81*: 250.

GRENSTED, L.W. 1949. *Lissodema cursor* Gyll. (Col., Pythidae) in Cheshire. *Entomologist's Monthly Magazine, 85*: 28.

GRIFFITH, C.F. 1947. *Gnorimus nobilis* L. (Col., Scarabaeidae); a recent capture. *Entomologist's Monthly Magazine, 83*: 276.

HAINES, F.H. 1942. *Agrilus viridis* L. (Col., Buprestidae) in the New Forest. *Entomologist's Monthly Magazine, 78*: 204.

HALLETT, H.M. 1951. The Coleoptera of Herefordshire. First supplement. *Transactions of the Woolhope Naturalists' Field Club, (1951)*: 279-82.

HALSTEAD, A.J. 1988. 1987 Annual Exhibition. Imperial College, London SW7 - 24 October 1987. Coleoptera. *British Journal of Entomology and Natural History, 1*:40.

HALSTEAD, A.J. 1989. 1988 Annual Exhibition. Imperial College, London SW7 - 19th November 1988. Coleoptera. *British Journal of Entomology and Natural History, 2*: 48.

HALSTEAD, A.J. 1990. 1989 Annual Exhibition. Imperial College, London SW7 - 28 October 1989. Coleoptera. *British Journal of Entomology and Natural History, 3*: 86.

HALSTEAD, D.G.H. 1963. Coleoptera: Sphaeritidae & Histeridae. *Handbooks for the Identification of British Insects, 4*(10): 1-16.

HAMMOND, P.M. 1959. *Crepidodera impressa* F. (Col., Chrysomelidae) in Essex. *Entomologist's Monthly Magazine, 95*: 250.

HAMMOND, P.M. 1963. New county records for Carabidae (Col.). *Entomologist's Monthly Magazine, 99*: 79.

HAMMOND, P.M. 1968. *Harpalus frolichi* Strum (Col., Carabidae) in Norfolk and Essex. *Entomologist's Monthly Magazine, 104*: 90.

HAMMOND, P.M. 1969. A note on wing-development, etc., in British *Bradycellus* s. str. (Col., Carabidae). *Entomologist's Monthly Magazine, 105*: 155-156.

HAMMOND, P.M. 1971. *Rypobius ruficollis* Jacqu. (Coleoptera, Corylophidae); a genus and species new to Britain. *Entomologist's Gazette, 22*: 241-243.

HAMMOND, P.M. 1979. Beetles in Epping Forest. *in*: *The Wildlife of Epping Forest. Essex Naturalist*, No. 4: 43-60.

HAMMOND, P.M. 1982. *Cymindis macularis* (Fischer v. Waldheim) (Col., Carabidae) - apparently a British species. *Entomologist's Monthly Magazine, 118*: 37-38.

HAMMOND, P.M. *and others*. 1989. Some recent additions to the British insect fauna. *Entomologist's Monthly Magazine, 125*: 95-102.

HARDE, K.W. 1966. *In: Die Kafer Mitteleuropas. Freude, H., Harde, K.W. & Lohse, G.A. Volume 8: 7-94. Krefeld. Goecke & Evers.*

HARDING, P.T. 1978. A bibliography of the occurrence of certain woodland Coleoptera in Britain, with special reference to timber-utilising species associated with old trees in pasture-woodlands. Unpublished Institute of Terrestrial Ecology report to the Nature Conservancy Council. (Institute of Terrestrial Ecology project number 405).

HARDING, P.T. *and others*. 1986. *Provisional distribution maps of Coleoptera: Coccinellidae (Ladybirds)*. Abbots Ripton, Biological Records Centre, Institute of Terrestrial Ecology.

HARDING, P.T. & ROSE, F. 1986. *Pasture-woodlands in lowland Britain*. Monks Wood, Institute of Terrestrial Ecology.

HARDY, J.R. & STANDEN, R. 1917. Notes on local captures of longhorn beetles during 1917. Fourth annual report. *Report. Lancashire and Cheshire Fauna Committee*, No. 84: 226-230.

HARE, D. & JEFFREY, P. 1953. *Strangalia aurulenta* Fab. in Devon. *Entomologist's Record and Journal of Variation, 65*: 301.

HARMAN, T.W. 1977. *Odontaeus armiger* Scop. (Col.: Scarabidae) in Kent. *Entomologist's Record and Journal of Variation, 89*: 231.

HARWOOD, P. 1921. *Anthonomus cinctus* Kollar in Britain. *Entomologist's Monthly Magazine, 57*: 226-227.

HARWOOD, P. 1922. *Pterostichus angustatus* Dufts., and *Anchomenus quadripunctatus* De G., etc. in Kent. *Entomologist's Monthly Magazine, 58*: 249-250.

HARWOOD, P. 1928. *Longitarsus nigerrimus* Gyll. in Dorset. *Entomologist's Monthly Magazine, 64*: 11.

HARWOOD, P. 1929. *Trox perlatus* Goeze in Dorset: an addition to the British coleopterous fauna. *Entomologist's Monthly Magazine, 65*: 171.

HARWOOD, P. 1930. *Myrmecoxenus subterraneus* Cherv., a myrmecophilous beetle new to Britain. *Entomologist's Monthly Magazine, 66*: 153.

HARWOOD, P. 1944. *Elater tristis* L. (Col., Elateridae) near Aviemore, Inverness-shire. *Entomologist's Monthly Magazine, 80*: 17.

HARWOOD, P. 1947. *Cryptocephalus coryli* L. and *C. punctiger* Payk. (Col., Chrysomelidae) in Inverness-shire. *Entomologist's Monthly Magazine, 83*: 88.

HARWOOD, P. 1956. Reappearance of *Diaperis boleti* L. (Col., Tenebrionidae). *Entomologist's Monthly Magazine, 92*: 17.

HENDERSON, C.W. 1944. *Bagous lutosus* Gy. (Col., Curculionidae) and *Cryptocephalus frontalis* Marsh. (Col., Chrysomelidae) in Leicestershire. *Entomologist's Monthly Magazine, 80*: 24.

HENDERSON, J.L. 1951. *Ceuthorrhynchus turbatus* Schultze (Col., Curculionidae), an addition to the British list. *Entomologist's Monthly Magazine, 87*: 309-310.

HENDERSON, J.L. 1957. British records of *Lagria atripes* Mul. et Guil. (Col., Lagriidae). *Entomologist's Monthly Magazine, 93*: 236.

HENDERSON, J.L. 1961. *Chaetocnema aerosa* Letzner (Col., Chrysomelidae) in Britain. *Entomologist's Monthly Magazine, 97*: 259.

HENDERSON, M.K. 1988. 1987 Annual Exhibition. Imperial College, London SW7 - 24th October 1987. Coleoptera. *British Journal of Entomology and Natural History, 1*: 40.

HENDERSON, M.K. 1989. 1988 Annual Exhibition. Imperial College, London SW7 - 19th November 1988. Coleoptera. *British Journal of Entomology and Natural History, 2*: 48.

HIGNETT, J. 1940. *Corymbites angustulus* Kies.; an Elaterid new to the list of British Coleoptera. *Entomologist's Monthly Magazine, 76*: 14.

HOBBY, B.M. 1939. *Odontaeus armiger* Scop. (Col., Scarabaeidae) in Hants. *Entomologist's Monthly Magazine, 75*: 174.

HOBBY, B.M. 1955. *Cryptocephalus primarius* Harold (Col., Chrysomelidae) in Berkshire. *Entomologist's Monthly Magazine, 91*: 173.

REFERENCES

HODGE, P.J. 1977a. *Agonum gracilipes* (Duft.) (Col., Carabidae) in East Sussex. *Entomologist's Monthly Magazine, 113*: 150.

HODGE, P.J. 1977b. *Zabrus tenebrioides* (Goeze) (Col., Carabidae) in East Sussex. *Entomologist's Monthly Magazine, 113*: 153.

HODGE, P.J. 1977c. *Amara famelica* Zimm. and *Agonum sexpunctatum* (L.) (Col., Carabidae) in East Sussex. *Entomologist's Monthly Magazine, 113*: 155.

HODGE, P.J. 1977d. Notes on some species of *Bembidion* Lat. (Col., Carabidae) taken in Sussex. *Entomologist's Monthly Magazine, 113*: 164.

HODGE, P.J. 1977e. *Aulonium trisulcum* (Fourc.) (Col., Colydiidae) and *Ctesias serra* (F.) (Col., Dermestidae) in East Sussex. *Entomologist's Monthly Magazine, 113*: 246.

HODGE, P.J. 1979. *Aphodius villosus* Gyll. (Col., Scarabaeidae) in Sussex, a recent capture. *Entomologist's Monthly Magazine, 115*: 163-164.

HODGE, P.J. 1980. *Ceuthorhynchus pervicax* Weise and *C. pectoralis* Weise in Kent and Sussex. *Entomologist's Monthly Magazine, 116*: 256.

HODGE, P.J. 1983. *Rhynchites olivaceus* Gyll. (Col., Attelabidae) in Sussex. *Entomologist's Monthly Magazine, 119*: 159.

HODGE, P.J. 1987. *Tychius quinquepunctatus* (L.) (Col.,Curculionidae) in South Wales. *Entomologist's Monthly Magazine, 123*: 48.

HODGE, P.J. 1989. *A survey of the insects of the Somerset Moors peat production zones with recommendations for their conservation.* Taunton, Nature Conservancy Council.

HODGE, P.J. 1990. 1989 Annual Exhibition. Imperial College, London SW7 - 28th October 1989. Coleoptera. *British Journal of Entomology and Natural History, 3*: 86-87.

HOLFORD, N.A. 1968. *Dasytes niger* L. (Col., Melyridae) in Savernake Forest, Wiltshire. *Entomologist's Monthly Magazine, 104*: 272.

HOLLAND, W. 1905. *Harpalus honestus* Duft. at Streatley, Berks. *Entomologist's Monthly Magazine, 41*: 255.

HOOPER, M. 1978. Final report to the Nature Conservancy Council, on survey of the Stanford Training Area. (Institute of Terrestrial Ecology project No. 465).

HORSFIELD, D. 1981. *Harpalus quadripunctatus* Dejean (Col., Carabidae) in northern England. *Entomologist's Monthly Magazine, 117*: 124.

HORSFIELD, D. 1987. *Phyllodecta polaris* Schneider (Col., Chrysomelidae) in Sutherland. *Entomologist's Monthly Magazine, 123*: 32.

HORSFIELD, D. 1988. *Nebria nivalis* (Payk.) (Col.) and other carabids from Beinn Spionnaidh, Sutherland and a second record of *N. nivalis* from Ben Hope. *Entomologist's Monthly Magazine, 124*: 199-200.

HORTON, G.A.N. 1980. *Pyrrhidium sanguineum* L. and *Criocephalus rusticus* L. (Col., Longicornia) in Monmouthshire. *Entomologist's Record and Journal of Variation, 92*: 52.

HORTON, G.A.N. 1989. 1988 Annual Exhibition. Imperial College, London SW7 - 19 November 1988. Coleoptera. *British Journal of Entomology and Natural History, 2*: 50-51.

HOUSTON, K. & COULSON, J.C. 1972. Fauna of the Cow Green area of Upper Teesdale. Teesdale Trust, Unpublished report.

HUGGINS, H.C. 1953. Some notes on *Strangalia aurulenta* Fab. *Entomologist's Record and Journal of Variation, 65*: 149-150.

HUNTER, F.A. 1951. Notes on collecting Longicorns in 1951. *Entomologist's Record and Journal of Variation, 63*: 224-225.

HUNTER, F.A. 1953. Collecting notes on Cerambycid Coleoptera, 1952. *Entomologist's Record and Journal of Variation, 65*: 60-63.

HUNTER, F.A. 1955. Notes on the distribution of *Osphya bipunctata* (F.) Col., Melandryidae. *Entomologist's Record and Journal of Variation, 67*: 121-122.

HUNTER, F.A. 1959. Collecting longhorn beetles in 1958. *Entomologist's Record and Journal of Variation, 71*: 122-126.

HUNTER, F.A. 1977. *Ecology of pinewood beetles. In: Native Pinewoods of Scotland.* Ed. by R.J.H.Bunce & J.R.Jeffers, 42-55. Proceedings of Aviemore Symposium, 1975. Cambridge, Institute of Terrestrial Ecology.

IMMS, A.D. 1927. *Ischnodes sanguinicollis* Panz. in Hertfordshire. *Entomologist's Monthly Magazine,* 63: 161.

IMMS, A.D. 1947. *Prionus coriarius* (L.) (Col., Cerambycidae) in Devon. *Entomologist's Monthly Magazine,* 83: 245.

JENNINGS, F.B. 1906. *Procas armillatus,* F., near Dartford. *Entomologist's Monthly Magazine, 42:* 138.

JESSOP, L. 1986. Coleoptera: Scarabaeoidae. *Handbooks for the Identification of British Insects,* 5(11): 1-53.

JOHNSON, C. 1962a. The Scarabaeoid (Coleoptera) fauna of Lancashire and Cheshire and its apparent changes over the last 100 years. *Entomologist,* 95: 153-165.

JOHNSON, C. 1962b. *Rhizophagus parvulus* Payk. (Col. Rhizophagidae): an addition to the British list. *Entomologist's Monthly Magazine,* 98: 231.

JOHNSON, C. 1963a. *Agonum quadripunctatum* (Deg.) (Col., Carabidae) in Scotland. *Entomologist's Monthly Magazine,* 99: 64.

JOHNSON, C. 1963b. *Amara montivaga* St. (Col., Carabidae) in Oxfordshire. *Entomologist's Monthly Magazine,* 99: 128.

JOHNSON, C. 1963c. *Pediacus depressus* Hbst. (Col., Cucujidae) in Kent. *Entomologist's Monthly Magazine,* 99: 209.

JOHNSON, C. 1965. Coleoptera report 1959-64. Thirty-fifth report and report of recorders. *Report. Lancashire and Cheshire Fauna Committee,* 47: 46-55.

JOHNSON, C. 1966a. Taxonomic notes on British Coleoptera: no. 4 - *Simplocaria maculosa* Erichson (Byrrhidae). *Entomologist,* 99: 155-156.

JOHNSON, C. 1966b. Coleoptera: Clambidae. *Handbooks for the Identification of British Insects,* 4(6a): 1-13.

JOHNSON, C. 1967. A sixth British and new county record for *Nebria nivalis* Payk. (Col., Carabidae). *Entomologist's Monthly Magazine,* 103: 72.

JOHNSON, C. 1975a. Synonymic and other notes on British Coleoptera. *Entomologist's Monthly Magazine, 111:* 111-113.

JOHNSON, C. 1975b. Nine species of Coleoptera new to Brtiain. *Entomologist's Monthly Magazine, 111:* 177-183.

JOHNSON, C. 1978. Notes on Byrrhidae (Col.), with special reference to, and a species new to, the British fauna. *Entomologist's Record and Journal of Variation, 90:* 141-147.

JOHNSON, C. 1982. *Phyllobius obtoni* Iniuelaen (Coleoptera, Curculionidae) new to Britain. *Entomologist's Gazette,* 33: 221-222.

JOHNSON, C. 1990. Data on recent captures of Sherwood Forest beetles, based on the Manchester Museum Collection. Unpublished report to Nature Conservancy Council.

JOHNSON, C., ROBINSON, N.A. & STUBBS, A. 1977. Dunham Park - A conservation report on a parkland of high entomological interest. Nature Conservancy Council, unpublished. (CST report no. 5).

JOHNSON, W.F. 1902. *Bembidion argenteolum,* Ahr., at Loch Neagh. *Entomologist's Monthly Magazine,* 38: 218.

JONES, R.A. 1977a. Scarce Cerambycidae (Col.) in West Sussex. *Entomologist's Monthly Magazine, 113:* 154.

JONES, R.A. 1977b. *Mononychus punctumalbum* (Herbst) (Col., Curculionidae) in Wiltshire. *Entomologist's Monthly Magazine, 113:* 246.

JONES, R.A. 1977c. *Ceuthorhynchus unguicularis* C.G. Thomson (Col., Curculionidae) in Sussex. *Entomologist's Monthly Magazine, 113:* 247.

JONES, R.A. 1984. 1983 Annual Exhibition, Chelsea Old Town Hall - 29 October 1983. Exhibit of Coleoptera and Hemiptera. *Proceedings and Transactions of the British Entomological and Natural History Society, 17:* 17.

JONES, R.A. 1987. *Synchita seperanda* Reitter - a third British locality. *Entomologist's Record and Journal of Variation,* 99: 43-44.

JONES, R.A. 1988. 1987 Annual Exhibition. Imperial College, London SW7 - 24th October 1987. Coleoptera. *British Journal of Entomology and Natural History, 1:* 40-41.

JONES-WALTERS, L.M. 1984. *Odontaeus armiger* Scop. (Col., Scarabaeidae) on Martin Down NNR, Hampshire. *Entomologist's Monthly Magazine, 120:* 144.

KEMP, S.W. 1902. Coleoptera caught in Ireland during May and June, 1902. *Entomologist's Monthly Magazine, 38:* 177-178.

REFERENCES

KENDALL, P. 1981. *Bromius obscurus* (L.) in Britain (Col., Chrysomelidae). *Entomologist's Monthly Magazine, 117*: 233-234.

KENWARD, H.K. 1990. A belated record of *Procas granulicollis* Walton (Col., Curculionidae) from Galloway, with a discussion of the British *Procas* spp. *Entomologist's Monthly Magazine, 126*: 21-25. (With a footnote by A.A. Allen.).

KEVAN, D.K. 1949. *Cantharis abdominalis* F. var. *cyanea* Curtis. (Col., Cantharidae), and other beetles, in Ross-shire. *Entomologist's Monthly Magazine, 85*: 263.

KEVAN, D.K. 1955a. *Badister anomalus* (Perris) (Col., Carabidae) new to Britain. *Entomologist's Monthly Magazine, 91*: 6.

KEVAN, D.K. 1955b. The identification of *Badister peltatus* (Pz.), *dilatus* Chaud., and *anomalus* (Perris) (Col., Carabidae, Licinini). *Entomologist's Monthly Magazine, 91*: 207-210.

KEVAN, D.K. 1959. *Ceuthorrhynchus chalybaeus* Germar (Col., Curculionidae) in Scotland and some observations with regard to its status, and to that of its allies *C. timidus* Weise and *C. moguntiacus* Schultz. *Entomologist's Monthly Magazine, 95*: 114-116.

KEVAN, D.K. 1960. Further Scottish records of *Phytonomus diversipunctatus* (Schrank) (= *elongatus* Paykull) and its comparison with allied species; also notes on the identification of *P. rumicus* (L.) and *P. adspersus* (F.) (Col., Curculionidae). *Entomologist's Monthly Magazine, 96*: 35-38.

KEVAN, D.K. 1961. Further observations on the status of *Ceutorrhynchus chalybaeus* Germar (Col., Curculionidae). *Entomologist's Monthly Magazine, 97*: 30-31.

KEVAN, D.K. 1963a. *Sitona brevirostris* Solari (Col., Cuculionidae) new to the British list. *Entomologist's Monthly Magazine, 99*: 39-41.

KEVAN, D.K. 1963b. The identity of *Cassida sanguinolenta* Müller and *C. prasina* Illiger (= *Chloris* Suffrian); *C. denticollis* Suffrian new to the British list (Col., Chrysomelidae). *Entomologist's Monthly Magazine, 99*: 168-174.

KEVAN, D.K. 1964. *Heptaulacus villosus* (Gyll.) (Col., Scarabaeidae) in Fife. *Entomologist's Monthly Magazine, 100*: 90.

KEVAN, D.K. 1967. The British species of the genus *Longitarsus* Latreille (Col., Chrysomelidae). *Entomologist's Monthly Magazine, 103*: 83-110.

KEY, R.S. 1981. *Nebria nivalis* (Payk.), (Col., Carabidae) on Scafell Pike, Cumbria. *Entomologist's Monthly Magazine, 116*: 160.

KEY, R.S. 1983. *Calosoma inquisitor* (L.) (Col., Carabidae) from Radnorshire and Brecknockshire. *Entomologist's Monthly Magazine, 119*: 248.

KEY, R.S. 1990. 1989 Annual Exhibition. Imperial College, London SW7 - 28th October 1989. Coleoptera. *British Journal of Entomology and Natural History, 3*: 87.

KEYS, J.H. 1911. *Barypithes duplicatus*, n.sp., and notes on other members of the genus. *Entomologist's Monthly Magazine, 47*: 128-132.

KEYS, J.H. 1916. *Anchonidium unguiculare* Aubé: A genus and species of Coleoptera new to the British list. *Entomologist's Monthly Magazine, 52*: 112-113.

KEYS, J.H. 1921. *Cathormiocerus attaphilus* Bris.: An addition to the British Coleoptera. *Entomologist's Monthly Magazine, 57*: 100-102.

KEYS, J.H. 1922. Coleoptera at the Lizard, in 1920 and 1921. *Entomologist's Monthly Magazine, 58*: 35-37.

KEYS, J.H. 1923. *Cathormiocerus*, etc., at the Lizard. *Entomologist's Monthly Magazine, 59*: 67-68.

KIDSON-TAYLOR, J. 1906. *Otiorrhynchus morio*, F., v. *ebeninus*, Gyll. in Sutherland. *Entomologist's Monthly Magazine, 42*: 272.

KIRBY, P. & LAMBERT, S.J.J. 1989. 1988 Annual Exhibition. Imperial College, London SW7 - 19th November 1988. Coleoptera. *British Journal of Entomology and Natural History, 2*: 49.

LANE, S.A. 1990. Nationally scarce and uncommon Chrysomelidae (Col.) in Warwickshire (VC 38). *Entomologist's Monthly Magazine, 126*: 7-8.

LAST, H. 1943. *Malthodes crassicornis* Mkl. (Col., Cantharidae) in Surrey. *Entomologist's Monthly Magazine, 79*: 113.

LAST, H. 1946. *Ptomaphagus varicornis* Ross. (Col., Cholevidae) in Surrey. *Entomologist's Monthly Magazine, 82*: 24.

LEE, J. 1980. *Aphodius villosus* Gyll. (Col., Scarabaeidae) in Yorkshire. *Entomologist's Monthly Magazine*, *116*: 84.

LEVEY, B. 1977. Coleoptera. Buprestidae. *Handbooks for the Identification of British Insects*, 5(1b): 1-8.

LINDROTH, C.H. 1960. On *Agonum sahlbergi* Chd. (Col., Carabidae). *Entomologist's Monthly Magazine*, *96*: 44-47.

LINDROTH, C.H. 1971. Taxonomic notes on certain British ground-beetles (Col., Carabidae). *Entomologist's Monthly Magazine*, *107*: 209-223.

LINDROTH, C.H. 1974. Coleoptera: Carabidae. *Handbooks for the Identification of British Insects*, 4(2): 1-148.

LINDROTH, C.H. 1985. *The Carabidae (Coleoptera) of Fennoscandia and Denmark*. Leiden, E.J. Brill. (Fauna Entomologica Acandinavica, Volume 15).

LLOYD, R.W. 1938. *Prionus coriarius* and *Sirex gigas* in Herefordshire. *Entomologist's Monthly Magazine*, *74*: 232.

LLOYD, R.W. 1943. *Molorchus umbellatarum* Schreib. (Col., Cerambycidae) in Herefordshire and Monmouthshire. *Entomologist's Monthly Magazine*, *79*: 201.

LLOYD, R.W. 1944. *Apion sorbi* F. (Col., Curculionidae) in Herefordshire and Monmouthshire. *Entomologist's Monthly Magazine*, *80*: 108.

LLOYD, R.W. 1945. *Molorchus umbellatarum* Shb. (Col., Cerambycidae) in Monmouthshire. *Entomologist's Monthly Magazine*, *81*: 215.

LLOYD, R.W. 1946. *Lissodema cursor* Gyll. (Col., Pythidae) in Herefordshire. *Entomologist's Monthly Magazine*, *82*: 16.

LLOYD, R.W. 1948a. *Carabus clathratus* L. (Col., Carabidae) in the Hebrides and Co. Mayo. *Entomologist's Monthly Magazine*, *84*: 33.

LLOYD, R.W. 1948b. *Apion laevigatum* Payk. (Col., Curculionidae) in Somerset and the Isle of Wight. *Entomologist's Monthly Magazine*, *84*: 283.

LLOYD, R.W. 1949. Coleoptera at Moccas Park, Herefordshire. *Entomologist's Monthly Magazine*, *85*: 22.

LLOYD, R.W. 1951. *Gynandrophthalma affinis* Ill. (Col., Chrysomelidae) rediscovered in Wychwood Forest, Oxon. *Entomologist's Monthly Magazine*, *87*: 286.

LUFF, M.L. (ed.) 1980. *Preliminary atlas of British Carabidae (Coleoptera)*. Abbots Ripton, Biological Records Centre, Institute of Terrestrial Ecology.

LUFF, M.L. 1974. Further records of *Tapinotus sellatus* F. (Col., Curculionidae), with a brief note on its biology. *Entomologist's Monthly Magazine*, *110*: 152.

LUFF, M.L. 1977. Interim maps of *Carabus* species. Newsletter (January) of the ground beetle distribution mapping scheme.

LUFF, M.L. 1978. Interim maps of *Cicindela*, *Calosoma* and *Cychrus* species. Newsletter (July) of the ground beetle distribution mapping scheme.

LUFF, M.L. & WALKER, M. 1981. *Triplax scutellaris* Charp. (Col., Erotylidae) and other interesting Coleoptera recently found in Northumberland. *Entomologist's Monthly Magazine*, *117*: 62.

MACKECHNIE-JARVIS, C. 1967. *Chaetocnema aerosa* Letz. (Col., Chrysomelidae) in Hampshire and Surrey. *Entomologist's Monthly Magazine*, *103*: 208.

MACKECHNIE-JARVIS, C. 1969. *Longitarsus rutilus* Ill. (Col., Chrysomelidae) on Scillonia. *Entomologist's Monthly Magazine*, *105*: 69.

MACKECHNIE-JARVIS, C. 1976. Field Meetings. Savernake Forest - 29th/30th May 1976. *Proceedings and Transactions of the British Entomological and Natural History Society*, *9*: 122.

MAJERUS, M. & KEARNS, P. 1989. *Ladybirds*. Slough. Richmond Publishing. (Naturalists' Handbooks 10).

MAJERUS, M.E.N. 1989. *Cocinella magnifica* (Redtenbacher): a myremcophilous ladybird. *British Journal of Entomology and Natural History*, *2*: 97-106.

MAJERUS, M.E.N., FORGE, H. & WALKER, L. 1990. The geographical distributions of ladybirds in Britain (1984-1989). *British Journal of Entomology and Natural History*, *3*: 153-165.

MAJERUS, M.E.N. & FOWLES, A.P. 1989. The rediscovery of the 5-Spot Ladybird (*Coccinella 5-punctata* L.) (Col., Coccinellidae) in Britain. *Entomologist's Monthly Magazine*, *125*: 177-181.

445

REFERENCES

MARRINER, T.F. 1938. The Cumberland Chrysomelidae. *Entomologist's Record and Journal of Variation, 50*: 63-67.

MARSH, R.J. 1988. *Aeletes atomarius* (Aubé), (Col., Histeridae) in Yorkshire. *Entomologist's Monthly Magazine, 124*: 166.

MASSEE, A.M. 1940a. *Lixus paraplecticus* L. (Col., Curculionidae) in Kent. *Entomologist's Monthly Magazine, 76*: 22.

MASSEE, A.M. 1940b. *Gnorimus nobilis* L. (Col., Scarabaeidae) in Kent. *Entomologist's Monthly Magazine, 76*: 107.

MASSEE, A.M. 1944. *Anisoxya fuscula* Ill. (Col., Melandryidae) and other beetles in Kent and Essex. *Entomologist's Monthly Magazine, 80*: 237.

MASSEE, A.M. 1945a. Abundance of *Labidostomis tridentata* L. (Col., Chrysomelidae) in Kent. *Entomologist's Monthly Magazine, 81*: 164-165.

MASSEE, A.M. 1945b. *Lissodema cursor* Gy. (Col., Pythidae) found at East Malling. *Entomologist's Monthly Magazine, 81*: 231.

MASSEE, A.M. 1946. *Gnorimus nobilis* L. (Col., Scarabaeidae) in London and Kent. *Entomologist's Monthly Magazine, 82*: 236.

MASSEE, A.M. 1947. Adundance of *Cryptocephalus 6-punctatus* L. (Col., Chrysomelidae) in Kent in 1947. *Entomologist's Monthly Magazine, 83*: 211.

MASSEE, A.M. 1949. *Baris scolopacea* Germ. (Col., Curculionidae) at Higham marshes, North Kent. *Entomologist's Monthly Magazine, 85*: 136.

MASSEE, A.M. 1950. *Acritus homoeopathicus* Woll. (Col., Histeridae) recorded in Kent. *Entomologist's Monthly Magazine, 86*: 360.

MASSEE, A.M. 1952. *Dryophilus anobiodes* Chev. (Col., Anobiidae) and other beetles associated with broom in Kent. *Entomologist's Monthly Magazine, 88*: 213.

MASSEE, A.M. 1955. *Longitarsus quadriguttatus* Pont. (Col., Chrysomelidae) found in Kent. *Entomologist's Monthly Magazine, 91*: 285.

MASSEE, A.M. 1956. *Saperda carcharias* L. (Col., Lamiidae) in Kent. *Entomologist's Monthly Magazine, 92*: 30.

MASSEE, A.M. 1958. *Mesosa nebulosa* F. (Col., Lamiidae) recorded in Kent. *Entomologist's Monthly Magazine, 94*: 155.

MASSEE, A.M. 1961. Further records of *Anthonomus cinctus* Redt. (Col., Curculionidae) in Kent. *Entomologist's Monthly Magazine, 97*: 123.

MASSEE, A.M. 1963. *Ceuthorrhynchus moelleri* Thoms. (Col., Curculionidae) recorded in Kent. *Entomologist's Monthly Magazine, 99*: 96.

MASSEE, A.M. 1967. *Rhopalodontus perforatus* (Gyll.) (Col., Cisidae) found at its old haunt in the Black Forest, near Rannock, Perthshire. *Entomologist's Monthly Magazine, 103*: 62.

MAY, A.H. 1933. An old record *Agelastica alni* L. from Essex. *Entomologist's Monthly Magazine, 69*: 190.

McCLENAGHAN, I. 1988. 1987 Annual Exhibition, Imperial College, London SW7 - 24 October 1987. Coleoptera. *British Journal of Entomology and Natural History, 1*: 41.

McCLENAGHAN, I. 1989. 1988 Annual Exhibition, Imperial College, London SW7 - 19 November 1988. Coleoptera. *British Journal of Entomology and Natural History, 2*: 50.

MENDEL, H. 1979. Additions to and comments on the Suffolk list of Anobiidae (Coleoptera). *Transactions of the Suffolk Naturalists' Society, 18*: 94-97.

MENDEL, H. 1980. Notes on Suffolk Carabidae (Coleoptera) including two species new to the county list. *Transactions of the Suffolk Naturalists' Society, 18*: 141-143.

MENDEL, H. 1982. *Hemicoelus nitidus* (Hebst) (Col., Anobiidae) new to Britain. *Entomologist's Monthly Magazine, 118*: 253.

MENDEL, H. 1985. *Trixagus brevicollis* (de Bonvouloir) (Col., Throscidae) in Britain. *Entomologist's Monthly Magazine, 121*: 95.

MENDEL, H. 1988a. *Provisional Atlas of the click beetles of (Coleoptera : Elateroidea) of the British Isles.* Monks Wood, Biological Records Centre, Institute of Terrestrial Ecology.

MENDEL, H. 1988b. *Drypta dentata* (Rossi) (Coleoptera: Carabidae) rediscovered in the Isle of Wight. *British Journal of Entomology and Natural History, 1*: 62.

MENDEL, H. 1989. Saproxylic beetles (Coleoptera) of the Icklingham Plains, an area of Suffolk Breckland with a remarkable dead-wood fauna. *Transactions of the Suffolk Naturalists' Society*, *25*: 23-28.

MENDEL, H. 1990a. The identification of British *Anthicus* Stephens (Coleoptera: Oedmeridae). *Entomologist's Gazette*, 41: 209-211.

MENDEL, H. 1990b. Suffolk click beetles of the Genus *Ampedus* including *A. balteatus* var. *adrastiformis* Reitter (Coleoptera: Elateridae) new to Britain. *Transactions of the Suffolk Naturalists' Society*, *26*: 30-32.

MENDEL, H. & OWEN, J.A. 1990. *Limoniscus violaceus* (Müller) (Col: Elateridae), the violet click beetle in Britain. *Entomologist*, *109*: 43-46.

MENZIES, I.S. 1946. *Agrilus sinuatus* Oliv. (Col., Buprestidae) in Surrey. *Entomologist's Monthly Magazine*, *82*: 44.

MENZIES, I.S. 1954. *Agrilus sinuatus* (Ol.) (Col., Buprestidae): some further Surrey records. *Entomologist's Monthly Magazine*, *90*: 102.

MENZIES, I.S. 1990. 1989 Annual Exhibition. Imperial College, London SW7 - 28th October 1989. Coleoptera. *British Journal of Entomology and Natural History*, *3*: 87-88.

MENZIES, I.S., NASH, D.R. & OWEN, J.A. 1991. *Colydium elongatum* (Fabricius) (Col. Colydiidae) in Wiltshire, Berkshire and Surrey. *Entomologist's Record and Journal of Variation*, *103*: 61-62.

MILES, P.M. 1942. *Copris lunaris* L. (Col., Scarabaeidae) in the Oxford district. *Entomologist's Monthly Magazine*, *78*: 240.

MITCHELL, A.V. 1927. *Strangalia aurulenta* F. in the Plymouth district. *Entomologist's Monthly Magazine*, *63*: 40.

MITFORD, R.S. 1923. *Lixus algirus*, L., at Fairlight. *Entomologist's Record and Journal of Variation*, *35*: 158.

MOORE, B.P. 1958. A further record for *Pediacus depressus* Herbst. (Col., Cucujidae). *Entomologist's Monthly Magazine*, *94*: 92.

MOORE, D. 1989. 1988 Annual Exhibition. Imperial College, London SW7 - 19 November 1988. Coleoptera. *British Journal of Entomology and Natural History*, *2*: 50.

MORGAN, I.K. 1988. Coleoptera. A summary of interesting beetle records from Carmanthenshire in 1987. *Dyfed Invertebrate Group Newsletter*, No. 9: 14-16.

MORGAN, I.K. 1989. Carmarthenshire Coleoptera records 1988. *Dyfed Invertebrate Group Newsletter*, No. 10: 13-16.

MORGAN, I.K. 1990. Coleoptera. Carmarthenshire Coleoptera records. *Dyfed Invertebrate Group Newsletter* No.16: 14-16.

MORGAN, M.J. 1980. *Saperda scalaris* (L.) (Col., Cerambycidae) in North Wales. *Entomologist's Monthly Magazine*, *116*: 221.

MORGAN, M.J. 1984. Chrysomelidae (Coleoptera) in North Wales. *Entomologist's Monthly Magazine*, *120*: 97.

MORLEY, C. 1904. The Coleoptera of Norfolk and Suffolk. *Transactions of the Norfolk and Norwich Naturalists' Society*, *7*: 706-721.

MORLEY, C. 1941. *Apion armatum* Gerst. (Col., Curculionidae), new to Britain, in the New Forest. *Entomologist's Monthly Magazine*, *77*: 133-134.

MORRIS, M.G. 1960. New records of *Mecinus janthinus* Germ. (Col., Curculionidae). *Entomologist's Monthly Magazine*, *96*: 191.

MORRIS, M.G. 1962. Further records of *Apion semivittatum* Gyll. (Col., Curculionidae) in Kent. *Entomologist's Monthly Magazine*, *98*: 12.

MORRIS, M.G. 1964. Preliminary notes on the biology of *Anthonomus chevrolati* Desbr. (Col., Curculionidae). *Entomologist's Monthly Magazine*, *98*: 95-96.

MORRIS, M.G. 1965a. *Otiorrhynchus ligustici* (L.) (Col., Curculionidae) in Shropshire, with notes on its recorded distribution in Britain. *Entomologist's Monthly Magazine*, *101*: 169-171.

MORRIS, M.G. 1965b. *Ceutorhynchus unguicularis* C.G.Thompson (Col., Curculionidae) new to the British Isles, from the Suffolk Breckland and the Burren, Co. Clare. *Entomologist's Monthly Magazine*, *101*: 279-286.

MORRIS, M.G. 1966. *Anthonomus rufus* Gyll. (Col., Curculionidae) in Lancashire. *Entomologist's Monthly Magazine*, *102*: 220.

MORRIS, M.G. 1968. *Ceuthorhynchus unguicularis* Thompson (Col., Curculionidae) in Wiltshire and Suffolk. *Entomologist's Monthly Magazine*, *104*: 45.

447

REFERENCES

MORRIS, M.G. 1969. Notes on the life-history of *Dorytomus hirtipennis* Bedel (Col., Curculionidae). *Entomologist's Monthly Magazine, 105*: 207-209.

MORRIS, M.G. 1970. *Phyllodecta polaris* Schneider (Col., Chrysomelidae) new to the British Isles from Wester Ross and Inverness-shire, Scotland. *Entomologist's Monthly Magazine, 106*: 48-53.

MORRIS, M.G. 1972. *Otiorhynchus auropunctatus* Gyll. (Col., Curculionidae) in Wester Ross, Scotland. *Entomologist's Monthly Magazine*, : 108: 45.

MORRIS, M.G. 1973a. *Rhynchaenus decoratus* Germar (Col., Curculionidae) in North Hampshire. *Entomologist's Monthly Magazine, 109*: 161.

MORRIS, M.G. 1973b. *Pyrrhidium sanguineum* (L.) (Col., Cerambycidae) as a breeding species in Britain. *Entomologist's Monthly Magazine, 109*: 163.

MORRIS, M.G. 1974a. A preliminary account of the weevils of Wiltshire (Coleoptera, Curculionoidea). *Wiltshire Archaeology and Natural History Magazine, 69*: 30-38.

MORRIS, M.G. 1974b. Wiltshire Coleoptera, with particular reference to *Limobius borealis* (Payk.) (Curculionidae), and a record from Gloucestershire. *Entomologist's Monthly Magazine, 110*: 222.

MORRIS, M.G. 1975. A note on the habits of *Rhinoncus albicinctus* Gyll. (Col., Curculionidae). *Entomologist's Monthly Magazine, 111*: 14.

MORRIS, M.G. 1976a. The British species of *Anthonomus* Germar (Col., Curculionidae). *Entomologist's Monthly Magazine, 112*: 19-40.

MORRIS, M.G. 1976b. Proceedings: 8th January 1976. *Proceedings and Transactions of the British Entomological and Natural History Society, 9*: 44.

MORRIS, M.G. 1978a. *Polydrusus sericeus* (Schaller) (Col.: Curculionidae) in South Wiltshire. *Entomologist's Record and Journal of Variation, 90*: 22-23.

MORRIS, M.G. 1978b. *Polydrusus sericeus* (Schaller) (Col.: Curculionidae): an additional note. *Entomologist's Record and Journal of Variation, 90*: 55.

MORRIS, M.G. 1982a. *Hypera meles* (F.) (Col., Curculionidae) in Dorset. *Entomologist, Monthly Magazine 118. 206.*

MORRIS, M.G. 1982b. *Tapinotus sellatus* (F.) (Col., Curculionidae) in Dorset. *Entomologist's Monthly Magazine, 118*: 100.

MORRIS, M.G. 1982c. *Apion difficile* (Col.: Apionidae) in Dorset and Worcestershire. *Entomologist's Monthly Magazine, 118*: 246.

MORRIS, M.G. 1983. Abundance of *Larinus planus* (F.) and *Rhinocyllus conicus* (Frölich) (Col.: Curculionidae) in Dorset and adjacent counties. *Entomologist's Monthly Magazine, 119*: 42.

MORRIS, M.G. 1986. *Tropideres niveirostris* (Col.: Anthribidae) in Dorset. *Entomologist's Monthly Magazine, 122*: 110.

MORRIS, M.G. 1987. *Rhynchaenus pseudostigma* Temperé in Britain - a preliminary note. *Coleopterists Newsletter, 29*: 5.

MORRIS, M.G. 1989. Some recent records of *Cassida nebulosa* L. (Col., Chrysomelidae). *Entomologist's Monthly Magazine, 125*: 168.

MORRIS, M.G. 1990. Orthocerus weevils. Coleoptera Curculionoidea (Nemonychidae, Anthribidae, Urodontidae, Attelabidae and Apionidae). *Handbooks for the Identification of British Insects, 5*(16): 1-108.

MORRIS, M.G. & COLLIER, M.J. 1989. *Ceutorhynchus querceti* (Gyll.) (Col., Curculionidae) in West Norfolk. *Entomologist's Monthly Magazine, 125*: 128.

MORRIS, M.G. & RISPIN, W.E. 1988. A beetle fauna of oolitic limestone grassland, and the responses of species to conservation management by different cutting regimes. *Biological Conservation, 43*: 87-105.

MOSELEY, K.A. 1979. *Nebria nivalis* (Payk.) and *Notiophilus aestuans* (Mots.) (Col., Carabidae) and other Coleoptera from the Isle of Skye. *Entomologist's Monthly Magazine, 114*: 221-222.

MURPHY, J.E. 1918. Re-occurrence of *Anchomenus* (*Agonum*) *sahlberghi* Chaud. in Scotland. *Entomologist's Monthly Magazine, 54*: 198-199.

NASH, D.R. 1971. Some Coleoptera found inhabiting the same beech stump in the New Forest, Hants. *Entomologist's Monthly Magazine, 107*: 191.

NASH, D.R. 1974a. *Dorcatoma serra* Pz. (Col., Anobiidae) : a notable addition to the Suffolk list. *Entomologist's Monthly Magazine, 110*: 240.

NASH, D.R. 1974b. A melanic specimen of *Harpalus* (*Ophonus*) *schaubergerianus* Puel (*rufibarbis* auctt.) (Col., Carabidae) from Suffolk. *Entomologist's Monthly Magazine, 110*: 252.

NASH, D.R. 1975. *Haplocnemus nigricornis* (F.) (Col., Dasytidae) in East Suffolk. *Entomologist's Monthly Magazine, 111*: 48.

NASH, D.R. 1977. *Crepidodera impressa* (F.) (Col., Chrysomelidae) in Suffolk. *Entomologist's Monthly Magazine, 113*: 94.

NASH, D.R. 1978. *Cossonus linearis* (F.) and *C.parallelepipedus* (Herbst) (Col., Curculionidae) inhabiting the same stump in the Suffolk Breck. *Entomologist's Monthly Magazine, 114*: 89.

NASH, D.R. 1979a. *Selatosomus bipustulatus* (Linnaeus) (Col., Elateridae) in Wiltshire and Suffolk. *Entomologist's Record and Journal of Variation, 91*: 135.

NASH, D.R. 1979b. *Bradycellus csikii* Laczó (Col., Carabidae) discovered in Suffolk. *Entomologist's Record and Journal of Variation, 91*: 279-280.

NASH, D.R. 1981. *Coccinella distincta* Faldermann (Col., Coccinellidae) in Wiltshire. *Entomologist's Monthly Magazine, 117*: 214.

NASH, D.R. 1982a. *Pityogenes trepanatus* (Nördl.) (Col., Scolytidae) new to Suffolk. *Entomologist's Monthly Magazine, 118*: 54.

NASH, D.R. 1982b. *Agonum gacilipes* (Dufs.) (Col., Carabidae) in Suffolk. *Entomologist's Monthly Magazine, 118*: 124.

NASH, D.R. 1982c. *Epierus comptus* (Erichson) (Col., Histeridae) new to Britain. *Entomologist's Record and Journal of Variation, 94*: 165-167.

NASH, D.R. 1990. *Trinodes hirtus* (F.) (Col.: Dermestidae) rediscovered in Suffolk. *Entomologist's Record and Journal of Variation, 102*: 186.

NATURE CONSERVANCY COUNCIL, 1988. 1987 Annual Exhibition. Imperial College, London SW7 - 24th October 1987. Coleoptera. *British Journal of Entomology and Natural History, 1*: 41.

NATURE CONSERVANCY COUNCIL, 1990. 1989 Annual Exhibition. Imperial College, London SW7 - 28th October 1989. Coleoptera. *British Journal of Entomology and Natural History, 3*: 88.

NELSON, J.M. 1978. Observations on the biology of *Chrysolina marginata* (L.) (Coleoptera: Chrysomelidae). *Entomologist's Gazette, 29*: 237-243.

NELSON, J.M. 1983. Insects from gravel patches on the shore line of two contrasting Scottish lochs. *Entomologist's Gazette, 34*: 133-134.

NEWBERRY, E.A. 1913. *Trachyphlaeus digitalis*, Gyll., An addition to the British list of Coleoptera. *Entomologist's Monthly Magazine, 49*: 126-127.

NEWBERY, E.A. 1902. A revision of the British species of *Bagous*, Schoen. *Entomologist's Record and Journal of Variation, 14*: 149-156.

NICHOLSON, G.W. 1916. *Lycoperdina succincta* L. in Suffolk. *Entomologist's Monthly Magazine, 52*: 253-254.

NICHOLSON, G.W. 1921. *Cryptocephalus biguttatus* Scop. on *Erica tetralix*. *Entomologist's Monthly Magazine, 57*: 36-37.

NICHOLSON, G.W. 1930. *Anthonomus chevrolati* Desbr. in Sussex. *Entomologist's Monthly Magazine, 66*: 132.

NICHOLSON, G.W. 1931. *Tropideres sepicola* F. and *Tomoxia biguttata* Gyll. in West Sussex. *Entomologist's Monthly Magazine, 67*: 237.

ONSLOW, N. 1989. A second Kent county record for *Tomoxia bucephala* Costa (Col.: Mordellidae). *Entomologist's Record and Journal of Variation, 101*: 159.

ORTON, P.D. 1988. Some interesting Coleoptera from Upper Strathglass, East Inverness-shire. *Entomologist's Monthly Magazine, 124*: 179.

OSBORNE, P.J. 1951. *Elater elongatulus* F. (Col., Elateridae) in Buckinghamshire. *Entomologist's Monthly Magazine, 87*: 214.

OSBORNE, P.J. 1953. *Calosoma inquisitor* L. (Col., Carabidae) in Oxfordshire. *Entomologist's Monthly Magazine, 89*: 250.

OSBORNE, P.J. 1957. *Agrilus sinuatus* (Ol.) (Col., Buprestidae) in Oxfordshire. *Entomologist's Monthly Magazine, 93*: 273.

OWEN, J.A. 1982. *Anaspis schilskyana* Csiki (Col., Scraptiidae) at Moccas Park, Hereford. *Entomologist's Monthly Magazine, 118*: 68.

REFERENCES

OWEN, J.A. 1983a. More about *Phyllodecta polaris* Schneider (Col., Chrysomelidae) in Britain. *Entomologist's Monthly Magazine, 119*: 191.

OWEN, J.A. 1983b. *Dryopthorus corticalis* Payk. (Col., Curculionidae) struggles to survive at Windsor. *Entomologist's Monthly Magazine, 119*: 224.

OWEN, J.A. 1984. *Bembidion virens* Gyll. (Col., Carabidae) in Easter Ross. *Entomologist's Monthly Magazine, 120*: 258.

OWEN, J.A. 1988a. 1987 Annual Exhibition. Imperial College, London - 24th October 1987. Coleoptera. *British Journal of Entomology and Natural History, 1*: 41-43.

OWEN, J.A. 1988b. *Phalacrus brunnipes* Brisout (Col., Phalacridae) in Scotland. *Entomologist's Monthly Magazine, 124*: 82.

OWEN, J.A. 1988c. A note on the life history of *Phyllodecta polaris* Schneider (Col.: Chrysomelidae). *Entomologist's Record and Journal of Variation, 100*: 90-91.

OWEN, J.A. 1988d. *Anaspis bohemica* Schilsky (Col.: Scraptiidae) at Coylumbridge. *Entomologist's Record and Journal of Variation, 100*: 191-192.

OWEN, J.A. 1989a. 1988 Annual Exhibition. Imperial College, London SW7 - 19th November 1988. Coleoptera. *British Journal of Entomology and Natural History, 2*: 51-53.

OWEN, J.A. 1989b. *Cyphon kongsbergensis* Munster (Col.: Scirtidae) in Wester Ross, Scotland. *Entomologist's Monthly Magazine, 125*: 156.

OWEN, J.A. 1989c. *Baris laticollis* (Marsham) (Col., Curculionidae) as a pest of cultivated cabbages. *Entomologist's Record and Journal of Variation, 101*: 49-50.

OWEN, J.A. 1990a. 1989 Annual Exhibition, Imperial College, London SW7 - 28 October 1989. Coleoptera. *British Journal of Entomology and Natural History, 3*: 88-90.

OWEN, J.A. 1990b. *Mordellistena nanuloides* Ermisch (Col.: Mordellidae) from the Isle of Grain, Kent. *Entomologist's Record and Journal of Variation, 102*: 24.

OWEN, J.A. 1990c. *Hemicoelus nitidus* (Herbst.) (Col.: Anobiidae) at Windsor. *Entomologist's Record and Journal of Variation, 102*: 274.

OWEN, J.A. & CARTER, I.S. 1988. Further Scottish records for *Cis dentatus* Mellié (Col.:Cisidae). *Entomologist's Record and Journal of Variation, 100*: 188.

PARMENTER, L. 1956. Society: London Natural History Society. *Entomologist's Monthly Magazine, 92*: 198.

PARRY, J.A. 1962. The genus *Apion* Herbst. and some other notable weevils in East Kent. *Entomologist's Record and Journal of Variation, 74*: 267-271.

PARRY, J.A. 1975. *Dyschirius angustatus* Ahrems (Col., Carabidae) in East Sussex. *Entomologist's Monthly Magazine, 111*: 160.

PARRY, J.A. 1978a. *Perileptus areolatus* (Creutz.) and *Thalassophilus longicornis* (Sturm) (Col., Carabidae) in North Wales. *Entomologist's Monthly Magazine, 114*: 215-216.

PARRY, J.A. 1978b. *Bradycellus distinctus* Dejean (Coleoptera: Carabidae) in Kent and Sussex. *Entomologist's Record and Journal of Variation, 90*: 305-306.

PARRY, J.A. 1979. Notes on the Donaciini (Col.: Chrysomelidae), with a list of recent East Kentish localities known to the author. *Entomologist's Record and Journal of Variation, 91*: 323-326.

PARRY, J.A. 1980a. *Bledius annae* Sharp (Col., Staphylinidae) in Kent. *Entomologist's Monthly Magazine, 116*: 197.

PARRY, J.A. 1980b. Notes on the Donaciini (Col: Chrysomelidae), with a list of recent Kentish localities known to the author. *Entomologist's Record and Journal of Variation, 92*: 9-12. (Concluded from volume 91: 323-326.).

PARRY, J.A. 1981. *Rhyncoenus populi* (F.) (Col., Curculionidae) new to Britain. *Entomologist's Monthly Magazine, 117*: 253.

PARRY, J.A. 1982. A weevil new to Britain: *Apion intermedium* Eppelsheim (Col., Curculionidae) in Kent. *Entomologist's Monthly Magazine, 118*: 227-229.

PARRY, J.A. 1983. *Anthicus scoticus* Rye (Col., Anthicidae) new to Kent. *Entomologist's Monthly Magazine, 119*: 38.

PARSONS, M.S. 1988. 1987 Annual Exhibition. Imperial College, London SW7 - 24th October 1987. Coleoptera. *British Journal of Entomology and Natural History, 1*: 43.

450

PARSONS, M.S. 1989. *Hister quadrimaculatus* Linnaeus on Dungeness, Kent. *British Journal of Entomology and Natural History*, 2: 100.

PAVETT, P.M. 1987. *Thermal jamming in tiger beetles* of the Amateur Entomological Society, 46: 159.

PAVETT, P.M. 1988. Some interesting beetle records from Carmarthenshire. *Dyfed Invertebrate Group Newsletter*, No. 10: 9.

PAYNE, K.G. 1977. *Synaptus filiformis* (F.) and *Cardiophorus erichsoni* du Buysson (Col., Elateridae) in South-west England. *Entomologist's Monthly Magazine*, 113: 206.

PEACOCK, E.R. 1977. Coleoptera: Rhizophagidae. *Handbooks for the Identification of British Insects*, 5(5a): 1-20.

PERKINS, R.C.L. 1926. Coleoptera on burnt areas in S. Devon and a note on *Strangalia aurulenta*. *Entomologist's Monthly Magazine*, 62: 288.

PERRING, F.H. & FARRELL, L. 1983. *British Red Data Books: 1. Vascular plants*. 2nd ed. Lincoln, Royal Society for Nature Conservation.

PHILP, E.G. 1965. Some additions to the list of Kentish Carabidae. *Entomologist's Monthly Magazine*, 101: 121.

PHILP, E.G. 1990. *Scaphisoma assimile* Erichson (Col: Scaphidiidae) in Kent. *Entomologist's Record and Journal of Variation*, 102: 116.

PICKARD-CAMBRIDGE, A.W. 1946. Entomology on the Sussex-Surrey border and in Hampshire. *Entomologist's Monthly Magazine*, 82: 199-200.

PLANT, C.W. 1985. *Apion semivittatum* Gyllenhal (Coleoptera: Apionidae) in South Essex. *Entomologist's Record and Journal of Variation*, 97: 25-26.

PLANT, C.W. & DRANE, A.B. 1988. A review of the records of *Acupalpus elegans* Dejean (Coleoptera: Carabidae) in Britain, with a note on its seperation from *Acupalpus dorsalis* (F.). *Entomologist's Gazette*, 39: 227-232.

PONTIN, A.J. 1959. Some records of predators and parasites adapted to attack aphids attended by ants. *Entomologist's Monthly Magazine*, 95: 154.

POPE, R.D. 1969. A preliminary survey of the Coleoptera of Redgrave and Lopham Fens. *Transactions of the Suffolk Naturalists' Society*, 14: 25-40 and supplement 14: 189-207.

POPE, R.D. 1973. The species of *Scymnus* (S.Str.), *Nephus* (Tullus) and *Nephus* (Col., Coccinellidae) occurring in the British Isles. *Entomologist's Monthly Magazine*, 109: 3-39.

POPE, R.D. 1988. A rare spider beetle (Col., Ptinidae) new to West Sussex. *Entomologist's Monthly Magazine*, 124: 76.

PORTER, D.A. 1987. *Platycis cosnardi* (Chevrolat) (Col., Lycidae) in West Sussex. *Entomologist's Monthly Magazine*, 123: 106.

PORTER, D.A. 1989. 1988 Annual Exhibition. Imperial College, London SW7 - 19 November 1988. Coleoptera. *British Journal of Entomology and Natural History*, 2: 53.

PRANCE, D.A. 1985. *Athous subfuscus* (Müller) (Col., Elateridae) in southern England. *Entomologist's Record and Journal of Variation*, 97: 96.

PRANCE, D.A. 1988. The Surrey record of *Triplax lacordairii* Crotch (Col.: Erotylidae). *Entomologist's Record and Journal of Variation*, 100: 237.

READ, R.W.J. 1980. Two interesting species of Curculionidae (Coleoptera) from west Cumbria. *Entomologist's Gazette*, 31: 162-164.

READ, R.W.J. 1981. *Notaris scirpi* (F.) (Col.: Curculionidae) in Cumbria with notes on three other species of the genus. (*Notaris acridulus* (Linnaeus); *Notaris aethiops* (Fabricius); *Notaris bimaculatus* Fabricius). *Entomologist's Record and Journal of Variation*, 93: 73-74.

READ, R.W.J. 1985. *Magdalis duplicata* Germar (Col., Curculionidae) new to England from Cumbria. *Entomologist's Monthly Magazine*, 121: 88.

READ, R.W.J. 1988. Records of local and uncommon Chrysomelidae (Coleoptera) from Cumbria. *Entomologist's Record and Journal of Variation*, 100: 99-100.

READ, R.W.J. 1989a. *Procas armillatus* (F.) (Curclionidae) from Cumbria. *Coleopterists Newsletter*, 35: 7.

READ, R.W.J. 1989b. Preliminary list of Curculionoidea from Cumbria, including VC 70 Cumberland. *Coleopterists Newsletter*, No. 36: 3-12.

REFERENCES

READ, R.W.J. 1989c. *Ceutorhynchus euphorbiae* Brisout (Col.; Curculionidae) in West Cumbria. *Entomologist's Monthly Magazine, 125*: 117.

REID, C.A.M. 1985. Distribution of *Bembidion schueppeli* Dejean (Coleoptera, Carabidae) in the British Isles. *Entomologist's Gazette, 36*: 197-200.

RICHARDS, O.W. 1927. Capture of *Cryptocephalus primarius* Harold in Berkshire. *Entomologist's Monthly Magazine, 63*: 161.

ROBERTS, M. 1959. *Prionus coriarius* L. (Col., Prionidae) in Hertfordshire. *Entomologist's Monthly Magazine, 95*: 108.

ROCHE, P.J.L. 1943. *Calosoma inquisitor* L. (Col., Carabidae) in Lancashire. *Entomologist's Monthly Magazine, 79*: 143.

ROCHE, P.J.L. 1944. Further additions to the Beds. list of Coleoptera. *Entomologist's Monthly Magazine, 80*: 30.

ROEBUCK, A. & BROADBENT, L. 1945. Some notes on *Cryptohypnus quadripustulatus* F. (Col., Elateridae) on grass fields in the Midlands. *Entomologist's Monthly Magazine, 81*: 8.

RUSSELL, H.M. 1951. *Macroplea appendiculata* Panz. (Col., Chrysomelidae) in Yorkshire. *Entomologist's Monthly Magazine, 87*: 146.

RUSSELL, H.M. 1953. *Saprinus virescens* Pk. (Col., Histeridae) in Yorkshire. *Entomologist's Monthly Magazine, 89*: 213.

SANKEY, J. 1957. *Platydema violaceum* (F.) (Col., Tenebrionidae) in Surrey. *Entomologist's Monthly Magazine, 93*: 278.

SANKEY, J.H.P. 1950. *Odontaeus armiger* (Scop.) (Col., Scarabaeidae) in Surrey. *Entomologist's Monthly Magazine, 86*: 264.

SANKEY, J.H.P. 1956. *Diaperis boleti* (L.) (Col., Tenebrionidae) in Hants. *Entomologist's Monthly Magazine, 92*: 405.

SAUNDERS, C.J. 1936a. A note on *Odontaeus armiger* Scop. *Entomologist's Monthly Magazine, 72*: 179.

SAUNDERS, C.J. 1936b. *Lissodema cursor* Gyll. in Sussex. *Entomologist's Monthly Magazine, 72*: 219.

SHARP, D. 1912a. *Strophosomus curvipes*: A coleopteran new to Britain. *Entomologist's Monthly Magazine, 48*: 150-151.

SHARP, D. 1912b. Notes on the British species of *Ophonus*. *Entomologist's Monthly Magazine, 48*: 181-185.

SHARP, D. 1913. *Bradycellus distinctus*, Dej., in England. *Entomologist's Monthly Magazine, 49*: 54.

SHARP, D. 1917a. Studies in Rhynchophora. *Entomologist's Monthly Magazine, 53*: 26-32.

SHARP, D. 1917b. Studies in Rhynchophora. *Entomologist's Monthly Magazine, 53*: 100-108.

SHARPE, J.S. 1946. A few Coleoptera of western Merionethshire. *Entomologist's Monthly Magazine, 82*: 203-205.

SHAW, H.K.A. 1949. *Chrysolina marginata* L. (Col., Chrysomelidae) near London. *Entomologist's Monthly Magazine, 85*: 168.

SHAW, S. 1954. Insects in 1950-1951, additions to the Lancashire and Cheshire lists. 30th report and report of the recorders. *Report. Lancashire and Cheshire Fauna Committee, 30*: 57-65.

SHIRT, D.B., ed. 1987. *British Red Data Books: 2. Insects*. Peterborough, Nature Conservancy Council.

SIDE, K.C. 1957. *Pterostichus (Feronia) kugelanni* (Panz.) (Col., Carabidae) new to East Sussex. *Entomologist's Monthly Magazine, 93*: 257.

SINCLAIR, M. 1985. *Coeliodes nigritarsis* (Hartmann) (Col., Curculionidae) in S.-W. Scotland. *Entomologist's Monthly Magazine, 121*: 61.

SKIDMORE, P. 1962. *Triplax lacordairii* Crotch in the New Forest with notes on other British species of *Triplax* (Col., Erotylidae). *Entomologist's Monthly Magazine, 98*: 271.

SKIDMORE, P. 1966. Miscellaneous notes on some insects in the Doncaster Museum. *Entomologist, 99*: 228-229.

SKIDMORE, P. 1972. Miscellaneous notes on some insects in the Doncaster Museum Collections (2). *Entomologist, 105*: 180-182.

SKIDMORE, P. 1973. *Chrysanthia nigricornis* Westh: (Col., Oedemeridae) in Scotland, a genus and species new to the British list. *Entomologist, 106*: 234-237.

SKIDMORE, P. 1985a. *Exochomus nigromaculatus* (Goeze) (Col., Coccinellidae) in Britain. *Entomologist's Monthly Magazine, 121*: 239-240.

SKIDMORE, P. 1985b. *Cyphon kongsbergensis* Munster (Col., Scirtidae) in Scotland. *Entomologist's Monthly Magazine, 121*: 249-252.

SKIDMORE, P. & HUNTER, F.A. 1980. *Ischnomera cinerascens* Pand (Col., Oedemeridae) new to Britain. *Entomologist's Monthly Magazine, 116*: 129-132.

SKIDMORE, P. & JOHNSON, C. 1969. A preliminary list of the Coleoptera of Merioneth, North Wales. *Entomologist's Gazette, 20*: 139-225.

SKIDMORE, P., LIMBERT, M. & EVERSHAM, B. 1985. The insects of Thorne Moors. *Sorby Record 23* (supplement) 1985. 152p.

SMITH, K.G.V. 1948. Notes on *Gnorimus nobilis* L. (Col., Scarabaeidae) in Worcestershire. *Entomologist's Monthly Magazine, 84*: 288.

SPEIGHT, M. 1968. A second record of *Agrilus sinuatus* (Oliv.), (Col., Buprestidae), in Oxfordshire. *Entomologist, 101*: 42.

SPEIGHT, M.C.D. 1989. *Saproxylic invertebrates and their conservation*. Strasbourg, Council of Europe. (Nature and Environment Series, No. 42).

STAINFORTH, T. 1944. Reed-beetles of the genus *Donacia* and its allies in Yorkshire (Col. Chrysomelidae). *Naturalist, 69*: 81-91.

STEELE, R.C. & WELCH, R.C. eds. 1973. *Monks Wood, a nature reserve record*. Huntingdon, Nature Conservancy.

STEPHENS, J.F. 1839. *A manual of British Coleoptera, or beetles*. London: Longman, Orme, Brown, Green and Longmans.

STOKES, H.G. 1952. *Agrilus sinuatus* Oliv. (Col., Buprestidae) in Salisbury, Wiltshire. *Entomologist's Monthly Magazine, 88*: 62.

STOTT, C.E. 1925. *Odontaeus armiger* Scop. (*mobilicornis* F.) in Surrey and Kent. *Entomologist's Monthly Magazine, 61*: 184-185.

STOTT, C.E. 1929. Re-occurrence of *Cryptocephalus decemmaculatus* in Staffordshire. *Entomologist's Monthly Magazine, 65*: 268-269.

STOTT, N. 1981. *Osphya bipunctata* F. (Col., Melandryidae) in Buckinghamshire. *Entomologist's Monthly Magazine, 117*: 138.

STRETTON, G.B. 1943. Some observations on the leaf-rolling habits of *Byctiscus populi* L. (Col. Curculionidae). *Entomologist's Monthly Magazine, 79*: 252-255.

SYMES, H. 1952. Some notes on *Prionus coriarius* Linn. *Entomologist's Record and Journal of Variation, 64*: 153-155.

TAYLOR, S. 1943. *Lycoperdina succicola* L. (Col., Endomychidae) new to Norfolk. *Entomologist's Monthly Magazine, 79*: 278.

THOMAS, J. 1972. Two records of *Leistus montanus* Steph. (Col., Carabidae). *Entomologist's Monthly Magazine, 108*: 30.

THOMPSON, M.L. 1923. *Sphaerites glabratus* in Yorkshire. *Entomologist's Monthly Magazine, 59*: 278.

THOMPSON, R.T. 1958. Coleoptera: Phalacridae. *Handbooks for the Identification of British Insects, 5*(5b): 1-17.

THORNLEY, A. & WALLACE, W. 1907. *et seq.* Lincolnshire Coleoptera. *Transactions. Lincolnshire Naturalists' Union, 1*: (1907) to *3*: (1915).

THOULESS, H.J. 1921. *Tapinotus sellatus* F. at Horning. *Entomologist's Monthly Magazine, 57*: 88.

TOMLIN, J.R.LE B. 1921. Notes on the Coleoptera of Glamorgan. *Entomologist's Monthly Magazine, 57*: 34-36.

TOMLIN, J.R.LE B. 1949. *Herefordshire Coleoptera. Part 1*. Hereford, Woolhope Naturalists' Field Club.

TOZER, D. 1939. Rediscovery of *Agrilus biguttatus* F. (Col., Buprestidae) in Sherwood Forest. *Entomologist's Monthly Magazine, 75*: 88.

TOZER, D. 1944. Notes on Midland Coleoptera for 1943. *Entomologist's Monthly Magazine, 80*: 21.

TOZER, D. 1953. *Licinus silphoides* Rs. (Col., Carabidae) in Northamptonshire. *Entomologist's Monthly Magazine, 89*: 135.

TURK, F.A. 1942. *Cassida nebulosa* L., (Col., Chrysomelidae) new to Cornwall. *Entomologist's Monthly Magazine, 78*: 72.

TURNER, H.J. 1921. Societies. The South London Entomological and Natural History Society: September 8th, 1921. *Entomologist's Monthly Magazine, 57*: 263.

REFERENCES

TURNER, M. 1976. 28th report of the Entomological section. *Report and Transactions of the Devonshire Association for the Advancement of Science, 108*: 183-185.

TWINN, D.C. 1952. *Anaglyptus mysticus* L. and *Harpalus obscurus* F. (Col.) in Cambridgeshire. *Entomologist's Monthly Magazine, 88*: 155.

TYLER, P.S. 1959. *Ludius ferrugineus* L. (Col., Elateridae) captured as an adult. *Entomologist's Monthly Magazine, 95*: 225.

UHTHOFF-KAUFMANN, R.R. 1947. *Obrium cantharinum* L. and *O.brunneum* F. (Col., Cerambycidae) in Great Britain. *Entomologist's Monthly Magazine, 83*: 77-78.

UHTHOFF-KAUFMANN, R.R. 1985. The genus *Obrium* (Col., Cerambycidae) in Great Britain: A re-appraisal. *Entomologist's Record and Journal of Variation, 97*: 216-223.

UHTHOFF-KAUFMANN, R.R. 1988. The occurrence of the Genus *Strangalia* Serville (Col.: Cerambycidae) in the British Isles. *Entomologist's Record and Journal of Variation, 100*: 63-71.

UHTHOFF-KAUFMANN, R.R. 1990. The Genera *Nathrius* Brethes and *Molorchus* F. in Great Britain (Col.: Cerambycidae). *Entomologist's Record and Journal of Variation, 102*: 239-242.

UHTHOFF-KAUFMANN, R.R. 1991a. The distribution and occurrence of the Tanner Beetle, *Prionus coriarius* L. (Col.: Prionidae) in Great Britain. *Entomologist's Record and Journal of Variation, 103*: 3-5.

UHTHOFF-KAUFMANN, R.R. 1991b. The genera *Lamia* F., *Mesosa* Latr. and *Leiopus* Serv., (Col: Lamiidae) in the British Isles. *Entomologist's Record and Journal of Variation, 103*: 73-77.

VAN EMDEN, F.I. 1945. Larvae of British beetles: V: Elateridae. *Entomologist's Monthly Magazine, 81*: 13-37.

VERDCOURT, B. 1944. *Prionus coriarius* L. and *Lucanus cervus* L. (Col.) in Berkshire. *Entomologist's Monthly Magazine, 80*: 248.

VERDCOURT, B. 1983. Persistence of *Elater* (=*Ludius*) *ferrugineus* L. (Col., Elateridae) in a suburban garden at Windsor. *Entomologist's Monthly Magazine, 119*: 210.

WAKELY, S. 1951. Foodplants of *Baris scolopacea* Germar. *Entomologist's Record and Journal of Variation, 63*: 97.

WAKELY, S. 1953. *Hippodamia 13-punctata* (L.) (Col., Coccinellidae) at Hastings, Sussex. *Entomologist, 86*: 10.

WALKER, J.J. 1894. *Smicronyx coecus*, Boh., at Portland. *Entomologist's Monthly Magazine, 30*: 210.

WALKER, J.J. 1900a. Coleoptera and Lepidoptera at Rannoch. *Entomologist's Monthly Magazine, 36*: 21-28.

WALKER, J.J. 1900b. The Coleoptera and Hemiptera of the Deal sandhills. *Entomologist's Monthly Magazine, 36*: 94-101.

WALKER, J.J. 1925. *Agrilus sinuatus* Ol. in Surrey. *Entomologist's Monthly Magazine, 61*: 183.

WALKER, J.J. 1927. The New Forest, May, 1927. *Entomologist's Monthly Magazine, 63*: 188-190.

WALKER, J.J. 1928a. *Meloe rugosus* Marsh, near Oxford. *Entomologist's Monthly Magazine, 64*: 89.

WALKER, J.J. 1928b. *Aphodius lividus* Ol. at Harpenden. *Entomologist's Monthly Magazine, 64*: 234-235.

WALKER, J.J. 1930a. *Procas armillatus* F. in abundance near Brighton. *Entomologist's Monthly Magazine, 66*: 231.

WALKER, J.J. 1930b. Habitat of *Rhizophagus aeneus* Richt. (*coeruleipennis* Sahlb.). *Entomologist's Monthly Magazine, 66*: 256.

WALKER, J.J. 1931. *Ceuthorrhynchus mixtus* Muls. and associated species of the genus. *Entomologist's Monthly Magazine, 67*: 208-209.

WALKER, J.J. 1932a. An annotated list of the Coleoptera of the Isle of Sheppey. *Transactions of the Entomological Society of the South of England, 7* (1931-1932): 81-140.

WALKER, J.J. 1932b. *Donacia impressa* Payk. and other species of the genus in the Oxford district. *Entomologist's Monthly Magazine, 68*: 165-166.

WALKER, J.J. 1932c. Three weeks in the New Forest. *Entomologist's Monthly Magazine, 68*: 172-175.

WALKER, J.J. 1933. Localities for *Gymnetron beccabungae* L. *Entomologist's Monthly Magazine, 69*: 189-190.

WALKER, J.J. 1936. *Lissodema cursor* Gyll.: a black variety at Oxford *Entomologist's Monthly Magazine*, 72. 209.

WALSH, G.B. & RIMINGTON, F.C. eds. 1953. *The natural history of the Scarborough district. Volume 2 - Zoology.* Scarborough, Scarborough Field Naturalists' Society.

WATERSTON, A.R. 1981. Present knowledge of the non-marine invertebrate fauna of the Outer Hebrides. *Proceedings of the Royal Society of Edinburgh*, 79B: 215-321.

WEAL, R.D. 1953. *Prionus coriarius* L. (Coleoptera, Prionidae), a teratological specimen. *Entomologist's Gazette*, 4: 12.

WELCH, R.C. 1968. *Acritus homoeopathicus* Woll. (Col., Histeridae) in Huntingdonshire. *Entomologist's Monthly Magazine*, 104: 122.

WELCH, R.C. 1975. Coleoptera from Kirkudbrightshire and Wigtownshire. Unpublished Institute of Terrestrial Ecology report to Nature Conservancy Council South-West Scotland region, a supplement to: Harding & Welch (eds.) 1975 Invertebrates survey. Galloway coast and woods, June 1974. Unpublished, as above.

WELCH, R.C. 1978. *Aphodius (Heptaulacus) villosus* Gyll. (Col., Scarabaeidae) in Angus. *Entomologist's Monthly Magazine*, 114: 209-210.

WELCH, R.C. 1980. *Nebria nivalis* (Payk.) (Col., Carabidae) from Mull, Skye and the Cairngorms, with a new character for its separation from *N. gyllenhali* (Schoen.). *Entomologist's Monthly Magazine*, 116: 166.

WELCH, R.C. 1981. New Coleoptera recorded for Woodwalton Fen NNR, collected during the Coleopterists' weekend meeting on the 14th May 1980. *Report. Huntingdonshire Fauna and Flora Society*, 34: 29-34.

WELCH, R.C. 1983a. *Nebria nivalis* (Payk.) (Col., Carabidae) from Ben Hope, Sutherland and the Paps of Jura. *Entomologist's Monthly Magazine*, 119: 16.

WELCH, R.C. 1983b. *Dirhagus pygmeus* F. (Col., Eucnemidae) from Kirkcudbrightshire, a second Scottish record. *Entomologist's Monthly Magazine*, 119: 80.

WHICHER, L.S. 1948. *Aphodius* spp. (Col., Scarabaeidae) in N. Hants. *Entomologist's Monthly Magazine*, 84: 182.

WHICHER, L.S. 1952a. *Aphanisticus emarginatus* Olivier (Col., Buprestidae) on the Hampshire mainland. *Entomologist's Monthly Magazine*, 88: 17

WHICHER, L.S. 1952b. *Hister quadrimaculatus* L. (Col., Histeridae) in N. Kent. *Entomologist's Monthly Magazine*, 88: 208.

WHICHER, L.S. 1952c. *Anisodactylus poecilloides* Steph (Col., Carabidae) in N. Kent. *Entomologist's Monthly Magazine*, 88: 212.

WHICHER, L.S. 1953. *Nebria livida* L. and other beetles from the Norfolk coast. *Entomologist's Monthly Magazine*, 89: 32.

WHITE, I. 1983. *Review of invertebrate sites in Scotland. Review of Dumfries and Galloway region.* Invertebrate Site Register report no. 21. Nature Conservancy Council, unpublished. (CSD report No. 637).

WHITEHEAD, P.F. 1989a. Observations on a population of *Meloe rugosus* Marsham (Coleoptera, Meloidea) at Broadway, Worcestershire, England. *Abstracts Int. Congress Coleopterology. Barcelona, 1989.* Asociacion Europea de Coleopterologia.

WHITEHEAD, P.F. 1989b. *Harpalus dimidiatus* Rossi (Col: Coleoptera) in Worcestershire. *Entomologist's Monthly Magazine*, 125: 118.

WHITEHEAD, P.F. 1990a. Analysis of a Coleoptera fauna from Broadway, Worcestershire. *Entomologist's Monthly Magazine*, 126: 27-32.

WHITEHEAD, P.F. 1990b. Rare beetles from an apple tree in Worcestershire. *Entomologist's Monthly Magazine*, 126: 236.

WHITTEN, A.J. 1990. Recovery: A proposed programme for Britain's protected species. Nature Conservancy Council, unpublished. (CSD report No. 1089).

WILKINSON, W. 1953. Box Hill, Surrey, a confirmed locality for *Odontaeus armiger* (Scop.) (Col., Scarabaeidae). *Entomologist's Monthly Magazine*, 89: 166.

WILLIAMS, B.S. 1924. *Anisoxya fuscula* Ill. at Harpenden. *Entomologist's Monthly Magazine*, 60: 63.

WILLIAMS, B.S. 1928a. *Aphodius lividus* at Harpenden. *Entomologist's Monthly Magazine*, 64: 234.

REFERENCES

WILLIAMS, B.S. 1928b. *Cryphalus asperatus* Gyll. at Bricket Wood. *Entomologist's Monthly Magazine, 64*: 235.

WILLIAMS, S.A. 1968. *Aegalia rufa* (F) (Col., Scarabaeidae) in Lancashire. *Entomologist's Monthly Magazine, 104*: 277.

WILLIAMS, S.A. 1972. Some interesting beetles taken in an autokatcher at Horton Kirby, W. Kent. *Entomologist's Monthly Magazine, 108*: 237.

WILLIAMS, S.A. 1975. *Cyrtotriplax bipustulata* (F.) (Col., Erotylidae) in Surrey. *Entomologist's Monthly Magazine, 111*: 48.

WILLIAMS, S.A. 1979. *Hypopycna rufula* (Er.) (Col., Staphylinidae) at Otford, W. Kent. *Entomologist's Monthly Magazine, 115*: 247.

WILLIAMS, S.A. 1984. *Cymindis macularis* (Fischer v. Waldheim) (Col., Carabidae) confirmed as a British species. *Entomologist's Monthly Magazine, 120*: 107.

WILSON, W.A. 1958. *Coleoptera of Somerset.* Somerset Archaeological and Natural History Society. Supplement to Proceedings, Volumes 101-102. Taunton. Pheonix Press.

WOLLASTON, V. 1843. Note on the capture of Coleoptera in Lincolnshire, in June, 1843. *Zoologist, 1*: 269-271.

WOOD, J.J. 1899. Coleoptera in the Lake District. *Entomologist's Monthly Magazine, 35*: 213.

WOOD, O.E. 1964. The Coventry Nature Reserve (Tile Hill Wood): report for 1963. *Proceedings of the Coventry and District Natural History and Scientific Society, 3*: 195-197.

WOOD, T. 1886. *Langelandia anophthalma*, Aubé, at St. Peter's, Kent; a species new to Britain. *Entomologist's Monthly Magazine, 23*: 93.

WOODROFFE, G.E. 1966. Some uncommon Coleoptera from the Oxfordshire Chilterns. *Entomologist's Monthly Magazine, 102*: 118.

ZATLOUKAL-WILLIAMS, R.G.Z. 1973. Further records of *Pyrrhidium sanguineum* (L.) (Col., Cerambycidae) in Herefordshire. *Entomologist's Monthly Magazine, 109*: 49.

INDEXES

Index to the names of invertebrates mentioned in the text

This index includes all family, generic and specific names of invertebrates mentioned in the text including synonyms of beetles listed in the data sheets. References to the page numbers of data sheets are listed first and are shown in **bold** type. References to page numbers of the three appendices are shown in *italic* type.

3-guttata = triguttata, Abdera	350
4-collis = latridioides, Hypocoprus	344
4-cornis = quadricorniger, Phytobius	288
4-fasciata = quadrifasciata, Abdera	349
4-guttatus = quadriguttatus, Longitarsus	198
4-guttatus = quadriguttatus, Mycetophagus	367
4-maculata = quadrimaculata, Dendroxena	409
4-maculatus = quadrimaculatus, Aphodius	388
4-maculatus = quadrimaculatus, Hister	339
4-maculatus = quadrimaculatus, Nephus	219
4-maculatus, Scymnus = quadrimaculatus, Nephus	219
4-nodosus = quadrinodosus, Phytobius	288
4-notatus = quadrinotatus, Hister	339
4-punctata = quadrimaculata, Dendroxena	409
4-punctata, Xylodrepa = quadrimaculata, Dendroxena	409
4-punctatum = quadripunctatum, Agonum	102
4-punctatus = quadripunctatus, Harpalus	135
4-punctatus = quadripunctatus, Notiophilus	143
4-pustulata = quadripustulata, Lissodema	380
4-pustulatum = quadripustulata, Lissodema	380
4-pustulatum = quadripustulatum, Bembidion	116
4-pustulatus, Cryptohypnus = quadripustulatus, Fleutiauxellus	323
4-pustulatus, Hypnoidus = quadripustulatus, Fleutiauxellus	323
4-signatus = quadrisignatus, Dromius	126
4-striatus, Saprinus = rugiceps, Hypocaccus	340
5-punctata = quinquepunctata, Coccinella	217
5-punctatus = quinquepunctatus, Tychius	309
5-punctatus, Aoromius = quinquepunctatus, Tychius	309
6-guttata = sexguttata, Leptura	160
6-maculata = sexmaculata, Judolia	158
6-punctatum = sexpunctatum, Agonum	103
6-punctatus = sexpunctatus, Cryptocephalus	185
8-maculatum = octomaculatum, Bembidion	115
10-maculatus = decemmaculatus, Cryptocephalus	182
10-notata = decemnotata, Phytodecta	205
12-guttata = duodecimguttata, Vibidia	220
12-guttata, Halyzia = duodecimguttata, Vibidia	220
12-spot ladybird	220
12-striatus = duodecimstriatus, Anommatus	169
13-punctata = tredecimpunctata, Hippodamia	218
13-spot ladybird	218

A

Aacupalpus	100
Abax	*31*
Abdera	**348-350**, *20, 23, 26, 42*
abdominalis = fowleri, Longitarsus	195
abdominalis = teutonus, Stenolophus	150
abdominalis, Ancistronycha	**94**, *25, 37*
aberratus, Rhantus	*17*
abietinus, Ebaeus = flavipes, Hypebaeus	359
abietis, Cryphalus	*53*
Abraeus	**335**, *23, 33*
absinthii = absynthii, Longitarsus	191
absynthii, Longitarsus	**191**, *23, 46*
Acalles	**230**, *27, 50*
Acalyptus	**230**, *27, 52*
Acanthocinus	**155**, *26, 44*
aceris, Phenacoccus	219
Acilius	*17*
Aclypea	**409**, *19, 23, 33*

Acmaeops	**155**, *20, 43*
Acrantus	421
Acritus	**336**, *21, 33*, 338
aculeata = holomelaena, Mordella	361
aculeata = leucaspis, Mordella	362
aculeata, Mordella	362
acuminata, Melanophila	*35*
Acupalpus	**99-100**, *19, 23, 25, 32*, 99
acuticollis, Mordellistena	**362**, *22, 42*
Aderidae	**61**, *15, 20, 26, 43*
Aderus	**61**, *20, 26, 43*
Adimonia	190
Adonia	**216**, *25, 40*
Adonis' Ladybird	216
Adoxus	172
Adrastus	**314**, *21, 36*
adustum = semipunctatum, Bembidion	117
adustum, Bembidium = semipunctatum, Bembidion	117
aedilis, Acanthocinus	**155**, *26, 44*
Aegialia	**382-383**, *19, 25, 34*, 384
Aeletes	**336**, *21, 33*
aenea, Chrysomela	*45*
aeneopiceus, Caulotrupodes	*50*
aenescens, Ilybius	*17*
aeneus = immundus, Saprinus	342
aeneus, Cyanostolus	**377**, *23, 39*
aeneus, Malachius	**360**, *21, 39*
aeneus, Ochthebius	*18*
aeneus, Paracymus	*18*
Aepopsis	101
Aepus	**100-101**, *24, 29*
aequalis, Orthoperus	**226**, *22, 40*
aerata, Batophila	*46*
aeratus, Gyrinus	*17*
aerea, Phyllotreta	**204**, *26, 45*
aerosa, Chaetocnema	**175**, *22, 46*
aeruginosus, Longitarsus	**192**, *20, 46*
aesthuans, Notiophilus	**142**, *24, 29*
aestuans = aesthuans, Notiophilus	142
aethiops, Elater = nigerrimus, Ampedus	316
aethiops, Notaris	**282**, *24, 51*
aethiops, Pterostichus	**147**, *24, 30*
Aetophorus	125
afer, Apion	*48*
affine, Apion	**72**, *23, 47*
affinis = nigroruber, Pyropterus	346
affinis, Abdera	**348**, *20, 42*
affinis, Berosus	*18*
affinis, Caenocara	**63**, *22, 38*
affinis, Chromoderus	**257**, *19, 49*
affinis, Dictyopterus = nigroruber, Pyropterus	346
affinis, Dorytomus	**262**, *21, 50*
affinis, Enochrus	*18*
affinis, Gynandrophthalma	**190**, *20, 45*
affinis, Olibrus	*40*
affinis, Plateumaris	**206**, *26, 44*
Agabus	*17*
Agapanthia	*44*
agaricola = agricola, Eledona	414
agaricola, Heledona = agricola, Eledona	414
Agelastica	**170**, *22, 45*
agilis, Longitarsus	**192**, *23, 46*
Agonum	**101-103**, *19, 22, 24, 31*

INVERTEBRATE INDEX

agricola, Eledona — 414, *25, 41*
Agrilus — **88-90**, *23, 25, 35*
Agriotes — 314, *21, 36*
albicinctus, Rhinoncus — 293, *20, 52*
albidus, Cleonus = affinis, Chromoderus — 257
albinus, Platystomos — 70, *26, 47*
algirus, Lixus — 276, *20, 49*
alienus, Lasius — 180
alismatis, Hydronomus — 270, *27, 50*
Allodactylus — 311
alni, Agelastica — 170, *22, 45*
alni, Cis — 211
alni, Dryocoetinus — 398, *24, 53*
alni, Phymatodes — 163, *26, 44*
Alophus — 49
alpina, Amara — 104, *21, 31*
alpinus, Cyrtonotus = alpina, Amara — 104
alpinus, Oreodytes — 17
alternans, Helophorus — 17
alternans, Trachyphloeus — 303, *26, 48*
Altica — **170-171**, *23, 26, 46*
aluta, Limnebius — 18
alvearius, Trichodes — 215, *19, 38*
Amara — **104-107**, *19-22, 24, 31*
ambiguus, Calathus — 119, *24, 31*
ambiguus, Sitona — 49
Ambrosia beetle — 404
Ammoecius — 383
Ampedus — **314-319**, *19-21, 23, 25, 35, 36*
Amphimallon — 383, *23, 34*
Amphimallus — 383
Anacaena — 18
Anaclyptus — 156
Anaglyptus — 156, *26, 44*
analis = variegata, Grammoptera — 158
analis, Baris — 240, *21, 52*
Anaspis — **405-407**, *19, 22, 23, 42*
Anchomenus — 101
Anchonidium — 231, *21, 50*
anchusae, Longitarsus — 192, *26, 46*
Ancistronycha — 94, *25, 37*
ancora, Tetratoma — 419, *26, 41*
anglicanus, Dryops — 18
angulosus, Ceutorhynchus — 246, *24, 51*
angustata, Feronia = angustatus, Pterostichus — 147
angustatus, Anthicus — 67, *26, 43*
angustatus, Dyschirius — 128, *21, 29*
angustatus, Hydrochus — 17
angustatus, Hylastes — 53
angustatus, Pterostichus — 147, *24, 30*
angusticollis, Ernobius — 37
angusticollis, Monotoma — 378, *21, 39*
angustimarginalis = marginellus, Malachius — 361
angustulus, Agrilus — 88, *25, 35,* 90
angustulus, Selatosomus — 328, *21, 36*
Anisandrus — 404
Anisodactylus — 108, *21, 22, 31*
Anisoxya — 350, *23, 42*
Anitys — 62, *25, 38*
annulipes = cineraceum, Apion — 73
Anobiidae — **62-66**, *2, 15, 20-23, 25, 37*
anobioides, Dryophilus — 65, *21, 37*
Anobium — 62, *25, 37,* 65, **213-215,** 225
anomalus, Badister — 109, *19, 32*
Anommatus — 169, *22, 23, 40*
anophthalma, Langelandia — 222, *21, 41*
Anoplodera — 160
Anoplus — 231, *26, 50*
Anostirus — 319, *20, 36*
Anthaxia — 90, *19, 35*
Anthicidae — **67-68**, *2, 15, 20, 21, 23, 26, 43*

Anthicus — **67-68**, *20, 21, 23, 26, 43*
Anthocomus — 358
Anthomus — 232
Anthonomus — **231-233**, *21, 22, 27, 52*
Anthophora — 355, 356
anthracina, Feronia = anthracinus, Pterostichus — 147
anthracinus, Pterostichus — 147, *24, 30*
Anthracus — 99
Anthrenus — 311, *19, 37*
Anthribidae — 2, 15, *20, 23, 26, 47,* 69
Anthribus — **69-71**, *23, 26, 47*
antirrhini, Gymnetron — 268
antracina, Feronia = anthracinus, Pterostichus — 147
Anurida — 101
Aoromius — 309
Apalus — 355, *20, 43*
Aphanisticus — **90-91**, *19, 25, 35*
Aphis — 219
Aphodius — **383-389**, *19, 21, 23, 25, 34*
Aphthona — 171, *23, 46*
apiarius, Trichodes — 216, *19, 38*
apicalis, Haliplus — 17
Apion — **72-83**, *20-24, 26, 47, 48,* 74, 75, *78-82*
Apionidae — **72-83**, 2, 15, *20-23, 26, 47*
Apis — 216
Aplocnemus — 357, *23, 25, 38,* 357
Apoderus — 47
appendiculata, Macroplea — 200, *21, 44,* 201
Apple bud weevil — 232
Apteropeda — 172, *20, 26, 46*
aquatica, Donacia — 186, *21, 44*
aquaticus, Notiophilus — 142
aquilonia, Formica — 378
arctica, Miscodera — 141, *24, 29*
arcticus, Bledius — 128
arcuatus, Ceuthorrhynchus = arquatus, Ceutorhynchus — 246
arcuatus, Clitostethus — 216, *20, 40*
arcuatus, Plagionotus — 163, *19, 44*
ardosiacus, Harpalus — 130, *24, 31*
arduus, Bagous — 234, *22, 50*
arenariae, Sibinia — 297, *27, 52*
arenarius = putridus, Aphodius — 387
arenarius, Plagiogonus = putridus, Aphodius — 387
areolatus, Perileptus — 145, *22, 29*
argenteolum, Bembidion — 110, *22, 30*
argillaceus, Bagous — 234, *21, 50*
Arhopalus — 43
arida, Chaetocnema — 46
aridula, Chaetocnema — 46
arietinus, Byrrhus — 92, *25, 35*
aristatus, Trachyphloeus — 303, *26, 48*
armatum, Apion — 72, *22, 48*
armeriae = rufitarsis, Omophlus — 416
armiger, Odonteus — 334, *23, 33*
armillatus = granulicollis, Procas — 292
armillatus, Procas — 292, *22, 51*
armoraciae = concinnus, Phaedon — 204
Aromia — 156, *26, 44*
arquatus, Ceutorhynchus — 246, *22, 51*
artemisiae = vespertinus, Phyllobius — 286
arundinis, Hypera — 271, *19, 49*
arvernicus, Helophorus — 17
Asaphidion — 108, *24, 30*
Asclera — 370
asellus, Cardiophorus — 320, *25, 36*
asper, Psammodius — 395, *23, 34*
asperatus, Trachyphloeus — 304, *26, 48*
asperatus, Trypophloeus — 404, *24, 53*
aspersa, Helix — 313
assimile, Apion — 80
assimile, Scaphisoma — 381, *22, 33*

astragali, Apion 72, 24, 48
Atactogenus 258
ater = longitarsis, Xyletinus 66
ater = parvulus, Longitarsus 197
ater = smartii, Cryptolestes 227
ater, Ocypus 118
ater, Philonychus 117, 26, 41
ater, Rhyncolus 50
ater, Salpingus 42
aterrima, Feronia = aterrimus, Pterostichus 148
aterrimus, Pterostichus 148, 19, 30
Athous 320, 21, 25, 36, 324
atomaria, Laria = atomarius, Bruchus 88
atomarius, Aeletes 336, 21, 33
atomarius, Bruchus 88, 26, 44
atomarius, Orthoperus 40
atomarius, Stilbus 373, 22, 40
atomus, Ceutorhynchus 246, 24, 51
atra = cruciferae, Phyllotreta 204
atratus, Anchomenus = nigrum, Agonum 102
atratus, Laccobius 18
atricapillus, Cercyon 18
atricillus, Longitarsus 196, 199
atripes, Lagria 415, 20, 41
atrocephalus, Laccobius 18
atropae, Epitrix 190, 26, 46
atroviolaceum = stomoides, Bembidion 117
atrovirens, Aphthona 46
attaphilus, Cathormiocerus 244, 20, 48
Attelabidae 83-87, 2, 15, 19, 21, 23, 26, 47
attelaboides, Cimberis 47
Attelabus 47
attenuata, Psylliodes 207, 20, 46
attenuata, Strangalia 167, 19, 44
Aulonium 220, 23, 41
Aulonothroscus 420, 21, 36
auratus, Carabus 29
auratus, Rhynchites 84, 19, 47
aureolus, Cryptocephalus 180, 26, 45
auriculatus, Dryops 18
auriculatus, Ochthebius 18
aurifer, Otiorhynchus 48
auritus, Scymnus 219
auropunctatus, Otiorhynchus 284, 20, 48
aurora, Dictyoptera 345, 25, 37
aurulenta, Strangalia 167, 23, 44
autographus, Dryocoetes 53
autumnalis, Meloe 355, 20, 43
axillaris, Cymindis 124, 23, 32
Axinotarsus 357, 20, 39, 358
azureus, Harpalus 130, 24, 31

B

bacchus, Rhynchites 84, 19, 47
Badister 109-110, 19, 22, 23, 25, 32
Baeckmanniolus 337, 25, 33
Bagous 234-240, 19-22, 24, 27, 50
bajulus, Hylotrupes 44
Balaninus 261
ballotae, Longitarsus 193, 26, 46
balteatus, Malthinus 95, 25, 37
barbata, Melandrya 352, 20, 42
barbicornis, Magdalis 277, 24, 50
Baris 240-242, 21, 24, 27, 52
barnevillei, Malachius 360, 21, 39
barnevillei, Trichosirocalus 305, 27, 51
Barynotus 242, 26, 49
Barypeithes 243, 22, 26, 49
Barypithes 243

Batophila 46
bearei, Longitarsus 193, 22, 46
beccabungae = veronicae, Gymnetron 260
beccabungae, Gymnetron 260, 24, 52, 260
Bee-eating beetle 215
Beet carrion beetle 409
Bembidion 110-118, 19, 21, 22, 24, 30
Bombidium 111
Berosus 18
betulae, Byctiscus 83, 26, 47
betulae, Curculio 261, 27, 52
betuleti = betulae, Byctiscus 83
betuleti, Rhynchites = betulae, Byctiscus 83
bicolon, Ochthebius 18
bicolor = vilis, Lixus 277
bicolor, Enochrus 18
bicolor, Taphrorychus 402, 24, 53, 228
bicolora, Donacia 186, 20, 44
bicornis, Sulcacis 213, 25, 40
bidentatus, Silvanus 411, 25, 39
Bidessus 17
bifasciata = biflexuosa, Abdera 349
bifasciatus, Anthicus 67, 26, 43
bifenestratus, Cercyon 18
biflexuosa, Abdera 349, 26, 42
biguttata = bucephala, Tomoxia 365
biguttatus = pannonicus, Agrilus 89
biguttatus, Agabus 17
biguttatus, Cryptocephalus 180, 20, 45
biguttatus, Notiophilus 143
bilineatus, Cryptocephalus 181, 26, 45
bilineatus, Graphoderus 17
bilineatus, Graptodytes 17
bimaculatus, Anthicus 67, 23, 43
bimaculatus, Laemophloeus = unifasciatus, Notolaemus 228
bimaculatus, Notaris 282, 27, 51
binodulus, Bagous 234, 19, 50
binodulus, Cryphalus = asperatus, Trypophloeus 404
binotatus = capucinus, Lepyrus 274
binotatus = nemorivagus, Anisodactylus 108
binotatus, Hallomenus 351, 26, 42
Biorhiza 262
Biphyllidae 87, 2, 25, 39
bipunctata, Osphya 353, 21, 42
bipunctatum, Bembidion 111, 24, 30
bipunctatus, Cryptocephalus 181, 26, 45
bipunctatus, Ellescus 265, 27, 52
bipustulata, Anacaena 18
bipustulata, Mycetochares = humeralis, Mycetochara 416
bipustulata, Tritoma 331, 23, 39
bipustulatus, Badister 109
bipustulatus, Panagaeus 144, 25, 32
bipustulatus, Selatosomus 328, 25, 36
bishopi = foveolatus, Rabocerus 381
bisignatus, Nephus 218, 19, 40
bispinus, Xylocleptes 228
bissexstriatus, Hister 338, 25, 33
bistriatus, Tachys 151, 24, 30
bisulcatus, Tachys 30
Black-headed cardinal beetle 376
Blaps 41
Blastophagus 403
Bledius 113, 128, 129
Blethisa 118, 24, 29
Blitophaga 409
Blue ground beetle 121
bohemica, Anaspis 405, 22, 42
boleti, Diaperis 414, 20, 41
boleti, Scaphisoma 382, 25, 33
Bolitophagus 413, 21, 41, 212
Bombardier beetle 118

INVERTEBRATE INDEX

Bombus	355
borealis, Limobius	274, 24, 49
borealis, Pelophila	145, 21, 29
Bostrichidae	87, 2, 15, 19, 38
Bostrichus	87, 19, 38
Bostrychus	87
bovistae, Caenocara	63, 21, 38
bovistae, Lycoperdina	330, 21, 40
braccata, Plateumaris	206, 23, 44
Brachinus	118, 19, 25, 32
Brachonyx	52
Brachynus	118
Brachysomus	243, 21, 26, 49
Brachytarsus	**69**
Bracteon	113
Bradycellus	119, 22, 23, 32
brevicauda, Mordellistena	363, 22, 42
brevicollis = melleti, Harpalus	133
brevicollis, Altica	170, 23, 46
brevicollis, Aulonothroscus	420, 21, 36
brevicollis, Malthodes	95, 20, 37, 96
brevicollis, Meloe	355, 20, 43
brevicornis, Aderus	61, 20, 43
brevis, Aphodius	383, 19, 34
brevis, Bagous	235, 20, 50
brevis, Hydrochus	17
Brindalus	389, 19, 34
britannicus, Cathormiocerus	244, 20, 48
britannus, Anthonomus	52
britteni, Altica	46, 171
Bromius	172, 20, 45
Brontes	229
Bruchela	47
Bruchidae	87-88, 2, 11, 20, 26, 44
Bruchidius	87, 20, 44
Bruchus	88, 26, 44, 87
brunnea = neuwaldeggiana, Mordellistena	364
brunneipes = brunnipes, Acupalpus	99
brunneum, Obrium	44
brunneus , Lasius	376
brunneus, Agabus	17
brunneus, Hylastes	53
brunneus, Lasius	225, 265, 335, 336
brunneus, Longitarsus	193, 26, 46
brunneus, Lyctus	215, 343
brunnipennis, Anthonomus	52
brunnipes, Acupalpus	99, 23, 32
brunnipes, Apion	73, 20, 48
brunnipes, Orthoperus	226, 21, 40
brunnipes, Phalacrus	372, 23, 40
brunsvicensis, Chrysolina	45
bucephala, Tomoxia	365, 23, 42
Buprestidae	88-92, 2, 15, 19, 20, 23, 25, 35
buprestoides = laevigata, Xylita	354
buprestoides, Melasis	333, 25, 36
buqueti, Thaneroclerus	38
buyssoni, Gnathoncus	337, 23, 33
Byctiscus	83, 21, 26, 47
Byrrhidae	92-93, 2, 19, 22, 23, 25, 35
Byrrhus	92, 25, 35, 141
Byturidae	2, 15

C

caelatus, Psammodius	34
Caenocara	63, 21, 22, 38
Caenopsis	244, 26, 48
Caenoptera	162
Caenorhinus	85
caerulea = cyanea, Ischnomera	369

caerulea, Ischnomera	368, 21, 43, 369
caeruleipennis, Rhizophagus = aeneus, Cyanostolus	377
caeruleus, Helops	415, 25, 41
caeruleus, Korynetes	213, 25, 38
caesus, Peltodytes	17
caesus, Pleurophorus	395, 19, 34
Calambus	328
Calathus	119, 24, 31
calceatus, Harpalus	31
calceatus, Rhynchaenus	294, 22, 53
caliginosus, Plinthus	289, 24, 50
Callidium	44, 163, 165
Callistus	120, 19, 32
callosum, Bembidion	30
Calomicrus	173, 23, 45
Calosoma	120, 22, 29
cambricus = cinerascens, Sitona	298
cambricus = puberulus, Sitona	299
cambricus, Sitona	49, 299
cambricus, Sitones = cinerascens, Sitona	298
cambricus, Sitones = puberulus, Sitona	299
cambricus? = cinerascens, Sitona	298
campestris, Ceutorhynchus	247, 27, 51
campyloides, Athous	320, 25, 36
canaliculata, Stenelmis	18
canaliculatus = linearis, Lyctus	347
canaliculatus, Acilius	17
canaliculatus, Phytobius	287, 27, 52
cancellatus, Carabus	29
Candidula	313
cantabricus, Hydroporus	17
Cantharidae	94-98, 2, 15, 20, 21, 23, 25, 37
cantharinum, Oberea	163
cantharinum, Obrium	162, 19, 44
Cantharis	94, 21, 25, 37, 98
cantiana, Monacha	313
cantianum = filirostre, Apion	75
canus = olivaceus, Bruchidius	87
capucina, Eucnemis	332, 20, 36
capucinus, Bostrichus	87, 19, 38
capucinus, Lepyrus	274, 19, 49
Carabidae	99-154, 2, 11, 15, 19-22, 24, 29
caraboides, Hydrochara	18
caraboides, Melandrya	352, 26, 42
caraboides, Platycerus	345, 19, 33
Carabus	120-121, 19, 22, 24, 29
carbonaria, Magdalis	277, 26, 50
carcharias, Saperda	166, 23, 44
cardinalis, Ampedus	314, 20, 35
Cardiophorus	320-321, 19, 20, 25, 36
cardui, Coeliodes = umbrinus, Stenocarus	301
Carida	348
carinata, Silpha	410, 19, 33
carinatus = elevatus, Tropiphorus	307
carinatus, Hydrochus	17
cariniceps, Hylis	333, 20, 36
carlinae = planus, Larinus	273
carpini, Acalyptus	230, 27, 52
carus, Paratillus	38
caspius = dimidiatus, Harpalus	131
caspius, Gyrinus	17
Cassida	173-174, 206, 20, 22, 23, 26, 47
castaneum, Tetropium	43
castaneus = brunneus, Longitarsus	193
castaneus, Anostirus	319, 20, 36
castaneus, Hypophlaeus = unicolor, Corticeus	413
castaneus, Hypophloeus = unicolor, Corticeus	413
castaneus, Pissodes	50, 289
Caterpillar-hunter	120
Cathormiocerus	244-245, 20, 21, 48
caucasicus, Ernoporus	399, 20, 53

Caulotrupodes	50
cavifrons, Rhynchites	84, 26, 47
cellaris, Oxychilus	313
Cepaea	313
cephalotes, Trachys	39
ceramboides, Pseudocistela	418, 26, 41
Cerambycidae	155-168, 2, 11, 15, 19-21, 23, 26, 43
cucumbyriformis, Judolia	44
Cerapheles	358, 23, 39
cerasi, Magdalis	278, 26, 50
cerasi, Orsodacne	44, 203
cerasorum, Balaninus = betulae, Curculio	261
Cercyon	18
cerdo, Apion	73, 26, 48
cerealis, Chrysolina	176, 3, 10, 20, 45
cerinus = ballotae, Longitarsus	193
Cernuella	313
Cerophytidae	36
Cerophytum	36
cervus, Lucanus	345, 25, 33
Cerylon	170, 25, 40
Cerylonidae	169-170, 2, 19, 22, 23, 25, 40
Cetonia	390, 25, 34
Ceuthorhynchidius	305
Ceuthorhynchus	246
Ceuthorrhynchidius	248, 251
Ceuthorrhynchus	246
Ceutorhynchus	246-257, 20-22, 24, 27, 51
Chaetarthria	18
Chaetocnema	175-176, 22, 23, 26, 46
chalcographus, Pityogenes	53
Chalcoides	176, 26, 46
chalcomera, Psylliodes	208, 26, 46
chalconatus, Agabus	17
chalybaeus, Ceuthorhynchus = pectoralis, Ceutorhynchus	250
chalybaeus, Ceuthorrhynchus = pectoralis, Ceutorhynchus	250
chalybaeus, Ceuthorrhynchus = thomsoni, Ceutorhynchus	255
championi = brunnipes, Phalacrus	372
championi, Ophonus = melleti, Harpalus	133
chevrolati, Anthonomus	231, 22, 52
chevrolati, Ceuthorhynchidius = barnevillei, Trichosirocalus	305
chevrolati, Ceuthorrhynchidius = barnevillei, Trichosirocalus	305
Chlaenius	122, 19, 25, 32
chloris = denticollis, Cassida	173
chlorizans, Baris	52
chlorocephala, Lebia	138, 25, 32
Choragidae	15
Choragus	69, 23, 47
Chromoderus	257, 19, 49
chrysanthemi, Ceuthorhynchus = campestris, Ceutorhynchus	247
chrysanthemi, Ceuthorrhynchus = campestris, Ceutorhynchus	247
chrysanthemi, Mantura	201, 23, 46
Chrysanthia	368, 20, 43
chrysocephala = luridipennis, Psylliodes	208
Chrysolina	176-179, 3, 10, 20, 23, 26, 45, 138
Chrysomela	179, 20, 45
chrysomela = pulchellus, Polydrusus	291
chrysomela, Polydrusus	48
Chrysomelidae	170-210, 2, 3, 8, 10, 11, 19-23, 26, 44
cicatricosa = cicatricosus, Meloe	356
cicatricosus, Meloe	356, 20, 43
Cicindela	123-124, 20-22, 24, 29
Cicones	221, 20, 23, 41, 225
Cillenus	113
Cimberis	47
cinctus = piri, Anthonomus	232
cineraceum, Apion	73, 23, 48
cinerascens, Ischnomera	369, 20, 43
cinerascens, Sitona	298, 22, 49
cinerea, Donacia	187, 26, 44
cinereus, Graphoderus	17
cinnabarinus, Ampedus	315, 21, 35
Cionus	257, 24, 49
circulatus, Macinus	279, 27, 52
circumcinctus, Dytiscus	17
circumdata = interrupta, Galeruca	190
circumflexus, Dytiscus	17
circumfuscus, Luperus = circumfusus, Calomicrus	173
circumfusus, Calomicrus	173, 23, 45
Cis	210-212, 21, 25, 40
Cisidae	210-213, 2, 21, 25, 40
Cissophagus	400
Cistela	418
cisti, Bruchidius	44
Clambidae	213, 2, 15, 22, 23, 34
Clambus	213, 22, 23, 34
clarki, Bembidion	111, 24, 30
clarus = longiseta, Longitarsus	195
clathratus = clatratus, Carabus	120
clatratus, Carabus	120, 22, 29
claudicans = collignensis, Bagous	235
clavicornis, Orthocerus	223, 25, 41
clavipes, Donacia	187, 26, 44
clematidis, Leptophloeus	228, 20, 39
Cleonus	258, 26, 49, 257
Cleridae	213-216, 2, 15, 19, 21, 25, 38
Clinocara	353
Clitostethus	216, 20, 40
Clivina	29
clypealis, Hydrovatus	17
Clythra	180
Clytra	180, 19, 45
Clytus	156, 163
cnemerythrus = tempestivus, Bagous	240
Cneorhinus	258, 26, 49
Cneorrhinus	258
coarctatus, Enochrus	18
Coccidula	40
coccinatus, Elater = cardinalis, Ampedus	314
coccinea, Pyrochroa	376, 26, 42
Coccinella	216-217, 21, 23, 40
Coccinellidae	216-220, 1, 2, 11, 15, 19-23, 25, 40
Codiosoma	293
coecus, Smicronyx	300, 21, 51
Coelambus	17
Coeliodes	259, 24, 27, 51, 252, 301, 311
Coenocara	63
coenosus, Aphodius	384, 25, 34
coerulea = caerulea, Ischnomera	368
coerulea = cyanea, Ischnomera	369
coeruleipennis, Rhizophagus = aeneus, Cyanostolus	377
coeruleus = caeruleus, Helops	415
coeruleus, Corynetes = caeruleus, Korynetes	213
coeruleus, Dasytes	358, 22, 38
collaris, Acmaeops	155, 20, 43
collaris, Clivina	29
collaris, Mecinus	279, 27, 52
collignensis, Bagous	235, 21, 50
collinum, Gymnetron	268, 24, 52
collinus = collinum, Gymnetron	268
Colliuris	143
Colobopterus	389
coluber, Cis	210, 21, 40
Colydiidae	220-225, 2, 16, 19-23, 25, 41
Colydium	222, 21, 41
comari, Phytobius	287, 27, 52
complanata = fusca, Amara	105
complanata, Nebria	141, 22, 29
complanatus, Laemostenus	31
comptus, Epierus	337, 22, 33
concinnus, Phaedon	204, 26, 45
concolor, Lamprosoma	45

461

conducta, Chaetocnema   175, *22, 46*
confluens, Polydrusus   **290**, *26, 49*
confusa, Chaetocnema   *46*
conicicollis, Monotoma   378
conicus, Rhinocyllus   **293**, *24, 49*
Coniocleonus   259-260, *19, 21, 49*
connexus, Polistichus   **146**, *20, 32*
Conopalpus   **351**, *26, 42*
conspersus, Agabus   17
conspersus, Anthonomus   **232**, *27, 52*
conspurcatus, Aphodius   **384**, *25, 34*
consputus, Acupalpus   **99**, *25, 32*
consputus, Aphodius   **384**, *21, 34*
constrictus, Ceutorhynchus   **247**, *27, 51*
consularis, Amara   **104**, *24, 31*
contractus, Ceuthorhynchus = insularis, Ceutorhynchus   249
contractus, Ceutorrhynchus = insularis, Ceutorhynchus   249
convexiusculus, Cercyon   18
Copris   **390**, *19, 34*
cordatus, Harpalus   **131**, *21, 31*
coriaceus = nigrescens, Orthoperus   226
coriarius, Prionus   **164**, *23, 43*
cornutus, Epiphanis   36
coronatus, Liparus   **275**, *26, 49*
corticalis, Dryophthorus   **265**, *20, 50*
Corticeus   **413**, *21, 41*
coryli, Apoderus   47
coryli, Cryptocephalus   **181**, *20, 45*
coryli, Triotemnus   **403**, *20, 53*
Corylophidae   226-227, *2, 21, 22, 25, 40*
Corymbites   319, 321, 328
Corynetes   213
cosnardi, Platycis   **346**, *22, 37*
Cossonus   **260**, *24, 27, 50*
Cossus   229
cossus, Cossus   229
crassicornis, Chrysolina   **176**, *20, 45*
crassicornis, Malthodes   **96**, *21, 37, 96*
crassicornis, Noterus   17
crassipes, Donacia   **187**, *26, 44*
crassirostris, Tychius   **307**, *22, 52*
crataegi, Otiorhynchus   48
crenatus, Dendrophagus   **227**, *25, 39*, 229
crenulatus, Georissus   17
Crepidodera   180, *23, 46*, 176, 202
crepitans, Brachinus   **118**, *25, 32*
cribratum, Trinophyllum   44
crinifer, Limnebius   18
crinitus = macularius, Sitona   298
crinitus, Sitones = macularius, Sitona   298
cristata, Feronia = cristatus, Pterostichus   148
cristatus, Pterostichus   **148**, *24, 30*
crotchi = salinus, Anthicus   68
cruciatus, Selatosomus   **328**, *19, 36*
cruciferae, Phyllotreta   **204**, *26, 45*
cruentatum, Apion   *48*, 80
cruxmajor, Panagaeus   **144**, *19, 32*
cruxminor, Lebia   **138**, *19, 32*
Cryphalus   *53*, 399
Cryptamorpha   39
cryptica, Cantharis   98
Crypticus   **413**, *25, 41*
Cryptocephalus   180-185, *19, 20, 23, 26, 45*
Cryptohypnus   323, 326
Cryptolestes   **227**, *23, 39*
Cryptophagidae   2, 11
Cryptorhynchidius   261
Cryptorhynchus   **261**, *27, 50*
csikii, Bradycellus   **119**, *22, 32*
Ctenicera   321, *23, 36*
Cteniopus   *41*

Ctesias   **311**, *25, 37*
Cucujidae   227-229, *2, 20, 23, 25, 39*
Cuethorrhynchus   255
cuprea, Cetonia   **390**, *25, 34*
cupreus, Harpalus   **131**, *19, 31*
cupreus, Rhynchites   **85**, *26, 47*
cupreus, Riolus   18
Curculio   261-262, *27, 52*
curculioides, Mycterus   **368**, *19, 42*
Curculionidae   230-311, *2, 8, 16, 19-22, 24, 26, 48*
curculionoides = curculioides, Mycterus   368
Curimopsis   92-93, *19, 23, 35*
cursitans, Amara   31
cursor, Lissodema   **380**, *23, 41*
curta, Amara   **104**, *24, 31*
curtirostre, Apion   78
curtisi, Apion   74, *24, 48*
curtisi, Haemonia = mutica, Macroplea   201
curtisii = curtisi, Apion   74
curtulum = curtisi, Apion   74
curtulum, Apion   74
curtum, Asaphidion   30
curtus, Longitarsus   **193**, *23, 46*
curvimanus, Barypeithes   49
curvipes = fulvicorne, Strophosoma   301
cuspidatus, Saprinus   **342**, *25, 33*
cyanea, Ischnomera   **369**, *26, 43*, 368
cyanella, Lema   45
cyaneus, Orobitis   52
cyanocephala, Lebia   **138**, *19, 32*
cyanoptera = sophiae, Psylliodes   209
Cyanostolus   **377**, *23, 39*
cylindrica, Phytoecia   **163**, *26, 44*
cylindricollis, Sitona   49
cylindricum, Sinodendron   336
cylindricus, Oxylaemus   **224**, *19, 41*
cylindricus, Teredus   **225**, *20, 41*
Cylindrinotus   **414**, *25, 41*
Cylindronotus   414
cylindrus, Bagous   **235**, *21, 50*
cylindrus, Platypus   **374**, *27, 53*, 222
Cymindis   124-125, *19, 23, 25, 32*
cynoglossi, Dibolia   **185**, *20, 46*
Cyphon   **396**, *23, 25, 35*
Cyrtonotus   104
Cyrtotriplax   331
czwalinai, Bagous   **236**, *20, 50*

D

dahli = nigrum, Agonum   102
dahli, Anchomenus = nigrum, Agonum   102
Dascillidae   2
Dasytes   358-359, *22, 23, 25, 38*
dauci, Hypera   **271**, *26, 49*
davisi, Oreodytes   17
dawsoni, Trichosirocalus   **305**, *27, 51*
decemmaculatus, Cryptocephalus   **182**, *20, 45*
decemnotata, Phytodecta   **205**, *26, 45*
decoratus, Hygrotus   17
decoratus, Rhynchaenus   **294**, *22, 53*
deflexicollis, Hydrocyphon   **397**, *25, 35*
Demetrias   **125**, *25, 32*
Dendrobium   65
Dendrophagus   **227**, *25, 39*, 229
Dendroxena   **409**, *25, 33*
dentata, Donacia   **187**, *23, 44*
dentata, Drypta   **128**, *19, 32*
dentatus, Cis   **211**, *21, 40*
denticolle, Anobium = denticollis, Hadrobregmus   65

denticolle, Dendrobium = denticollis, Hadrobregmus 65
denticollis, Cassida 173, 20, 47
denticollis, Hadrobregmus 65, 26, 37
denticollis, Rhinoncus = quadrinodosus, Phytobius 288
denticollis, Dryobius 186
depressus, Licinus 139, 25, 32
depressus, Pediacus 229, 25, 39
depressus, Potamonectes depressus 17
depressus, Pytho 377, 23, 42
Dermestidae 311-313, 2, 16, 19-21, 25, 37
dermestoides, Hylecoetus 347, 25, 39
dermestoides, Pediacus 229
Derodontidae 2, 37
Deronectes 17
desertus, Otiorhynchus 284, 26, 48
desjardinsi, Cryptamorpha 39
desmaresti, Tetratoma 419, 23, 41
desmarestii = desmaresti, Tetratoma 419
Diachromus 126, 19, 32
Diaperis 414, 20, 41
Diastictus 390, 20, 34
Dibolia 185, 20, 46
Dicheirotrichus 126, 25, 32
Dichirotrichus 126
Dicronychus 322, 20, 36
Dictyoptera 345, 25, 37
Dictyopterus 345
diecki, Anommatus 169, 22, 40
difficile, Apion 74, 23, 47
diffinis, Ophonus = ardosiacus, Harpalus 130
difforme, Apion 75, 26, 48
difformis = campyloides, Athous 320
difformis, Orthathous = campyloides, Athous 320
digitalis, Trachyphloeus 304, 24, 48
diglyptus, Bagous 236, 20, 50
dilatatus, Badister 109, 25, 32
dimidiatus = kugelanni, Pterostichus 149
dimidiatus, Baeckmanniolus 337, 25, 33
dimidiatus, Dytiscus 17
dimidiatus, Harpalus 131, 22, 31
dimidiatus, Poecilus = kugelanni, Pterostichus 149
Diplocoelus 87, 25, 39
Diplolepis 262
Dirhagus 332, 21, 36
Dirrhagus 332
discoideus, Harpus = smaragdinus, Harpalus 137
discus, Trechus 153, 24, 30
dispar, Apion 75, 21, 48
dispar, Thanatophilus 411, 19, 33
dispar, Xyleborus 404, 27, 53
dissectus, Plegaderus 342, 25
dissimile, Apion 75, 26, 48
dissimilis = borealis, Limobius 274
Dissoleucas 71
distincta = magnifica, Coccinella 216
distinctus, Aphodius 385, 25, 34
distinctus, Bradycellus 119, 23, 32
distinctus, Gyrinus 17
distinctus, Miarus 52
distinguendus = nigrofasciatus, Longitarsus 197
divaricata = magnifica, Coccinella 216
diversipunctata, Hypera 271, 21, 49
diversipunctatus, Phytonomus = diversipunctata, Hypera 271
Dolichosoma 359, 25, 38
domesticus, Xyloterus 222
Donacia 186-189, 20, 21, 23, 26, 44
Dorcatoma 63-64, 23, 25, 38, 214
Dorcus 336
dorsalis, Acupalpus 32, 99, 100
dorsalis, Helophorus 17
dorsalis, Longitarsus 194, 26, 46

Dorytomus 262-264, 21, 22, 24, 27, 50, 51
dresdensis, Dorcatoma 63, 23, 38, 214
Drilidae 313, 2, 23, 36
Drilus 313, 23, 36
Dromius 126-127, 19, 21, 23, 32
druparum, Anthonomus = rectirostris, Furcipus 266
Drupenatus 265, 27, 50
Drusenatus 265
Dryocaetes 398, 403
Dryoceotinus 398
Dryocoetes 53, 225, 398, 403
Dryocoetinus 398, 24, 53
dryographus, Xyleborus 405, 27, 53
dryophagus = dryographus, Xyleborus 405
Dryophilus 65, 21, 37
Dryophthorus 265, 20, 50
Dryopidae 18
Dryops 18
Drypta 128, 19, 32
dubia = barbata, Melandrya 352
dubia, Scraptia 407, 19, 42
dubia, Stenostola 167, 26, 44
dubius, Ptinus 38
dubius, Sphindus 412, 25, 39
duodecemstriatus = duodecimstriatus, Anommatus 169
duodecemstriatus, Anommatus 169
duodecimguttata, Vibidia 220, 22, 40
duodecimstriatus, Anommatus 169, 23, 40
duplicata, Magdalis 278, 24, 50
Dyschirius 128-129, 19-22, 24, 29
Dytiscidae 2, 16, 17
dytiscoides = violaceum, Platydema 417
Dytiscus 17

E

Ebaeus 359, 19, 38
Eccoptogaster 402
echinatus, Brachysomus 243, 26, 49
Ectinus 326
edmondsi, Tachys 151, 19, 30
edwardsi, Phloiophilus 373, 25, 38
edwardsii = edwardsi, Phloiophilus 373
Elaphrus 130, 22, 24, 29
Elater 322, 20, 36, 314
Elateridae 314-329, 2, 10, 19-21, 23, 25, 35
elateroides, Cerophytum 36
elateroides, Trixagus 420, 21, 36
Eledona 414, 25, 41
elegans = marginellus, Malachius 361
elegans, Acupalpus 99, 19, 32, 100
elevatus, Tropiphorus 307, 26, 49
Elleschus 265
Ellescus 27, 52, 265
Elmidae 2, 11, 16, 18
Elminthidae 15
Elodes 397, 22, 35
elongantulus, Ampedus 315, 23, 35
elongata = diversipunctata, Hypera 271
elongata, Elodes 397, 22, 35
elongata, Phytonomus = diversipunctata, Hypera 271
elongata, Rhagonycha 97, 23, 37
elongatulus, Elater = elongantulus, Ampedus 315
elongatulus, Hydroporus 17
elongatum, Colydium 222, 21, 41
elongatum, Nemozoma 421, 21, 36
elongatus, Hydrochus 17
elongatus, Lixus 49
elongatus, Tillus 215, 25, 38
emaaginatus = emarginatus, Aphanisticus 90

emarginatus, Aphanisticus 90, 19, 35
emarginatus, Spercheus 17
Endomychidae 330, 2, 20, 21, 25, 40
Endophloeus 222, 20, 41
Enochrus 18
Ephimeropus 239
ephippium, Bembidion 111, 22, 30
Epierus 337, 22, 33
Epiphanis 36
Epipolaeus 289
Epitrix 190, 26, 46
equestris, Amara 105, 24, 31
equiseti, Dicronychus 322, 20, 36
equiseti, Grypus 267, 27, 51
ericeti, Agonum 101, 24, 31
ericeti, Altica 171, 26, 46
erichsoni, Cardiophorus 321, 20, 36
erichsoni, Laricobius 37
erichsoni, Oulema 203, 20, 45
erichsonii, Lema = erichsoni, Oulema 203
Erirrhinus 282, 302
Ernobius 37
Ernopocerus **399-400**
Ernoporus 20, 24, 53, 399
Eros 345
erosus, Orthotomicus 53
Erotylidae 331, 2, 21, 23, 39
erratus, Calathus 119
erythroleucos, Coeliodes 259, 27, 51
erythroleucus = erythroleucos, Coeliodes 259
Eryx 417
Eubria 374, 21, 35
Eubrychius 266, 27, 51
Eucinetidae 2, 34
Eucinetus 34
Eucnemidae 332-334, 2, 20, 21, 25, 36
Eucnemis 332, 20, 36
Euheptaulacus 391, 19, 23, 34
Euophryum 50
euphorbiae, Ceutorhynchus 247, 24, 51
Europhilus 103
Eurynebria 141
exaratus, Atactogenus = plumbeus, Cneorhinus 258
exaratus, Cneorrhinus = plumbeus, Cneorhinus 258
exaratus, Ochthebius 18
exiguus, Acupalpus 100, 25, 32
exiguus, Cryptocephalus 182, 20, 45
exiguus, Zacladus 311, 27, 51
Exochomus 217, 22, 40
exsculptus, Ochthebius 18
extensus, Dyschirius 128, 19, 29

F

faber, Strophosoma 301, 26, 49
fabricii, Teretrius 343, 19
fagi, Cerylon 170, 25, 40
fagi, Diplocoelus 87, 25, 39
fagi, Ernoporus 399, 24, 53
fagi, Silvanoprus 411, 20, 39
famelica, Amara 105, 21, 31
farimairei = melanarius, Prionychus 417
fasciata, Mordella = villosa, Variimorda 366
fasciatus = affinis, Chromoderus 257
fasciatus = arietinus, Byrrhus 92
fasciatus, Anthribus 69, 23, 47
fasciatus, Aphodius 385, 23, 34
fasciatus, Trichius 34
fasciculata = dauci, Hypera 271
fasciculatus, Phytonomus = dauci, Hypera 271

fasciculatus, Pogonocherus 164, 26, 44
fastuosa, Pilemostoma 206, 23, 47
femoralis, Scymnus 219, 25, 40
femorata, Oncomera 370, 26, 43
fenestratus, Ilybius 17
fergussoni, Bledius 129
Feronia 147
ferrea = dubia, Stenostola 167
ferrugatus, Elater = pomorum, Ampedus 316
ferruginea = ferrugineum, Ostoma 371
ferruginea, Zilora 354, 26, 42
ferrugineum, Ostoma 371, 20, 38
ferrugineus = parallelepipedus, Cossonus 260
ferrugineus, Elater 322, 20, 36
ferrugineus, Hetaerius 338, 22, 33
ferrugineus, Hydroporus 17
ferrugineus, Longitarsus 194, 20, 46
festivus, Cis 211, 25, 40
fibulatus, Malthodes 96, 25, 37
figurata, Cantharis 37
filiformis, Synaptus 329, 20, 36
filirostre, Apion 75, 26, 48
filirostris, Dorytomus 262, 27, 50
fissirostris, Caenopsis 244, 26, 48
Five-spot ladybird 217
flavescens, Drilus 313, 23, 36
flavicollis = squamulatus, Tychius 310
flavicollis, Acupalpus 100, 23, 32
flavicollis, Zeugophora 210, 20, 44
flavicornis, Dorcatoma 64, 25, 38
flavicornis, Olibrus 371, 22, 40
flavimanum, Apion 76, 23, 48
flavipes = puncticollis, Dasytes 359
flavipes, Graptodytes 17
flavipes, Hypebaeus 359, 20, 39
flavipes, Luperus 200, 26, 45
flavipes, Polydrusus 290, 26, 49
flavipes, Zorochros 36
flavoguttatus, Malthodes 37
Fleutiauxellus 323, 23, 36
flexuosa, Abdera 349, 26, 42
flexuosa, Phyllotreta 45
floricola = cuprea, Cetonia 390
fluviatile, Bembidion 112, 24, 30
foetidus = fasciatus, Aphodius 385
foliorum, Rhynchaenus 294, 24, 53
Formica 180, 216, 217, 223, 338, 340, 344, 378, 390
formicetorum = angusticollis, Monotoma 378
foveatoscutellatum = soror, Apion 81
foveolatus, Raboceras 381
foveolatus, Rabocerus 381, 23, 41
foveolatus, Salpingus = gabrieli, Rabocerus 381
fowleri, Longitarsus 195, 23, 46
fracticornis, Onthophagus 393, 22, 34
fraxini, Corticeus 41
frit = collignensis, Bagous 235
frit = subcarinatus, Bagous 240
frit, Bagous 236, 21, 50
froelichi, Harpalus 132, 20, 31
froelichii = froelichi, Harpalus 132
frolichii = froelichi, Harpalus 132
frontalis = schmidti, Scymnus 220
frontalis, Cryptocephalus 182, 23, 45
frontalis, Malthinus 95, 25, 37
frontalis, Rhantus 17
fronticornis, Rhopalodontus = bicornis, Sulcacis 213
fulgidicollis, Helophorus 17
fuliginosus = umbrinus, Stenocarus 301
fuliginosus, Lasius 376
fuliginosus, Pycnomerus 41
fullo, Polyphylla 395, 19, 34

fulva, Amara 105, *24, 31*
fulva, Leptura 159, *21, 43*
fulvicollis, Mycetophagus 366, *10, 11*
fulvicorne, Anobium 214
fulvicorne, Strophosoma 301, *21, 49*
fulvicornis = fulvicorne, Strophosoma 301
fulvus, Berosus 18
fulvus, Cyrtonotus = fulva, Amara 105
fulvus, Trechus 153, *24, 29*
fumigatum, Bembidion 112, *24, 30*
funesta, Oxythyrea 34
furcatus, Haliplus 17
Furcipus 266, *27, 52*
fusca, Amara 105, *19, 31*
fusca, Cantharis 94, *21, 37*
fusca, Formica 338
fuscicornis = lutea, Rhagonycha 98
fuscicornis, Podagrica 207, *26, 46*
fuscipes, Podagrica 207, *23, 46*
fuscocinerea, Hypera 271, *26, 49*
fuscocinereus, Phytonomus = fuscocinerea, Hypera 271
fuscula = testacea, Scraptia 408
fuscula, Anisoxya 350, *23, 42*
fuscula, Scraptia 408, *20, 42*
fusculus, Heterocerus 334, *25, 35*
fuscus = ambiguus, Calathus 119
fuscus = linearis, Lyctus 347
fuscus, Telophorus = fusca, Cantharis 94

G

gabrieli, Raboceras 381
gabrieli, Rabocerus 381, *26, 42*
Galeruca 190, *20, 45*, 138
Galerucella *45*, 139
Galerucs 138
ganglbaueri, Longitarsus 195, *23, 46*
Gastrallus 65, *20, 37*
gemellatus, Sitona 298, *20, 49*
geminus, Hydroglyphus 17
geniculatum, Bembidion 30
genistae, Apion 76, *23, 48*
geographicus, Ceutorhynchus 248, *27, 51*
Georissus 17
Geotrupes 334, *23, 25, 33, 34*, 387, 392
Geotrupidae 334, *2, 16, 23, 25, 33*
germanica, Cicindela 123, *21, 29*
germanicus, Nicrophorus 33
germanus = palliatus, Ptinus 375
germanus, Diachromus 126, *19, 32*
germanus, Liparus 275, *21, 49*
germanus, Rhyssemus 396, *19, 34*
gibbus = tenebrioides, Zabrus 154
gigas, Ernobius 37
gilvipes, Bembidion 112, *24, 30*
glabra, Hydrothassa 45
glabratus, Sphaerites 412, *21, 32*
glabrirostris = lutulentus, Bagous 238
glabrirostris, Bagous 237, *27, 50*
glabriusculus, Hydroporus 17
glaucus = hollbergi, Coniocleonus 259
glaucus, Cleonus = hollbergi, Coniocleonus 259
Globicornis 312, *20, 37*
globosa, Apteropeda 172, *26, 46*
globula = globulus, Rutidosoma 296
globulosus = punctatus, Onthophilus 341
globulus, Rutidosoma 296, *24, 51*
Gnathoncus 337, *23, 33*
Gnorimus 391-392, *19, 20, 34*
goettingensis = violacea, Chrysolina 179

goettingensis, Chrysomela = violacea, Chrysolina 179
Gonodera 418
Gracilia 157, *20, 44*
gracilipes, Agonum 101, *22, 31*
gracilis, Nanophyes 83, *21, 48*
gracilis, Phloeophagus 286, *19, 50*
gracilis, Pterostichus 140, *01, 30*
gramineus, Cardiophorus 321, *19, 36*
graminis, Chrysolina 177, *23, 45*
graminis, Miarus 280, *27, 52*, 281
Grammoptera 157-158, *21, 23, 43*
Grammostethus 338, *25, 33*
granarius, Cercyon 18
granularis, Graptodytes 17
granulatus, Trypophloeus 404, *19, 53*
granulicollis, Procas 292, *22, 51*
granulum, Abraeus 335, *23, 33*
Graphoderus 17
grapii, Rhantus 17
Graptodytes 17
griseostriatus, Potamonectes 17
grisescens, Galerucella 45
griseus, Dryops 18
griseus, Helophorus 17
Gronops 266-267, *22, 26, 50*
Grypidius 267
Grypus 267, *27, 51*
guttatus, Panspoeus 36
guttifer, Malthodes 97, *25, 37*
guttiger, Ilybius 17
gyllenhali, Apion 77, *26, 48*
gyllenhali, Nebria 142
gyllenhalii = gyllenhali, Apion 77
Gymnetron 267-270, *24, 27, 52, 53*
Gynandrophthalma 190, *20, 45*
Gyrinidae 2, *16, 17*
Gyrinus 17

H

Hadrobregmus 65, *25, 37*
haematocephalus, Pachytychius 285, *20, 51*
haematodes, Apion 80
Haemonia 200
haemoptera, Chrysolina 177, *26, 45*
haemorrhoidalis = marginata, Lebia 139
Halacritus 338, *22, 33*
halensis, Scarodytes 17
Haliplidae 2, *16, 17*
Haliplus 17
Hallomenus 351, *26, 42*
halophilus, Enochrus 18
Haltica 170
Halyzia 40, 220
hannoverana = hannoveriana, Hydrothassa 191
hannoveriana, Hydrothassa 191, *21, 45*
Haplocnemus 357
Harminius 324, *25, 36*
Harpalus 130-137, *19-22, 24, 25, 31*
Harpus 137
harwoodi = longiceps, Rhynchites 85
Hazel leaf roller 83
heasleri = czwalinai, Bagous 236
hederae, Kissophagus 400, *27, 53*
Hedobia 66
Heledona 414
Helicella 313
Helix 313
Helmidomorphus 239
Helobium 118

INVERTEBRATE INDEX

Helochares	18
Helophorus	17
helopioides, Oodes	144, 25, 32
Helops	415, 25, 41, 414
helveticus = flavicornis, Olibrus	371
Hemicoelus	66, 22, 37
hemipterus, Phosphaenus	344, 20, 37
hemisphaerica, Cassida	173, 23, 47
Henbane flea beetle	208
hepaticus, Ceutorhynchus	248, 22, 51
Heptaulacus	392, 19, 34, 391
herbigrada, Aphthona	46
Hetaerius	338, 22, 33
Heteroceridae	335, 2, 16, 21, 25, 35
Heterocerus	335, 21, 25, 35
heydeni, Haliplus	17
hippocastani, Melolontha	393, 22, 34
Hippodamia	218, 22, 40
hirtipennis, Dorytomus	262, 24, 50
hirtulus, Ceutorhynchus	249, 27, 51
hirtus, Brachysomus	243, 21, 49
hirtus, Trinodes	313, 21, 37
hispidulus, Heterocerus	335, 21, 35
hispidus = perlatus, Trox	420
hispidus, Trachodes	303, 27, 50
Hister	338-339, 19, 22, 25, 33
Histeridae	335-343, 2, 16, 19, 21-23, 25, 32
hollbergi, Coniocleonus	259, 19, 49
holomelaena, Mordella	361, 22, 42
holomelina, Grammoptera	43
holosericeus = tristis, Chlaenius	122
holsaticus, Longitarsus	46
Homaloplia	393
homoeopathicus, Acritus	336, 21, 33
honestus, Harpalus	132, 19, 31
hookeri, Apion	75
Hop flea beetle	207
Hop root weevil	289
Horned dung beetle	390
horridus, Trichosirocalus	306, 24, 51
hortensis, Chaetocnema	175
hospes = luridipennis, Psylliodes	208
humerale, Bembidion	113, 19, 30
humeralis = binotatus, Hallomenus	351
humeralis, Anthonomus	232, 22, 52
humeralis, Mordellistena	363, 22, 42, 364
humeralis, Mycetochara	416, 23, 41
humeralis, Synchita	224, 25, 41, 225
huttoni, Pentarthrum	50
hybrida = maritima, Cicindela	123
hybrida, Cicindela	123, 20, 29
Hydaticus	17
Hydraena	18
Hydraenidae	2, 16, 18
Hydrochara	18
Hydrochus	17
Hydrocyphon	397, 25, 35
Hydroglyphus	17
Hydronomus	270, 27, 50
Hydrophilidae	2, 16, 17
Hydrophilus	18
Hydroporus	17
Hydrothassa	191, 21, 45
Hydrovatus	17
Hygrobiidae	2, 16
Hygrotus	17
Hylastes	53
Hylastinus	227
Hylecoetus	347, 25, 39
Hylesinus	377
Hylis	333, 20, 21, 36
Hylobius	270, 20, 49
Hylophilus	61
Hylotrupes	44
hyoscyami, Psylliodes	208, 20, 46
Hypebaeus	359, 20, 39
Hypera	271-273, 19-22, 24, 26, 49
Hyperaspis	218, 25, 40
hyperici, Chrysolina	138
Hypnoidus	323, 326
Hypocaccus	339-340, 21, 23, 25, 33
Hypocassida	191, 19, 47
hypochaeridis, Cryptocephalus	45, 180
Hypocopridae	344, 2, 22, 39
Hypocoprus	344, 22, 39
Hypophlaeus	413
Hypophloeus	413
Hypulus	351, 20, 42

I

ictor, Dorytomus	263, 27, 50
ignavus = honestus, Harpalus	132
ignicollis, Hydrochus	17
illigeri, Hister	339, 19, 33
Ilybius	17
immaculatum, Scaphium	382, 22, 33
immarginatus, Gastrallus	65, 20, 37
immundus, Saprinus	342, 25, 33
imperialis, Demetrias	125, 25, 32
imperialis, Ptinomorphus	66, 25, 37
impressa, Crepidodera	180, 23, 46
impressa, Donacia	188, 23, 44
impressus, Haplocnemus = pini, Aplocnemus	357
impressus, Selatosomus	329, 25, 36
impunctipennis, Dyschirius	129, 24, 29
inaequalis = longicollis, Pterostichus	149
inaequalis, Gronops	266, 22, 50
incarnatus, Bruchidius	44
indagator = inquisitor, Rhagium	165
inexspectatum, Anobium	62, 25, 37
infima, Amara	106, 22, 31, 104
inquinatus = distinctus, Aphodius	385
inquisitor, Calosoma	120, 22, 29
inquisitor, Rhagium	165, 26, 43
insignis = vectensis, Dromius	127
insignis, Orthochaetes	283, 27, 51
insularis, Ceutorhynchus	249, 20, 51
intermedium, Apion	77, 24, 48
interpunctatus, Rhynchites	85, 26, 47
interrupta, Galeruca	190, 20, 45
interruptus, Nicrophorus	409, 25, 33
intersecta, Candidula	313
intricatus, Carabus	121, 19, 29
intrudens, Syagrius	8, 50
inversus = ulmi, Anthonomus	233
iota, Rhynchaenus	295, 27, 53
Ips	53, 213
iricolor, Bembidion	30
iridis, Lixus	49
Ischnodes	324, 23, 36
Ischnomera	368-370, 20, 21, 26, 43
isotae, Enochrus	18
itala, Helicella	313

J

jacobaeae = tabidus, Longitarsus	199
jacquemarti, Cis	212, 25, 40
janthinus, Mecinus	279, 24, 52
Judolia	158, 23, 44

juglandis = humeralis, Synchita ... 224
jungermanniae, Smicronyx ... 300, 27, 51

K

kiesenwetteri = difficilis, Apion ... 71
kirbyi = cursor, Lissodema ... 380
Kissophagus ... 400, 27, 53
kluki = brunnipes, Orthoperus ... 226
kongsbergensis, Cyphon ... 396, 23, 35
Korynetes ... 213, 25, 38
kugelanni, Pterostichus ... 149, 19, 30
kutscherae, Longitarsus ... 198

L

labiatus, Agabus ... 17
Labidostomis ... 191, 20, 45
labile, Gymnetron ... 52
Laccobius ... 18
Laccophilus ... 17
Laccornis ... 17
Lacon ... 324, 19, 35
lacordairei = lacordairii, Triplax ... 331
lacordairii, Triplax ... 331, 21, 39
Laemophloeus ... 228, 20, 39
Laemostenus ... 31
laevicolle, Apion ... 77, 24, 48
laevigata, Xylita ... 354, 23, 42
laevigatum = brunnipes, Apion ... 73
laevigatum, Apion ... 78, 22, 48
laevigatus = immarginatus, Gastrallus ... 65
laeviuscula, Clytra ... 180, 19, 45
Lagria ... 415, 20, 41
Lamia ... 159, 20, 44
laminatus, Haliplus ... 17
Lamprohiza ... 344, 19, 37
Lamprosoma ... 45
Lampyridae ... 344, 2, 19, 20, 37
Lampyris ... 344
Langelandia ... 222, 21, 41
lapathi, Cryptorhynchus ... 261, 27, 50
lapidosus = fulvus, Trechus ... 153
lapponicus, Dytiscus ... 17
lapponicus, Elaphrus ... 130, 22, 29
Large fruit bark beetle ... 402
Laria ... 87
Laricobius ... 37
Larinus ... 273, 26, 49
Lasiorhynchites ... 84
Lasiotrechus ... 153
Lasius ... 180, 219, 225, 265, 335, 336, 376
latecincta = crassicornis, Chrysolina ... 176
laterale, Bembidion ... 113, 24, 30
lateralis, Cillenus = laterale, Bembidion ... 113
Lathridiidae ... 2, 16
lathridioides = latridioides, Hypocoprus ... 344
laticollis, Baris ... 241, 24, 52
laticollis, Helophorus ... 17
laticollis, Trachyphloeus ... 304, 24, 48
laticornis, Agrilus ... 88, 25, 35, 90
latifrons = coluber, Cis ... 210
latipalpis = thoracica, Anaspis ... 407
latirostris = conicus, Rhinocyllus ... 293
latirostris, Platyrrhinus = resinosus, Platyrhinus ... 70
latridioides, Hypocoprus ... 344, 22, 39
latus, Deronectes ... 17
latus, Symbiotes ... 330, 25, 40
Lebia ... 138-139, 19, 25, 32
Leiodidae ... 2

Leiosoma ... 273-274, 21, 26, 49, 50
Leistus ... 139, 22, 29
lejalui, Ochthebius ... 18
Lema ... 45, 203
lemoroi, Apion ... 78, 22, 47
lenensis, Ochthebius ... 18
Leperisinus ... 400, 27, 53, 421
lepida, Feronia = lepidus, Pterostichus ... 149
lepidii, Baris ... 241, 24, 52
lepidus, Pterostichus ... 149, 24, 30
lepidus, Stictonectes ... 17
Leptinidae ... 2
Leptophloeus ... 228, 20, 39
Leptura ... 159-161, 19, 21, 23, 43, 158, 222
Lepyrus ... 274, 19, 49
lethifera, Blaps ... 41
leucaspis, Mordella ... 362, 22, 42
leucogaster, Litodactylus ... 276, 27, 52
leucophthalmus, Sphodrus ... 31
lichenum, Ptinus ... 375, 21, 38
lichtensteini, Pityophthorus ... 401, 21, 53
lichtensteinii = lichtensteini, Pityophthorus ... 401
Licinus ... 139-140, 23, 25, 32
ligustici, Otiorhynchus ... 284, 20, 48
lilii, Lilioceris ... 45
Lilioceris ... 45
limbatum, Omophron ... 143, 19, 29
limbatus, Scymnus ... 220, 25, 40
limbatus, Thymalus ... 371, 25, 38
Limexylon ... 348
Limnebius ... 18
Limnichidae ... 344, 2, 23, 35
Limnichus ... 344, 23, 35
Limnoxenus ... 18
Limobius ... 274-275, 20, 24, 49
limonii, Apion ... 78, 26, 47
Limoniscus ... 325, 10, 19, 36
limosus, Bagous ... 237, 27, 50
linariae, Gymnetron ... 268, 24, 53
lineare, Dolichosoma ... 359, 25, 38
linearis, Corticeus ... 41
linearis, Cossonus ... 260, 24, 50
linearis, Lyctus ... 347, 25, 38, 214, 215, 343
lineatocribratus, Cis ... 212, 25, 40
lineatulus, Tychius ... 308, 24, 52
lineatus, Xyloterus ... 53
lineellus, Sitona ... 49
lineola, Orsodacne ... 203, 26, 44
Lionychus ... 140, 21, 32
Liosoma ... 273
Lipara ... 126
Liparus ... 275, 21, 26, 49
Lissodema ... 380, 23, 26, 41
Litodactylus ... 27, 52, 266, 276
litorale, Bembidion ... 113, 24, 30
litoralis = littoralis, Pogonus ... 146
littoralis, Pogonus ... 146, 24, 30
livens, Agonum ... 101, 24, 31
livescerum = reflexum, Apion ... 79
livida, Nebria ... 142, 22, 29
lividus, Aphodius ... 385, 19, 34
lividus, Helochares ... 18
Lixus ... 276-277, 20, 22, 49
lloydi, Gymnetron = potentillae, Sibinia ... 297
lobatus, Sphinginus ... 361, 22, 39
longiceps, Dromius ... 126, 23, 32
longiceps, Rhynchites ... 85, 26, 47
longicolle, Melasoma = tremula, Chrysomela ... 179
longicollis, Cionus ... 257, 24, 49
longicollis, Pterostichus ... 149, 24, 30
longicornis, Hydroporus ... 17

longicornis, Thalassophilus — **152**, *22, 29*
longiseta, Longitarsus — **195**, *22, 46*
longitarsis, Bagous — **237**, *20, 50*, 234
longitarsis, Helophorus — *17*
longitarsis, Xyletinus — **66**, *20, 37*
Longitarsus — **191-199**, *20, 22, 23, 26, 46*
longulus, Hydroporus — *17*
lonicerae, Rhynchaenus — *53*
Lucanidae — **345**, *2, 16, 19, 25, 33*
Lucanus — **345**, *25, 33*
lucens, Lipara — 126
lucida, Amara — **106**, *24, 31*
Ludius — 322
lugens, Agonum — *31*
lugens, Megapenthes — **326**, *19, 36*
lugubris, Formica — 217, 223, 340, 378
lunaris, Copris — **390**, *19, 34*
lunatum, Bembidion — **114**, *24, 30*
lunatus, Callistus — **120**, *19, 32*
lunatus, Gronops — **267**, *26, 50*
Luperus — **200**, *26, 45*, 173
luridipennis, Pogonus — **146**, *21, 30*
luridipennis, Psylliodes — **208**, *8, 20, 47*
luridus, Acupalpus — 100
luridus, Berosus — *18*
lutea, Rhagonycha — **98**, *25, 37*
luteola, Galerucella — 139
luteola, Psylliodes — **209**, *22, 47*
luteorubra, Platynaspis — **219**, *23, 40*
lutosus, Bagous — **238**, *20, 50*
lutulentus, Bagous — **238**, *27, 50*
lutulosus, Bagous — **238**, *24, 50*
Lycidae — **345-346**, *2, 22, 23, 25, 37*
Lycoperdina — **330**, *20, 21, 40*
lycopi, Longitarsus — **196**, *26, 46*
lycopi? = fowleri, Longitarsus — 195
Lyctidae — **347**, *2, 16, 25, 38*
Lyctus — **347**, *25, 38*, 214, 215, 343
Lymantor — 403
Lymexylidae — **347-348**, *2, 16, 20, 25, 39*
Lymexylon — **348**, *20, 39*
Lymnaeum — 115
Lyperosomus — 147
Lythraria — **200**, *26, 46*
lythropterus, Elater = cinnabarinus, Ampedus — 315
Lytta — *43*

M

macer, Pterostichus — *30*
Macrocephalus — *70*
Macronychus — *18*
Macroplea — **200-201**, *21, 23, 44*
macularis, Cymindis — **124**, *19, 32*
macularius, Sitona — **298**, *26, 49*
maculosa, Simplocaria — **93**, *22, 35*
Magdalis — **277-278**, *24, 26, 50*
magnifica, Coccinella — **216**, *23, 40*, 217
majalis, Dorytomus — **263**, *22, 50*
major, Oulimnius — *18*
Malachius — **360-361**, *21, 25, 39*
mali, Scolytus — **402**, *27, 53*
Malthinus — **95**, *25, 37*
Malthodes — **95-97**, *20, 21, 25, 37*
Mantura — **201-202**, *23, 26, 46*
Margarinotus — 338, 341
marginalis = sanguinolenta, Chrysolina — 178
marginalis, Axinotarsus — 358
marginalis, Chrysomela = sanguinolenta, Chrysolina — 178
marginata, Chrysolina — **178**, *23, 45*

marginata, Lebia — **139**, *19, 32*
marginatus, Grammostethus — **338**, *25, 33*
marginatus, Hydroporus — *17*
marginatus, Polydrusus — **290**, *20, 49*
marginellus, Malachius — **361**, *25, 39*
marinus, Aepus — **100**, *24, 29*
marinus, Ochthebius — *18*
maritima, Anurida — 101
maritima, Cicindela — **123**, *24, 29*
maritimum, Bembidion — *30*
maritimus = dimidiatus, Baeckmanniolus — 337
maritimus, Cathormiocerus — **245**, *21, 48*
maritimus, Fleutiauxellus — **323**, *23, 36*
maritimus, Pachylopus = dimidiatus, Baeckmanniolus — 337
maritimus, Saprinus = dimidiatus, Baeckmanniolus — 337
markovichianus, Endophloeus — **222**, *20, 41*
Masoreus — **141**, *23, 32*
matthewsi, Mantura — *46*
maurus, Malthodes — **97**, *25, 37*
Mecinus — **279**, *24, 27, 52*
Megachile — 215, 216
Megapenthes — **326**, *19, 36*
megaphallus, Hydrochus — *17*
Megatoma — **312**, *25, 37*
melanarium, Gymnetron — **268**, *27, 53*
melanarius = melanarium, Gymnetron — 268
melanarius, Agabus — *17*
melanarius, Prionychus — **417**, *20, 41*
melancholicus, Harpalus — **132**, *19, 31*
melancholicus, Selatosomus — *36*
Melandrya — **352**, *20, 26, 42*
Melandryidae — **348-354**, *2, 16, 20-21, 23, 26, 42*
melanocephalus, Enochrus — *18*
melanocephalus, Longitarsus — 198
Melanophila — *35*
melanostictus = distinctus, Aphodius — 385
melanostoma, Anaspis — **406**, *22, 42*
Melanotus — **326**, *19, 36*
melanura, Odacantha — **143**, *25, 32*, 125
Melasis — **334**, *25, 36*
Melasoma — 179
meles, Hypera — **272**, *24, 49*
melleti, Harpalus — **133**, *22, 31*
Meloe — **355-356**, *19-21, 26, 43*
Meloidae — **355-356**, *2, 19-21, 26, 43*
Melolontha — **393**, *22, 34*
melolontha, Melolontha — 393
Melyridae — **357-361**, *2, 19-23, 25, 38*
memnonia, Magdalis — *50*
mendax, Saprosites — *34*
menthastri, Chrysolina — *45*
menyanthidis — 187
menyanthidis = clavipes, Donacia — 187
menyanthis — 187
menyanthis = clavipes, Donacia — 187
meridionalis, Badister — **109**, *22, 32*
meridionalis, Eucinetus — *34*
Merophysiidae — *2*
Mesites — **280**, *27, 50*
Mesosa — **161**, *21, 44*
metallicum, Scaphidema — **418**, *25, 41*
metallicus, Corymbites = nigricornis, Selatosomus — 329
metallicus, Hypocaccus — **339**, *21, 33*
Metallites — 290
Metoecus — *43*
Miarus — **280-281**, *21, 22, 27, 52*
micans = mollis, Polydrusus — 291
micans, Agonum — 103
micans, Orchesia — **353**, *26, 42*
Microlomalus — 341
Micromalus — 341

Microrrhagus 332
micros, Maurus 280, 21, 52
micros, Tachys 151, 00, 30
micros, Trechus 30
millefolii, Olibrus 377, 25, 40
millum = cineraceum, Apion 73
minutus = quercicola, Ampedus 317
minimum, Apion 79, 21, 48
minor, Molorchus 44
minor, Orchesia 353, 26, 42
minor, Tomicus 403, 21, 53
Mint flea beetle 194
minuta, Elodes 35
minuta, Gracilia 157, 20, 44
minuta, Platycis 346, 25, 37
minuta, Trachys 91, 20, 35
minutissima, Hydraena 18
minutissimus, Bidessus 17
minutus = minuta, Platycis 346
minutus = minuta, Trachys 91
minutus = nigrellus, Clambus 213
minutus = pallidulus, Clambus 213
minutus, Gyrinus 17
Mire Pill Beetle or Bog Hog 92
Miscodera 141, 24, 29
misellus = maurus, Malthodes 97
misellus, Acalles 50
mixtus, Ceutorhynchus 249, 27, 51
mixtus, Limobius 275, 20, 49
Mniophila 202, 26, 46
mobilicornis, Odontaeus = armiger, Odonteus 334
Moccas beetle 359
moelleri, Ceutorhynchus 250, 22, 51
moguntiacus, Ceuthorhynchus = thomsoni, Ceutorhynchus 255
moguntiacus, Ceuthorrhynchus = thomsoni, Ceutorhynchus 255
molleri, Ceutorhynchus = moelleri, Ceutorhynchus 250
molleri, Ceutorrhynchus = moelleri, Ceutorhynchus 250
mollina, Omiamima 283, 24, 48
mollinus, Omias = mollina, Omiamima 283
mollis, Opilo 214, 25, 38
mollis, Polydrusus 291, 26, 49
Molorchus 162, 23, 44
Monacha 313
monilis, Carabus 121, 24, 29
monilis, Laemophloeus 228, 20, 39
Mononychus 281, 24, 51
monostigma, Demetrias 125, 25, 32
Monotoma 378, 21, 39
Monotomidae 16
montanus, Leistus 139, 22, 29
monticola, Bembidion 114, 24, 30
moraei, Cryptocephalus 45
Mordella 361-362, 366, 22, 42, 363
Mordellidae 361-366, 2, 9, 16, 22, 23, 26, 42
Mordellistena 362-365, 22, 42, 43
morio, Otiorhynchus 285, 22, 48
mortisaga, Blaps 41
moschata, Aromia 156, 26, 44
mucronatus, Haliplus 17
muelleri, Agonum 102
multipunctata, Blethisa 118, 24, 29
multipunctatum, Helobium = multipunctata, Blethisa 118
muralis, Apalus 355, 20, 43
muricatus, Phytobius 287, 24, 52, 288
murina = fuscocinerea, Hypera 271
murinus, Phytonomus = fuscocinerea, Hypera 271
murinus, Porcinolus 93, 25, 35
Murmidius 170, 19, 40
murraea, Cassida 47
muscorum, Mniophila 202, 26, 46
muscorum, Otiorrhynchus = desertus, Otiorhynchus 284

Musk beetle 156
mutator, Geotrupes 25, 33, 392
mutilatus, Mesocoelopus 201, 23, 44, 200
muticus = clavicornis, Orthocerus 223
mutilatus, Salpingus = foveolatus, Rabocerus 381
Mycetochara 416, 23, 41
Mycetochares 410
Mycetophagidae 366-367, 2, 16, 19, 23, 25, 41
Mycetophagus 366-367, 19, 23, 25, 41, 62, 324
Mycteridae 368, 2, 16, 19, 42
Mycterus 368, 19, 42
Myelophilus 403
Myrmechixenus 223, 21, 22, 41
myrmecophilus, Cathormiocerus 245, 21, 48
Myrmecoxenus 223
Myrmetes 340, 25, 33
mysticus, Anaglyptus 156, 26, 44
mysticus, Malthodes 37

N

Nanophyes 83, 21, 48
nanuloides, Mordellistena 364, 22, 42
nanus, Helophorus 17
nanus, Ochthebius 18
nasturtii, Drupenatus 265, 27, 52
nasturtii, Longitarsus 196, 26, 46, 199
natator, Gyrinus 17
navale, Lymexylon 348, 20, 39
Nebria 141-142, 22, 29
nebulosa, Cassida 174, 22, 47
nebulosa, Mesosa 161, 21, 44
nebulosus, Anthribus 69, 26, 47
nebulosus, Coniocleonus 260, 21, 49, 259
Necrobia 38
Necrophorus 409
Negastrius 326-327, 20, 36
neglectus, Harpalus 31
neglectus, Hydroporus 17
neglectus, Xylophilus = brevicornis, Aderus 61
neglectus, Xylophilus = populneus, Aderus 61
Nemonychidae 2, 16, 47
nemoralis, Aphodius 386, 23, 34
nemoralis, Cepaea 313
nemorivagus, Anisodactylus 108, 22, 31
Nemosoma 421
Nemozoma 421, 21, 38
Nephus 218-219, 19, 20, 40
neuwaldeggiana, Mordella 363
neuwaldeggiana, Mordellistena 364, 22, 42
newaldeggiana = neuwaldeggiana, Mordellistena 364
Nicrophorus 409-410, 23, 25, 33
nigellus = brevicollis, Malthodes 95
niger, Aphodius 386, 19, 34
niger, Dasytes 358, 23, 38
niger, Lasius 180, 219
niger, Limnoxenus 18
niger, Otiorhynchus 48
nigerrimus, Ampedus 316, 19, 35
nigerrimus, Longitarsus 196, 20, 46, 198
nigra = braccata, Plateumaris 206
nigra, Strangalia 168, 23, 44
nigrellus, Clambus 213, 23, 34
nigrescens, Orthoperus 226, 25, 40
nigriceps, Aphthona 171, 23, 46
nigriceps, Perigona 31
nigricorne, Bembidion 114, 24, 30
nigricornis, Aplocnemus 357, 23, 38
nigricornis, Chlaenius 122, 25, 32
nigricornis, Chrysanthia 368, 20, 43

nigricornis, Selatosomus — 329, *21, 36*
nigrinus, Ampedus — 316, *25, 35*
nigrinus, Ernobius — *37*
nigrinus, Scymnus — *40*
nigripes, Globicornis — 312, *20, 37*
nigrirostris, Hypera — *272*
nigrita = tyrolensis, Silpha — *411*
nigrita, Curimopsis — 92, *19, 35*
nigrita, Hydraena — *18*
nigritarsis = lutulentus, Bagous — *238*
nigritarsis, Cionus — 257, *24, 49*
nigritarsis, Coeliodes — 259, *24, 51*
nigrofasciatus, Longitarsus — 197, *23, 46*
nigrofasciatus, Luperus = circumfusus, Calomicrus — *173*
nigrolineatus, Coelambus — *17*
nigromaculatus, Exochomus — 217, *22, 40*
nigropiceum, Bembidion — 115, *22, 30*
nigroruber, Pyropterus — 346, *23, 37*
nigrum, Agonum — 102, *24, 31*
nitens, Attelabus — *47*
nitens, Carabus — 121, *24, 29*
nitens, Normandia — *18*
nitida, Amara — 106, *22, 31*
nitidicollis, Hydrochus — *17*
nitidula, Anthaxia — 90, *19, 35*
nitidula, Chalcoides — 176, *26, 46*
Nitidulidae — *2*
nitidulus = aequalis, Orthoperus — *226*
nitidulus = cuspidatus, Saprinus — *342*
nitidulus = subnitescens, Saprinus — *343*
nitidulus, Chlaenius — 122, *19, 32*
nitidulus, Cryptocephalus — 183, *20, 45*
nitidulus, Dryops — *18*
nitidulus, Rhizophagus — 378, *25, 39*
nitidus = cylindricus, Teredus — *225*
nitidus, Dyschirius — 129, *22, 29*
nitidus, Hemicoelus — 66, *22, 37*
nitidus, Limnebius — *18*
nivalis, Nebria — 142, *22, 29*
niveirostris, Tropideres — 71, *20, 47*
nobilis, Cassida — 174, *26, 47*
nobilis, Gnorimus — 391, *20, 34*
noctiluca, Lampyris — *344*
nodulosus, Bagous — 239, *20, 50*
Normandia — *18*
normannum, Bembidion — *30*
Northern Cockchafer — *393*
Notaris — 282, *24, 27, 51*
Noteridae — 2, 16, *17*
Noterus — *17*
Notiophilus — 142-143, *24, 29*
Notolaemus — 228, *23, 39*
novemlineatus, Coelambus — *17*
nubila = nebulosa, Mesosa — *161*
nuchicornis, Onthophagus — 394, *23, 34*
nutans, Onthophagus — 394, *19, 34*

O

Oak pin-hole borer — *374*
Oberea — 162, *20, 44*, 163
obliquum, Bembidion — 115, *24, 30*
obliteratus, Longitarsus — *46*
oblongicollis, Rhizophagus — 379, *20, 39*
oblongiusculus, Scybalicus — *32*
oblongopunctata, Feronia = oblongopunctatus, Pterostichus — *150*
oblongopunctatus, Pterostichus — 150, *24, 30*
oblongulum, Leiosoma — 273, *26, 49*
oblongus, Laccornis — *17*
Obrium — 162, *19, 44*

obscura, Cantharis — 94, *25, 37*
obscura, Donacia — 188, *23, 44*
obscura, Silpha — 410, *20, 33*
obscurus, Bromius — 172, *20, 45*
obscurus, Dyschirius — 129, *20, 29*
obscurus, Harpalus — 133, *19, 31*
obscurus, Helochares — *18*
obscurus, Hylastinus — *227*
obscurus, Paralister — 341, *19, 33*
obscurus, Telophorus = obscura, Cantharis — 94
obsoletus, Dicheirotrichus — 126, *25, 32*
obsoletus, Hydroporus — *17*
obtusata, Mantura — 202, *26, 46*
obtusus, Tropiphorus — 307, *24, 49*
occidentalis, Bledius — *128*
ochraceum = ochraceus, Amphimallon — *383*
ochraceus, Amphimallon — 383, *23, 34*
ochroleucus, Longitarsus — 197, *26, 46*
ochropterus, Enochrus — *18*
Ochrosis — 202, *21, 46*, 200
ochrostoma = nitidulus, Cryptocephalus — *183*
Ochthebius — *18*
octomaculatum, Bembidion — 115, *19, 30*
oculata, Oberea — 162, *20, 44*, 162
oculatus = plumbeus, Dasytes — *359*
oculatus, Aderus — 61, *26, 43*
Ocypus — *118*
Odacantha — 143, *25, 32*, 125
Odontaeus — *334*
Odonteus — 334, *23, 33*
Oedemera — 370, *20, 43*
Oedemeridae — 368-370, 2, 16, *20, 21, 26, 43*
oelandica, Adimonia = interrupta, Galeruca — *190*
Oil beetle — *356*
olexai, Hylis — *21, 36*
Olibrus — 371-372, *22, 25, 40*
olivaceus, Bruchidius — 87, *20, 44*
olivaceus, Rhynchites — 85, *23, 47*
olivieri = asperatus, Trachyphloeus — *304*
olssoni, Phytobius — 288, *21, 52*
Omaloplia — 393, *25, 34*
Omiamima — 283, *24, 48*
Omias — *283*
Omophlus — 416, *20, 41*
Omophron — 143, *19, 29*
Oncomera — 370, *26, 43*
ononidis, Hypera — 272, *22, 49*
ononidis, Sitona — 299, *26, 49*
Onthophagus — 393-395, *19, 22, 23, 25, 34*
Onthophilus — 341, *22, 33*
Oodes — 144, *25, 32*
opaca, Aclypea — 409, *23, 33*
opacus, Bledius — *128*
opacus, Gyrinus — *17*
Opatrum — 416, *25, 41*
Ophonus — *130*
ophthalmicus = olivaceus, Rhynchites — 85
ophthalmicus, Lasiorhynchites = olivaceus, Rhynchites — 85
Opilio — *214*
Opilo — 214, *25, 38*
orbicularis, Scirtes — 398, *23, 35*
Orchesia — 353, *26, 42*
Orchestes — *294*
Oreodytes — *17*
oricalcia, Chrysolina — 178, *26, 45*
orichalcia = oricalcia, Chrysolina — *178*
orichalcia, Chrysomela = oricalcia, Chrysolina — *178*
orni, Leperisinus — 400, *27, 53*
Orobitis — *52*
Orsodacne — 203, *26, 44*
Orthathous — *320*

Orthocaetes 283
Orthocerus 223, 25, 41
Orthochaeter 283, 27, 51
Orthoperus 228, 21, 22, 25, 40
Orthotomicus 53
Osmia 215, 216, 355, 356, 375
Osphya 363, 21, 42
Ostoma 371, 20, 39
Otiorhynchus 284-285, 20, 22, 26, 48
Otiorrhynchus 284
Oulema 203, 20, 45
Oulimnius 18
ovalis, Murmidius 170, 19, 40
Oxychilus 313
Oxylaemus 224, 19, 21, 41
Oxythyrea 34

P

Pachylopus 337
Pachyta 155, 158
Pachytychius 285, 20, 51
palliatus, Ptinus 375, 23, 38
palliatus, Tanymecus 302, 26, 49
pallida = nigriceps, Aphthona 171
pallida, Biorhiza 262
pallida, Cantharis 37, 98
pallidipenne, Bembidion 115, 24, 30
pallidulus, Clambus 213, 22, 34
pallidus, Cylindrinotus 414, 25, 50
pallipes, Apion 47
pallipes, Asaphidion 108, 24, 30
paludosum, Bembidium = litorale, Bembidion 113
palustris, Eubria 374, 21, 35
palustris, Hydraena 18
Panagaeus 144, 19, 25, 32
pannonicus, Agrilus 89, 23, 35
Panspoeus 36
papposus, Limnebius 18
Paracymus 18
paradoxus, Metoecus 43
Paralister 341, 19, 33
parallelepipedus, Cossonus 260, 27, 50
parallelepipedus, Dorcus 336
parallelepipedus, Paromalus 341, 19, 33
parallelogrammus, Coelambus 17
parallelopipedus = parallelepipedus, Cossonus 260
parallelopipedus = parallelepipedus, Paromalus 341
parallelopipedus, Micromalus = parallelepipedus, Paromalus 341
parallelus = melleti, Harpalus 133
parallelus, Abax 31
parallelus, Harpalus 133, 21, 31
parallelus, Platypus 53
parallelus, Platypus 374, 22
parallelus, Tychius 308, 24, 52
paraplecticus, Lixus 276, 20, 49
Paratillus 38
Paromalus 341, 19, 33
parvula = acuticollis, Mordellistena 362
parvula = nanuloides, Mordellistena 364
parvula = parvuloides, Mordellistena 365
parvula, Mordellistena 364, 22, 42
parvuloides, Mordellistena 365, 22, 43
parvulus, Ceutorhynchus 250, 21, 51
parvulus, Cryptocephalus 183, 26, 45
parvulus, Longitarsus 197, 23, 46
parvulus, Rhizophagus 379, 21, 39
parvulus, Tachys 152, 24, 30
pastinacae, Hypera 273, 20, 49
patricia = equestris, Amara 105

Patrobus 145, 24, 29
patruelis = nigrofasciatus, Longitarsus 197
nautillus, Rhynchites 86, 21, 47
paykulli, Aphodius 380, 33, 34
paykulli, Gyrinus 17
Pear weevil 277
pectinicornis, Ctenicera 321, 23, 36
pectinicornis, Ptilinus 215
pectinicornis, Schizotus 376, 23, 42
pectoralis, Ceutorhynchus 250, 24, 51
Pediacus 229, 23, 39
pedicularius = conspersus, Anthonomus 232
pedicularius, Ebaeus 359, 19, 38
pellucens, Sibinia 52
pellucidus, Longitarsus 46
Pelophila 145, 21, 29
peltatus, Badister 110, 23, 32
Peltidae 371, 2, 20, 25, 38
Peltodytes 17
Pentarthrum 50
perforatus, Rhopalodontus 212, 21, 40
Perigona 31
Perileptus 145, 22, 29
Peritelus 48
perlatus, Trox 420, 19, 33, 421
pervicax, Ceutorhynchus 251, 24, 51
Peryphus (Bembidion) 112, 114, 117
petro, Bagous 239, 19, 50
Phaedon 204, 26, 45, 343
Phalacridae 371-373, 2, 16, 22, 23, 25, 40
Phalacrus 372, 23, 40
Phenacoccus 219
phlegmatica, Magdalis 278, 24, 50
Phloeophagia 293
Phloeophagus 286, 19, 24, 50
Phloeophilus 373
Phloeosinus 53
Phloeotrya 354
Phloiophilidae 373, 2, 25, 38
Phloiophilus 25, 38, 373
Phloiotrya 354, 26, 42
Phosphaenus 344, 20, 37
Phyllobius 286, 26, 48
Phyllobrotica 45
phyllocola = varians, Anthonomus 233
Phyllodecta 204, 21, 45
Phyllotreta 204-205, 23, 26, 45
Phymatodes 163, 26, 44
Phytobius 287-289, 21, 24, 27, 52, 266, 276
Phytodecta 205, 26, 45
Phytoecia 163, 26, 44
Phytonomus 271
piceae, Cryphalus 53
piceus = edmondsi, Tachys 151
piceus, Hydrophilus 18
piceus, Mycetophagus 366, 25, 41, 62, 324
piceus, Myrmetes 340, 25, 33
picicornis, Baris 242, 27, 52
picipennis = vernalis, Harpalus 137
picipes = fabricii, Teretrius 343
picipes = ganglbaueri, Longitarsus 195
picipes, Rhizophagus 379, 23, 39
piger, Cleonus 258, 26, 49
Pilemostoma 206, 23, 47
pillumus, Pseudostyphlus 293, 24, 51
pilosellus, Ceutorhynchus 251, 21, 51
pilosus, Polydrusus 49
pilosus, Xylechinus 53
pilumnus = pillumus, Pseudostyphlus 293
pimpinellae, Anthrenus 311, 19, 37
Pine chafer 395

INVERTEBRATE INDEX

Pine sawyer 159
pineti, Brachonyx 52
pini, Aplocnemus 357, 25, 38
pini, Ernobius 37
piri, Anthonomus 232, 22, 52
pisorum, Bruchus 44
Pissodes 289, 21, 50
Pityogenes 401, 24, 53
Pityophthorus 401, 21, 53
plagiatus, Aphodius 387, 25, 34
plagiatus, Stenolophus 32
Plagiogonus 387
Plagionotus 163, 19, 44
planata, Uleiota 229, 23, 39
planatus, Brontes = planata, Uleiota 229
plantaginis, Hypera 272
plantagomaritimus, Longitarsus 198, 26, 46
plantarum, Miarus 281, 22, 52
planus, Larinus 273, 26, 49
Plateumaris 206, 23, 26, 44
Platycerus 345, 19, 33
Platycis 346, 22, 25, 37
Platydema 417, 20, 41
Platyderus 146, 24, 31
Platynaspis 219, 23, 40
Platypodidae 374, 2, 16, 22, 27, 53
Platypus 374, 22, 27, 53, 222
Platyrhinus 70, 26, 47
Platyrrhinus 70
Platystomos 70, 26, 47
Platystomus 70
Plegaderus 342, 25, 32
Pleurophorus 395, 19, 34
Plinthus 289, 24, 50
plumbeus, Cneorhinus 258, 26, 49
plumbeus, Dasytes 359, 25, 38
plumipes, Anthophora 355
Podagrica 207, 23, 26, 46
Poecilium 163
poeciloides, Anisodactylus 108, 21, 31
Poecilus 149
Pogonochaerus 164
Pogonocherus 164, 26, 44
Pogonus 146, 21, 24, 30
polaris, Phyllodecta 204, 21, 45
poligraphus, Polygraphus 53
Polistichus 146, 20, 32
polita, Chrysolina 138
politus = picipes, Rhizophagus 379
politus, Dyschirius 29, 128
Polydrosus 290
Polydrusus 290-291, 20, 24, 26, 48, 49
Polygraphus 53
polylineatus, Tychius 309, 22, 52
Polyphylla 395, 19, 34
Polystichus 146
Pomatinus 18
pomonae = quercicola, Ampedus 317
pomorum, Ampedus 316, 25, 35
Pontania 79
ponticus, Laccophilus 17
Poophagus 265
Poplar borer 166
Poplar leaf roller 83
populi, Byctiscus 83, 21, 47
populi, Chrysomela 45
populi, Mycetophagus 367, 23, 41
populi, Rhynchaenus 295, 22, 53
populnea = populneus, Aderus 61
populnea, Xylophila = populneus, Aderus 61
populnea, Xylophilia = populneus, Aderus 61

populneus, Aderus 61, 26, 43
porcatus, Otiorhynchus 48
porcicollis, Brindalus 389, 19, 34
Porcinolus 93, 25, 35
porcus, Aphodius 387, 25, 34
posthumus, Ceuthorhynchus = pumilio, Ceutorhynchus 252
posthumus, Ceuthorrhynchidius = pumilio, Ceutorhynchus 252
posthumus, Coeliodes = pumilio, Ceutorhynchus 252
Potamonectes 17
Potamonectes depressus 17
potentillae, Sibinia 297, 27, 52
Potosia 390
poweri, Ochthebius 18
praetermissa, Amara 107, 24, 31
praeusta = ustulata, Grammoptera 157
praeustus, Ampedus 35
praeustus, Elater = cardinalis, Ampedus 314
prasina, Cassida 174, 26, 47
prasinus, Polydrusus 49
pratensis, Rhynchaenus 295, 27, 53
primarius, Cryptocephalus 184, 20, 45
primita = primitus, Sibinia 297
primitus, Sibinia 297, 27, 52
Prionocyphon 397, 25, 35
Prionus 164, 23, 43
Prionychus 417, 20, 25, 41
Procas 292, 22, 51
Procraerus 327, 21, 36
proxima, Pontania 79
pruni = mali, Scolytus 402
Psammobius 389, 395
Psammodius 395, 23, 34, 389
Psammoporus 383
Pselactus 293, 27, 50
Pselaphidae 2, 16
Psephenidae 374, 2, 21, 35
pseudacori = punctumalbum, Mononychus 281
Pseudocistela 418, 26, 41
pseudopumila, Mordellistena 365, 22, 42
pseudopustulata, Hyperaspis 218, 25, 40
pseudosardous = marginellus, Malachius 361
pseudostigma, Rhynchaenus 296, 22, 53
Pseudostyphlus 293, 24, 51
Psylliodes 207-209, 8, 20-22, 26, 46, 47
Pterostichus 147-150, 19, 24, 30
Ptiliidae 2, 9, 11
Ptilinus 215, 225
Ptinidae 375-376, 21, 23, 25, 38
ptinoides, Acalles 230, 27, 50
Ptinomorphus 66, 25, 37
Ptinus 375-376, 21, 23, 25, 38
puberulus, Sitona 299, 22, 49, 298
pubescens = cavifrons, Rhynchites 84
pubescens, Apion 79, 26, 48, 79
pubescens, Cyphon 396, 25, 35
pulchella, Hydraena 18
pulchellus, Negastrius 326, 20, 36
pulchellus, Polydrusus 291, 26, 49
pulchellus, Scymnus = quadrimaculatus, Nephus 219
pulicarius, Axinotarsus 357, 20, 39, 358
Pullus 220
pulvinatus, Ceutorhynchus 251, 24, 51
pumila = pseudopumila, Mordellistena 365
pumila = scrobiculatus, Trachys 92
pumilio, Ceutorhynchus 252, 24, 51
pumilus, Malthodes 96
punctatulus, Harpalus 134, 22, 31
punctulatus, Licinus 140, 23, 32, 139
punctatum, Anobium 63
punctatus, Helochares 18
punctatus, Ochthebius 18

472

punctatus, Onthophilus   341, *22, 33*
puncticeps, Harpalus   *31*
puncticollis, Dasytes   *359, 25, 38*
puncticollis, Harpalus   *134, 21, 31*
punctiger, Ceutorhynchus   252, *27, 51*
punctiger, Cryptocephalus   184, *23, 45*
punctipennis, Cyphon   396, *23, 35*
punctolineatus, Melanotus   326, *19, 36*
punctulata = aerea, Phyllotreta   204
punctulata = serra, Dorcatoma   64
punctulatus, Cis   40
punctum, Halacritus   338, *22, 33*
punctumalbum, Mononychus   281, *24, 51*
pusillus = aesthuans, Notiophilus   142
pusillus = rachifer, Adrastus   314
pusillus, Aphanisticus   91, *25, 35*
pusillus, Ochthebius   18
pusillus, Tychius   309, *27, 52*
putridus = fasciatus, Aphodius   385
putridus, Aphodius   387, *25, 34*
Pycnomerus   41
pygmaea = oculatus, Aderus   61
pygmaea, Hydraena   18
pygmaea, Xylophila = oculatus, Aderus   61
pygmaeus = femoralis, Scymnus   219
pygmaeus = pusillus, Tychius   309
pygmaeus = schmidti, Scymnus   220
pygmaeus, Dirhagus   332, *21, 36*
pygmaeus, Hylophilus = oculatus, Aderus   61
pygmaeus, Limnichus   344, *23, 35*
pygmaeus, Olibrus   372, *25, 40*
pygmaeus, Xylophilus = oculatus, Aderus   61
pyrenaeum, Leiosoma   274, *21, 50*
pyrenaeus, Barypeithes   243, *22, 49*
pyrenaeus, Geotrupes   *23, 33*
pyri, Anthomus = piri, Anthonomus   232
Pyrochroa   376, *26, 42*
Pyrochroidae   376, *2, 16, 23, 26, 42*
Pyropterus   346, *23, 37*
Pyrrhidium   165, *20, 44*
Pythidae   377, *2, 16, 23, 42*
Pytho   377, *23, 42*

Q

quadricollis = latridioides, Hypocoprus   344
quadricorniger, Phytobius   288, *24, 52*
quadricornis = quadricorniger, Phytobius   288
quadridens, Pityogenes   401, *24, 53*
quadrifasciata, Abdera   349, *23, 42*
quadrifasciata, Strangalia   44
quadrifoveolata, Monotoma   39
quadriguttatus = quadripunctatus, Notiophilus   143
quadriguttatus, Longitarsus   198, *23, 46*
quadriguttatus, Mycetophagus   367, *23, 41*
quadriguttatus, Notiophilus   143
quadrillum, Lionychus   140, *21, 32*
quadrimaculata, Dendroxena   409, *25, 33*
quadrimaculata, Phyllobrotica   45
quadrimaculatus, Aphodius   388, *19, 34*
quadrimaculatus, Hister   339, *22, 33*
quadrimaculatus, Nephus   219, *20, 40*
quadrinodosus = muricatus, Phytobius   287
quadrinodosus, Phytobius   288, *24, 52*
quadrinotatus, Hister   339, *19, 33*
quadripunctata = quadrimaculata, Dendroxena   409
quadripunctata, Clytra   45
quadripunctata, Silpha = quadrimaculata, Dendroxena   409
quadripunctata, Xylodrepa = quadrimaculata, Dendroxena   409

quadripunctatum, Agonum   102, *19, 31*
quadripunctatus, Anchomenus = quadripunctatum, Agonum   102
quadripunctatus, Enochrus   18
quadripunctatus, Harpalus   135, *22, 31*
quadripunctatus, Notiophilus   143, *24, 29*
quadripustulata, Lissodema   380, *26, 41*
quadripustulatum = quadripustulata, Lissodema   380
quadripustulatum, Bembidion   116, *24, 30*
quadripustulatus = bipustulatus, Panagaeus   144
quadripustulatus, Fleutiauxellus   323, *23, 36*
quadrisignatus, Dromius   126, *19, 32*
quadristriatus, Saprinus = rugiceps, Hypocaccus   340
quadristriatus, Tachys   30
quadrituberculatus = olssoni, Phytobius   288
quadrituberculatus, Macronychus   18
quadrituberculatus, Phytobius   288
quenseli, Amara   107, *22, 31*
quenselii = quenseli, Amara   107
querceti, Ceutorhynchus   252, *21, 51*
querceti, Cryptocephalus   184, *20, 45*
querceus, Lacon   324, *19, 35*
quercicola, Ampedus   317, *25, 35*
quercicola, Ceutorhynchus   253, *24, 51*
quercinus, Hypulus   351, *20, 42*
quercus, Trypodendron = signatus, Xyloterus   405
quinquelineatus, Hygrotus   17
quinquepunctata, Coccinella   217, *21, 40*
quinquepunctatus, Tychius   309, *21, 52*
quisquilius, Crypticus   413, *25, 41*

R

Raboceras   381
Rabocerus   381, *23, 26, 41, 42*
rachifer, Adrastus   314, *21, 36*
radiolus, Apion   82
Rainbow leaf beetle   176
rapae, Ceutorhynchus   253, *27, 51*
ratzeburgi, Scolytus   402, *27, 53*
raucus, Otiorhynchus   285, *26, 48*
rectirostris, Furcipus   266, *27, 52*
reflexum, Apion   79, *24, 48*
reichei = reichi, Smicronyx   300
reichei, Longitarsus   46
reichi, Smicronyx   300, *21, 51*
reppensis = pseudopustulata, Hyperaspis   218
resedae, Ceutorhynchus   253, *27, 51*
resinosus, Platyrhinus   70, *26, 47*
reticulata, Silpha = undata, Aclypea   409
reticulatus, Bolitophagus   413, *21, 41, 212*
retusa, Anthophora   355
revestita, Strangalia   168, *20, 44*
Rhagium   165, *26, 43*
Rhagonycha   97-98, *23, 25, 37*
Rhantus   17
Rhinocyllus   293, *24, 49*
Rhinoncus   293, *20, 52, 288*
Rhipiphoridae   *2, 16.43*
Rhizophagidae   377-379, *2, 16, 20, 21, 23, 25, 39*
Rhizophagus   378-379, *20, 21, 23, 25, 39, 377*
Rhizotrogus   383
rhododactylus = putridus, Aphodius   387
Rhopalodontus   212, *21, 40*
Rhopalomesites   280
Rhynchaenus   294-296, *21, 22, 24, 27, 53*
Rhynchites   83-86, *19, 21, 23, 26, 47*
Rhynchosinapis   8
Rhyncolus   50, 286
Rhysothorax   382
Rhyssemus   396, *19, 34*

INVERTEBRATE INDEX

Rhytidosoma 296
Rhytidosomus 296
Riolus 18
Risophilus 125
rivularis, Oulimnius 18
rivularis, Trechus 153, 21, 30
robini, Aepus 101, 24, 29
robinii = robini, Aepus 101
robinii, Aepopsis = robini, Aepus 101
roboris, Acalles 230, 27, 50
roboris, Anoplus 231, 26, 50
rosae, Diplolepis 262
rosenmulleri = textor, Lamia 159
rostellium = rostellum, Gymnetron 269
rostellum, Gymnetron 269, 24, 53
rotundicollis = ardosiacus, Harpalus 130
rotundicollis, Ophonus = ardosiacus, Harpalus 130
rubens, Anitys 62, 25, 38
rubens, Apion 80
rubens, Trechus 154, 24, 30
ruber, Coeliodes 259, 27, 51
rubidus, Curculio 261, 27, 52
rubiginosum, Apion 80, 21, 48
rubiginosus = ferrugineus, Longitarsus 194
rubra, Leptura 43
rubromaculatus = femoralis, Scymnus 219
rudis, Bagous 50
rufa, Aegialia 382, 19, 34
rufa, Formica 216, 217, 223, 340, 344, 378
rufa, Leptura 43
rufa, Osmia 375
rufibarbis = schaubergerianus, Harpalus 136
rufibarbis, Harpalus 136
rufibarbis, Ophonus = schaubergerianus, Harpalus 136
ruficeps, Ampedus 317, 19, 35
ruficollis, Cardiophorus 321, 19, 36
ruficollis, Necrobia 38
ruficollis, Platyderus 146, 24, 31
ruficollis, Rypobius 227, 22, 40
ruficollis, Silis 98, 25, 37, 99
ruficorne, Aulonium 41
rufifrons, Hydroporus 17
rufinasus, Stenopelmus 50
rufipennis, Ampedus 318, 20, 35
rufipes 205
rufipes = decemnotata, Phytodecta 205
rufipes = erichsoni, Cardiophorus 321
rufipes = vaudoueri, Phloiotrya 354
rufipes, Bruchela 47
rufipes, Bruchus 44
rufipes, Hydraena 18
rufipes, Phloeotrya = vaudoueri, Phloiotrya 354
rufipes, Thanasimus 214, 21, 38
rufitarsis, Harpalus 132
rufitarsis, Omophlus 416, 20, 41
rufocincta = praetermissa, Amara 107
rufovillosum, Xestobium 213, 214
rufulus, Trichosirocalus 306, 24, 51
rufum, Euophryum 50
rufus, Anthonomus 233, 21, 52
rufus, Rhysothorax = rufa, Aegialia 382
rugiceps, Hypocaccus 340, 23, 33
rugifrons, Hypocaccus 340, 25, 33
rugosa = rugosus, Meloe 356
rugosus, Meloe 356, 21, 43
rupicola, Harpalus 135, 24, 31
rupicoloides, Ophonus = melleti, Harpalus 133
ruricola, Omaloplia 393, 25, 34
rustica, Mantura 202, 26, 46
Rutidosoma 296, 24, 51
rutilus, Longitarsus 199, 23, 46

ryei, Apion 80, 26, 48
Rypobius 227, 22, 40

S

sabuleti, Aegialia 383, 25, 34, 384
sabulicola, Harpalus 135, 21, 31
sabulicola, Negastrius 327, 20, 36
sabulosum, Opatrum 416, 25, 41
sabulosus, Trox 421, 23, 33
sahlbergi, Agonum 102, 19, 31
sahlbergi, Chaetocnema 175, 23, 46
sahlbergii = sahlbergi, Chaetocnema 175
salicariae, Lythraria 200, 26, 46
saliceti, Orchestes = foliorum, Rhynchaenus 294
salicinus, Dorytomus 263, 27, 50
salicis, Dorytomus 264, 24, 51
salinus, Anthicus 68, 23, 43
Salpingidae 380-381, 2, 16, 23, 26, 41
Salpingus 42, 381
saltuarius, Cryphalus 53
sanguinea, Formica 180, 338
sanguineum = rubiginosum, Apion 80
sanguineum, Pyrrhidium 165, 20, 44
sanguineus, Ampedus 318, 19, 36
sanguinicollis, Ischnodes 324, 23, 36
sanguinicollis, Ischnomera 370, 26, 43
sanguinolenta = crassicornis, Chrysolina 176
sanguinolenta = prasina, Cassida 174
sanguinolenta, Chrysolina 178, 23, 45
sanguinolenta, Chrysomela = crassicornis, Chrysolina 176
sanguinolenta, Leptura 159, 21, 43
sanguinolentus, Ampedus 318, 23, 36
sanguinolentus, var. = bipunctatus, Cryptocephalus 181
sanguinosa, Cassida 47
Saperda 166, 23, 44
Saprinus 342-343, 19, 22, 25, 33, 337, 339
Saprosites 34
sartor = textor, Lamia 159
Sawyer beetle 164
saxatile, Bembidion 116, 24, 30
scaber, Otiorhynchus 285, 26, 48
scabricollis, Lixus 276, 22, 49
scalaris, Saperda 166, 23, 44
scalesianus, Hydroporus 17
scaliae, Aphis 219
scanicus, Ellescus 52
Scaphidema 418, 25, 41
Scaphidiidae 381-382, 2, 22, 25, 33
Scaphisoma 381-382, 22, 25, 33
Scaphium 382, 22, 33
Scaphosoma 381
scapularis, Lebia 139, 19, 32
Scarabaeidae 382-396, 2, 19-23, 25, 34
Scarce Seven-spot ladybird 216
Scarodytes 17
schaubergerianus, Harpalus 136, 25, 31
schilskyana, Anaspis 406, 22, 42
Schizotus 376, 23, 42
schmidti, Scymnus 220, 25, 40
schneideri, Tychius 310, 27, 52
schoenherri = squamosus, Barynotus 242
schoenherri, Apion 80, 24, 48
schonherri = schoenherri, Apion 80
schonherri = squamosus, Barynotus 242
schrankii = nitidulus, Chlaenius 122
schueppeli, Bembidion 116, 22, 30
schuppeli = schueppeli, Bembidion 116
schuppeli, Bembidium = schueppeli, Bembidion 116
scirpi, Notaris 282, 27, 51

scirrhosus, Thryogenes 302, *27, 51*
Scirtes 398, *33, 35*
Scirtidae 396-399, *8, 22, 23, 25, 35*
scitulum, Agonum 103, *22, 31*
scitulus, Anchomenus = scitulum, Agonum 103
scitulus, Europhilus = scitulum, Agonum 103
sclopeta, Brachinus 118, *19, 32*
scolopacea, Baris 242, *21, 52*
Scolytidae 398-405, *2, 11, 16, 19-21, 24, 27, 53*
Scolytus **402**, *27, 53*, 220, 330, *377*
scoticus, Anthicus **68**, *21, 43*
Scraptia 407-408, *19-21, 42*
Scraptiidae 405-408, *2, 19-23, 42*
scrobiculatus, Trachys **92**, *23, 35*
scrofa, Aphodius **388**, *19, 34*
scrophulariae, Anthrenus **311**, *19, 37*
scropulariae = scrophulariae, Anthrenus 311
scutellaris, Orchestes = testaceus, Rhynchaenus 296
scutellaris, Paracymus 18
scutellaris, Tachys **152**, *22, 30*
scutellaris, Triplax **331**, *21, 39*
scutellata, Coccidula 40
scutellata, Leptura **160**, *23, 43*, 222
Scybalicus 32
Scydmaenidae 2
Scymnus 25, 40, 216, **219-220**
secalis, Trechus 30
sedecimguttata, Halyzia 40
sedi, Apion **81**, *26, 47*
Selatosomus 328-329, *19, 21, 25, 36*
sellatus, Tapinotus **302**, *24, 52*
seminiger, Hydaticus 17
seminulum, Chaetarthria 18
semipunctatum, Bembidion **117**, *22, 30*
semistratus = cuspidatus, Saprinus 342
semistriatus = subnitescens, Saprinus 343
semivittatum, Apion **81**, *23, 47*
senecionis = ganglbaueri, Longitarsus 195
seniculus, Apion 74
separanda, Synchita **225**, *21, 41*, 221
sepicola, Tropideres **71**, *20, 47*
septempunctata, Coccinella 217
septentrionalis, Anaspis **407**, *19, 42*
septentrionis, Otiorrhynchus = scaber, Otiorhynchus 285
septentrionis, Oulema 45
septentrionis, Patrobus **145**, *24, 29*
sericeus = olivaceus, Rhynchites 85
sericeus, Lasiorhynchites = olivaceus, Rhynchites 85
sericeus, Polydrusus **291**, *24, 49*
sericeus, Rhynchites *47*
seripilosus = reichi, Smicronyx 300
serra, Ctesias **311**, *25, 37*
serra, Dorcatoma **64**, *23, 38*
serricornis, Prionocyphon **397**, *25, 35*
serripes, Harpalus **136**, *25, 31*
servus, Harpalus **136**, *25, 31*
setiger, Orthochaetes **283**, *27, 51*
setigera, Curimopsis **93**, *23, 35*
setosus, Ceuthorrhynchus = atomus, Ceutorhynchus 246
sexdentatus, Ips *53*
sexguttata, Leptura **160**, *21, 43*
sexmaculata, Judolia **158**, *23, 44*
sexpunctatum, Agonum **103**, *22, 31*
sexpunctatus, Anchomenus = sexpunctatum, Agonum 103
sexpunctatus, Cryptocephalus **185**, *20, 45*
sexpunctatus, Ptinus **375**, *25, 38*
sharpi, Bradycellus 119
sheppardi, Choragus **69**, *23, 47*
Sibinia 297-298, *24, 27, 52*
sigma, Dromius **127**, *23, 32*
signata = primitus, Sibinia 297

signaticollis, Berosus 18
signatum = signatus, Xyloterus 106
signatum, Trypodendron = signatus, Xyloterus 106
signatus, Xyloterus **405**, *27, 53*, 222
Silis 98, *23, 37*, 99
Silpha 410-411, *19, 20, 25, 33*
Silphidae 409-411, *2, 16, 19, 20, 23, 25, 33*
silphoides = punctatulus, Licinus 140
Silvanidae 411, *2, 20, 25, 39*
Silvanoprus **411**, *20, 39*
Silvanus **411**, *25, 39*
silvicola 107
silvicola = quenseli, Amara 107
similaris, Dryops 18
similis, Onthophagus 393, 394
similis, Silvanus = fagi, Silvanoprus 411
simplex = oblongicollis, Rhizophagus 379
Simplocaria **93**, *22, 35*
simulatrix, Laccobius 18
Sinodendron 336
sinuata = vittata, Phyllotreta 205
sinuatus = illigeri, Hister 339
sinuatus, Agrilus **89**, *23, 35*
sinuatus, Laccobius 18
Sitaris 355
Sitona 298-299, *20, 22, 26, 49*
Sitones 298
skrimshiranus, Stenolophus **150**, *23, 32*
smaragdinus, Harpalus **137**, *25, 31*
Smicronyx **300**, *21, 27, 51*
socius, Cathormiocerus **245**, *20, 48*
sodalis, Badister 32
sodalis, Sibinia **298**, *24, 52*
sophiae, Psylliodes **209**, *21, 47*
sorbi = laevigatum, Apion 78
sordidus, Agriotes **314**, *21, 36*
sordidus, Aphodius **388**, *23, 34*
soror, Apion **81**, *23, 47*
spadix, Pselactus **293**, *27, 50*
sparganii, Donacia **189**, *23, 44*
spartii, Cryptolestes **227**, *23, 39*
Spercheus 17
Sphaeriestes 381
Sphaeriidae 2
Sphaerites **412**, *21, 32*
Sphaeritidae 412, *2, 16, 21, 32*
sphaeroides, Peritelus *48*
Sphindidae 412, *2, 25, 39*
Sphindus **412**, *25, 39*
Sphinginus **361**, *22, 39*
Sphodrus 31
spinimanus, Trachyphloeus **305**, *26, 48*
spinulosus = markovichianus, Endophloeus 222
splendida, Apteropeda **172**, *20, 46*
splendidula, Lamprohiza **344**, *19, 37*
spreta, Amara **107**, *24, 31*
squamosus, Barynotus **242**, *26, 49*
squamulatus = asperatus, Trachyphloeus 304
squamulatus, Tychius **310**, *27, 52*
Stag beetle 345
staphylea, Chrysolina 138
Staphylinidae 2, 9, 11, 16
Stenelmis 18
Stenocarus **301**, *27, 51*
Stenolophus **150**, *23, 25, 32*, 99
Stenopelmus 50
Stenostola **167**, *26, 44*
stephensi, Bembidion 30
stephensi, Tychius 52
stercorarius = obscurus, Paralister 341
stercorarius, Geotrupes 387

stercorarius, Hister = obscurus, Paralister 341
stercorarius, Margarinotus = obscurus, Paralister 341
Stereocorynes 286
sternalis, Cercyon 18
Stictonectes 17
stierlini, Asaphidion 30
stigma, Rhynchaenus 296
Stilbus **373**, *22, 40*
stolidum, Apion **82**, *26, 48*
stomoides, Bembidion **117**, *24, 30*
Strangalia **167-168**, *19, 20, 23, 44*
strenua, Amara **107**, *21, 31*
striatellus, Dryops 18
strigifrons, Helophorus 17
striolata, Trichia 313
striolatus, Agabus 17
Strophosoma **301**, *21, 26, 49*
Strophosomus 301
sturmi, Bembidium = octomaculatum, Bembidion 115
subaeneus, Ilybius 17
subcarinatus, Bagous **240**, *24, 50*
subcoerulea, Chaetocnema **176**, *26, 46*
subferruginea, Hypocassida **191**, *19, 47*
subfuscus, Athous **320**, *21, 36*
subglossa = affinis, Caenocara 63
subglossa, Coenocara = affinis, Caenocara 63
subnitescens, Saprinus **343**, *19, 33*
subnotatus, Trechus **154**, *19, 30*
subpilosus, Ptinus **376**, *25, 38*
subquadratus = azureus, Harpalus 130
subspinosa, Zeugophora 44
substriatus, Pomatinus 18
subterraneus, Aphodius **389**, *19, 34*
subterraneus, Bledius 128
subterraneus, Myrmechixenus **223**, *22, 41*
subviolaceus, Riolus 18
succincta, Lycoperdina **330**, *20, 40*
suffriani, Gyrinus 17
Sulcacis **213**, *25, 40*
sulcatus = punctatus, Onthophilus 341
sulcicollis, Psammobius = asper, Psammodius 395
sulcifrons, Barypeithes **243**, *26, 49*
sulcirostris = piger, Cleonus 258
sulphureus, Cteniopus 41
sus, Euheptaulacus **391**, *19, 34*
sutor = textor, Lamia 159
suturalis = limbatus, Scymnus 220
suturalis = ononidis, Sitona 299
suturalis = parallelus, Platypus 374
suturalis, Longitarsus **199**, *26, 46*
suturalis, Orthotomicus 53
suturalis, Rhantus 17
suturellus, Ceuthorhynchus = pervicax, Ceutorhynchus 251
suturellus, Ceuthorrhynchus = pervicax, Ceutorhynchus 251
suturellus, Longitarsus 196
Syagrius **8**, *50*
sycophanta, Calosoma 29
sylvatica, Cicindela **124**, *22, 29*
Symbiotes **330**, *25, 40*
Synaptus **329**, *20, 36*
Syncalypta 93
Synchita **224-225**, *21, 25, 41*, 221
syrites, Ceuthorhynchus **254**, *20, 51*
Systenocerus 345

T

tabidus, Longitarsus **199**, *26, 46*
Tachypus 108
Tachys **151-152**, *19, 22, 24, 30*

tanaceti, Galeruca *45*, 138
tanaceti, Galerucs 138
Tanymecus **302**, *26, 49*
Taphrorychus **402**, *24, 53*, 228, 405
Tapinotus **302**, *24, 52*
tardii, Mesites **280**, *27, 50*
tardus = froelichi, Harpalus 132
tardyi = tardii, Mesites 280
tardyi, Rhopalomesites = tardii, Mesites 280
Tarsostenus **214**, *19, 38*
taurus, Onthophagus **394**, *19, 34*
Telophorus 94
tempestivus = czwalinai, Bagous 236
tempestivus, Bagous **240**, *27, 50*
tenebrioides, Zabrus **154**, *22, 31*
Tenebrionidae **413-418**, *2, 16, 20, 21, 23, 25, 41*
tenebroides = tenebrioides, Zabrus 154
tenebrosus, Harpalus **137**, *22, 31*
tenellus = fasciatus, Aphodius 385
Teredus **225**, *20, 41*
Teretrius **343**, *19*
terminatus, Cerapheles **358**, *23, 39*
terminatus, Ceutorhynchus **254**, *27, 51*
terrestris, Bombus 355
terricola, Laemostenus 31
terricola, Tropiphorus **307**, *26, 49*
tessulatus = paykulli, Aphodius 386
testacea, Hydraena 18
testacea, Scraptia **408**, *21, 42*
testaceum, Bembidion **117**, *24, 30*
testaceus, Conopalpus **351**, *26, 42*
testaceus, Rhynchaenus **296**, *21, 53*, 294
testudinarius, Heptaulacus **392**, *19, 34*
tetrastigma, Phyllotreta 45
Tetratoma **419**, *23, 26, 41*
tetratoma, Clinocara = minor, Orchesia 353
Tetratomidae **419**, *2, 16, 23, 26, 41*
Tetropium 43
teutonus, Stenolophus **150**, *25, 32*
textor, Lamia **159**, *20, 44*
thalassina, Donacia **189**, *26, 44*
Thalassophilus **152**, *22, 29*
Thanasimus **214**, *21, 38*
Thanatophilus **411**, *19, 33*
Thaneroclerus 38
thapsi = nigritarsis, Cionus 257
thapsus = nigritarsis, Cionus 257
thomsoni, Ceutorhynchus **255**, *22, 51*
thoracica, Anaspis **407**, *23, 42*
thoracica, Cantharis 37
thoracicus = gramineus, Cardiophorus 321
thoracicus, Dyschirius 29
Throscidae **420**, *2, 21, 36*
Throscus 420
Thryogenes **302**, *27, 51*
thujae, Phloeosinus 53
Thymalus **371**, *25, 38*
tibialis, Amara 104
tibialis, Procraerus **327**, *21, 36*
tibialis, Tychius **310**, *24, 52*
tigrina = pastinacae, Hypera 273
tiliae, Ernoporus **400**, *20, 53*
Tilloidea **215**, *19, 38*
Tillus **215**, *25, 38*
Timberman 155
Tiresias 311
titillator 159
titillator = textor, Lamia 159
tobias, Anthicus 43
tomentosus = terricola, Tropiphorus 307
tomentosus, Rhynchites **86**, *26, 47*

Tomicus	403, *21, 53*
tomlini = longitarsis, Bagous	237
Tomoxia	365, *23, 42*
Trachodes	303, *27, 50*
Trachyphloeus	303-305, *24, 26, 49, 244*
Trachys	91-92, *20, 23, 35*
translucida, Rhagonycha	98, *25, 37*
transparens = clarki, Bembidion	111
transversalis, Hydaticus	*17*
transversovittatus, Hylobius	270, *20, 49*
Trechus	153-154, *19, 21, 24, 29, 30, 152*
tredecimpunctata = tredecimpunctata, Hippodamia	218
tredecimpunctata, Hippodamia	218, *22, 40*
tremula, Chrysomela	179, *20, 45*
tremulae = tremula, Chrysomela	179
tremulae, Dorytomus	264, *27, 51*
tremulae, Melasoma = tremula, Chrysomela	179
trepanatus, Pityogenes	401, *24, 53*
triangulum, Ceutorhynchus	255, *27, 51*
Trichia	313
Trichius	*34*
Trichodes	215-216, *19, 38*
Trichosirocalus	305-306, *24, 27, 51*
tricornis, Bledius	113
tridentata, Labidostomis	191, *20, 45*
triguttata, Abdera	350, *23, 42*
triguttatus, Alophus	*49*
trimaculatus, Ceutorhynchus	255, *27, 51*
Trinodes	313, *21, 37*
Trinophyllum	*44*
Triotemnus	403, *20, 53*
Triplax	331, *21, 39*
tristis = coenosus, Aphodius	384
tristis, Ampedus	319, *20, 36*
tristis, Anthicus	68, *20, 43*
tristis, Arhopalus	*43*
tristis, Cercyon	*18*
tristis, Chlaenius	122, *19, 32*
tristis, Silpha	*33*
trisulcatum = trisulcum, Aulonium	220
trisulcum, Aulonium	220, *23, 41*
Tritoma	331, *23, 39*
Trixagus	420, *21, 36*
Trogidae	420-421, *2, 16, 19, 23, 33*
troglodytes, Liosoma = pyrenaeum, Leiosoma	274
troglodytes, Oulimnius	*18*
troglodytes, Trachys	*35*
Troglops	*39*
Trogossitidae	421, *2, 21, 38*
Tropideres	71, *20, 47*
Tropiphorus	307, *24, 26, 49*
Trox	420-421, *19, 23, 33*
truncorum, Phloeophagus	286, *24, 50*
Trypodendron	405
Trypophloeus	404, *19, 24, 53*
tuberculatus, Helophorus	*17*
turcica = scapularis, Lebia	139
turneri, Zeugophora	210, *23, 45*
Turnip flea beetle	204
Two-spot wood-borer	89
Tychius	307-310, *21, 22, 24, 27, 52*
typographus, Ips	213
tyrolensis, Silpha	411, *25, 33*

U

Uleiota	229, *23, 39*
uliginosus, Agabus	*17*
uliginosus, Elaphrus	130, *24, 29*
ulmi, Anthonomus	233, *27, 52*

umbellatarum, Molorchus	162, *23, 44*
umbrinus, Stenocarus	301, *27, 51*
uncinatus = tomentosus, Rhynchaenus	*80*
uncinatus, Otiorhynchus	*48*
undata = undatus, Cicones	*221*
undata, Aclypea	409, *19, 33*
undata, Megatoma	312, *25, 37*
undatus, Cicones	221, *20, 41, 225*
undulatus, Agabus	*17*
undulatus, Harminius	324, *25, 36*
unguiculare, Anchonidium	231, *21, 50*
unguicularis, Agabus	*17*
unguicularis, Ceutorhynchus	256, *21, 51*
unicolor = olivaceus, Bruchidius	87
unicolor = translucida, Rhagonycha	98
unicolor, Corticeus	413, *21, 41*
unicolor, Laria = olivaceus, Bruchidius	87
unifasciata, Tilloidea	215
unifasciatus, Notolaemus	228, *23, 39*
unifasciatus, Tilloidea	215, *19, 38*
unipunctatus = monostigma, Demetrias	125
unipustulatus, Badister	110, *25, 32*
unistriatus, Bidessus	*17*
univittatus, Tarsostenus	214, *19, 38*
urinator, Gyrinus	*17*
Urodontidae	2, *16, 47*
urticae, Ceutorhynchus	256, *21, 51*
urticarium, Apion	*47*
ustulata, Grammoptera	157, *21, 43*
ustulatus, Cercyon	*18*

V

vacca, Onthophagus	395, *25, 34*
validirostris = ictor, Dorytomus	263
validirostris, Pissodes	289, *21, 50*
vaporariorum, Cymindis	125, *25, 32*
vaporariorum, Myrmechixenus	223, *21, 41*
variabilis = pubescens, Cyphon	396
variabilis = punctipennis, Cyphon	396
variabilis, Gnorimus	392, *19, 34*
varians, Anthonomus	233, *27, 52*
varians, Chrysolina	45, *138*
variegata = humeralis, Mordellistena	363
variegata = variegatus, Cicones	221
variegata = variegatus, Meloe	356
variegata, Adonia	216, *25, 40*
variegata, Grammoptera	158, *23, 43*
variegata, Mordella	363
variegatus = nebulosus, Anthribus	69
variegatus, Cicones	221, *23, 41*
variegatus, Haliplus	*17*
variegatus, Meloe	356, *19, 43*
Variimorda	366, *26, 42*
Varimorda	366
variolosus, Oxylaemus	224, *21, 41*
varipes, Apion	82, *26, 48*
varius, Brachytarsus = nebulosus, Anthribus	69
varius, Haliplus	*17*
varius, Leperisinus	421
vaudoueri, Phloiotrya	354, *26, 42*
vectensis, Dromius	127, *21, 32*
velatus = velutus, Eubrychius	266
velatus, Litodactylus = velutus, Eubrychius	266
velatus, Phytobius = velutus, Eubrychius	266
velutus, Eubrychius	266, *27, 51*
ventralis, Ochrosis	202, *21, 46*
venustus = parallelus, Tychius	308
vernalis, Geotrupes	334, *25, 34*
vernalis, Harpalus	137, *22, 31*

INVERTEBRATE INDEX

veronica = veronicae, Gymnetron — 269
veronicae, Gymnetron — **269**, *27, 53*, 267
verrucatus, Ceutorhynchus — **256**, *21, 51*
versutum, Agonum — **103**, *24, 31*
versutus, Anchomenus = versutum, Agonum — 103
verticicornis = nutans, Onthophagus — 394
vesicatoria, Lytta — *43*
vespertinus, Phyllobius — **286**, *26, 48*
vestigator, Nicrophorus — **410**, *23, 33*
vestitus, Cis — 211
Vibidia — **220**, *22, 40*
viciae = atomarius, Bruchus — 88
viciae, Laria = atomarius, Bruchus — 88
vicinum, Apion — **82**, *26, 48*
viduatus, Ceutorhynchus — **257**, *27, 51*
vilis, Lixus — **277**, *20, 49*
villifrons, ?Taphrorychus = dryographus, Xyleborus — 405
villosa, Variimorda — **366**, *26, 42*
villosa, Varimorda — 366
villosoviridescens, Agapanthia — *44*
villosulum, Gymnetron — **270**, *27, 53*
villosulus = villosulum, Gymnetron — 270
villosus, Curculio — **262**, *27, 52*
villosus, Dryoceotinus — 398
villosus, Euheptaulacus — **391**, *23, 34*
viminalis, Phytodecta — *45*
violacea = violaceus, Meloe — 356
violacea, Chrysolina — **179**, *26, 45*
violaceum, Callidium — *44*
violaceum, Platydema — **417**, *20, 41*
violaceus, Cryptocephalus — **185**, *19, 45*
violaceus, Limoniscus — **325**, 10, *19, 36*
violaceus, Meloe — **356**, *26, 43*
Violet click beetle — 325
virens, Bembidion — **118**, *21, 30*
virens, Leptura — **161**, *19, 43*
virescens, Oedemera — **370**, *20, 43*
virescens, Saprinus — **343**, *22, 33*
virgata, Cernuella — 313
viridis, Agrilus — **90**, *23, 35*
viridis, Ochthebius — *18*
vittata, Cassida — **174**, *47*
vittata, Phyllotreta — **205**, *23, 45*
vittatus, Acrantus — 421
vittatus, Polystichus = connexus, Polistichus — 146
vulneratus, Diastictus — **390**, *20, 34*
vulneratus, Malachius — **361**, *21, 39*
vulneratus, Plegaderus — *32*

W

walkerianus, Tachys — **152**, *19, 30*
waltoni, Apion — *48*, 74
waltoni, Caenopsis — *48*
waltoni, Phytobius — **289**, *27, 52*
waterhousei = ferrugineus, Longitarsus — 194
waterhousei, Sitona — **299**, *26, 49*
wetterhali = wetterhalli, Masoreus — 141
wetterhalii = wetterhalli, Masoreus — 141
wetterhalli, Masoreus — **141**, *23, 32*
wetterhallii = wetterhalli, Masoreus — 141
Willow beetle — 261
Wood tiger beetle — 124
wrightii, Rhynchosinapis — 8

X

Xestobium — 213, 214, 225
Xyleborus — 404-405, *27, 53*, 377
Xylechinus — *53*
Xyletinus — **66**, *20, 37*
Xylita — **354**, *23, 42*
Xylocleptes — 228
Xylodrepa — 409
Xylophila — 61
Xylophilia — 61
Xylophilus — 61
Xyloterus — **405**, *27, 53*, 222

Z

Zabrus — **154**, *22, 31*
Zacladus — **311**, *27, 51*
zenkeri, Aphodius — **389**, *25, 34*
Zeugophora — **210**, *20, 23, 44, 45*
zigzag = parallelus, Harpalus — 133
Zilora — **354**, *26, 42*
zonatus, Graphoderus — *17*
zonatus, Trichius — *34*
Zorochros — *36*

Index to scientific names of plants mentioned in the text.

A

Acacia | 405
acanthium, Onopordum | 258, 306
acanthoides, Carduus | 258, 273
Acer | 30, 107, 408
acetosa, Rumex | 72, 202
acetosella, Rumex | 72, 80, 201, 301, 304
Achillea | 173, 174, 177, 178, 191, 255, 269, 305
Acinos | 82
acre, Sedum | 81
acutiformis, Carex | 282
Aegopodium | 178
aequalis, Alopecurus | 236
Ajuga | 172
alba, Melilotus | 272, 307
alba, Nymphaea | 187
alba, Reseda | 254
album, Chenopodium | 257, 266
album, Sedum | 81
Alchemilla | 287
Alisma | 188, 270
Alliaria | 247, 253, 254, 255
aloides, Stratiotes | 234, 239
Alopecurus | 236
alsine, Stellaria | 271
Althaea | 82, 207, 276
altissima, Melilotus | 308
altissimum, Sisymbrium | 241, 254
amara, Cardamine | 250, 251
amara, Iberis | 247
amphibia, Rorippa | 252
amphibium v. terrestre, Polygonum | 288
amphibium v. natans, Polygonum | 294
Anagalis | 203
anagallis-aquatica, Veronica | 267, 269, 270
Anemone | 273
Angelica | 275
anglica, Genista | 76
anglicum, Sedum | 81
angustifolia, Galeopsis | 185
angustifolium, Chamaenerion | 172
annua, Mercurialis | 81
Anthemis | 75, 78, 82, 269, 305
Anthriscus | 178, 275
Anthyllis | 181, 284, 308, 310
Antirrhinum | 179
Apium | 254
aquatica, Mentha | 82, 177, 194
aquatica, Oenanthe | 271, 276
aquatica, Scrophularia | 199
aquaticum, Myosoton | 271
Aquilegia | 273
Arabidopsis | 247, 249
Arabis | 247, 256
arboreus, Lupinus | 62
Arctium | 258, 302
Armeria | 298
Armoracia | 204, 253
Artemisia | 173, 178, 190, 192, 291, 364
articulatus, Juncus | 90
arundinacea, Phalaris | 282
arvense, Cerastium | 271
arvense, Cirsium | 173, 258, 273, 276, 293, 362
arvense, Equisetum | 267
arvense, Thlaspi | 250
arvense, Trifolium | 75, 80, 82, 309
arvensis, Acinos | 82
arvensis, Anagalis | 203
arvensis, Mentha | 82
arvensis, Myosotis | 194, 248

arvensis, Sinapis | 254
arvensis, Spergula | 174, 267, 297
arvensis, Stachys | 257
ascendens, Chaenorhinum | 196
Aster | 193
Astragalus | 72, 299
Atriplex | 6, 16, 207, 233, 308
Atropa | 190
aureum, Trifolium | 79
auricula-judae, Auricularia | 417
Auricularia | 417
australis, Phragmites | 187, 206, 282
autumnalis, Leontodon | 371

B

baccifer, Cucubalus | 173
Ballota | 178, 185, 193
Barbarea | 241, 250
beccabunga, Veronica | 267, 269, 270
bella-donna, Atropa | 190
bellidifolium, Limonium | 78
Berula | 271
Beta | 257, 276, 291, 302
betulinus, Piptoporus | 211, 212, 213, 226, 414
binervosum, Limonium | 78
Blackstonia | 300
Boletus | 412
Bovista | 63, 330
bovista, Lycoperdon | 63, 330
Brassica | 204, 205, 241, 253, 254
bulbosus, Ranunculus | 283
Butomus | 189, 239

C

caelatum, Lycoperdon | 330
Cakile | 204, 241
Calamintha | 92, 196
Calluna | 114, 134, 259, 260
Calmintha | 76
Caltha | 191
Campanula | 280, 281
campestre, Lepidium | 250, 253
campestre, Trifolium | 79, 309, 310
campestris, Artemisia | 173, 190
canadensis, Elodea | 239
canadensis, Solidago | 297
canina, Peltigera | 223
cannabinum, Eupatorium | 192
caprea, Salix | 90
Cardamine | 250, 251
Cardaria | 254
Carduus | 174, 208, 255, 258, 273, 276, 293, 306
Carex | 91, 186, 189, 282
Carlina | 273
carota, Daucus | 254, 275
carota ssp. gummifer, Daucus | 273
cataria, Nepeta | 82, 194, 196
catenata, Veronica | 270
caucalis, Anthriscus | 275
Centaurea | 72, 273, 276, 293, 296, 302
Centaurium | 300
Cerastium | 267, 271, 297
Ceratocystis | 227
Ceratophyllum | 240
Chaenorhinum | 279
Chaerophyllum | 275
chamaedrys, Veronica | 269
Chamaenerion | 172

cheiranthoides, Erysimum — 241
cheiranthos, Rhynchosinapis — 241, 248
Cheiranthus — 253
cheiri, Cheiranthus — 253
Chenopodium — 174, 242, 257, 266
Chrysanthemum — 178
Cicuta — 178
cicutarium, Erodium — 271, 274, 275, 277
cinerea, Erica — 171
cinerea, Salix — 90
Cirsium — 173, 182, 208, 255, 258, 273, 276, 293, 306, 362
claviculata, Corydalis — 249, 292
Clematis — 228
Clinopodium — 76
Cochlearia — 204, 247, 249, 254
Collybia — 224, 419
concentrica, Daldinia — 70, 224, 334
confluens, Pyronema — 336
Conium — 178, 254
Convolvulus — 191, 302
conyza, Inula — 206
Coriolus — 414
corniculatus, Lotus — 285, 299, 304, 309, 310
coronopus, Plantago — 177, 244, 245, 279, 291, 303, 306
Corticium — 373, 419
Corydalis — 249, 292
cotula, Anthemis — 75
cracca, Vicia — 73, 77, 88, 272, 299
Crambe — 254
Crepis — 372
crispum, Petroselinum — 254
crocata, Oenanthe — 276
Cucubalus — 173
Cuscuta — 300, 301
Cynoglossum — 185, 192, 198
Cytisus — 65, 67, 173, 227, 290, 300, 308

D

Daldinia — 70, 224, 334
Datura — 190
Daucus — 254, 273, 275
dealbata, Acacia — 405
decidua, Larix — 168
Dentaria — 251
Descurainia — 209, 249, 251, 253, 254
Digitalis — 202
Diplotaxis — 255
Dipsacus — 195
dissectum, Cirsium — 273
Draba — 247
draba, Cardaria — 254
drucei, Thymus — 3, 176
dubium, Trifolium — 79, 80, 309
dulcamara, Solanum — 203, 209
dysenterica, Pulicaria — 206, 240

E

Echium — 192, 194, 196, 198, 248
Eleocharis — 175, 189
Elodea — 239
emersum, Sparganium — 189
epithymum, Cuscuta — 300
Equisetum — 235, 238, 267
erecta, Berula — 271
erectum, Sparganium — 186, 200, 238, 282
Erica — 171, 181, 280
Erodium — 271, 274, 275, 277
Erophila — 249

Erysimum — 241
erythraea, Centaurium — 300
Eupatorium — 192
Euphorbia — 280
europaea, Cuscuta — 300
europaeus, Lycopus — 177, 196, 246
europaeus, Ulex — 173, 290, 300

F

faba, Vicia — 309
falcata, Medicago — 272
fasciculare, Hypholoma — 61
Festuca — 301
Filago — 73, 269, 372
filiformis, Potamegeton — 200
Fistulina — 367
fistulosa, Oenanthe — 276
flammula, Ranunculus — 235
flavum, Glaucium — 256
flos-cuculi, Lychnis — 173
fluitans, Glyceria — 186, 203, 236
fluviatile, Equisetum — 235, 238
foetidissima, Iris — 281
fomentarius, Fomes — 63, 212, 413, 414
Fomes — 63, 212, 413, 414
forsteranum, Sedum — 81
fragiferum, Trifolium — 74, 309
fruticosus, Rubus — 65, 172
Fumaria — 249, 253
fusipes, Collybia — 224

G

gale, Myrica — 295
Galeopsis — 185, 246
Galium — 179
gallica, Filago — 73, 372
gallii, Ulex — 173
Ganoderma — 63
gemmatum, Lycoperdon — 330
Genista — 74, 76, 173, 290, 308
Gentianella — 301
Geranium — 171, 271, 274, 311
germanica, Gentianella — 301
giganteus, Meripilus — 414
glabra, Turritis — 256
glandulifera, Impatiens — 92, 93, 103, 111-117, 140, 145, 148, 151-153, 213, 217, 323, 327, 383, 397
Glaucium — 256
glaucum, Chenopodium — 257
Glaux — 175
Glechoma — 92, 172, 179, 195, 248
glomerata, Campanula — 280
Glyceria — 186, 203, 236
glycyphyllos, Astragalus — 72, 299
Gnaphalium — 73
graveolens, Apium — 254

H

Halimione — 242, 291
hastata, Atriplex — 257
Hedera — 62
hederacea, Glechoma — 92, 172, 179, 195, 248
helenioides, Cirsium — 182
Helianthemum — 180, 181, 184, 303, 304
helix, Hedera — 62
hepatica, Fistulina — 367

Heracleum 275
herbacea, Salix 204
heterophyllum, Lemnum 230
Hieracium 180, 250, 301, 361
hirsuta, Arabis 247, 256
hirsuta, Vicia 299
hispidus, Polyporus 353
hispus, Inonotus 64
Honkenya 174
Hordeum 301
Humulus 208, 289
hybridum, Chenopodium 257
Hydrocharis 239
hydropiper, Polygonum 289
Hyoscyamus 92, 190, 208
Hypholoma 61

I

Iberis 247
Impatiens 92, 93, 103, 111-117, 140, 145, 148, 151-153, 213, 217, 323, 327, 383, 397
inodorum, Tripleurospermum 82, 178
Inonotus 64
Inula 206
Iris 281
islandica, Rorippa 250, 252

J

jacea, Centaurea 273
jacobaea, Senecio 248
japonica, Torilis 178
Jasione 280
Juncus 90, 91

K

kali, Salsola 257

L

lacustris, Scirpus 188
ladanum v. canescens, Galeopsis 185
Laetiporus 351, 366, 414
laevigatum, Taraxacum 251
Lamium 172
lanceolata, Plantago 177, 244, 279, 305, 306
lapathifolium, Polygonum 288
lappa, Arctium 258, 302
Larix 168
Lathyrus 73, 298, 299, 309
latifolia, Campanula 280
latifolia, Typha 282
latifolium, Lepidium 241
latifolium, Sium 271, 276
Leontodon 250, 371, 372
Lepidium 204, 241, 250, 253
Leucanthemum 82, 181, 247
Limonium 78, 180
Linaria 176, 177, 179, 268, 279
lingua, Ranunculus 186
Linum 198
Lithospermum 199
littoralis, Atriplex 242, 257
Lotus 272, 285, 298, 299, 304, 309, 310
lucens, Potamogeton 200
Lupinus 62

lupulina, Medicago 76, 298, 299
lupulus, Humulus 208, 289
Luronium 270
lutea, Nuphar 187, 189
lutea, Reseda 178, 242, 254
luteola, Reseda 242, 254
luteus, Boletus 412
Lychnis 173
lychnitis, Verbascum 258
Lycoperdon 63, 330
Lycopus 177, 196, 246
Lysimachia 200, 302
Lythrum 83, 200, 270, 287

M

maculatum, Conium 178, 254
Magnolia 404
major, Plantago 172, 177, 279
majus, Antirrhinum 179
Malva 207, 276
marina, Spergularia 267, 297
marina, Zostera 201
maritima, Armeria 298
maritima, Artemisia 192, 291, 364
maritima, Beta 291
maritima, Cakile 204, 241
maritima, Crambe 254
maritima, Glaux 175
maritima, Plantago 176, 177, 178, 198, 204, 279, 306
maritima, Suaeda 242
maritima, Triglochin 204
maritimum ssp. inodorum, Tripleurospermum 293
maritimus, Scirpus 200
Marrubium 92, 185, 193
Matricaria 75, 78, 203, 269, 293
media, Plantago 172, 177, 196, 279
media, Spergularia 267, 297
Medicago 76, 272, 298, 299, 307, 308
medium, Trifolium 308, 309
Melilotus 272, 307, 308
Mentha 76, 82, 92, 177, 185, 194, 196, 246
Mercurialis 81, 307
Meripilus 414
merismoides, Phlebia 373
mesenterica, Auricularia 417
Mespilus 203
millefolium, Achillea 173, 174, 178, 191, 255, 305
minor, Ulex 173, 290
minus, Chaenorhinum 279
mite, Polygonum 289
molle, Geranium 271, 274
montana, Jasione 280
montanus, Lathyrus 309
morisonii, Spergula 297
morsus-ranae, Hydrocharis 239
moschata, Malva 207
moschatum, Erodium 271
murinum, Hordeum 301
Myosotis 194, 248
Myosoton 271
Myrica 295
Myriophyllum 200, 237, 266, 276, 287

N

napus, Brassica 254
Nasturtium 205, 241, 247, 249, 265, 343
natans, Luronium 270
nemorosa, Anemone 273

nemorum, Lysimachia	200
Nepeta	82, 194, 196
nepeta, Calamintha	92
niger, Hyoscyamus	92, 190, 208
nigra, Ballota	178, 185, 193
nigra, Centaurea	293, 296, 302
nigrescens, Bovista	330
nigricans, Schoenus	91
nigrinus, Polyporus	212
nigrum, Solanum	196
nigrum, Verbascum	258
nodosa, Scrophularia	258
nudicaulis, Teesdalia	247, 252
nummularium, Helianthemum	184, 303
Nuphar	187, 189
nutans, Carduus	255, 258, 273, 293, 306
Nymphaea	187

O

obtusifolius, Rumex	202
Oenanthe	271, 276
officinale, Cynoglossum	192, 198
officinale, Nasturtium	205, 241, 247, 249, 265, 343
officinale, Sisymbrium	241, 248, 249, 253, 254
officinale, Taraxacum	252
officinalis, Althaea	82
officinalis, Cochlearia	249, 254
officinalis, Fumaria	249, 253
officinalis, Melilotus	272, 308
officinalis, Sanguisorba	287
officinalis, Veronica	269
oleracea, Brassica	253, 254
oleraceum, Cirsium	258, 273
Onobrychis	77, 79, 87, 299
Ononis	272, 298, 299
Onopordum	258, 306
Origanum	76
ovina, Festuca	301

P

palustre, Cirsium	273, 276, 293
palustre, Equisetum	267
palustris, Caltha	191
palustris, Eleocharis	175, 189
palustris, Potentilla	287
palustris, Stachys	177, 246, 256, 257
Papaver	301
parviflora, Fumaria	249
Pastinaca	273, 275
patula, Atriplex	257
pectinatus, Potamogeton	201
pectinatus x filiformis, Potamegeton	200
Peltigera	223, 236
Peplis	288
peploides, Honkenya	174
perenne ssp. anglicum, Linum	198
perennis, Mercurialis	81, 307
perfoliata, Blackstonia	300
perlatum, Lycoperdon	63, 330
persicaria, Polygonum	287
petiolata, Alliaria	247, 253, 254, 255
Petroselinum	254
Phaeolus	371
Phalaris	282
Phlebia	373
Phragmites	187, 206, 282
Phyteuma	281
phyteuma, Reseda	254

pilosa, Genista	76
pilosella, Hieracium	301
piperata, Mentha (x)	194
Piptoporus	211, 212, 213, 226, 414
Pisum	299, 309
Plantago	172, 176, 177, 178, 196, 198, 202, 204, 244
	245, 246, 269, 279, 291, 303, 304, 305, 306
plantago-aquatica, Alisma	270
Pleurotus	331, 382
plicata, Glyceria	236
plumbea, Bovista	63
podagraria, Aegopodium	178
Polygonum	75, 202, 287, 288, 289, 294, 302
Polyporus	63, 64, 212, 225, 337, 349, 351, 353, 382, 414
Populus	86, 178, 210, 261, 296, 404
Poria	337
portula, Lythrum	83
portula, Peplis	288
portulacoides, Halimione	242, 291
Potamogeton	188, 200, 201, 237
Potentilla	287
pratense, Geranium	171, 274
pratense, Trifolium	80, 82, 272, 299, 308, 309
pratensis, Cardamine	250, 251
pratensis, Festuca	301
pratensis, Lathyrus	73, 298, 299
prostrata, Atriplex	266
Prunella	73, 172, 196
Prunus	84, 180, 203
pseudacorus, Iris	281
ptarmica, Achillea	177
Pulicaria	206, 240
Pulmonaria	194
purpurea, Linaria	179
pyrenaicum, Geranium	274
pyriforme, Lycoperdon	330
Pyronema	336
Pyrus	203

Q

quercinum, Corticium	373, 419

R

radiatus, Polyporus	349, 353
radicata, Collybia	419
radicicola, Ceratocystis	227
Ranunculus	186, 235, 273, 283, 360
rapa, Brassica	204, 205
raphanistrum, Raphanus	254
Raphanus	204, 241, 254
recutita, Matricaria	293
repens, Linaria	268
repens, Ononis	272, 298, 299
repens, Ranunculus	273
repens, Trifolium	74, 77, 272
Reseda	178, 242, 254
Rhacomitrium	142
rhoeas, Papaver	301
Rhynchosinapis	209, 241, 248
robertianum, Geranium	274
Rorippa	241, 250, 252
Rosa	203
rosea, Althaea	276
rostrata, Carex	189
rotundifolia, Campanula	280
rotundifolium, Geranium	271
rubra, Spergularia	267, 297
Rubus	65, 172

Rumex	72, 80, 201, 202, 301, 304
rupestris, Spergularia	297
ruralis, Spergularia	297
rusticana, Armoracia	204, 253

S

Sagina	297
Sagittaria	186, 188, 240, 270
sagittifolia, Sagittaria	188, 240, 270
salicaria, Lythrum	200, 270, 287
Salicornia	175, 242
Salix	62, 79, 85, 86, 90, 156, 164, 180, 181
	190, 191, 204, 230, 264, 274, 295, 296
Salsola	257
Salvia	185
Samolus	374
sanguineum, Geranium	274
Sanguisorba	287
sativa, Medicago	76, 272, 299, 308
sativa ssp. falcata, Medicago	76
sativa, Onobrychis	299
sativa, Pastinaca	273, 275
sativa, Vicia	88, 299, 309
sativa ssp. nigra, Vicia	309
sativum, Pisum	299, 309
sativus, Raphanus	204, 241, 254
scabiosa, Centaurea	273, 302
Schoenus	91
schweinitzii, Phaeolus	371
Scirpus	188, 200
scoparius, Cytisus	65, 67, 173, 227, 290, 300, 308
scorodonia, Scrophularia	199
scorpioides, Myosotis	248
Scrophularia	192, 197, 199, 258, 269
scutellata, Veronica	267, 269, 270
Sedum	81
semidecandarum, Cerastium	267
Sempervivum	81
Senecio	194, 195, 197, 206, 248
sepium, Vicia	77, 88, 309
serpyllifolia, Veronica	269
Silene	173
Sinapis	254
Sisymbrium	241, 248, 249, 253, 254
Sium	271, 276
Solanum	196, 203, 209
Solidago	297
sophia, Descurainia	209, 249, 251, 253, 254
Sorbus	85, 203, 231, 277, 280
soulangiana, Magnolia	404
Sparganium	186, 187, 189, 200, 238, 282
speciosa, Galeopsis	246
Spergula	174, 267, 297
Spergularia	267, 297
Sphagnum	92, 100, 101, 103, 121, 122, 125, 148,
	151, 152, 196, 236, 237, 285, 396
sphondylium, Heracleum	275
spicata, Mentha	194
spicatum, Myriophyllum	200, 266, 276
spinosa, Ononis	299
Spiraea	311, 312
squalidus, Senecio	197
squamosus, Polyporus	64, 414
Stachys	172, 177, 185, 246, 256, 257
Stellaria	271, 297
Stereum	211, 419
stramonium, Datura	190
Stratiotes	234, 239
striatum, Trifolium	309, 310

Suaeda	242
submersum, Ceratophyllum	240
subhirsuta, Ononis	190
subterraneum, Trifolium	74, 77, 79
sulphureus, Laetiporus	351, 366, 414
sulphureus, Polyporus	223
sylvatica, Stachys	246, 256
sylvestris, Angelica	275
sylvestris, Anthriscus	178, 275
sylvestris, Malva	276
sylvestris, Rorippa	241
Symphytum	192, 194, 196

T

Tanacetum	173, 177, 305
Taraxacum	251, 252
Teesdalia	247, 252
telephium, Sedum	81
telmateia, Equisetum	267
tenuifolia, Diplotaxis	255
tenuis, Lotus	299
tetrahit, Galeopsis	246
tetralix, Erica	171, 181
tetrasperma, Vicia	299
Teucrium	172, 202
thaliana, Arabidopsis	247, 249
Thalictrum	193
thapsus, Verbascum	257, 258
Thlaspi	250
Thymus	3, 176, 195
Tilia	400
tinctoria, Genista	74, 76, 173, 308
tinus, Viburnum	216
Torilis	178
trachelium, Campanula	280
Tramete	351
Trametes	332
Trichaptum	354
Trifolium	74, 75, 77, 79, 80, 82, 272, 299, 308, 309, 310
Triglochin	204
Tripleurospermum	75, 78, 82, 178, 293
tripolium, Aster	193
tuberosus, Lathyrus	299
Turritis	256
Typha	187, 282, 373

U

Ulex	173, 290, 300
uliginosus, Lotus	298, 299
umbellatus, Butomus	189, 239
Urtica	208
usitatissimum, Linum	198
Ustulina	221
Utricularia	196, 239

V

valerandi, Samolus	374
Verbascum	197, 199, 257, 258
verna, Erophila	249
Veronica	172, 248, 267, 269, 270
versicolor, Coriolus	414
verticillatum, Myriophyllum	266, 276
Viburnum	155, 157, 158, 161, 216
Vicia	73, 77, 88, 272, 299, 309
viciifolia, Onobrychis	77, 79, 87
virosa, Cicuta	178

PLANT INDEX

vitalba, Clematis	228
vulgare, Cirsium	273, 293, 306
vulgare, Clinopodium	76
vulgare, Echium	192, 194, 196, 198, 248
vulgare, Leucanthemum	82, 181, 247
vulgare, Limonium	78, 180
vulgare, Marrubium	92, 193
vulgare, Origanum	76
vulgare, Tanacetum	173, 177
vulgaris, Alchemilla	287
vulgaris, Aquilegia	273
vulgaris, Barbarea	241
vulgaris, Beta	257, 302
vulgaris ssp. *maritima, Beta*	276
vulgaris, Calluna	259, 260
vulgaris, Carlina	273
vulgaris, Filago	73, 372
vulgaris, Linaria	177, 179, 268, 279
vulgaris, Lysimachia	200, 302
vulgaris, Prunella	73
vulgaris, Ustulina	221
vulgaris, Utricularia	196, 239
vulneraria, Anthyllis	181, 284, 308, 310

W

Wistaria	66
wrightii, Rhynchosinapis	209

Z

Zostera	201

484